WITHDRAWN

*American
Steamships on
the Atlantic*

American Steamships on the Atlantic

Cedric Ridgely-Nevitt

Drawings by the author

Newark
University of Delaware Press
London and Toronto: Associated University Presses
1981

© 1981 by Associated University Presses, Inc.

Associated University Presses, Inc.
4 Cornwall Drive
East Brunswick, New Jersey 08816

Associated University Presses
69 Fleet Street
London EC4Y 1EU, England

Associated University Presses
Toronto M5E 1A7, Canada

Library of Congress Cataloging in Publication Data

Ridgely-Nevitt, Cedric.
    American steamships on the Atlantic.

    Bibliography: p.
    Includes index.
    1. Steam-navigation—United States—History.
2. Steam-navigation—Atlantic Ocean—History.
3. Steamboats—United States—History. 4. Atlantic
Ocean—Navigation—History. I. Title.
VM623.R54   387.5′0973   78-66835
ISBN 0-87413-140-5

Printed in the United States of America

Contents

List of Appendixes	8	
Preface	9	
1 Inventing and Perfecting the Steamboat	13	
2 The *Robert Fulton*	52	
3 Auxiliary Steamships, Part 1	58	
4 Steamboats on the High Seas	73	
5 Auxiliary Steamships, Part 2	83	
6 The First Ocean Steamships, 1846-1850	98	
7 The Ocean Steam Navigation Company, 1847-1857	128	
8 The *United States*	140	
9 The Collins Line, 1850-1858	149	
10 The Havre Line, Part 1, 1850-1861	172	
11 Screw-Propelled Steamships, 1850-1855	187	
12 Experimental Craft	208	
13 The Vanderbilt European Line, 1855-1860	222	
14 The North Atlantic Steamship Company, 1859-1860	250	
15 Occasional Atlantic Liners, Part 1	257	
16 Iron Screw Steamships	281	
17 The Baltimore and Liverpool Steamship Company, 1865-1868	291	
18 The Havre Line, Part 2, 1865-1867	296	
19 The New York Mail Steamship Company, 1863-1867	301	
20 Ruger Brothers, 1865-1870	311	
21 The New York and Bremen Steamship Company, 1867	314	
22 The American Steamship Company, 1867-1871	316	
23 William H. Webb, 1868-1870	322	
24 Occasional Atlantic Liners, Part 2	324	
25 The End of the Side-Wheel Era	347	
Appendixes (A-W)	350	
Glossary of Marine Terms	369	
Bibliography	374	
Index of Names of Vessels	378	
Index of Owners, Operators, Shipbuilders, and Engine Builders	382	

List of Appendixes

A	Fulton-Livingston Steamboats	350	M	Vanderbilt Steamships, 1850-1865	359
B	Fulton-Livingston Engines	350	N	Vanderbilt Steamship Engines, 1850-1865	359
C	Early Steamboats	351	O	Wood Screw Steamships	360
D	Coastwise Steamboats	351	P	Wood Screw Steamship Engines	361
E	Steamboat Engines	352	Q	Iron Screw Steamships	362
F	Auxiliary Steamships	352	R	Iron Screw Steamship Engines	363
G	Coastwise Steamships, 1820-1850	353	S	Coastwise Steamships, 1861-1865	363
H	Coastwise Steamship Engines, 1820-1850	354	T	Pacific Mail Steamships, 1861-1869	364
I	Coastwise Steamships, 1851-1860	355	U	Walking-Beam Engines, 1861-1869	365
J	Coastwise Steamship Engines, 1851-1860	356	V	Converted Steamships	366
K	Atlantic Liners, 1847-1857	357	W	First Atlantic Sailings (in 2 parts)	367
L	Atlantic Liner Engines, 1847-1857	358			

Preface

What was a steamship like, and what happened to her? Why did she have certain engineering features, what accepted ideas were derived from previous vessels, and what new elements were carried on by her successors? The author has tried to answer these questions for American steam-powered, ocean-going craft from their origin in Robert Fulton's day until the wooden-hulled sidewheelers were replaced, about 1873, by iron-screw steamers powered by multiple expansion engines.

The passenger-carrying mail liners built to cross the Atlantic have always captured attention as the largest and fastest vessels afloat, but, because less than a score were built between 1847 and 1867, coastwise vessels must also be considered. Many made voyages to England, France, Ireland, and, in a few cases, to a Mediterranean port. In the late 1840s coastwise developments led directly to the design features of the Atlantic liner. Then, from 1850 to 1860, there was an engineering interchange between the two; each affected the other's characteristics.

Throughout the text the author has expressed engineering opinions and has drawn conclusions, sometimes based on limited contemporary evidence. These are a result of four decades of experience as a naval architect; during that time he has observed successes—widely recorded—and failures—frequently concealed—of twentieth-century steamships. There were similar successes and failures in the 1800-1870 epoch, and clear statements that steamers behaved badly were, and are, infrequent. The author has tried to draw likely conclusions from recorded events, not from enthusiastic predictions by some newspaper reporter whose attitude was, all too frequently, that anything American was a revolutionary advance beyond a previous European vessel. He admits reading between the lines and takes full responsibility for damning some steamships and praising others.

Because the author has been engaged, since 1936, in locating the very small number of plans of early steamships that still survive, he is indebted to many helpful individuals and institutions. Most wooden commercial vessels, sail or steam, were built without plans of any kind. Since those original drawings which do exist would be unreadable if reproduced on a small scale, he has either redrawn or retraced them. A considerable number of line drawings were prepared long ago for his 1939 undergraduate thesis.

A very large part of the text is based on contemporary newspaper accounts of the steamships, their sailings, and their unfortunate appearances in the disaster columns of daily papers and weekly or semiweekly shipping periodicals.

Much of this information came from the New York Historical Society's Library; its magnificent newspaper collection is not duplicated elsewhere. The New York *Evening Post* is a particularly good source concerning the early decades of the nineteenth century. The ocean steamship and the Atlantic liner are best reported in the New York *Herald, Commercial Advertiser,* and *Journal of Commerce*. Arrivals and departures are chronicled in the *New York Shipping and Commercial List*. Roger N. Mohovich, newspaper librarian, his predecessors, and his colleagues have been extremely helpful. They have exhumed between two and three tons of bound volumes from their stacks. Concerning the years after 1869, the Maritime Association of the Port of New York, formerly the New York Maritime Exchange, possessed an almost complete set of the New York *Maritime Register*. Vincent Filosa of this group allowed the author to consult those ponderous tomes over a period of years.

For other newspapers the author has relied upon the Library of Congress, the New York Public Library, the Boston Public Library, the Enoch Pratt Library at

Baltimore, the Maryland Historical Society, and the Boston Athenaeum.

The Peabody Museum of Salem, Massachusetts, has provided plans, illustrations, and information, chiefly from the Bradlee collection. In turn, the author has been assisted by L. W. Jenkins, Walter Muir Whitehill, Ernest Dodge, and P. C. F. Smith. The Mariner's Museum at Newport News, Virginia, has copied plans and photographs, mostly from its Eldredge collection; John L. Lochhead and Harold S. Sniffin have been most accommodating. The Birmingham, England, Public Library supplied copies of original Boulton and Watt drawings.

Many early works on shipbuilding, naval architecture, and marine engineering were collected by the eminent New York shipbuilder William H. Webb. These, together with the surviving plans and records of his building yard (incomplete, as a result of two different shipyard fires) are now a valued part of the library at Webb Institute of Naval Architecture. It was Webb who founded and endowed this college to provide to deserving young Americans a free education in the design of ships and their engines. The author has imposed upon Phyllis C. Blood, its first librarian, her successors, and Professor Fred Forrest, its present librarian. Many line drawings were re-created from dimensions recorded in Webb's Book of Offsets. Corresponding small-scale spar plans are, however, only conjectures, but all are based on contemporary illustrations. In those cases where actual mast and spar dimensions survive, they are followed. The chief reason that so many Webb steamers are used as illustrations is that other builders' designs have not survived.

Marian Virnelson steered the author to drawings inherited from John W. Griffiths, her grandfather; Frank A. Taylor, of the Smithsonian's U.S. National Museum, provided copies.

The late Howard I. Chapelle lent drawings from his private collection, as well as encouragement. The late Frank C. Bowen was equally encouraging and made material available from the extensive collection he maintained at Gravesend, England. His lists of transatlantic arrivals and departures, compiled by B. W. Ginsberg, were most useful.

The author has consulted three divisions of the U.S. National Archives: the photograph collection, when it was first made available by Dr. Vernon Tate; the naval plan collection; and the ship registers and enrollments. Dr. Forrest R. Holdcamper and Kenneth R. Hall have located and photostated hundreds of ship documents.

The New York Historical Society's manuscript collection includes Robert Fulton, John Ericsson, Richard Meade, and Lawrence Stryker material, as well as logs of several steamships plying the Aspinwall route. The New Jersey Historical Society has a splendid collection of Fulton drawings, as well as the only accurate contemporary view of John Fitch's *Experiment*. The Fitch notebooks (and all Fitch information comes from them) were originally deposited, by him, with the Library Company of Philadelphia. They are now at the Pennsylvania Historical Society. Thomas J. Dunnings, Jr., Alan Frazer, and Peter Parker from these three institutions have been of assistance.

The San Francisco Maritime Museum is noted for a photograph collection that, strangely enough, includes rate items taken on the East Coast. Matilda Dring presided over it and welcomed viewers as if they were family members, rather than strangers. Perhaps all the individuals seriously involved in maritime history do form a family group.

Many illustrations of marine engines came from contemporary technical works in New York's Engineering Societies Library. Their files of the *Journal of the Franklin Institute* provided a considerable part of the detail dimensions and technical descriptons of steamships and their engines.

Many members of the Steamship Historical Society of America have answered questions and have steered the author toward other sources, both animate and inanimate. Word-of-mouth data from informed individuals provided rapid solutions to many nagging questions.

One individual who has been particularly helpful, over a nine-year period, is N. R. P. Bonsor, resident of Jersey, that delightful island off the Channel shores of France. He has been preparing a long-awaited second edition of *North Atlantic Seaway* simultaneously with the present author's actual writing of this work. The interchange of information and ideas has been both enjoyable and useful. Not only has he been amiable enough to consult periodicals not available on this side of the Atlantic, he has also made portions of his own records available.

The published works of two present-day maritime historians have been consulted and quoted extensively. Professor E. Kenneth Haviland has written the definitive history of American steamships in Chinese waters. Professor John Haskell Kemble's works on gold-rush steamers, transpacific steamships, and the Pacific Mail Steamship Company are standard reference material. The author wishes to emphasize his indebtedness to both.

Sadye Cohen has cheerfully typed three different drafts of the text. Further manuscript assistance has been tendered by D. Trouillon. Many chapters have been read by, and helpful questions and comments received from, H. T. Fitzpatrick, Jr., miniature model maker, and the late Randall J. LeBoeuf, Jr., whose extensive collection of Robert Fulton material is now at the New York Historical Society.

To the few mentioned above, and the large number of others who assisted the author, he offers his heartfelt thanks.

American Steamships on the Atlantic

1
Inventing and Perfecting the Steamboat

THE question of who invented the steamboat has created an extensive body of printed matter that has been growing steadily for the past two centuries. Every country has claimed the real inventor as its own: the Marquis de Jouffroy in France, William Symington in England, and John Fitch in the United States are a few of the better-known examples. Possible dates are equally varied. Jonathan Hulls, in England, proposed a sternwheel steamship in 1736.[1] Even this date can be pushed farther back if we consider Denis Papin, who operated a paddle-wheel boat on the river Fulda at the German city of Kassel in 1707.[2] Although it was propelled by manpower, Papin did publish other material dealing with the possibility of powering vessels by steam.

If we restrict our interest to the actual development of American steamers as a practical means of carrying passengers, there is still no general agreement as to priority. There have been partisans who insist that John Fitch, James Rumsey, Colonel John Stevens, and even, possibly, Robert Fulton invented the steamboat. Other favorites include Samuel Morey, Elijah Ormsbee, Nicholas Roosevelt, and Oliver Evans.[3] Each pioneer has had enthusiastic support, not only from certain contemporaries, but also from later authors who were to write biographies long after their heroes were dead.

The whole idea of invention for a complicated entity such as a steamboat is unrealistic. Such an assemblage of complex parts did not suddenly spring from the inventor's mind and glide into the water with steam up, smoke streaming from its stack, froth churning behind its paddle wheels, and an admiring crowd gathered on shore. To invent the hull; to invent the steam engine; to invent paddle wheels, the screw propeller, or the jet pump; and furthermore, to make them work together as a harmonious whole, is a little too much to expect of one individual. Yet some of the pioneers tried to do all these things at once, did them very badly, and, even though the result may have finally operated, contributed little or nothing toward the further development of the steamboat. What was really needed was an engineer who could take existing elements and combine them in an intelligent way (with the minimum amount of trial and error) in order to produce a workable and useful craft. If others did not want to utilize the result or pay for its services, it could never endure save on the printed page, and it could never lead to improved craft that would solve further transportation problems. If a success were to be achieved, the technology of the inventor's time would have to be advanced enough to manufacture all the necessary components going into his steamboat. Engine parts had to be cast, forged, turned, drilled, finished, and made to fit together with airtight and steam-tight joints before a steam engine could become possible. Boilers had to be made up of formed and riveted plates, again with pressure-tight joints. Wood hulls would have to be designed and built to support the heavy machinery and to resist its dynamic forces. These elements form the minimum basic requirements for any self-propelled vessel.

From a marine engineering standpoint, the steamboat first became a real possibility in 1776 when James Watt's and Mathew Boulton's newly formed partnership started manufacturing steam engines on a commercial basis. Their factory was at Soho, near Birmingham, England. The technology and the required machine tools were created there to produce the necessary motive power. The year 1782 might be an even better date, for in that year the firm produced an engine with a rotational motion and, simultaneously, Watt obtained a patent for using steam on both sides of its piston to make it double-acting.[4] From this date on, both the necessary knowledge and the skilled workmen to apply it would gradually emigrate from England to other countries and, with time, would establish the foundries, machine shops, and large-scale forges necessary to build land and marine engines.

It was a complete lack of such facilities that hampered steamboat pioneers in the United States. Up until 1800 the only available metal workers were the local blacksmiths. What they could not hammer to shape in iron could not be attained elsewhere.

A boat's hull was much less difficult. One of the major reasons for Great Britain's commercial interest in the American Colonies was the availability of forests to supply badly needed shipbuilding timber. Skilled shipwrights came from England and Scotland, set up shipyards, and had an extensive shipbuilding industry established long before the Revolutionary War. Native shipbuilders served apprenticeships under experienced masters from abroad and, in their turn, became master builders—who were their own designers—in New England, New York, on the Delaware River, and in the Chesapeake Bay area. By the end of the eighteenth century all these regions were capable of constructing wooden ships of considerable size.

John Fitch

One of the most original characters to produce eighteenth-century steamboats was John Fitch, a Connecticut farm boy, Revolutionary War soldier, backwoods surveyor, self-taught engraver, button founder, and silversmith, who pursued the idea of propelling a boat by steam with the ardor of a religious zealot. He lacked adequate mechanical ability and was completely devoid of engineering judgment; furthermore, his unfortunate personality prevented him from enlisting the assistance of the patrons he needed to provide financial or influential aid. To assist him, in fact to do most of the engineering and mechanical work involved, he acquired Henry Voight, a Philadelphia clock and watch maker.[5]

This intrepid pair undertook the invention of the steamboat in the broadest sense of the word. They knew little or nothing of James Watt's work, had never seen a steam engine or a boiler. They had read little on the subject and, even worse, Fitch was too independent to take advice from those who had. Starting in 1786 they first produced a model steam boiler and engine with a one-inch diameter cylinder that failed to work. Undaunted, they advanced to a three-inch diameter cylinder and ordered a small skiff to use for propulsion experiments. These were simultaneous activities; the steam engine was never installed in the boat. Several arrangements of paddles, operated by manpower, were tried and, finally, one with oars proved successful in July of 1786. Apparently the engine also operated about the same time. As a result, many authorities, starting with Westcott, infer July 27, 1786, as the date when this craft was actually driven by the steam engine. Fitch's statement was: "I have now invented a simple and practical way of rowing a Boat, applicable to an Engine."[6] At no time did he specifically claim to have installed his model engine in the boat or to have propelled her by anything but muscle power.

Satisfied that a steam engine could be produced and that power could be efficiently used to propel a boat by oars, Fitch and Voight proceeded to construct an actual steamboat. Her details were published in the *Columbian Magazine* in December of 1786 together with an illustration bearing the premature date of 1786. Although written as if the vessel existed, this was a description of their intentions and we do not know whether the boat was actually built as described or illustrated. There was to be a horizontal cylinder of 12" bore by 3'-0" stroke that drove twelve oars—more correctly called paddles, for they did not row the vessel in the normal sense but operated like canoe paddles—six of them dipping vertically into the water as the other six were lifted out. Their fore-and-aft travel was 5'-6" for each stroke of the piston.

The engine was completed in May of 1787 and promptly demonstrated that wood, as a material for the cylinder heads, was a mistake. There were further difficulties with both the boiler and the condenser that kept the inventors busy throughout the summer, but eventually the 45' long by 12" beam craft was ready for a public trial on the Delaware River on August 22, 1787. Because the Constitutional Convention for the United States was meeting in Philadelphia, its distinguished delegates were invited to attend the trials.[7] Unfortunately, the boat proved extremely slow. Fitch and his considerable number of financial backers, most of them tradesmen of very limited means, realized that she could not be put to any commercial use; their expenditures to date were just a first installment. A better and faster vessel was imperative. Nevertheless, this was the first vessel in American waters actually to move under steam on a number of occasions; she took out invited guests and was in use until the middle of December of 1787.

Now the weaknesses of Fitch and Voight were to become apparent. They had achieved an operating, if barely so, steamboat after two years of trial and error. Being unable to assess the good features of this design, they blundered along to make even more mistakes in everything that followed. Both were intrepid tinkerers with no real technical understanding. The essential element of intuition that would guide later, more gifted experiments was missing.

As a first step, in order to increase speed, an iron casting for an 18″ diameter cylinder was ordered, but it was of such poor quality that the foundry broke it up for scrap rather than deliver it. This, of course, was an example of the limitations of the technology of the era. There were then only three steam engines in the United States,[8] the largest being a pre-Watt pumping engine of British manufacture. American iron works lacked any experience in producing the sound, thin-walled castings needed.

To try to circumvent the problem, the inventors resolved to use the cylinder they had, despite its proved lack of power. To reduce the boat's resistance, the original hull, 45′ by 12′, would be replaced by a new one 60′ long by 8′ in beam. Furthermore, to improve the propulsion, a new scheme had three rectangular paddles operating behind the stern of the vessel to replace the many oars installed along the sides. These were sensible things to do. But the old engine, as installed in the new boat, had to be changed; there were boiler difficulties that including an embarrassing breakdown during a public trial held on the Delaware from Philadelphia to Burlington, New Jersey. After many modifications everything worked well and several completed trips to Burlington were made. On October 12, 1788, she carried thirty passengers on the route and made the journey in three hours and ten minutes.[9] This was still too slow for commercial operations.

More modifications were necessary. Because the new hull and stern paddles had improved the speed over that of the 1787 boat, these were retained, but it was decided that a new engine—and with it a new boiler, new air pump, new condenser, and all the other auxiliary parts of a steam power plant—was essential. With new financial backing, however, came a gentleman of ideas and wealth, Dr. William Thornton, who threw himself into the boat and machinery design with great enthusiasm. Once again that quality greatly exceeded his mechanical ability. When completed, practically nothing worked. Numerous condensers were tried and discarded. The air pump was replaced. There were always boiler problems. By the end of 1789 the new boat still was not operable. After a December trial she caught fire and had to be sunk to save the hull. When spring arrived she had been raised; each problem was solved, the eighth condenser really opereated properly[10] and, finally, on April 16, 1790, a reasonable speed was attained during a completely successful trial. A passenger cabin, especially designed by Thornton, was added and on June 14 the steamboat was advertised in the *Federal Gazette*, to take passengers from Philadelphia up the Delaware River to Trenton, New Jersey, on Monday, Wednesday, and Friday, returning the next day, with intermediate stops at Burlington, Bristol, and Bordentown. The fare, all the way to Trenton, was five shillings, or $1.00. At other times, during the summer, she went down river to Chester, Pennsylvania, and to Wilmington, Delaware. Westcott counted twenty three advertisements in various Philadelphia papers covering thirty one trips, adding up to a distance of 1,380 miles.

Fitch, by an enormous expenditure of energy, had overcome endless engineering and financial difficulties and achieved the first steamboat in the world to carry paying passengers. Speed trials had been made over a carefully measured mile along Front Street in Philadelphia and showed eight miles per hour, a remarkable value; in service the speed was considerably lower. If the 1790 drawing of the vessel owned by the New Jersey Historical Society is a correct, rather than a diagrammatic, illustration, the passenger cabin was at the bow, no boiler or stack was shown (the author suspects it was on deck amidships with coal or wood piled alongside) and the engine was located well aft. The vertical piston, through a complicated ratchet arrangement, drove, by a sprocket chain, a transverse crankshaft supported on projecting brackets extending aft of the transom stern. There were three cranks and three "oars" whose basic motion was still similar to that of a canoe paddle. The rudder, steered by a tiller, was forward of the paddles. The scale on the drawing indicates a length of 60′. Fitch's French Patent of 1791, now in the Smithsonian's Watercraft Collection, shows an identical drawing except that the scale is omitted.

At this point eighteenth-century reality must be injected into what seemed a successful conclusion. The steamboat *Experiment*, to apply the name appearing on the drawing, took longer than the stage coaches did between Philadelphia and Trenton. Although her fares were lower, the number of passengers who chose to ride her was limited. Novelty had little or no commercial value. Throughout her one summer of operation she lost money.

Although one additional boat, appropriately named *Perseverance*, was started, she was soon abandoned and nothing further came of this enterprise. None of the individuals involved ever achieved anything additional to improve the design or construction of steamboats. No satisfactory craft was ever again propelled by the unique stern-paddle system. Both the boiler and steam engine installed were so crude that neither influenced any later marine engineering development. In short, the saga of

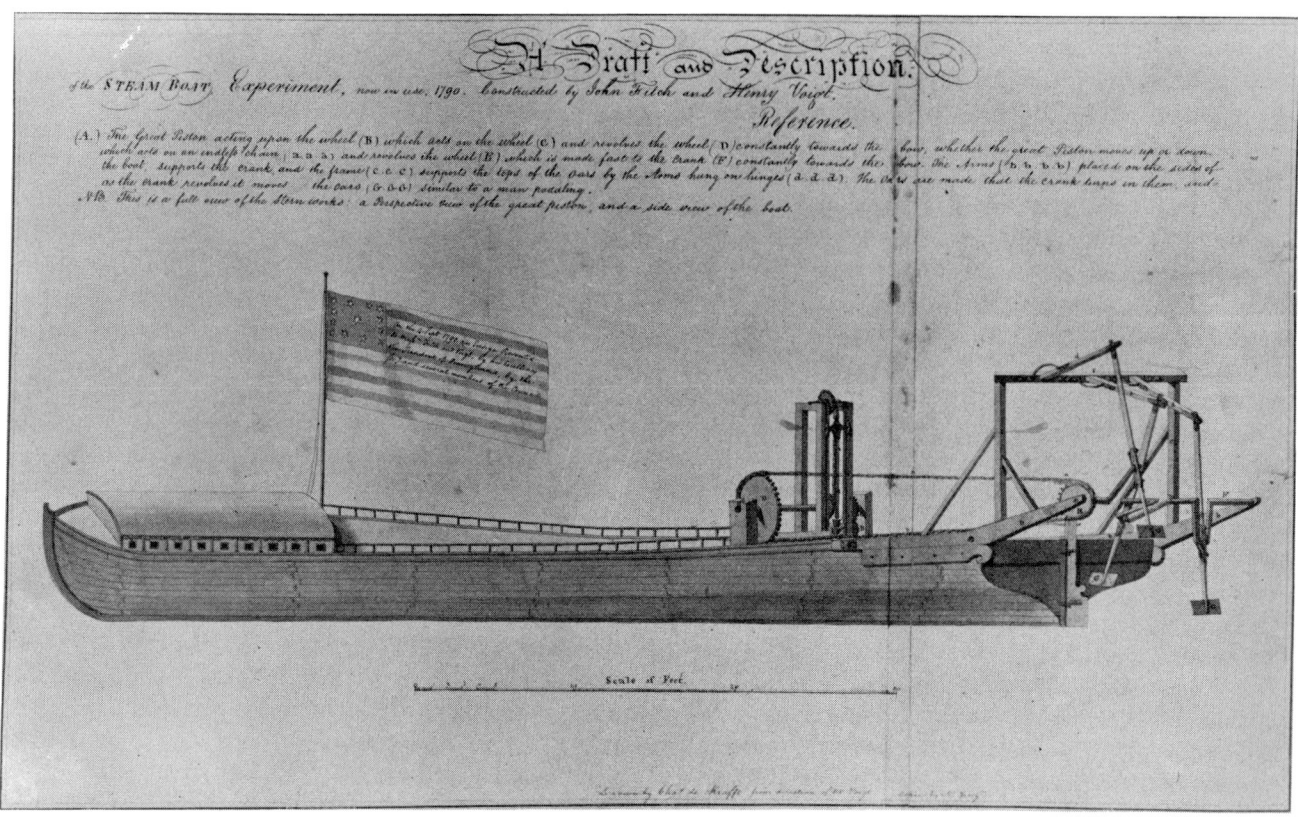

The Fitch-Voight steamboat of 1790. *Courtesy of New Jersey Historical Society.*

the three Fitch steamboats is a neat and complete element in itself. It was not influenced by the design of others and did not pass anything on to posterity except two historical "firsts." The 1787 boat with twelve oars was the first to make a series of trips under steam power in the United States. The third boat, of 1790, was the first steamboat to carry passengers for hire on regularly scheduled trips anywhere in the world. The first was too slow and the last a commercial failure.

James Rumsey

In some respects James Rumsey resembled John Fitch; he was an equally intrepid amateur and an exact contemporary. Both were born in 1743, both conceived of the idea of steam propulsion about the same time, both produced a first crude steamboat in 1787. Thereafter they were bitter enemies, each trying to prove his own priority, often by stretching truth a little beyond its region of elasticity and by deliberately confusing dates and events. In important ways, however, they were very different.

Rumsey, as a housebuilder and an innkeeper on the banks of the Potomac River at what was then Bath, Virginia (now Berkeley Springs, West Virginia), possessed the admirable features of charm, good manners, and a ready tongue. He was able to impress the wealthy, the cultured, and the intellectuals whom he met. They were willing to listen to him and to assist him, whereas they had been repelled by Fitch's crude demands for attention. As a mechanic or an engineer, Rumsey was even less well equipped than his rival. Nevertheless, he was an enthusiastic inventor and conjured up new, and usually impractical, ideas in a number of fields.

Rumsey's first appearance in recorded history was in 1784, when he demonstrated a "stream" boat, or, more exactly, a model of one, to George Washington, who was visiting Bath. This was a catamaran with a water wheel between the two hulls. The wheel, turned by the current, drove a series of poles that pushed against the stream bed (apparently streams were to have flat bottoms of uniform depth) and walked the assemblage upstream. Although a lifting-oneself-by-one's-own-bootstraps device, the model actually did work in a small rectangular sluice with water fed into it from a stream. Washington seems to have been unduly impressed and publicly praised Rumsey's engineering ability.

This stream boat had nothing to do with subsequent steam-powered vessels. The reason for discussing it is that garbled accounts have said that Rumsey demonstrated a *steam*boat before General Washington. Furthermore, Rumsey, in his claims for priority, did keep on mixing his *steam* and *stream* boats together, trying to make his reader think they were identical and not just concurrent projects.

The real work of producing a boat propelled by steam started in May of 1785 when Rumsey[11] employed a mechanic, Joseph Barnes, at £5 per month. Because this was a large wage for that era, Barnes must have been a highly skilled individual. He traveled, in order to locate sources for castings, as far away as Baltimore and, during Rumsey's long absences during 1785 and 1786 when he was serving as a rather inept Chief Engineer of the Potomac Company, which was planning a canal along the river, Barnes had complete charge of the steamboat project. The hull was built near Bath and the machinery installed by December of 1785; the boat was then floated down the river to Shenandoah Falls where Rumsey and his canal crew were at work. About that time ice set in and the engine was removed and stored ashore for the winter. A second boiler was constructed (there is no evidence that the first had even been tried) and trials began in April of 1786.

Many modifications were necessary, the first boiler was installed and failed, just like the second, and, by the end of the year, the vessel was still inoperable. During its second winter the hull was badly damaged by ice. After repairs there were further experiments that included a third boiler failure. But, finally, after two-and-one-half years of development, a public trial was held on December 3, 1787.[12] Horatio Gates, late Major General, Continental Army, and the Reverend Robert Stubbs, the teacher at the Academy in nearby Shepherds-Town, plus lesser citizens of neither the clergy nor the military, provided written certificates to verify the date. Further trials were held on December 11 and a speed of four miles per hour was estimated. Considering the engine and the type of propulsion, such a speed could be obtained only going downstream with considerable aid from the current. Rumsey's success followed that of Fitch and his oared steamboat's public trial on August 22, 1787.

Since neither contemporary illustrations nor dimensions of this launch exist, we can only draw upon a Rumsey proposal[13] of a year later for a larger boat and assume that she was basically similar.

Jet propulsion was utilized and it involved drawing water in through an opening in the bottom of the boat about one-third of its length from the bow; it was then pumped aft through a rectangular trunk on top of the keel and expelled through the stern. The engine consisted of a single-cylinder pump, set vertically just above the bottom opening that admitted the water. On top of the pump came a single-cylinder steam engine of the same stroke. The two pistons, one for steam and one for water, were attached to the same piston rod. It was therefore a simple piece of machinery with the absolute minimum of moving parts. The upward stroke, drawing in the water to the bottom side of the pump cylinder, was accomplished by steam acting on the under side of the upper piston. Then, as the steam was led to a condenser, atmospheric pressure forced both pistons down, squirting the water astern and driving the boat forward. Just as Fitch and Voight had great difficulty getting satisfactory materials and making them fit, so did Rumsey and Barnes. To try to seal the steam piston against leakage, water was pumped on top of it—a system tried and discarded by Watt much earlier. This kept the whole cylinder cool, condensed a great deal of steam in the cylinder when it should have been pushing the piston up, and produced an extremely uneconomical engine. Fitch, to his credit, had used a double-acting cylinder, a separate condenser, and an air pump; in other words, he and Voight had developed, with limited knowledge of their predecessors' work, an engine with the same basic elements that James Watt had incorporated in his designs. In terms of thermodynamic efficiency, Rumsey's engine was successful only in that it operated for two trials. It was the most inefficient design possible.

Jet propulsion has been a favorite amateur project for years. The uninformed see a fish taking in water by mouth and expelling it aft through its gills and conclude that nature has developed this as an ideal way of propelling a boat. It has been reinvented innumerable times. Benjamin Franklin proposed it and hence was extremely sympathetic to Rumsey's plans. He was equally unsympathetic to the ones used by Fitch. In the light of present-day knowledge, we know that this is the least efficient way possible to propel a ship. True, it is very satisfactory for a high-speed airplane, but at 5, 10, or 15 knots it justs wastes horsepower in any land, air, or water vehicle.

Driven by an engine of very low efficiency and hampered by a propulsive system of equally wasteful characteristics, Rumsey's designs could have no possible future. They form, once again, an isolated piece of history that led nowhere. Rumsey, happily unaware of all this, raised considerable sums of money from the Rumseian Society, formed in Philadelphia in 1788, sailed to England, and ordered a really large vessel from a Dover shipyard in 1789. Once more nothing worked properly, despite the more advanced English technology available. Rumsey's death, as a result of a stroke that incapacitated him while he was delivering a lecture to the Society of Arts in London, left her unfinished. The ship was tried by his surviving partners and is said to have made four knots on the Thames in February 1793.[14] She added nothing to the glory of her designer. Legal proceedings in connection with Rumsey's estate declared her a failure.[15]

To recapitulate, Rumsey produced two vessels. The first was a small launch, probably no more than an enlarged rowboat, on the Potomac River that made two successful trials in December of 1787. She had taken over two years of continuous modifications before this limited success was achieved. The posthumous boat made one known trial on the Thames after three years of development work. The author wonders if James Rumsey deserves the 75' tall column erected at Shepherdstown, West Virginia, in 1915 to commemorate his achivements on the Potomac.[16]

There were other experimenters, all on a smaller scale, before 1800. Since none of them achieved even so great a success as the two just discussed in detail, their adventures are not recapitulated here. For many years there have been innumerable claims as to who invented the steamboat, and there will be others. Everyone has a right to his individual hero. The present author's reason for reviewing the Fitch and Rumsey claims is to try and show them in their proper engineering perspective. To a certain extent the idea of a ship genealogy is instructive. The steamship of today is a descendant, improved by innumerable designers, engineers, and shipbuilders, of that of a century ago. And these predecessor craft were not unique; they descended from even earlier steamboats. A really successful design of any era was based on the past elements that worked well, plus correction of ones that worked badly. Both the Fitch and the Rumsey designs were so unsatisfactory in both the commercial and the engineering sense that they had no descendants. Inventive they may have been. They were also sterile.

Robert Fulton and the First Successful Steamboat

The genesis of the first practical American steamboat took place in Paris; it resulted from the meeting of two widely different individuals, each of whom had part of the necessary talent not only to invent one steamboat, but also to create a whole transportation system consisting, eventually, of a steamboat line, a marine-engine works to build machinery, and a repair yard to maintain the boats in running order.

Robert R. Livingston, former Chancellor of the State of New York, Minister Plenipotentiary of the United States to France, arrived in Paris in November 1801. A scion of a wealthy family, a lawyer with many connections and extensive influence, particularly in the New York State Legislature at Albany, he had long yearned to create a steam-propelled vessel. It is to be regretted that, like many others, he had absolutely no engineering ability and was completely unaware of his shortcomings. In 1797 he had joined forces with his brother-in-law, Colonel John Stevens of Hoboken, and Nicholas J. Roosevelt, who had established a foundry at Belleville, New Jersey, on the Passaic River. John Stevens was another gentleman of wealth with a wide range of interests, one of which was the use of steam in any form. While less impractical than Livingston, he was unable to carry out his own ideas—which teemed forth in large numbers. In short, Roosevelt, the mechanic of the trio, was to do the actual work while bombarded by endless letters, each with imperative new instructions. Needless to say, when things went wrong, the two wealthy gentlemen blamed their hard-working mechanic. Obviously he had not carried out their instructions. Roosevelt, freed of their partnership, went on to build, in 1799, two important pumping engines for the Philadelphia Water Works. Designed by Benjamin Latrobe, these were the first large steam engines produced in America. They were also entirely successful.

Robert Fulton, having started his career as a miniature painter in Philadelphia, traveled to London to study under the eminent American artist Benjamin West. A man of great charm, Fulton managed to live abroad for many years, usually without visible or gainful employment. During that time he came in contact with persons having scientific or inventive interests and published a treatise on canals in 1796. A year later, in the midst of the French Revolution, Fulton traveled to France, where he became interested in torpedoes (today they would be called mines) for blowing up ships, and in submarines.

For almost three years Fulton worked on the planning and construction of a submarine, the *Nautilus,* without beng able to elicit financial help from the French Navy under the First Consul, Napoleon Bonaparte. This frustrating episode was completed and the iron hull, which had corroded badly in salt water, scrapped by the end of 1801.

Shortly thereafter Fulton and Livingston met and soon discussed steam navigation. By June of 1802 Fulton had a 3' long model boat built in Paris complete with two clockwork motors and a series of means of propulsion including rotating paddle wheels; an endless chain of paddles, like the caterpillar treads on a tank; and oars or paddles. In addition, various shapes of bows could be fitted to the model. On October 10 Fulton and Livingston signed a partnership deed[17] that would lead in five years to an operating steamboat providing better transportation than any ever before available to paying passengers; best of all, their venture would operate at a profit.

One of the parts of the agreement covered the construction of an experimental steamboat by Fulton in England at a cost of not over £500 to be advanced by Liv-

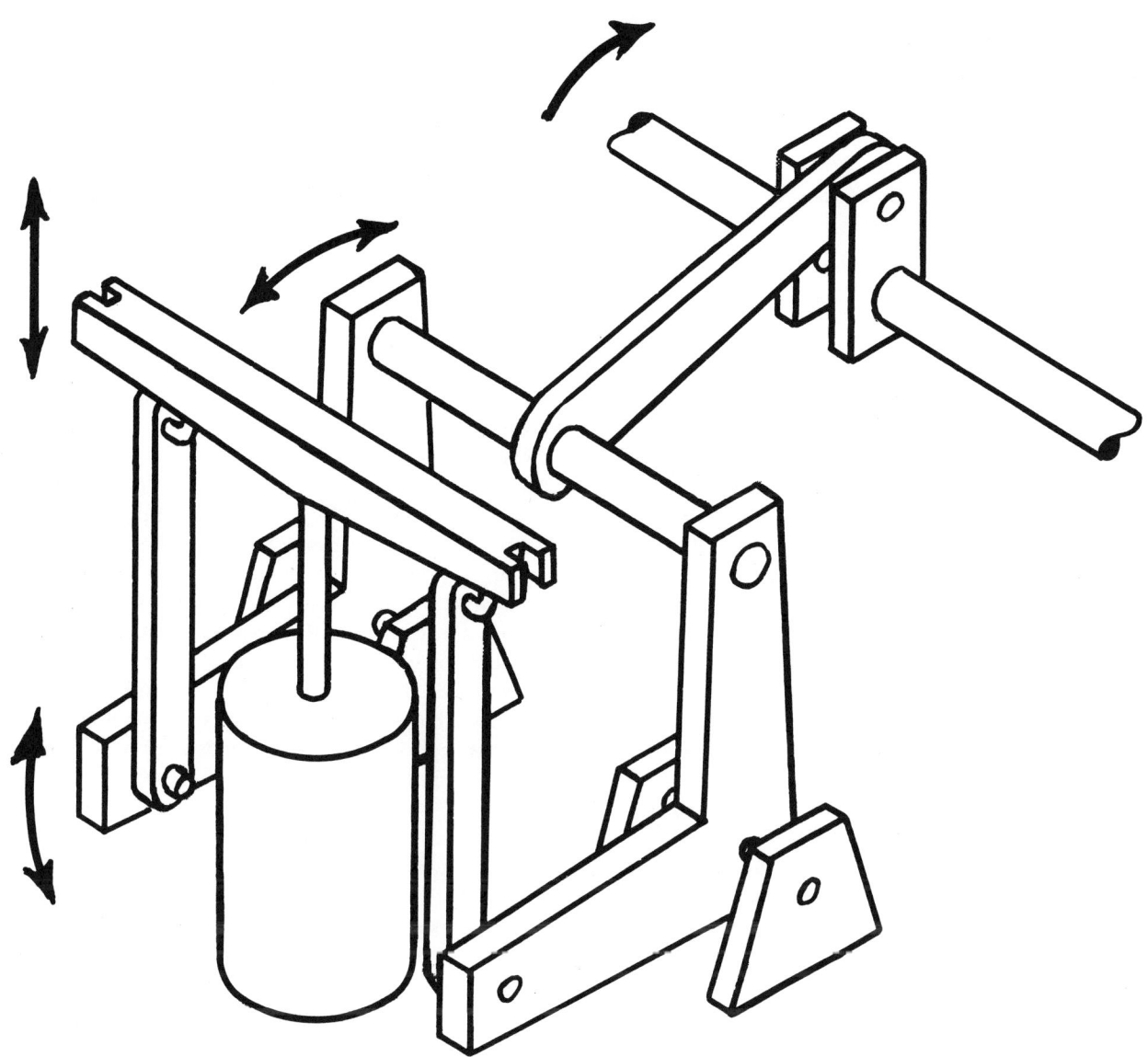

Diagrammatic arrangement of a bell crank engine.

ingston. Livingston, always careful in money matters, saw to it that a further clause provided for Fulton to return £250 over a two-year period plus 7% interest if the craft failed. Livingston already held a monopoly, granted by the State of New York, that gave him the sole rights for operating steamboats on its waters.

The experimental boat was undertaken in France rather than England. Here, at once, we have an engineering approach to the problem. We have already seen the use of a model boat constructed without delay by a professional instrument maker. Experiments[18] with it led to the choice of paddle wheels for propulsion. As a second means of arriving at rapid results, an existing steam engine was hired from Perier Brothers, machinery manufacturers, who had been associated with Fulton's submarine experiments. Fulton took advantage of a country whose technology was more advanced than his own, thereby avoiding the many disappointments of his predecessors who had to design and build their own engines, usually with untrained workmen and a complete lack of adequate tools. The boiler and other machinery items required were constructed by Etienne Calla, the builder of the clockwork model.

If the surviving drawing in the Conservatoire des Arts et Métiers in Paris correctly shows the installation, there is a boiler aft, then a vertical cylinder forward of it with a transverse crosshead on top of the piston rod and bell cranks on either side of the cylinder, with the paddle shaft driven by a horizontal connecting rod from the vertical legs of the bells.[19] To assist the engine to pass through dead center (top or bottom of stroke) a pair of flywheels were geared up to turn twice as fast as the paddle wheels. The wheels are forward of the engine about one-third of the boat's length from the bow. Unfortunately, the drawing dates from January of 1803 and precedes the construction of the vessel; no evidence is available to show how exactly it was followed.

There were difficulties; a virtuoso boiler of complex design proved unsatisfactory. It was replaced by a simpler one of well-tried type. The great error was inadequate strength for the hull which sank to the bottom of the Seine under the weight of its machinery. The latter was salvaged, undamaged, and a new hull of greater size and strength, 74.6' long by 8.2' beam by 3.2' depth constructed. By the end of July Fulton was inviting his friends for trials. These were satisfactory and on August 9, 1803, a public demonstration for a group of distinguished French scientists was made and a speed equivalent to that of a "rapid pedestrian" attained, despite her towing two other boats and going against the current in the Seine.[20]

The first full-scale demonstration took place only ten months after the Fulton-Livingston partnership was signed. Although intended as a preliminary step, the boat was longer than Fitch's Delaware River steamer of 1790, and much larger than Rumsey's only American boat, each of which took several agonizing years to evolve. In one respect it did not live up to its designer's anticipations; it did not achieve the speed that his spring-driven model had led him to expect. This proved a valuable experience, for it led Fulton to make conservative predictions for all subsequent steamers, which in general were to make better-than-predicted speeds. No one is ever blamed for exceeding his promises.

Even before the August trials, the partners were sure of success. Livingston, always sanguine, wrote his friends in the New York Legislature, who passed a twenty-year renewal of his steamboat monopoly provided that an operating steamboat could be produced in two years. Fulton, on August 6, 1803, wrote the firm of Boulton and Watt requesting the price of a 24-horsepower, double-acting engine of 4'-0" stroke[21] and enclosed a drawing showing the way he wanted the cylinder and condenser arranged. Because the governmental license that was required to export an engine to the United States could not be obtained, the firm declined the order.

The lukewarm reactions of Napoleon and his ministers to the long series of submarine and mine experiments negated any financial return to Fulton. Word of these, however, had reached England, and Lord Stanhope, his friend and correspondent for many years, persuaded the Admiralty to negotiate with him to return to England and continue his work to their mutual advantage. After discussions with a British secret agent visiting France under the name "Mr. Smith," an agreement was reached. Although not specifically covered, the subject of exporting a steam engine was, no doubt, a matter of considerable interest to both parties. At any rate, an American calling himself "Mr. Francis" arrived in London in May of 1804 and immediately undertook the procurement[22] of a steam engine from Boulton and Watt.

The latter quoted £386 for a 24" bore by 4'-0" stroke cylinder, its covers, piston, piston rod, valve gear, condenser, and air pump, together with the connecting piping. Fulton asked for two essential changes from their normal construction, a relocation of the condenser away from its usual position and, since the boat would travel through salt water, a brass air pump (to avoid corrosion). The manufacturers rearranged the condenser in a considerably less expensive way than his drawings proposed, provided the brass air pump, and increased the price to £548, an exorbitant charge entirely out of line with a material substitution that made for simpler casting and easier machining. The younger members of the Boulton and Watt families, who were now in charge of the works, were obviously sharp businessmen. Most of the engine parts were based on their standard 20 horsepower engine with its stroke reduced from 5'-0" to 4'-0". They also supplied a sketch of a boiler that was used to order a copper one from the London firm of Cave and Son, for £476-11-2. The engine builder advised that it should be 16'-0" long with a stack of 23" to 26" in diameter. As with brass for the air pump, copper was used to avoid corrosion; it had the great advantage of being easier to bend, roll, or rivet than the rather poor grade of iron plate available at that time. It was less likely to crack and could be soldered to assure tightness.

While this engine was under construction, Fulton proceeded to Portsmouth Dockyard to supervise the construction of "torpedoes" to use against the French fleet; he did not see the engine when it was temporarily assembled by the builders at their Soho works but accepted its parts crated for shipment. It could not be delivered until February of 1805 when ice-bound canals thawed. This time no difficulties were raised and an export license was issued on March 22. There is no record of when or by what vessel either it or the boiler came to New York.

Fulton's work for the Admiralty delayed any progress on the eventual steamer for almost two years. Not until October of 1806 did he sail from Falmouth and it was December before he reached New York. A contract must have been placed promptly, however, with Charles

Brownne, who had just established a shipyard of his own in New York City. Because drawings[23] of Fulton's second steamer, *Rariton*, and third, *Car of Neptune*, still survive, it is almost certain that he supplied the builder with dimensions, arrangements, shapes, and, especially, the location of the heavy keelsons and other timbers required to support the engine, the paddle wheels, and the boiler. Certainly Brownne had no knowledge of these matters and no idea of the weight of all the propulsion machinery. After his disappointment in the speed of the Paris boat, Fulton spent much time and effort on the powering problem—that is, what is the resistance of a hull moving through the water, how much power will be lost in the paddle wheels, and, finally, what horsepower and size of engine must be installed to make a given speed? As a sound, if self-taught, engineer, he made use of what was published and developed a powering method of his own based on models of submerged geometrical objects tested by Colonel Mark Beaufoy[24] for a British organization called the Society for the Improvement of Naval Architecture.

The data from a series of rectangular models with wedge-shaped ends were used and are explained in his 1809 Patent Specification.[25] Those of Fulton's detractors who claim that he did not invent the paddle wheel, the steam engine, or the combination of the two, should note that he was the first designer of ships to regularly use the results of model tests to power his vessels. No other designer for the first two-thirds of the nineteenth century dared try this approach. Moreover, his vessels, after Paris, never failed to meet their predicted speeds.

The powering method strongly affected the design of his hulls. They were given shapes of the same general type as one particular family of Beaufoy models. The steamers had flat bottoms, flat sides sloping out so that the deck was wider than the bottom, and the only ship-shaped portions were the bow and the stern, which did merge smoothly with the rest of the hull. But, for a considerable length amidships, there was no curvature; the flat side met the flat bottom with a sharp corner or chine. This had the advantage of providing ample floor space for the cabins, as well as making the hull both light in weight and easy to build. Here, once again, Fulton, the innovator, had to furnish a drawing to show the shipbuilder the unusual shape he desired.

No contract for or plan of this first Fulton-Livingston steamer survives. It was under construction by March 16, 1807, for Fulton mentions it in a letter of that date to Livingston. She must have been launched soon thereafter, for an account book shows that the engine was carted from a Mr. Barker's warehouse on April 23 to the ship for installation. Since the machinery in the Paris boat worked well, the same arrangement was planned. But the crosshead, its guides, bell cranks, connecting rods, bearings, flywheels, gears, shafting, and paddle wheels all had to be designed and built. Fulton set up a shop near the Manhattan end of the Powles, or Paulus, Hook (now Jersey City) ferry, and the whole project proceeded with great speed for on July 4, 1807, he wrote his partner, "I have the wheels up, they turn admirably."

In six short months a large hull had been built, men and tools assembled, engine parts designed and constructed, and everything fitted together successfully. Preliminary tests[26] on August 9, 1807, tried out two different sizes of floats on the paddle wheels to see which was more effective. Ten days later the craft was ready for public trials.

Before continuing with this first, as yet unnamed, Hudson River steamboat, it may be well to review what we know about her. With regard to her hull and appearance, the answer is very little. Yet numerous artists have painted her and a full-size replica was built for the 1909 Hudson-Fulton Celebration. They are all based on sheer guesswork. Numerous incorrect dimensions have been published, even though they come from excellent sources such as a Fulton letter[27] or an 1817 biography by a personal friend[28] and business colleague.

The really essential document, the vessel's Custom House Enrollment, had been lost for years[29] and its meager information lost with it. It was finally found[30] during the early 1960s. Dated September 3, 1807, and signed by Peter A. Schenck of the Port of New York, it gives dimensions measured on the actual hull of 142' length by 14' width by 4' depth, the latter being measured inside the hull from the top of the floors to the under side of the deck. Her tonnage (approximately the volume of the hull in tons of 95 cubic feet) was 78-71/95, her stern was square, there was neither figurehead nor quarter galleries, one deck was installed, and two masts fitted. No contemporary pictures or drawing exists. No newspaper account of her own day gives any description, but there are three 1850 sources that the author feels give reasonable information. John W. Griffiths, an American naval architect of reknown and the son of a New York shipwright of Brownne's day, gives detail sizes[31] of the ship's structure. The bottom was formed of 1 1/2" thick tongue and groove, yellow pine planks with white lead to seal the joints. Floor timbers were laid across it at 24" intervals and were of 8" x 8" oak in way of the engine and 4" x 8" spruce elsewhere. The side frames were spruce 6" x 8" where they met the chine log (at the junction of the flat side and bottom) and tapered to 4" x 4" at the gunwale. Griffiths mentions an exterior flywheel 10' in diameter, says the paddle wheels extended below the bottom of the boat, and gives the paddle shaft size as 4 1/2" diameter. Later drawings show square shafts (Fulton always called them "axels"). Richard V.

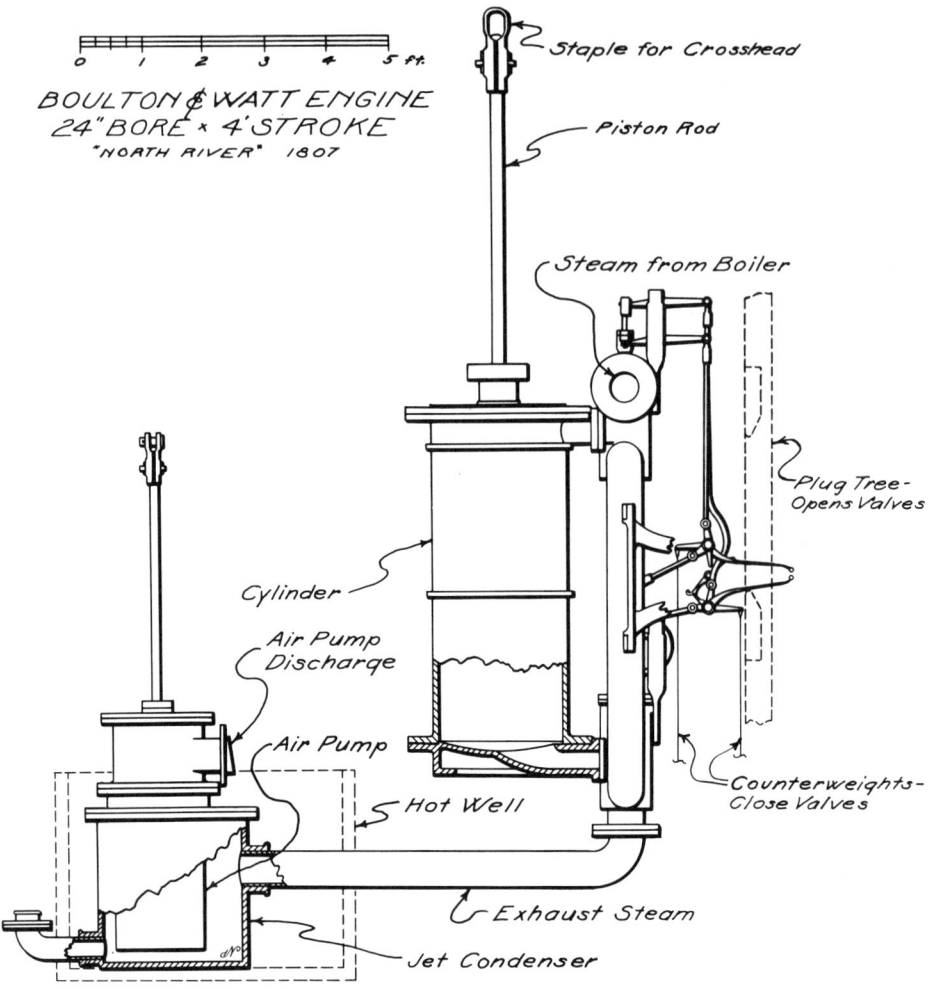

Boulton and Watt engine 24″ bore by 4′-0″ stroke for the *North River*.

DeWitt, who had seen the boat as a youth, confirms in his 1858 description that there were outside flywheels just forward of the paddle wheels and that they were so large that they were partly immersed. Fulton, in his 1810 Patent Specification, refers to external flywheels and says that he once propelled the ship, when her regular wheels were damaged, by mounting paddles[32] on the flywheels. Unfortunately, he gives an 1808 date for the event, which cannot be correct; early that year the exterior flywheels had been replaced by one or more interior ones. The paddle wheels overhung the sides, had no protection around them, and, in conjunction with the faster-turning flywheels, must have thrown up clouds of spray.

Professor James Renwick, an eminent engineer who served for many years on the faculty of Columbia University, visited the vessel during her first year of operation and, long afterward, described her as follows:

> It was decked for a short distance at the stem and stern. The engine was open to view, and from the engine aft a house like that of a canal-boat was raised to cover the boiler, and the apartment for the of-

ficers. In these, by the addition of a few berths, the passengers were accommodated. There were no wheel guards. The rudder was of the shape used in sailing vessels, and moved by a tiller. The boiler was of the form usual in Watt's engines, and was set in masonry. The condenser . . . stood . . . in a large cold-water cistern.[33]

Although our knowledge of the hull is woefully inadequate, two Boulton and Watt drawings give an exact picture of the cylinder and the air pump. The present author has traced portions of the two drawings to show a single view of the assembled engine. While we have it available, the first definite, detailed drawings of an actual steamer's engine, it may be desirable to consider the parts, the reason for their inclusion, and the terminology that will be continuously used in describing later engines.

At this early period steam was supplied only slightly above atmospheric pressure, possibly at two or three pounds per square inch (abbreviated psi). The force developed on the piston could be increased if the steam were applied at the top end of the cylinder and, simultaneously, a vacuum (theoretically 14.7 psi below atmosphere) on the bottom. For that reason, the condenser, which created a vacuum, was an essential element. After steam from the boiler had been admitted by a valve and used to force the piston to the end of its stroke, the boiler supply was cut off and the remaining steam admitted to the condenser by an exhaust valve. The condenser was just a closed, airtight container with a pipe connected to the sea admitting a jet of cold water to condense the steam and thereby create a vacuum. It was descriptively entitled *jet condenser*. The cylinder was double-acting, since both the top and bottom ends were alternately connected to the steam supply and to the condenser, forcing the piston up and down. Rumsey, as we have already seen, used a simpler form of engine with only the under side of the piston being alternately connected to the boiler and condenser; it was, therefore, single-acting. The top end of his cylinder was open to the air.

Although the condensation of steam forms a vacuum, there is always some leakage of air through the space around the piston rod, the operating rods for the valves, and any joints in the system not completely tight. Therefore an *air pump* is essential to clear out such leakage and maintain a continuous vacuum. This pumped the water out of the condenser as well. Since the water was still hot, the chamber (Renwick called it the cistern) into which it was discharged was named the *hot well*. This first engine was unique in combining all these elements. The hot well surrounded the jet condenser, and this, in turn, surrounded the air pump. Finally, another pump, which was not supplied by Boulton and Watt, took part of the water from the hot well and forced it back into the boiler to replace what had been evaporated into steam. This was given the name of *feed pump* or *force pump*. Since sea water was mixed in with the condensed steam, the boiler gradually accumulated the salt left behind by evaporation and, from time to time, had to discharge it overboard (called *blowing down* the boiler) to keep it from caking out as a solid deposit. Excess water from the hot well drained overboard. No drawings of the many parts built by Fulton are known to exist. These parts had to transmit the force from the top of the vertical piston rod to the paddle shaft and must also have included the levers and rods used to work the air pump and the feed pump; they also had to operate the two supply and two exhaust valves controlling the motion. Wheels, shafting, cranks, and bearings all had to be manufactured in the United States. Fulton always used geared flywheels with his engines, which meant further machine work.

Let us now return to the chronology of the vessel. The hull was not entirely complete when the final, and conclusively successful, trial began on Monday, August 17, 1807, with about forty invited guests aboard, most of them friends and relatives of Robert R. Livingston.[34] The vessel, still undocumented and unnamed, sailed from a berth on the North River in the Greenwich Village section of today's New York City at 1 P.M. Exactly twenty-four hours later[35] she arrived at Chancellor Livingston's estate, Clermont, having averaged just under five miles per hour. After an overnight stay she sailed at 9 A.M. and made the forty miles to Albany in eight hours. The next day, Thursday, she steamed, this time with the addition of five paying passengers, to Clermont in nine hours, departed from there at 7 P.M. and reached New York on Friday at 4 P.M. Fulton notes that no sails were used. In command was Captain Andrew Brink, formerly of the Hudson River sloop *Maria*. The engineer is unknown, but he was discharged at Albany (tradition says he was a Scot and celebrated too much). His assistant, Charles Dyke, replaced him[36] and was to serve his new employers for many years thereafter.

After this triumphant demonstration, the boiler and engine were closed in, two cabins with twelve berths in each were erected, and the vessel was enrolled at the Custom House. Her name was *North River Steam Boat* and never during her lifetime was she called *Clermont*. Without any waste of time she was put into commercial service, starting from a Cortlandt Street dock with sixteen passengers, including Fulton, at 6:42 A.M. on Friday, September 4. Advertisements in the *New York Evening Post* and the *Albany Gazette* provided the following information:

New York to:	Fare	Time
Newburgh	$3.00	14 hours
Poughkeepsie	4.00	17 hours
Esopus	4.50	20 hours
Hudson	5.00	30 hours
Albany	7.00	36 hours

Service was soon interrupted by a collision that damaged one of the exposed paddle wheels. After repairs she became an immediate success as far as the traveling public was concerned. On October 1 she arrived at New York with sixty passengers and departed[37] the next day with ninety. Having only twenty four berths available, she was obviously crowded. In other respects this was also a very unsatisfactory trip. Sailing at 10 A.M. against waves, a strong tide, and a gale, she had to anchor and await improved conditions. Next morning she proceeded, but soon collided with an anchored sloop that carried away one paddle wheel. Undaunted, she continued on the other and reached Albany October 4. Again she was repaired, returned to service, and carried capacity crowds. Her luck on Friday, November 13, again left her; she suffered from a broken paddle shaft, but it was replaced within forty eight hours. Twice-a-week service to Albany continued until the end of November, when ice forming in the upper river ended her first season. Yet, with less than three months of service, Fulton reported a profit of $1,000 based on rather optimistic accounting practices.

This is the point where the steamboat became a new and useful means of transportation. It was running between important cities where both commercial and passenger traffic were already extensive. Since the stage coach had to traverse difficult roads and mountainous country, a trip from New York to Albany was long, slow, uncomfortable, and very expensive. For greater comfort the summer traveler went by Hudson River sloop. In addition, these sturdy sailing craft carried all the freight but had to depend on unpredictable winds and could never be scheduled. The *Steam Boat*, as Fulton called her, presented a new mode of travel faster and cheaper than the stage. Undaunted by accidents and delays, she sailed at definite times and dates, as advertised in the local press. And, above all, she proved a pleasant way to travel. Chancellor Livingston had the funds and the influence to smooth her way. Because the scheme led to a financial return, he was doubly interested and willing to make further investments. At last he had the steamboat he had longed for. For the third time he obtained an extension of a monopoly in New York waters for Fulton and himself or their heirs. This was particularly important because New York claimed the entire Hudson River all the way to the New Jersey shore line.

Fulton, after years of impecunious living abroad, turned out to be a successful businessman and a great organizer. To him were left all the practical details. In addition to the design of the vessel and most of its engine parts, he had to acquire shops, find skilled mechanics, and get all the essential items made and installed in the boat. Furthermore, he had to train engineers in her operation, set up the steamboat line, and keep its accounts. Finally, he was to prove a most successful press agent; the newspapers did very little actual reporting of events. They would, however, publish what was sent them and Fulton wrote the necessary newsy letters as well as preparing the paid advertisements covering sailing dates.

It is about at this point, the end of 1807, that we can say that the steamboat had evolved, rather than that it had suddenly been invented. Robert Fulton had played a major share in this evolution. He applied useful knowledge from the past, developed theories of his own, and designed craft that worked without endless periods of adjustment, modification, patching, and improvising. When he made mistakes, he recognized the difficulties and rectified them. In addition, when elements proved successful, he used them over again as a way of avoiding new problems. The arrangement of machinery in his Paris boat of 1803 worked well; he carried it over to the *Steam Boat* of 1807 and would use it again in later craft.

Fulton, then, may not be the inventor of the steamboat; he was, however, the first scientific marine engineer to assemble one. Furthermore, because the contemporary shipbuilders had no background for designing what he needed, he had to assume the task of being his own naval architect. His first full-size boat worked, made money, and served as the ancestor from which all future steam-propelled vessels in the United States were to descend, largely by a continuing process of evolution.

Rebuilding the *Steam Boat*

The continuing development toward better and larger craft was to start at once. The first season had shown a number of defects in the *Steam Boat* needing immediate attention. Fulton wrote Livingston that the hull was too weak and should be scrapped.[38] Then, since the replacement would be heavier and have increased resistance, the boiler should be replaced by one of greater capacity operating at 5 psi higher pressure in order to increase the engine's horsepower. Because of the numerous paddle-wheel casualties, the cast iron

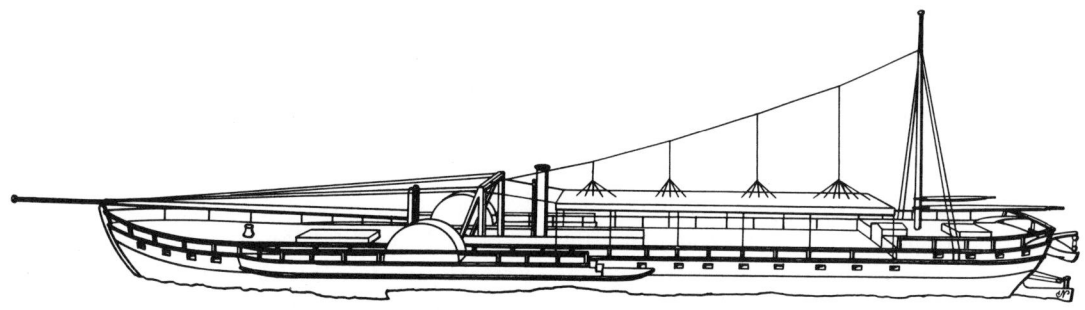

The *Steamboat* as rebuilt in 1808. The foremast is lowered and resting on the stem.

"axels" should be replaced by stronger wrought iron ones, if they could be obtained.

All this was taken in hand and a new boat emerged approximately twice the size of the old one. Livingston chose to assert his inventive genius and insisted on the new boiler being of wood with metal flues and furnaces, a scheme that had worked at a very low steam pressure at the Philadelphia waterworks. The exterior flywheels were replaced with a smaller one placed inside the hull. The side wheels were covered by paddle boxes and protected from damage by heavy timber guards extending outboard beyond the deck. Since an outer bearing for the paddle shaft could now be supported by the guard (in addition to an inboard one at the side of the hull) the likelihood of failure was lessened; furthermore, clutches were installed so that either wheel could be disconnected from the engine to improve maneuverability. The guards were most useful additions; they were led aft and ended in steps leading down to serve as gangways for embarking passengers from small boats alongside. Water closets were installed adjacent to the paddle boxes and firewood for fuel was stored on the guards to keep the deck free for the passengers.[39]

Communication was achieved by mounting a steering wheel on a platform above the engine and just forward of the engineer's station so that the pilot and engineer could see and hear each other. Fulton devised the rig as well as the hull details. He mounted a large mast with yards for a square sail and a topsail forward and a small one aft setting a spanker. In addition there were jibs and lower studding sails. The topsail was small and was set flying without braces. All this was to provide a large deck area amidships, unencumbered by rigging but protected by an awning to keep sun, soot, and rain off the passengers.

When not in use the forward mast was hinged and could be let down on deck with its topmast extending ahead of the bow. When lowered there was less windage and the helmsman had a better view ahead. It might be remarked that this unusual sail plan, the guard arrangement, the steering wheel forward, a steering gear without any tiller encroaching into passenger quarters—all were actual inventions; they had never appeared before on sail-, oar-, or steam-propelled craft.

The enlarged vessel was needed to take care of a reasonable number of passengers on an overnight trip and every effort was made to provide them as much luxury as possible. Aft was a ladies cabin, with twelve berths under a raised quarter deck; then, proceeding forward, a "Great Cabin" with twenty-four berths and, forward of that, another with sixteen more.[40] In addition there were "sophas" in each cabin taking care of a 50% overflow after the berths were filled and making space for a total of seventy-eight sleeping passengers. The great cabin served as the dining saloon and connected with the galley, pantry, and larders, which were situated port and starboard outboard of the boiler room. Needless to say, there was a bar for the gentlemen.

The three contemporary drawings of the rebuilt Steam Boat that survive consist of a Fulton drawing at the U.S. Naval Academy labeled "View of the Steamboat passing the Highlands," a slightly different version of the same vessel in the same locality made by a draftsman as one of the illustrations for Fulton's 1809 Patent Specifications, and an uncompleted, unlabeled sketch, probably by Fulton, bound with some damaged copies of the same patent drawings in an incorrectly dated binding at the New York Historical Society. The first two show the foremast lowered. The third has the mast raised and the vessel at anchor. There was a crutch on the stem to support the mast in its lowered position and to serve as a lead for the anchor cable over the bow. One wonders about anchor-

25

ing when the mast was lowered. The author has traced a copy of the Patent Specification drawing belonging to the American Society of Mechanical Engineers and has corrected the view's erroneous perspective.[41] It is the most detailed of the three.

The revised machinery drawings, if they were ever made, have not survived. It is to be regretted that numerous authors have published the patent drawings, several copies of which survive in both the United States and England, or generally similar ones from the New Jersey Historical Society, and claimed that these represent the engines of the so-called *Clermont*. None of these shows the unique air pump, within a condenser, surrounded by a hot well, which Boulton and Watt supplied. Instead, the cylinder is shown on top of a condenser, and the other items are completely separate, not combined units. The author believes that the 1809 drawings were for the *Car of Neptune*, which was being engined at the time they were delineated. Making all the elements individual and separate led to simpler casting and machining, most important considerations when Fulton was manufacturing his own machinery without the experienced foundrymen that Boulton and Watt had available. The other parts, however—the vertical cylinder, an overhead crosshead operating on guides supported by a wood frame, bell cranks with counterweights to balance the piston's weight, and a geared flywheel—are still similar to the Paris steamer and it is reasonable to assume that the *Steam Boat* had a like arrangement. Fulton, as already stated, tended to retain any trouble-free elements in his designs.

The new hull, new boiler, new interior, rearrangement of flywheels and related machinery items were accomplished with remarkable speed and a new steamer with the original cylinder and air pump was enrolled at the Custom House on May 14, 1808. Since, however, the first document was surrendered and a new one issued on replace it, the legal fiction was established that it was the same boat. It was not. The name of *North River Steam Boat* remained, but "of Clermont," the new port of registry, was added. Posterity has reversed all this and insists on the vessel being named *Clermont*. Fulton called her the *Steam-Boat*, the *Steamboat*, or the *Steam Boat* as long as she was the only such craft on the Hudson. After the *Car of Neptune* came out, he referred to her in his advertisements and accounts as the *North River*. A comparison of figures from the two enrollments shows that no single dimension of the 1807 *Steam Boat* carried over into the 1808 version.

Date	Length	Beam	Depth	Tonnage
1807	142'	14'	4'	78-71/95
1808	149'	17'-11"	7'	182-48/95

The first voyage after rebuilding was a fiasco. The wood boiler was leaking steam at every seam; it took 48 hours to reach Albany and 56 hours returning.[42] Nevertheless, the disaster had beneficial results; from this point on Livingston left engineering matters to his partner. One cannot but be impressed by the speed with which Fulton could act. The design, construction, and installation of a copper boiler took only two weeks; sailings for Albany resumed on June 1. Perhaps Fulton had managed to start the new boiler secretly. Thereafter the *North River* was scheduled to sail from New York on Wednesdays and from Albany on Saturdays with regular stops at West Point, Newburgh, Poughkeepsie, Esopus, Red Hook, Catskill, and Hudson. All of the way passengers were landed or picked up in the river by rowboat. The through fare of $7.00 included the use of a berth, sixty pounds of baggage, and four meals served at specified hours. The passage time for the 145-mile trp was conservatively quoted as thirty five hours at a speed just above four miles per hour. In practice, a favorable wind and tide occasionally led to one as short as twenty four or six miles per hour. Normal times were in the thirty-to-thirty-five-hour range. When the river froze and the steamboat season ended, the stages took over for the winter months; their fare was higher, $10.00, and the trip lengthened to fifty seven hours or more.

Successors to the *Steam Boat:* The *Phoenix*

Colonel John Stevens of Hoboken, once the *North River* had proved a success, undertook the construction of a steamboat of his own for commercial, rather than experimental, purposes. Like Livingston, he was a wealthy gentleman-scientist interested in a wide range of things from horticulture to transportation. He corresponded widely, expressed his scientific views freely, and considered himself a highly successful inventor and innovator, particularly in the field of steam engines, pumps, and boilers, and had patented numerous "improvements" in these. It should be strongly emphasized that most of them were paper improvements and could not be achieved with the materials and methods available to him. Since his ideas were expressed and preserved on paper, posterity has rated him as a great, pioneering engineer. In reality, he was somewhat like Livingston in that he provided the financial backing that would lead to the development of a successful steamboat once he could find a competent engineer to incorporate some of his sounder ideas into a functioning object and to ignore the far more numerous impractical ones. He did, however, propose more realistic solutions than did Livingston, his brother-in-law and rival.

The man who served as a "Fulton" to John Stevens was his son, Robert Livingston Stevens. He was only seventeen in 1804 when the first successful family project, a launch called *Little Juliana* (after Robert's sister), steamed around New York harbor.[43] He was her engineer; his brother, John Cox Stevens, was her helmsman. She was highly original and extremely simple. There was a vertical engine of 4 1/2" bore by 9" stroke driving twin screws geared together. It operated at a high pressure, 20-40 psi. The engine arrangement eliminated the need for crosshead guides and resulted in a small number of working parts. The high-pressure steam produced a large force on its small piston without the necessity of a vacuum. As a result, the cylinder could exhaust to the air and did not require the installation of a condenser, air pump, or hot well. The boiler, one of the elder Stevens's too-advanced ideas, had very little space for water but made steam rapidly in a series of projecting tubes. Unfortunately, the brackish harbor water deposited salt and chemical scale in these at the high temperatures associated with the elevated steam pressure and troubles soon developed. Both the engine and boiler were preserved, rebuilt, and operated again in a new hull in 1844; today they survive in the Smithsonian Institution in Washington, D.C. There were other small-boat experiments, both before and after this, and a workshop had been set up on the Colonel's estate at Hoboken, New Jersey, to assemble their engines from parts built by others.

When it came to a real steamer, not just a launch, none of the *Little Juliana*'s features was retained. Instead, based on our limited knowledge of the eventual vessel, she seems to have been given most of the elements of the *North River* as rebuilt in the winter of 1807-08. It is likely that Robert Stevens was responsible for this very intelligent choice; it had previously been uncharacteristic of his father to agree with, adopt, or follow any one else's ideas. The talented son must have exerted a restraining influence, for the craft appeared as a two-masted vessel driven by paddle wheels and engines that were supplied steam from a low-pressure boiler. Since neither drawings nor detailed descriptions of the hull or machinery are available, our knowledge of her leaves much to be desired. Furthermore, there are extensive gaps during her first two years of existence; we do not know whether she was running, tied up, or being altered. Many accounts have erroneous dates; those who back Stevens as the inventor of the steamboat have had her started as early as 1806 and in operation a week or two after the *North River*.[44] Others have had her carrying passengers out of Philadelphia a year before she ever arrived there. These dates are fabrications.

The contract for the hull, eventually to be named *Phoenix*, was dated January 7, 1808, and specified a length on keel of 100', a molded beam of 16' and a depth from top of floor timbers to the underside of deck planking of 6'-4". This makes her considerably smaller than the original *Steam Boat*. It was to be completed in three months' time for the small sum of $900.00, with the elder Stevens supplying all the materials.[45] Nathaniel Sayre, William Gailer, Joseph Morgan, and William Williams were the builders and John Floyd supervised the work. No joiner work (interior partitions or furniture) or wood carver's work was included in the price.

Like many vessels before and after her, the *Phoenix* was not even launched by the specified delivery date. Colonel Stevens advertised in the *New York Evening Post* inviting the public to attend her Hoboken launching on April 9, 1808. He not-too-modestly calls her a "beautiful steamboat" and says, "for model and workmanship, she is allowed by judges to be the superior of any in the country." There was only one other steamboat and it belonged to Fulton and Livingston.

In a letter to Livingston on July 23, 1808, Stevens noted that the hull and machinery were nearly finished and asked for permission to operate her from New York to Albany.[46] Because the chief asset of the *North River*'s owners was their monopoly in New York waters, nothing but acrimonious correspondence followed and no agreement between the brothers-in-law, each wanting to be publicly acknowledged as the man really responsible for producing the steamboat, was ever possible.

Since the vessel was never enrolled at the New York Custom House, we have no official completion date, no set of dimensions, no official description. The one report of her operation comes from a Stevens advertisement trying to raise funds to build another steamboat (actually to pay for the one already built) and start a service, partly by water, from New York to Philadelphia. In this the Colonel describes a trip on September 27, 1808, from Hoboken to Perth Amboy, New Jersey.[47] Because of several delays for "adjusting various parts of the machinery," it sounds as if it must have been one of her early trials. There were also adverse winds and tides that made it a very slow passage of 6 hrs. 45 min. Returning things went better and the estimated 30 miles to Powles Hook, New Jersey, was completed in 5 hrs. 26 min. at 5 1/2 miles per hour. Apparently this was considered a satisfactory demonstration of completeness, for the vessel was, at long last, given the name of *Phoenix*.

From that September 1808 date until the late spring of 1809 nothing is known of her career. One of Stevens's letters says that she had been operated for several hundred miles during the summer but does not mention any date or route.[48] The long silence leads one to suspect that her owner was involved with other of his many affairs and tied her up to save the costs of continued operation. Although he constantly claimed the right to operate on

New York State waters despite the legislative grant to Livingston and Fulton, he did not choose to risk bringing the matter before the courts for a legal test.

About the time of the completion of the *Rariton*, second vessel to be constructed under Fulton's design and supervision, the Stevens steamer, without fanfare, left the Hudson for the Delaware River. Since this involved a considerable voyage on the open sea down the New Jersey coast before the shelter of Delaware Bay was reached, the project was approached with the utmost caution. A schooner was hired as an escort and there were extensive delays waiting for the best possible weather. Robert L. Stevens was on board as both captain and engineer. His log, a very laconic document, survives.[49] Moses Rogers, later master of the *Savannah*, has been credited with her command, but the Stevens family papers do not record his presence until after her arrival at Philadelphia.[50]

On Saturday, June 10, 1809, she sailed from Hoboken but after two hours anchored at Quarantine, off Staten Island, where she remained until Monday noon. When she did get underway, a heavy wind and sea damaged the starboard paddle wheel. Next day, while passing Sandy Hook, the port wheel gave way and she had to be warped into Cranberry Inlet for repairs. Both injuries appear to have been to the wooden buckets (or paddles) attached to the wheel arms and were easily repaired. Bad weather kept the *Phoenix* at anchor until the 17th, a whole week after leaving Hoboken. The first real progress at sea proved to be a fifteen-hour passage to Barnegat Bay. Here she stayed for four days sheltering from a gale. On Wednesday the 21st she again steamed for almost fifteen hours and reached Cape May, where she anchored for the night. Thursday she proceeded to Newcastle, Delaware, and on Friday ended the trip at Philadelphia thirteen days after leaving Hoboken. The schooner escort proved useless and was left behind.

To the *Phoenix* belongs the honor of being the first steamboat to make a voyage by sea, even though it was a very slow and cautious one. Since the distance from Sandy Hook to Cape May, approximately one hundred twenty five land miles, was completed in thirty hours of steaming, she achieved a speed of four miles per hour under favorable weather conditions. At Philadelphia Moses Rogers became her master and Robert Stevens her engineer. A letter by Henry Voight, former partner of John Fitch but now chief coiner of the United States Mint at Philadelphia, shows that she made an eight-hour trip from Philadelphia up the river to Trenton, New Jersey, on July 9 and returned, also in eight hours, the next day.[51] Cabins were installed, the ship fitted out for passengers, and in August a regular service was undertaken.

The *Phoenix* became the Delaware River successor to Fitch's and Voight's 1790 craft, but was not her descendant in any way. Her hull, machinery, boiler, and side wheels were all developed independently. In one respect there was a similarity. She was not a commercial success. Excellent roads existed over the flat country from Philadelphia to upriver ports, and stage coaches could make the thirty-five-mile trip to Trenton more rapidly than the steamboat; as a result, she did not offer the great advantages that the *North River* could when compared with land competitors who had to face streams, mountains, and a long, dismal road trip.

During her first season, 1809, the *Phoenix* started from Philadelphia on Monday, Wednesday, and Friday for Trenton and returned the next day.[52] On Sunday she went to Bristol, Pennsylvania, via Burlington, New Jersey, and returned the same day. Advertisements appeared only once a week and the schedule was an awkward one with a different sailing time each day. It is concluded that this was to take advantage of the tide and current. The fact that such an unusual arrangement was adopted implies an acceptance on the owner's part that his steamer was not fast enough to cope with the normal forces of nature.

After the start of this service on August 9, Burlington, Bristol, and Whitehill were added as way stops. Breakdowns must have occurred for there were no advertisements on August 29 or September 19 among the Tuesdays regularly employed for announcing schedules. Occasionally she headed down river to Chester, Pennsylvania, on Sunday excursions. On October 11, when the advertisements ceased, it is assumed that the service ended. It would seem that 1808 and 1809, her first two years, saw only two actual months of commercial operation. The very brief newspaper announcements gave no fares, no passage times, no information about meals. Certainly there was very little to attract the traveler. The Stevens family were inept as press agents. Like many a wealthy individual, John Stevens was remarkably parsimonious about small expenditures. Robert L. Stevens did not dare to prepare or insert an advertisement without permission from his father.

It is one of the ironies of history that a Stevens boat, the *Phoenix*, and a Livingston one, the *Rariton*, had to combine their operations and that Stevens had to adjust his sailing schedule to suit those of the *Rariton* before either could achieve commercial success.

The *Rariton*

The second American steamboat to be designed by Fulton was owned by two members of the Livingston family, John R. and Robert J. Livingston, who agreed to pay Fulton and Robert R. Livingston one-sixth of the gross income from the vessel in return for permission to operate on the waters of New York State. The design and construction were carried out just as if she were for the senior partners; the Chancellor advanced the money and Fulton handled all the details until the boat was finished. When completed, she was transferred to her final owners before her enrollment on July 6, 1809, at the Custom House. Named *Rariton* (not *Raritan*), she was 124'-0" long by 21'-0" beam by 6'-8" depth of hold, of 163 tons, built by Charles Brownne, and ornamented with an eagle figurehead. Fortunately, an original plan of her hull survives at the New-York Historical Society; it is the earliest authentic steamship drawing extant.

She had many of the unique Fulton characteristics. The cross section shows a trapezoidal form with a flat bottom, sharp chines, and sloping sides. Also indicted is a semicircular, trough-shaped keel, which was used to collect the bilge water and lead it to a pump. There is a ladies' cabin aft, an entranceway with stair, then the main cabin, going forward, with its bar. The boiler, as the heaviest single item in the ship, is nearly amidships. Forward of that is the cylinder and then the paddle wheels. The forward mast comes down right in the middle of the engine between the port and starboard bell cranks. Only someone like Fulton, who planned both the hull and the machinery, would have chosen such a location. In later ships the different elements supplied by the shipbuilder and the engine builder would have been kept apart to avoid interference. Forward are the crew's quarters and a "kitchen" in an out-of-the-way and inconvenient location. The plan shows quarter galleries aft but her official document states she had none. The total cost was around $26,000.[53]

The *Rariton* was intermediate in size between the *Phoenix* and the enlarged *North River*. Her engine was of the same type as the latter's, this time of American manufacture, but of unknown size. Major suppliers, as noted in the surviving account book, are Messrs. McQueen and Smallman for castings and iron work.[54] Payments of $3,000 to the former and $2,450 to the latter are listed. For the boiler, $4,500 is listed to Paul Revere for copper and rivets and an additional $1,200 to a coppersmith named Bennett for labor.

The *Rariton* made her maiden voyage from the Battery to Amboy, New Jersey, on July 8, 1809, just one day ahead of the *Phoenix*'s first trip up the Delaware to Trenton.[55] Going, she took five and one half hours and, returning, four hours, corresponding to speeds of four to five miles per hour. Thereafter, her route was extended around the western side of Staten Island to Elizabethtown Point (later Elizabethport), then Perth Amboy, and up the Raritan River to Brunswick (now New Brunswick). Trips were made on Mondays, Wednesdays, and Fridays, followed by a return the next day. On Sundays the run shortened to a round trip to Perth Amboy. The fare was 12 shillings, or $2.40, all the way to New Brunswick.[56]

Fulton's account book lists 2,404 passengers carried in 1809, half of them to Perth Amboy, with the total of $2,675.25 taken in. Evidently this sum was not sufficient, for John Livingston, in a letter to Mahlon Dickinson in 1813,[57] wrote that he offered the *Rariton* for sale after being nearly "annihilated" but no one would buy her. Only after she was made a part of a connecting service with the *Phoenix* did she prove profitable. The two Livingstons tried much harder than Stevens did to attract the public. Advertisements of considerable size, in both New York and Philadelphia papers, gave fares; offered to take children at half price; listed breakfast, dinner, and tea on board "more agreeable than in many road houses." As a convenience, the *Rariton* sailed at the same time every day with a passage time of six to eight hours from New Brunswick to New York. Not only was the reader informed, he was also enticed by the "elegant and easy style of travelling."[58] From omission of any advertisement, service seems to have been suspended for the winter about the 19th of December. The *Rariton*'s 1809 season lasted five months to the *Phoenix*'s two; she was a far superior vessel.

The *Rariton*, perhaps to achieve fame, suffered the first recorded boiler explosion on a passenger-carrying steamer. At first thought this would seem more disastrous than it actually was. She had stopped for passengers at Perth Amboy while returning from her second trip to New Brunswick.[59] Several passengers and one of her proprietors, John Livingston, were standing by the boiler when clouds of steam started rising around it. The actual failure was a small rupture, near the bottom, and the chief result was that boiling water ran out into the hull. It must be remembered that, since steam pressures were a few pounds per square inch, no catastrophic hazard with pieces of metal flying through the air was likely.

Only one person was below deck at the time, a seaman discharged for drunkenness, who was sleeping it off in the crew's quarters. He would have been safe had he stayed put but, instead, he jumped out of his berth into the boiling water, fell down, and was so severely scalded that he died several days later. The other casualty was the engineer, James Law, who was on deck but, going below to invetigate, had his feet burned. Livingston blamed Law for not lifting the safety valve as soon as the engine was stopped and allowing the steam to blow off. The engineer admitted his fault and was replaced, the boiler was quickly repaired, and service resumed. Today the blame seems unfair. Either the safety valve did not function as it should have, or the boiler was defective. Yet, for many years to come it was standard practice to lift safety valves manually and blow off steam whenever a vessel stopped.[60]

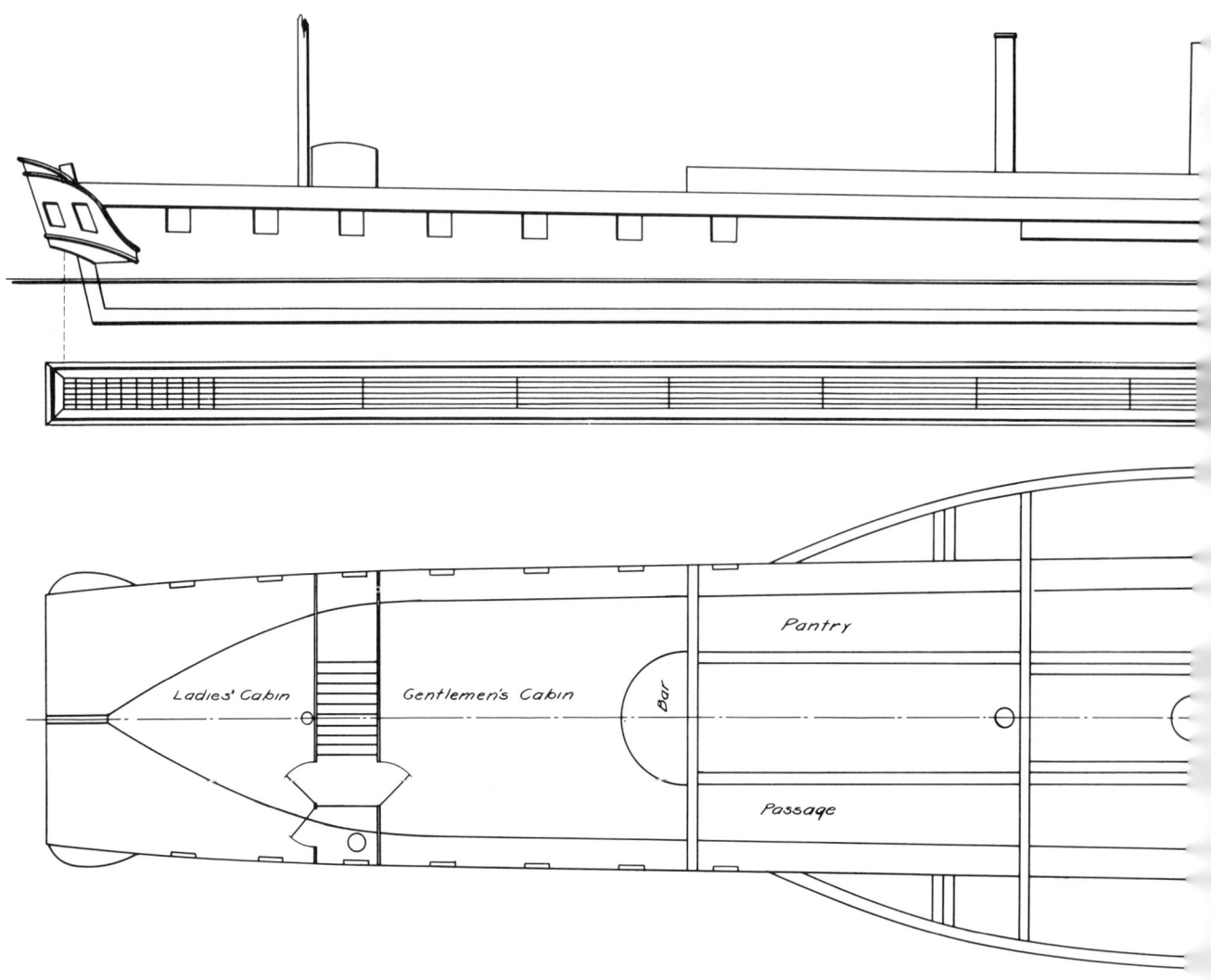

Arrangement of the *Rariton* of 1809.

Alternate Routes from New York to Philadelphia

From November of 1809 to May of 1810 there were no advertisements of the *Phoenix* in the *Philadelphia Aurora*. The winter was utilized for extensive reconstruction of the boat in almost every respect. The original two-cylinder engine was replaced by a single-cylinder Rogers, to Colonel Stevens in March of 1810 shows that a new boiler was awaiting installation at Bordentown.[62] It was a complex design, having two water drums side by side and a third on top, all with internal flues but an external furnace of fire brick. Since the father and the son were unhappy with the hull, it may have been considerably altered. Certainly there were changes amidship, for there is mention of mounting the paddle shafts at a higher level. There was considerable strengthening to try to avoid the serious vibrations that had been evident. A portrait at the Mariners Museum shows her after the rebuilding and bears Captain Rogers's name painted on the paddle boxes.

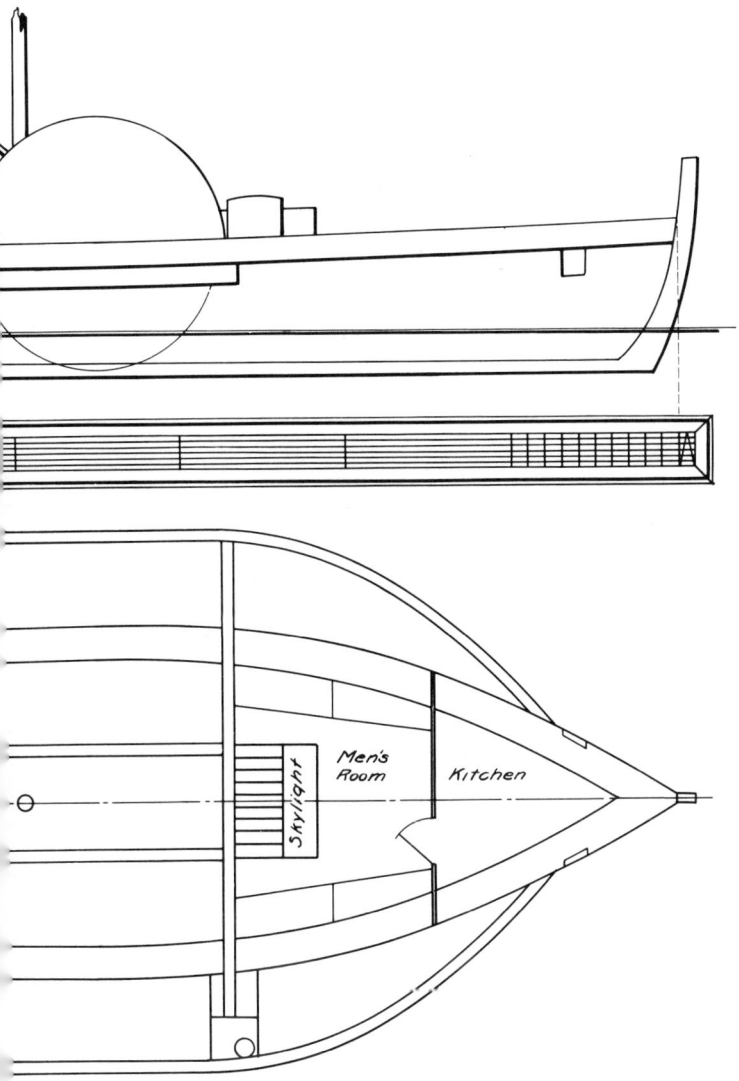
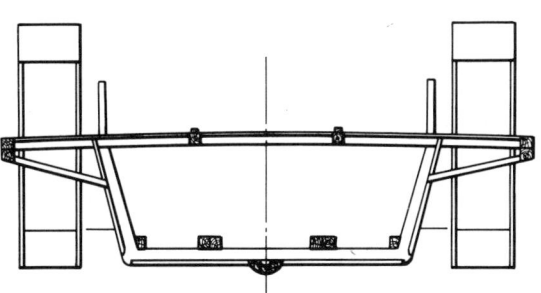

—Rariton—
1809

After a drawing by Robert Fulton
Length between Perpendiculars 119'-9"
Breadth Moulded 21'-0"
Depth of Hold 6'-8"
Cedric Ridgely-Nevitt   January 1941

In appearance she was still a smaller edition of the *North River*. The engine was forward of the boiler, the tall forward mast had a large square sail, and the smaller mizzen had a spanker. The passengers were surrounded by an open rail and protected from weather and sparks by an awning. The guards, although much smaller than those on a Fulton steamer, were still used for the storage of firewood. Only her complicated three-drum boiler illustrated John Stevens's propensity for something radically new.

The second engine and boiler enabled her to steam on a fixed schedule and no longer did she have to wait on the tide and a favorable current. Unfortunately, the Stevenses became more reticent than ever about announcing her sailings. The first advertisement mentioning her as being in operation is one in the *Philadelphia Aurora* giving the *Rariton*'s schedule. The latter steamer had started service near the end of April. On May 4, 1810, she was noted as leaving New York from the Battery every day at 6 A.M. for Elizabethport and Perth Amboy. Stages would take passengers on to Philadelphia via Trenton or to Bordentown where they could proceed via the "steamboat line" to Philadelphia. This was, of course, the *Phoenix*.

John Stevens's *Phoenix* after extensive alterations and the installation of a second engine. *Courtesy of The Mariners Museum, Newport News, Va.*

In June the *Rariton* went back to her 1809 schedule of steaming all the way to New Brunswick three days a week and returning the next day. From New Brunswick there was a "new and elegant" line of stages to Trenton and Philadelphia.

The possibility of collaboration between boat and stage impressed the owners of the latter for, on July 24, 1810, there appeared a joint announcement bearing the names of seven different stage lines listing three alternate ways of reaching New York, all starting with the *Phoenix*, sailing from Philadelphia at 2:30 P.M., Monday, Wednesday, and Friday for Bordentown, where the coaches would await her arrival in the evening. For a through fare of $4.25, a passenger could continue by road to New Brunswick, traveling all night, to catch the *Rariton* to New York. Less expensive routes went to Elizabethport and South Amboy, where the passengers boarded sailing packets for the city. Everyone advertised except the owner of the *Phoenix*.

A month later, on August 9, a new advertisement was finally inserted by Stevens, listing sailings at 7 A.M. three days a week to Bordentown, with stages to New Brunswick connecting with the *Rariton*. The new departure times made it far more convenient for the traveler. He could have an overnight rest at New Brunswick, northbound, and Bordentown, southbound, to recover from the twenty-five-mile stage journey. Neither vessel had to sleep passengers aboard. The *Rariton*'s owners were still more lavish with their announcements of the same service and advised their customers to avoid "dangerous ferry, mud, dust and musquetoes" by means of this "elegant and easy style of travelling."

August 1810 marks the beginning of the Stevens-Livingston cooperation. At last one could journey between the two largest cities of the United States in reasonable comfort and at a reasonable cost. The combination of two boat owners plus the intervening stages, all scheduled for their mutual benefit, proved a success. For the second time on American waters, steamboats wee in their element, providing a new and useful service on a well-traveled route. Starting in 1810, Albany, New York, and Philadelphia were brought closer together by four steamboats, the *Phoenix, Rariton, North River,* and the *Car of Neptune.*

Developments on the Hudson

When we followed the *Phoenix* on her 1809 coastwise voyage to the Delaware we left the *North River* at the end of her second summer. She had continued as long as conditions permitted and, when the Hudson froze, was laid up for a winter overhaul. But, since the 1808 season had been longer and even more successful than her first, a companion vessel was ordered. Both the boat and her engine were under way simultaneously with the *Rariton*.[63] The hull, as usual, was ordered from Charles Brownne, on October 1, 1808.

Her specifications give a length on keel of 157', length on deck 163', extreme beam 22', and breadth of bottom 18'. The flat bottom and three feet up on the sloping sides were to be planked with 3" tongue and groove pine; the upper planking was reduced to 2" thick. Every third or fourth frame was to be white oak, the intermediate ones pine, and all heavily stressed parts supporting the engine were of oak. The cost, including all the cabin fittings, capstan, rudder, and steering gear was $8,200. The plan supplied the builder has not survived, but Fulton's earlier design for her is preserved in two unlabeled drawings in the New Jersey Historical Society's collection. Here the dimensions are 176' on deck, 167' on the keel, with respective beams of 22' and 18' at deck and bottom. The ship, as finally built, had dimensions of 169' by 25'-2" by 7'-3" depth, with a tonnage of 295. Fulton had second thoughts on the subject of beam and increased it.

The detail arrangements shown resemble the rebuilt *North River*. Going from aft forward we have a ladies' cabin with sixteen berths, an entryway with the captain's room to port and water closets to starboard, then a large cabin for gentlemen with twenty berths extending as far forward as the boiler room. There is an extensive galley on the starboard side of the machinery space equipped with ovens, stew pans, a tea boiler, a buttery, and a serving table. On the port side are the coal bin (for the stove, not the ship's boiler) and the steward's room. Forward of the boiler are the fireman's and engineer's station, the cylinder, the bell cranks, and the paddle shafts.

On centerline, ahead of all the machinery, is the crew's mess room, flanked by their berths port and starboard. At the bow comes another cabin with twenty berths and a semicircular bar having ample space for bottle storage under the stair leading down to it, and, in the triangular-shaped bow, a second ladies' cabin containing eight berths. All berths are two high, except in the captain's room. Additional sleeping space is provided by twenty four "sophas" in the various cabins. All this adds up to eighty eight spaces for passengers, eleven for the crew, or a grand total of ninety nine. There are large skylights over all but the ladies' cabin in the bow, which must have proved dark and noisy from the slap of waves on the water side of the thin planking.

For machinery the ever-busy Fulton had completed work shops near Powles Hook (Jersey City), not only to build engines from his own design, but also to do the necessary repair work during winter lay-ups. A former employee of Boulton and Watt, Charles Stoudinger, was engaged as foreman. He had also worked for Nicholas Roosevelt. Iron castings were obtained from Robert McQueen and brass ones from James P. Allaire. The author is convinced that the bell-crank engine delineated in Fulton's 1809 patent drawings represents the engine arrangement for this, the second Hudson River steamboat.

In comparison with the *North River*, the cylinder's diameter was increased from 24" to 33" and its stroke from 4'-0" to 4'-4"; the horsepower would be about doubled. The boiler shell consisted of 1/4" thick sheets of copper bought from Paul Revere. James Bennet constructed it from Fulton's designs, which gave dimensions of 18' long by 8' wide.[64]

All the above material demonstrates that the *Car of Neptune*, completed in the fall of 1809, was an enlarged *North River* in every respect. She even emulated her predecessor in one unsatisfactory feature. The *North River*, as we have seen, had a series of accidents to her paddle shafts, the largest and most difficult elements to manufacture without heavy, steam-powered machinery. The *Car of Neptune*, less than a month after her completion, reached New York on October 8, 1809, with a passenger list of seventy, to be temporarily withdrawn for replacement of defective shafts.[65] The same problem waa to plague paddle steamers throughout their heyday and would reappear years later in screw steamers that would suffer propeller-shaft failures.

As a larger and faster steamer than the *North River*,

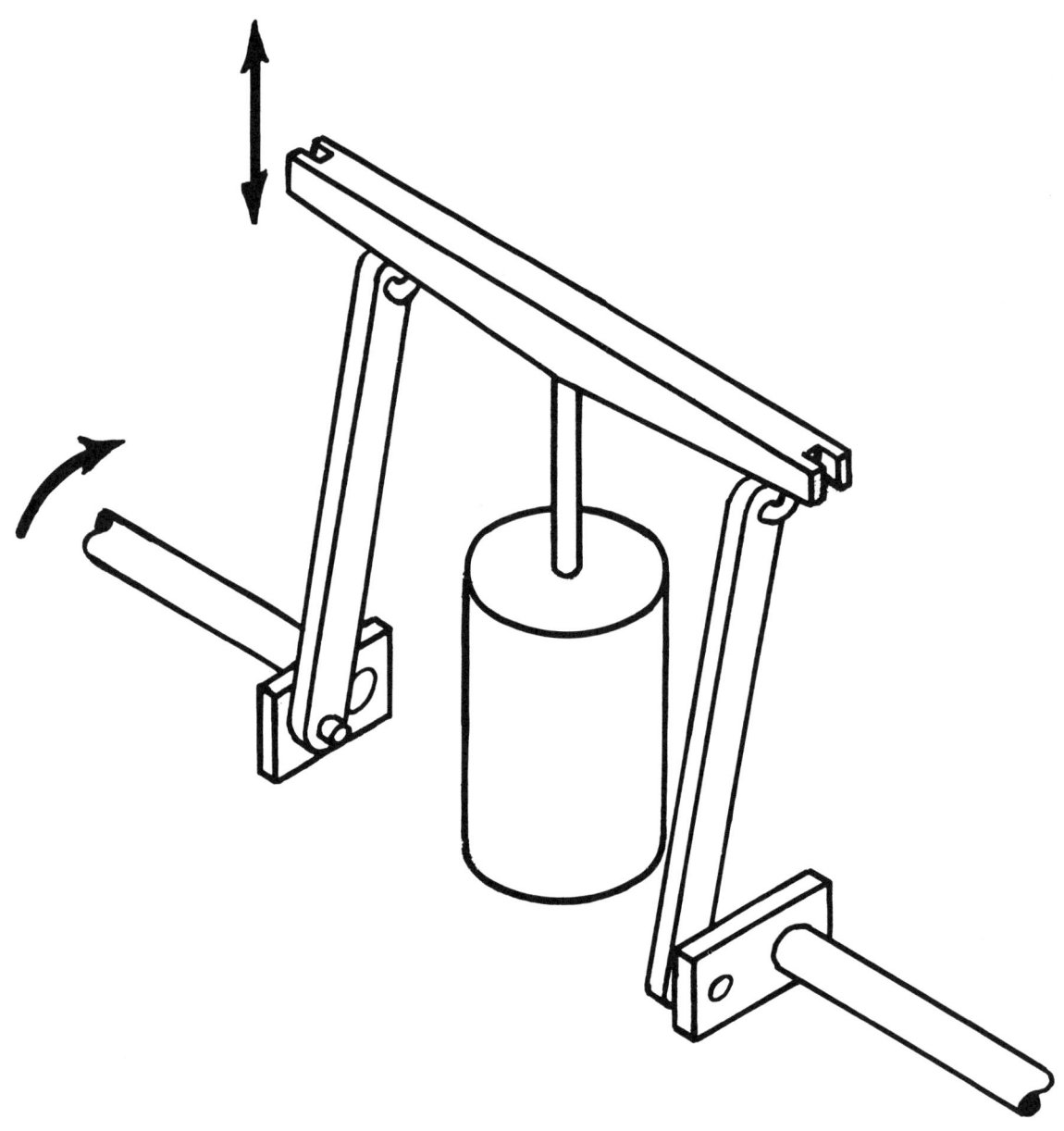

Diagrammatic arrangement of a crosshead engine.

the *Car* immediately became popular with her passengers and retained their favor for many years. Under the best of conditions she could make the trip to Albany in twenty hours at seven miles per hour. In spite of being a long steamboat, the longest ever designed by Fulton, she represented the first break away from Beaufoy's model tests. In this and subsequent vessels the hull's proportions were revised to make them continuously wider and, after the *Car*, shorter. The changes proved successful from both a speed and horsepower standpoint. Fulton was conducting full-scale experiments but proceeding cautiously as he gradually modified his designs a little at a time.

The *Paragon* of 1811 continued the new trend; being some 20" wider and 2'-0" shorter than the *Car of Neptune,* her ratio of length to beam was 6.2 instead of 6.7. The 1808 *North River* had 8.3 and her first version an all time high of 9.9. The increased width made for less cramped cabins and more workable galleys. In addition, it created a stronger, less flexible hull.

The *Paragon*'s engine was simplified by eliminating the bell cranks between the cylinder and the crank shaft. The result was a cross head engine. John and Robert Stevens had already tried this arrangement when they reengined the *Phoenix.* Such an omission saved weight and cost and was carried over into subsequent vessels.

The year 1811 was to prove an interesting one on the Hudson River. When the ice broke up in the spring, the *Car of Neptune* and the *North River* began their regular runs. The *Paragon* was nearing completion but, like most craft, was not ready on time and did not join the Fulton-Livingston fleet until the fall. Despite legislative protection, serious competition appeared in the form of two new vessels named *Hope* and *Perseverance*, which were built in Albany and engined in New York by Robert McQueen, who had previously supplied iron castings to both Fulton's engine works and the Stevens's shops. The first hull, launched[66] at Albany by Moses Kenyon on March 19, 1811, was 149' long by 20'-8" beam by 7'-7" depth or almost the same size as the *North River*. The woodcut used to advertise these steamers shows that they were direct copies of the *North River*, with a bowsprit added, the topmast eliminated, and the after guards shortened. An undesirable feature was that Captain Elihu S. Bunker, who supervised their building, allowed Kenyon to use such heavy construction that it increased their draft and adversely affected speed. The *Hope* had a score of owners, of whom James Van Ingen seems to have been the most active.

Under Captain Bunker the *Hope* steamed upriver for the first time on June 22, 1811, while a band on board serenaded her fifty passengers with "Washington's March" and other patriotic airs. Unfortunately, the time of thirty eight hours to Albany was not very good. The *North River* usually did it in thirty two. Thereafter, an anonymous contribution (obviously by Fulton) appeared in the *New York Evening Post:*

An Ode On Steamboats
While the Car of old NEPTUNE still glides o'er the waves,
And boastings of Hope she contemptuously braves,
And calls on her sister, swift footed NORTH RIVER,
To see the poor mimic creak, rattle, and quiver;
Or dare in the race of fair science to run,
And enter the list of the fleet RARITON;
Like a poor wounded snake, she drags her slow length;
As defective in speed, as deficient in strength;
Til defeated, abash'd, she returns in disgrace,
Condemned by all virtues that witness'd the race.[67]

Legal, as well as verbal, action started, but while the lawyers filed papers the *Perseverance* was completed. With two lines and four boats operating, the Albany traveler could leave on the Fulton steamers from Cortlandt Street on Tuesdays and Saturdays or on the Van Ingen ones from nearby Liberty Street on Wednesdays and Saturdays. On July 27 Captain Bartholomew of the *North River* sailed from Albany just after the *Hope*. Taking advantage of his speed he overtook the later, rammed her on the starboard quarter, kept his engine running, and pushed her aground.[68] With his lighter draft, he had no difficulty in backing clear and proceeding on his way. Apparently no physical damage was done.

Eventually the legal situation was settled during the winter, when the New York Court of Error upheld the existing monopoly and turned both the *Hope* and *Perseverance* over to Fulton and Livingston.[69] No attempt was made to utilize the two craft; they were broken up. The *Perseverance's* engine, however, was installed in a second steamer named *Phoenix* built in 1815 on Lake Champlain.[70] Captain Bunker did not, however, allow this setback to interfere with his enthusiasm for propulsion by steam. Temporarily, he headed South to replace Moses Rogers as master of the Delaware River *Phoenix* but, as we shall see, he would return to New York and start anew.

Without opposition, the partners started the 1812 operation with three departures a week for Albany. A fourth vessel of their own, the *Fire-Fly*, was completed in September. In spite of her small size, only 81'-6" long by 14' beam, she was to prove a very useful addition. While it was first intended that she extend the service up the river from Albany to Troy, she was actually placed on a local operation between New York and Newburgh or Poughkeepsie. As a result, the *Paragon, Car of Neptune,* and *North River* could be used in express service with fewer stops.

With their own needs temporarily under control, Fulton and Livingston started a steamer for others who were willing to cooperate with them technically and financially. As in the case of the *Rariton*, the hull was ordered from Charles Brownne, the engine from Fulton's Jersey City establishment. Such a craft was the *Richmond*, intended for the James River in Virginia. By the time of her completion, July of 1814, however, the United States was at war, New York was blockaded by a British fleet, and any attempt at delivery would involve great risk. The *Richmond* was therefore taken over for the Hudson service and, as a newer, larger, more luxurious vessel, became the replacement for the *North River*. For the latter's epitaph we have the following advertisement:

Steam Boat Notice
The Public are informed that the old North River Steam Boat is laid aside, and the staunch new Boat called the RICHMOND, with handsome accommodations substituted in her stead, this Boat in consequence will take the day of the North River, leaving New-York on Thursday and Albany on Tuesday as heretofore advertised.

Bartholomew, Master

The *Richmond*, with dimensions of 154'-6" by 28'-6" by 9'-0" continued the shorter-length, wider-beam trend

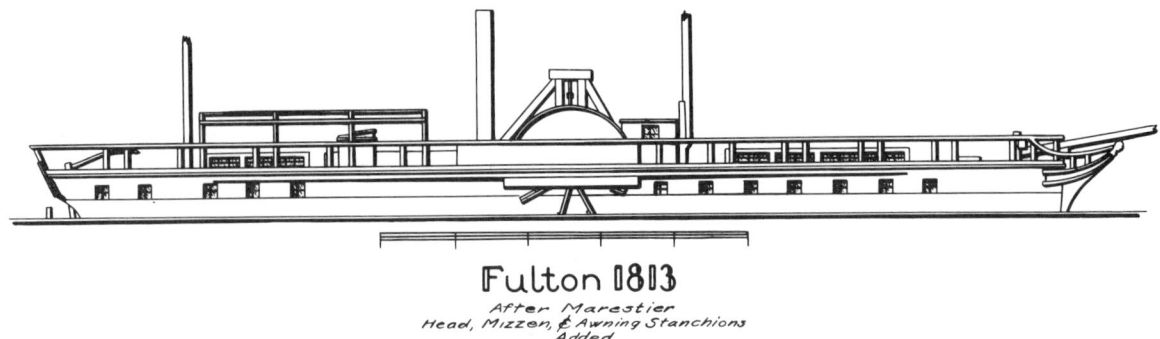

Outboard profile of the *Fulton* of 1813.

already discussed and was a better match in size and engine power to the *Car of Neptune* and *Paragon*. With the *North River's* replacement, seven years after steam navigation on the Hudson began, the line had a well-balanced fleet of three large, generally similar, long-run steamers and one small one for short trips and special uses. They had acquired, in Captains Wiswall, Roorback, and Bartholomew, three trained and trustworthy operators for their principal boats. An engine works under Charles Stoudinger had been built. Furthermore, next to the engine works, a dry dock was installed to maintain and repair the steamers of the fleet. When the *North River* entered it on October 14, 1813, she was claimed to be the first vessel to enter a dry dock in the United States.[72]

In the same year that saw the passing of the *North River*, another steamer of advanced design made her debut. Like the *Richmond*, she was built for owners operating within the framework of the New York monopoly. Cadwallader D. Colden, friend and biographer of Fulton, was the chief financial backer and Captain Elihu Bunker was responsible for the principal ideas incorporated into the vessel. The latter insisted on departing from Fulton's standard flat-bottom, flared-side design and placed the hull order with Adam and Noah Brown, well-known shipbuilders in New York City. Furthermore, since she was intended for Long Island Sound service, the hull was of increased scantlings (greater strength) when compared with Hudson River craft. The interior arrangements were conventional, but the whole machinery installation was moved farther aft than it had been on the *Paragon* or *Car of Neptune*. As a result, the forward cabin with twenty eight berths and a curving grand staircase at its after end became the largest public space on board. Aft was a smaller saloon with twelve berths and the ladies' cabin, right at the stern, with eight. She had the usual ketch rig of a Fulton steamboat with a hinged mast forward and a smaller mizzen aft.

Captain Bunker was loyal to the Hudson River sloops he had sailed earlier and steered her from a tiller aft instead of a raised wheel forward. The latter soon replaced the former when he realized its advantages. The engine, from Fulton's works, had a 36" bore and a 4'-0" stroke, making it the most powerful built to date. It was a typical crosshead type whose design features had been established for the *Paragon*.

Named *Fulton*, the round-bottomed hull was only 134' long, 30' 9" wide, and 8'-9" deep. Her draft of 6'-3" was considerably higher than the *Paragon*'s 4'-1", but the increase was an advantage in the rougher water expected on Long Island Sound.

Like the *Richmond*, she was caught by the War of 1812 and could not venture outside New York harbor for fear of capture by anything from a ship of the line, such as H.M.S. *Superb*, 74 guns, to the British privateer *Lively* that was cruising nearby.[73] To avoid laying her up it was decided to put her into the Albany service under a special agreement whereby her owners paid Fulton and Livingston $3.00 per passenger for the privilege. One admires Captain Bunker for returning to the Hudson with a newer, faster, better boat only three years after being forced to abandon the *Hope*. He squarely faced the problem of charging a $10.00 fare, to his rivals' $7.00. In his advertisements he listed the advantages of speed and luxury; he would take only sixty passengers.[74] There would be no crowding on the *Fulton*, which was scheduled to make one trip a week to Albany, starting in May of 1814, and, in between times, entertained passengers on day excursions as far as Sandy Hook (but never outside) for a sum of $3.00 including dinner. Captain Bunker proved to be the happy combination of a far-sighted businessman, one of the earliest to understand the possibilities of a steamboat service, and a skillful master who could get the most out of his boat and crew.

A third steamboat for others, the 186-ton *Washington*, was also completed in 1814, somewhat smaller than either the *Richmond* or the *Fulton* and of considerably lower power. She was intended for the Potomac River, and, unlike the other two, was laid up. After 1815, when the war ended and she reached Washington, she was the only steamboat on the Potomac river for the next five years.

Long Island Sound Service

The year 1815 and the lifting of the British blockade of the Atlantic Coast enabled the *Fulton* and her energetic master to begin the New York to New Haven, Connecticut, service for which she had been expressly built. Because Hell Gate, at the junction of the East and Harlem Rivers, was notorious for its tidal curents, rocks, and whirlpools, there was a considerable question as to whether it could be passed on any regular schedule by a steamboat. Sailing packets had to await favorable wind and water conditions before they dared venture through.

On Tuesday, March 21, 1815, the *Fulton* sailed from Beekman Slip on the East River, passed safely through these perilous waters, and reached New Haven eleven hours later without the use of sails.[75] Fog delayed her return until Friday. There were several stops en route because of low visibility, but no problems at Hell Gate despite a three-or four-knot current; the return passage, including delays, took fifteen hours.

Two round trips a week were undertaken, and, before the end of March, a fast passage of eight hours was achieved with a following wind at a speed above nine miles per hour. The New York newspapes began to report passenger lists as high as 204; it was fortunate that sailings were by day and that the limited number of berths aboard did not need to be used. Blessed with immediate acceptance by the public, the *Fulton* was put on a three-round-trip-per-week schedule by June. As on the Hudson, the steamboat had proved, at a fare of $5.00 per passenger, to be both cheaper and more comfortable than the stage and considerably faster as well. Her cost of $87,000, a third of it for her copper boiler[76], had been high, but she was destined to be a most satisfactory steamer for years to come.

In 1814 her harbor excursions had proved a great attraction to the public. They were continued, and on July 4, 1815, she made three separate trips[77] starting at 9 A.M., 1 P.M., and 8 P.M. to entertain the holiday throngs. The evening one, of course, included music and fireworks as a fitting conclusion to that glorious day. Tickets were a dollar apiece.

Fulton's death, on February 23, 1815, left the Hudson River operation without a guiding hand or a desire to build newer, larger steamers for its own service or for others under license. It was fortunate that at least three important vessels had already been planned and would be carried along to actual completion.

The one of immediate interest was almost a duplicate of the *Fulton*, with the engine power increased by giving the cylinder a 36″ bore by 5′-0″ stroke instead of 36″ x 4′-0″. She had been started at the yard of Adam and Noah Brown and was supposed to be named *Emperor of Russia*, perhaps to further negotiations with the Imperial Court for the introduction of steamboats to Russia waters, plus, it need not be said, exclusive rights to their use. Her name gave rise to many garbled rumors, both in the United States and Europe, that the vessel was intended for either transatlantic service or operation abroad.[78]

Whatever the intent, Fulton was far too intelligent an engineer to send a 135′ steamer of 31′ breadth and 10′-7″ depth across the Atlantic. She was incapable of carrying enough wood to make even a fraction of the crossing. But, without his direction, the Livingston heirs did not know just what to do with the craft. Since Captain Bunker and his financial backers were anxious to expand their booming Long Island Sound business, they arranged to take her over. Renamed *Connecticut*, she had a number of owners, including Cadwallader Colden; Adam Brown, her builder; John Mott; Richard S. Williams; and Josiah Ogden. Elihu S. Bunker and William Comstock, both of them shipmasters, became shareholders at a later date.

Upon completion, in September of 1816, Captain Bunker moved to the new boat and turned the *Fulton* over to Richard Law. The latter remained on the New York-New Haven service, sailing three times a week, and the former undertook a longer trip to New London, via New Haven, twice a week, as an overnight run. The *Connecticut* arrived back from her maiden voyage on October 1 and was "said to surpass any boat that has ever been built in beauty and strength."[79] Outbound she had managed to overtake the *Fulton*, despite several miles lead, and pass her.

Upon the advent of winter the *Fulton* was laid up, but the more powerful *Connecticut* continued between New York and New Haven as long as ice conditions permitted.

The real question in the minds of the professionals in the marine field was, how would a steamboat perform in heavy weather? There are two letters by Captain Bunker dealing with this problem.[80] They were written to J. Bronson in August and September of 1815. The first states in part:

> I have never had so much sea since I have been in operation as yesterday, and although there was never a single doubt on my mind with regard to crossing the

Atlantic with a Steam Boat with safety and ease, I consider yesterday's passage a further confirmation of my opinions, and indeed I consider it almost an experiment. I am satisfied that a Steam Boat of the length of the FULTON, drawing 8 or 9 feet, will scud better than any other vessel in a gale of wind, and that her engine may at all times be kept in operation, without being interrupted by the swell, whether great or small.

Portions of the second letter follow:

Yesterday I started from New Haven as usual. When I got out on the Sound found a heavy gale from the eastward; and it being the last of the ebb-tide, and running to windward, had raised a heavy sea. This way by far the strongest gale that I have known on the Sound, and more than double the sea running. The wind veering from E. to E.S.E. could not make a lee under either shore of the broad Sound.

The boat yesterday, notwithstanding the heavy sea, was perfectly manageable with one man at the wheel, although she was running at the rate between 13 and 14 knots having made the passage in 6 hours and 40 minutes—No other sail set on the Sound but the square sail on the FULTON'S forward mast.

In 1817 the Sound route was extended to New London, with the *Connecticut* operating from New York to New Haven and the *Fulton* from New London to New Haven. Their sailings were timed to meet at the latter port where through passengers were exchanged. It is believed that the division of the run into two parts was adopted because the *Connecticut's* more powerful engine enabled her to cope with the difficult navigation through Hell Gate on a regular schedule. Next year the pair, under the name of "Sound Steam Boat Line," started service March 9 for New London, but on the 17th advertised the extension of the *Fulton's* route to Norwich, instead of New Haven, stopping at New London on the way.

The New York-Norwich operation with passenger stages from there to Boston continued from 1818 until the summer of 1824. When occasions warranted, there were excursions for special purposes. On July 15 of 1823 the *Connecticut* was chartered to tow a new light ship from Manhattan Island out to its station at Sandy Hook. Captain Bunker promptly advertised a "sailing party" for the occasion, and offered to take passengers along for a fare of $1.50 that would include a "collation."[81]

The Later Careers of The *Rariton* and the *Phoenix*

The New York-to-Philadelphia route was last considered in 1810, after the *Phoenix* was rebuilt and the joint service with the *Rariton* established. The author does not intend to follow the Delaware end of the operation in detail, for the vessels there were not directly in the line of evolution from the river steamboat toward the ocean-going steamship. The *Phoenix* continued running between Philadelphia, Trenton, and way points, except in the winter months. There were further alterations[82] made to try and better her performance; Robert Stevens oversaw the changes and his father sent advice from Hoboken.

In 1812 a new and much improved vessel, the 139' long *Philadelphia*, was built in Kensington and engined at Trenton to replace the *Phoenix*. Work proceeded so slowly that the new steamer was not completed until October of 1813; for the rest of the year the *Phoenix* was laid up and the *Philadelphia* took over.[83] Early in 1814 the *Philadelphia's* shafts showed defects and had to be replaced. The *Phoenix* was reactivated to tow her down river to Philadelphia on March 30.[84] The fall of 1813 is the last available date for regular operation of the *Phoenix* on the important New York-Philadelphia route. It is the year of her effective, if not her final, demise. She saw but five seasons of useful operation. As a result, her record is considerably less satisfactory than that of the *North River*, completed a year before her and in continued employment for a year afterward. Both were rebuilt extensively and, although the *North River's* changes are better recorded, those of the *Phoenix* were just as extensive. She had a new engine, a new boiler, and many hull changes.

The northern branch of the Philadelphia service remained unchanged until 1816 when the *Olive Branch* joined the *Rariton* and sailings could be increased from three times per week to six. Since the New York-New Brunswick and Trenton-Philadelphia steamboats guaranteed the least travel by land, they were preferred, but they were no longer the exclusive water route for through travelers. Another—via Elizabethport, a stage to Bristol or Bordentown, then boat again—to Philadelphia had appeared in 1815, starting with a small steamer called the *Sea Horse* at the New York end and the *Bristol* on the Delaware. Other competition was to develop;[85] no doubt it spurred on the Livingstons and the Stevenses toward pleasing their patrons.

Constructed by Noah Brown, the *Olive Branch* was the second New Brunswick steamboat planned by Fulton,[86] but it was not completed until after his death. She was considerably wider and deeper than, but of the same length as, the *Rariton*, and had a more powerful engine. The builder's name implies that she was ship-shaped, like the *Fulton*.

The newer boat replaced the *Rariton* after 1816.[87] Since the latter's enrollment was not abandoned until 1820, she probably saw occasional use as a spare boat. The last event noted in her career was a more serious repetition of the first. In 1818 the braces supporting the front head of her boiler, which faced the engine and the firing station, became so corroded that the whole head blew out, scalding her engineer to death.[88]

The *Olive Branch* continued on the New Brunswick route until 1824, but her experience during her later years was too close to the first year of service by the *Rariton* for comfort. New competitors had moved in; profits existed, but they were not liberal enough to warrant building new craft for the run. John Livingston was seeking a more satisfactory route, but he could not make a change without further grants by the successors to Fulton and Livingston. The latter had no intention of permitting anyone to share their profitable Hudson River operation.

The *Chancellor Livingston*

Under the original Fulton-Livingston partnership agreement, each member would pass on his share to his individual heirs. After Livingston's death, at the age of 67, in 1813 and Fulton's, at 50, in 1815, the Fulton offspring were young children and his wife was one of the extensive Livingston clan. In order to manage the Hudson River enterprise, a company[89] was planned under Edward P. Livingston, Robert L. Livingston, and Dominick Lynch, Jr., with a capital of $820,000—the excessive value estimated for the boats, patent rights, and other properties—under the name North River Steamboat Company. It was eventually formed, but with much less capital.

The first two vessels completed under this new management, the *Connecticut* and *Olive Branch,* appeared in 1816. These, as we have seen, were of moderate size and presented no great changes in the gradual development of the Fulton type of craft. A third, however, represented a bold break with the past. She was, appropriately, named after Chancellor Livingston, whose belief in Fulton and steam navigation and consequent financial backing had been so essential to the development of the steamboat. The principal characteristics of, and the decision to build, a large, advanced type of vessel had been made before Fulton's death. She had been scheduled for completion in 1816. The *Chancellor Livingston* proved to be a posthumous memorial to the engineering partner, as well as to the financial one, but she was not finished until 1817.

Her hull was ordered from a new source, Henry Eckford, the most original and successful of all American shipbuilders of the 1812-30 period. No doubt Eckford was the person most responsible for the final design. Isaac Webb, first his apprentice and later his partner, was in charge of the hull's construction. The Fulton engine works, now that no partner was left to oversee it, was first leased by James P. Allaire, whose name has already been mentioned as the owner of a brass foundry that had supplied casting for Fulton's engines. For a brief time Charles Stoudinger was a partner, but very shortly his death left the management in Allaire's hands. Allaire has always been listed as the builder of the *Chancellor Livingston*'s engines, but it is likely that the design came from Stoudinger, the foreman who had operated the works so long under Fulton's direction. In 1816 the facilities were transferred from Jersey City to Allaire's foundry at Cherry Street in Manhattan Island.[90] Allaire, with the *Chancellor*'s engine to his credit, achieved immediate recognition as the principal engine builder in the country; by virtue of his further accomplishments he continued as such until he retired in 1850.

The basic arrangement of the *Chancellor Livingston* was still that of a forward passenger cabin and an after one, each with upper and lower berths along the sides of the ship. The machinery installation, however, was completely reversed; the earlier steamers had the boiler amidships, since it was the heaviest item in the ship, with the engines and paddle wheels forward. Here the boiler was well forward and the lines of the ship drawn to provide a full bow and a fine stern so as not to disturb the trim. Instead of locating the machinery to suit the hull, the hull was designed to suit the machinery. This placed the engine aft and the paddle wheels amidships. An attractive sailing-ship bow with an overhanging stem, curved head rails, and ornamental trail boards gave the figurehead an appropriate setting. There were quarter galleries aft, for purely aesthetic reasons. The hull was painted black and buff and had a port strake along the cabin windows in a style copied from British warships. There was a bowsprit forward, again from the standpoint of appearance rather than usefulness, but, for the first time on the Hudson, no masts and no sails.[91] Steam propulsion had become reliable enough to justify the omission of any emergency aid to propulsion.

A third innovation was the installation of a large cabin on deck, the ladies' cabin, and above it an open promenade deck covered by an awning extending from the stern almost as far forward as the paddle boxes. This represents the first departure from sailing-ship practice, in which all cabins were below decks, toward the characteristic American steamboat of the 1840 period, where all cabins and passenger spaces would be located in superstructures built on and above a shallow hull.

It is most fortunate that Jean Baptiste Marestier, a French engineer, was able to obtain plans of the ship, which were reproduced in his 1824 *Mémoire*. A Swedish lithograph printed for another foreign visitor, the Swedish Baron Klinkowstrom, has almost identical details with the exception of the paddle box shape.[92] The former shows them to be semicircular, with no houses fore and aft. The second shows extensive houses of such height that they completely cover the paddle wheels. An American engraving shows that reality was about halfway

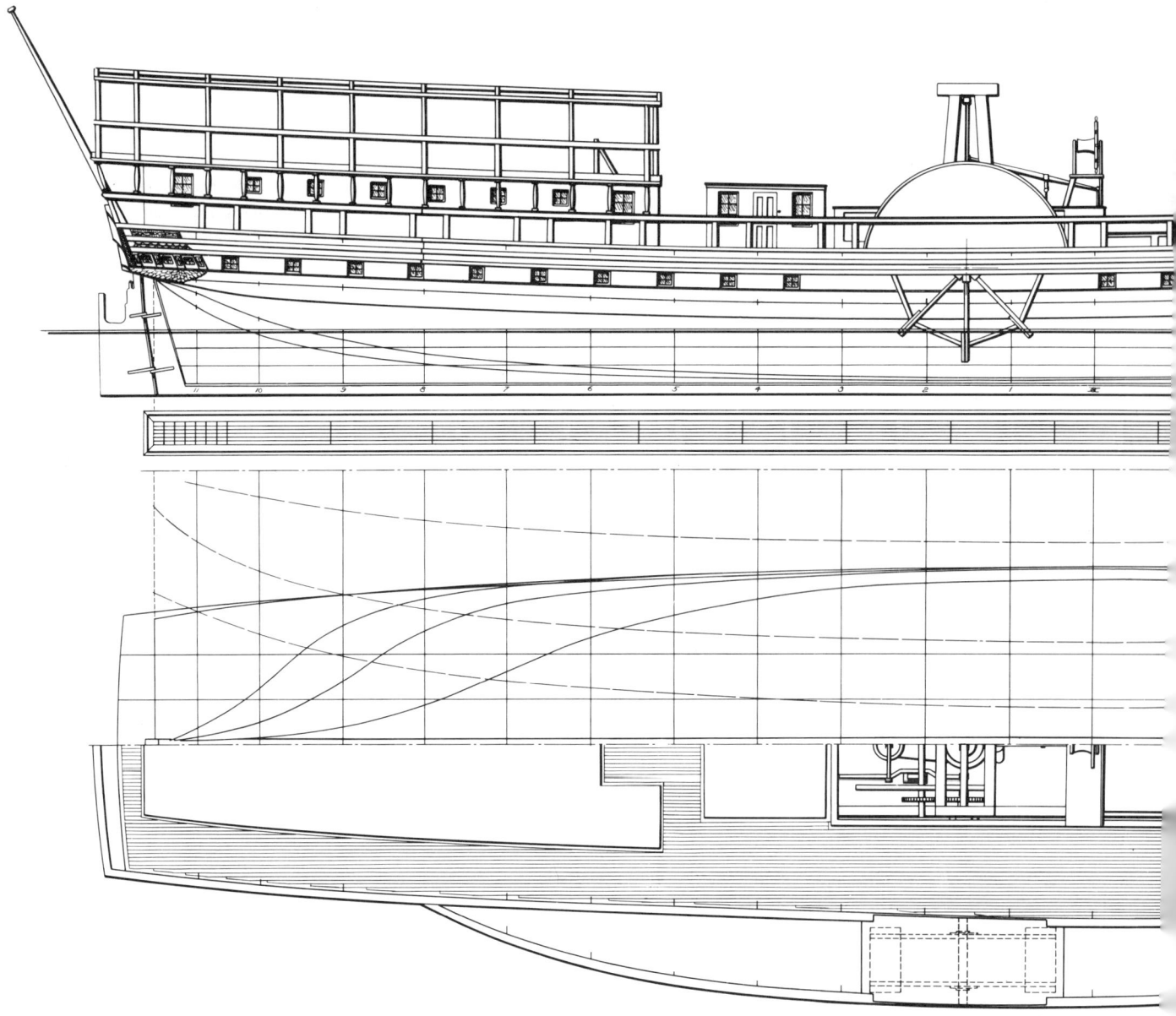

Chancellor Livingston, 1817, lines and deck.

between the two.[93] There were curved boxes and short houses attached forward and aft for the water closets. This was the traditional location that started with the *North River* and carried on for almost a century. The paddle wheels splashed up lots of water and the ventilation was excellent.

The interior arrangement placed a long main cabin aft with thirty eight berths; forward was a short one with wings extending aft along each side of the boiler. Moreover, there was enough beam so that extra tiers of berths along the center could be installed, making fifty six in all. On deck the ladies' cabin contained twenty four; the total was one hundred eighteen passenger berths. For the crew, engineers, cooks, a pilot, and some extra spaces, twenty-three more were provided, or a grand total of one hundred forty-one.[94] As usual, settees in the after cabin, which was also the dining saloon, gave additional sleeping space. Galleys, stores, coal bunkers, and service spaces filled the midship region of the hull either side of the engine. There were several deckhouses for the captain, officers, crew, and three

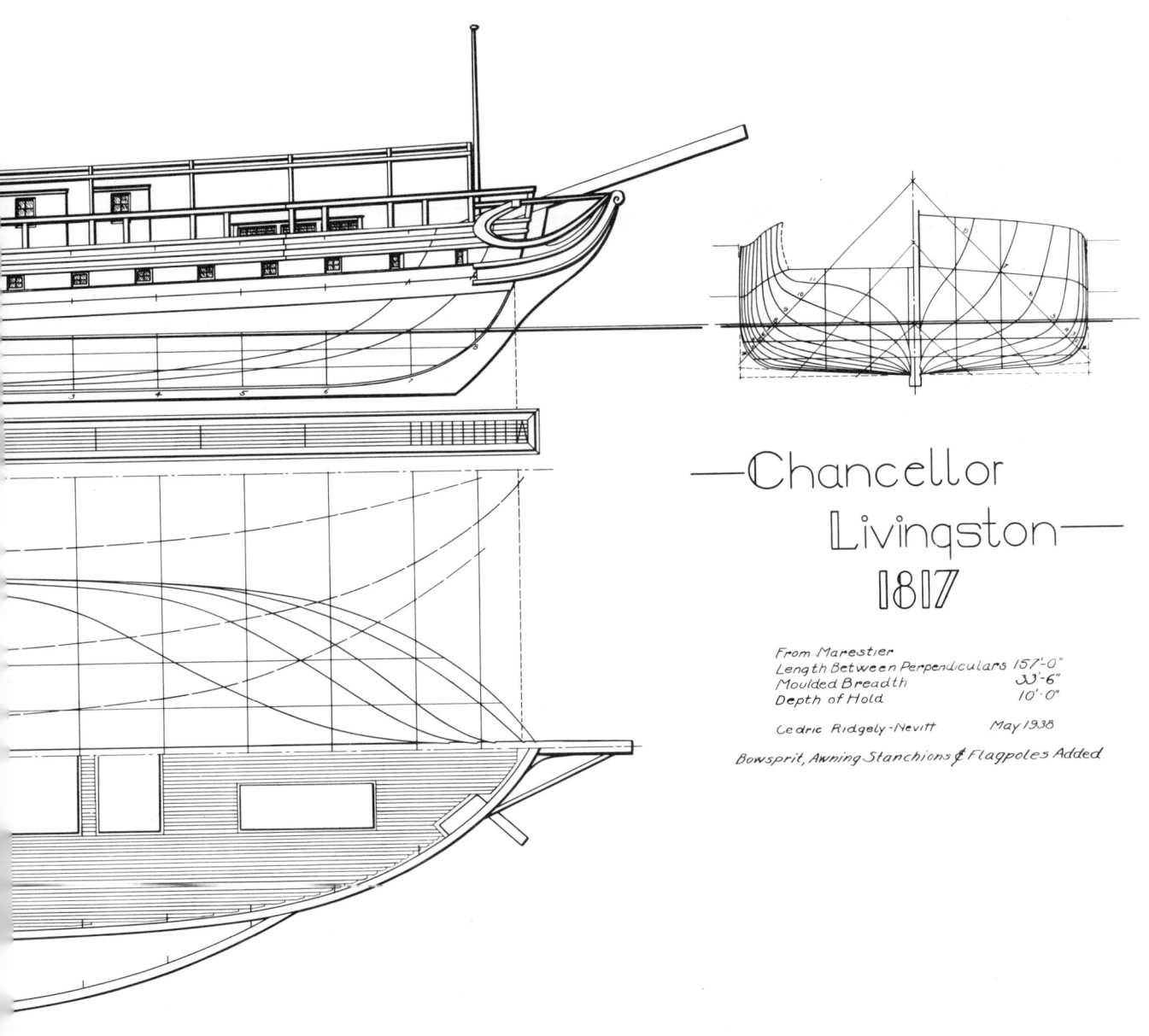

—Chancellor Livingston—
1817

From Marestier
Length Between Perpendiculars 157'-0"
Moulded Breadth 33'-6"
Depth of Hold 10'-0"

Cedric Ridgely-Nevitt   May 1938

Bowsprit, Awning Stanchions & Flagpoles Added

large entryways to the staircases going below. All the interior joiner work was constructed by David Cook, who probably built the deck houses as well.

Guards projecting beyond the hull extended almost three-quarters of the ship's length and, as before, provided the space needed to stow the very large amount of wood required when it was used instead of coal. The steering wheel was on a raised platform just above the control station for the engine; the boiler arrangement in Marestier would have put a single stack directly ahead of it to block the view. As completed, two stacks were fitted, side by side, and the helmsman could see between them. Later on, a flagpole was put between the stacks; the pilot had more obstructions than ever.

The *Richmond* of 1814, with a tonnage of 370, was the largest previous steamer. The *Livingston*'s value of 495 shows her to be 34% bigger. If we go back to the final version of the *North River*, the *Livingston* represents a two-and-one-half-times-larger craft with the size increase showing up in larger beam, depth, and draft but only a small length change. Her official dimensions were 157' long, 33'-6" beam, and 10'-3" depth of hold. Various

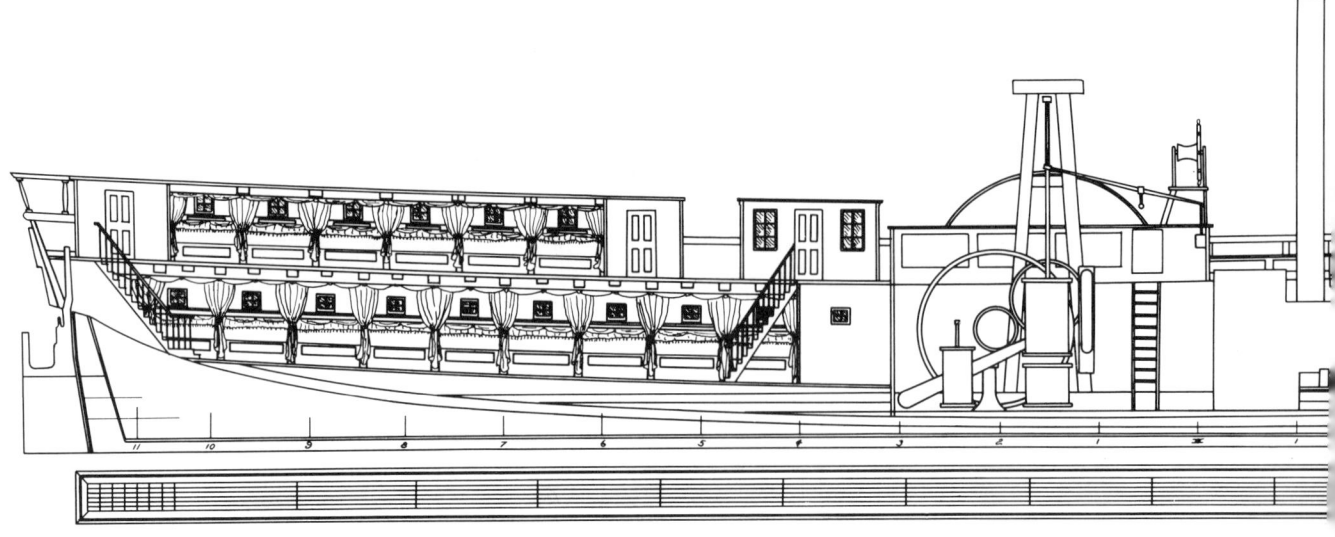

Chancellor Livingston, inboard profile.

sources give drafts from 6'-0" to 7'-3". Although the dimensions are difficult to check, for Mariestier's scale, labeled "pieds," is incorrect for American feet, the extreme beam over guards is about 49' and the draft shown on the plans corresponds to the 6'-0" value.

The large engine and its even larger copper boiler took a whole year to construct. It was a 40" bore by 5'-0" stroke crosshead engine,[95] having two geared flywheels 14'-0" in diameter and turning 18'-0" wheels with eight floats 3'-0" deep by 5'-10" wide, and was capable of producing about 75 Indicated Horse Power.

The engine drawing in Marestier is most complete and shows all the essential parts missing from the Boulton and Watt example given earlier for the *North River.* The cylinder and valve gear are exactly similar save in size. Everything is driven from the overhead crosshead: there are connecting rods port and starboard driving the cranks on the shafts, another pair of rods down to pivoted, counterbalanced levers which, in turn, operate a second transverse crosshead of shorter stroke driving the air pump. Extending forward is a lever to operate the piston of the feed pump drawing part of the condensed water from the hot well and forcing it back into the boiler. Its long vertical rod also serves as the "plug tree" or valve-operating device. Pins projecting sideways from it engage the slightly flexible, curved springs which, through further levers, open the four steam and exhaust valves. The valve levers have small weights hanging on leather straps to close the valves again.

The weight of the piston, piston rod, crosshead, and all the attached connections is neatly balanced by the heavy ends of the air-pump levers, so that there is no tendency for the engine to stop in any fixed position. This is the best and most complete early steamboat engine drawing extant.

The copper boiler, about 25'-3" long, 12'-3" wide, and 10'-6" high, had a single wide furnace at the firing end from which two circular flues extended forward, reversed themselves, and returned again. They carried the heated gases through the water and increased the rate at which steam could be produced. The technical name *return-flue* boiler is an apt description. The Marestier drawing shows the flues joining, but, as built, each had its individual stack. The dry weight was about twenty tons and the water, at its working level 6" above the top of the flues, added another 27 tons. Note that here and afterwards tons of weight used in describing ships are always 2,240 pounds. The American short ton of 2,000 pounds is a modern invention unknown throughout the rest of the world and never applied to seagoing ships anywhere.

A final innovation was the use of grate bars and a furnace design that permitted the use of anthracite coal as fuel instead of wood. Allaire experimented with such an installation on the *Car of Neptune* and even served on board her as a fireman until he was assured of successful performance before he would make a permanent installation on the *Livingston.*[96] In 1820 Marestier noted her as the only steamer he saw that burned coal. This was not always the case for, on a later trip in 1822, the fuel was wood.

We have no record of the steam pressure carried. Comparisons based on contemporary vessels having low-pressure copper boilers indicate that it was no more than

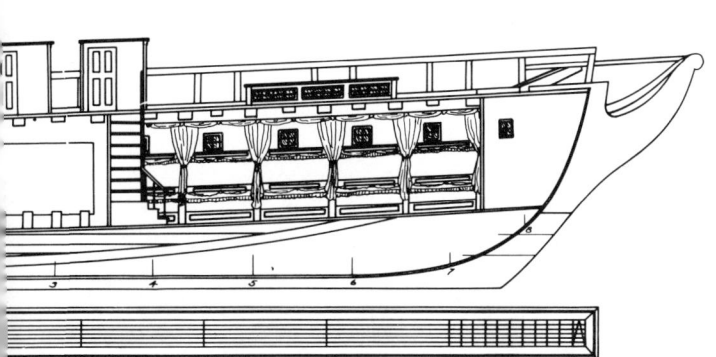
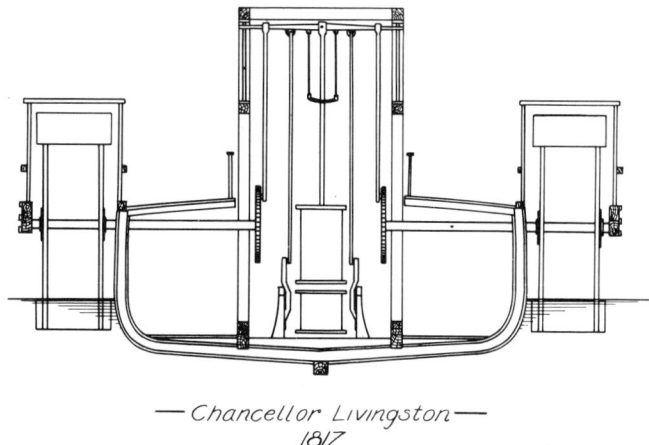

—Chancellor Livingston—
1817

six to eight pounds per square inch and could be accurately measured by a mercury column. There were, by this time, engines and boilers of far higher pressure, particularly the ones built by Oliver Evans on the Delaware River, where fresh water waa available and therefore no scale deposited inside the boilers. These had pressures of 150 psi and consequently required iron boilers to resist the larger stresses. They did not have a very satisfactory safety record; boiler failures and violent explosions were altogether too frequent. The more conservative owners found it advantageous to advertise vessels of the "Boulton and Watt" or "Fulton-Livingston" principle with "low-pressure steam" and "copper boilers." They had to pay considerably more for them, for copper was extremely expensive and larger engines were necessary. Iron plates of good quality were not available in this country; they had to be imported from England.

The *Chancellor Livingston* was a progressive vessel in a number of features and, at the same time, conservative in retaining the reliable type of machinery that had served her predecessors so well. Her cost, $110,000, made her the most expensive American vessel ever built for private ownership. With her advent, the *Car of Neptune*, now eight years old, was retired. Unfortunately, this left the North River Steamboat Company with a remarkably unbalanced fleet of the *Paragon*, blessed with reasonable speed but cramped quarters, and the *Richmond*, a somewhat larger boat with such deliberate movements that she had been christened *Lady Richmond* by the traveling public. For odd jobs there was the diminutive *Fire-Fly*. None could compare with the newest addition.

The *Chancellor Livingston* started with a trial run to Newburgh and back, making the upriver trip in 9 hours[97] and returning in 8 hours 15 minutes at speeds of 6.67 and 7.27 miles per hour respectively. By the summer of her first year she had made the record time of 18 hours from Albany to New York.[98] If stops were excluded, the speed came to 8.5 miles per hour. The one design change not a success was the increased draft, which prevented her from passing over the shallows in the Hudson eight miles below Albany. The passengers had to be transferred to stages, but they were quite willing to put up with a few miles by road in return for the pleasure of traveling on such a splendid craft.

In June of 1817 New York used its steamboats lavishly during the visit of President James Monroe. The *Richmond* was sent on a special trip to bring him from Staten Island to the Battery with the sloop of war *Saranac* and the local revenue cutter in attendance. The *Connecticut* was made available for an inspection trip covering the fortifications in the harbor from Sandy Hook to Hell Gate and, for the climax of the occasion, the *Chancellor Livingston* bore the president upriver to West Point.

Two years later, very near the start of the spring season, the *Chancellor Livingston*, bound up the river in the vicinity of Red Hook, was completely disabled by a broken crosshead.[100] As we have already seen, every part of the machinery was driven from this transverse beam. It was fortunate that the piston rod was not bent, nor was the cylinder damaged before the engine could be stopped. Captain Wiswall transferred his passengers to a sailing sloop for delivery to Albany and sent a messenger on horseback galloping to New York to order a spare and to send another steamer to Albany so that the next south-

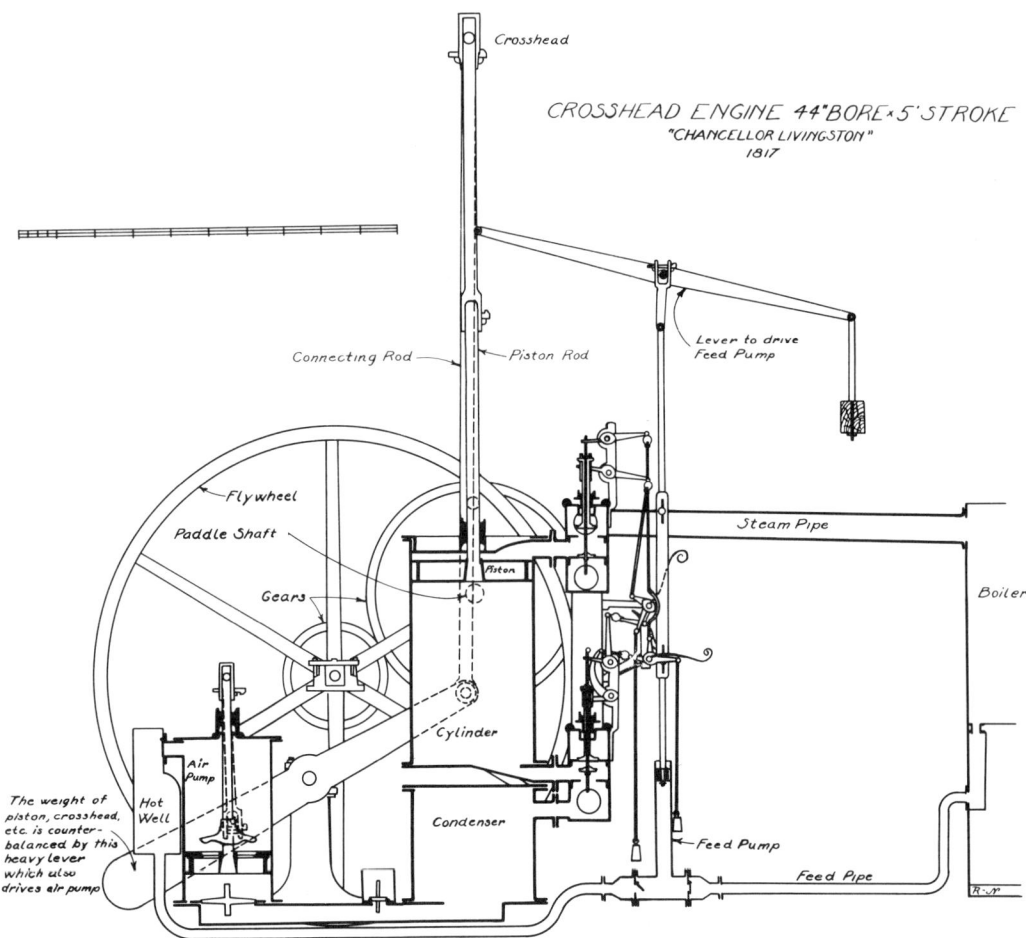

Crosshead engine 44″ bore by 5′-0″ stroke for the *Chancellor Livingston*.

bound sailing would suffer as little delay as possible.

The year 1822 was unusually busy for the North River Steamboat Line's vessels. First there was a yellow fever epidemic in the city. It is difficult for the present-day visitor to visualize the streams, ponds, and marshes then present on Manhattan Island, all capable of breeding clouds of mosquitoes. As a consequence, all who could deserted the city, overloading every available means of transport. The Hudson River steamboats kept going but could no longer operate from Cortlandt Street. The North Battery, in the village of Greenwich, outside the limits of the city, was used as a temporary terminal.

A second event inspired by the epidemic was a series of Camp Meetings organized at Haverstraw by the Committee of the Methodist Society. It was arranged that the steamers should stop at the camp grounds for half an hour on both up and down river trips for the convenience of the participants.[101] Unfortunately, the boats were unable to cope with all who wished to attend the evangelistic festivities.

One traveler has left an account of an eventful trip on the *Livingston* under these special circumstances.[102] Starting at Belleville, New Jersey, at 11 A.M. on September 10, he crossed the Hudson by the Stevens-owned, horse-driven ferry from Hoboken and reached the North Battery by 3:30 P.M. There he and the other passengers for the *Chancellor Livingston* found only the little *Fire-Fly* awaiting them. The *Livingston*, crowded with 500 to 600 participants from the camp meeting, was drawing far more water than usual and grounded in Haverstraw Bay.

Her would-be passengers, aware only of her absence, were loaded on the *Fire-Fly*, which headed upstream, searching for the larger boat, and, not finding her, returned to State Prison Wharf, also in the Greenwich

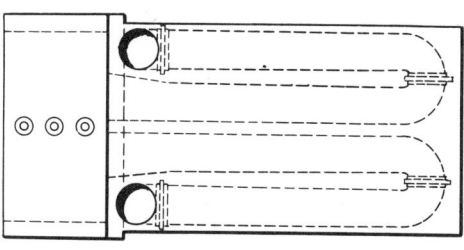
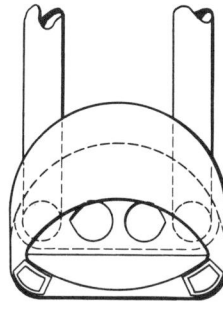
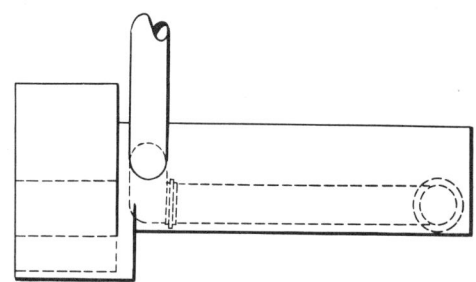

Return-flue boiler for the *Chancellor Livingston*.

Village area. At 8:00 P.M., to try and entertain the restive travelers, the *Fire-Fly* headed out for a sunset sail. During this she nearly crushed a rowboat full of men and women. Despite this diversion, some of her passengers were so dissatisfied that she had to return and put them ashore. With the more resigned ones still on board, she sailed again at 9:15, met the *Livingston* at 10 P.M. and returned with her to the North Battery. Off State Prison Wharf, just before landing, a man and woman fell overboard, whether from a state of religious fervor or purely because of the overcrowded conditions is unknown, but were safely rescued, to everyone's gratification.

Back at the original starting point, the revival group was landed, the Albany passengers loaded, some of the crew of the *Fire-Fly* taken over to replace those crewmen allowed ashore, and the *Livingston* sailed about 11:00 P.M., some seven hours late. To solve the fuel problem a sailing sloop was lashed alongside and the crew worked during the night moving her load of wood onto the steamer.

The Reverend Stryker found his berth hot and the machinery noisy (probably because of the gearing); he was unable to sleep, and returned to the deck about 3 A.M., where he walked up and down with another insomniac passenger. Haverstraw was passed—without a stop—at 5 A.M. and, later in the morning, three cadets appeared on the dock at West Point, where they entertained the ship's company with "Hail Columbia" and other marches played on the bugle. At 9:30 P.M. the bar below Albany was reached after a long, slow voyage lasting over 22 hours. No doubt the sloop alongside contributed to the low speed and made for poor steering until it could be cast off.

With the engine stopped, Stryker slept well during his second night on board. At 8 A.M. stages started arriving and loading the passengers and their baggage to deliver them to Albany.

The masters of the line's steamers were the chief executives of the company at this time. The fact that Captain Samuel Wiswall could land hundreds of passengers, acquire fresh hands from the *Fire-Fly*, lash a loaded sloop to his guards, and get underway in an hour's time testifies to his being well worth the high salary—for 1822—of $1,500 per year. The trust placed in him by the absentee owners was amply justified. As had the original Chancellor, Robert L. Livingston, as president of the company, conducted much of his business from Clermont, upriver and out of touch with its day-to-day operations.

In the late fall when the pestilence abated, there was again a shortage of space for New York residents returning to the city. Not only were the three available steamboats overloaded, but also every sloop on the river. In October the *Chancellor Livingston* carried 130 passengers southbound; in November there were 250. By November 19 the *Fulton* had been chartered from Captain Bunker, and the North River Company's fleet, for the first time in its history, was sailing every day.[103] At a time when service was normally interrupted by ice, everything was still running until the *Richmond* suffered an engine breakdown on December 10 and had to be towed to New York by the *Livingston*.

All this activity justified the order for a new steamer,

the first of the line not planned by Robert Fulton. And, because that optimistic engineer was not involved, it resulted in a backward, rather than a forward, step. Instead of being larger and more commodious when compared with the *Chancellor Livingston*, the new boat became a pallid copy of smaller size. With a length of 135′, a beam of 31′-6″, and a depth of 9′, her tonnage was only 346; in short, she was very little larger than the *Paragon*, which had been sunk in 1820 and never replaced. Her builders, Blossom, Smith, and Dimon, were new to the line, but both Stephen Smith and John Dimon had served apprenticeships under Henry Eckford, builder of the *Livingston*. The engine supplier, the West Point Foundry located at Cold Spring on the Hudson, was also new, but the machinery was a conventional crosshead, or, as it was often called, "square" engine of 4′-0″ bore. It was supplied steam by a massive copper boiler weighing 30 tons that cost almost one-third the price of the entire ship.[104] A trial speed of ten miles per hour was attained.

The name chosen, *James Kent*, was to honor the retiring Chancellor of the State of New York and she bore his likeness, not at the bow, but as part of the carving ornamenting the transom stern. Like the *Livingston*, she had a ship's bow and bowsprit, no masts, two stacks side by side, and a large deck house aft furnished as a drawing room for the feminine passengers.

The year 1824 was a decisive but unhappy one for the North River Steamboat Company, for it saw the end of their sole right to navigate the waters of New York. During the lifetime of Fulton and Livingston, large amounts of time and money had to be expended in legal actions defending this monopoly. We have specifically considered only the *Hope* and *Perseverance* encroachment in 1811, the first of many such cases. Not only was the North River Company involved in litigation; so were all the steamboat owners who operated under licenses. John Livingston, who had built the *Rariton* and later replaced her by the *Olive Branch,* was preyed upon by competitors seeking—and taking over—part of the lucrative New York-New Brunswick-Philadelphia trade. The most difficult and least reputable of these was the young Captain Cornelius Vanderbilt, who was operating unlicensed steamers, financed by Thomas Gibbons, from New York waters to New Jersey ports on the Kill van Kull.[105]

From 1818 on, there were rate wars, misleading advertisements, court proceedings, and attempts to arrest captains and detain boats, with often chaotic conditions for the unwary passenger who might board a steamer in New Brunswick only to find it unable to land in New York and would have to be transferred in the middle of the harbor to another that could. But this one might not deliver him until it had made its scheduled trip all the way to Staten Island and back.

Vanderbilt was a shrewd young man who made money for himself and his employer, circumvented the law, deceived his passengers, ruined his competitors, and somehow persuaded later generations to look upon him with awe and respect. The legal proceedings between Gibbons, his employer, and Ogden, a licensed competitor, with Daniel Webster as counsel for Gibbons, eventually reached the Supreme Court of the United States. In March of 1824 Chief Justice Marshall delivered the Court's opinion that the Fulton-Livingston grant by New York State was unconstitutional insofar as it applied to any navigation between states.

John Livingston acted promptly to test the case further and shifted his *Olive Branch* from New Brunswick to the Hudson River. His April 21 advertisement announced sailings three times a week to Albany,[106] noted her draft as being shallow enough to pass the bar below the city, and quoted a capacity of fifty three berths together with a number of settees "with backs." The through fare was cut to $5.00 and deck passengers could travel for half price.

The North River Company promptly applied in the New York courts for an injunction to prevent her operation. This was denied by the Chancellor of the State in June of 1824,[107] because the *Olive Branch* had landed two boxes of merchandise and taken on passengers in the State of New Jersey after leaving New York City and before reaching Albany. This showed that, in the legal sense, she navigated between states and therefore the Supreme Court's ruling was applicable.

Thus the successors to Fulton and Livingston were not directly involved in the critical law suit to establish their monopoly rights before the highest court of the United States. One of their licensees, who had tried to defend his own (and in principle their) steamboat route against deadly competition, sued and lost. To make matters worse, one of the Livingston clan, the first person ever to operate under a license, became the first competitor to establish a legal right to navigate the Hudson, his previous operations having been encroached upon by both parties involved in the Supreme Court action. John Livingston's anxiety to find a more profitable route for the *Olive Branch* can be appreciated.

With the monopoly matter settled in 1824, further opposition moved in during the 1825 season, bringing faster vessels and lower rates. The conservatism of the North River Steamboat Company put it in a poor defensive position and forced it to withdraw in 1826, the same year that Robert L. Stevens brought his family name back to the Hudson River with the highly successful *New Philadelphia*. This Philadelphia-built steamboat made the passage from New York to Albany in 12 hours 23 minutes at 11 1/2 miles per hour.[108] The *Kent*'s best time had been 14 hours 30 minutes. Competition and progress went hand in hand; in 1826 there were sixteen steamers on the river as opposed to three the year the monopoly was declared unconstitutional.

It is not the author's intention to follow the river steamboat further, for it proceeded in a different direction from the oceangoing vessel. Nevertheless, the development up to and including the *Chancellor Livingston* was part of the direct line of progress from the *North River* of 1807 to the first oceangoing steamship, the *Robert Fulton* of 1820.

The Three Fulton Steamers on Long Island Sound

The *Fulton* and the *Connecticut*, planned partly by Fulton, partly by Bunker, and operated under a license granted by the North River Steamboat Company, were fortunate in having neither competitors of consequence nor legal tests from 1816, the year the *Connecticut* began running, until June of 1822, when the State of Connecticut denied any New York steamer the right to use her waters. New Haven, New London, and Norwich could no longer be served by the two boats mainly owned by Colden but operated by Bunker.

Starting on June 4, 1822, the *Connecticut* sailed from New York at 4 P.M., was detained at Sands Point, Long Island, by a gale for 8 1/2 hours, arrived off Fishers Island at 8 P.M. on the 5th, where she remained another 3 hours, rounded Point Judith at 2 A.M. on the 6th, touched at Newport, and arrived at Providence, Rhode Island, at 8 A.M.[110] There were two major problems posed by this long route bypassing Connecticut waters. First was the large amount of firewood required in order to steam twice the distance previously covered. A boatload had to be picked up somwhere during the trip. The second matter for concern was the craft's seaworthiness once she passed outside the shelter of Long Island and was exposed to the open sea while doubling Point Judith. Because further voyages proved possible, the *Fulton* joined the *Connecticut* to provide two sailings a week connecting with stages from Providence to Boston. Fares were $9.00 to Newport and $10.00 to Providence. Because of the short land journey, the new service appealed to Boston-bound passengers; the net result of Connecticut's punitive action was to encourage long-distance travel. The citizens of New Haven and New London suffered since the replacement steamer, *United States*, owned in Connecticut but unlicensed for New York operation, had to land her passengers at Byram Cove, twenty five miles by stage from New York City.

Both the *Fulton* and *Connecticut* must have been very strongly built. In August of 1822 the latter sailed from New York on the 25th with 45 passengers. After a minor collision with a sloop, in which she lost a boat, she steamed along until 2 A.M. next morning when, about 15 miles east of New Haven, she managed to run into the *Fulton* in a head-on collision.[111] Passengers were thrown out of their berths but no one was injured. Although both lost their bowsprits, the chief damage was above the waterline. Captain Bunker, unwilling to put into a Connecticut port, took advantage of the fair weather prevailing and headed both back to New York steaming at about half their normal speeds. On a temporary basis, the little steamboat *Enterprize*, commanded by his son, R. S. Bunker, left her local New York to Oyster Bay run and substituted for her larger colleagues. The *Enterprize* was later to become the first steam vessel on active service in the U.S. Navy.

Two years later, when Connecticut complied with the Supreme Court's antimonopoly ruling, the Bunker pair tried returning to Connecticut ports but soon droppped them for the more profitable Rhode Island ones.[112] One physical change had been made since the boats were new; the *Connecticut* had an upper deck added (like the *Chancellor Livingston's*) to protect the passengers from the sun and rain—nothing as unpleasant as soot or sparks was ever mentioned in Captain Bunker's advertisements. When neither the view nor the open air could tempt the voyager, he could consult a 500-volume library to while away the time.[113] Ice conditions made year-round service impossible, but, as on the Hudson, the winter lay-up was used to overhaul the vessels and their machinery. As a rule the first sailing was early in March and the last sometime in December.

Just as the Hudson River saw the arrival of competition after 1824, so did Long Island Sound. Unlike the North River Steamboat Company, the Rhode Island line had an aggressive management, which, instead of giving up, countered by adding new boats of superior speed and comfort. In turn came the *Washington* in 1825, which had extensive improvements made in 1827; the *Benjamin Franklin* in 1828; and the *President* in 1829. Captain Bunker was intimately involved in their planning and invariably took over as master of the newest addition.

The *Fulton* and *Connecticut* were further remodeled from time to time and by 1826 *Boston Courier* advertisements stated that ladies' cabins had been added on deck and all the lower cabins enlarged and refurnished, and that the engines had been increased in power although the new boilers were still of the safe, low-pressure type built of copper. Pressures must, however, have been increased in order to increase speed and horsepower.

When the *Washington* appeared, she reopened the New London service that had been abandoned in 1822. The *Fulton* and *Connecticut* continued the through route to Newport and Providence. In 1826, however, all three were scheduled, at various times, from New York to Providence with sailings three times a week and to New London twice.

Most of the newer, rival lines concentrated on New Haven, Hartford, Bridgeport, and the Connecticut ports not served by Captain Bunker. The occasional one that tried to break into his lucrative Providence operation soon retired in defeat.

There is always an exception to any rule and on the Sound it proved to be the *Chancellor Livingston*. After the collapse of her original owners in 1826, she was extensively rebuilt and reengined to make her faster. The hull was altered, the row of rectangular windows replaced by circular ports to make her more seaworthy, solid bulwarks substituted for open rails to keep spray and water off her deck, and a three masted schooner rig added and supplemented by a square sail on the foremast. A new boiler gave her three stacks athwartships and a new crosshead engine, 56" bore by 6'-0" stroke, replaced the original 40" by 5'-0" one. Like the original, it was built by James Allaire.

The *Washington* was 137'-6" long with a tonnage of 339; the *Benjamin Franklin*, 142'-6" and 410. Despite their double walking beam engines, again by Allaire, the rebuilt *Livingston* still outclassed them in size, speed, and luxury. As a one-boat line she was able to hold her own from the spring of 1828, when she started under Captain Charles Coggeshall, until 1832. Because her appearance forced the New York to Providence fare down from $10.00 to $6.00, it is evident that the Bunker steamboats had heretofore been remarkably profitable. They still made money despite the 40% cut.

Near the end of her first season there was a race, October 9, 1828, between the *Livingston* and *Franklin*,[114] both setting out from Providence at the same time. The latter reached Newport 15 minutes ahead, but the former's piston was known to be cracked and cautious operation had prevailed. The defect was not immediately remedied and led, a few weeks later, to severe engine damage, the cylinder casting being broken. Her sails brought her home to Newport. Some sources refer to her having had two engines, others three. It is believed that the major repairs necessary at the end of 1828 led to the idea that a third engine was installed. Instead, it was an extensive replacement of the damaged parts of the second. She may have had three different cylinders, but the last two were identical in size.

Next year the *Fulton*, *Benjamin Franklin*, and *Washington*, for Captain Bunker, and the *Chancellor Livingston*, in opposition, continued on the Sound. Two events of importance were the retirement of the *Connecticut*, at the age of 13 years, and the unpopular decision of the owners of the *Livingston* to banish decanters of free whiskey and brandy from her dinner tables. Red wine was substituted. The *President* appeared in October.[115] With a length of 158', a tonnage of 518, berths for 137 passengers, and a speed permitting a 16-hour passage from New York to Providence, she became a most successful replacement for the older *Fulton*. At long last the *Chancellor Livingston* was exceeded in every respect. The *President*'s machinery[116] duplicated the *Washington*'s and the *Franklin*'s and had two beam engines with the wheels completely independent of each other. The *President*'s interior design included a completely new feature; she had individual staterooms replacing the long rows of upper and lower berths curtained off from the public cabins.

The *Fulton* was retired from the Providence run in 1830 and, by 1831, only the *Chancellor Livingston* remained on the Sound as the last representative from the Fulton era of steam navigation in the United States. She chose to demonstrate her disdain for newer craft on May 14. The *Washington*, under Captain Tomlinson, was rammed by the *Livingston*, under Captain Comstock, around midnight off New Haven.[117] The former was headed for Newport and the latter for New York. The *Chancellor* cut into the *Washington* just forward of the paddle box but was able to take off all fifty two passengers and her entire crew before the latter sank. Next morning she met the *President* and transferred the rescued passengers so that they could continue their journey. Captain Comstock was asleep at the time of the collision and an inexperienced pilot, who had just joined the vessel, was blamed for the collision.

The year 1832 was a disastrous one for all steamboat lines. A cholera epidemic in New York resulted in, first, the quarantine of passengers before they could land at Providence and, later, in complete disruption of the New York to Rhode Island and Boston service. This, together with the fact that she was now outclassed, led to the sale of the *Chancellor Livingston*, in April of 1833, to Amos H. Cross of Portland, Maine, and Cornelius Vanderbilt. Already in Maine waters was the *Connecticut*, which had been purchased in April of 1829 by Messrs. Seward and Samuel Porter.[118] These two remaining elements of the Fulton fleet brought regular down-east service from Boston to Portland and Bath for the first time. Previous operations in this region had been sporadic and short-lived, with very small craft, like the *Patent*, which had reached Maine waters in 1823; the steam brig *New York*, in 1824; and the *Legislator* in 1826. The *Chancellor Livingston*, under Captain William Porter, sailed from Boston twice a week to Portland. The *Connecticut*'s schedule was similar.

On July 23, 1833, however, the *Livingston* made an excursion around Boston's island-dotted harbor under John Stubbs, one of the local pilots. There were two hundred passengers aboard and there was a brass band from Philadelphia to serenade them.[119] About 6 P.M. she struck Hangman's Ledge bow on at full speed and drove one-third of her length aground. The sloop *Glance* of Cohasset came alongside and offered to take off the passengers for $25.00, but the evening steamer to Hingham, the *General Lincoln* and its Captain Beal, intervened as a less commercial Samaritan and, after putting ashore both her own and the *Livingston*'s passengers, returned to try to help her off. Low water showed no actual holes in the hull although some deck beams were broken and bout two feet of water had leaked in. She was pulled off on July 25.

Although the Boston *Transcript* reported that there was "no probability she will go to sea again as a steamer" and a New York paper added that the keel was broken in two places, the real damage was less than reported.[120] For the remainder of the year the steamer *McDonough* (or *MacDonough*) was brought from Hartford to take her place.

Not only was the *Livingston* repaired for the 1834 season, but the even older *Connecticut* was sent back to New York where she was completely rebuilt and the single boiler from 1826 replaced by two new ones mounted, not in the hull, but on the guards in the latest Hudson River style.[121] Under Captain William Porter she returned to the Boston to Portland run by day and continued on to Bath and Gardiner, Maine.

The *Chancellor Livingston* and the *MacDonough* both made overnight runs to Portland. That many sailings a week proved to be more than the route could stand. On August 12 the *MacDonough* was advertised as taking the place of the *Chancellor Livingston* for the rest of the season. It was the end of an enviable seventeen-year record. In her early years she reigned supreme on the Hudson. A decade later, rebuilt and equipped with a new engine, she became the queen of all the steamers on Long Island Sound. And, as a dowager, she spent her last two years providing service of unprecedented luxury from Boston to Portland. After the *North River* she is the most important steamboat to appear in American history, and she reflects the greatest credit on Fulton, who conceived her; Henry Eckford and his foreman, Isaac Webb, who designed and built her; Charles Stoudinger, designer of her first engine; and James Allaire, who built both the original engine and its later replacement. Even when dismantled in 1835, she ceased to exist only in part. A new hull was constructed for the Cumberland Steam Navigation Company and her engine transferred to it. Measuring 163'-6" long by 27'-2" beam by 10'-7" deep, of 445 tons, the new boat was named *Portland* and achieved a certain amount of fame as the first Maine steamer to burn anthracite coal instead of wood, just as the *Livingston* was the first coal burner on the Hudson.

After twelve years on the Maine coast the *Portland* was acquired by Charles Morgan in 1847,[122] just in time for an extremely profitable charter to the U.S. Army during the Mexican War. She carried troops and supplies from New Orleans to Vera Cruz and Tampico, Mexico. Despite the onus of throwing overboard 100 horses to keep from sinking on one stormy voyage, she was chartered again in 1848, at the excessive rate of $8,000 per month. She later reverted to Morgan's New Orleans, Galveston, Rio Grande operations until dismantled in 1854.

Although the original steamer saw only seventeen years' service, her second engine, despite its 1828 catastrophe, lasted for twenty-seven.

The exact end for either the *Fulton* or the *Connecticut* is in doubt. The last date available for the former is the spring of 1831 when she was engaged on local service out of New Haven.[123] The Lytle List[124] shows that her papers were marked *abandoned in 1838*, but this could be a meaningless date—the vessel may have been scrapped years before with no notice ever getting back to the Custom House where she was enrolled. She may perhaps have been converted to a barge, laid up, and have existed for many years after such official notice was received. The *James Kent*, for example, was marked *abandoned in 1842*, but continued to exist, first as a coal barge on the Delaware and Hudson Canal, and then as a stake boat in New York harbor for twenty years. She was finally broken up in 1895.[125]

A month after the *Chancellor Livingston* was withdrawn, the *Connecticut* set out from Providence, Rhode Island, for New York. There she was advertised to sail for Charleston, South Carolina; Key West; New Orleans; Matamoras; and Tampico, Mexico, still under Captain Samuel Porter. She would depart November 18, 1834, and offered to take 160 passengers and light freight.[126] She did reach the Gulf of Mexico, for the New York "Commercial List," in November of 1835, reports that she had put into Havana in distress and was condemned. This date is in fair agreement with the Lytle List's report that she was sold to foreign owners in 1836.

The *Connecticut* was the first American steamer to travel from New York all the way to Maine waters and then to return, head South around stormy Cape Hatteras, and make a long winter voyage to the Gulf and, if the advertisement is correct, clear to Mexican waters. It would be of great interest if it were possible to find out more about a three-thousand-mile voyage by a wood-burning steamer with many stops along the way for fuel. Perhaps the wild rumor of 1815 that she was to cross the Atlantic is almost justified by her 1834 exploit and her 1835 operation in the Gulf of Mexico. With her passing, the Fulton-Livingston era of steam navigation finally ended. The steamboat had been invented, developed, and adapted to operate in waters far from the Hudson River. In the *Fulton, Connecticut,* and *Chancellor Liv-*

ingston a remarkably successful transitional type of craft halfway between a steamboat and an oceangoing steamship had evolved.

Notes

1. Jonathan Hulls, *Description and Draft of a Newly-Invented Machine for Carrying Vessels or Ships out of and into any Harbor*, (London, 1737).
2. H. Philip Spratt, *The Birth of the Steamboat*, (London, 1958), pp. 23-27.
3. James T. Flexner, *Steamboats Come True*, (New York, 1944), presents the only account that is both accurate and unbiased in its treatment of American claimants.
4. R. J. Law, *The Steam Engine*, Science Museum, (London, 1965), pp. 13-14.
5. Thompson Westcott, *Life of John Fitch*, (Philadelphia, 1857), pp. 157-59.
6. Ibid., pp. 159-60.
7. Ibid., pp. 132-33.
8. Charles Frederick Partington, *An Historical and Descriptive Account of Steam Engines*, (London, 1822), p. 46; Carroll W. Pursell, Jr., *Early Stationary Engines in America*, (Washington, D.C., 1969), pp. 5-9.
9. Westcott, *Fitch*, pp. 255-56.
10. Ibid., pp. 281-84.
11. Deposition by Joseph Barnes dated December 10, 1787, in James Rumsey, *A Short Treatise on the Application of Steam*, (Philadelphia, 1788).
12. James Rumsey, *Short Treatise*.
13. James Rumsey, *Explanation of a Steam Engine, and the Method of Applying It in a Boat* (Philadelphia, 1788).
14. *Gentleman's Magazine*, February 1793, p. 182.
15. Flexner, *Steamboats*, p. 213.
16. Alexander Crosby Brown, "James Rumsey—Steamboat Inventor," *Steamboat Bill* (1969), pp. 134-38.
17. Alice Crary Sutcliffe, *Robert Fulton and the "Clermont"* (New York, 1909) and H. W. Dickinson, *Robert Fulton, Engineer and Artist* (London, New York, and Toronto, 1913) published the essential documents and correspondence covering the development of the vessel that was never named *Clermont* in her lifetime. The Fulton-Livingston agreement is in Sutcliffe, *Fulton*, pp. 117-22.
18. Robert Fulton, 11 page MS, January 9, 1803, New York Historical Society.
19. *L'Illustration, Histoire de la Marine* (Paris, 1934), pp. 324-25.
20. Dickinson, *Fulton*, pp. 155-59.
21. Ibid., pp. 160-64.
22. Ibid., pp. 168-78 quotes the correspondence. The originals are in the Boulton and Watt Collection of the Birmingham, England, Public Library.
23. Drawings at the New York Historical Society and the New Jersey Historical Society, respectively. A later proposal for the *Car of Neptune*, still unlabeled, is at the Philadelphia Maritime Museum.
24. Mark Beaufoy, *Nautical and Hydraulic Experiments* (London, 1834), is the final report on these. Fulton used partial reports available before he left England.
25. Dickinson, *Fulton*, pp. 289-312.
26. Ibid., p. 216.
27. Sutcliffe, *Fulton*, p. 192.
28. Cadwallader D. Colden, *The Life of Robert Fulton* (New York, 1817) table of steamboat dimensions following the text.
29. Records of the Bureau of Marine Inspection and Navigation, Record Group 41, U.S. National Archives.
30. Donald C. Ringwald, *First Steamboat to Albany*, The American Neptune, (July 1964), pp. 157-71.
31. John W. Griffiths, *A Treatise on Marine and Naval Architecture* (London, 1856), p. 149.
32. Dickinson, *Fulton*, p. 320.
33. James Renwick, *Reminiscences of the First Introduction of Steam Navigation*, The Historical Magazine (August 1858), p. 227.
34. Ringwald, *The American Neptune*, p. 158.
35. Sutcliffe, *Fulton*, pp. 222-223.
36. *Nautical Gazette*, September 5, 1907, pp. 162-63. Samuel W. Staton, ed., wrote 126 installments of *A History of The First Century of Steam Navigation* that were published 1907 through 1910. Many original documents and newspaper reports are extensively quoted throughout the series. The actual period covered was 1807 through 1825.
37. *Nautical Gazette*, October 3, 1907, pp. 226-27.
38. Sutcliffe, *Fulton*, pp. 259-64.
39. Deposition by Charles Brownne, shipbuilder, dated July 23, 1811, *Nautical Gazette*, January 23, 1908, p. 64.
40. Regulations for the North River Steam-Boat reproduced in Sutcliffe, *Fulton*, pp. 268-69.
41. Frederick D. Herbert, *Robert Fulton's Original Drawings*, Transactions, Society of Naval Architects and Marine Engineers 1934, pp. 21-39.
42. Renwick, *Reminiscences*, p. 228.
43. Archibald Douglas Turnbull, *John Stevens* (New York and London, 1928), pp. 185-93.
44. Geo. Henry Preble, *A Chronological History of the Origin and Development of Steam Navigation* (Philadelphia, 1883), and H. Philip Spratt, *Birth of the Steam Boat*, are two examples of standard works that have important Fitch and Stevens dates advanced one year ahead of actuality.
45. Turnbull, *John Stevens*, pp. 236-37.
46. Ibid., p. 258.
47. *New York Evening Post*, October 20, 1808.
48. Turnbull, John Stevens, p. 259.
49. Ibid., pp. 275-78.
50. Moses Rogers letter, July 1, 1809, to John Stevens. New Jersey Historical Society.
51. Turnbull, John Stevens, p. 280.
52. *Philadelphia Aurora and General Advertiser*, August through October 1809.
53. John H. Morrison, *History of American Steam Navigation* (New York, [1903], p. 168.
54. Livingston-Fulton Account Book. New York Historical Society.
55. *New York Evening Post*, July 10, 1809.
56. *Philadelphia Aurora and General Advertiser*, 1809-10.
57. J. R. Livingston letter, November 17, 1813, New Jersey Historical Society.
58. *Philadelphia Aurora* advertisement, June 11, 1810.
59. John Livingston letter printed in the *New York Evening Post*, July 15, 1809.
60. J. H. Ward, *Steam for the Million* (New York, 1862), p. 89.
61. Morrison, *Steam Navigation*, p. 37.
62. Turnbull, *John Stevens*, pp. 286-87.
63. John H. Morrison, *History of New York Ship Yards* (New York, 1909), pp. 31-32.
64. Contract, Robert Fulton and James Bennett, dated May 12, 1809, New York Historical Society.
65. *New York Evening Post*, October 9, 1809.
66. *Nautical Gazette*, January 23, 1908, p. 63.
67. July 1, 1811.
68. *Nautical Gazette*, January 30, 1908, p. 93.
69. *Nautical Gazette*, February 27, 1908, p. 155.
70. Francis B. C. Bradlee, *Steam Navigation in New England* (Salem, 1920), pp. 140-41.
71. New York Evening Post, July 5, 1814.
72. Livingston-Fulton Account Book, New York Historical Society.
73. *Nautical Gazette*, May 7, 1908, p. 311.
74. *New York Evening Post* advertisement, May 11, 1814.
75. *Nautical Gazette*, April 9, 1908, p. 262.
76. Jean Baptiste Marestier, *Mémoire sur les bateaux à vapeur des Etats-Unis d'Amérique*, (Paris, 1824).
77. *New York Commercial Advertiser*, advertisement, June 30, 1815.
78. *Lancaster* (Pennsylvania) *Intelligencer*, August 10, 1816; Pierre Andriel, *Coup d'oeil historique sur l'utilité des batiments-à-vapeur dans le royaume des Deux-Siciles*, (Naples, 1817).
79. *New York Evening Post*, October 2, 1816.
80. *Nautical Gazette*, September 3, 1908, p. 113.
81. *New York Commercial Advertiser*, July 11, 1823.
82. Turnbull, *John Stevens*, pp. 314-15.
83. *Philadelphia Aurora*, advertisements October 30, 1813, through January 5, 1814.
84. Robert L. Stevens letter, March 31, 1814, to John Stevens. New Jersey Historical Society.
85. *Nautical Gazette*, July 9, 1908, p. 17.
86. Ibid., August 13, 1908, pp. 76-77.
87. *New York Evening Post* advertisements March 24 through November 13, 1817.
88. Morrison, *Steam Navigation*, p. 168.
89. *New York Evening Post*, November 27, 1815.
90. *Commercial Pathfinder*, September 1868. (Clipping in a scrapbook

of William H. Webb.) Webb Institute of Naval Architecture.
91. Enrollment dated March 29, 1817. U.S. National Archives.
92. Axel Klinkostrom, *Bref om de Forenta Staterna forfattade under en resa till Amerik* (Stockholm, 1824).
93. Landing of General Lafayette at Castle Garden, drawn by Imbert, engraved by Sam'l. Maverick. New York Public Library.
94. *Nautical Gazette,* June 4, 1908, pp. 358-59.
95. Marestier, *Memoire,* trans. Sidney Withington, Marine Historical Association (Mystic, Conn., 1957), pp. 13-17. This translation is a partial reprint of the 1824 French edition.
96. *Commercial Pathfinder,* September 1868.
97. *New York Evening Post,* March 31, 1817.
98. Ibid., July 10, 1817.
99. *Nautical Gazette,* October 1, 1908, p. 161.
100. Ibid., April 8, 1909, p. 291.
101. Robert L. Livingston Letterbook, 1817-26, August 15, 1822. New York Historical Society.
102. Rev. Herman Stryker Diary, September 10 to September 16, 1822. New York Historical Society.
103. *Nautical Gazette,* September 30, 1909, p. 254.
104. Ibid., October 28, 1909, p. 327.
105. Wheaton J. Lane, *Commodore Vanderbilt,* (New York 1942), pp. 29-43.
106. *New York Evening Post,* April 21 and May 19, 1824.
107. *Nautical Gazette,* November 11, 1909, pp. 363-64.
108. Morrison, *Steam Navigation,* p. 50.
109. *Nautical Gazette,* February 3, 1910, p. 87.
110. Ibid., August 19, 1909, pp. 146-47.
111. *New York Evening Post,* August 27, 1822.
112. *Nautical Gazette,* November 18, 1909, p. 381.
113. *New York Evening Post,* August 3, 1821.
114. Henry Whittemore, *The Past and the Present of Steam Navigation on Long Island Sound,* (Providence and Stonington Steamship Company, 1893), pp. 39, 42.
115. Ibid., p. 42.
116. Morrison, *Steam Navigation,* p. 266.
117. Whittemore, *Past and Present,* p. 43.
118. Francis B. C. Bradlee, *Some Account of Steam Navigation in New England* (Salem, 1920), pp. 54-56.
119. *Boston Transcript,* July 24, 1833.
120. *New York Shipping and Commercial List,* July 27, 1833.
121. *Boston Transcript,* July 1, 1834.
122. James P. Baughman, *Charles Morgan* (Nashville, Tenn., 1968), pp. 47, 50, 96.
123. Melancthon W. Jacobus, *The Connecticut River Story* (Hartford, Conn., 1956), p. 48.
124. Forrest R. Holdcamper, ed., William M. Lytle, comp., *Merchant Vessels of the United States, 1807-1868,* (Mystic, Conn., 1952).
125. G. W. Murdock, annotated picture albums, New York Historical Society.
126. *New York Shipping and Commercial List,* November 12, 1834.

2
The *Robert Fulton*

THE *Chancellor Livingston*'s success, starting in 1817 on the Hudson, has been dealt with in considerable detail because this vessel led directly to an even more important one, the *Robert Fulton* of 1820. The latter was the first oceangoing steamship, both in concept and execution, to appear on American waters. David Dunham, merchant of New York City, who had managed coastwise packets for some years, was the leader of the enterprise that ordered her, Henry Eckford was both builder and part owner, and James P. Allaire served as engine builder and owner. During her construction Jasper Lynch, formerly employed by Fulton, supervised the engine work[1] and, in the newspapers, was erroneously credited with ownership. Thus the shipbuilder and engine builder who had combined their talents so successfully on the *Livingston* joined forces a second time to produce the largest steamship in the world and placed her, under the management of an experienced shipowner, on the New York to Charleston and New Orleans route. Both New York and New Orleans were major ports and the cargoes carried between them formed an essential part of the country's coastwise trade.

Her original register, of April 22, 1820, although difficult to decipher, seems to show dimensions of 159″ in length, 33′-5″ beam, and 17′-3″ depth. There was a transom stern with quarter galleries and a bust as a figurehead. Her hull was constructed, in Eckford's normal manner, of the best possible materials: live oak, locust, cedar, and Georgia pine, with copper fastenings throughout the bottom and the bilges, and covered with sheet copper below the water[2] to prevent the growth of barnacles that would slow her down. It would also protect her from the marine borers called teredos. The length and beam dimensions were close to the *Chancellor Livingston*'s, but, as befitted a seagoing ship, the depth was considerably increased, as was the draft, the latter being about 10′. Since there were no dredged channels into Charleston, the draft was lower than that of a contemporary sailing packet.

The engine was an enlargement of the *Livingston*'s, still of the single-cylinder, crosshead type, with the bore enlarged from 40″ to 44″ in diameter and the stroke retained at 5′-0″. The flywheels, 12′-0″ in diameter with 6″ wide rims and 7″ spokes, were turned at twice the speed of the engine by means of 36 tooth gears meshing with 72 tooth ones on the paddle shafts. A colossal copper boiler 30′-10″ long by 12′-10″ by 8′-10″ high[3] had four return flues and two stacks side by side. As a coal burner, the *Fulton* was the first designed to use this fuel exclusively. The required volume of firewood made any long sea voyage impossible. The wheels were 18′ in diameter with ten floats 6′-6″ wide and were mounted on 8″ square shafts.

All the machinery was surrounded by wooden bulkheads lined with lead to make them watertight; the space could be flooded to put out a fire without sinking the ship. The rig, as originally planned, was a small one consisting of lug sails, a most unusual departure from normal American practice. Such a radical idea was typical of Eckford and shows that he considered the sails a purely auxiliary means of power. When they were furled everything was lowered on deck, leaving only bare poles and shrouds with the minimum of wind resistance. The *Robert Fulton* was a very large vessel for her time. The cost, $130,000, was equally large and represented a major financial risk assumed by the three owners. Both hull and machinery required an extensive building period; although launched in May of 1819, a whole year was needed to install the machinery and to finish and outfit her properly. There were extensive rumors, just as there had been during the *Connecticut*'s construction. One was that she was intended to operate from New York to Liverpool or London[4].

About the middle of April in 1820 a trial was made.[5] She steamed down the East River, rounded the Battery into the North River for a ways, then turned around and headed down the harbor as far as Gravesend Bay before returning. The distance was estimated to be 22 to 24 miles and the time two and one-half hours, all without sails. A journalist's reaction was: ". . . truly one of the wonders of the present age."

The first departure of the *Robert Fulton*, on April 25, 1820, for New Orleans, via Charleston and Havana, was celebrated by a sailing party for several hundred invited guests,[6] including Cadwallader Colden, who was doubly honored as a steamboat owner and as Mayor of New York. The well-wishers went down the harbor to Sandy Hook on her, dined aboard, and then transferred to the *Connecticut*, under Captain Elihu Bunker. During the return a series of flowery resolutions were passed by the guests praising the steamship, her owners, New York City, Captain Bunker, and any other subject brought to their attention; everyone had a wonderful time.

Under Captain Inott or Mott (the author is not sure whether there were two different masters on the *Robert Fulton*'s first two voyages or only the latter with a sometimes misspelled name) the subject of all the excitement steamed smoothly along. The only untoward event as she headed South was that some of the paddles worked loose, fouled the guards, and were broken. The engine had to be stopped to permit replacement but was in continuous use during all the rest of the trip. Upon reaching Havana, permission was requested to enter the harbor without the payment of port dues.[7] Not only was this granted, but the Governor, the principal dignitaries of the town, and the merchants of note all came out to inspect their unusual guest. On May 10 she left for New Orleans, steamed from there on May 28, from Charleston June 11, and arrived New York June 14, 1820. In all she was 17 days 12 hours from New Orleans to New York, including stops of two days at Havana and four at Charleston. On board were sixty to seventy passengers. Once more there was a series of resolutions, which included the following statement:

> Having experienced on our passage, several heavy blows, we have had good opportunities to judge whether any good cause existed for the apprehension at first entertained as to her safety in rough weather, and do not hesitate to declare them groundless. Being unincumbered with that heavy weight of spars and rigging, she consequently rolls and labours much less than a vessel rigged in the usual way, and propelled only by sails; and the manner in which she contended for several days with a heavy head sea, and wind, convinces us that the duration of her passage will never be materially affected by the weather she has to encounter.[8]

The *Robert Fulton*'s second trip was less of a triumph. Sailing from New York on July 2, she was south of the Florida Keys nine days later when a flywheel gear gave way, overloaded one side of her transmission system, and broke the crosshead.[9] She put about and sailed back to New York, arriving off Sandy Hook on July 20. A 1,200-mile trip in eight to nine days under a limited rig indicates that, despite her design as a steamship, she had a hull with excellent sailing characteristics.

Repairs occupied all of August. An excursion trip around Long Island was advertised for September 11 or 12 but was called off because of lack of patronage.[10] It was followed by a third departure, on October 10, this time for Charleston only. New Orleans had been canceled because of a yellow fever outbreak. Despite a head wind that developed almost into a hurricane, Charleston was reached in five days and a quick return made. After this Captain Mott decided that more sails would be an asset, perhaps to save coal, perhaps to steady the ship and reduce rolling.

In the extremely short time of two weeks, new masts were made, stepped, rigged, and the *Robert Fulton* became a three-masted, square-rigged ship with topgallants on all masts and large spencers on the fore and main in addition to the usual jibs, staysails, and a spanker.[11] This was a flexible arrangement; she could send the yards down and operate as a three-masted schooner if the wind were ahead of the beam, or set all her square sails if it were aft. The lower masts were made taller than usual and the topmasts shorter in order to facilitate this novel operation.

Cold weather helped abate the yellow fever outbreaks; on November 5 the whole itinerary was undertaken—Charleston, Havana, New Orleans, and an extra stop at Savannah, Georgia. Southbound there was such a strong following wind that it carried away the fore and main yards soon after leaving New York; the new square sails were eliminated by the whims of nature. The steamer carried no freight other than gold or silver. Slaves and servants were in limbo; they were not listed as passengers. The slaves, like cargo, had to be declared at the Custom House before they could board.

Piracy was still a serious problem in the Caribbean in the 1820s and the coast of Cuba was a favorite haunt from which pirates emerged to prey on the steady stream of traffic between East Coast cities and New Orleans. Returning from that port, the *Robert Fulton*, on March 30, 1821, was chased and fired on by a "piratical" three-

Newspaper cut of the *Robert Fulton*. Courtesy of The New-York Historical Society, New York City.

masted schooner; her paddle wheels enabled her to escape with ease.

The U.S. Navy was assigned the task of clearing up the situation and decided that a steam-powered vessel of small size and light draft was essential for the work.[12] For this they acquired the steamboat *Enterprize,* formerly captained by R. S. Bunker, son of the master of that name whom we have seen on many occasions. She had been built in Hartford, Connecticut, in 1818 and ran on Long Island Sound from New York to Mamaroneck and Oyster Bay in 1821 and 1822 before being purchased in December 1822 by Commodore David Porter.[13] She was refitted, renamed *Sea Gull,* and served as the flagship of a fleet of schooners under Porter's command in West Indian waters during 1823 and 1824. She was unique in having a rotary engine, of which nothing is known of its details. Before joining the navy she had neither masts nor sails.

A much grander approach had been advocated by a New York paper,[14] which advised the purchase and arming of the *Robert Fulton* for the work. There was even a discussion of a whole fleet of steam frigates to be copied after her.

Following her escape from pirates, the *Fulton* made two more voyages to New Orleans in 1821. Then, in August, a grand excursion was offered that would take three days and two nights.[15] She was to sail from New York on Thursday, August 9, head for Sandy Hook, and, subsequently, stop to allow her passengers to fish for sea bass. She would take on additional passengers at Long Branch, New Jersey. She would then circumnavigate Long Island. One hundred and seventy-five passengers paid the $10 fare and proved so enthusiastic that a second "party of pleasure" was announced from New York and New Haven to Newport and Providence and return for $12.50 for adults and at half the price for children and servants.[16] This time she carried a more comfortable load of eighty persons. One of these, John Quincy Adams, then Secretary of State, left her at Providence to continue on his way toward Boston and Quincy.

The trip was equally successful and each landing found the whole waterfront crowded with the curious, gathered to see and welcome this fabulous craft.[17] The passengers hurried ashore to see the sights and the visitors rushed aboard with the identical purpose. At Newport, in an attempt to keep down the throngs, only ladies were permitted aboard. Some six hundred of the fair sex took advantage of the opportunity.

A fairly detailed account of her next voyage, one going only to Charleston and Savannah, is available.[18] Also, for the first time, advertisements gave fares: $40 to Charleston, $50 to Savannah, children and servants at half price, with beds, bedding, and meals included. Liquors, however, were extra. When the *Robert Fulton* got under way at 10:35 A.M. on September 30, 1821, a strong south wind was blowing. As she steamed toward Sandy Hook all the yards were lowered and the topmasts struck so that windage would be reduced. At 3:20 in the afternoon she rounded the Hook, put her visitors on board the pilot boat, and headed into a heavy sea and a rising gale. Both were severe enough to prevent progress; it must be remembered that her engine could produce less than 100 horsepower and roughly one-half of this would be consumed by the friction in the engine and the fluid losses in the side wheels. After two hours of attempting the impossible, Captain Barnard put back into the shelter of Sandy Hook to ride out the storm. Anchor was weighed next day at 3 P.M. and the steamer plowed along into heavy swells. When the wind veered to a favorable direction the topsails were set, the wheels disconnected—this report is the only indication that it was possible to do so—the fires drawn, and the *Robert Fulton* continued on course entirely under sail from 2 A.M. until 8 A.M. on October 2. At times during the night a speed of 11 knots was attained. As the wind died, the sails were taken in, steam raised, and a 7-knot speed under power was achieved. The seasick passengers emerged from their misery as wind and wave diminished. The next night a head wind arose; once more the spars and running rigging were stripped from the masts. On October 4 the wind was favorable again and sail hoisted. When squalls appeared, she proceeded under steam and reefed topsails.

It is clear that Captain Barnard took every possible advantage of the wind and used his seamen in the best sailing ship tradition to set sails, furl them, strike topmasts as necessary, and see that the ship sped along as rapidly as possible with the least expenditure of fuel. On October 6 the Charleston pilot was picked up and the dock reached by 10 A.M. Once the passengers were landed, gold and silver shipments were taken aboard and she headed out at 3 P.M. After striking the bar with her keel on three separate occasions, she made sail for Savannah, took on the pilot there, and anchored at the river mouth waiting for a favorable tide. She did not reach the city until twenty four hours after leaving Charleston. On the 8th at 3 P.M. she got under way but the Savannah pilot proved even worse than the Charleston one and put her on an oyster bed, where the falling tide left her hard aground until 5 A.M. the next day.

The wind being adverse, the *Fulton* beat her way under fore-and-aft sails plus steam to Charleston, docking there at 6 P.M. on the 10th. Two days were spent coaling and preparing for the return. At 11 A.M., October 12, she headed out to sea at 8 to 10 knots. Since two ministers were aboard, one conducted the Sunday services while the other preached the sermon. Northbound the wind was variable; at times she set sail; at others she steamed into a heavy sea under bare poles. On Wednesday, October 17 the New York pilot came aboard. Quarantine was passed without delay and the Battery reached at 3 P.M. The round trip took 17 days, including all stops and detentions. The newspaper account quotes 1,500 nautical miles in 12 1/2 days, 120 miles per day, or 5 knots speed, much of it against adverse wind and sea conditions. It continues:

> She is beyond any question, the safest, most expiditious, and most comfortable of any species of naval architecture that has ever floated on the ocean; and experience has uniformly extracted the substance of this statement from any passenger who ever sailed on her.

The modern reader should realize that, for 1821, this is a completely true statement. The only through route from New York to the major cities of the Southeast, Charleston and Savannah, was served by small sailing packets, well below 100' long. Moreover, the usual passage past stormy Cape Hatteras resulted in an average speed below 5 knots. No sailing ship of her size ran to New Orleans and none had her passenger facilities. The open deck space when the weather was good, and the extensive cabin, galley, and steward's staff to look after creature comforts when it was bad, did not exist on any other seagoing vessel.

After four voyages in 1820 and five in 1821, the *Robert Fulton* was put up for sale and purchased by David Dunham on December 28, 1821. The *Commercial Advertiser* stated that it was "merely to quiet some old claims made by others." She continued to Charleston, Havana, and New Orleans with occasional additions, such as Pensacola or Savannah, and sometimes deletions, usually to avoid quarantine and delay when yellow fever epidemics were rampant.

Her twelfth voyage, in the summer of 1822, is also recorded in some detail, starting with her departure from New Orleans on June 28, 1822,[19] with 106 cabin passengers who paid $100 each to go to New York. The sea was smooth, the weather warm; the engine and boiler made it hot below. The crew rigged up an awning over almost the whole deck, strung up eighty hammocks, and moved settees and mattresses up for the others. Even the ladies slept in the open to take advantage of the cooling breezes. On June 30 they breakfasted and dined there as well. When nearing Cuba, the ship's guns were got ready for action and a hose rigged to throw hot water on any enemy who ventured too close. No pirates appeared, but a Dutch galliot headed for Havana was nearly run down. Havana harbor was reached on July 4 and a salute was fired by the U.S. naval sloop *Hornet* anchored there. The local bumboat fleet surrounded the *Fulton*, offering pineapples, coconuts, oranges, limes, bananas, and other produce. Twenty more passengers boarded. When she sailed she had with her three parakeets, fruit enough "to kill the passengers," and baggage "to fill a good sized ship." The author of the account was pleased that no monkeys were included. On the 6th, the day after leaving Havana, the weather became squally and the sea rough, and on July 7 torrents of rain drove the passengers below decks for the first time. Next morning everyone returned topsides, where they remained until Charleston was reached on July 9. Because there was one sick passenger, the ship was quarantined and no one was allowed ashore except to transfer to the fenced-off quarantine grounds.

When the *Fulton* headed out on the 11th, the ship *Atticus* was crossing the bar to windward of her and drifted down on her, carrying away the bowsprit and jibboom. Despite this, she spent all day under sail on the 12th while some repairs were made to the engine. The fact that she logged 10 knots, despite missing headsails, is another example of her exceptional sailing qualities.

Off the New Jersey coast, the last night at sea was celebrated by hot whiskey punch and suitable toasts. New York was reached on July 15 but a four-day quarantine was established because of sickness; everyone had to stay aboard. In this case, with good weather and sea conditions, a 1,800 nautical-mile trip had been made in thirteen days at sea at an average speed of about 6 knots.

Thereafter, the *Fulton* was withdrawn for four months while the hull and rigging were repaired, overhauled, and a new boiler installed.[20] Under a new master, Captain Chase, she started on a voyage that was very long in time. Leaving New York near the end of November, she did not get back until February 3, 1823.

For over a month before her return she was advertised as sailing beyond New Orleans to Vera Cruz, Mexico, provided enough passengers appeared.[21] In fact, when she left New York on February 9, it was for her usual ports. During her absence David Dunham was lost overboard from a Hudson River sloop and an auction sale was arranged at the Tolentine Coffee House by the administrators of his estate.[22]

Despite the sale she returned to her regular route, still managed by Dunham and Company, leaving New York on June 8, 1823, and not returning again until October 4th. Without the senior Dunham in charge, the sailings became irregular and no real attempt seems to have been made to keep her running. Yet, on her arrival just after his death, she had a quite acceptable passenger list of fifty-two from New Orleans, eleven from Havana, and five from Charleston. There is no reason to believe that she had been unprofitable. The year 1823 saw only three voyages and 1824 two, the first marred by a boiler leak that forced her to return to New York and spend four days repairing it.[23]

August 16 of 1824 saw the arrival of the Marquis de Lafayette, the Revolutionary War hero, returning to America. The U.S. Navy put two hundred sailors aboard the *Robert Fulton* to man her yards and fire a fifteen-gun salute as she led a grand procession of vessels steaming up the harbor.[24] The Marquis was aboard the *Chancellor Livingston*, which followed the *Robert Fulton*. The smaller *Bellona* and *Nautilus* came next and towed the ship *Cadmus*, which had brought Lafayette across the Atlantic. Finally, on either flank, came the *Oliver Ellsworth* and the *Connecticut*. A memorable event it was.

On her final arrival, January 10, 1825, she brought not only passengers from New Orleans, but also a cargo of hides and cotton. Prior to 1823, only specie had been carried.

In all the *Robert Fulton* made eighteen voyages in five years of active service, thirteen of them under David Dunham's direction in her first three years. The public was regularly informed by advertisements of her plans, usually well in advance, and any delays or changes were announced as soon as the information became available. Starting in 1821, a rather large cut showing the steamer in her ship rig was inserted to catch the reader's attention. The surviving watercolors of her, at the Peabody Museum, New York Historical Society, and Mariners Museum, all appear to be later reconstructions by Richard V. deWitt, based on the cut with further details added by the artist.

In 1827 she was sold to the Brazilian Navy for 152,000 milreas, and converted to a 24-gun sailing sloop; as late as 1838 she was still operating and some time thereafter was wrecked on the Brazilian coast.[25]

The *Robert Fulton* as a steamship, and her short commercial life, 1820-25, are somewhat difficult to place in their proper perspective. One point of view is to consider her the culmination of the James Watt, Robert Fulton, Hudson River type of craft. Without doubt she was just this. Her hull was a direct development from the *Chancellor Livingston*, with the necessary changes to make her seaworthy, fast under sail, and an admirable craft for the unprotected waters of the Atlantic. Her crosshead engine was, the year it appeared, the ultimate step in a direct development starting with the *North River*'s Soho-built cylinder, air pump, and valve gear. The boiler was an improvement, since it had internal furnaces and flues, but was still a large, low-pressure, copper affair containing tons of boiling sea water. The chief engineering change was the exclusive use of coal instead of wood, an essential feature to enable her to stay at sea for the 1,250 nautical miles from Charleston to New Orleans without refueling at Havana.

The *Robert Fulton* was the first successful American oceangoing steamship. Although we have no financial reports, the fact that she continuously operated, that

New York newspapers listed anywhere from seventy to one hundred passengers on her arrival (a further group of servants and slaves should be added), and that the income warranted a number of improvements in rig and interior, and a boiler replacement, proves that David Dunham had the profits to finance the changes and was convinced that she would pay off any additional investment.

All this is a far cry from the *Savannah,* of 1819, which sailed for Europe without a passenger, could not find a buyer abroad, returned to the United States under sail, was sold to pay her debts, and had her engine removed, all within eighteen months.

Yet the *Fulton*'s career changed abruptly after Dunham's death. There were few sailings and long delays at New York or New Orleans, all without explanation. She obviously was not a success in her last two years, for which no clear reason emerges except the apparent loss of interest of her managers, Dunham and Company, once the senior member of the firm was no longer present. As in Fulton's case, Dunham's enthusiasm and continuous attention to the details of operation were the essential part of his steamship's success. Without his personal attention she sank into obscurity; when she became a sailing ship even her new name was lost; she just vanished into South American waters without any reported sailing date. Worst of all, there were no direct descendants. When she stopped running, the only oceangoing steamship in American waters ceased to exist. For two decades there was no real successor; Americans devoted their energies to filling the rivers, bays, sounds, and lakes with lightly built steamboats of ever-increasing size and speed, but long-distance travel by sea was put aside until the late 1840s, when the seagoing steamship had to be redeveloped by another generation of shipbuilders.

Notes

1. *New York Evening Post,* March 4, 1820.
2. Ibid., March 2, 1820.
3. Jean Baptiste Marestier, *Mémoire sur les bateaux à vapeur des Etats-Unis d'Amérique* (Paris, 1824; partial English translation by Sidney Withington, Marine Historical Association, Mystic, Conn., 1957), pp. 8, 9, 14, 34, 58.
4. *New York Evening Post,* March 4, 1820.
5. *New York Commercial Advertiser,* April 17, 1820.
6. Ibid., April 26, 1820.
7. *New York Evening Post,* May 29, 1820.
8. Ibid., June 15, 1820.
9. Ibid., July 22, 1820.
10. *New York Commercial Advertiser,* September 5 and 11, 1820.
11. Ibid., October 25, 1820.
12. Francis B. C. Bradlee, *Piracy in the West Indies and its Suppression* (Salem, 1923), pp. 33-36.
13. Frank M. Bennett, *The Steam Navy of the United States,* (Pittsburgh, 1896), p. 16.
14. *New York Evening Post,* December 13 and 14, 1822.
15. *New York Commercial Advertiser,* August 2, 1821.
16. Ibid., August 16, 1821.
17. Preble, *History,* p. 109. The footnote identifying the vessel as the Long Island Sound steamer *Fulton* is incorrect.
18. *New York Commercial Advertiser,* October 13, 1821.
19. Ibid., July 15, 1822.
20. Ibid., October 4, 1822.
21. Ibid., December 20, 1822.
22. Ibid., April 30, 1823.
23. Ibid., May 28-June 3, 1824.
24. *Nautical Gazette,* January 6, 1910, p. 15.
25. John H. Morrison, *History of American Steam Navigation* (New York, 1903), p. 437; Henry Howe, *Memoirs of the Most Eminent American Mechanics* (New York, 1846), p. 213; *New York Evening Post,* December 21, 1827.

3
Auxiliary Steamships Part 1

The *Savannah*

ONE path leading to the American oceangoing steamship has already been chronicled above in considerable detail. Starting with a steamboat, the *North River* of 1807, it proceeded to the Long Island Sound steamers, *Fulton* and *Connecticut* of 1813 and 1816, then to the queen of the Hudson River fleet, the *Chancellor Livingston* of 1817, and, finally, to the first successful coastwise steamship, the *Robert Fulton* of 1820. These were all craft designed, built, and engined by the most talented men of their day. None of them was a visionary radical; instead, all were practical individuals, successful in their respective fields, and their major interest was to develop a workable, reliable, steam-driven vessel that would be of service to the traveling public and would at the same time prove to be a profitable investment.

Each example cited was, for its time, the largest, the most expensive, the most powerful craft to appear. Because costs were high, great faith was necessary on the part of their financial backers. And, in every case, the venture involved starting a service where a demand existed for better transportation than that previously available.

There were many other steamers that did not fit into the successful category just outlined. Backers could be carried away by the idea of newness and would try unsuitable routes. Being either unwilling or unable to risk the requisite capital, they could skimp by employing less experienced designers and builders who would produce cheaper boats and engines. An obvious way to do this was to take an existing type of ship instead of one especially designed for the purpose, install an engine of low power to keep the cost down, and count on using the wind as much as possible, rather than burning expensive fuel. This was the approach of the timid and the unimaginative. Instead of a real steamship, a sailing ship with auxiliary steam power resulted. The classic example was the *Savannah* of 1819, about which a great deal has been written. Her only claim to fame was that she was the first auxiliary steamship to cross the Atlantic. Historical firsts are always of interest, whether they are successes or failures. The *Savannah*, like Fitch's and Rumsey's steamboats, was a historian's success and a businessman's failure.

Moses Rogers's name must always be linked with the *Savannah*'s. He had commanded the Stevens steamer *Phoenix* on the Delaware River in 1809, and later transferred to the *Eagle* in those same waters, as master and part owner until 1815, when he took her to Chesapeake Bay.[1] His third steamboat was the *New Jersey*, which ran with the *Eagle* on the upper reaches of the Chesapeake. Like Captain Elihu Bunker, he made himself into an experienced master in steam during its infancy and became a great advocate of its use. Unlike Bunker, he did not have the vision to undertake only services where large numbers of prospective passengers were available and immediate fiscal rewards possible. Legend has seemed to magnify his reputation. Upon his death he was reported to have both served with Fulton and commanded the *Phoenix* on her venture by sea from New York to Philadelphia in 1809. Since the Fulton-Livingston account books do not mention him, the first claim is incorrect. The second is questionable.

Rogers traveled South in 1817 to take over a new steamboat that was to operate from Charleston to Savannah and was to be named *Charleston*.[2] Once she started service, in December, her master came into contact with the maritime circles of Savannah. The two seemed to interact favorably and soon a plan for setting up a steamship company was formed, funds raised, and an Act of Incorporation signed by the Governor of Georgia on December 19, 1818, for the Savannah Steam Ship Company.[3] Specifically, the purpose was "to attach, either as auxiliary or as principal, the propulsion of steam to sea vessels, for the purposes of navigating the Atlantic and other oceans, and that they have provided a ship for that purpose, which is now in a sufficient state of forwardness to afford sanguine expectations of the experiment being tested. . . ." Of the large number of stockholders, the most important were William Scarborough, wealthy planter, merchant, and leading citizen of Savannah, and Joseph P. McKinne, his partner in a shipping firm. Scarborough, as president of the company, aided by Moses Rogers, as a stockholder, was the guiding spirit. McKinne lent commercial and shipping experience. But if we consider the funds announced for forming the company, $50,000, and compare this cost with the $42,000 cost of the *Washington*, the last steamboat planned and supervised by Fulton, that sum could not go very far. The *Fulton*, for Long Island Sound service, ran to $87,000 and the *Chancellor Livingston* to $110,000.

The Savannah Steam Ship Company seems to have been dazzled by the success of a few early steamboats operating in local waters; it carefully avoided realities: Savannah was an unsuitable terminus for an Atlantic service and the available capital was insufficient.

A sailing packet under construction at the New York shipyard of Samuel Crockett and William Fickett was purchased during the summer of 1818. She had been started as a Havre packet.[4] The only major change in her hull was to eliminate the second deck forward of the mainmast to make room for an engine and a pair of boilers. Her official dimensions were 98'-6" long, 25'-10" beam, 14'-2" depth of hold, and a tonnage of 319.[5] There was a square stern, a man burst as a figurehead, and a conventional three-masted ship rig. The launching took place on August 22, 1818, and the *New York Evening Post* of the day before reported, "It is believed that this ship, as to beauty of model and excellence of workmanship has not been surpassed by any ever built in this city. . . ." In fact, she was a typical packet ship of her era. It should be emphasized that only because of her standard characteristics could Messrs. Rogers, Scarborough, and their associates afford to buy her. Because she was the product of a reputable shipyard in New York, where the highest standards of shipbuilding were maintained, they were assured of quality in both her workmanship and materials.

To power the ship a number of unrelated contractors were brought together, with no one of them completely responsible for the work involved. It is as tribute to Captain Rogers that he manged to extract workable machinery from the many hands engaged in its design and construction. James Allaire's foundry supplied the air pump and cylinder castings. Daniel Dod, of Elizabeth, New Jersey, constructed two copper boilers and provided the drawings for the engine. Stephen Vail, owner of the Speedwell Iron Works near Morristown, New Jersey, constructed the engine and the metal parts for the paddle wheels. Nothing remains to show who was responsible for the square paddle shafts, criticial items in the performance of the ship; already recorded has been Fulton's continuous difficulty in obtaining "axels" of adequate strength.

The Vail account book shows that work started on the engine three weeks ahead of the launching date.[6] There were difficulties with Dod's drawings and both the air pump and cylinder piston rods had to be lengthened. The fact that they were patched up, rather than discarded and replaced, indicates that rigid economies were practiced.

In late October of 1818 the *Savannah* left New York for Elizabeth to have her machinery installed.[7] It is probable that the copper boilers, the largest items, went in first. The Vail accounts show that the engine had not advanced very far. By December 28 she had been towed back to New York to avoid being frozen in or damaged by ice on the Raritan River. Work continued, both in the Vail shop and on board the ship, throughout the winter. Early in March of 1819 she was reported almost ready to sail for Savannah, but she was delayed by a series of trials.[8] The *New York Gazette* of March 20 states: "She was again tried yesterday. She works to advantage, and her velocity is such to give entire satisfaction." A further report is more specific and gives a steam pressure of one inch of mercury (1/2 psi) and notes that she went from Fly Market Wharf, on the East River, to Staten Island and back in one hour and fifty minutes, part of it against wind and tide.[9]

The *Savannah*, like the *North River*, has suffered from numerous reconstructions, almost all of them based upon artistic imagination. It is particularly unfortunate that two incorrect reconstructions have produced models prominently displayed in the Smithsonian Institution's Water Craft Collection for every school child to see.[10]

Savannah, 1819, sail plan.

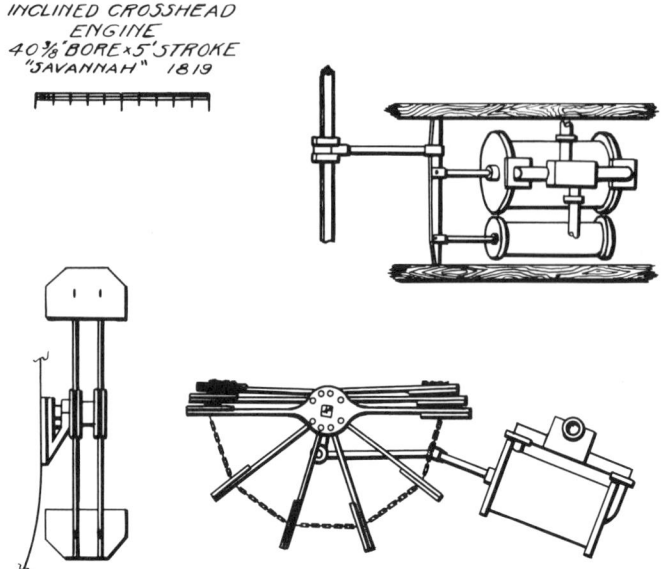

Inclined crosshead engine 40 3/8″ bore by 5′-0″ stroke for the *Savannah*.

The only contemporary evidence of high quality is the survey of the ship and her machinery made by Marestier after her return from Russia. He printed an outboard profile of the ship, two views of the engine, and two of the paddle wheels.[11] In no respect do these disagree with any of the rather sketchy newspaper accounts nor is there any variance that would conflict with events in her log book. The one obvious error is that the foremast and mainmast are shown as identical in height instead of the latter being taller. If one measures the profile drawing in the original French edition and assumes the scale to be 1/200 of full size (a common fraction for a metric drawing) the length dimension and the paddle-wheel diameter both check with quoted sizes. The Marestier dimensions are very close to the Custom House ones, the only problem being that his draft is probably intended to be depth. A 14′ draft is too large for a packet of this size.

Source	Length	Beam	Depth
Register	98′-6″	25′-10″	14′-2″
Marestier	100′	26′	14′

The present author has enlarged the Marestier drawing and shortened the foremast in the accompanying drawing. There is no difficulty in fitting the machinery, as delineated, in the hull. Unless new material is discovered, he sees no reason for departing from the available evidence printed by a trained engineer who actually saw the ship.

The author, as a naval architect, has made some engineering estimates of weight and draft to try to find what would be reasonable values. His figures lead to a light draft of 9′-6″ to bottom of keel; this corresponds to the waterline drawn by Marestier. Then, if we add seventy-five tons for coal, fifty-five tons for wood, crew, water, and stores, and one-hundred tons for cargo, the loaded draft would be 11′-6″. All these would be reasonable for the shaft position shown; in the light condition the wheels would dip into the water about two feet, which is very low, and, at the loaded condition, four feet, a high value.

The engine is the item of greatest interest; it had both good and bad features. The cylinder, 40 3/8″ in diameter by 5′-0″ stroke, is extremely large. In fact, it is identical in size to that of the *Chancellor Livingston* of 1817, the largest marine engine in the United States. The air pump scales 18″ to 20″ diameter from the drawings. Both are inclined, a departure from most of the steamboat engines of the era. An unusual element was the use of a single transverse crosshead attached to the piston rods for both the cylinder and pump, with a connecting rod on the port end driving the paddle shaft. Steam was admitted to and exhausted from the cylinder by a simple slide valve.

Looking at this arrangement, the modern engineer is appalled. It has practically everything wrong with it. It shows that intrepid amateurs can sometimes succeed against all the odds, for it worked. Let us consider the difficulties in detail.

1. The crosshead acting as a lever with various forces from the cylinder, pump, and connecting rod

would rarely be in equilibrium. It would always be twisting so that the starboard guide at one end would lead or lag the port one at the other. Fortunately, all these must have been far larger than necessary and did not fail under the moderate forces applied.

2. The valve on top of the inclined cylinder provided no means for water condensing inside the cylinder to escape. At very low steam pressures there must have been large amounts of liquid collecting. As a result, drains would have been installed and left open, otherwise the water would be trapped between the piston and the lower cylinder head.

3. The air pump was much too big for the job and its inclined piston would result in poor performance of the valves installed in the piston.

4. The fact that two boilers had to be installed in a limited space and still leave enough room to transport coal to them, fire the furnaces, haul the ashes, and operate the engine indicates that they must have been very small. The *Chancellor Livingston*'s single one was 25'-3" long by 12'-4" wide by 10'-10" high.[12] Because no equivalent volume could be installed in a 100-foot ship, the *Savannah* did not have the steaming capacity necessary to keep her overly large cylinder supplied. Furthermore, the open drains and the excess air pump size wasted a lot of whatever power was developed by the cylinder. The author regards the 1/2 psi pressure reported on trials as significant; that is probably all she could produce. With allowance for losses through valves and piping, steam would have reached the piston at less than atmospheric pressure. Air would leak into the cylinder instead of steam leaking out.

With all these drawbacks the installation was about the equivalent of a Boulton and Watt land-based pumping engine of the 1790 period and its actual efficiency must have been well below that of the *North River*'s engine.

Let us turn now from the dark to the lighter side. There were some advanced elements. The paddle wheels were entirely original and could be collapsed at sea and taken aboard when the ship operated under sail. They were 16' in diameter, with 10 buckets 4'-8" wide by 2'-9" deep, with the corners cut off at a 45° angle. One master pair of spokes extended all the way across the wheel and were part of the center casting fastened to the paddle shaft. The other spokes were attached to the center by chains connecting all the pivoted spokes to the master pair. To stow them the engine was stopped, with the master spoke horizontal and the top chain disconnected so all the upper paddles collapsed onto it. The crew removed the bolts, picked up the spokes, and stowed them on deck. Thereafter the wheels were revolved one-half revolution to bring the remaining spokes on top; their chain was disconnected, and, as a rule, the rest were lashed down to the horizontal master spoke rather than being taken completely off. The ship's log shows that the whole thing could be accomplished in under thirty minutes.

Previous steamboats had cast iron centers to provide sockets for wheel arms, but the latter, as well as the paddles, were always made of wood. The *Savannah*'s arms and pivots had to be of wrought iron. They had sufficient weight so that the engine's stopping at dead center was unlikely. As a result, the gears and flywheels of the Fulton-Allaire crosshead engines were eliminated. Cost, weight, and space were saved.

Another advanced feature was the use of coal as the main source of fuel. The *Savannah* was the first steamer after the *Chancellor Livingston* to have grates and furnaces designed for it. The limited space aboard permitted no other choice.

There is considerable disagreement among contemporary sources as to speed, capacity for fuel, and possible cruising range. Once again the author has resorted to engineering computations to try to ascertain reasonable values. Using 16 rpm (revolutions per minute) for the wheels and atmospheric steam pressure, the actual horsepower developed by the steam in the cylinder is about 60. Technically, this is called *Indicated Horse Power*, commonly abbreviated IHP. In light of the efficiency of the machinery, paddle wheels, and the hydrodynamics of the hull, that amount should drive the ship at about five knots without the aid of sails, but including the rather large wind resistance of the spars and rigging. Then, assuming that seventy-five tons of coal were available and a fuel consumption of ten pounds per IHP per hour (about half the thermal efficiency of the Cunard liner *Britannia* of 1840) she could travel 1,150 miles under steam. Since, in addition to the coal, there were twenty-five cords of wood (the quantities quoted by Stevens Rogers in 1838), another two-hundred miles might be added.[13] An optimistic estimate, then, might be 1,400 miles, a pessimistic one 1,000, depending on how many times fires had to be lit, steam raised, used, and then the plant shut down.

These figures clearly show that a ship of this size with a poorly designed engine of low efficiency simply could not stow enough fuel to cross the Atlantic. Moses Rogers and the other owners were well aware of her limitations. They had chosen to accept an auxiliary steamship that relied primarily on the wind. The folding paddle wheels were a very successful solution to the problem of steaming, when possible, and did not interfere with sailing qualities whenever winds were fair or fuel was short.

Finally, having considered the hull, the engines, and the paddles, we come to what the visitors to the ship saw and admired—the passenger quarters. These were typical for a sailing packet. It is to be regretted that so many varied and conflicting reports were written. One New York paper reported:

> There are 32 berths all of which are staterooms. The cabin for ladies is entirely distinct from that intended for gentlemen, and is admirably calculated to afford

the retirement which is so rarely found on board of passenger ships.[14]

An English source says:

> The State Cabin which is entirely wainscotted with mahogany is twenty-four feet long, and perhaps twelve in width—on the sides are the cots, twenty in number, and admirably calculated for comfort and accommodation. The floor is painted to imitate a tessellated pavement—looking glasses are let into the mast which divides the cabin and this arrangement gives a pleasing effect to the tout ensemble.[15]

International disagreement was further compounded when a Russian report stated:

> In the cabin are 40 small compartments with beds on each of which stands a number.[16]

In 1819, as today, newspaper reports were often inaccurate. Yet they may be the only source of information available. That is the reason why the author, from time to time, offers opinions and interpretations. A combination of engineering judgment and history can sometimes lead to a better understanding of the actual ships than a strict summation of conflicting documents.

Not until the eve of sailing for Savannah were there any advertisements. With only two days' notice for a projected March 27 departure, it is not surprising that neither passengers nor freight appeared.[17] On March 28, 1819, the *Savannah* sailed; the *Gazette* listed her as a "ship," with no mention of steam power. Moses Rogers was in command and served as his own chief engineer. His future brother-in-law, Stevens Rogers, was first mate and kept the log, which still survives at the Smithsonian Institution, U.S. National Museum, in Washington. The two Rogerses were unrelated, although they had sailed together before and were both natives of New London, Connecticut.

The engine was not put to work until the second day out, but, when the wind became favorable, the wheels were soon stowed again.[18] The ship arrived at Savannah on April 6, after a nine-day passage with about forty-two hours under steam, the longest period being seventeen hours on April 2 and 3. The *Robert Fulton*, as a fully powered steamship, would make the same distance in five days. The *Savannah* was welcomed with great enthusiasm, saluted by the revenue cutter *Dallas*, and met by many citizens.

President James Monroe was on one of his jaunts around the country and was scheduled to visit Charleston and then Savannah. Scarborough and his friends decided to send the *Savannah* to their rival port with an offer to transport the President in suitable style. She was coaled, her sailing advertised, and, with eight paying passengers, got under way on April 14. The short trip took far too long. She went down the Savannah River as far as Tybee light, met fog, then, at the insistence of the pilot, anchored for the night. That worthy went home to sleep in a comfortable bed, came out the next day, and took her out to sea. In the afternoon she made Charleston light but, because the tide was falling, she had to anchor a second time. Finally, on the third day, she did reach Charleston. The light-draft local steamboats took only a day. President Monroe was too good a politician to antagonize the many Charleston voters by traveling on a ship from a much smaller city and promised to inspect her later, at her home port. After two weeks in Charleston, the *Savannah* steamed back in twenty-seven hours with seven passengers on board.[19] This time there was better scheduling; she did not have to await favorable tides at either bar. Her draft was a serious problem; there were no dredged channels, the sand bars shifted regularly, and the pilots, wisely, had to see the available landmarks to be sure of the ship's position.

Back at her home port she was advertised to sail to New York "should a sufficient number of passengers offer. . .," but was unable to find enough to make the voyage worthwhile.[20] Happily, the President's arrival turned everyone's attention to patriotic ventures rather than commercial ones. No one in Savannah seems to have been aware of New York's 1817 welcome, which had enabled President Monroe to travel on the larger, faster, and far more elegant *Fulton* and *Chancellor Livingston*. On Tuesday, May 11, the President; John C. Calhoun, his Secretary of War; and assorted generals, judges, ministers, and local dignitaries were all awakened at a very early hour, taken aboard the *Savannah* at 7 A.M. and transported under steam and sail down the river to Tybee light, where lunch was served.[21] On the way back, since the tide was falling, the whole party had to transfer to a local steamboat, the *Altahama*. The *Savannah* had to be left behind. Nevertheless, the President expressed enough interest in the ship to make the owners think of the U.S. Navy as a possible purchaser. For the time being, the plan to send her across the Atlantic still held.

At about this time the Savannah Steam Ship Company must have got over the self-induced intoxication that accompanied their original fund-raising and the contracting for a ship and engine even before their official incorporation. They now had a ship of moderate size, with an engine of very moderate power and a quite limited cruising range. Word-of-mouth advertising produced neither passengers nor freight for Liverpool and a belated advertisement announcing the sailing did no better.[22] It was clear that there was no immediate future for a transatlantic steamship between a small coastal port, Savannah, and a major shipping metropolis, Liverpool.

Faced with these unfortunate realities, the owners had another attack of wishful thinking. They would send her to Russia to demonstrate to its all-powerful and incredibly wealthy ruler their wonderful new invention. If he were to purchase the *Savannah*, their financial investment would be recouped—probably at a large profit—and the stockholders could undertake some new and, it was hoped, equally exciting venture. The ship cleared at the Custom House on May 20, 1819, for St. Petersburg instead of Liverpool.

Actually, she did not sail until the 22nd. In the early morning of the 20th John Weston, a crew member, fell overboard when coming aboard, was drowned, and caused the delay. Throughout history, sailors returning from shore leave have found their steps somewhat unsteady.

Steam was used down the river; then, with the open sea ahead, the *Savannah* reverted to sailing ship practice, anchored, and awaited a favorable wind. This is the first clue that the owners had been niggardly in supplying coal. It was an expensive item in Charleston, for it was mined in England and shipped across the Atlantic. Not until 5 A.M. on the 24th was steam raised and sails set. As soon as she had crossed the bar and dropped the pilot, the wheels were shipped and she continued under sail. Thereafter steam was used sparingly for various periods from four to eighteen hours at a time.[23] Off the coast of Ireland on June 17 she was sighted by the revenue cutter *Kite*, which investigated to see if she were on fire. Next day the log reports that no coal remained. There must, however, have still been some wood, for steam was raised again on June 20, the day the *Savannah* steamed up the Mersey to Liverpool with her sails furled. When there was an audience, she was a steamship.

The *Savannah*'s twenty-nine-day crossing in the late spring of the year was neither slower nor faster than one would expect from a well-designed sailing packet. She left Savannah under steam, arrived at Liverpool in the same condition, and steamed for a total of eighty to 100 hours during the crossing. Many years later Stevens Rogers mentioned seventy-five tons of coal and twenty-five cords of wood, which would take her less than 1,200 miles at five knots and permit 240 hours under power. Allowing for intermittent use, she could go for perhaps 180 hours under steam. Since she did only half of this, she must have been very poorly supplied for the crossing. The forced economy may have been advantageous, because the limited use of the engine kept the awkward cylinder, air pump, and crosshead design from either gradually tearing itself apart or causing such excessive wear that the guides or the piston rod's stuffing box failed.

Various sources, contemporary or after the fact, have grossly exaggerated the times spent under steam.[24] *Niles New York Register* for August 21, 1819, states that "she worked her engine eighteen days." Stevens Rogers, in 1856, listed fourteen days under steam.[25] Like many stories, this one improved with the telling. Passage time went down to twenty-two days (instead of twenty-nine) and steaming time went up.

Once Liverpool was reached and a cheap supply of coal became available, the ship's bunkers were filled. Stevens Rogers's seventy-five tons of coal and twenty-five cords of wood were probably loaded at Liverpool rather than Savannah.

On the day she left Liverpool, July 23, bound for Russia, two other steamers headed down the Mersey—the packet *Waterloo*, bound for Belfast, and a local tug towing the ship *George Canning* out of the harbor.[26] The former, as a full-powered steamer, was much faster than the *Savannah*. In Danish waters the latter was held up from August 9 to the 14 at Elsinore and did not reach Stockholm until the 22d.

Although Great Britain had produced a considerable number of powered craft by 1819, they were still a great rarity in Sweden and Norway, which at that time were under a single monarch. The *Savannah*, therefore, received a state visit from Crown Prince Oscar. On September 1 an excursion was organized for the diplomatic corps and a pleasant trip to the island of Vaxholm completed. Further fuel was loaded and, best of all, King Charles XIV offered to barter hemp and iron, but not cash, for the ship.[27] The offer was declined; Captain Rogers was sure he could do better.

The *Savannah* left Stockholm on September 5, 1819; Kronstadt, the Russian naval base, was reached four days later and Saint Petersburg on the 13th. At least 240 hours, or ten full days out of the thirty-three under way since leaving Liverpool, were under steam. This corresponds to about 1,200 miles and would fit in with the quantity of coal mentioned by Stevens Rogers.

For the first and last time since her Charleston visit, the *Savannah* carried passengers. Thomas Graham, Lord Lynedoch, the retired General who had served under Wellington in Spain, persuaded Captain Rogers to take him and his nephew, Robert Graham, from Stockholm to the Russian city of palaces.

Unlike Stockholm, St. Petersburg already had two steamboats in operation, the *Elizabeth*, built and engined there in 1815, which had operated between St. Petersburg and Kronstadt, and a successor constructed two years later.[28] No one was discourteous enough to mention them to the American seafarers. Two different steaming parties were arranged for distinguished visitors. Representatives of the naval forces inspected the ship and

Captain Rogers was entertained ashore. In later years the story that the Czar visited the *Savannah* was told; it was another fairy tale.

In the meantime fall and bad weather had arrived. The *Savannah* was shifted to an anchorage off Kronstadt where she took on fuel from September 25 through 28.

Financial affairs did not come to their expected conclusion. Neither the Czar nor his naval advisers wanted the *Savannah*. There was no reason why they should; she was unsuitable for arming with any reasonable weight of battery. Furthermore, winter would soon close this far northern port. Taking advantage of a turn in the weather, Captain Rogers steamed out from Kronstadt on October 10, 1819. Once past the guard ship, the wheels were unshipped and the *Savannah* became just one among eighty sail heading down the Gulf of Finland; all had been awaiting a favorable wind.

A fair wind it was, and Copenhagen was reached in four days, three were then spent at anchor. On October 17 sail was set and she continued on her way with a brief pause at Elsinore for clearance formalities. Once the northern tip of Denmark was passed, adverse gales forced her to make for the Norwegian port of Arendal. A day later, with improved weather, she ventured out into the Skagerrak and thence north of Scotland to the Atlantic and the long voyage home entirely under sail. The coal aboard was useful as ballast. On November 30, 1819, Tybee light was seen and steam raised for the first time since leaving Kronstadt to bring her up the Savannah River. Fifty-one days from Russia, including all stops, was, once more, a creditable performance for any sailing packet less than 100' long. The *Savannah*, throughout her career, proved a fast, able, well-designed ship under sail.

Whether this was any consolation to her owners is questionable. They had built and engined her for a price of more than $50,000. Thereafter she had been equipped, supplied, and operated for eight months. She had been a sensation in some of her European ports of call, but there was no remuneration in sight to reward her pioneering exploits. The bill for her engine was still unpaid.

After three days in her home port, she was sent off again in search of a purchaser, this time to Washington in the hope that President Monroe's diplomatic expressions of interest were real ones. It was felt that she would be of value to the U.S. Navy for the suppression of Cuban pirates. This last idea was typical of the *Savannah's* owners. The Navy had been entirely disinterested in its one steam warship, *Demologos* (or *Fulton*). During that vessel's trials she had always had her machinery manned by workmen from Fulton's engine works.[29] Thereafter, her single paddle wheel never turned again and no one was ever trained to operate her engine or fire her boiler.

The *Savannah* was not designed with enough strength to support a gun battery on her deck; she was not at all suitable for conversion to a warship. Even if she had been, her speed was too low and her draft far too great for the specific purpose of chasing pirates in their small, fast, light draft schooners. When the Navy did undertake their extermination, it did so with a "mosquito fleet" of just such schooners, with the *Sea Gull*, ex-*Enterprize*, as an equally small steamboat of little draft, to support them.

Leaving Savannah on December 3, the *Savannah* steamed down the river, then headed for Washington, weathered a winter gale off Cape Hatteras, and reached Chesapeake Bay in ten days, partly under power. Thereafter, the engine was used all the way up the Potomac River. When she reached Washington, her fires were hauled for the last time on December 16, 1819.[30]

The sanguine Savannah Company, which at first had thought that a minor Southern port with 7,000 citizens (Charleston, practically next door, had 24,000) could support a transatlantic steamship, had then decided that the Czar of all the Russians needed their creation, and, as a last resort, had then sent her to the seat of the United States government to dazzle the eyes of all in authority, was, finally, bereft of ideas. In March of 1820 she was advertised for sale; in August she was acquired by Captain Nathan A. Holdridge of New York City. Leaving Washington on September 1, she worked her way downstream, sailed from Hampton Roads on the 6th, reached New York on the 10th; there her engine was sold to James Allaire for $1,600. His interest was sentimental; he retained the cylinder and displayed it as a piece of historical interest.

Holdridge extended the second deck for the length of the ship; as a result she had to be remeasured by the proper authorities. Her new document shows a vessel 98'-6" by 25'-10" by 12'-11" with a tonnage of 291 as opposed to the original 98'-6" by 25'-10" by 14'-2" and 319 tons. Under the official rules, different formulae were used to compute the tonnage for single- and two-decked hulls. The second variation, the depth, does not mean that the ship was cut down, thereby lowering the whole main deck. It is a fictitious value, one-half the beam. The ship was unchanged in size; the differences are mere technicalities that tend to confuse reports of how big a ship really was. All this is emphasized because modern sources have badly misinterpreted the official dimensions and their changes.[32]

With the machinery gone and a tween deck throughout her length, the *Savannah* reverted to a standard three-masted packet, rather larger than most vessels of that type in the coastwise trade. Under her new owner and master she sailed from New York on October 26, 1820, for Savannah with, for the first time in her career, both paying cargo and passengers. Only five days later she reached Savannah; it was a rapid passage, shorter than her first, partly under steam, which had taken twice as long. At last the *Savannah* was a commercial success. She carried cotton in season northbound, merchandise south, and passengers in both directions for the next year. Then disaster struck.

On Monday morning, November 5, 1821, at 3 A.M., with an east-north-east wind and low visibility, she went aground at the Fire Place on Fire Island south of the present Long Island town of Bellport.[33] At 9 A.M. the ship was upright in the sand, not leaking, and Captain Holdridge went for help, hoping that the 250 bales of cotton on board could be saved. Her crew and three passengers came ashore and the cargo was unloaded over a period of time, but additional gales blew the wreck further ashore, started leaks, and salvage became impossible.

The *Savannah*, to sum up, was a sailing ship converted by the addition of a low-powered and poorly designed steam plant of minimum cost. Her unique feature was a pair of side wheels whose iron spokes could be folded up and removed when under sail. In service her engine was operated for brief periods that add up to about three weeks of steaming during the nine-month period from her original departure from New York until she was tied up at Washington. Although there were no reported breakdowns, it is doubtful that the peculiar cylinder with its unbalanced air pump and crosshead could have been kept running for long periods at full power.

Her captain, Moses Rogers, seems to have been something of a combination of Fitch and Rumsey. Like the former, he was able to get machinery built that would run. Like the latter, he could interest others in his project and persuade them to invest. But the Savannah Steam Ship Company seems to have been wholly impractical in the commercial sense. It created an auxiliary steamship and did not know what to do with her. It put her on a route that, at the time, did not support any regular sailing packet line and then expressed surprise when neither passengers nor shippers trusted the craft to deliver themselves or their cargoes across the Atlantic. The *Savannah* was the wrong type of ship at the wrong place at the wrong time.

Her real interest is as a curiosity, for she did cross the Atlantic and used steam for about four days' total time during a twenty-nine-day passage from Savannah to Liverpool. There was much newspaper excitement, but nothing permanent was achieved. Neither her unique engine arrangement nor her folding paddle wheels were ever copied in a subsequent steamship.

The one temporarily successful survivor of the experiment was Daniel Dod of Elizabeth, New Jersey, designer of her engine and builder of her boilers. He continued as an engine builder until May 9, 1823, when the boiler of the steamer *Patent* exploded during her trials, killing Dod and four others. His later engines were considerable improvements over those of the *Savannah*. But can the same be said of his boilers? Three of them, on the *Atalanta*, *Powhatan*, and *Patent*, exploded.[34]

The *Fidelity*

Two years after the *Savannah* and one after the *Robert Fulton*, an existing two-masted schooner named *Fidelity* was engined with the intention of carrying passengers and freight between New York and Norfolk at weekly intervals.[35]

A diminutive craft, 88' long by 19'-4" beam by 9'-0" depth, she registered 139 tons. Launched by the newly organized shipyard of Brown and Bell in 1820, she was owned by George Suckley and commanded by Richard Leech. Nothing is known about the engine. She was advertised to leave New York on Monday, May 7, 1821, with a fare of $15.00 to Norfolk, and to return on the 10th. She immediately found that weather, rather than a schedule, was the key feature affecting her arrival. On the way down she met strong head winds, burned up all her wood, and arrived on the date she was supposed to leave.[36] After docking at 8 A.M., she hastily fueled and did manage to sail at 5 P.M. Northbound, she reached Sandy Hook on the 12th, but her boiler started leaking so badly that the fires had to be drawn before she could reach her pier.

No further Norfolk sailings were undertaken. The next reference to the *Fidelity* was an advertisement announcing her departure from New York on December 1 for St. Augustine, Florida; Pensacola; and Mobile, Alabama. This was a delivery voyage for future operation on the Gulf of Mexico. Boiler troubles continued, for she had to put into Norfolk on December 11 in distress, with one of her firemen dead.[37] Her fifteen passengers were landed and repairs delayed her until January 22. After that she made a creditable passage of four days to Savannah, where she took on coal, rather than wood, and continued on to the Gulf. There, in more sheltered waters, she seems to have operated in passenger service. The Lytle List shows her enrollment as abandoned in 1828.

The *New York*

Because the *Savannah* made but one round trip between Savannah and Charleston with paying passengers and the *Fidelity* did no better on her intended route between new York and Norfolk, neither succeeded in the commercial sense. The first ocean steamer to do so, after

Newspaper cut of the auxiliary half-brig *New York*. Courtesy of The New-York Historical Society, New York City.

the *Robert Fulton*, was the *New York*, which did carry passengers and freight on a regular schedule between New York and Norfolk for over a year until she ran aground near Cape Henry. She was an auxiliary steamer regularly called a "steam brig" in the newspapers. The best illustrations available, two charming cuts used to head her sailing announcements, show her to be square rigged on the foremast and fore and aft on the main; she was really a hermaphrodite (or half) brig.

Her hull was built at Norfolk by William F. Hunter under the direction of Captain Richard Churchward.[38] Both were part owners, together with George Rowland, Charles Rowland, and John Allmand. Captain Churchward had been master of the packet schooner *Tell-Tale*, which had been sailing regularly between New York and Norfolk. Several of the Rowland family were merchants at Norfolk with shipping connections. The assemblage of an experienced captain, a shipbuilder, and several businessmen, all acutely aware of local conditions, was a far more hardheaded group than the visionaries of the Savannah Company.

The *New York*, rated at 281 tons, was similar to the *Savannah* in size, but was an extremely large brig for the 1820s. Since she was designed for steam, the choice of a two-masted rig was reasonable, for it provided a clear space for the machinery between the masts. Her dimensions were 105' by 27'-6" by 11'. She had a single deck, a square stern, and a billet head. The engine was by Daniel Dod, who had moved from New Jersey to New York in 1820, and it represents a considerable advance over his designs for the *Savannah*.[39] The awkward inclined air pump driven by an unbalanced lever was discarded. Instead, the cylinder was vertical and the paddle wheels were driven by a walking beam. This was originally called a "working beam," but usage changed it. Because the wheels were no longer folding, they were constructed with wooden spokes. Consequently, Dod did not trust their momentum to assure regular motion, and, following Allaire, he adopted a geared flywheel.

The exact arrangement is somewhat of a mystery. The first illustration of the brig, in the *New York Evening Post* for October 22, 1822, shows the port side with a walking beam on that side and the boiler and stack to starboard. The owners were unsatisfied with this cut. It showed a

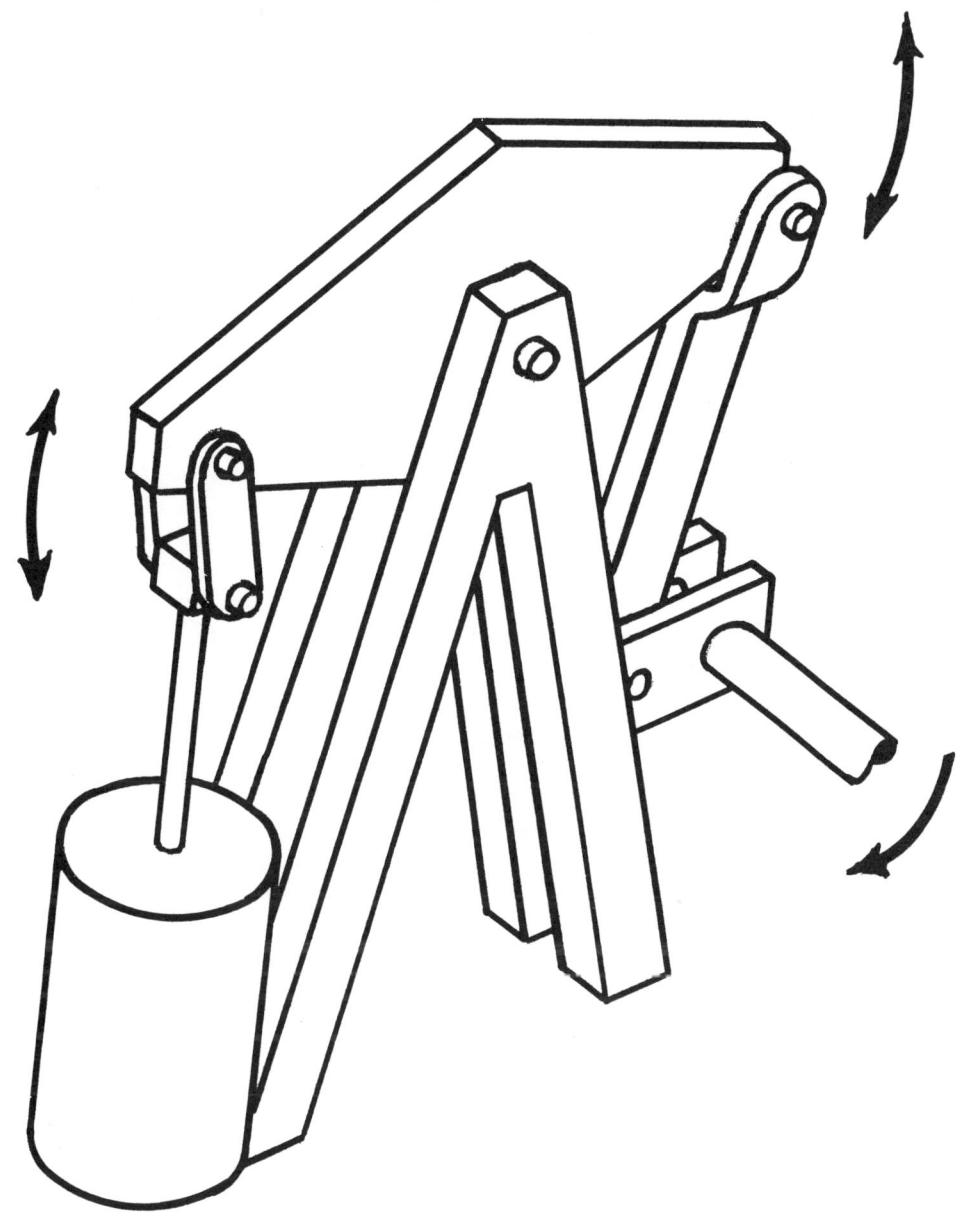

Diagrammatic arrangement of a walking beam engine.

thick, clumsy bowsprit and had the fore topgallant yard crossed abaft the mast in a most unseamanlike manner. They replaced it by a better view with a reasonable bowsprit, the yard crossing forward of the mast, and the stern much lower so that an awning protecting the passengers appeared clearly. The difficulty is that the starboard side is now shown and the beam for the engine is now on that side, with the stack and boiler again behind it. Because the stack was in line with the center of the paddle wheels, the paddle shaft had to be in two parts, for it could not pass through the boiler to which the stack attached. Probably the "balance wheel" (flywheel) shaft was geared to the port and starboard paddle shafts and passed abaft the boiler. There are several references to the "engine" (in the singular) with horsepowers of 50 or 60 quoted, but, whether the boiler was on the port side and the engine to starboard or vice versa is uncertain.

With regard to accommodations, there was a forward cabin with twelve berths and an after one with sixteen. ". . .elegantly finished, and afford a new and imposing evidence of the superior taste of our townsmen, Messrs.

Lemuel and Robert Langley."[40] Settees were added to bring the total accommodations up to fifty. Unlike the *Savannah*, the *New York* was not taken away for the installation of her machinery; all the work seems to have been done at Norfolk under the supervision of a Mr. Sutcliffe, who became her chief engineer. On August 31, three months after her launch, a short trial was made down the Elizabeth River toward Hampton Roads.[41] This was followed by a longer excursion to Old Point Comfort on September 10.

On Saturday, September 14, 1822, the *New York* left Norfolk for Jersey City. It has already been noted that yellow fever was widespread in New York that year; the New Jersey port was substituted to avoid quarantine delays. The engine broke down and practically the whole 3½ day passage was under sail. Repairs kept her in port until October 1. When they were completed, her return took fifty hours, thirty-five under steam and sail and fifteen lying to in a heavy gale.[42] Thereafter a fortnightly service was scheduled since, as an auxiliary of limited power, more frequent sailings could not be assured. As winter approached they became less frequent. The shortest passage reported was thirty-six hours, New York to Cape Charles, at the entrance to Chesapeake Bay, and the longest six days from Norfolk to New York with three spent at anchor because of head winds. After her arrival at New York on January 4, 1823, she was laid up until March 14, when she headed South. In 1823 she was advertised to sail on three voyages a month carrying passengers and light freight. Published reports indicate that passengers used her regularly but that cargoes were chiefly carried southbound. The largest list of passengers found gives twenty-seven first class and nineteen steerage. Captain Churchward and her owners had succeeded in establishing a new steam route and travelers were taking advantage of it.

Then, in September of 1823, the *New York*'s good fortune began to desert her. An onshore gale forced her to take shelter in Delaware Bay during a northbound passage.[43] Conditions continued so adverse that she headed upriver and discharged her passengers at Philadelphia on September 13 instead of at New York. Both weather and luck continued their onslaught and, although she did reach New York on her next voyage, when she returned with thirty passengers and a cargo of merchandise, she ran hard aground in a thick fog on October 4, 1823, about half a mile south of the Cape Henry lighthouse.[44] There was no immediate damage; she was about one hundred feet from the beach and her hull was tight. While the cargo was being taken out, northerly winds forced her further aground. To lighten her, coal and ballast were unloaded and the machinery was dismantled and put ashore. Only the shafting and the flywheel remained in place. About the end of the month she was hauled off and towed into Norfolk by the steamboat *Hampton*.

The owners, daunted by their difficulties, decided to sell the brig in her disassembled condition. An auction was set for January 14, 1824, with the engine, furniture, and vessel to be sold separately.[45] The advertisements for the sale give a little more information about the *New York*. The hull was of first-rate construction, for not only was it copper fastened, but even the deck planks were copper nailed. The bottom was covered with twenty-six-ounce copper plates. In addition to the twenty-eight berths mentioned previously, there must have been twenty-two settees to account for the fifty hair mattresses listed.

The next date available is an advertisement to sail from New York to Boston on May 5, 1924.[46] It is believed that the reinstallation of her machinery was done in New York, where experienced mechanics were available. The *New York*'s departure, with fares of $7 to Newport and $10 to Boston, is the first known steamer sailing taking passengers from New York all the way around Cape Cod to Boston. She arrived there on May 7, forty-five hours from New York and twenty-two from Newport, with twelve passengers.[47] The trip was a delivery one; she was scheduled to sail from Boston on the 9th for Portland and other Maine ports. Upon her arrival she became the second steamer running regularly from Boston down east, the first having been the little steamboat *Patent* of 98 tons, which started service in 1823. It was on the latter's machinery trials that Daniel Dod met his death.

The *New York*'s first trip was a memorable one, according to a letter from her master published in the *Boston Courier* for May 15. After she passed Cape Ann, the wind increased to gale force and carried away the fore topmast. Being unable to make headway, she put about toward Provincetown, sheltered by the tip of Cape Cod. Two days after leaving Boston she set out from Provincetown. Some twenty miles south of Cape Ann she encountered the schooner *Superior* floating on her side with both topmasts gone. Captain Bartlett stopped, fastened a hawser to one of the masts, and tried to haul her upright, without success. Because fuel was short, he could not tow her on her beam ends and had to leave the wreck behind. Portland was reached on May 12 and, after five hours in port, the *New York* continued on to Eastport. Returning, she took three days from Portland to Boston, much of the time at anchor or groping through fog. The *Patent* made better time; she sailed four hours later and arrived a day earlier.[48] Captain Porter, of the latter steamer, had long experience in Maine waters and had a somewhat faster boat of lighter draft.

For the summer of 1824 the *New York* left Boston about twice a month, always to Portland, but then to various ports such as Lubec, Eastport, Belfast, Bangor, and Owl's Head. By prior arrangement passengers could land at other stops. Upon occasion excursions were made in the vicinity of Boston. One was advertised for August 20, 1824, up and down the bay for $1.50, which included a "cold collation." As soon as it was over she loaded her Maine passengers and left that same evening for Portland and Eastport. A second harbor jaunt was planned for September 14 and a trip to Salem on the 16th.

Although a departure for Maine on September 23 was announced, the *New York* headed South instead. The occasion was the Marquis de Lafayette's tour of the United States and his impending visit to Yorktown, Virginia. The *New York* became the first American cruise ship, for she set out from New York on August 14[49] for the Revolutionary War battlefield via Norfolk and would provide passengers from both ports with food and lodging during the extensive celebrations. Her trip from New York was a slow one, four days to Norfolk with a passenger list of twenty five; it must have been considerably swelled by Virginia sightseers. Off Yorktown there was a grand procession of eight steamboats from the Baltimore, Richmond, Washington, Norfolk area.[50] These were the *Virginia*, with Lafayette and his son aboard; the *United States*; the *Petersburg*; the *Mount Vernon*; the *Potomac*; the *Richmond*; the *Eagle*; and a second *Virginia*. As an outsider, the *New York* was not part of the official festivities. On October 25 she left Norfolk with her gratified onlookers.

Back in New York it was decided to try breaking into the lucrative New York, Newport, Providence trade originated and dominated by Captain Bunker's *Fulton* and *Connecticut*. Their protection, under license from the Fulton-Livingston monopoly, had just been declared void, but an auxiliary steamer with cramped passenger cabins was hardly a competitor in speed or comfort. To offset these drawbacks a cut-rate fare of $6.00 was quoted for the *New York*, with sailings from New York on Fridays and from Providence on Tuesdays, touching at Newport in both directions. Service started November 13, 1824, but temporarily ceased a month later when the boat was withdrawn for machinery alterations.[51] It was resumed and continued until sometime in January of 1825.

Three months in winter weather competing with the long-established *Fulton* and *Connecticut* operation proved to the *New York*'s latest owners, as it would to other competitors, that the New York-Providence service was not practical. The *New York* then returned to her original route, New York to Norfolk, but extended it up the James River to Richmond. She left New York April 15, 1825, made Norfolk in forty-eight hours, two hours shorter than her first passage back in 1822, spent two days there, and left for the state capital.[52] Four voyages were completed, the last reaching New York June 11 or 12. On this final trip she was advertised to carry horses and carriages, as well as passengers. A further sailing was announced for June 15, 1825, but was not made. Instead she returned to Boston and made a trip to Portland and beyond in August and a second in September.

When next heard from, almost a year had elapsed, but during that time a new and larger engine of 100 horsepower had been installed. The *New York* was to sail under a new master, Captain Benjamin Harrod, from Lloyd's wharf in Boston, on Monday, August 21, 1826, for Portland, Bath, Belfast, and Eastport.[53] Nothing seems to have worked satisfactorily on this trip. Its events indicate that Captain Harrod had been a particularly poor choice. First he ran her aground going up the Kennebec River and had to await the next tide before she floated off. She continued on to Bath; then to Belfast; thereafter, off Owl's Head, she ran into the *Patent* and disabled her so badly she had to tow her back to Belfast.[54] At 9 P.M. the same evening, August 24, the *New York* caught fire near the stack and burned so rapidly that passengers and crew barely escaped in her boats. At the time of the fire the engine room was deserted and no officer was at his post; only a helmsman and a lookout were on deck, no fire buckets were available, and it was impossible to stop the engine. Since the baggage had to be abandoned, one of the passengers, a Mr. Thayer, was understandably upset over the loss of $1,800 in his trunk.[55] It was indeed fortunate that all on board reached Petit Menan light safely. Oddly enough, there seems to have been no censure of Captain Harrod; in those early days steamships were considered dangerous items that could be expected to burn, explode, or otherwise destroy themselves.

The *New York* was a slightly smaller successor to the *Savannah* built by Norfolk interests who wished to improve the sailing schooner service then existing to and from New York. As a half-brig supplied with a low-powered engine, she was unable to steam continuously on her intended route and occasionally had to anchor when there were head winds. Yet she provided regular service for a year until she stranded when trying to enter Chesapeake Bay in a thick fog. She survived this catastrophe, was sold, refitted, and thereafter became a steamer in search of a profitable route under a series of owners and masters. She tried Boston to Portland to Eastport operations for a time, made a special excursion to Yorktown in connection with Lafayette's return there, attempted for three months to compete with the *Fulton* and the *Connecticut* on Long Island Sound, and finally perished in a holocaust in Maine waters. Although generally ignored, she was a far more successful craft than the *Savannah*, and it is believed that she was a financial success prior to running aground.

But, like the *Robert Fulton*, she had no immediate successors. The auxiliary steamship just ceased to exist, outside of the U.S. Navy, for almost two decades. It would be redeveloped in a new form in the 1840 period by New England interests.

Notes

1. John C. Emmerson, Jr., *The Steamboat Comes to Norfolk Harbor* (Portsmouth, Va., 1949), pp. 3-17.
2. Malcolm Bell, Jr., *Savannah Ahoy!* (Savannah, Ga., 1959), pp. 16-17.
3. Ibid., pp. 21-22.
4. Stevens Rogers in J. Elfreth Watkins, *The Log of the Savannah*, Report of the National Museum for 1890, p. 637.
5. Certificate of Registry dated March 27, 1819. Record Group 41, U.S. National Archives.
6. Watkins, Log, pp. 618-22.
7. Frank O. Braynard, *S.S. Savannah* (Athens, Ga., 1963), pp. 49, 52.
8. *New York Gazette and General Advertiser*, March 3, 20, 25, 1819.
9. *New York Mercantile Advertiser*, March 27, 1819.
10. Model of the Steamship *Savannah*, described in Carl W. Mitman, Catalogue of the Watercraft Collection, United States National Museum Bulletin 127 (Washington, D.C., 1923), pp. 61-63; Howard I. Chapelle, *The Pioneer Steamship Savannah: A Study for a Scale Model*, United States National Museum Bulletin 228 (Washington, D.C., 1961).
11. Jean Baptiste Marestier, *Mémoire sur les bateaux à vapeur des Etats-Unis d'Amérique* (Paris, 1824; partial English translation by Sidney Withington, Marine Historical Association, Mystic, Conn., 1957), Plate III, fig. 10, and Plate VII, figs. 32-34.
12. Marestier, *Mémoire*, p. 33.
13. Watkins, Log, p. 628.
14. *New York Mercantile Advertiser*, March 27, 1819.
15. *Cheshire Chronicle*, August 7, 1819, quoted in Braynard, *S.S. Savannah*, p. 66.
16. *Northern Post of the New Paper of Sankt-Peterburg*, October 29, 1819, quoted by Braynard, *S.S. Savannah*, p. 68.
17. *New York Gazette and General Advertiser*, March 25, 1819.
18. Watkins, Log, pp. 629-30.
19. Braynard, *S.S. Savannah*, p. 98.
20. *Columbian Museum and Savannah Gazette*, May 5-8, 1819, quoted in Braynard, *S.S. Savannah*, p. 101.
21. Bell, *Savannah Ahoy!*, pp. 36-37.
22. *Columbian Museum and Savannah Gazette*, May 19, 1819.
23. Watkins, Log, p. 633.
24. Preble, *Chronological History*, pp. 98-99.
25. Watkins, Log, p. 637.
26. *Cheshire Chronicle*, August 7, 1819.
27. Braynard, *S.S. Savannah*, pp. 153-60.
28. H. Philip Spratt, *Transatlantic Paddle Steamers*, 2d ed. (Glasgow 1967), pp. 98-99; *New York Herald*, September 18, 1851.
29. Frank M. Bennett, *The Steam Navy of the United States* (Pittsburgh, Pa., 1896), p. 16.
30. Braynard, *S.S. Savannah*, pp. 185-87.
31. Ibid., p. 195.
32. Ibid., p. 197, says that the ship was cut down; it was not. Chapelle, *Study for a Scale Model*, fig. 6, p. 71, completely ignored the evidence of the first document in reconstructing plans for the Smithsonian model. As a result, he made many unnecessary changes from the Marestier drawings of the *Savannah*.
33. *New York Daily Advertiser*, November 7, 1821.
34. Francis B. C. Bradlee, *Some Account of Steam Navigation in New England*, (Salem, Mass., 1920), pp. 49-50; Greville Bathe and Dorothy Bathe, *Oliver Evans* (Philadelphia, 1935; reprint ed. 1972), pp. 252-53.
35. *Nautical Gazette*, July 15, 1909, pp. 50-51.
36. Emmerson, *Steamboat*, pp. 184-87, 211.
37. *New York Evening Post*, December 17, 1821.
38. *Nautical Gazette*, September 16, 1909, p. 219.
39. Bradlee, *Steam Navigation*, p. 53.
40. Emmerson, *Steamboat*, p. 259-51.
41. Ibid., pp. 253-54.
42. *New York Evening Post*, October 9, 1822.
43. Emmerson, *Steamboat*, p. 298.
44. Ibid., pp. 299-301.
45. *New York Commercial Advertiser*, December 24, 1823.
46. *Nautical Gazette*, September 23, 1909, pp. 236-37.
47. *Boston Courier*, May 8, 1824.
48. Ibid., May 21 and 22, 1824.
49. *Nautical Gazette*, September 23, 1909, p. 237.
50. Emmerson, *Steamboat*, pp. 335-42.
51. *New York Commercial Advertiser*, December 15, 1824.
52. Emmerson, *Steamboat*, pp. 369-70.
53. *Boston Courier*, August 19, 1826.
54. Bradlee, *Steam Navigation*, pp. 53-54.
55. *Boston Courier*, August 30, 1826.

Massachusetts, sail plan.

4

Steamboats on the High Seas

The Charleston Route

THE *Robert Fulton* (1820-25) was the first true oceangoing steamship in the United States. She was large, heavily built, and especially designed for service from New York to Charleston and New Orleans. Her contemporary, the *New York* (1822-26), became the first successful auxiliary steamship in ocean service. With the hull of a sailing brig and an engine of limited power she operated on two occasions between New York and Norfolk, the route for which she was designed, but often had to anchor when there were adverse winds. While the fuel consumption was low, her passage times were markedly affected by wind and wave.

Neither vessel had a direct successor. Instead, the three decades from 1820 to 1850 were responsible for two other outstanding American developments in marine transportation. The first was the creation of many sailing packets operating at regular departure dates in both transatlantic and coastwise service. Fast, well-built, seaworthy ships were launched by the hundreds to carry seaborne passengers and cargoes.[1] In the coastwise trade sometimes brigs, or even schooners, could do the work whenever a larger, three-masted, square-rigged ship was not justified.

Meanwhile, a second phenomenon, the steamboat, was being developed and began to appear on every river, bay, sound, or lake that could boast towns of any size. It was a fast, high-powered, shallow-draft, very lightly built vessel eminently suited for smooth, protected waters. An essential part of making steamboats fast was to construct them as lightly as possible. In general they moved far away from the heavily built *Chancellor Livingston* and became a new type, quite different from all the examples considered so far in these pages. When used in fresh water, their engines could be designed for high pressures and the whole complication of condensers and air pumps done away with. Steam was exhausted directly to the air. The result was a light, cheap, powerful engine.

The hull became long, thin, and shallow. It was so flexible that large trusses had to be added port and starboard. These used wood in compression and iron rods in tension, and their stiffening effect permitted heavy weights, such as the engines and boilers, to be installed without the hulls bending excessively and working out of shape. Another means of distributing concentrated loads was to install heavy vertical poles, like masts, and then to hang decks, guards, and cabins from iron tie rods leading to their tops.

In time, extensive guards overhung the hull from bow to stern and on them were built large, but very light, superstructures with many cabins and saloons for passengers as well as enclosures to protect the freight, all of the latter being stowed on the main deck rather than inside the hull. It must be realized that the hog trusses and poles were unique American inventions and that the enormous wood steamboats on the Hudson, the Great Lakes, the

Mississippi, and Long Island Sound could never have been created without these essential load-carrying and load-distributing members. Robert L. Stevens, son of John Stevens, is credited with the invention of the hog truss for the Hudson River steamboat *North America* of 1827.[2]

Like the sailing packets, the steamboats were built in enormous numbers as the westward-bound settlers moved inland along the waterways of the United States. During these years of expanding trade it was not difficult to build a sailing vessel and profit from its operation along the coast or to acquire a steamboat and put it to work carrying supplies and passengers on any sheltered body of water. The railroad, as a serious competitor, was a long way off. It was an era of canals connecting inland rivers and lakes. The ocean steamer was ignored completely. It would be large, heavy, costly, slow, and would require vast quantities of fuel. Why should prospective owners undertake such a risk?

A few reckless individuals began to think of a cheaper solution to ocean transportation than by means of a vessel especially designed for the service. Many steamboats had been built in New York, or Philadelphia, or Baltimore, or wherever an engine works was available, and then sent by sea on a long delivery voyage. Under a careful master, with good weather and special precautions to protect vulnerable openings, most of them had arrived safely in ports up and down the eastern seaboard of the United States and all around the Gulf of Mexico. One steamboat, the *Braganza*, ex-*Swift*, went all the way to Rio de Janeiro in 1820.

It seems strange that James P. Allaire, with his long experience and direct personal involvement in creating a highly successful steamship like the *Robert Fulton*, should be the entrepreneur to set up a second Charleston service, this time with an entirely unsuitable steamboat. Allaire was a master iron founder and the most successful of the early steam engine designers and builders, but he totally lacked an appreciation of naval architecture or ship construction. He was instrumental in extending steam navigation and invested in many steamboats, as either sole or part owner, but left their hull design to others, sometimes capable, sometimes considerably less than that.

One example was the *David Brown*, 130'-11" by 18'-1" by 8'-4" with a tonnage of 190, which was built by Brown and Bell in 1832 to run from Red Bank, New Jersey, to New York. Although Allaire's marine engine works was in the latter port, he also owned the Howell Works in New Jersey, which produced innumerable cast-iron stoves, kettles, pots, griddles, grates, and other domestic items in iron or brass. The *Brown* was built to deliver these goods to New York and did so for several months.

Late in 1832, however, Allaire reopened the Charleston route, which had been served by sailing packets for the last seven years, and he did so with the *David Brown*. Going around Cape Hatteras was a far cry from turning Sandy Hook. Nevertheless, despite adverse winter weather, the *David Brown* sailed from New York on December 22, 1832, for Charleston and made a second voyage starting January 5 of 1833. Then, we do not know why, she was withdrawn until the spring. One would like to think that Captain James Pennoyer advised caution and that Allaire concurred.

On April 20 the *Brown* resumed her sailings and continued at two-week intervals to the end of 1833, aside from two interruptions.[3] The first, in July, consisted of boiler difficulties off Cape Lookout. After the fires were drawn and the boilers drained, sail was set and Captain Pennoyer headed for Beaufort, North Carolina. Although he grounded her several times while crossing the bar, no damage resulted. A vessel called the *Cygnet* was chartered at Newbern to deliver the passengers.[4]

The second accident came shortly afterward. The *David Brown* sailed from Charleston on August 17, 1833, with fifty passengers. In a heavy gale off Hatteras she lost her rudder and had to put into Norfolk for repairs on the 23rd.[5] She was a lucky steamer, in spite of being "well calculated for speed, by light and narrow construction. . . ." In three years of ocean service she completed forty-five round trips from New York to Charleston. Fortunately, she was laid up for the winter months. her last trip from New York was December 8, 1835. It is interesting that the *Brown*, an out-and-out steamboat, was the coastwise successor to both the *Robert Fulton*, to Charleston, and the *New York*, to Norfolk, and that she resembled neither in concept or design. Her crosshead engine was an advance over the *Fulton*'s, for it eliminated the geared flywheel.

The first year's operation of this local steamboat in ocean waters proved so profitable that a second one was built to join her in 1834. This was the *William Gibbons* of 294 tons, a hundred tons larger, built by Brown and Bell. Her hull had overhanging guards, in way of the wheels, and a deckhouse extending from just forward of the stack to the stern. The engine was of the crosshead type and atop the wood frame for its guides was a large bell. At this time and long afterward the steam whistle was not used at sea; instead, a bell was the standard signaling device, inherited from the sailing ship. There was a two-masted schooner rig and a square sail on the foremast. The only good reproduction of the vessel is the diminutive newspaper cut used to advertise the New York and Charleston Steam Packet Company. Not only was "Steam Packet W. Gibbons" painted on the paddle box, but there was "Wm. Gibbons" on a swallowtail pennant from her main topmast.[6]

On her first trip to Charleston, March 1, 1834, she steamed away from Pier 32 on the East River under the command of Captain Pennoyer, formerly of the *Brown*.[7] A large party of invited guests accompanied her on the latter craft as far as the Narrows. They must have outnumbered the twenty-three passengers. A newspaper account gave her appearance as "beautiful" and commented on the total lack of vibration from the machinery. The *David Brown* followed her to Charleston a week later. From then until the end of November a weekly service was maintained by the two steamboats except for occasional interruptions for repairs.

Starting in September, their formal management was given to Charles Morgan, who had been increasingly active in both ownership and operation of coastwise sailing packets since 1819. Morgan became a part owner of the *Brown* in 1832 and of the *Gibbons* just after her completion. Allaire, his brother-in-law John Haggerty, and Morgan were the principal owners, but others held shares from time to time.[8] The *Brown* was laid up for the winter of 1834 but the larger, more powerful *Gibbons* carried on throughout this hazardous season.

The owners' luck continued. They ordered a third steamboat, the *Columbia*, another Brown and Bell product. She was 164'-6" long, 22'-6" wide, and 11'-10" deep, of 423 tons, an enlarged *Gibbons* with, for the first time, a small hog frame installed to stiffen the long, shallow hull. The usual crosshead engine had a 56" bore and a 6'-0" stroke, and, in the event of failure, there was still a two-masted schooner rig. An innovation in the passenger quarters was the inclusion of four private staterooms in addition to the usual curtained berths lining her saloons.[9]

The *Columbia* headed South on her maiden voyage[10] on March 21, 1835, and alternated with the *William Gibbons* thereafter. At first, the *David Brown* was withdrawn, had a new boiler installed, and was tried out on the Norfolk-New York route for two months, but, beginning June 17, she was added to the Charleston service, leaving New York in the middle of the week while the others sailed on Saturdays.

The *William Gibbons*, about to complete what would normally be her last trip of the 1835 season, was heading for the Narrows in New York harbor when her steam chimney gave way.[11] This was an extension of the steam space at the top of the boiler projecting up around the base of the stack. Its purpose was to dry out all moisture from the steam before it entered the main steam line leading off the chimney at its top. Although the boiler was low in the ship, the steam chimney extended up through the cabins. The actual blowout occurred between the main and promenade decks and the escaping steam killed the bartender and two passengers in the forward cabin. Two firemen died and the Second Engineer was badly scalded when the inner wall gave way and steam escaped from the furnace doors into the fire room. Captain Halsey, standing on deck immediately over the damage, was uninjured. Once again, as in the *Rariton* and the *Patent*, a boiler explosion on a boat having a copper boiler and a steam pressure of a few pounds per square inch did not destroy the vessel or do damage except to those directly in the path of escaping steam and boiling water. The *Gibbon*'s superstructure was slightly damaged. She was towed up the harbor and quickly repaired. During the winter the *Columbia* carried on but the smaller *David Brown* was sold; she started in 1836 on a new route connecting Key West, Havana, and New Orleans.

March 1836 saw the *William Gibbons* and the *Columbia* engaged in their usual weekly sailings to Charleston. For two successive years the New York and Charleston Steam Packet Line had operated regularly and profitably. For most of its third year Providence continued to smile upon these frail craft as they hurried up and down the East Coast of the United States. But all good things come to an end. On October 8, 1836, the *Gibbons* left New York for her fifty-ninth voyage without her usual master, who had been attacked by a sudden illness. Morgan, as agent for the line, had asked Edward L. Halsey, her former captain, to make the trip, but it is not clear whether he was in complete command or, as he stated afterwards, on board in an advisory capacity and as a social figurehead, with the first officer, Joshua Andrews, in actual charge. Later, Halsey placed all the blame for subsequent events on Andrews, while twenty-eight passengers printed a statement blaming Halsey for their woes.

Regardless of who was in charge, there were 128 passengers on board when, in the early hours of October 10, the *Gibbons* was thought to be past Cape Hatteras and her course changed accordingly.[12] She was actually twenty-five miles north of the Cape and ran aground in the breakers off New Inlet. She tried backing off without success, then went ahead and forced herself over the bar, but damaged her rudder and became unmanageable. Moreover, a severe gale made her condition doubly perilous. She was deliberately beached on Boddy's Island, North Carolina. From this point on, discipline broke down. While some of the crew were landing the passengers by boat, others were raiding the liquor supplies; opening the mail, looking for any money enclosed; and, finally, breaking open the passengers' baggage and pocketing their valuables. In particular, Andrews himself, an unnamed second engineer, and five firemen were involved in looting and, eventually, when the weather improved, they commandeered one of the boats and headed for Elizabeth City, North Carolina. The abandoned passengers spent two days in the open with little food or water before they were able to escape in local sailing craft to Norfolk and nearby ports. It was a group on the schooner *South Boston*, which had been chartered to carry them on to Charleston, that published the resolution condemning Captain Halsey. To them there was no question as to who was actually in command. Andrews and three of his cohorts were apprehended in New York and sent to Raleigh for trial on a charge of "gross outrages committed on the property of the passengers."[13]

Despite this discreditable incident, a replacement was ordered, a boat generally similar to the *Columbia* but somewhat smaller, 365 tons to the latter's 423, that was named *New York*. Morgan and Haggerty were now the sole owners of the line.

There was no service to Charleston during January and February of 1837; the *Columbia* opened the year's operations on March 11, continuing at two-week intervals. Later she was joined by the *New York* on June 15; thereafter they alternated with weekly sailings.

James Allaire had disposed of his shares in the *William Gibbons* at the proper moment, just three weeks before her loss.[14] Regrettably, he chose to return to the Charleston trade and contracted for a new steamboat, intended to be the largest and most luxurious on the route.[15] She was a long, very narrow boat, 210'-9" in length by 22'-6" beam by 11'-6" depth of hold, with the usual crosshead engine of the 56" bore by 9'-0" stroke. Brown and Bell built the hull and Allaire the machinery. Accommodations for 120 passengers were provided and her total cost came to about $90,000.[16] The fact that she could be built for less than $100,000 in 1837 while the *Robert Fulton* had cost $130,000 in 1820 is a clear indication that a steamboat was a much less substantial craft than an ocean steamer. Despite the reputation of her builders, there were serious errors in her design.

Because the engine and boilers were improperly located, she drew too much water aft and too little forward. To correct this, a large amount of ballast had to be placed under the forward cabin floor. Then, during her first trip with this excessive weight on board, two of the iron stay rods installed to distribute the load broke and the whole hull was strained, with the ends drooping.[17] After she returned to New York the ballast was removed (apparently as a lesser of two evils) and the stay rods were replaced, but the hull remained out of shape.

The managing agent for the *Home* was not Charles Morgan, as many sources have implied, but, as listed in the *Shipping and Commercial List*, was Allaire's Howell Works Company. Morgan did, however, sell tickets for her. The maiden voyage, with its structural failures, started from New York on September 7, 1837, and was followed by a second on September 23. Both were well patronized. Newer, faster boats always drew passengers.

A new speed record to Charleston, sixty-four hours at 9.8 knots, was made.

With about ninety passengers and a crew of forty-five she steamed down New York's Upper Bay on Saturday, October 7, just 364 days after the *William Gibbon*'s last departure. Although a pilot was aboard, he was dismissed by Captain White who, shortly afterwards, ran her aground while still in the Lower Bay. There she stuck for several hours until a second pilot was procured who got her off and out beyond Sandy Hook. There was no apparent damage.

On Monday, in an increasing gale, the hull began to work so badly that the superstructure came apart. Doors were twisting off, panels falling out, and decks over the staterooms leaking, and the ends of the ship could be seen flexing up and down. To make matters worse, she nearly grounded on Wimble Shoals, north of Cape Hatteras. When sails were set to help her escape the latter hazard, they put a further strain on her frail hull; the *Home* was caught in a beam sea, rolling became excessive, and serious leaks started. Simultaneously, Captain White was intoxicated and incapable of action. Moreoever, the other officers proved incompetent and unwilling to take over. Instead, one of the passengers, Captain John Salter, was asked to assume command. This he did, aided by two other experienced mariners, Alfred Hill and Andrew A. Lovegreen, who had taken passage.

Although every available man was put to work bailing, water steadily rose in the hull, the added weight burying her wheels so deeply that they became less and less effective. Rising water drowned out the fires about 7 P.M. on Monday. Proceeding under sail, Captain Salter ran her on the beach near Ocracoke Inlet, at 10:30 P.M. to keep her from sinking.[18] Because no boat could be launched in the violent surf, all the passengers were sent to the upper deck. When the bow broke off, seven of them were washed ashore with the topgallant forecastle. As the *Home* disintegrated, twenty passengers and twenty of the crew managed to get through the breakers alive.[19] Included were both Captains Salter and White, two women, and two lucky souls who owed their survival to having donned india-rubber life preservers. These were not furnished on board but had been purchased ashored by timid folk before sailing.

The *Home* tragedy, coupled with the incompetence of her officers, was widely reported and followed by extensive recriminations. A special committee was formed in Charleston to investigate the disaster. It heard testimony from the survivors and reported "that the steam packet, 'the HOME', was most unfaithfully built — and never was seaworthy, and that when she left the port of New York, on that most ill-fated destination to the City of Charleston, she was entirely unfit for the safe conveyance of passengers." Just to be sure their sentiments were clear, they censured the captain, the owner, and the shipbuilder. Needless to say, the same Charleston citizens had previously welcomed the *Home* and all the earlier steamboats coming from New York, had highly praised them, their owners, their builders, and their captains as long as they stayed out of trouble.

A second steamboat managed to survive the same storm although she came very close to the fate of the *Home*. This was the *Charleston* of 569 tons, built in Philadelphia in 1836 to operate from that city to Charleston. She steamed down the Delaware River on October 7, reaching the open sea about 9 P.M. This would put her ahead of the *Home*, which was then hard aground in New York harbor. She met with damage on Monday when her bulwarks were washed away and the shutters protecting the lower cabin windows stove in. The crew had excellent direction and nailed up boards and canvas to protect the openings.[20] Then, as conditions deteriorted, deck houses became damaged and the waves, breaking under the sponsons, which were fitted below the guards, forced up the deck and flooded the interior. As the *Charleston* sank deeper in the water, the passengers were put to work pumping and bailing.

By Monday noon the fore hatch had been stove in and the fires in one of the two boilers put out.[21] Sail was set and her master tried to run her ashore, but, luckily, the wind would not permit it. She finally was driven around Cape Lookout on Monday night and managed to anchor in more sheletered waters. There she rode out the storm about fifty miles from the spot where the *Home* was going to pieces. On October 10 she made nearby Beaufort for temporary repairs and a day later continued on to Charleston. The *Charleston* was a better boat; the underwater portion of the hull stayed tight during the storm, and she was obviously manned by real seamen.

With these two reports, a disaster and a near one, widely publicized, together with the comment that not a single square-rigged vessel was lost or driven ashore by the same storm, confidence in steamboats to South Carolina declined. Within a month Morgan withdrew the *Columbia*, sent her to the Gulf of Mexico, and put her to work between New Orleans and Galveston.[22]

The *New York* continued to Charleston on a fortnightly schedule through the rest of 1837, made one voyage per month in January and February of 1838, and then started fortnightly sailings again until October, when Morgan ended his entire East Coast operation and sent her to join the *Columbia* in his increasingly profitable Texas line.

One final steamboat was especially built for the New York-Charleston route, this time by a group of Charleston citizens with Captain James Pennoyer, formerly of the *David Brown* and the *William Gibbons*, as a part owner, supervisor of construction, and, eventually, master. She was the largest and best of this special breed of steamboat. Named the *Neptune*, she was built in New York by

Neptune, original appearance with two stacks. *Courtesy of The Mariners Museum, Newport News, Va.*

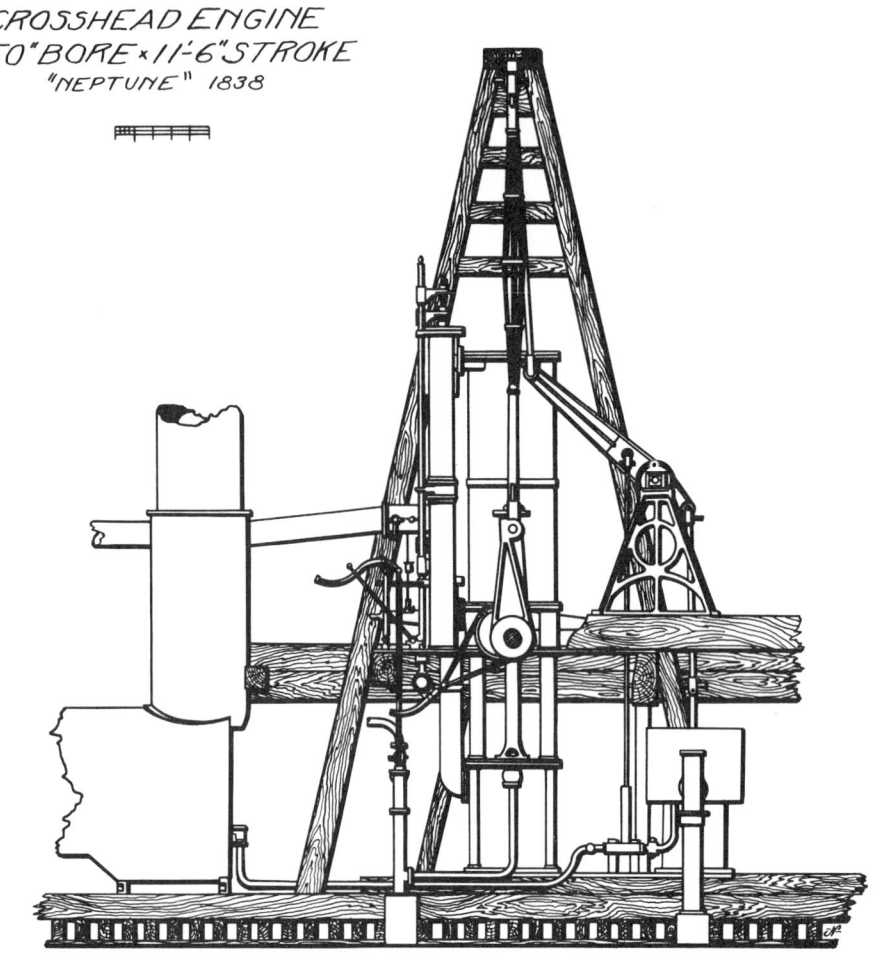

Crosshead engine 50″ bore by 11′-6″ stroke for the *Neptune*.

Lawrence and Sneeden[23] and engined by Allaire with the usual single-cylinder crosshead engine. She was 215′ in length with a beam of 25′-4″ and a depth of 14′. Her tonnage of 745, for the first time, exceeded the *Robert Fulton*'s 702 of eighteen years earlier. The 50″ diameter piston had a very long stroke, 11′-6″, and was said to develop 250 Indicated Horsepower. She was a far better design than the *Home* and her cost ran to $137,000. Originally she had two stacks on her single boiler, but later changes resulted in one. The paddle wheels were quite large, necessarily so to go with the engine's long stroke, 25′ in diameter. After the *Home* catastrophe special consideration was given to life-saving equipment, and all her mattresses, 150 in number, were designed to float and act as life preservers.[24] Four boats with a total capacity of 140 people were also supplied.

Her first sailing was in late January of 1838, and continuing voyages were made about once a month. Because passengers and freight had become scarce, the owners were soon in financial difficulties; the *Neptune* did not help matters when she was caught in a Saint Patrick's Day gale that wrecked one of her stacks, damaged her engine, and forced her into Norfolk for repairs.[25]

After only nine voyages to New York, she made one to Baltimore and, while returning, touched ground at the mouth of the Cape Fear River, where she put in to get fuel.[26] She got off, but then lay idle for a year before being acquired for the New Orleans-Galveston trade. After some modifications at New York, she sailed from there October 26, 1839, for New Orleans.

The early success of the Allaire-Morgan steamboats led to the establishment of additional routes from Philadelphia, Norfolk, Baltimore, and Washington to southern ports. Captain Pennoyer, after leaving the *William Gibbons*, acquired a very small boat, the *Dolphin* of 133 tons. Originally intended to carry passengers between Charleston and St. Augustine, Florida, she was used in May of 1835 to test the possibility of running between the former port and Norfolk in the sheltered sounds behind Cape Lookout.[27] Although only 115'-6" long with a draft below 5'-6", she grounded so frequently that the dangerous outside route had to be accepted. She was then joined by a new and much larger boat, the *South Carolina*, of 466 tons and a length of 172'; the two provided about five sailings per month between Norfolk and Charleston, connecting with existing lines to Philadelphia or Washington.

The adventurous Captain Pennoyer was soon seeking new waters to conquer. In August of 1835 he announced the extension of services from Charleston to Saint Augustine, Cape Florida (now Cape Canaveral), Key West, and Havana. The fare quoted was $40 and the Charleston *Courier* commented: "The DOLPHIN is an entirely new boat, of elegant model, with superior engine and heavy copper boilers, copper fastened and coppered—handsomely fitted for fifty passengers, and well provided in every respect. The commander is too well known to the public to need any additional recommendation. He was the first to introduce, *successfully*, Atlantic Steam Navigation, and a polished gentleman." The "superior engine" was, of course, an Allaire product.[28]

The *Dolphin* steamed in and out of Charleston for the next fourteen months, at first to Havana, later to Baltimore; sometimes on the Norfolk run; and, finally, to Florida ports. On December 17, 1836, she stopped off the Saint John's River to take a pilot for Jacksonville. When she started again, the boiler (previous accounts listed "boilers") exploded, killing half of the thirty persons aboard and sinking the boat.[29] Captain Pennoyer was not on board at the time. There is no record of the steam pressure carried, but values had been rising with time and explosions were becoming more hazardous.

The bad reputation of the coastwise steamboats was not always a matter of weather or seamanship; the *Dolphin*'s demise was probably a result of unsatisfactory operation by the engineer on watch. Because of the small number of deaths, it did not create any great stir; it was, however, the prelude to a much more serious event of the same kind.

When the *Dolphin* was withdrawn from the Norfolk-Charleston service, she had been replaced by the *Georgia* of 551 tons, built at Baltimore in 1836, a much better running mate for the *South Carolina*. The combination was so successful that a third boat, the *Pulaski*, was built in Baltimore and the three made twice-weekly sailings in 1838 between Savannah, Charleston, and Norfolk under the name of the Atlantic, Savannah, and Charleston Steam Packet Company, with fares of $25 to Norfolk, $28 to Baltimore, and $30 to Philadelphia by way of connecting steamboats.[30] The trio were said to be especially reinforced for ocean service.[31]

The *Pulaski*, an enlarged version of the *South Carolina* and *Georgia*, of 687 tons, was built by John A. Robb in Baltimore and was 206'-1" long by 25'-2" beam by 13'-7" depth of hold, powered by a 200 horsepower walking beam engine, provided by Watchman and Bratt, that had a 48" bore and a 12'-0" stroke.[32] She had two copper boilers. One hundred and sixteen passengers were carried in three public cabins and, in addition, four staterooms were available for families. The line advertised that their masters and engineers were men of skill and experience who had been hired without regard to cost. Advertisements cost very little.

The *Pulaski* arrived in Charleston on October 21, 1837, under Captain Dubois and seems to have behaved admirably for the next eight months. On Thursday, June 14, 1838, she departed from Charleston with a crew of 37 and some 150 to 160 passengers. She had left Savannah the previous day and was bound for Baltimore. Friday, about forty miles south of Wilmington, North Carolina, at 11 P.M. the starboard boiler exploded with great violence, shattering the hull all the way to the waterline.[33] Relieved of its weight of water she heeled to port, but since there was a gale blowing, kept rolling the hole in her side under and taking the sea through it. Forty minutes later the hulk broke in two, the stern disintegrated and sank, but the bow with its foremast remained afloat.

The two quarter boats were tight and usable; the two wood ones stowed on deck were so dried out that they filled immediately. Those who did get off were forced to land through heavy surf onto the barrier island nearby, and further fatalities resulted when the boats capsized. A group on the bow drifted for four days before they were taken off by the schooner *Henry Camerdon*. Others escaped by clinging to a portion of the after promenade deck and pieces of floating wreckage. About sixty people survived, a little less than one-third of the total number on board.

The engineer on watch was blamed for the explosion. It was reported that the steam pressure had been fluctuating between twenty-six and twenty-nine inches of mercury (12 1/2 to 14 pounds per square inch), a quite normal value, but, when one of the gage cocks was opened to check the water level, steam escaped, showing that the water was far too low. This was, and is, a great hazard; if the liquid level is allowed to fall in such a boiler the furnaces emerge, become red hot, and give way. Salt water boilers were particularly prone to such difficulties. As the salt accumulated in the boiler water, it had to be blown overboard, and, if the process were not stopped in time, the water level dropped too far. Steam pressures, while still low, had reached values that could cause a catastrophe.

The *Pulaski*'s violent end was the last of the train of events that, rightfully, damned the steamboat's performance on the Atlantic. Morgan's *New York* was transferred three months later and Pennoyer's *Neptune* followed soon afterward. The *Charleston* had departed for the Gulf early in 1838. Unsatisfactory vessels and unsatisfactory operation, combined with the gales found off Cape Hatteras, convinced passengers that the steamboat should stay in sheltered waters where it belonged. Owners, whether optimistic, ignorant, or rapacious, had to bow to economic pressures.

The New York-Charleston service started in December of 1832 with the *David Brown* and ended six years later with the last sailing of the *Neptune* in October of 1838. James Pennoyer had commanded both. There were two serious wrecks, the *William Gibbons* in October of 1836 and the *Home* in October of 1837. Furthermore, two boiler explosions led to sinkings, the *Dolphin* in December of 1836 and the *Pulaski* in June of 1838.

The *New York* and *Neptune* survived for a while in the New Orleans to Texas trade, but both were lost in storms, the former near Galveston on September 7, 1846, and the latter, as a Mexican War transport, on November 25 of the same year.[34] High seas and shallow-draft steamboats just did not go together.

Americans learned by bitter experience that the frail steamboat, although a magnificent engineering achievement, could not solve every water-transportation problem. It had clearly proved unseaworthy on the stormy Atlantic.

The *Natchez*

A large vessel built in Baltimore in 1838 belongs to a special case, somewhere in between a steamboat and a steamship. She was a far more seaworthy craft than the *William Gibbons, Columbia, New York,* and *Home* series that had been developed by Brown and Bell from 1834 to 1838. Named the *Natchez*, she was intended to make the long voyage from New York to the Gulf of Mexico by sea and then steam up the Mississippi River, first to New Orleans and then to Natchez, Mississippi. She was therefore the first successor to the *Robert Fulton* of 1820 to steam as far as New Orleans. In size, 792 tons, she was larger than the *Fulton* or the *Neptune*, the best of the ocean steamboats. She was 206'-9" long by 28'-2" beam and 14'-1" deep. Her hull was by Baltimore's Rogers, Brown, and Cully, and her 56" bore by 10'-0" stroke engine by Charles Reeder and Sons.[35] It was a walking-beam engine, an early example of the type to go to sea. Others had been installed in the steam brig *New York* in 1822 and the steamboat *Charleston* in 1836. The *Natchez* was heavily built and had two full decks and a three-masted rig. Her draft soon proved to be too large for regular operation up the Mississippi.

She was completed in Baltimore for George Brown and immediately sent to New York, arriving there on July 6, 1838. Thence she sailed on the 18th or 19th. From Natchez on August 23, via New Orleans on the 26th, she reached New York September 5 with almost fifty passengers. She departed again on the 24th, but, after loading at Natchez, ran aground five miles below that city.[36] In order to float her the cargo was taken off, but, simultaneously, the river was falling so rapidly that she was left high and dry.[37] Presumably she came off during the spring floods in 1839, but no further voyages were reported until October 26, when she left New York for New Orleans. From then until April of 1842 she ran regularly between New Orleans and Havana but, once a year, made the longer trip to New York for docking and overhaul.

Her last New Orleans departure was April 30, 1842, under her regular captain, Joseph Swiler, for Havana.[38] On May 27, however, she arrived at New York with a new master, consigned to a new agent, J. B. Lasala and Company. She was scheduled to sail for Charleston and Havana on June 15, 1842, but nothing further was reported. It is probable that she had been sold—hence the new agent and captain. Morrison says that she was bought by an unnamed South American country and the Lytle List indicates that she was lost in a collision on November 21, 1842. The author has been unable to confirm either of these statements.

The *Bangor*

Despite the unsuitability of the shallow-draft, lightly constructed steamboat for continuous rough water service, many a one was sent on a long ocean trip as the only possible way to deliver it to a distant buyer. The first American vessel to actually steam across the Atlantic was a case in point.

Just before the conclusion of the *Chancellor Livingston's* service in Maine waters, a group of Boston merchants formed the Boston and Bangor Steamship Company.[39] They contracted with Brown and Bell for a boat of the same length as the *Chancellor Livingston*, but of smaller beam and depth, the *Bangor* of 385 tons. She was 156' long, 25'-10" in beam, and 10'-1" depth of hold, powered by a crosshead engine 36" bore by 9'-0" stroke. There were still cabins in the hull, as in her illustrious predecessor, but also deck ones for two-thirds of her length and overhanging guards from bow to stern. Amidships was a short hog frame to reinforce the hull. There was a figurehead and a two-masted schooner rig. Because the pilot house was immediately abaft the foremast, it is doubtful that her one square sail was ever set; it would block the helmsman's view. The *Bangor* was to leave

Boston July 25, 1834, under Captain George Barker, for a day trip to Portland. Tying up overnight, she would sail the next morning for Owl's Head, Belfast, Bucksport, and Frankfort, and would reach Bangor in the evening. Fares ran from $3, for Portland, to $6, for Bangor.[40] After this maiden voyage there were occasional delays for repairs, but the *Bangor* continued on and soon proved a very satisfactory boat. At last Maine had a fine new steamer especially designed for her waters instead of an old vessel taken from some other service. It is significant that the *Chancellor Livingston* and the *Connecticut* had to be withdrawn after the *Bangor* made her debut.

Because she burned wood, it was convenient to have so much local timber available. She consumed twenty-five cords per one way trip,[41] a figure corresponding to a pile four feet wide by four feet high and two-hundred feet long. There must have been much sweat and toil loading, and later shifting logs during a trip. Both timber and labor had to be cheap, and in Maine they were.

The *Bangor* was laid up for the winter after the first week in December.[42] In succeeding years she continued between Boston, Portland, Bangor, and other Maine ports. She provided regular service except for one serious accident in November of 1836, when she was leaving Bucksport for Boston.[43] First the paddle shaft broke, inside one of its bearings, and, as a consequence, the beam driving the air pump gave way and the whole engine was thrown out of line. Fortunately, the 1836 season was about to close and it was possible to use the winter lay-up to make the necessary replacements.

By 1842 competition, partly because of an almost-completed railroad from Boston to Portland and partly because of the encroachment of other steamboats, became severe. The *Bangor* was transferred in the spring to the Portland-Calais route, via Belfast, and was soon thereafter sold to the Turkish government. In preparation for

Bangor, ready to cross the Atlantic. Deckhouses have been removed and windows boarded up. *Courtesy of Peabody Museum of Salem.*

her ocean voyage, the passenger spaces in the hull were converted to coal bunkers and most of her deck cabins removed, both to compensate for the extra fuel weight Boston, on August 16, 1842, she headed North to Nova Boston, on August 15, 1842, she headed North to Nova Scotia (where coal could be bought more cheaply than in Boston); then to Fayal, Madeira; Gibraltar; Malta; and Constantinople, coaling at each port.

Under a new name, *Sudaver*, she became a passenger ferry running between the Ottoman capital and Princes' Islands in the Sea of Marmora. About 1850 she seems to have been extensively rebuilt.[44] Later she was sold to a private firm, the Idarei Feraide Company, and lasted well into the 1880s.[45] The *Bangor*, with over fifty years of service to her credit, shows that Brown and Bell did build successful steamboats even though the *Home* did not last out three voyages. The desires of the individual owners, together with their willingness to insist on, and pay for, strength and quality, had a determining influence on the final product.

Notes

1. Carl C. Cutler, *Queens of the Western Ocean* (Annapolis, Md., 1961), devotes 177 pages of Appendices to lists of packets sailing regularly from American ports from 1817 to 1860.
2. John H. Morrison, *History of American Steam Navigation* (New York, 1903), p. 51.
3. *New York Shipping and Commercial List*, 1832-35.
4. John C. Emmerson, Jr., *Steam Navigation in Virginia and North Carolina Waters, 1826-1836* (Portsmouth, Va., 1949), p. 232; hereafter cited as Emmerson II.
5. Ibid., pp. 217, 237.
6. *New York Shipping and Commercial List*, May 23, 1835.
7. *New York Commercial Advertiser*, March 3, 1834.
8. James P. Baughman, *Charles Morgan* (Nashville, Tenn., 1968), p. 251.
9. Morrison, *Steam Navigation*, p. 437.
10. *New York Shipping and Commercial List*, 1834-37.
11. Emmerson II, pp. 341-42.
12. Ibid., pp. 367-75.
13. Ibid., pp. 385, 387.
14. Baugman, *Charles Morgan*, p. 251.
15. Morrison, *Steam Navigation*, pp. 437-41.
16. Baughman, *Charles Morgan*, p. 17.
17. *Charleston Courier*, October 30, 1837. Morrison, *Steam Navigation*, pp. 439-40, contains a quotation from W. C. Redfield, one of the *Home*'s engineers, that implies, incorrectly, that the broken "braces" were a trivial matter. They were major strength members.
18. *Charleston Courier*, October 30, 1837.
19. *New York Commercial Advertiser*, October 17, 1837.
20. Ibid., October 19, 1837.
21. S. A. Howland, *Steamboat Disasters and Railroad Accidents in the United States*, 2d ed. (Worcester, Mass., 1840), pp. 38-44.
22. Baughman, *Charles Morgan*, p. 252.
23. Alexander Crosby Brown, "An Early American 'Neptune,'" *The American Neptune* (1952), pp. 148-52.
24. *Charleston Courier*, November 6, 1837.
25. *New York Shipping and Commercial List*, March 24, 1838.
26. Ibid., March 31, 1838.
27. *Baltimore American and Commercial Daily Advertiser*, May 22 and July 19, 1835. Emmerson II, pp. 293, 300, 304, 305.
28. Ibid., p. 326.
29. Ibid., pp. 386-87.
30. *Charleston Courier*, March 28 and June 25, 1838.
31. *Baltimore American*, October 19, 1837.
32. *Charleston Courier*, October 23, 1837.
33. Ibid., June 20 and 23; July 2, 1838.
34. *New Orleans Picayune*, September 10 and December 6, 1846.
35. Morrison, *Steam Navigation*, pp. 442-43.
36. *New York Commercial Advertiser*, November 9 and 21, 1838.
37. *New York Shipping and Commercial List*, November 14 and 28, 1838.
38. *New Orleans Price Current and General Intelligencer*, December 1839 through May 1842.
39. *New York Shipping and Commercial List*, May 11 through June 18, 1842.
40. *Boston Transcript*, July 24, 1834.
41. Francis B. C. Bradlee, *Some Account of Steam Navigation in New England* (Salem, Mass., 1920), pp. 87-89.
42. *Boston Transcript*, November 11 and December 6, 1834.
43. *Baltimore American*, November 9, 1836.
44. The *New York Herald* of September 18, 1851, states that she "renewed her youth."
45. Morrison, *Steam Navigation*, pp. 394-95.

5

Auxiliary Steamships Part 2

The *Clarion*

AFTER the loss of the steam brig *New York* in 1826, fifteen years elapsed before another auxiliary steamship appeared. An even smaller vessel, the *Clarion*, was a three-masted sailing bark built at Philadelphia in 1838 by Joseph P. Vogels for himself and her Captain, John H. Young. She measured a mere 226 tons, was 91.9' long with a beam of 23.5'. How a full-length lower deck could be installed in a depth of 11.75' is a question, but, since she had a raised quarterdeck, there may have been standing room in her after cabin. If the *New York* was an oversized brig, the *Clarion* was an undersized bark. She traded to Havana and New Orleans.

In 1840 she was purchased by Captains Russell and Stephen Glover, who had witnessed some of John Ericsson's screw-propeller experiments in England. They brought her to New York and contracted for the installation of engines to convert her into a twin-screw steamship at a reported cost of $15,000.[1] The use of two propellers avoided having to modify the whole stern to make room for a centerline one. Ericcson, after arriving in the United States in 1839, had designed engines for three small screw steamboats prior to starting on the *Clarion*, his first seagoing effort. Each propeller was driven by an independent engine of 28" bore and 2'-8" stroke. They were to turn at 45 rpm and the combined horsepower was estimated at 70.[2] The conversion took much longer than anticipated; a sailing to Havana was announced for February 5, 1841, but the shafting installation kept her high and dry on dock until the 10th, when she was lowered into the water.[3] Not until the first week of April could trials be carried out. When they were, a speed of 7 1/4 knots without sails at a very low coal consumption of 300 pounds per hour was reported. An opportunity to show off her speed came when she was steaming back through the Narrows and was able to pass the packet *Alabama* sailing in under topsails and a main topgallant.[4]

On April 14, when she departed for Havana under Captain Edward Dunn, she was the first of three steam-powered vessels to make that run. Two larger New York steamers, the *Lion* and the *Eagle*, had just been finished for the Spanish naval authorities in Cuba. Their American officers, after delivering them, returned on the *Clarion* and arrived back at New York on May 20. She cleared again for Havana on June 16 and, returning, made a northbound trip of nine days loaded with a cargo of sugar and cigars.[5] This was a good passage for so small a sailing vessel, which suggests that part of it must have been under steam. She could not, however, carry enough coal for extended use of her engines; her advertisements stated that power would not be used unless calms or head winds were encountered.

In August she made a trip to Nantucket under charter to some branch of the United States government; afterward she had machinery alterations before returning to the Havana trade. Starting in 1842, the Virgin Island port of St. Thomas was added to her itinerary.

Late in March or early in April of 1842, the *Clarion*,

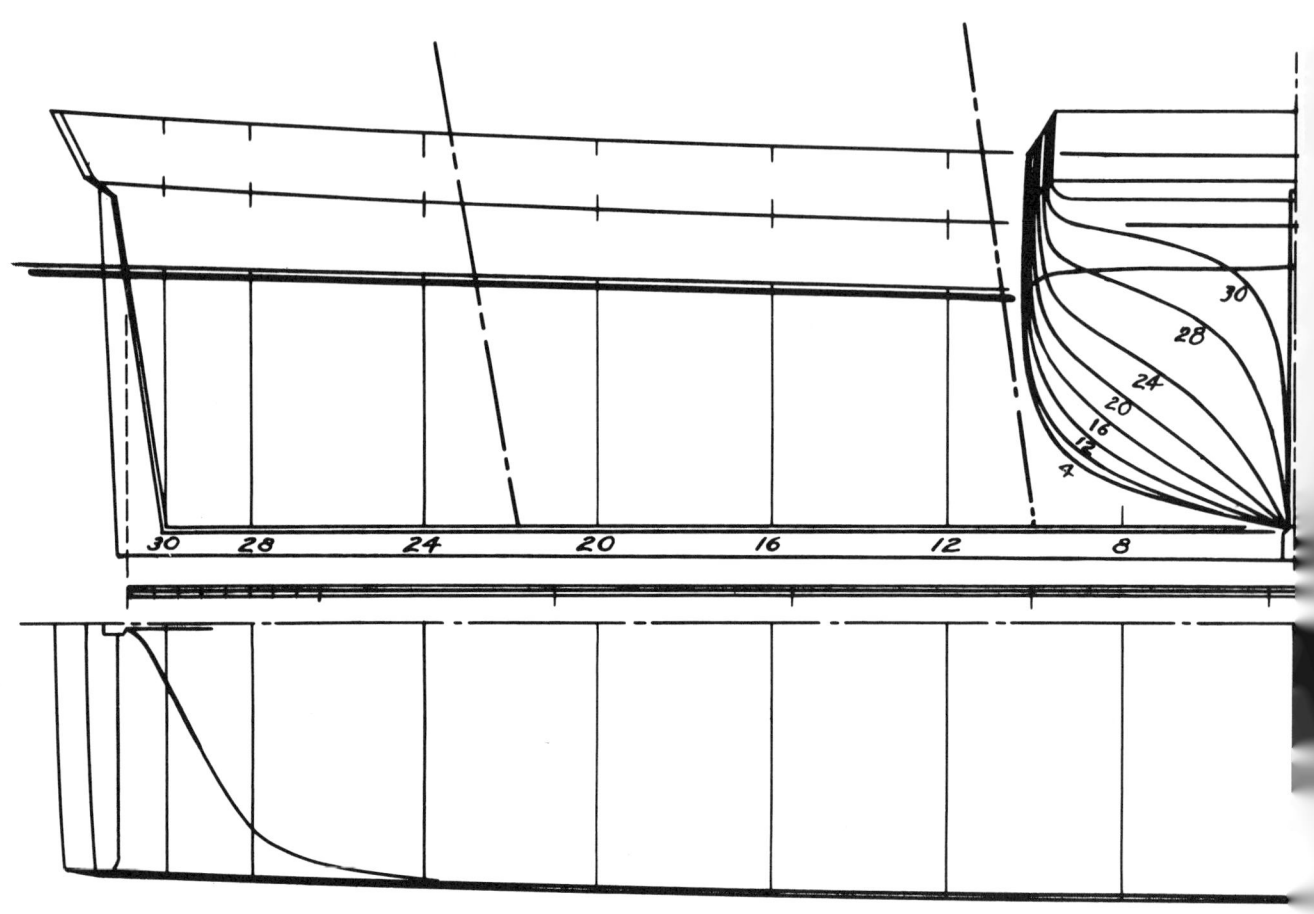

—Clarion—
1838

Mld. Length on L.W.L.   88'-2"
Mld. Beam   23'-0"
Depth of Hold   13'-9"
Cedric Ridgely-Nevitt   April 1975

From a photostat - Not faired

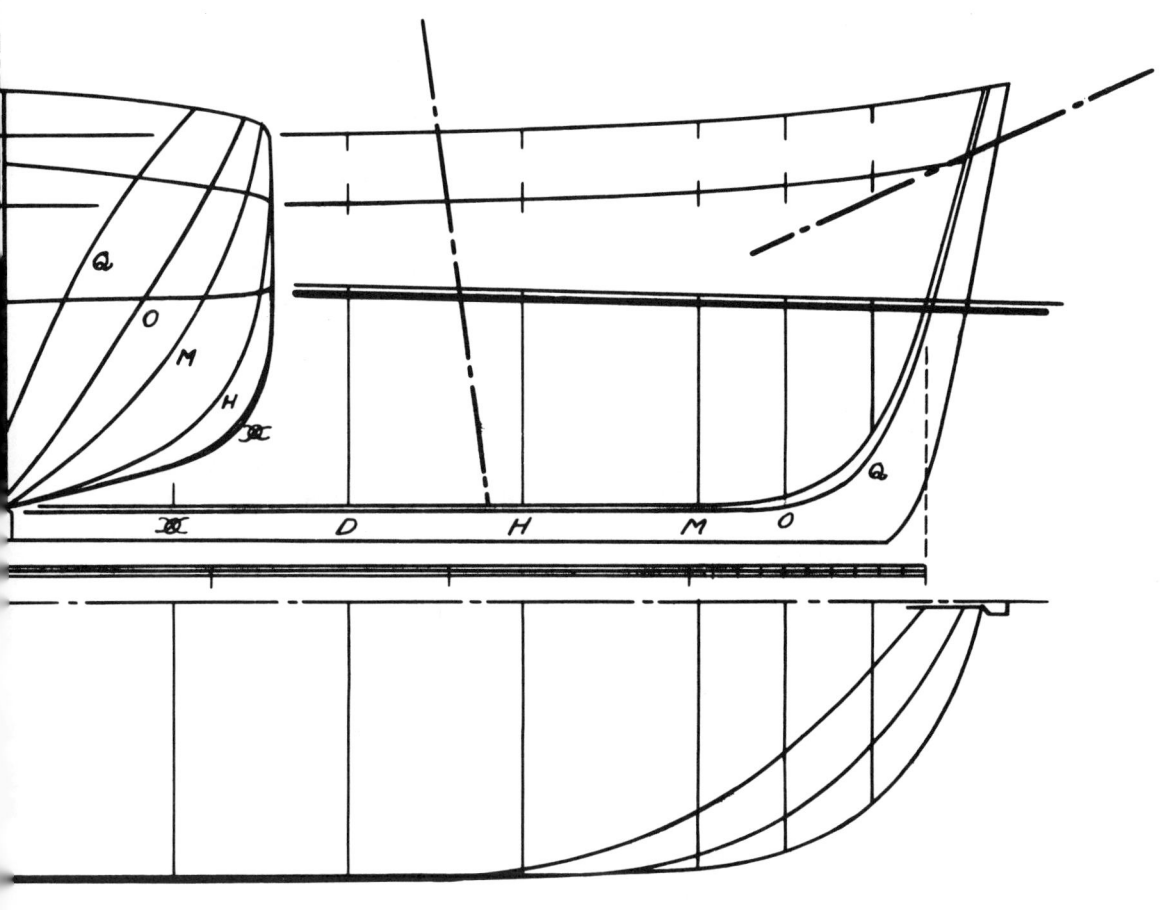

after stopping at St. Thomas, was passing through Old Bahama Channel bound for Havana when she piled up on a reef under sea conditions so bad that no boat could be launched. Her masts were cut away, the hull was holed and full of water, but the crew managed to stay aboard the wreck for three days until the waves went down and a passing sloop manned by turtle hunters took off as many as she had room for and set out for Nassau, 150 miles away.[6] The overloaded rescuer began to leak badly but, by continuous bailing, was kept afloat.

The *Clarion*'s brief career, less than a year, as an auxiliary steamship has been ignored by historians. It is easy to do this because so little is known about her and there is no portrait of her.[7] Yet she, like the *Savannah*, is another historic first. The *Clarion* was the first steam-powered screw steamer under the American flag to ply regularly on an ocean route. Not only that, her twin-screw engines were the first to operate on the high seas. She led the way to such later auxiliaries as the *Marmora* and the *Midas* in 1844. Her lines drawing survives and shows that she had a particularly attractive hull with masts raking well aft.

The *Midas*

Aside from the *Clarion*, the auxiliary steamship idea lay dormant for two decades. But with the prospect of sending steamships from the United States to China in the offing, the idea was revived and, implemented by the enthusiastic espousal of Boston's Robert B. Forbes, several vessels of this type were produced.

Forbes was, in succession, a shipmaster, shipowner, and partner in Russell and Company, the foremost American firm in the China trade. He was keenly interested in any means of improving ships; he personally developed the first successful double-topsail rig in order to make sail handling easier and reduce the number of crew members needed to man a square-rigger. He and John Ericsson were firm friends, despite the fact that his investments in the latter's projects, in particular the unfortunate steamboat *Iron Witch* on the Hudson River, resulted in large monetary losses.

In 1844 Forbes contracted for a small steamer to be built by Samuel Hall of East Boston for use on the China coast. Like the *New York*, she was a half-brig and, like the *Clarion*, she was given twin screws with machinery designed by Ericsson and built by Hogg and Delamaster in New York. J. M. Forbes and W. C. Hunter were part owners.[8] She was a very small vessel of 186 tons, 100′ by 21′-7″ by 9′-5″, with a single deck, a half-height forecastle, and a raised quarterdeck with a well aft around the steering wheel.[9] Forbes introduced his double-topsail rig on the foremast, stepping a combined topmast and topgallant abaft the masthead. When housed, it slipped down between the lower mast and the spencer mast mounted behind it. Lacking square sails on the main, its topmast was stepped forward of the lower mast in the usual manner.

The screws were of the Ericsson type which had the inner two-thirds formed of either plates or cast blades surrounded by a cylindrical ring. Then, outside the ring, separate blades made of plates rolled to shape were riveted on. With this construction, the outer blades need not be exactly in line with the inner ones and Forbes's own model of the *Midas* shows five blades outside the ring and four inside. The two shafts came through the hull and were exposed to the sea. Just forward of each propeller was a two-armed shaft bracket with a bearing that had, of course, to be lubricated by salt water. Inside the hull the shafts could be disconnected from the engines, thereby permitting the propellers to rotate freely when under sail. On her first trial, in Boston's harbor on October 28, 1844, she was said to have made seven knots against the wind.[10]

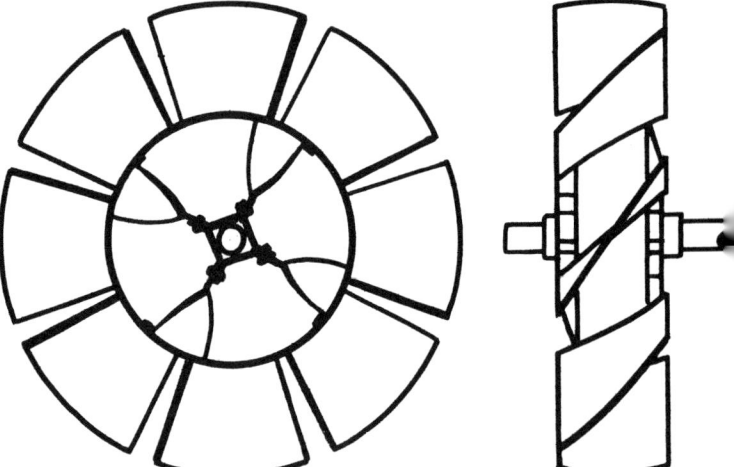

Ericsson propeller.

The *Midas* sailed from New York on November 4 with her screws in place but disconnected from the engines. She became the first American steamer to cross the South Atlantic and to round Africa's Cape of Good Hope. Unfortunately, she sailed most of the way. Thereafter she put into Mauritius with the shaft stuffing boxes leaking badly. Leaving there on March 22, 1845, she reached Hong Kong, via Singapore, on May 21. Steam was used occasionally and with such little understanding that her boilers accumulated extensive scale. The propellers, windmilling for many thousands of miles, had practically worn out their bearings by the time she arrived. It might be remarked that Ericsson always had trouble with his bearings. During and after the Civil War, his monitors had endless problems with them.

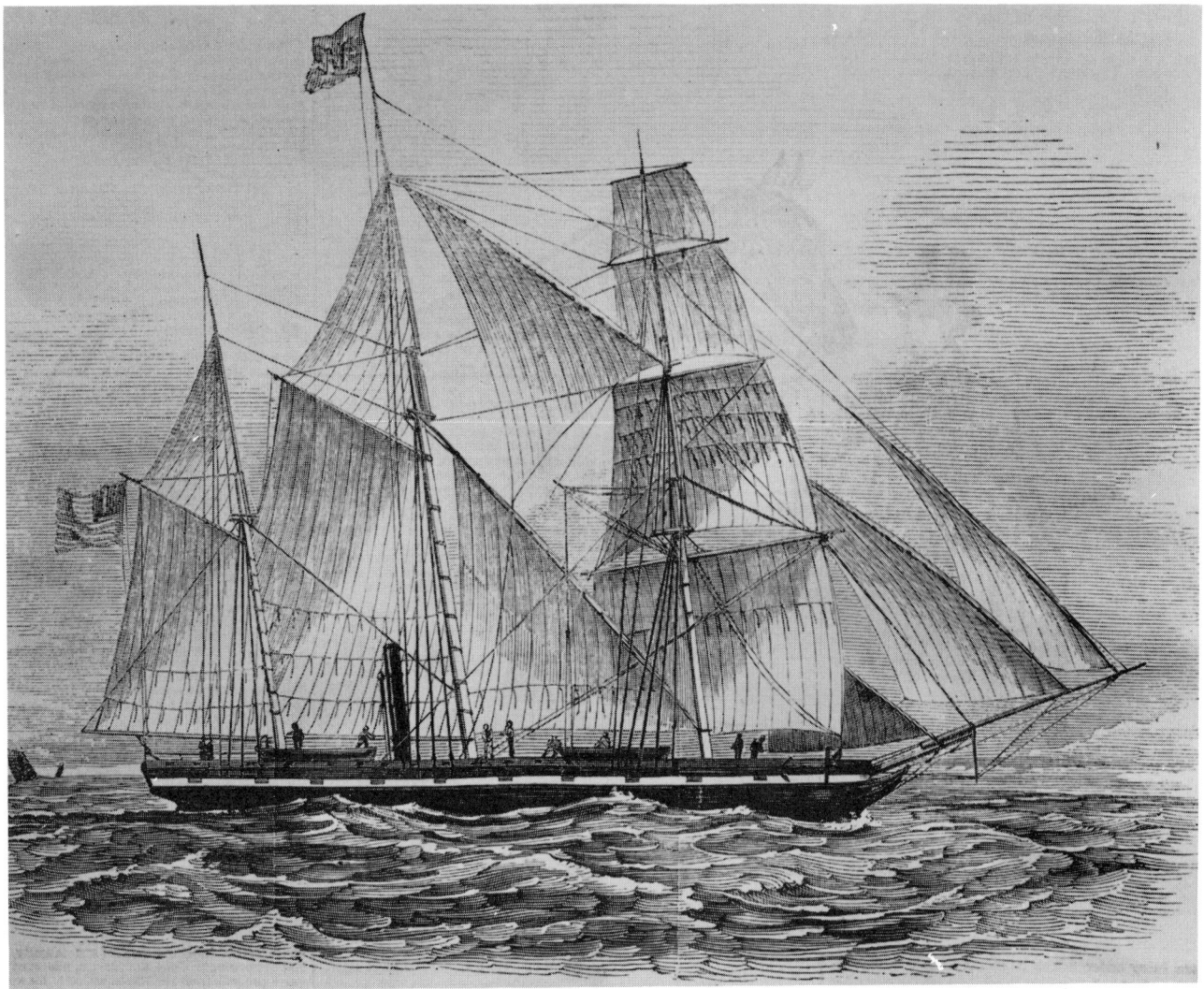

Marmora, first twin-screw steamer to cross the North Atlantic. *Courtesy of The Mariners Museum, Newport News, Va.*

Forbes wrote, "She arrived in China with her brass bearings much injured, her boiler nearly ruined by neglect and bad engineering, and her reputation damned." She was put to work running from Hong Kong to Canton twice a week, with salvage and towing work as added duties when the occasion arose.[11] In the spring of 1846 she went as far afield as Manila in the Philippines. But after a year and a half her boiler did become useless. In 1847 her propellers were removed and stowed in her hold, and she returned to New York under sail by way of Rio de Janeiro. She was sold to Padelford and Fay of Savannah, who took out her engines. Beginning in September she sailed as a topsail schooner, no longer encumbered by the drag of shafts, struts, and spinning wheels. She operated between Savannah and Rio de Janeiro for a number of years.

The *Marmora*

While the *Midas* was being built at Boston, another auxiliary steamship was under way in Maine. To her credit, she was the first American steam-powered vessel to take paying passengers across the Atlantic. And, to make the passage even more memorable, she was the first twin-screw vessel ever to cross that broad ocean. Named *Marmora*, she was built by Johnson Rideout and enrolled at Bath on August 13, 1844.[12] She was 143'-3" in length, 24'-10 1/2" in beam, and 11'-4" in depth with a tonnage of 380, and had three masts and a lion figurehead. She was built for Captain William R. Page and Rufus Hallowell.

The machinery was designed by Ericsson and built by the Phoenix Foundry in New York. It is presumed that the parts were shipped to Bath for installation. The choice of screw propulsion seems to have resulted from an agreement made in October of 1843[13] whereby Page and Hallowell were appointed agents to persuade the governments of any country other than France to install Ericsson engines in and propellers on vessels trading to the Mediterranean or the Black Sea.

The *Marmora*, then, was another example of a small steamer propelled by a pair of screws rather than by paddle wheels. They were driven by two tiny walking-beam engines set transversely in the ship, with the beams just under the deck. The compact arrangement took up little space.[14]

She was a handsome vessel with a straight sheer and a painted white port strake along her black hull. Rigged as a barkentine, the mainmast was very tall, the mizzen diminutive, but the thin stack, just abaft the main, must have interfered badly with sail handling. Perhaps it could be lowered.

Nothing is known of operations in Maine waters, but, like the *Bangor*, she was sold to the Turkish government. The agreement between her owners and Ericsson before building, and the choice of name imply that she was intended for export.

In August of 1845 the *Marmora* was in New York and was advertised as a "remarkably fast sailer. . . ." One expects a certain amount of exaggeration but the statement, "Having crossed the Atlantic several times, passengers may safely count on reaching Liverpool in 14 days"[15] was an outright lie. After a number of delays, the *Marmora*, under the command of William Page, sailed from New York on September 2. On board were sixteen cabin and twenty-four steerage passengers. Her lofty rig proved essential. Shortly after leaving New York, the fragile propellers with their thin outer blades were damaged by seas; Captain Page had to sail his steamer the rest of the way, dragging his useless screws behind him.[16] Instead of the promised fourteen days, the *Marmora* took twenty-four to Liverpool.[17]

After repairs she left Liverpool October 31, 1845, but, when a fire was discovered in her coal bunkers, she put into Cork, Ireland, where it was extinguished. The original coal supply was discharged and replaced before the journey resumed. Misfortune continued to follow the *Marmora*, for she was wrecked on the coast of Morocco before reaching her destination.[18]

The *Edith*

Never a person to do things by halves, the energetic Mr. Forbes ordered a second auxiliary steamer before the *Midas* was even finished. A much larger boat intended for the opium trade from India to China, she was named *Edith* after Forbes's daughter. Of 407 tons, she was 121' by 26'-3" by 14', with the hull by Samuel Hall, the machinery by Hogg and Delamater, and a three-masted bark rig designed by Forbes.[19] Ericsson's engine arrangement was a great improvement over that of the *Midas* or the *Marmora*. There was a single screw with its shaft far enough off centerline so that it missed the sternpost. Once again, Ericsson avoided any change in normal hull construction. When the *Edith* was under sail her propeller could be disconnected from the shaft and lifted completely out of the water.

The *Edith* started her career in a spectacular way when her launching ways collapsed and left her heeled over with her port bilge resting on the mud. Her galley, a separate house that had not been fastened down as yet, slid across the deck and pinned the yard foreman, Mr. Delano, against the rail. He was extricated in a bruised and lacerated condition and with a broken arm. A day later, November 14, 1844, she still lay in her ignominious position, with the shipyard wanting to drag her off, scraping across the bottom, and with Forbes insisting that she be jacked up and launched properly in order to avoid further damage.[20] It is not known who prevailed, but, despite the accident, she was quickly completed.

With her hold filled with coal, she sailed from New York for Hong Kong, via the Cape of Good Hope, on January 18, 1845, under Captain George W. Lewis. Because she was heavily loaded, Forbes's intention was to use steam extensively during the early part of the voyage until she was lightened enough to sail well. From then on, sails were to be used. Unfortunately, there was a leak in the steam piping that permeated the insulation under the cabin floor and made the quarters damp and soggy. It also seems to have terrified everyone in authority. As a result, she sailed on, plowing through the water at slow speed, and not until she neared India was the trouble located and steam raised again.

Oriental traders had doubts about trusting a valuable cargo like opium to a steamer, but eventually the *Edith* was loaded and made a fast passage, just under twenty-five days, from Bombay to Hong Kong via Macao, using steam whenever the wind fell. She arrived on September 11, 1845, making the passage from New York in just under nine months. For some reason, possibly prejudice, she was diverted from the lucrative opium trade and sailed for Shanghai on October 8.[21] She met bad weather and had to return to port twice for repairs. The second time her overdue cargo was unloaded and sent on by a sailing ship. Her engine was so weak that she could make no headway against rough seas or adverse winds. Forbes's business associates on the spot lost patience with her; she was sent home with a valuable cargo of tea and silk from Wampoa on December 8, 1845, for Boston, via Rio de Janeiro, entirely under sail.[22]

The *Edith*, to date, was a successor to the *Savannah* in that she was a low-powered auxiliary steamer in search of a commercial reason for existence. The next step in her career proved to be different; when the *Savannah* went to

Washington, the naval authorities did not want her. But when the *Edith* came home, the Mexican War's outbreak had created a demand for transports of all kinds. The War Department first chartered, then bought her; used her in the Gulf of Mexico; and, once peace arrived, turned her over to the U.S. Navy. The latter's steam warships were all auxiliaries; none could carry enough coal to steam all the way to Africa, the Mediterranean, or the Orient. An auxiliary steamer would fit in nicely.

The *Edith*, was sent around South America, sailing from New York on November 7, 1848, but her West Coast operations as part of the Pacific Squadron were of brief duration. She was wrecked on August 24, 1849, two days out of San Francisco, off Cape Conception.[23] At the time, she was headed for Santa Barbara and San Diego to pick up delegates who were to attend California's state constitutional convention at Monterey.

The *Massachusetts*

The largest American auxiliary steamship for commercial use was the *Massachusetts* of 1845. She was ordered about the time the *Edith* sailed for India and was, first, an enlarged *Edith*, and, second, a reincarnation of the *Savannah*. Like the latter, she had a packet-ship hull; furthermore, with a hoisting screw, she could sail unencumbered by underwater appendages. Like both, she had engines of limited power that could drive her at a low speed in smooth water. Her owners were Robert B. Forbes, John M. Forbes, John S. Gardner, and Edward King. Samuel Hall built her hull in East Boston; the machinery, designed by John Ericsson, was by Hogg and Delameter of New York; the double topsail rig was of the Forbes pattern.

She is the first American transatlantic liner to have plans that still survive. In addition there are several lithographs, a good oil painting, a number of logs, and excellent descriptions of her machinery. The plans, logs, and trial reports represent the ship as rebuilt in the Norfolk Navy Yard during 1853 and 1854, but the changes from the original were minor.

She had a length on deck of 161'-3", an extreme beam of 31'-8", a depth of hold of 20'-0", and a tonnage of 750. There were two full decks, a transom stern, and a billet head. Forward, a topgallant forecastle extended almost to the foremast; aft, a full height poop was fitted from the mainmast to the transom. Amidships were the usual packet's high bulwarks and a long deckhouse that housed the gallery. Joining the forecastle and the poop on either side were gangways level with the rail.[24] The lines show a typical sailing hull with a full bow, a fine stern, moderate dead-rise, a large bilge radius, and the maximum section well forward. When fully loaded she would draw about 16' but, in naval operation, averaged 14' to 15'. At the lower figure she had a displacement (or actual weight) of 1,168 tons in salt water and, for every additional inch of draft, could carry another 9.64 tons of coal, stores, or cargo.[25] This would make the fully loaded displacement 1,400 tons. For comparison, the *Savannah*'s was about 380.

The machinery was designed for a much higher steam pressure than we have considered thus far. The simplest way to increase horsepower and, simultaneously, reduce space, weight, and cost, was to increase the pressure. As it goes up, so does the boiling point of water. And, if higher temperatures prevail in the boilers, insoluble scale is formed whenever salt water is used. The scale acts as insulation, the metal parts are overheated until they deteriorate and collapse under the high pressure. All these problems were discovered early in the history of steamboats operating on salt water; as a result, low-pressure boilers and massive engines were retained from Fulton's time through the 1830s.

The difficulties could be avoided if fresh water, instead of salt, were used in the boilers. River and lake steamers took advantage of it at an early date. But the jet condenser injected whatever water the boat was floating in to condense the steam. Fresh water for boilers of seagoing ships required that the simple jet condenser be eliminated and the more complex surface condenser be substituted. In the latter the exhaust steam from the engine passed through tubes around which cold sea water was dripped or circulated. The condensed water, therefore, remained fresh when pumped back into the boilers.

All this is very simple in theory; the invention of the surface condenser goes back to James Watt. In practice the problem was anything but simple. Galvanic corrosion, the type found when dissimilar metals are in contact in a liquid capable of carrying an electric current (such as salt water) results in one of the metals corroding badly and soon developing holes through it. The manufacture of thin-walled tubes, in the first part of the nineteenth century, was difficult. Even if they could be obtained, corrosion soon destroyed the boundaries separating the steam from the salt water. From that point on the salt water did get into the fresh, was pumped into the boilers, and all the basic problems of sea water in the boilers returned. Although many patented types of surface condensers were developed, entirely reliable ones were not produced prior to 1870. Up until that time the major ocean steamers had low-pressure engines, jet condensers, and, therefore, large, heavy power plants with high fuel consumption.

John Ericsson was never one to be daunted by mere matters of practicality. He "invented" his own surface condenser (which had no original features)[26] and

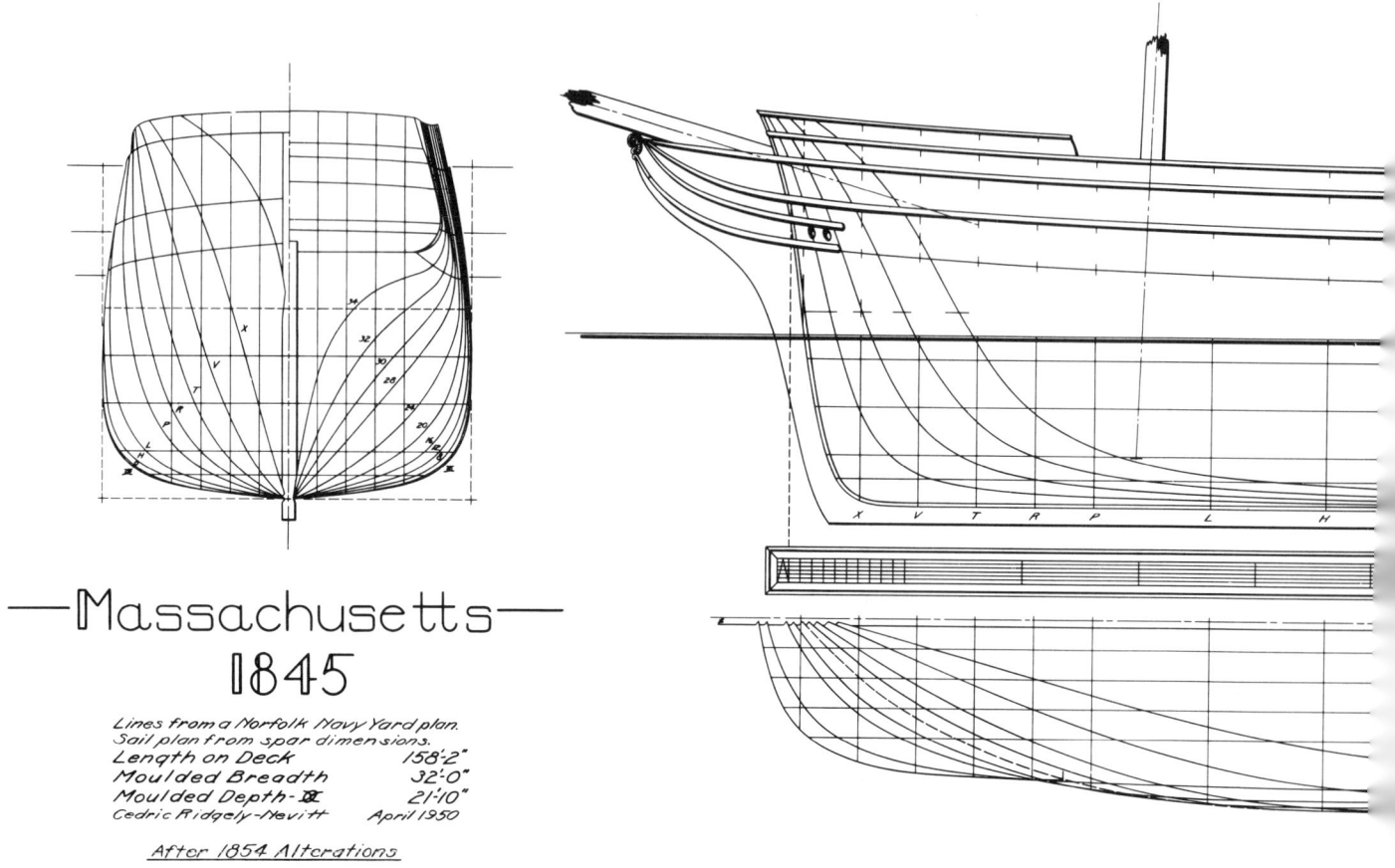

Massachusetts, 1845, lines.

assumed that it would continue to function regardless of any corrosion problem.

A second item that had to be changed, when pressures were increased, was the boiler material. While copper did not corrode and was easy to work, it was too weak and too easily deformed to withstand high pressures. As technology advanced, wrought iron plates of increased sizes and improved quality became available and were imported from England. Since wrought iron was a much stronger material, and a considerably cheaper one, it had, by 1840, become the standard for boiler construction despite its greater corrosive rate.

The *Massachusetts* had two iron boilers 14'-8" long and 7'-0" in diameter, each with two furnaces.[27] Flues ran from the furnaces to an internal combustion chamber at the other end of the boiler, with a series of horizontal tubes returning the hot gases to the furnace end before they passed out and up the stack. It was, in principle, almost a "scotch" boiler that would become the standard type some forty years later. Her pair were designed for anthracite coal and were intended to operate at 40 psi (pounds per square inch). A telescoping stack could be lowered when the fires were out. The engine was located well aft and consisted of two inclined cylinders in a Vee-arrangement turning a single crankpin on the shaft. They were 24 7/8" bore by 3'-0" stroke and were almost at right angles to each other.

The propeller shaft extended out through the port side clear of the sternpost and had a hollow tube inboard, then a smaller-diameter solid shaft outboard. The latter part could be pushed aft to fit into a square hole in the propeller hub, or pulled forward, telescoping into the tube and disconnecting the propeller. The propeller and its bearing were carried by an arm supported from a

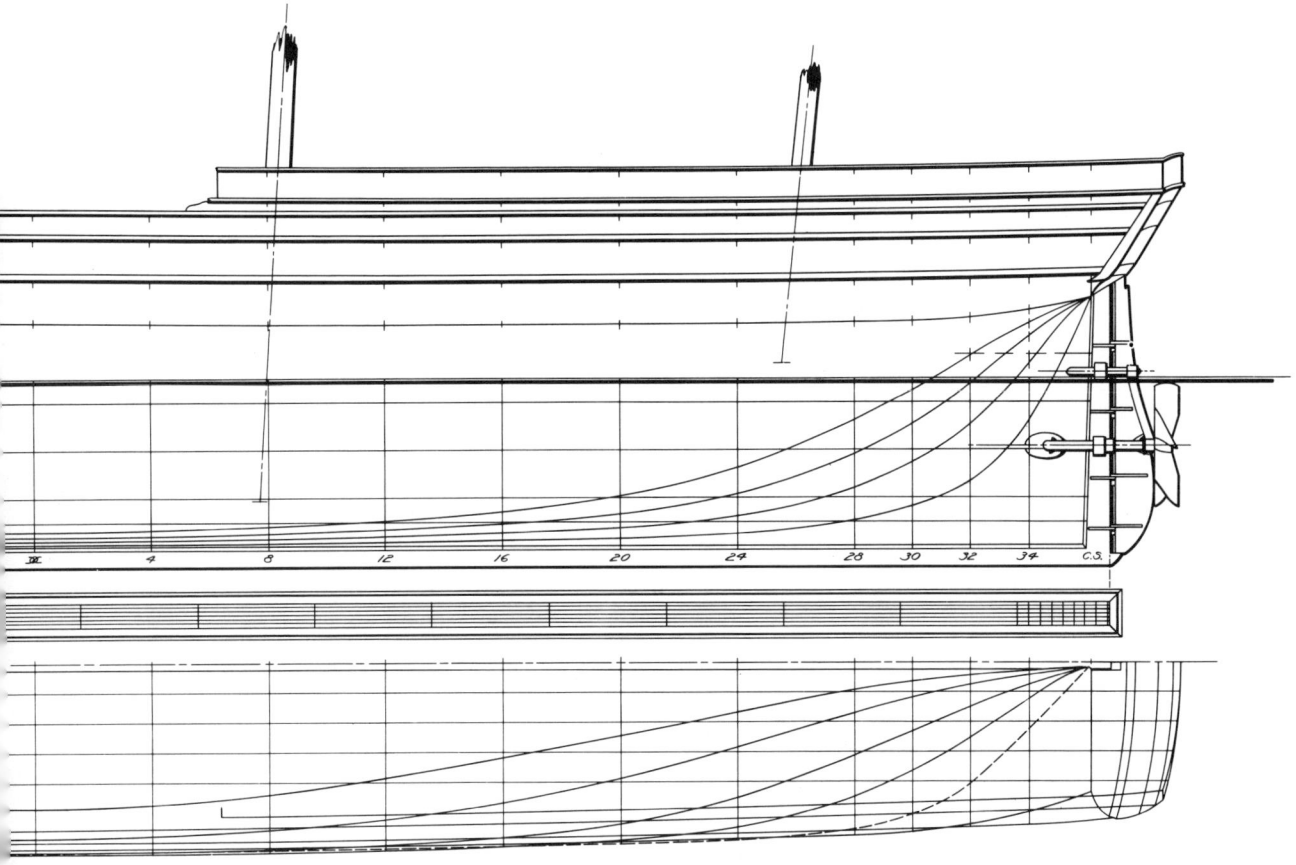

second shaft piercing the hull at about the water line. The latter could be rotated through half a revolution, either by a hand capstan or a small steam engine. In the lower position it held the propeller in line with the driving shaft, which could be moved aft to engage the socket in the hub. When the propeller shaft was pulled forward and disengaged, the upper shaft was rotated to bring the propeller out of the water and hold it behind the transom, where it could be secured in place by chains. A smaller version had been tried out on the *Edith*.

All this was highly original and produced a compact steam plant costing only $24,000, compared to the hull's cost of $56,000.

There were, however, several important drawbacks. The problem of keeping a surface condenser tight has already been mentioned. If it were not kept so, the use of return tubes—rather than the larger flues found in earlier craft—inside the boilers would lead to rapid scale formation and eventual failure. The boiler was not designed with any condenser malfunction in mind. The screw-hoisting gear was clever and worked well in smooth water, but was intended for a ship that did no backing. As long as the *Massachusetts* was going ahead, the thrust from the propeller pushed against the shaft and transferred the load to a thrust bearing inside the ship. But when she backed, the propeller tried to pull away from the shaft and put all the load on the supporting arm. This badly overloaded the bolts holding it to the sternpost. The whole arrangement was suitable only for a ship of low power; even so, she had to be handled with care to avoid wrenching the whole installation off when the engine was stopped or backed. And it was impossible to secure the propeller out of water when seas were running.

The Ericsson propeller, 9 1/2' in diameter, was located aft of the rudder, which had a notch in it so that it could be turned to port without being stopped by the

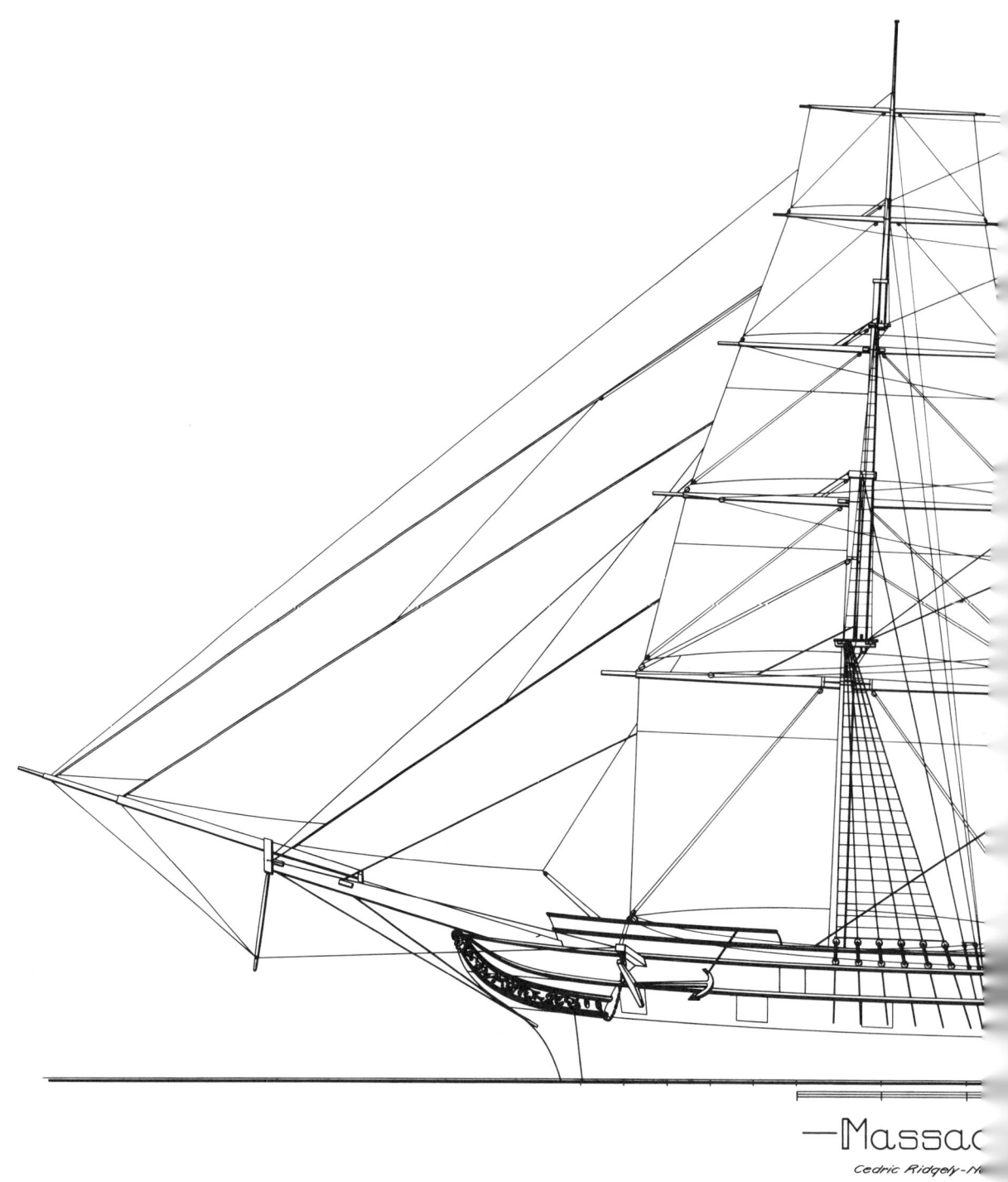

Massachusetts, sail plan.

ts 1845—
June 1950

shaft. The cylindrical ring of Ericsson's earlier propellers was omitted and diagonal iron braces used to strengthen the six cast brass blades. These must have interfered with the flow of water and reduced the screw's efficiency.

The Forbes rig was the most successful innovation; it halved the size of the topsails, the largest square sails on the ship and the ones most frequently in use. It also made it easier to shorten sail without sending men aloft to man the yards and tie reef points. Spars could be quickly sent down and there was considerable standardization in their sizes. The crossjack, fore lower topsail, and main upper topsail yard, for example, were identical in size, 47' long by 14" diameter. Similarly, the mizzen lower topsail, fore upper topsail, and main topgallant yard were interchangeable.[28]

The *Massachusetts* was reported as making 9 miles per hour with 170 IHP and 8 tons of coal per day when first completed. Ericsson always exaggerated. But, after she was reboilered in 1854 with a new (and probably superior) five-bladed propeller 10'-6" diameter, her maximum speed under steam in smooth waters was 6.9 knots.[29] At the time, the ship was fully loaded to a 16'-3" draft and the steam pressure was 37 psi, practically the same that Ericsson had designed for. The measured Indicated Horsepower was 241 at 54 revolutions per minute. In service she averaged 4.86 knots without sails. It is unlikely that a higher rate could have been attained with the original boilers and screw.

All this shows just how underpowered she was; the *Massachusetts* was little faster than the *Savannah*, despite the quarter of a century that had elapsed. On the other hand, being a larger boat with a more economical power plant, she could carry the coal required to drive her a much longer distance.

An advertisement in the *New York Commercial Advertiser*, starting August 8, 1845, indicates that thirty first-class passengers could be carried in cabins in the long poop. On the second deck were spaces for fifty in the second cabin, plus an unstated number in the steerage. To assure the timid, the advertisement noted that the ship was modeled, arranged, and rigged in precisely the same way as the most approved packets but could also be moved by steam in calm and moderate winds. However, the latter ability "will not interfere with efficiency as a sailing packet."

Forbes had planned to send her from Boston to Liverpool, but very sensibly selected New York as a city with a larger potential for passengers.[30] Arriving there on September 2, 1845, she took four long days from Boston. The engine cannot have had much use. Scheduled to sail on September 1, the date was put off to the 6th, then to the 10th, and, finally, the 15th.[31] All these delays imply machinery adjustments. She headed out to sea under the command of A. H. White, bearing with her Forbes, his family, and friends—a total of eight first-cabin passengers—plus thirty others in the steerage. The *Massachusetts* arrived at Liverpool on October 3. It was no faster a crossing than that made by the crack packet *Yorkshire*, which sailed on September 16 and reached Liverpool October 4. Returning, the *Massachusetts* left on October 22 and reached New York either November 22d or 23d, via Holmes Hole, in Martha's Vineyard, on the 19th. She had been aground twice in that vicinity and had jettisoned fifty bags of salt from her cargo. Steam was used extensively on the eastbound passage and sparingly when returning. The Welsh coal loaded at Liverpool melted the grate bars in her boiler furnaces.

On her second voyage the *Massachusetts* left New York December 11, making a nineteen-day passage to Liverpool. Coming home she departed January 21, 1846, arrived at the Grand Banks, off Newfoundland, by the 14th of February, but was held up by adverse winds and did not get to New York until March 4. A forty-two-day crossing, even in midwinter, was a very poor showing.

The *Massachusetts*, as a single ship operation, could not compete with the many regularly scheduled American sailing packets or the small number of full-powered British steamships crossing the Atlantic.

A third trip was advertised for May of 1846 but abandoned. Forbes by this time was forced to admit that his enthusiasm for the auxiliary steamship was making great inroads into his own, as well as his friends', capital. They were all fortunate that the War Department, spurred on by the Mexican War, was anxious to obtain transports with steam power, regardless of their speeds. That agency bought the *Massachusetts* and the *Edith* to carry the troops and supplies needed to attack Mexico's Gulf ports. In March of 1847 she took General Winfield Scott and his staff to Vera Cruz, where a major amphibious landing was successfully achieved by the Army and Navy working harmoniously together. Over 8,000 troops were put ashore on March 9, 1847.[32]

The *Massachusetts*' two Atlantic voyages and year as a transport were enough to develop the difficulties inherent in Ericsson's installation. The condenser was no longer tight, the boilers were deteriorating; when she reached New York on September 25, 1847, she took nineteen days from Vera Cruz and had very little use of her engine on the way. Work was done on the boilers and the engine, and a completely new condenser, along with a new circulating pump for it, was installed. The condenser and pump cost $2,271.53.[33] The overhaul must have been extensive, for the *Massachusetts* was unable to get back to sea until December 31.

Then, four days out of New York, the propeller shaft broke inside the stern tube, and again she was without power. Instead of returning, she sailed on. A new shaft was ordered in Cincinnati, where it could be sent to New

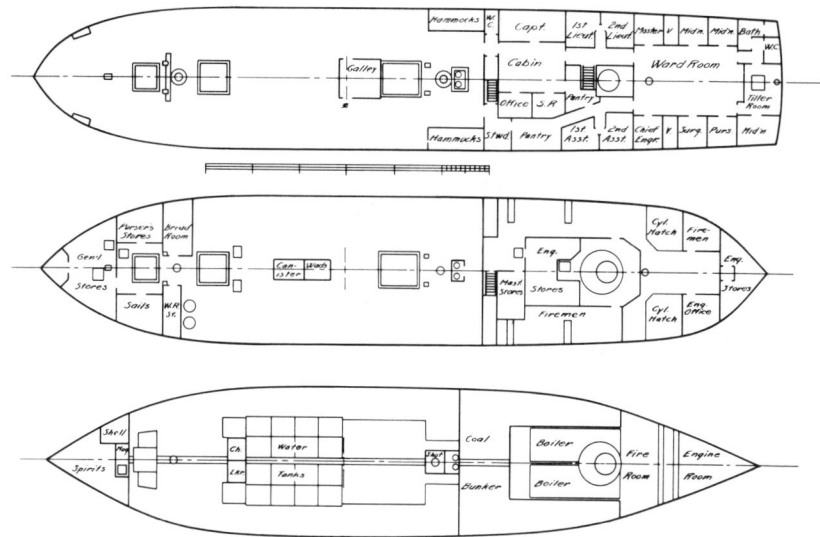

Massachusetts, 1854 deck plans.

Orleans by steamboat.[34] A faster delivery was possible by river than by sea.

The *Massachusetts* returned again to New York on August 14, 1848, and this time made a much better passage, thirteen days from Vera Cruz. Now that the war had ended, she headed for the West Coast, leaving New York on November 11, 1848, still under Army control but intended for naval use. The actual transfer took place on August 1, 1849, at the Mare Island Navy Yard near San Francisco.[35]

The Navy found her a most useful ship. She could sail long distances but could get in and out of close waters under steam with, because of her low horsepower, a limited expenditure of coal. The latter, transported all the way from England by sailing ship, was an expensive commodity. In November of 1849 the *Massachusetts* cruised to the Hawaiian Islands and, after her return, was sent exploring the West Coast picking sites for lighthouses and buoys. By 1852 repairs for both the hull and the machinery were again in order. Sailing from San Francisco on August 13, she reached the Norfolk Navy Yard on March 17, 1853. On the way she spent twelve days at Punta Arenas, Chile; fourteen at Valparaiso; and eight at Rio de Janeiro. Since there was power to aid her, the Strait of Magellan was chosen rather than the stormy passage around Cape Horn.

Although the *Massachusetts* lay at Norfolk undergoing repairs for over a year, the actual work involved did not justify such a long stay. Bureaucracy was at work. The chief item accomplished was the replacement of the worn-out boilers by new ones 19' long by 9' wide, somewhat larger than the original ones, but of more rugged design, with flues to provide the heating surface rather than tubes. The stack was increased in diameter to improve the draft. Ericcson's had been a very tall, thin one with a fan inside. A five-bladed bronze screw of 10'-6" diameter was substituted for the original one with its awkward braces. The carpenters removed the long deckhouse forward of the mainmast, substituted a much smaller one, and cut four gunports through the bulwarks port and starboard.

Three Norfolk drawings, dated from July 27, 1853, to July 5, 1854, survive in the Naval Plan Collection at the National Archives. These show the lines and the deck arrangements from the main deck down. The author has used these and the published spar dimensions[36] to produce the accompanying drawings of the *Massachusetts*. The rigging is based on two lithographs by the eminent artist Fitz Hugh Lane showing her as a packet. The author could not resist indicating the original white port strake on the outboard profile; it may have been painted out by 1854.

On July 5, 1854, she put to sea on a circuitous route back to the Pacific. Nineteen days later she touched at Madeira to put ashore her Chief Engineer, B. F. Isherwood, who was severely ill. On September 9 she arrived at Rio de Janeiro, sailed on the 21st, lost her mainmast in a storm the next day, and put back. After fifty-two days in port being rerigged, she sailed from Rio and reached the Strait of Magellan a month later. All these times indicate that the voyage was made under sail, but, in the Strait, her engine enabled her to come to the rescue of the unfortunate U.S. sloop *Decatur* which had been trying to get through from October 13 to December 18.[37] The two vessels met in Boy's Bay; the *Decatur* was, by then, eighty-eight days out of Rio. The sloop was sometimes towed, sometimes abandoned when the weather became too severe, sometimes lost in the fog, but was always relocated and, finally, delivered into the Pacific on January 4, 1855. Steam triumphed over sail; the *Decatur* had spent twelve weeks in the Strait of Magellan!

The *Massachusetts* reached San Francisco May 8, 1855, ten months and three days from Norfolk. The rest of the year she spent cruising up and down the West Coast, as far north as Portland, Oregon, and as far south as Acapulco, Mexico.

Starting in 1856 she spent sixteen months as a floating fortress, protecting the settlers of the Oregon Territory who were trying to establish themselves around Puget Sound and its adjacent waterways. It was one of the rare cases where the Navy, rather than the Army, was keeping the local Indians under control. There were landings and shore skirmishes; even the ship's guns and crews to man them were sent ashore from time to time. She was decommissioned at San Francisco on June 17, 1857; the Navy did not consider itself the proper agent to continue protecting the white inhabitants from the natives. Instead, the *Massachusetts* was laid up for a while, but then turned over to the Army on May 5, 1859. She went back to Puget Sound and was kept there until January of 1862, when the Navy reclaimed her.

Because the Navy had, in the meantime, acquired another steamship named *Massachusetts*, the packet's name was changed to *Farralones*.[38] It was at this time that her engine was removed and her rig reduced to a bark before she was put to work as a stores ship for the Pacific Squadron. That service began in January of 1863 and continued for the next four years. After the Civil War ended, she was twenty years old and no longer needed. On May 15, 1867, she was sold for $15,000 in gold, not depreciated paper currency, to G. A. Moore and Company at San Francisco. After alterations and repairs, she became the sailing bark *Alaska*. In 1870 she was as far away as Liverpool,[39] but seems to have traded mainly in the Pacific. Her long career ended when she was driven ashore at Callao, Peru, on July 24, 1871.[40]

The *Massachusetts* might be considered America's first semipractical, steam-powered Atlantic liner. She made two voyages and carried paying passengers and cargo in both directions. But, as an auxiliary steamship, she could not compete with the scores of sailing packets or with the seven fully powered British liners then operating; they could command high fares because of their speed and regularity.

It would be better to classify the *Massachusetts* as the last trial made in the experimental era during which American craft of various, and mostly unsuitable, types demonstrated that the North and the South Atlantic could be crossed by vessels that had steam power, whether or not it was actually used. If we neglect the coastwise *Clarion*, there were five of these auxiliary steamships: the *Savannah* to Liverpool in 1819, the *Marmora* to the Mediterranean in 1845, the *Midas* and *Edith* around the Cape of Good Hope the same year, and the *Massachusetts*, with two voyages to Liverpool in 1845 and 1846.

As a contrasting, and considerably less seaworthy type, there was also one steamboat that crossed the Atlantic bound for Constantinople. The *Bangor*, propelled by paddle wheels that interfered with her sailing qualities, was the first American steam vessel to cross largely under power, in 1842.

To review the means of steam propulsion: there were two vessels driven by side wheels, two by single screws, and two by twin screws. Three of these, the ones that were really sailing packets with machinery added, the *Savannah*, *Edith*, and *Massachusetts*, all had special means of stowing their paddles or screws to prevent any interference with their sailing qualities.

The withdrawal of the *Massachusetts* from transatlantic service in 1846 signaled the abandonment of the auxiliary steamship as a useful type for Atlantic operation under the flag of the United States.

Notes

1. William Conant Church, *The Life of John Ericsson* (New York, 1890), 1: 109-10.
2. *New York Herald*, November 23, 1840; January 15 and 30, 1841.
3. Ibid., February 6, 15, 17, and March 5, 1841.
4. Ibid., April 5-7, 1841.
5. *New York Commercial Advertiser*, July 27, 1841.
6. Ibid., April 14, 1842.
7. The watercolor at the Peabody Museum catalogued as the *Clarion* bears no resemblance to the actual vessel. It has two masts, one deck, and an incredibly crude means of driving twin screws from a single engine. It must be some amateur inventor's proposal, not an actual craft. There is a photostat of her lines in the museum's files, but the location of the original is unknown.
8. Robert B. Forbes, *Personal Reminiscences*, 2d ed., (Boston, 1882), pp. 208-10.
9. A rigged model of the *Midas* made by Forbes belongs to the Massachusetts Historical Society and was, at one time, on display at the Peabody Museum at Salem, Massachusetts. The description of the vessel is based on it.
10. *Boston Transcript*, November 2, 1844.
11. Edward Kenneth Haviland, "American Steam Navigation in China," pt. 1, *The American Neptune* (July 1956), pp. 160-61.
12. William Avery Baker, *A Maritime History of Bath, Maine* (Bath, Me., 1973), 1: 405, 2: 828.
13. Church, *Ericsson*, 1: 156.
14. *Mechanics Magazine*, September 27, 1845, p. 224.
15. *New York Herald*, August 28, 1845.
16. *Mechanics Magazine*, October 4, 1845, p. 240.
17. *London Times*, September 27, 1845.
18. John H. Morrison, *History of American Steam Navigation* (New York, 1903), p. 394.
19. Forbes, *Reminiscences*, pp. 210-13.
20. *Boston Transcript*, November 13-15, 1844.
21. Haviland, in *American Neptune*, (July 1956), pp. 161-62.
22. *U.S. Nautical Magazine* (February 1856), p. 354.
23. *New York Herald*, February 1 and November 2, 1849.
24. *Mechanics Magazine*, October 18, 1845, pp. 257-62.
25. B. F. Isherwood, *Engineering Precedents for Steam Machinery* (New York, 1859), 2: 213-22.

26. J., C. Cresson and J. H. Alexander, "Report on Marine Condensers; and on Corrosion and Deposite in Steam Boilers" (Philadelphia, 1852), MS in author's collection, contains drawings of two Ericsson condenser designs.
27. B. H. Bartol, *A Treatise on the Marine Boilers of the United States* (Philadelphia, 1851), pp. 78-79.
28. Charles B. Stuart, *Naval and Mail Steamers of the United States*, 2d ed. (New York, 1853), pp. 63-66.
29. Isherwood, *Precedents* 2: 221.
30. Forbes, *Reminiscences*, p. 214.
31. *New York Herald*, September 16 and October 29, 1845.
32. K. Jack Bauer, *Surfboats and Horse Marines* (Annapolis, Md., 1969), pp. 75-81.
33. John Ericsson Letters, 1831-62. New York Historical Society.
34. David Wood letter, February 22, 1848, to John Ericsson, Ericsson Letters.
35. Navy Department, Office of the Chief of Naval Operations, Naval History Division, Ship's History Section, *History of Ships Named Massachusetts*, updated September 21, 1965.
36. Stuart, *Naval and Mail Steamers*, p. 64, and *Mechanics Magazine*, October 18, 1845, pp. 257-62.
37. Richard W. Meade II, U.S.N., Abstract Log of the *Massachusetts*. New York Historical Society.
38. Official Records of the Union and Confederate Navies in the War of the Rebellion, ser. 2, (Washington, 1921) 1: 82, 138; hereafter cited as Official Records.
39. Forbes, *Reminiscences*, p. 216.
40. *New York Maritime Register*, September 6, 1871.

6

The First Ocean Steamships, 1846-1850

The *Southerner* and *Northerner*

PAUL Spofford and Thomas Tileston were agents for a line of six sailing sloops operating between Boston and New York as early as 1824.[1] Over the years, as coastwide trade grew, they became a prosperous firm of New York merchants operating under the name of Spofford, Tileston, and Company. By 1840 they were particularly active in the New York, Charleston, and Havana trade. Charleston interests, in 1844, discussed with them the possibility of reopening a badly needed steamship service to that port and, in 1845, a group of New Yorkers and Charlestonians ordered a suitable steamship. William H. Brown was selected as the builder and Messrs. Stillman and Allen of the Novelty Iron Works were to design and construct the machinery. The association was to prove a happy one. Brown was one of the most talented shipbuilders in New York and had already constructed a large steam warship named the *Kamschatka* for the Russian Navy. He also produced some of the largest and fastest Hudson River and Long Island Sound steamboats. The engine builders had been working together since 1842, although the firm they managed could trace its origin back to 1833, when a boiler was built for the steamboat *Novelty*.[2] The combination of experienced shippers led by Spofford and Tileston, of an imaginative shipbuilder, and of a talented firm of land and marine engine builders produced the second actual oceangoing steamship in the United States, the *Southerner*. A similar group had created the *Robert Fulton* in 1820.

When completed in September of 1846, the *Southerner* was clearly a new type; in fact, her general arrangement set the pattern for ocean and coastwise steamers for the next decade. True, she had a rather full bow and transom stern that had been standard on sailing packets. The rig, however, was limited—three masts with a square topsail and a topgallant on the foremast. The hull was very heavily built; there were to be no more casualties such as those met by the *Gibbons* and the *Home*. An unusual feature was that the paddle boxes were not just added to the hull; instead, the side planking curved out and around the guards to increase their strength and protect them from the sea. Forward was a topgallant, or full-height, forecastle.[3] Abaft this was a short well deck, but the bulwarks were so high that it cannot be seen on any portrait of the steamshp; between the forecastle and the poop was the all-important "caboose"—in modern terminology, the galley. Aft of the well was a raised pilot house, which was adopted from the American steamboat and notably absent from foreign steamships, which still had their steering wheels located aft, sailing ship fashion. In the latter the helmsman has to see the sails; in a steamship it was more important to see ahead. From the pilot house aft was a raised deck with a solid bulwark. The

Southerner. Courtesy of The Mariners Museum, Newport News, Va.

combination of forecastle, high bulwarks, and poop produced a clear, unbroken, sheer line extending from bow to stern. Below this the black hull had a wide white strake along it but dispensed with the painted imitation gunports that had appeared on the *Massachusetts*. The poop bulwarks were white, instead of black, to suppress the appearance of excessive depth. The single stack, just forward of the paddle boxes, had a rather extensive cowl around its top. A number of Brown's contemporaries criticized her appearance;[4] nevertheless, they copied her when they received steamship orders.

She measured 191'-3" long on deck with a beam of 30'-8", a depth of 14'-1", and a tonnage of 785. This made her longer and larger than the *Robert Fulton*.

The *Southerner's* most unusual feature was her engine, which departed completely from previous American practice. Most of the vessels discussed so far had crosshead engines, pioneered by Fulton and further developed by Allaire. Another standard variety was the walking beam engine perfected by Robert L. Stevens and particularly popular on the Hudson River. This type had been used on Long Island Sound in the later Bunker boats, such as the *Washington* and *Benjamin Franklin*, in the 1820s. Neither was considered suitable for an oceangoing steamship.

The side lever engine had been universally used on British transatlantic liners. Canadian-built machinery of this type, by Bennet and Henderson of Montreal,[5] had been installed in the *Royal William* in 1831. Two years later she became the first vessel to use steam power all the way across the Atlantic. In 1838, when real passenger service began, both the *Sirius* of 703 tons and the larger *Great Western* of 1340, were propelled by side lever engines. A further reason for choosing such engines was their excellent performance in the first four Cunard steamships, *Britannia*, *Acadia*, *Caledonia*, and *Columbia*, which had been running between Liverpool and Boston since 1840. Cunard Line never tried anything else until it shifted from side wheels to propellers.

The side-lever engine was a compact unit that put all the moving parts well down in the ship and made them readily available to the engineers on watch. The cylinder was vertical and had a crosshead attached to the top of its piston rod. From each end of the crosshead rods led down

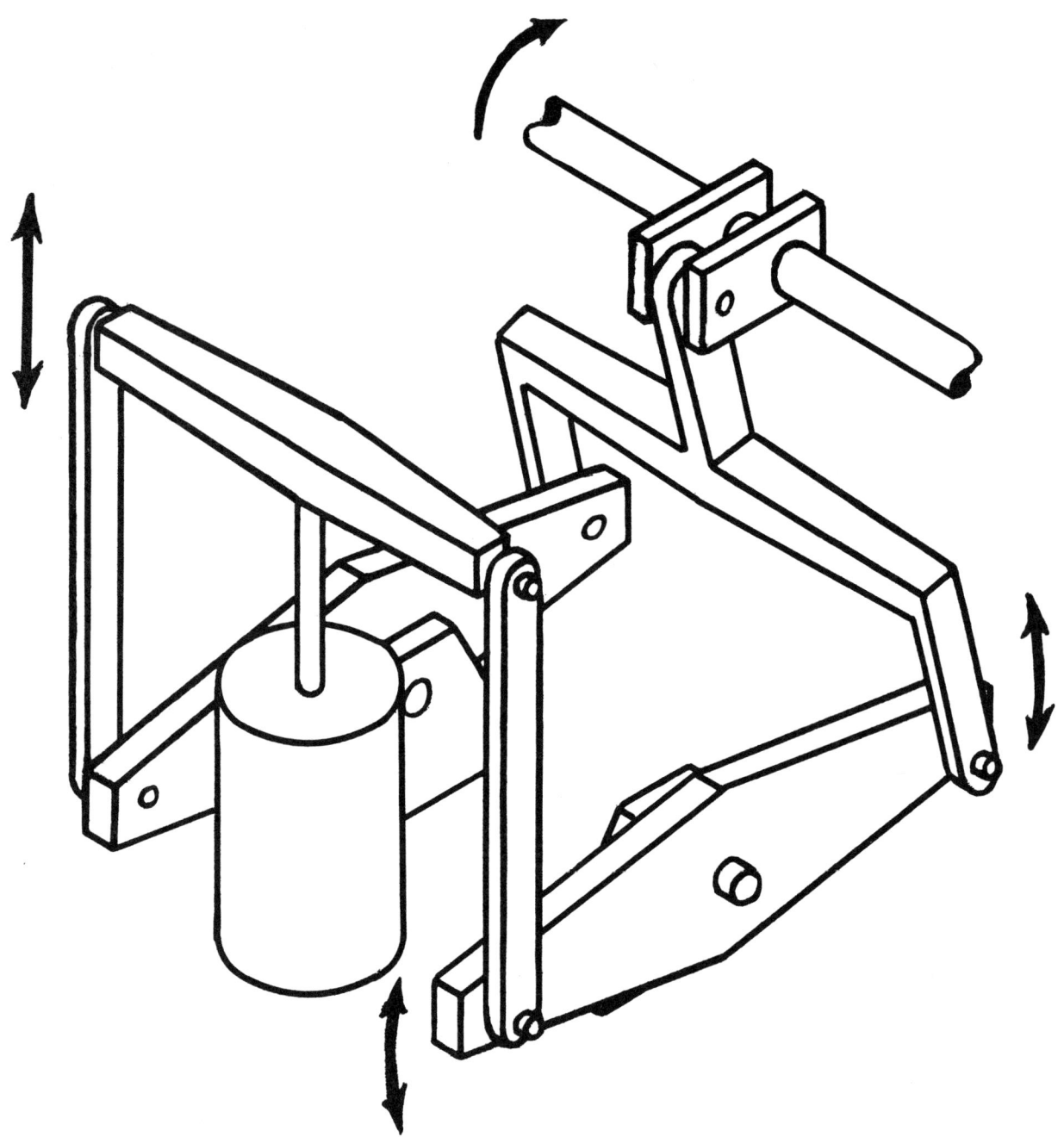

Diagrammatic arrangement of a side lever engine.

to a pair of large cast iron levers, similar to walking beams but mounted low down on each side of the cylinder rather than high above the ship. The other ends of the levers were connected by a "crosstail," and from its center a connecting rod reached up to turn the crank on the paddle shaft in effect, a compact variation of the bell-crank engine used by Fulton on the *North River*. The bell cranks were straightened out to form the two side levers and the connecting rod changed from horizontal to vertical. The whole assembly, with its air pump, hot well, and jet condenser, was mounted on a heavy bedplate that held all the working parts rigidly in line.

The United States Navy was the first American agency to install side lever engines. Two paddle-wheel frigates planned in 1839 and completed in 1842, the *Missouri* and the *Mississippi*, had similar hulls but tried different engine arrangements. The *Missouri* was fitted with inclined engines and the *Mississippi* with a pair of side lever ones of 75" bore and 7'-0" stroke. Both were designed by Charles W. Copeland, who had been employed as a consulting engineer.[6] Previously he had been superintendent of the West Point Foundry at Cold Spring on the Hudson River.

The *Mississippi* is best known as Commodore M. C.

Perry's flagship during his expedition to Japan in 1853. Charles B. Stuart, Engineer in Chief of the Navy, wrote of her, "She has...been altogether the most useful and economical side-wheel steamer in the Navy, and has, it is believed, steamed a greater distance than any war steamer now afloat."[7]

The Novelty Iron Works produced the first commercial side-lever engine of American manufacture for the *Southerner*. It was of 67" bore and 8'-0" stroke, and was considered very massive. The iron bedplate weighed 15 tons and each side lever 5-1/2. It proved so reliable that it became the standard type for the first-class ocean steamers that followed the *Southerner*. The paddle wheels were 31' in diameter and turned about 15 rpm.

The interior arrangement placed the crew in the forecastle and the first-class passengers in the poop, sailing ship fashion. Passengers had staterooms either side of a grand saloon running fore and aft. Each cabin had an airport—popularly called a porthole—through the side of the ship, and a fixed deadlight overhead. The former was an unusual feature not often found on sailing packets, which had only the latter—very small, thick, glass prisms fitted flush in the deck planking, which passed little light. These were four to six inches long and an inch or two wide. Sailing-ship quarters, even of the highest class, had always been dark, ill-ventilated spaces sealed off from the weather. This was essential, especially so when the ships heeled over when reaching or tacking. But a paddle-wheel steamer had to stay fairly upright for propulsion reasons. Hence the possibility of admitting outside air, without water or spray, was a great boon to her patrons.

The berths were against the inner stateroom bulkhead and were decorated with rich tapestry "in the most approved packet fashion." Forward of the boilers, on the lower deck, was a second cabin with a large number of berths. Quarters for 100 to 150 passengers were provided in two classes.[8]

On September 8, 1846, under the command of Captain Michael Berry, who had both invested in her and supervised her construction, a final trial trip was held with invited guests to celebrate her completion. Because of fog she did not go out to sea. The important event was a dinner and speeches by dignitaries from as far away as Georgia, with Mr. Tileston acting as host. Newspapers published exaggerated technical figures: the *Southerner* was said to have made 13 knots with the wheels turning at 17 1/2 rpm, and the engine to have produced 650 horsepower. A realistic value would be 400 IHP and perhaps 11.5 knots under the very best of conditions. The rpm was the only reasonable figure.

On September 12, 1846, she sailed from New York for Charleston and ten days later was back again after a highly successful maiden voyage. Her return took sixty hours at an average speed of 10.5 knots in good weather. Sailings continued every other Saturday. She met such a gale in November that she had to put into Delaware Bay for shelter on November 17. Coming out a day later, she left 140 sail at anchor behind the Delaware Breakwater, unable to go to sea because of adverse winds. Here was clear evidence of a real oceangoing steamship's ability to get off a lee shore and continue on her way. Even though the trip took four days, it was a triumphal return for the *Southerner* and Captain Berry.

Passengers, the United States mails, and valuable cargoes in both directions were attracted by her speed, comfort, and regularity. Northbound she carried cotton, rice, and molasses; and southbound, luxuries and manufactured goods from home and abroad. Her usual passage times were sixty hours in good weather, seventy in bad, with fifty-five hours, at over 11 knots, as the best reported. Earlier British liners without subsidies had abandoned transatlantic service to New York. As a result, the *Southerner* was left as the only oceangoing steamship regularly plying in and out of the harbor from September of 1846 until February of 1847 when an auxiliary screw steamer, the *Sarah Sands*, arrived from Liverpool. The *Southerner* was the only American steamship in ocean service anywhere for the first nine months of her career. Then she was joined by the liner *Washington*, which made her first trip from New York to Bremen in June of 1847. Although considerably smaller than the *Washington* or the *Sands*, it should be emphasized that she could leave both behind if a race were started; she was much faster.

Because the *Southerner* became an immediate success, her proprietors ordered a second steamship, the *Northerner*. The keel was laid in Brown's shipyard in March of 1847, about the time the *Southerner* was starting on her tenth voyage. Construction proceeded rapidly; the frames were up by July, and delivery was made in late September. She was a slightly enlarged version of the *Southerner*, 203'-6" by 32'-4' by 21'-7 1/2", with a tonnage of 1,012. The well deck between the forecastle and poop was covered over by a continuous spar deck, thus increasing the depth. The same barkentine rig and a white strake along the side make her appearance remarkably similar, but a light-colored stack instead of a dark one, and a narrow white wale just below the wider strake, distinguished her from the earlier craft.

The engine was a second Stillman and Allen side-lever one, with the cylinder bore increased from 67" to 70" and the stroke of 8'-0" retained. If there were no change in steam pressure, this would result in 9% more horsepower. The 31'-0" diameter wheels had 24 paddles, 7'-6" wide by 2'-6" deep and, at a draft of 12'-0", they dipped 4'-9" in the water.

Steam at a pressure of 18 psi was generated by two iron

return-flue boilers, each with two furnaces and an extensive steam chimney extending well up around the base of the stack. To assist combustion there was a forced draft fan discharging air into the ash pits below the grates.[9] In service about one ton of anthracite coal per hour was consumed; if the fuel rate were in the order of five pounds per indicated horsepower per hour, this would correspond to 450 IHP in service.

The *Northerner*, under Captain T. S. Budd, joined the *Southerner* thirteen months after the latter's reopening of the Charleston route. Sailing from New York on October 2, 1847, she reached Charleston on the 5th, put to sea again on the 9th, and made her first return voyage in fifty-five hours. With two completely successful steamers available, Spofford and Tileston set up a weekly service with Saturday sailings in each direction.

The *Northerner* and the *Southerner* were the pioneers in a rapidly expanding coastwise service that, during the 1850s, was to replace the sailing packets that had operated to Savannah, Mobile, New Orleans, Key West, and Havana. Although ocean steamship development proceeded at somewhat less than a snail's pace from 1820 to 1846, it was about to accelerate and suddenly produce new types of greatly increased sizes in the short four years from 1846 to 1850. Two factors led to the rapid production of many new craft, some very good, a few extremely bad, in such a brief period. One was Congressional action setting up payments for the carriage of the mails on a few essential routes and specifying large, fast, regularly scheduled steamships for them; the other was the 1848 discovery of gold in California, which created a completely new trade route with an insatiable demand for ships to deliver hordes of treasure seekers and all the tons of food, eqipment, and supplies they would need to live and work in an almost undeveloped and hitherto inaccessible region.

The gold rush and its demand for steamers for the California trade disrupted many plans for coastwise lines and transferred vessels from their original routes to new ones. The success of so many of these transplanted craft testifies to the skill and foresight of their designers, who had built great adaptability into their steamships just as they had, in prior years, into their sailing craft.

The *Northerner* was an early example to depart for other waters. On March 1, 1849, she sailed from New York to Chagres, on the Isthmus of Panama, touching at Charleston and Havana on the way, then returning to her regular route. A year later she was purchased by J. Howard and Son for their Empire City Line. It might seem strange that Spofford and Tileston chose to part with the newer and larger of their two steamers, but the offer of $180,000 was too tempting. Under Captain R. H. Waterman, famous for driving sailing ships and their crews to new speed records, the *Northerner* was to sail from New York for San Francisco by way of Rio de Janeiro, Buenos Aires, Valparaiso, and Panama (the old Spanish town on the Pacific side of the Isthmus of Panama). She was cleared at the New York Custom House on April 9, 1850, and reached Panama on July 16, ninety-six days out of New York.[10] The time under way, forty-seven days and one hour, was the shortest for any steamer that had yet made the trip. On August 15 she reached San Francisco with a passenger list of 415, most of whom had embarked at Panama.[11] The number was at least double her original capacity.

Despite the long voyage, she must have been in good condition for she set out for Panama two weeks later on August 31 with about two-hundred passengers and a large amount of gold dust on board. While leaving San Francisco she struck a rock, carried away her bowsprit, and sustained other damage, but no serious delay resulted. She stopped at Acapulco, Mexico, and Realejo, Nicaragua, on the way.[12] After three voyages to Panama, she was bought by the Pacific Mail Steamship Company, the established mail line on the route. Her size and her engine exactly matched the line's *California*, *Oregon*, and *Panama*.

The *Southerner* spent a profitable decade steaming between the two seaports for which she was designed. She seems to have been remarkably free of accidents and was rarely mentioned in the newspapers except for routine arrivals and departures. Nevertheless, there were occasional moments of crisis.

Nearing New York at 3 A.M. on November 16, 1847, she was nearly rammed by the schooner *Eliza Ann Thompson*, outbound from New York to Newbern, North Carolina.[13] The *Southerner* backed her engine fast enough so that she did the ramming, losing her jibboom and cutwater. She had slowed so much that no serious damage was done either vessel. Two of the schooner's terrified crew jumped aboard the steamer. Repairs were started and she sailed again, on her regular schedule, four days later.

Three years later came another collision, also off the New Jersey coast. This time the bark *Isaac Mead*, bound from New York for Savannah, was struck by the *Southerner* northbound, at 2 A.M. on October 4, 1850, and sank in five minutes. The collision was most unusual, directly head on, with the bow of the *Southerner* striking the *Mead*'s bowsprit and splitting her hull down the middle.[14] The *Southerner*, as the larger and higher vessel, lost her bobstay, cutwater, head rails, and flying jibboom, but retained her own bowsprit.

Three life boats of the very successful Francis type were launched into heavy seas and picked up seven crew and two passengers of the thirty-three on board the bark. Because the boats were made out of sheet metal, rather than wood, they proved easy to handle and watertight. Wood boats tended to dry out and leak badly. It was fortunate that these admirable craft had been developed and that the Novelty Iron Works was manufacturing them at such an early date.

The *Southerner*, as a mail steamer, did carry running lights, although no legal requirement for them existed in the United States. She was said to have a white light at the masthead, a red one on the starboard paddle box, and a blue one on the larboard one. The report had the sides confused.

For the second time, the *Southerner* was quickly repaired and made her Charleston departure without delay. These two collisions bear our earlier reports of sturdy construction. She had been overhauled in the summer of 1850 and had had new boilers installed.

As soon as the *Northerner* was sold, Spofford, Tileston, and Company ordered an even larger steamship, the *Union*, of 1,200 tons and a smaller one of 900, the *Marion*. Both were delivered in 1851. Following them came the *James Adger* in 1852 and the *Nashville* in 1854. As these joined the *Southerner*, Charleston sailings increased to twice a week. From Charleston a connecting line to Florida ports and, via Key West, to Havana was established in late 1848 with a Baltimore-built steamer, the *Isabel*.

The *Southerner* continued until 1857, when she was replaced by another newcomer, the *Columbia*. The *Southerner* was chartered to S. L. Mitchell, who had four steamships running to Savannah. Mitchill advertised her as a "new and splendid" steamship, a slight exaggeration, before she sailed from New York on September 9, 1857. She ran into the hurricane that sank a much larger steamer, the *Central America*, ex-*George Law*, off the South Carolina coast on September 12 and had to put into Charleston in distress, with her stack, paddle boxes, and boats gone, part of her cargo thrown overboard, and six feet of water in her hold.[15] She was patched up to return to New York without passengers or cargo, arriving September 29 to be sold for breaking up. She was dismantled in 1858.

The *Northerner* ran more or less regularly on the San Francisco-Panama route from August of 1851 to August of 1853,[16] but, as larger steamships came out to the Pacific, she was first demoted to occasional operation as a spare boat and then relegated to the northern mail service from San Francisco to the Oregon Territory.

On her final sailing, she left San Francisco on January 4, 1860, northbound for the Columbia River and Puget Sound ports. The next afternoon, at full speed with all sail set, she struck Blunts Reef, near Cape Mendocino, and started such a serious leak that the pumps could not keep it under control.[17] She was beached about three-and-a-half miles north of Cape Fortunas in heavy seas that broke over the ship. Boats were launched, but two out of three capsized. One of her engineers managed to get a line ashore, by which the master, the pilot, and the purser escaped. Six other passengers were rescued from a section of the deck that floated clear. Most of the women aboard were in the one boat that made shore safely.[18]

The *Southerner*, with eleven years of active service, and the *Northerner*, with over twelve, were the first direct successors of the *Robert Fulton* as well designed oceangoing steamships. After years of flirting with inexpensive solutions and trying obvious side issues, such as unsuitable steamboats or auxiliary sailing ships of inadequate speed, operators had started in the right direction again. William H. Brown; the Novelty Iron Works; and Spofford, Tileston, and Company had, with their very first trial, produced an outstanding steamship in the *Southerner* of 1846 and had improved on her in the larger *Northerner* of 1847.

These two were the only immediate predecessors of the first American transatlantic liners, the *Washington*, in 1847; the *Herman*, 1848; and the *United States*, also 1848. Nevertheless, the later, and more successful, liners of 1850 and 1851, the *Atlantic*, *Baltic*, *Arctic*, and *Pacific* of the Collins Line, and the *Franklin* and *Humboldt* of the Havre Line, benefited materially from the experience gained in the design, construction, and operation of other coastwise steamships built in the 1848-50 period. For this reason, these coastwise liners are discussed here as both logical successors to this early pair and as important engineering predecessors of later Atlantic liners. Their successes and failures determined the course of American steamships from 1850 until the outbreak of the Civil War.

The *California, Oregon,* and *Panama*

In 1846, when the disputed western boundary line between the United States and Canada was settled by treaty, the present state of Oregon became an official part of United States. President Polk sent a message to Congress asking that a formal method of government be set up and that a mail route be established. Two acts of Congress dated March 3, 1847, resulted, one proposing a mail service from an Atlantic or Gulf port to Havana, Chagres, across the Isthmus of Panama, and then to the mouth of the Columbia River, by way of Monterey and San Francisco, once every two months.[19] One-hundred-thousand dollars per year for the Pacific part of the service was too low to attract any contractor. The second act empowered the Navy Department, rather than the Post Office, to contract for the monthly delivery of mails by either sailing ship or steamer. The ships were to be built under the supervision of the Navy, were to be suit-

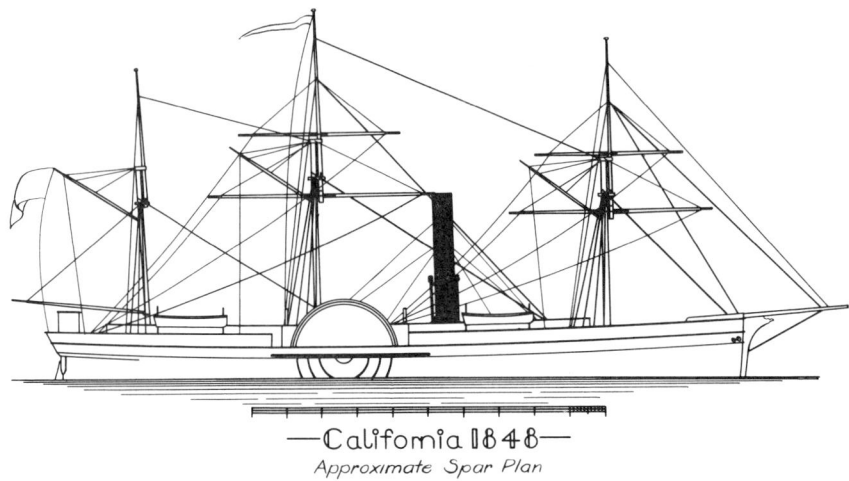

California, approximate spar plan.

able for conversion to warships, and could be taken over in the event of hostilities.[20]

The Secretary of the Navy advertised for bids to carry the mails from Panama to Astoria and eventually awarded a contract to Arnold Harris for ten years at $199,000 per annum. He proposed three steamers, two of at least 1,000 tons and one of not less than 600. The contract was signed November 16, 1847, but three days later was assigned to William H. Aspinwall, who was head of a successful firm of merchant traders, Howland and Aspinwall, which had many years of shipping experience to Europe and the Orient.

The venture, in 1847, was a speculative one. The Mexican War was still on and the United States had no clear title to California. That region was sparsely settled while the Oregon Territory, the terminus of the line, was devoid of towns of any size; the white settlers were widely scattered farmers. There was no indication that a steamship, even at a monthly interval, would attract many passengers or much in the way of freight. The Pacific Mail Steamship Company was incorporated by William H. Aspinwall, his partner Gardiner G. Howland, and Henry Chauncey. Three steamships were ordered, but, instead of two large and one small one, all were to be of 1,000 tons. Two went to William H. Webb, the eminent New York shipbuilder. Because the Webb yard had many other contracts, the third was placed with one of the oldest yards in New York, Smith and Dimon. Stephen Smith and John Dimon had, like Webb's father, been apprenticed to Henry Eckford; the *James Kent*, successor to the *Chancellor Livingston*, had been constructed by them as long ago as 1823. The Webb pair were to be named *California* and *Panama* and the Smith and Dimon one, *Oregon*. Oddly enough, their side lever engines were also split up, but not in the same way as the hulls. The *California* and *Oregon* had machinery by the Novelty Works while the *Panama*'s came from the venerable Allaire establishment.[21] No doubt the matter of delivery and, particularly, the actual installation of machinery influenced the choice. The major iron works were the only places in New York with cranes large enough to handle engines and boilers. After launching, the hulls were towed away from the shipyard and, as a rule, all subsequent work was at the engine builder's wharf, even though the woodwork was being done by the builder, the interior and the cabins by a firm of joiners, and the masting and rigging by other specialists. Fitting out had to be carefully planned to follow the machinery installation, which was always the most critical element involved. The decks could not be completed until the massive boilers and the engine castings had been lowered into place. Cabins, in turns, could not be built until the decks were completed.

The *California*'s keel was laid January 4, 1848; she was launched May 19 and delivered in October. The *Panama* followed closely; her keel was laid on February 21; she was launched July 29 and delivered in November.[22] The *Oregon*, in the yard next door, was launched August 5 and delivered within a week of the *Panama*.[23]

All three hulls appeared identical except for ornamental scrollwork at the bow and fancywork on the paddle boxes. Because a fire at Webb's shipyard destroyed his mold loft and most of the molds for the forward frames that had already been used for the *California*, Webb took the opportunity to improve the bow shape of the *Panama*, making it a little sharper and giving it more flaring sections. The two sterns, however, were identical. It is unlikely that Smith and Dimon used Webb's lines; the *Oregon* must have been somewhat different in shape, but only the most expert eye could discern the changes. Their register dimensions were:

	Length	Beam	Depth of Hold	Tonnage
California	199'-2"	33'-6"	20'-0"	1,057
Panama	200'-4"	33'-10 1/2"	20'-2"	1,087
Oregon	202'-0"	33'-10"	20'-0"	1,099

The trio were about the same size as the *Northerner*, built the previous year, a little beamier, a little shallower, but with a somewhat larger sail area. They were three-masted barks instead of barkentines. Because they had to operate on a desolate shore without shipyards, docks, or machine shops, a definite ability to sail, in the event of a breakdown, was essential. There was a short forecastle, a short poop, and a series of deckhouses from foremast to mizzen. Two elements were of particular interest.

First was the whole bow shape which, for the first time, broke away from sailing-ship practice. The stem was vertical and free of forward rake, cutwater, headrails, trail boards, and all the accustomed elements so appealing to the eye. The combination of bowsprit, jibboom, and dolphin striker was replaced by a short bowsprit supported by a plain knee to the stem. True, there was a carved gryphon on the *California* to camouflage the utilitarian simplicity and serve as a figurehead, but a real change had been made to do away with the complications developed by centuries of tradition.

A second unusual item was even more important, but was completely buried within the hull's timbers. These were the first steamships discussed by the present author to be fitted with diagonal iron strapping to strengthen their wooden hulls.[24] Perhaps the greatest single problem confronting the nineteenth-century shipbuilder was that of creating ever longer and larger ships from a relatively weak, flexible material like wood. Prior to 1847 a ship of two-hundred feet in length and over one-thousand tons was an extreme rarity, but the next decade would see oceangoing craft over three-hundred feet long and above three-thousand tons.

The major strength members in a wooden ship, in the lengthwise direction, are the two layers of planking, one outside the frames and the other, properly called "ceiling," inside. These, together with the uppermost deck, create a box-shaped girder whose ends tend to droop or "hog" if supported by a wave crest amidships with troughs at the bow and stern. The opposite deflection, in which the midship region tends to "sag," was met when there was a wave crest at each end and a hollow under the center of the ship. In both cases the individual planks tend to slide lengthwise with respect to each other and such movement was resisted by the fastenings of the planks to the frames and the friction of the caulking inserted between them in order to make them watertight. Many hulls did deform badly with time, so that the keel line formed a curve; the ends usually hogged.

Steamboat builders had solved the problem by intro-

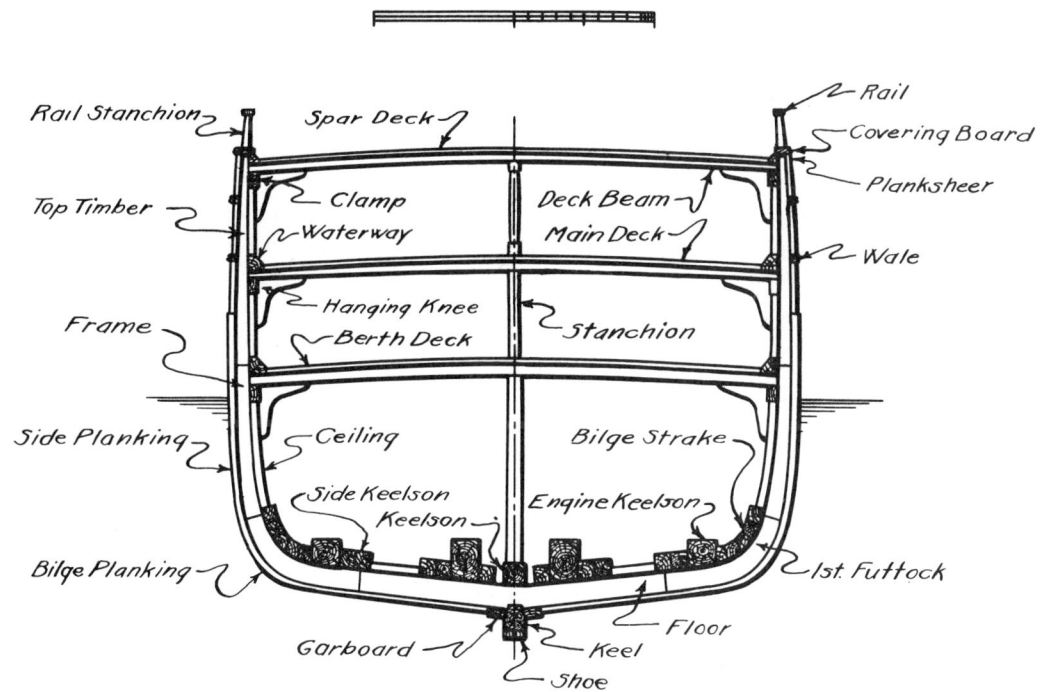

Hull structure of a wood steamship.

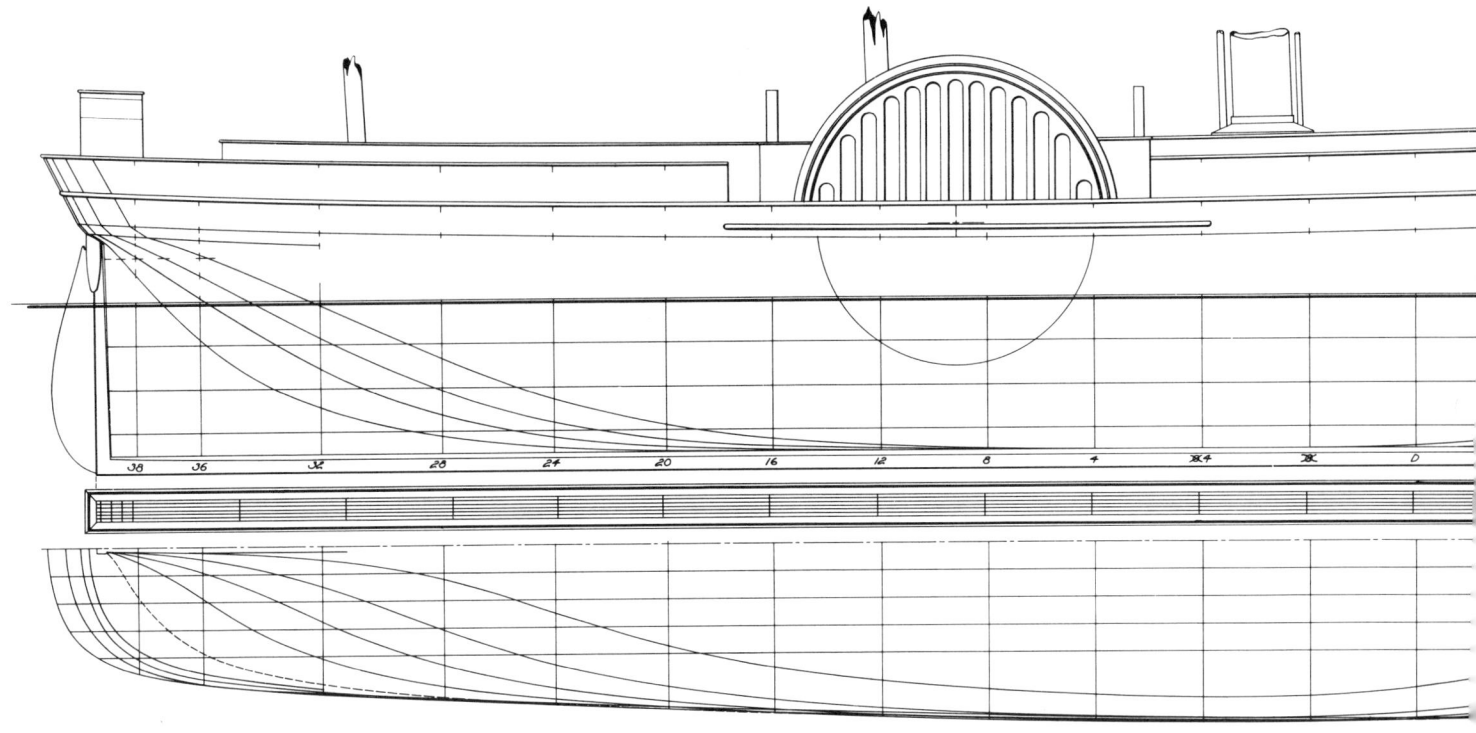

California, 1848, lines.

ducing extensive hog frames to hold their long, slender hulls in shape. A second means was the use of diagonal iron strapping which, while it did not act as a truss, did provide what is technically termed *sheer strength* and prevented sliding or working of the ceiling or planking. The date of strapping's first use in this country is uncertain; the first plan showing its application in a steamboat hull was published in 1839.[25] It had been used before that by the British Admiralty in ships of the line whose heavy guns tended to make them lose their original shape rapidly.[26] As applied to American steamships, and, much less frequently, sailing ships, it consisted of flat iron bars running from the head of the floor timbers at a 45° angle up the level of the clamps, the heavy ceiling strakes supporting the ends of the deck beams, at the highest deck. The straps were notched into the frames. A companion set at 45° in the opposite direction to the first, were notched into the planking. The two formed a series of X-braces for most of the ship's length. Where they crossed clear of the frames, the bars were riveted or bolted together. On the frame, a metal bolt was run from the planking through the crossing, the frame, and the ceiling, tying everything together. All bars had to be installed and riveted before the ceiling and side planking could be installed. In the *California*'s case, the bars were 4" wide by ¾" thick and spaced 3½' to 4' apart. Straps were expensive to fit, but without doubt contributed to the long lives of these remarkably well-built steamers. The *California* lasted for forty-six years, the *Oregon* for thirty-two, and the *Panama* for over twenty.

Because the *California*'s and *Oregon*'s single side lever engines came from the Novelty Works and, moreover, were identical in size—70" bore by 8'-0" stroke—to the engine previously supplied the *Northerner*, it is likely that they were little different from, or were even identical to, the latter's. The boilers, two in number, were of the return-flue type, similar to, but somewhat larger than, hers.[27]

For their size they were expensive ships, but justifiably so, for they had the finest possible workmanship and came from two of the best shipyards in New York. The *Oregon* cost $198,504 and the *Panama* $211,356.[28]

The author has redrawn the lines of the *California* from Webb's offsets, which show that—at some time be-

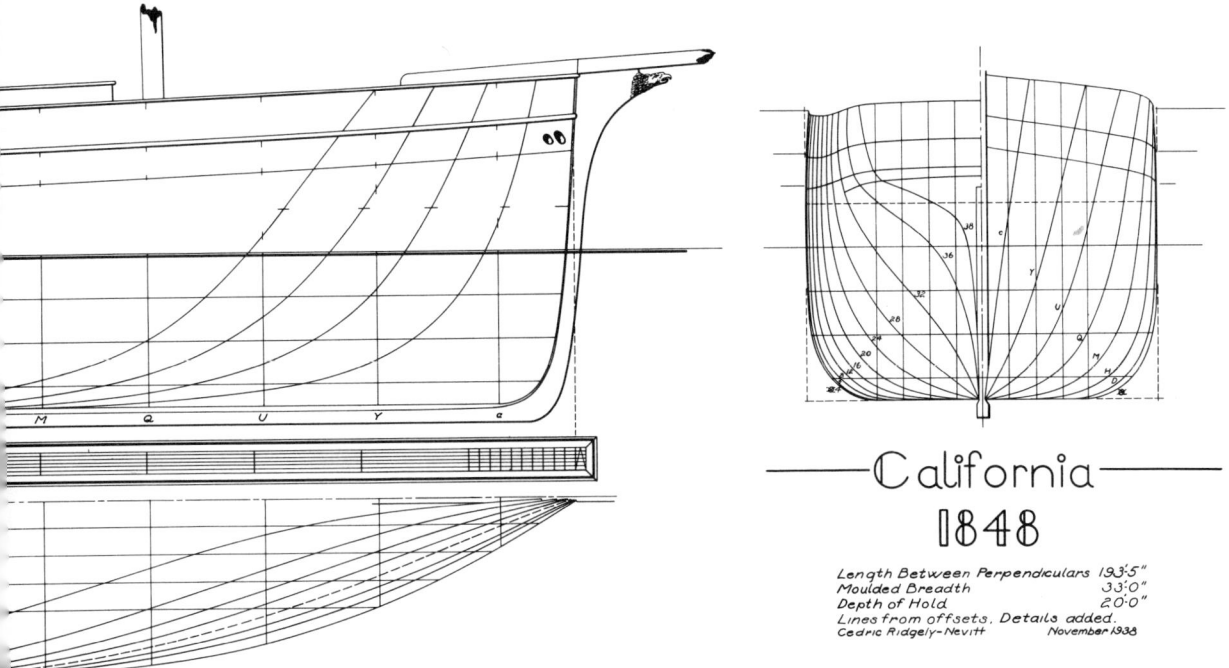

California
1848

Length Between Perpendiculars 193'-5"
Moulded Breadth 33'-0"
Depth of Hold 20'-0"
Lines from offsets. Details added.
Cedric Ridgely-Nevitt   November 1938

tween making her half model and the full-size fairing of the lines on the mold loft floor—the hull was lengthened by inserting four extra frames at 2'-6" spacing just aft of the original dead-flat section. It is likely that, while the ship was being constructed, two or three more were added. The length on deck from the offsets, 193'-5", is too short in comparison with the official value, 199'-2", or Webb's 200'-0" in his *Certificate Book*. The paddle boxes, deckhouses, and all details on the drawing are conjectures from a lithograph. The *Panama*, presumably, looked exactly like the *California*, but the *Oregon* could be distinguished by a circular decoration on the paddle box and a stack somewhat farther forward. All three were rather plain, austere-looking craft, purely utilitarian in appearance.

The *California* was completed in late September of 1848, then inspected and approved by a board of naval officers. Because she was headed for a distant shore devoid of repair facilities, she was loaded down with spare parts for her engine as well as 500 to 550 tons of coal and enough provisions for one year's operation. The latter were badly underestimated, for no large passenger lists were then expected. Quarters for fifty to sixty first-cabin and one-hundred to one-hundred-and-fifty steerage passengers were installed.[29]

In command was Captain Cleveland Forbes who had, in turn, been a shipyard apprentice, a sailor before the mast, and master of the steamboat *John Potter*. When she left New York on October 6, it was with one-hundred guests who traveled somewhat beyond Sandy Hook and enjoyed a festive dinner aboard before transferring to the steamboat *Orus*. A few passengers were left on board for South American ports but no one, the owners, the guests, or the crew, had any idea of the chaotic conditions that would exist by the time the *California* had circumnavigated South America and arrived at Panama.

The first port, Rio de Janeiro, was reached November 2 after some bad navigation; the *California* had gone too far and had to turn back. The average speed was low, 7 1/2 knots; on a long voyage no attempt was made to attain the designed speed, for a slower one conserved coal. A crack appeared in the crosshead of the engine shortly after sailing, but it was decided to proceed with a wary eye on it and a prayer that it would not increase.[30] A more serious matter was Captain Forbes's illness, which forced him off the ship at Rio de Janeiro. After a partial recovery, he resolved to retain command, coaled, and continued. The anchor was weighed November 25 and the Strait of Magellan entered December 7.[31] Some of the problems met in forcing treacherous passages through the strait have already been recorded for the *Massachusetts*. The *California* preceded her and was the first American steamship ever to pass through. She almost came to grief when she dragged anchor and touched a rock, but she was kedged off without damage. Her actual steaming time was forty-one-and-a-half hours in the strait, but 108 more were spent at anchor when visibility or weather was too poor to permit progress. She arrived at Valparaiso, Chile, on December 16, 1848, with her master seriously ill. To aid him, a mariner familiar with the Pacific coast, Captain Marshall, was taken on from a sailing ship, exchanged for one of the *California*'s mates. Having replenished her coal, the *California* sailed six days later and on December 27 stopped at Callao, where she loaded seventeen cabin and about eighty steerage passengers.

The news of the discovery of California gold had been slowly working its way across the continent and did not attract the general public's attention in the Eastern United States until December of 1848, two months after the *California*'s departure. Simultaneously, the information was gradually moving southward and this is what had caused the unexpected influx of Peruvian fortune hunters.

From Callao on January 10, 1849, via Paita, on the 14th, the *California* arrived at Panama on January 17 to receive a tumultous welcome from some five-hundred prospective passengers. The United States Mail steamer *Falcon* had sailed from New York on December 1, 1848, with a few passengers for California and Oregon, but had picked up a large number, who had just heard the news, when she stopped at New Orleans on the way. The *Crescent City* left New York two days before Christmas with 130 passengers. Two small steamboats, the *Isthmus* and *Orus*, and an assortment of sailing vessels had delivered more intrepid souls to the mercies of the Isthmus of Panama, its local canoes, and its tropical jungle trails. They had fought their way across and had been waiting at Panama for periods varying from several days to several weeks. Few of them had bothered to purchase through tickets and the local agent for the Pacific Mail was literally overwhelmed.

After protest meetings and heated discussions, the eminently fair decision was to honor the through tickets issued in New York or New Orleans; to retain the Peruvian passengers, whom everybody considered foreign interlopers that should never be allowed near American gold; and then to take as many others as possible in the order in which they had applied for passage. Temporary berths were installed in the ship's hold. Thus, on January 30, 1849, there were about 365 passengers and a crew of 36 crammed into the ship and overflowing onto the deck and the housetops.[32] Early next morning the *California* sailed. Fresh water was taken on at the nearby Isle of Taboga. From then on the voyage was a miserable one with poor food, mostly from the limited stores on board the steamer; little was available at Panama. Captain Forbes was unable to assert himself and Captain Marshall had little control over his loudly protesting passengers.

The arrival at Acapulco, the old Spanish port to which the Manila galleons had sailed for a century and a half, was a relief to everybody. At least some of the passengers could get ashore, buy provisions of their own, and escape the crowding for a few hours. Leaving on the 11th, the *California* touched at San Blas and Mazatlan. Arriving at the latter on the 15th, several ringleaders in a mutiny by the firemen were put ashore and Mexican substitutes shipped. Although a stop at San Diego was planned, it was abandoned when the coal supplies became dangerously low. Water and provisions followed suit. About 300 miles south of Monterey an attempt was made to remove the floats from the paddle wheels, but they were badly rusted in place. Instead of sailing as intended, it was decided to continue under sail and steam. To do this all the available wood in the ship—spare spars, berths, transoms, anything else not necessary—was to be cut up for the furnaces. Fortunately, the removal of wood dunnage in the hold uncovered spare bags of coal stowed below. The combined wood and coal got her to Monterey, California, on February 23. For four days wood was cut ashore and, luckily, a little more coal unearthed from beneath the cargo. On February 28, 1849, the *California* headed into San Francisco Bay, the first steamship ever to do so. The naval squadron that was anchored there saluted her with seventeen guns.

Not only did the passengers hurry ashore as fast as the boats could land them; so did the officers and crew. A week later only Captain Forbes and an apprentice from the engine-room staff, James DeKay, were left on board. Worst of all, the shipload of coal being sent from England had not arrived. The *California* swung at anchor with the ailing Forbes looking after the ship and his apprentice boy tending the machinery. The latter's work was recognized by the owners once the word got back to them. He was promoted to Assistant Engineer and awarded a gold watch.[33] Because he was a minor and a long way from home, the watch was sent not to him but to his father at Oyster Bay, New York. It would be pleasant to ascertain whether the young hero and his watch were ever united.

The second Webb steamer, the *Panama*, was approved by the naval inspection board in November and headed out to sea on December 1, 1848. Only five days out the engine was wrecked by a piece of wood left somewhere in the steam piping during its original assembly at the Allaire works.[34] The fact that trials had been completed and that the ship had proceeded on her way without difficulty must be recorded as a miracle. When the block did work its way into the cylinder, the piston, top cylinder head, and cylinder casting were all so badly damaged that the steamer had to sail back to New York. She anchored off South West Spit on the day after Christmas.[35] New castings had to be made and machined.

In the meantime the *Oregon* was completed and inspected, and she sailed for the Pacific Coast on December 8. Captain R. H. Pearson had served as first mate on the early clipper ship *Sea Witch*, owned by Howland and Aspinwall, under the celebrated Captain Waterman. During the construction of three steamers, he had been the owners' superintendent. As master of the *Oregon* he was a far more resolute character than either Forbes or his semireplacement, Marshall. The *Oregon* arrived in Rio de Janeiro on January 3, 1849, making a slightly faster passage than the *California*. There were minor engine problems; the piston rod's stuffing box leaked so badly that the engine had to be stopped to repack it.[36] The great complaint was that the fireroom was poorly ventilated. Furthermore, sleeping quarters for the firemen and engineers, adjacent to the machinery space, proved equally hot and unlivable. After leaving Rio on January 14, the Strait of Magellan was, for once, undergoing a rare spell of good visibility. The *Oregon* passed through in just thirty-six hours. But the Pacific was belying its name with heavy seas that rolled her guards completely under water. Valparaiso was reached on February 2 for a four-day fueling stop. As the ship headed toward the equator, the heat drove the firemen out of their quarters. New ones had to be improvised forward, but the fireroom continued at a debilitating temperature despite the rigging of canvas ventilators called "wind-sails." Stopping at Callao, February 12 to 18, and at Paita on the 19th, the *Oregon* reached Panama late on February 23. Here, for the first time, the *Oregon*'s complement learned that the *Panama*, which should have preceded them at every port, had not been lost.

About three weeks had elapsed since the *California*'s departure. Most of the travelers left behind had gone on by sailing ship, but others had arrived to refill the ranks of clamorous would-be passengers. Faced with inadequate food supplies and cooking facilities, Captain Pearson and the line's agent assigned fewer spaces; somewhere between 250 and 300 were on board on March 13 when she headed for Taboga to take on water. Left behind were 800 gold hunters at Panama and 530 more at Chagres.[37] There was no Acapulco stop, but one was necessary at San Blas for four days to take on coal of very poor quality. It had to be laboriously loaded into the *Oregon*'s boats, rowed out, and hoisted aboard. She left on March 25, brief stops being made at San Diego and Monterey to deliver and pick up the mail. San Francisco was reached April 1, 1849. There Captain Pearson anchored next to the line-of-battle ship *Ohio* and arranged with Commodore Jones to help him retain his crew. No one was allowed ashore. After some protests nine crew members were put in irons aboard the flagship. No boats were allowed to approach or leave the *Oregon*; the passengers boarded a barge and then pulled themselves ashore.

After unloading, the *Oregon*, despite some desertions, had a crew; but now she faced the same problem as the *California*: the collier from England had still not arrived. The opportunity was seized to send her spare crosshead over to replace the *California*'s cracked one. Captain Pearson proposed sailing the *Oregon* down to San Blas, the nearest port known to have coal available. He convinced the local Pacific Mail representative that this was necessary, raised the seamen's wages from $12 per month to $112 to compete with shore wages, shipped new men to replace the deserters, reloaded the ones sent onto the *Ohio* as mutineers, and with what coal was left steamed through the Golden Gate on April 13. Four days later the paddles were removed from the wheels and she continued under sail. Winds were favorable and San Blas was reached on the 21st, a remarkably fast passage. The dirty task of rowing out the coal was repeated; after stowing 150 tons, she steamed out on the 26th, touched at Acapulco, and reached Panama on May 5, 1849. Thus the *Oregon*, last of the trio, completed the first round trip from Panama to California and brought down the mails, seventy-five passengers, and over $165,000 of gold dust.[38] Captain Pearson had learned his trade well under R. H. Waterman.

While all this was going on, the *Panama*, commanded by David D. Porter, Lieutenant, U.S.N., left New York for the second time on February 18, 1849, the only one of the three steamers to have any measurable number of passengers, fifty-seven, for the long voyage to California. She made Rio in twenty-six days, had a relatively uneventful (or perhaps unrecorded since, as the third vessel, public interest decreased) voyage, and reached the Bay of Panama in May a few days before the *Oregon* anchored there after returning from California. Although there were some 2,000 persons awaiting passage, the *Panama* took 290 north on May 18 and arrived at San Francisco June 4.

In the meantime the coal supply had arrived, Captain Forbes had hired carpenters at exorbitant rates to repair the stripped quarters on his *California*, and a crew had been assembled at equally inflated wages. She left San Francisco on May 1 with fifty-four passengers and just under $350,000 in gold, or, as it was called in the papers, specie. Regrettably, no stop was made at San Blas and, for a second time, berths, bulkheads, boats, and fittings fed her fires to enable her to reach Panama. The *Panama*, northbound, had also run out of fuel, but made San Francisco on the loose timber and barrels aboard; she did not have to gut herself.

Once the first trips to San Francisco were completed and coal depots established there, at Acapulco, and at Panama, the three steamers began to operate on a monthly schedule in either direction. Temporarily, the Oregon part of the mail contract was maintained by sailing ships, but before the year was out other steamships

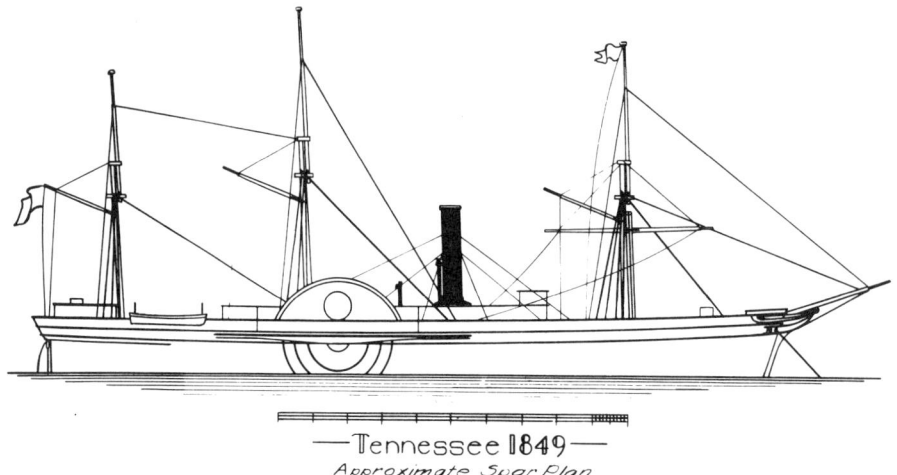

Tennessee, approximate spar plan.

were ordered and purchased to assure the regularity of that service as well.

These three Pacific steamships mark the coming of age of the American oceangoing side-wheel steamship. They were of original design, very heavily built, and fitted with reliable engines. They were able to make delivery voyages of 13,000 miles under steam. Their sails were to aid the motive power but not supplant it. Their own crews coped with whatever problems arose on the way. After brief stays in port, they proceeded to operate successfully on the route for which they were intended. What delays resulted were owing to the slowness of the sailing ships in bringing out the necessary coal, not to any defects or breakdowns in the steamships themselves.

The fact that they operated on a distant shore, free of shipyards or engine works until the Pacific Mail could establish its own repair facilities, was a remarkable achievement for its day. Webb and Smith and Dimon, the shipbuilders, were the finest designers and craftsmen in the field. The many men involved at both the Novelty and the Allaire Works deserve the highest credit for the dependability of their engines and boilers.

While the author does not intend to follow them in detail, it might be remarked that two of the three, as time went on, had their rigs reduced. There were new boilers installed in the *California* in 1856 and in the *Oregon* in 1858.[39] It was probably at this time that they became brigs and had their bowsprits removed. For some reason the *Panama* remained a bark as late as 1860 although she, too, had been reboilered. By the date of the boiler work and the concurrent rebuilding, all three had been considerably outclassed by newer, larger steamships on the main San Francisco-Panama run. The *Panama* and *Oregon* were transferred to the northern route, to Puget Sound and the Columbia River. The *California* was retained as a spare steamer. In 1861 the *Panama* was sold to Messrs. Holladay and Flint for local service,[40] and in 1868 fitted out as a revenue cutter and transport for the Mexican government and renamed *Juarez.*

The *Oregon,* sold at the same time to the same buyers, had her engine removed in 1869, was rigged as a sailing bark for the lumber trade. She was so badly damaged by a collision with the bark *Germania* in the spring of 1880 that she was condemned at San Francisco.[41]

The *California* passed to the Oregon and Mexico Steamship Company, returned to the Pacific Mail house flag from 1872 to 1874 and, under her original owner, celebrated her twenty-fifth birthday with special ceremonies. She then went to Goodall, Nelson, and Perkins, who kept her steaming along the coast for another year. In 1875 she was owned by Nicholas Richard, converted to a sailing bark and continued her useful career until lost in late 1894 or early 1895 on the Peruvian coast.[42] A final indication of the excellent design of many early steamships was that, whenever they were converted to sail, they remained satisfactory vessels. The *Oregon* and *California* demonstrated their excellent hull design regardless of the means of propulsion.

The *Cherokee* and *Tennessee*

As soon as a successful steamship line existed from New York to Charleston, its rival port, Savannah, sought to follow suit. Savannah had since 1819 grown rapidly to become an important city and now had sufficient trade to justify the operation of steamships to and from northern trade centers. The New York and Savannah Steam Navigation Company was set up with Samuel L. Mitchill as its president and New York agent. An order was placed with William H. Webb for two steamships.

William Webb was the son of Isaac Webb, who had served as foreman under Henry Eckford for the construction of the *Chancellor Livingston* and the *Robert Fulton*. Upon Isaac Webb's death in 1840, the son took over his fasther's yard and soon established himself as a talented designer and builder of sailing ships of every kind and size. With the advent of steamships, he was one of the first to undertake their construction. Simultaneously with these two Savannah steamers, to be named *Cherokee* and *Tennessee*, Webb was building the *California* and *Panama*.

The *Cherokee*'s keel was laid February 14, 1848; she was launched in June and delivered in September. The *Tennessee*'s keel was laid eight days after the *Cherokee* was launched; she, in turn, was launched October 25, 1848, and delivered in March of 1849.[43] This was an example of Webb's prompt and well-organized building practice. In spite of the fire in his shipyard that destroyed the mold loft, the half models, and some of the full-size molds for these steamers, no delay in delivery ensued.[44] The stern shapes of the *Cherokee* and *Tennessee* were identical. Since, however, the engine, boilers, and wheels were moved seven feet forward in the latter, the bow was filled out. Webb noted, after completion, that it should have been filled a little more. The official dimensions were:

	Length	Beam	Depth of Hold	Tonnage
Cherokee	210'-8"	35'-4"	15'-2"	1,244
Tennessee	211'-10"	35'-8"	22'-0"	1,275

Once again it is necessary to emphasize that such documented dimensions are only the product of the individuals, careful or careless, who measured them. The *Cherokee*'s depth is to a different deck; the Webb Certificate Book gives 22'-3" for the *Cherokee* and 22'-0" for her sister. They were probably identical. So were the beams.

The earliest detailed plans of a completely successful American ocean steamship are those of the *Tennessee*.[45] The author has redrawn these for publication and has used the original mold-loft offsets for the lines. Offsets are dimensions lifted from the frames of the ship as drawn full size on the mold-loft floor in order to construct patterns, or, in shipbuilding terms, *molds*, for cutting the ship's timbers to the exact curve needed.

Although a clipper bow with headrails and trail boards still indicates the *Tennessee*'s sailing-ship ancestry, the rest of the lines do not. The packet ship's flat transom has vanished and has been replaced by a counter stern. One unique feature introduced by Webb was the completely flat bottom, a great asset for a steamship, for it allowed the boilers and engines to be mounted low in the hull and permitted the shallowest possible draft, a particularly desirable feature when most American harbors were protected by natural sand bars and ships had to await high tides and favorable sea conditions before they could enter.

The steamer's deck plan is a great rarity; most merchant ships throughout the entire wood-shipbuilding era were built without any plans. Only a half model was carved out to develop the hull shape; it was expanded to full size in the mold loft, and, from that point on, arrangements were worked out directly on the ship as her construction proceeded. The skilled shipwright was a designer and a builder both. An inspired one was a great artist as well. The only essential items were very simple sketches to send to the sparmaker and the sailmaker; very few have survived. The *Tennessee* plans show the stack diameter too large; it should be 4'-9" diameter.[46] The latter value is shown on the approximate spar plan.

Forward was a pump-brake windlass, followed by mooring bitts on the deck (although the necessary openings in the bulwarks are omitted), circular scuttles each side of the pilot house for coaling, and guards located well above the water to prevent their rolling under. The deckhouse top over the boilers is a grating for both light and ventilation. The two deckhouses aft serve as entryways to the grand saloon below. In addition, the after one contained the smoking room. Accommodations for approximately two-hundred passengers were provided by means of staterooms opening from the after saloon and berths in the steerage space forward.[47]

The Novelty Iron Works supplied a side-lever engine of 75" bore by 8'-0" stroke and two iron return-flue boilers producing steam at 16 psi. The paddle wheels had twenty-four paddles each 15" deep by 8'-0" wide and 31'-4" in diameter, turning about 15 rpm. Forced draft was supplied to the fires and the fuel consumption came to 2,400 lbs. of anthracite coal per hour. A moderate spread of sail on a three-masted barkentine rig was fitted.

Under Captain Thomas Lyon the *Cherokee* sailed from New York October 5, 1848, made an extremely rapid turn around at Savannah, and was back again on the 11th. Northbound she made the passage in seventy hours at an average speed of 10 knots. Further sailings continued at two-week intervals with northbound times reported from sixty-one hours in good weather to seventy-eight in bad. Corresponding speeds are 11.5 and 9.0 knots. The cabin fare was $25.

The local railroads from Savannah to Macon and

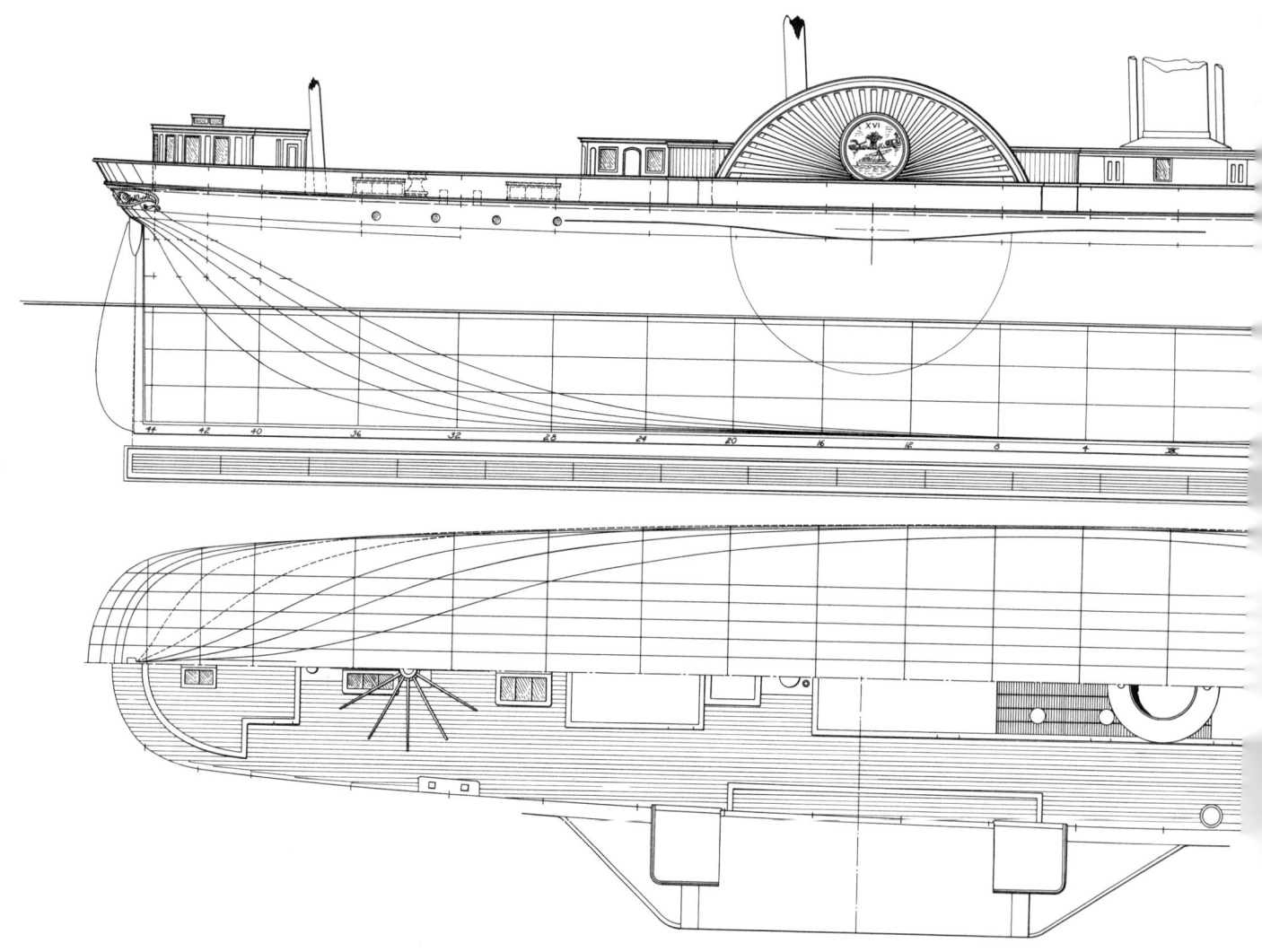

Tennessee; 1849, lines and deck.

Atlanta cooperated and made up special trains for the *Cherokee*'s passengers. Also connecting was the mail steamer *Isabel*, which started from Charleston but touched at Savannah on her way to Key West and Havana.

The *Tennessee*, under Captain John Collins, followed the *Cherokee* on March 21, 1849. The two steamships sailed every Wednesday and operated regularly without delays or postponements. The success of a steamship can be judged by the lack of news about her. In this sense both Savannah liners were eminently successful: they did not run aground, collide with other craft, catch fire, or make the disaster columns of the newspapers. Neither were there reported delays for machinery adjustments or repairs. The first unusual event of any kind was the withdrawal of the *Tennessee* after her fifteenth round trip.

The Pacific Mail Steamship Company, whose origin has been outlined in the preceding section, found that its first three ships were not adequate for the unexpected gold rush of 1849. First, they needed additional tonnage, preferably larger, for their Pacific service from Panama to San Francisco. Second, the mail line from New York to Chagres, on the Gulf of Mexico side of the Isthmus of Panama, was far behind schedule in its shipbuilding program and was not providing a satisfactory Atlantic connection. The Pacific Mail therefore decided to set up its own Atlantic line.

The *Tennessee* was purchased for $200,000[48] for the Pacific and, shortly afterward, the *Cherokee*, as the first of several vessels for the Atlantic. The former was trans-

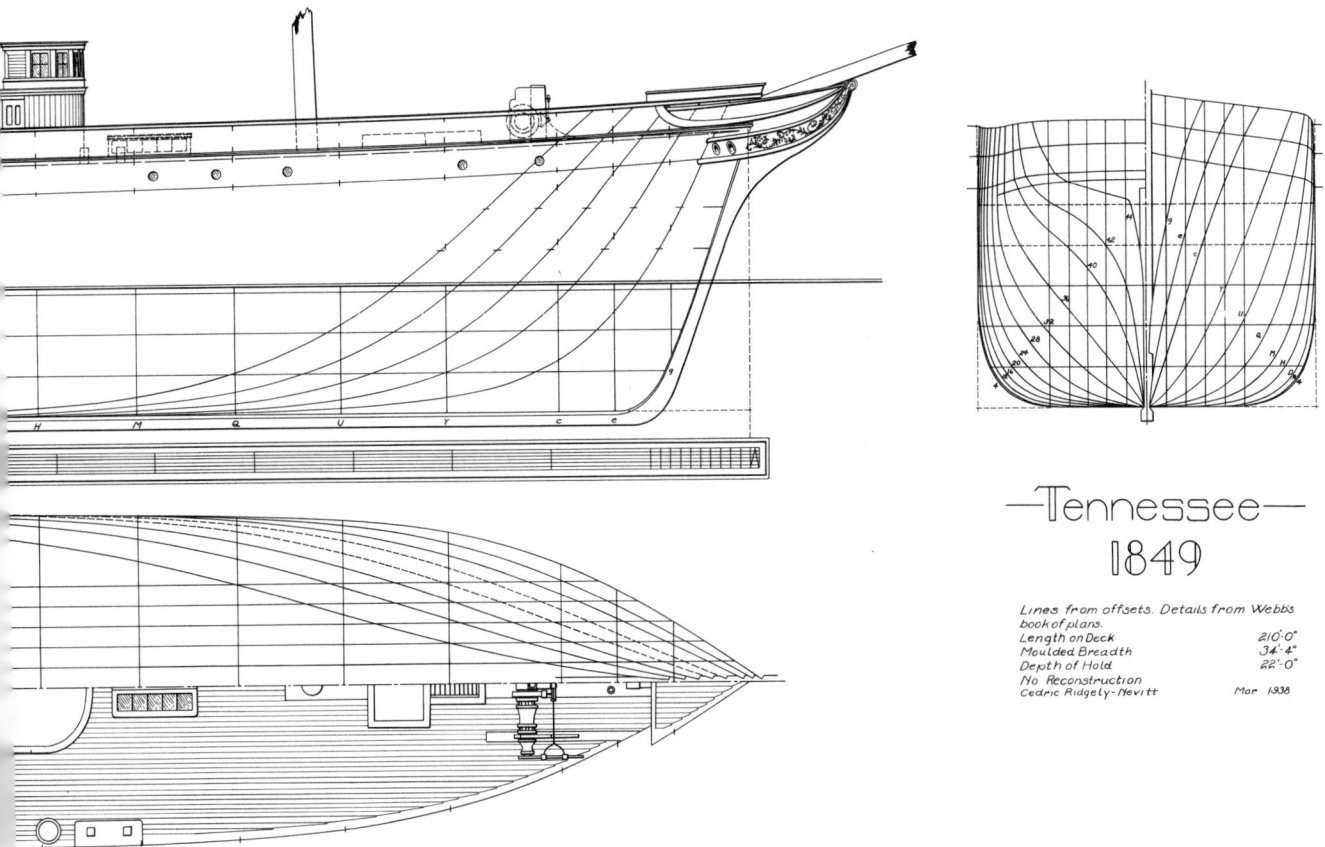

ferred after reaching New York on October 21, 1849. She was considerably altered to provide extra passenger berths for her intended service.[49] It is axiomatic that larger coal bunkers were necessary, as were additional water tanks. A lithographic portrait printed after the changes shows that an extra deck was added over the original houses from just aft of the forecastle all the way to the stern. Abaft the wheels, the sides were extended up to the new deck. The pilot house disappeared in the process. While providing for three times as many passengers, the alterations changed the *Tennessee* from a handsome ship into an ugly one.

The *Cherokee*, on the other hand, did not receive immediate changes. She had just completed her thirty-first voyage, arriving in New York on December 8, 1849, and sailed again, for her new owner, to Chagres on December 13.

The *Tennessee* steamed South from New York on December 6, 1849. Commanded by George A. Cole and carrying fifteen passengers, she made Rio de Janeiro in twenty-four days, where she remained for fourteen more coaling. Sailing from Rio she reached the Strait of Magellan in fourteen days and encountered some difficulties threading her way through the dangerous passages. On January 29, 1850, while anchored off Cape Possession, seas broke over her bow throughout the night and damaged the windlass, breaking one of its pawls. Three days later, when raising the anchor in Possession Bay, the other pawl broke, rendering the windlass useless. Fortunately, the weather cleared for two days, permitting repairs to be made. Fog, snow, and gales further hampered her, but she reached Valparaiso on February 16. Leaving there on the 25th, she reached Panama March 11, 1850, ninety-five days out of New York, fifty-seven of them under steam.[50] Coming into San Francisco Bay on April 14, 1850, she carried the largest passenger list, 551, recorded to that date. Six days later she set out for Panama with 114 passengers and $551,000 in specie.[51] Being larger and faster than the three steamships then in service, she was a most useful addition to the Pacific Mail fleet.

As for the *Cherokee*, she made her first Pacific Mail sailing from New York to Chagres on December 13, 1849. Northbound she stopped at Kingston, Jamaica, for coal and arrived at New York January 13, 1850. Her subsequent sailings were usually monthly, on the same route, until November, when she was joined by two more steamships, the *Crescent City* and *Empire City*, which increased sailings to twice a month. During this time the Pacific Mail's Atlantic operations were competing with the United States Mail Steamship Company's vessels to Havana, New Orleans, and Chagres, while the latter had sent some ill-assorted steamships to the Pacific to run against the Pacific Mail. Early in 1851 the lines decided to cooperate; the Pacific Mail took over all the steamers on the West Coast, the U.S. Mail all those on the East, and both agreed to work in concert rather than fight each other.

The *Cherokee* carried the mails from New York to New Orleans for the U.S. Mail on May 26, 1851. She was utilized on both the Chagres route and from New York to New Orleans via Havana. In 1853 she was spending most of her time running to New Orleans, because newer and larger steamers had taken over the Isthmus operation. On August 26 of that year she lay at her Warren Street wharf in New York loaded with a particularly valuable cargo, including French silk for southern dressmakers. She was ready to sail next day, but no one was aboard except two watchmen. Suddenly a fire broke out in the cargo hold, where some chemicals had been stowed, and spread to the sails and rigging.[52] By the time volunteer fire companies arrived, she was beyond saving. The firemen pumped water through her side ports and down her hatches, and cut holes in the bow to sink the hull. Next morning two lower masts and the stack were left standing, but everything else had burned to the waterline, about at the second-deck level. The steamer was valued at $175,000 and the cargo at $350,000.

Five days after the fire, William Ellis of the Atlantic Insurance Company had the hulk pumped out and afloat.[53] Because there was little damage to the machinery, there was talk of rebuilding her. The author has found no evidence that this was done. She was not dismantled, for she was still reported at out-of-the-way piers in New York until August of 1858.[54] But she never sailed again.

The year 1853 was equally unlucky for the *Tennessee*. In the early morning of March 6, groping through a dense fog, she ran aground about four-and-a-half miles north of the Golden Gate. The sea was smooth; after the ship swung broadside onto the beach, the passengers landed safely. Although attempts were made to haul her off, they proved unsuccessful; within twenty-four hours breakers of increasing size began pounding the ship, eventually breaking her back. The following report was published in San Francisco: "Her officers and crew feel as if they were attending the funeral obsequies of a dear and valued friend. She was a favorite craft and one of the best sea boats that plowed the Pacific Ocean. She was the home, the pride and refuge of her officers and crew. . . ."[55]

Like the two William H. Brown steamers, the *Northerner* and *Southerner*, the two William H. Webb ones, the *Cherokee* and *Tennessee*, were remarkably successful coastwise steamships; it is to be regretted that a fire and the hazards of the sea were to bring them to such untimely ends.

The *Ohio* and *Georgia*

The same legislation that led to the formation of the Pacific Mail Steamship Company also provided for the delivery of the United States mails from New York and other East Coast ports to Havana, New Orleans, and Chagres on the Caribbean side of the Isthmus of Panama. A contract was first let to Albert G. Sloo of Cincinnati, Ohio, who proposed to build five steamships of at least 1,500 tons each. Two routes were agreed on, the first from New York via Havana to New Orleans, with stops outside the bars at Charleston and Savannah, and the second from Havana to Chagres for the mails bound to the Pacific. Both were to have sailings twice a month with a total payment of $290,000 per year once everything was in operation. The steamships were to be built under naval supervision, were to be suitable for conversion to warships, and were to be commanded by officers of the U.S. Navy. Mr. Sloo had been active as a lobbyist trying to push the establishment of mail-carrying steamship routes through Congress. The contract, dated April 20, 1847, required two ships to be ready by October of 1848 and the whole operation by October of 1849. Sloo accomplished nothing himself, but four months later, on August 17, transferred the contract to George Law, Marshall O. Roberts, and Bowes R. McIlvaine of New York, who founded the United States Mail Steamship Company, popularly called the "Law Line" after its president and majority stockholder.

Law had started as a laborer in canal construction, had worked his way up to become a contractor, and had achieved fame for his work on the Croton Aqueduct and the large stone bridge bringing drinking water across the

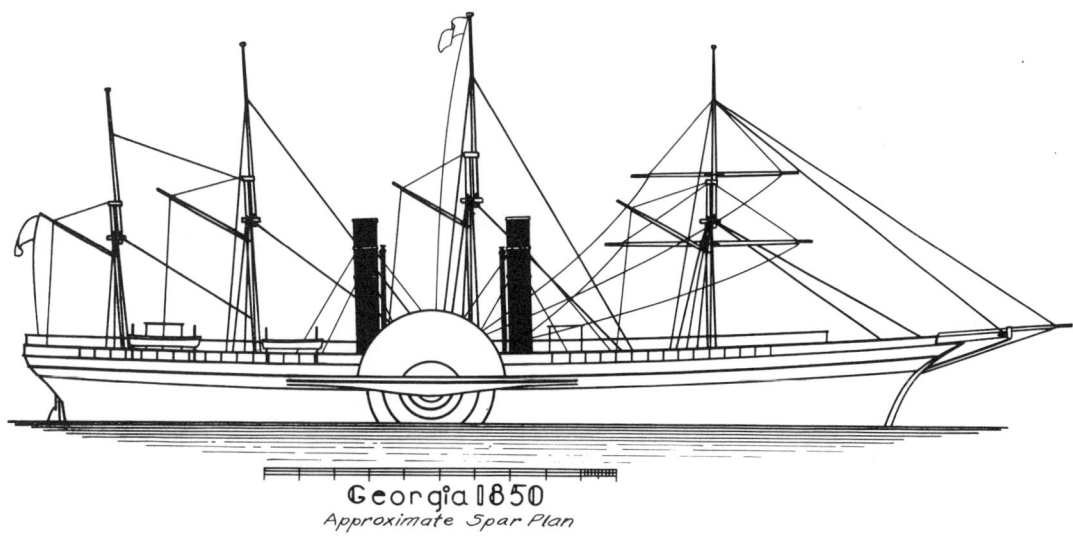

Georgia, approximate spar plan.

Harlem River to Manhattan Island. High Bridge in 1840 was considered one of the country's engineering marvels. Law further expanded his interests into banking and railroads. Unfortunately, he was not well informed in the shipbuilding field and chose a flamboyant solution rather than one based on sound experience.[56] Instead of starting five steamers of moderate size to meet the contract requirements, he ordered only two. These, however, were extremely large and were to be fitted with more powerful engines than any immediately available. In addition, instead of specifying fastenings, scantlings, and hull details of the highest quality, as Howland and Aspinwall had for the *California, Panama,* and *Oregon,* Law economized by adopting moderate timber sizes and limited fastenings and, worst of all, omitting diagonal strapping. One order, for the *Ohio,* was placed with Joseph Bishop and Jeremiah Simonson, the other, for the *Georgia,* with Smith and Dimon. Even at the start it was obvious that such large vessels could not be completed in time. To make matters worse, the engines went to T. F. Secor and Company, whose establishment was already building engines for the transatlantic liner *United States* and the coastwise steamers *Crescent City* and *Empire City.* The creation of four additional side lever engines of unprecedented size was to delay final deliveries of the *Ohio* and *Georgia* until after the contract date when five were supposed to be in service.[57]

The *Ohio* was 246' long by 46' beam by 32'-9" deep, with a draft fully loaded of 17'-2" and a tonnage of 2,432, about twice that of the *Tennessee*. She had a forward-curving stem terminating in a figurehead of a dragon with a long serpent's tail curling along the side of the ship for a distance of 40'.[58] Nothing so simple as the *California*'s vertical stem was considered suitable. After the *Ohio* was launched in August of 1848, a whole year later found her still tied up at Secor's wharf. She was not completed until September of 1849.

Captain William Skiddy, the naval constructor who supervised her building, described her structural details as follows:

> The ship OHIO is well timbered and well fastened —a good model except the projecting bow. There is not much copper used, which would be useless expenditure, considering the probable duration of the ship in sound state not to exceed ten years. She has no diagonal iron or wooden trussing on the frames, but has substituted a diagonal hog-bracing from the keelson to the beams, and from the beams to the bilge. This being an experiment on sea-steamers, it remains to be tested to prove its efficiency.[59]

It is a most revealing paragraph, especially with regard to the use of iron fastenings, which corroded rapidly because of galvanic action with the copper sheathing and led to a short life for the hull. Furthermore, a cheap solution to longitudinal strength was tried. Although the vessel had three full decks, there was very little height available for the 3" diameter round iron trusswork, which extended only from the keelson to the main (second) deck beams.[60] Both the shallow depth and the location, independent of planking or ceiling, would render it ineffective. The good features comprised the bolting of the ceiling strakes edgewise as well as to the frames, and filling in the bottom between the floors to provide a strong foundation for the engines and boilers. Lengthwise there were seven built-up oak keelsons 24" wide by 36" deep and, in way of the engines, their depth was almost doubled. In short, the bottom of the ship was made as rugged as wood construction permitted, but the hull as a whole was neglected; the ceiling, the main strength member keeping it from hogging or sagging, was only 7" thick.

There were two side lever engines of 90" bore by 8'-0" stroke, and these seemed to have raised problems for the engine builder, even though previous side-lever engines of 9'-0" stroke but smaller bore had been constructed for the *Crescent City* and *Empire City*. Each engine's bedplate weighed 34 tons and was anchored down to the wood keelsons by 60 enormous bolts 5' long and 2 1/2" in diameter. The side levers were 19'-0 3/4" long and turned on massive pivots 15" in diameter. The great wrought iron paddle shafts were 21" in diameter. When two cranks were added, the rotating assembly amounted to 51 tons of iron, without the weight of the wheels. There were four massive iron drop-flue boilers, two forward of the engine room and two aft. With a diameter of 14' and a length of 21'-6" they added up to 160 tons empty and held 120 more tons of water when filled. The designed steam pressure was 15 psi and the anticipated coal consumption 50 tons per day. The paddle wheels were excessively large for the engines, 36' in diameter with double buckets 10'-6" wide and 15" deep, 32 pairs being mounted on each wheel.[61] Because of their size, they turned very slowly, only one revolution every five seconds or 12 rpm.

The interior of the ship had staterooms for 250 first-cabin passengers, each room having three berths 7'-0" long.[62] Considerable exaggeration is suspected in so liberal a dimension. Mr. Law was an excellent, but not very truthful, press agent. There was a ladies' cabin aft on the uppermost deck and a social hall with a bar forward for the gentlemen. Well up in the bow was a "second-class" cabin with one-hundred berths. For the first-cabin passengers there was a barber shop and showers with hot and cold water—obviously salt. In the forward and after saloons the dining tables could seat all the cabin passengers at a single setting. The crew was said to number 125, but this again seems an exaggeration. Externally, the appearance was marred by an ungainly rig of four masts, with the one between the two stacks being particularly prominent. It would appear to have been a fire hazard. Happily, it was later removed to produce a conventional three-masted barkentine rig.

Trials were run on September 15, 1849, over two years after Law and his associates had taken over the contract.[63] They were more social than technical; in fact, the fires were lighted for the first time a few days earlier. Steam was produced at a very low pressure and a mere eight knots achieved. Hundreds of invitations had been sent out and everyone who received one boarded the ship. Instead of an anticipated guest list of 800, there were 1,500 people trying to find places to stand. When dinner time came around, the 400 available seats were filled by the ladies. They began to enjoy the food and were particularly delighted that they could at last sit down to rest their aching feet. Nor were they in any hurry to leave the comfort they had found and make way for the starving gentlemen, all of whom were beginning to feel most unchivalrous. The less restrained males made a spectacle of themselves by tracking down and appropriating food and drink wherever they could find it.

Five days later, September 20, 1849, the *Ohio* sailed for New Orleans under the command of Lieutenant James F. Schenck, U.S.N. She was the largest vessel, under sail or steam, ever completed in the United States and had the highest horsepower yet installed. The cost was about $400,000, of which $235,000 was for the engines and boilers. According to the advertisements, "the size, strength, and power of the *Ohio* are far beyond those of any sea steamer afloat."[64] The iron-screw steamer *Great Britain* of 1845 was larger than the *Ohio*, but, being British, was ignored.

Regardless of the fanfares, the supervisory naval officers refused to accept her because she did not comply with the mail contract. According to it the ship had to be coppered, a sound requirement since it prevented worms from boring through her timbers and, in addition, kept barnacles from foulding her bottom and slowing her down. It was a common practice to nail on the copper sheets after the maiden voyage; should there be small leaks or unsatisfactory caulking, the difficulties could be easily located and repaired. But because there was no dry dock large enough to take the *Ohio*, nothing could be done until one under construction was completed. The United States Mail management frequently overlooked contract requirements and seemed to expect others to do so too.

Because she was a year overdue, the fares were drastically cut to entice travelers away from other services that had been set up by the Pacific Mail Company and the Empire City Line. They were $100, to Chagres, in the after saloon, $80 in the forward one, including a stateroom berth, and $50 for the steerage (which had previously been termed "second class"). At New Orleans those continuing on to Chagres and the Pacific would transfer to the *Falcon*, a rather unusual steamship that had been first chartered and later purchased by the U.S. Mail Line to begin its operation once it became clear that vessels of its own design would not be ready. She started in December of 1848, carrying the mails to Chagres as a token part of the company's contract. She had been the first steamship to deliver passengers to the Isthmus of Panama to connect with the *California* on the latter's original voyage to San Francisco.

The *Ohio*'s maiden trip was a slow one that took seventy-eight hours to Charleston compared to the *Southerner*'s usual sixty. She did not enter the harbor but picked up the mails outside the bar. A similar stop off Savannah was made. At New Orleans the *Falcon* was not there to meet her. Steaming down the Mississippi on October 4 with only forty-nine passengers, the *Ohio* did meet the *Falcon* and stopped alongside so that the mails from Chagres could be put on board her for New York.[65] The *Ohio*'s passengers for the *Falcon* were, however, waiting at New Orleans.

The second voyage was also a frustrating one, although the passenger list of 450, most of whom were bound for California, must have gladdened the line's financial agents. The speed improved; her best average was 10.9 knots from Havana to Balize, the pilot station on the Mississippi. Leaving New Orleans under a local pilot, she spent sixteen hours hard aground before two towboats, the *Persian* and the *Hercules*, managed to free her. There were more delays both entering and leaving Havana, but, worst of all, the tardy *Falcon* missed another rendezvous scheduled there. Consequently, the *Ohio* reached New York without her California mails and had only thirty New Orleans passengers on board. There was some cargo, 350 bales of cotton and $75,000 in gold.

By January of 1850 the second steamship, the *Georgia*, was ready. While her appearance was similar, and her engines and boilers identical to the *Ohio*'s, considerable differences did exist. Smith and Dimon employed John W. Griffiths, who had achieved great success with his two early clipper ships, *Rainbow* and *Sea Witch*, to draw her lines. Griffiths teemed with ideas and was always certain of their success. Two of his theories were tried out on the *Georgia*. The first was to make the bow very long and sharp and the stern considerably shorter and fatter. These elements were completely the reverse of the sailing-packet hull that we have examined in the *Massachusetts*. In the latter the maximum beam was well forward; in the *Georgia* it was far aft of the half-length. Griffiths also reasoned that the beam of a steamer should be as large as possible to reduce the change in draft as coal was consumed.[66] He did not realize that high beam leads to excess stability and this, in turn, results in a quick, uncomfortable roll that can build up to large angles. The *Ohio* had been given a larger than normal beam in order to keep her draft down and the *Georgia* exaggerated it further. With a length of 248' and a beam of 48'-8", the ratio of length to beam was almost 5 to 1. William H. Webb's *Tennessee* was 6 to 1 and William H. Brown's very successful *Crescent City*, which will be discussed later, had a 7 to 1 ratio. Webb wrote a revealing comment on the lithograph of the *Georgia* in his possession: "And a beast she was."

The excess beam increased her tonnage to 2,727 over the *Ohio*'s 2,432. The depth, 33'-0", was almost identical. The maximum draft, however, was reduced to 16'-3". Captain Skiddy, the naval constructor for both, compared her structure to that of the *Ohio* as follows: "although of similar materials, has a lighter frame and is not so thoroughly fastened." There was neither diagonal strapping nor any substitute for it.[67]

The *Georgia* was launched on September 6, 1848, three weeks after the *Ohio*, but her delivery came in January of 1850, three months behind her semi-sister. Once again no preliminary trials were reported. One suspects that Mr. Law's communications to the press suppressed the dock trials that any prudent engine builder would insist on. At first she was intended to accompany the *Ohio* out to sea on the latter's fourth departure from New York, but she was not ready in time. A second date was set for trials on January 22, 1850; these were to include a race with the Cunard liner *Canada* when the latter should leave her Jersey City pier bound for Liverpool. Lieutenant David D. Porter, U.S.N., held his ship off Governor's Island until the *Canada* passed, then started after her and was well ahead off Coney Island. Whether the *Canada*'s master was competing or not cannot be ascertained. The *Georgia*'s wheels turned up to 14 rpm at a steam pressure of 17 1/2 psi, both exceptionally high values.[68] If Mr. Law wanted a public show, David Porter was just the man to put it on. Unfortunately, the *Georgia* zigzagged along the course, but the blame was placed on the inexperience of the helmsman and stretching tiller ropes. Actually, it indicated a basic defect in her hull design that will be discussed later. There was a luncheon for a limited number of guests and a quick return to the U.S. Mail dock.

The *Georgia*'s table of offsets was printed in several of Griffith's wordy treatises on shipbuilding and naval architecture.[69] The present author has drawn up a body plan from them. It is a very attractive one with a typical Griffith's sailing-ship bow and a large amount of flare to keep the deck dry. Had the vessel been built on a length of 300' instead of 250', it would have been extremely handsome.

On January 28, 1850, the *Georgia* steamed out of New York carrying 180 passengers, 200 tons of freight, and 700 tons of coal. With her fine bow and full stern she was drawing 16' at the bow and 15' at the stern, and was still steering badly. Charleston bar was reached in fifty hours, but seventeen were spent there and off Savannah awaiting boats from shore. From New York to New Orleans, a distance of 2,010 nautical miles, the running time was 6 days 23 hours, giving a speed of 12 knots—extremely satisfactory; but the consumption of 449 tons of anthra-

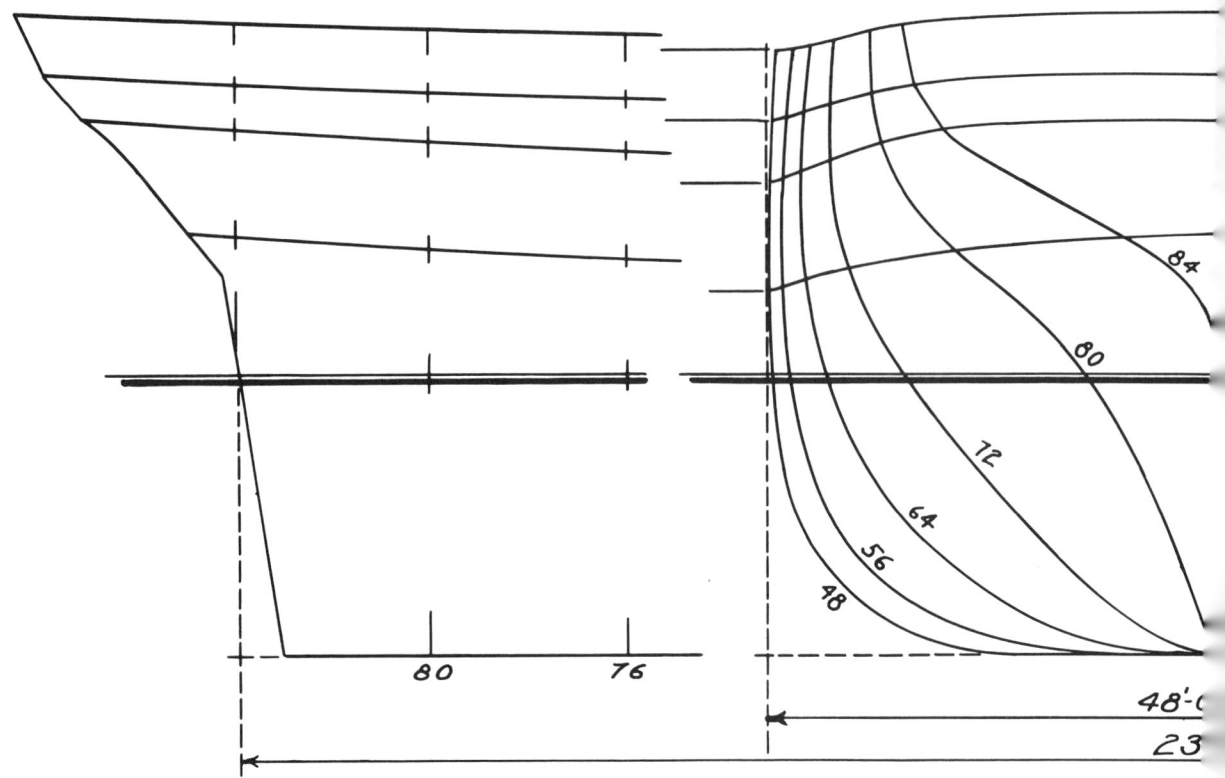

Georgia, 1850, lines.

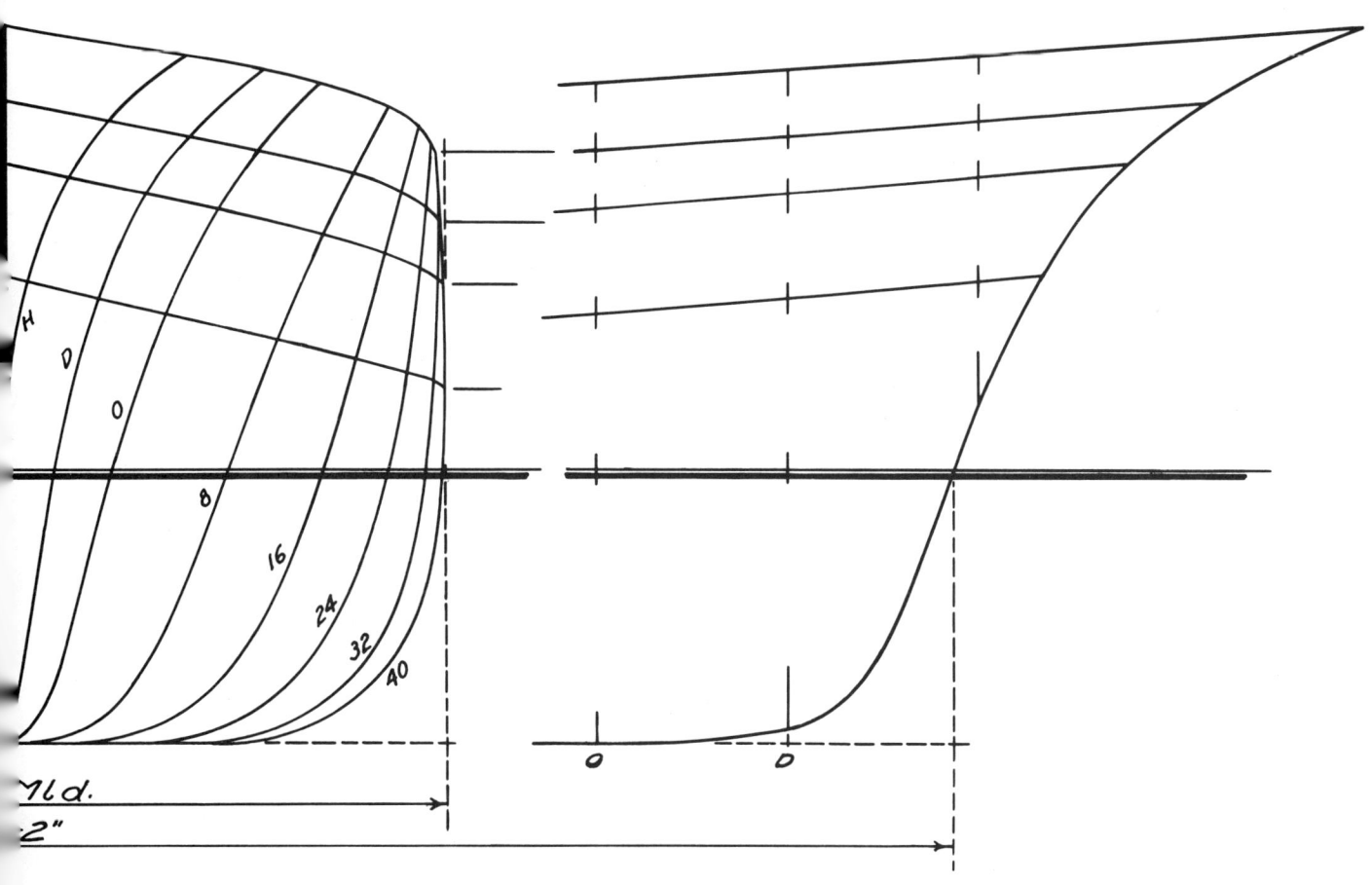

cite coal (64 tons per day) was both excessive and expensive. With the maximum section and the machinery well aft, much of the coal bunker space must have been forward, resulting, when it was filled, in trim by the bow. Whenever this happened, steering became erratic, just as it had on her trial. From New Orleans with 430 passengers, she headed for Chagres, sailed from there on the 26th of February, coaled at Havana, and arrived off Sandy Hook the evening of March 8. There were a number of very speedy parts of the maiden voyage, but, for unknown reasons, the total elapsed time was excessive. The author does not believe Captain Porter's statement: "At times (northbound from Chagres) she ran sixteen knots an hour, and averaged fourteen knots up to Havana." On the other hand, he accepts without question the continuation: "The last day's run we had a heavy sea on the quarter, and, being light, the ship rolled considerably. . . ."[70] Rolling and steering always plagued the *Georgia*.

Starting in April of 1850, the *Ohio* remained on the New York-Havana-New Orleans route sailing on the 27th or 28th of each month. The *Georgia*'s path was from New York to Havana and Chagres, with departures on the 13th of the month. The *Falcon*, as a third and much smaller steamer, followed a complex operation between New Orleans, Havana, and Chagres transshipping passengers and mails to and from the larger vessels. A knowing passenger carefully chose the steamship taking him all the way to his destination; the less experienced one might find himself, after a pleasant four days on a large steamer with ample room, jammed into an overcrowded small one for the latter half of his trip.

The *Georgia*'s interior must have been somewhat different from the *Ohio*'s. The advertisements for her second voyage, which, like the first, made both New Orleans and Chagres, quoted $100 in a stateroom berth and $80 in a standee berth all the way to Chagres.[71] The second category indicates that former public rooms had been filled with extra berths. They were needed; 650 passengers were on board leaving New York and 1,254 when she left New Orleans on March 22, 1850.

Once the Pacific Mail and the United States Mail reached a working agreement in January of 1851, the latter obtained three first-class steamers, the *Cherokee, Crescent City*, and *Empire City* from the former's new York-to-Chagres fleet, also the smaller *Philadelphia*. With these additions a twice-monthly service to both Chagres and New Orleans was finally achieved. Very early in their operation the *Ohio* and *Georgia* had demonstrated that the way stops off Savannah and Charleston were entirely unsatisfactory. In bad weather there were frustrating delays and often they made no shore contacts at all. The postal and naval authorities agreed to their elimination.

Once the two very large, very expensive steamships were in service, there were no published criticisms of them; their masters were their most enthusiastic champions. James F. Shenck wrote of the *Ohio* in August of 1850:

> The character of her machinery, the general strength of the ship, her capacity, the certainty with which she performs, may be judged by the fact that from the moment she commenced running she has not lost a day, has never been stopped to adjust or repair her machinery; has passed the last winter, in which she has encountered several severe storms and hurricanes with great steadiness, without the loss of a spar and without detention or injury of any sort.[72]

Yet, after less than four years of service, both were unobtrusively withdrawn and tied up never to move again under their own steam! They must, in reality, have been unsatisfactory, unprofitable craft. They were replaced by the *Illinois*, of 2,123 tons, built by Smith and Dimon in 1851, and the *George Law*, 2,141 tons, by William H. Webb in 1853. Once these two splendid steamships were available, the *Georgia*'s and *Ohio*'s sailings became less frequent and stopped completely in 1854. The *Ohio* was the last to make the run from New York to Chagres on March 6 of that year.

The *Ohio* met her first severe storm in late 1850 and very nearly sank. She started out from Havana for New York on December 18 with 350 passengers and $1,000,000 in specie, but as the wheels made their first revolution, one cylinder head was blown off. After the damaged engine was disconnected, she sailed again on the 19th with the remaining cylinder driving both wheels. Three days out she met a heavy gale[73] and tried to lay to with her bow to the sea, but the one engine could not hold her in position. When she turned away, on the 23d, her motions became so violent that the hull was strained, leaks started, and, on December 24th, the water rose high enough to put out her fires. All the passengers and crew were put to work on the hand pumps and hoisting water out of the hold in barrels. After twenty-four hours the wind decreased and control was gained over the leakage. The fires were lit again on Christmas day; Captain Schenck wanted to head for New York but the passengers petitioned him to make the nearest port. They prevailed, and the bailing had to continue until she reached Norfolk. Most of the saloon furniture had been destroyed as it slid around the decks, and some of the iron braces in the unusual hog trusses were broken.

Since there was no dry dock at Norfolk capable of receiving her, she was sent to New York. There she was put on the new sectional dry dock for examination by Captain Skiddy, who glossed over the damage:

> I examined her thoroughly and could not find any

indication of her being strained in the least; every butt and joint was as perfect as when first put together. . . . I am satisfied that the principal leakage arose from the inefficient fastenings of the discharge-valves on the sides. The ship rolling in the trough of a heavy sea at an angle of thirty degrees each way, with two or three feet of water rolling over the keelsons, washing coals, cinders, and ashes up in the wings, might easily alarm persons unaccustomed to sea casualties.[74]

To the passengers who bailed for their lives, and the engineers who had their fires drowned out by ten feet of water in the hold, this statement was an inadequate description of what took place. Why did the hog frames let go if the hull were not strained?

The author considers this event clear evidence that diagonal strapping was needed to keep a wood hull of this length and breadth in shape and prevent it from working and leaking in a seaway. The chief items done in dry dock were to recaulk the bottom; to remedy the discharge valve leaks, so conveniently blamed; to copper the bottom; and, at long last, to obtain the naval board's approval as a mail steamer on January 21, 1851.[75] The fact that caulking was necessary shows that the hull was not in the best of condition.

The *Georgia* escaped the perils of the sea for a time, but in 1853 she met a severe storm and duplicated the *Ohio*'s misadventure. She was southbound with 400 passengers, having sailed from New York on September 5. Off Cape Hatteras she ran into a tropical disturbance of such force that she became uncontrollable.[76] The water rose rapidly in her hold, putting out all the fires. Again tackles were rigged and casks used to help out the hand pumps. As the sea diminished, so did the leakage; fires were rekindled on September 8. Bailing, however, continued for two more days until Norfolk was reached. A little brig, the *Lady Chapman* of Bermuda, undamaged and undaunted by the storm, stood by to assist the far larger *Georgia*. It was a humiliating situation.

It is rare when two very similar steamers get into identical difficulties. If they do, it is likely that they suffer from similar design errors. The second near-loss confirms the fact that both ships suffered the same structural weakness. They were too large for the type of wood construction that had previously been satisfactory for hulls of smaller tonnage.

There were a series of defects in the *Georgia* and *Ohio* that led their owners to lay them up.

1. They were great coal eaters and very costly to run. High speed was possible but the cost prohibitive.
2. The excessive beam made them roll in a severe and uncomfortable fashion. The *Georgia* was worse than the *Ohio*.
3. The high beams also led to steering problems, which contributed to their difficulties. In heavy weather they became uncontrollable. The technical name of this is *directional instability*. The *Georgia*'s erratic steering in smooth water confirms its presence.
4. The wood hulls without diagonal iron strapping did not have adequate longitudinal or shearing strength. They deflected badly and started leaking in heavy seas.

It is easy for the twentieth-century naval architect to criticize his predecessors. They did not realize the existence of directional stability, a phenomenon not systematically investigated in the marine field until the 1940s. Most American shipwrights in 1850 lacked the theoretical background necessary to handle stability, much less rolling. The whole problem of a ship's strength was not explored until after the iron hull was well developed. Yet, somehow, the best builders did produce correct, intuitive solutions based on experience with their own creations and those of their contemporaries. George Law and his partners wanted large, unusual craft to catch the attention of both the Congressmen who appropriated money and the travelers who would patronize their steamers. The Law Line received what it demanded from the shipbuilders, and the ships' sizes and proportions were far enough away from conservative practice to create severe technical problems. In the case of the *Georgia*, John W. Griffiths, as a sometimes brilliant, sometimes erratic designer, must share the blame. This was his first steamship and one that should have made him more cautious. Alas, it did not.

In 1859, when the U.S. Mail Steamship Company's mail contract, which had been extended to eleven years instead of ten, expired, the two steamers were still moldering away at unused piers in New York harbor. While the line's affairs were being wound up, they were disposed of. In July and August of 1860 they were both reported as at the foot of 13th Street and the East River, near the Novelty Iron Works.[77] Probably the engines were being removed and broken up for scrap. The *Georgia*'s name was last listed in August and the *Ohio*'s in October.

The *Crescent City* and *Empire City*

After the consideration above of the difficulties developed by the *Ohio* and *Georgia* when they proceeded too far and too fast from past experience, it is salutory to return to William H. Brown, whose ships progressed regularly and consistently, each one profiting from its predecessor and increasing in size, strength, and overall success. Already discussed have been his Charleston pioneers, *Southerner*, 785 tons, of 1846, and *Northerner*, 1012 tons, of 1847. These were followed by two steamers designed for the New Orleans trade, *Crescent City*, 1,291 tons, in 1848, and *Empire City*, 1,751 tons, in 1849. After them were to come a pair of Atlantic liners, *Atlantic* and *Arctic* of over 2,800 tons. Brown, previously known for river and sound steamers, was perhaps the greatest single contributor to the development of the American ocean-going steamship; regrettably, no one has ever publicly recognized his genius. It might be remembered that the famous sailing yacht *America*, which made George Steers's name famous as a designer, was actually built in Brown's shipyard and that it is Brown's name, not Steers's, that appears on official records as her builder. Little is known about him except that some of his less successful contemporaries, such as Griffiths, disliked him. He was unusual in that he was never engaged in packet ship or clipper ship construction before or during his steamship activities. In short, he was an early specialist in steam-powered vessels and, in addition, one of the first to use diagonal iron strapping; he had introduced it into the Long Island Sound steamer *Narraganset* as early as 1836.[78]

In 1847 a Mr. Isaac Newton, who had considerable success with steamboats on the Hudson River, joined J. Howard and Son, New York shipping agents, to establish a service from New York to New Orleans. A steamship 233'-7" long by 33'-11" beam by 22'-8" depth was ordered from the Brown shipyard and a side lever engine 80" diameter by 9'-0" stroke from T. F. Secor and Company, who had been building walking-beam engines since 1838;[79] this, however, was their first side-lever engine and, unlike the later ones for the *Georgia* and *Ohio*, was delivered without delay. Under Captain Charles Stoddard she was advertised to leave New York June 1, 1848 for New Orleans, via Havana, with fares of $75 in the first cabin or $60 in the second for the entire trip.[80] The voyage was a most profitable one; she brought back 120 cabin and 60 steerage passengers when she entered the Lower Bay on June 22. Havana had been omitted because of a legal technicality, but would be included on all further trips. The second voyage brought $162,000 in gold and 463 passengers to New York on July 23.[81] On the third trip, in calm weather but aided by the current of the Gulf Stream, she made a remarkable day's run of 326 miles at a speed of 13.6 knots. In a heavy gale, however, her speed was down to 6.5 and a reasonable average might be around 10 knots. In October she ran aground southbound from New Orleans toward the Gulf and could not be removed even with six river steamboats struggling to pull her off. Eventually, all the cargo and fuel had to be unloaded before success was achieved; the delay amounted to eight days. This was not the only untoward event. North of Havana she ran short of coal and had to put into Wilmington, North Carolina. There was only 40 tons available there, but a schooner load of wood was bought. Further delays were encountered awaiting tides high enough to permit crossing the bar at the mouth of the Cape Fear River. A traveler who sailed from New Orleans on October 20 did not reach New York until November 9, but he could regale his friends with his adventures on board the only steamship then running between these ports.

After five voyages the *Crescent City* arrived back on December 9, 1848, just as the news of California gold was being recognized in New York. To take advantage of the new development she was scheduled to depart for Chagres on December 23, 1848. The *Falcon*, the first steamer chartered by the U.S. Mail Steamship Company, had sailed from New York on December 1, but did not acquire gold-hunters until she docked at New Orleans. Some 130 passengers left on the *Crescent City* while a hundred less adventurous souls came to the waterfront to cheer them on.[82] These, together with the *Falcon*'s New Orleans contingent, were part of the excited mob at Panama awaiting the first arrival of the *California*.

While she was away, the *Crescent City* was purchased—together with the second ship for the line, the *Empire City*, which was still on the stocks at the Brown shipyard—by Charles Morgan, and John and Joseph Howard. Since 1837, when Morgan transferred his steamboat activities from the stormy East Coast to the calmer Gulf of Mexico, he had assembled a profitable fleet for the New Orleans to Texas route. Now, twelve years later, he was to return to operating vessels out of New York. This time, instead of cheap and unsuitable steamboats, he took over strongly built ocean steamships planned by wiser operators. Furthermore, they were left under the experienced management of the Howards. As a result there was no change in the *Crescent City*'s schedule. She continued to sail from New York to Havana and Chagres about once a month. Northbound a stop at Kingston, Jamaica, was made for coal.

The second steamer, the *Empire City*, had her frames erected by October of 1848.[83] With dimensions of 238'-8" by 39'-4" by 24'-4" and a tonnage of 1,751, she was a much larger ship and is said to have cost sums varying from $220,000, which seems low, to $300,000 which is probably the opposite.[84]

A new feature meeting the approval of all who inspected her was a light hurricane deck providing a "grand" promenade from bow to stern and side to side.[85] All doors, ports, and windows in the deckhouses were protected from sun and rain by its overhang. She was a smart-looking steamship with little sheer, three nicely raking masts, the conventional barkentine rig, a stack well forward, and side wheels well aft. There was a clipper bow, a forecastle, as a continuation of the hurricane deck, and one narrow painted stripe at the level of the guards, which made her look long and lean.

Her interior was the most ornate of any steamship on the Atlantic. The saloon for the ladies in the after house had purple and gold damask upholstery on its rosewood furniture. Every stateroom door opening from it was decorated with a Hudson River scene. Below was the main saloon with two-berth staterooms on both sides; here the center panel of each door bore a scene from Washington Irving's *Sketch Book*. The forward bulkhead had murals of New York and New Orleans. Ornamental columns had gilded Corinthian capitals and these were set off by satin and zebra wood. Forward was a smoking room, labeled "Social Hall," and below it the dining saloon, both lined with staterooms.

Practical matters were considered in addition to the ornamental. The decks were covered with oilcloth patterned to look like carpet. It must have been much easier to clean when motions and passengers interacted unfavorably.

The 85" bore by 9'-0" side lever engine and its two very large iron boilers were built and installed by Secor and Company. Trials were held on July 10, 1849, when the *Empire City* headed out to sea for a seven-hour trip with many invited and some uninvited guests who slipped aboard. A special dinner was held four days later during which Mr. Morgan awarded a silver pitcher to Archer Guion for supervising the engine work.

On July 17 the *Empire City* departed from New York for Chagres with 193 passengers. Fares were $150 in the after saloon, $125 in the upper saloon, $115 in the foreward saloon, $100 in the lower cabin, and $80 steerage. Gold would be taken for 1% of its value and freight at the high rate of 70 cents per cubic foot. She left Chagres on August 8, took on coal and water at Kingston, and reached New York on the 12th with a mere 12 passengers; no one was returning yet from California. On her second voyage she brought back $603,000 in gold and the $6,030 charge made it a very profitable cargo.

Having both the *Empire City* and *Crescent City* available, their owners assumed the title of Empire City Line and provided fast, regular sailings twice a month direct to Chagres. Private enterprise, without a mail subsidy, anticipated regular service by the U.S. Mail Line by over six months. It was operating to the Isthmus by the shortest possible route and had two splendid steamships to its credit. During most of 1849 the U.S. Mail had only the *Falcon*, a small, bastard-type steamer designed around a pair of rebuilt engines from John Ericsson's disastrous Hudson River steamboat, the *Iron Witch*.

In December of 1849, just when everything was going well, the *Crescent City* chose to break down at the worst possible moment. There were plans afoot for creating a short route to California by way of Nicaragua instead of Panama. Cornelius Vanderbilt wished to visit the country for a personal inspection. Charles Morgan, always interested in any new transportation scheme, decided to emulate the "Commodore," a courtesy title awarded to Vanderbilt as a result of his many successful steamboat ventures. Both of them sailed on the *Crescent City* on December 13, 1849.[86] On the 16th a crosstail break led to a bent connecting rod, a bent piston rod, and the immobilization of the engine. Paddles were removed and sails set, which just flapped in the lightest of breezes. Progress was too slow to suit either of the eminent shipowners. Nearby, however, were small sailing craft. Morgan acted first by chartering the schooner *Sarah A. Smith*, buying her cargo of lumber on the spot, throwing it overboard, and sending the mails and thirty passengers on to Chagres. Vanderbilt, in turn, hired the brig *Roscoe* to take Morgan, himself, and his party to Havana, where they could board a Royal Mail steamship for New Orleans, Chagres, or Vera Cruz. The brig *Marcia* took a few passengers to Havana, leaving about fifteen on the *Crescent City*. She sailed slowly toward Charleston and was eventually picked up on December 27 and towed to that city by the *Governer Dudley*, a local steamboat. The extensive engine repairs having been completed, she left there on February 15, 1850, and reached New York about the 18th with Morgan again on board. Neither he nor Vanderbilt had ever reached Nicaragua. Later they would be, at various times, either rivals or partners in steamship operations to that country.

The fact that the Empire City Line established a Pacific branch has already been reported. Since there were few suitable steamships afloat, the line made do with the auxiliary screw steamer *Sarah Sands*, built of iron in England in 1847; the *New Orleans*, of 761 tons; and the *Northerner*, from the Charleston line, as their one satisfactory steamer. As soon as they arrived on the West Coast, in the summer of 1850, the Pacific Mail started negotiating for their purchase; by October they had taken over the Empire City fleet on both the Atlantic and Pacific. The Atlantic branch kept its own name and the same agent, Howard and Son, but its sailing dates were revised to connect with the Pacific Mail's Panama to San Francisco steamships. Then, only three months later, when the Pacific Mail and the U.S. Mail reached an agreement to stay on opposite sides of the Isthmus of Panama, the former's Atlantic steamers, *Crescent City*, *Empire City*, *Cherokee*, and *Philadelphia* were sold to the U.S. Mail. By this settlement three excellent vessels were acquired to round out a badly unbalanced assembly of unrelated craft. From this point on, the *Ohio* and *Georgia*, with the largest passenger capacity, could operate to Chagres, and the two *Cities* to New Orleans, usually in conjunction with the *Cherokee*. Either one could replace a Chagres boat when repairs were under way, and a rebuilt and enlarged *Philadelphia* could substitute wherever necessary.

By sheer chance the *Empire City* and *Crescent City* returned in 1851 to the service for which they were designed, and they continued to carry passengers, freight, and mails regularly and economically between New Orleans, Havana, and New York.

The *Crescent City*'s career ended late in 1855. She had sailed from New York on December 3 and was scheduled to arrive at Havana five days later. On the way she ran hard aground on Little Bahama Bank on the 7th, opening up her hull to the sea.[87] Although boats were lowered and her spars lashed together to make a raft, the vessel remained safely upright on the reef. One boat was dispatched to Sandy Cay, forty-three miles away, to summon the salvage schooner *Defiance,* but the only assistance she could render was to convey some passengers to Nassau. The *Crescent City*'s master, Captain A. G. Gray, chartered a passing brigantine, the *Alma,* to take his remaining passengers and the mails on to Havana.

The *Empire City* in 1853 had two new boilers installed, large ones weighing forty tons apiece.[88] Few of the iron boilers that used salt water lasted very long; they either corroded or collected scale and became overheated, with permanent damage when they were carelessly fired. Replacement was a normal, not an unusual, event, even though it involved removing cabins and opening up the decks.

The September 1857 hurricane that sank the U.S. Mail's *Central America,* ex-*George Law,* and damaged the *Southerner* almost claimed the *Empire City.* She followed the *Central America* out of Havana on September 8, both bound for New York. She may have been fortunate in that she could not keep up with the newer and larger steamer, which passed out of sight before sundown. From September 9 to the 13th she met increasing wind and sea. Near Cape Lookout a dismasted bark was seen, but waves were so high that nothing could be done to aid her. On the 14th the *Empire City* found herself off Cape Hatteras in a critical condition: the seas had broken over her decks; the starboard paddle box was gone; the forward houses on the guards were swept away. Moreover, since a number of airports were broken in, water was swirling through the cabins. All sails had been blown out and the main gaff had fallen to the deck.[89] The ship was running out of coal and had begun to burn her own woodwork, starting with awning stanchions and steerage berths. On the 15th she limped into Norfolk, but, because of fever at one of her ports of call, was quarantined and not allowed to dock. An inspection showed that the rolling and pitching had been so violent that bolts holding the boilers in place were broken and that the starboard one had a hole punctured in its bottom. After coaling and making temporary repairs, she left for New York on September 18. She met the Norwegian bark *Ellen* coming in with forty-one survivors of the *Central America*. These were taken aboard and a successful search made for the American brig *Marine,* badly overloaded by other rescued passengers. In all, ninety-two were taken over by the *Empire City* and delivered to New York on September 20, 1857.

Extensive rebuilding kept the battered ship out of service until January 2, 1858, when she returned to the New York, Havana, New Orleans route.

She continued on as part of the U.S. Mail fleet until it was disbanded in 1859, but was then acquired by Marshall O. Roberts, who had succeeded George Law as both the principal stockholder and managing agent of the line. Roberts carried on the service from New York to New Orleans and continued to advertise under the U.S. Mail name, even though the original company no longer existed. Soon afterward the Civil War requisitioned every available passesnger steamship for transporting troops. The *Empire City* was chartered, first by the War Department, later by its Quartermaster Corps, at rates varying from $1,000 to $775 per day. On her first charter, in March of 1861, when war seemed imminent, she embarrassed the U.S.S. *Mohawk*, which was to escort her from New York to Indianola, Texas. The *Mohawk* was left far behind throughout the trip and never caught up with the far faster *Empire City*.[90] The latter brought home troops that had been stationed in Texas.

In October of 1861 she was a transport on Admiral DuPont's expedition to Port Royal and in 1862 carried troops bound for Fernandina, Florida. A year later she went up the Mississippi and passed besieged Vicksburg, again as a transport. In 1864 she delivered men to General Butler during the invasion of the North Carolina Sounds. Just before the war ended she was sold outright to the War Department for $225,000, only a little below her first cost, and, for unknown reasons, was not disposed of when hostilities ceased. In 1866 she was considered for use as a floating hospital at the New York Quarantine Station,[91] but never put to use. The *Empire City* was laid up at Red Hook until May of 1869. Sometime thereafter she was broken up.[92]

She and the *Crescent City* were William H. Brown's intermediate examples of ocean steamships, bridging the gap between the early ones, the *Southerner* and *Northerner* of 1846 and 1847, and the Collins liners completed by him in 1850. They were intermediate in time, in design, and in size, but were never so in terms of success. From the point of view of profitable investments they were superior to the steamers that came before them as well as those that followed. Except for needed upkeep, they were always in demand and each successive owner kept them hard at work carrying passengers and freight.

Empire City as a Civil War transport. Shear legs have been erected to lift parts of her engine. *Photo by U.S. Army Signal Corps.*

The Early Ocean Steamships

Every coastwise steamship considered in this chapter was built and engined in New York City, the shipbuilding and engine-building center of the United States. Only one other steamship of over 1,000 tons was completed in the 1846-49 period. It was the *Isabel*, of 1,115 tons, a Maryland-built steamer of about the same size and general appearance as the *Cherokee* and *Tennessee*. Her builder was S. H. Duncan and her machinery was constructed by Charles Reeder, both of Baltimore. She has been mentioned as the mail steamer from Charleston via Savannah and Key West to Havana that was operated by Charleston's Mordecai and Company. Occasionally she made a voyage all the way to New York instead of turning around at Charleston. In regular service she connected with the Spofford and Tileston boats. During the Civil War she became a blockade runner and was renamed *Ella Warley*.

If she is included, there were a total of twelve large coastwise steamships constructed from the start of 1846 to the end of 1849. Four were by William H. Brown, gradually increasing in size from 785 tons on the *Southerner* to 1,751 for the *Empire City*. These led up to two Brown-built Collins liners for Liverpool service in 1850. Four were by William H. Webb, two for Savannah and two for San Francisco trade, but all were eventually engaged in the carriage of mail and passengers from the East to the West Coast. Webb was unusual in starting his steamship career with a transatlantic liner, the *United States*, and following it with these coastwise craft. Two further steamships, one small, the *Oregon*, for the Pacific and one large, the *Georgia*, for the Atlantic were by the old and respected firm of Smith and Dimon. The *Georgia*'s near-sister, the *Ohio*, by Joseph Bishop and Jeremiah Simonson, and the *Isabel*, by S. H. Duncan, complete the list. Ten of these were successful ships with the three for Pacific Mail being very rugged craft that lasted many years. The two largest, however, the only ones over 2,000 tons, were not adequate from a structural viewpoint, and both came extremely close to sinking when caught in heavy storms. The *Georgia*, in addition, had hull proportions that made her particularly difficult to steer.

Every one had side lever engines and return-flue boilers. The *Georgia* and *Ohio* had two engines and four boilers each to insure high speed, but, simultaneously, they became great coal burners and were far too costly to operate. The others had a single engine and a pair of boilers. The Novelty Iron Works provided machinery for six of the vessels and seems to have been somewhat more successful than their closest rival in avoiding delays. T. F. Secor engined four, the Allaire Works one, and Charles Reeder one. Only one of these organizations had had any previous experience with the side lever engine. Like the shipbuilders, they were producing newer and larger products than they ever had before. There were breakdowns of various parts in service, but the surprising thing is that there were not more of them. Both large castings and heavy forgings were difficult elements to produce free from concealed defects that might lead to cracks and failures after considerable operation. Such engines and boilers provided the starting point from which larger ones were to be developed to power the Atlantic liners of the 1850s.

The experience with ever-larger hulls and ever-increasing engine power helped to continue the dominance achieved by New York and its many skilled artisans during the height of wooden steamship building in the 1850-65 period. Baltimore built steamboats for the Chesapeake Bay and its adjoining waters, Philadelphia supplied steamboats for the Delaware River River and pioneered in the development of the screw-propelled vessel, but the large steamships for ocean and coastwise service were created in New York.

Notes

1. Carl C. Cutler, *Queens of the Western Ocean* (Annapolis, Md., 1961), pp. 173, 415.
2. Fred Irving Dayton, *Steamboat Days* (New York, 1925), p. 382.
3. *New York Herald*, September 9, 1846.
4. *Merchants Magazine and Commercial Review* 20 (January-June 1849): 519-23.
5. H. Philip Spratt, *Transatlantic Paddle Steamers*, 2d ed. (Glasgow, 1967), pp. 26-27.
6. Frank M. Bennett, *The Steam Navy of the United States* (Pittsburgh, Pa., 1896), pp. 35-36.
7. Charles B. Stuart, *Naval and Mail Steamers of the United States*, 2d ed. (New York, 1853), p. 36.
8. *New York Herald*, September 9, 1846; *New York Commercial Advertiser*, September 9, 1846.
9. B. H. Bartol, *A Treatise on the Marine Boilers of the United States* (Philadelphia, 1851), pp. 26-27.
10. *New York Tribune*, August 7, 1850.
11. *San Francisco Alta California*, August 16, 1850.
12. *New York Tribune*, October 7, 1850.
13. *New York Commercial Advertiser*, November 16, 1847.
14. *New York Herald*, October 5, 1850.
15. Ibid., September 21 and 30, 1857.
16. John Haskell Kemble, *The Panama Route, 1848-1869* (Berkeley and Los Angeles, Calif., 1943), p. 238; *San Francisco Alta California* (1850-53).
17. E. W. Wright, ed. *Lewis and Dyden's Marine History of the Pacific Northwest* (1895; reprint ed. New York, 1961), pp. 95-96.
18. Ernest A. Wiltsee, *Gold Rush Steamers* (San Francisco, 1938), pp. 275-76.
19. John Haskell Kemble, "The Genesis of the Pacific Mail Steamship Company," *California Historical Society Quarterly* 13, nos. 3 and 4 (1934): 241-43.
20. *Merchants Magazine and Commercial Review* 16 (January-June 1847): 419-20.
21. *New York Tribune*, October 20, 1848.
22. William H. Webb, *Certificate Book*, vol. 1. Webb Institute of Naval Architecture.
23. Kemble, *Panama Route*, p. 239.
24. William Skiddy to M. C. Perry, letter dated New York, January 21, 1849, House Executive Document 124, 32d Congress, 1st sess., pp. 140-41; Hereafter cited as House Executive Document 124.
25. Lauchlan McKay, *The Practical Shipbuilder* (New York, 1839), Plate 7.
26. Andrew Murray, *The Theory and Practice of Shipbuilding* (Edinburgh, Scotland, 1861), p. 91, fig. 36.
27. Stuart, *Naval and Mail Steams*, p. 148; Bartol, *Marine Boilers*, p. 26.
28. Kemble, *Panama Route*, pp. 239, 242.
29. *New York Tribune*, October 9, 1848.
30. Victor M. Berthold, *The Pioneer Steamer* California, *1848-1849* (Boston and New York, 1932), p. 27.
31. Ibid., pp. 28-31.
32. Ibid., p. 49.
33. *New York Commercial Advertiser*, July 23, 1849.
34. *New York Herald*, January 6, 1849.
35. *New York Commercial Advertiser*, December 26, 1848.
36. Kemble, "Genesis," p. 398.
37. *New York Commercial Advertiser*, April 7, 1849.
38. Kemble, "Genesis," pp. 401-3.
39. American Lloyds' Registry of American and Foreign Shipping, 1863.
40. Kemble, *Panama Route*, p. 242.
41. *New York Maritime Register*, April 21, 1880.
42. Kemble, *Panama Route*, p. 218.
43. William H. Webb, *Certificate Book*, vol. 1.
44. William H. Webb, penciled notes in his *Offset Book*. Webb Institute of Naval Architecture.
45. William H. Webb, *Plans of Wooden Vessels*, vol. 1, privately printed and distributed by the author (New York, ca. 1895).
46. Bartol, *Marine Boilers*, pp. 40-41.
47. Kemble, *Panama Route*, p. 248.
48. Ibid., p. 39.
49. *New York Herald*, November 8, 1849.
50. Ibid., April 6 and 25, 1850.
51. *San Francisco Alta California*, April 15 and 20, 1850.
52. *New York Tribune*, August 27, 1853; *New York Commercial Advertiser*, August 27 and 29, 1853.
53. *New York Herald*, September 1, 1853.
54. *New York Shipping and Commercial List*, August 28, 1858, has the last listing. The *Cherokee* was at Hunter's Point on the East River.
55. *San Francisco Alta California*, March 9, 1853.
56. Kemble, *Panama Route*, pp. 15-16.
57. House Executive Document 124, p. 39.
58. *New York Herald*, September 20, 1849.
59. Letter, William Skiddy to Commodore M. C. Perry, January 21, 1849, House Executive Document 124, pp. 140-41.
60. *Journal of the Franklin Institute* 18, 3d ser. (July-December 1849): 446-48.
61. Bartol, *Marine Boilers*, pp. 34 and 35. The author has quoted this source, which, as a technical one, he considers more trustworthy than the *Herald* which gives 37' by 10' wheels with 28 double paddles.
62. *New Orleans Picayune*, October 2, 1849.
63. *New York Commercial Advertiser*, September 17, 1849.
64. Ibid., September 20, 1849.
65. Ibid., October 13, 1849.
66. John W. Griffiths, *A Treatise on Marine and Naval Architecture*, new ed. (London, 1856), p. 158.
67. House Executive Document 124, p. 140.
68. *New York Herald*, January 24, 1850.
69. Griffiths, *Marine and Naval Architecture*, p. 163.

70. Letter, David D. Porter to Commodore M. C. Perry, March 9, 1850, House Executive Document 124, pp. 120-22.
71. *New York Commercial Advertiser*, March 12, 1850.
72. Letter, Jas. Findlay Schenck to Hon. T. J. Rusk, August 24, 1850, House Executive Document 124, pp. 130-31.
73. *New York Herald*, December 28, 1850; *New York Commercial Advertiser*, December 28, 1850.
74. Letter, William Skiddy to Commodore M. C. Perry, January 16, 1851, House Executive Document 124, pp. 175-76.
75. Letter, W. D. Salter, Charles H. Bell, and William Skiddy to Hon. W. A. Graham, January 21, 1851, House Executive Document 124, p. 175.
76. *New York Commercial Advertiser*, September 14, 1853.
77. *New York Shipping and Commercial List*, January-October, 1860.
78. John H. Morrison, *History of American Steam Navigation* (New York, 1903), p. 274.
79. James P. Baughman, *Charles Morgan* (Nashville, Tenn., 1968), p. 242.
80. *New York Commercial Advertiser*, June 22, 1848.
81. Ibid., July 24, 1848.
82. *New York Herald*, December 24, 1848.
83. *New York Daily Tribune*, October 20, 1848.
84. Kemble, *Panama Route*, p. 224; Baughman, *Charles Morgan*, p. 62.
85. *New York Herald*, July 16, 1849.
86. *New York Tribune*, January 5, 1850.
87. *New York Commercial Advertiser*, December 19, 1855.
88. *New York Herald*, September 1, 1853.
89. Ibid., September 17, 18, and 21, 1857.
90. *Official Records*, ser. 1, 4: 94, 101, and 106.
91. *New York Herald*, April 29, 1866.
92. *New York Shipping and Commercial List*, March 17 through May 15, 1869.

7
The Ocean Steam Navigation Company, 1847-1857

ON the last day of President John Tyler's term of office, March 3, 1845, he signed a bill authorizing the postmaster general to contract for the transportation of mails to Great Britain and the continent of Europe. Cave Johnson, the postmaster general appointed by Tyler's successor, James Polk, was a man of great energy who did his best to expand the services he directed. He advertised for bids to establish routes from New York to Liverpool, Bristol, or Southampton; to Antwerp, Bremen, or Hamburg; and to Havre, Brest, or Lisbon. A number of bidders responded with varied proposals and costs, but that of Edward Mills quoting $300,000 per annum to Havre appeared most attractive.[1] Johnson, however, preferred Bremen as a terminus with a stop at Cowes, or nearby Southampton, on the way. After considerable negotiation and the promise of favorable treatment by the City of Bremen and the adjoining Kingdom of Hanover, Mills agreed to provide four steamers with sailings twice a month to Bremerhaven, the port for Bremen, for a period of five years, at a cost of $400,000 per year. As an alternative, once two ships were in operation at monthly intervals, he could send the second pair to Havre instead of Bremen and the payment would be reduced to $350,000 because of the shorter distance.

It was not until March of 1846 that the agreement was sent to Congress for final approval; there the whole project was delayed, not only by the Mexican War, but also by other proposals submitted by Edward K. Collins of New York and Robert B. Forbes of Boston, both of whom tendered different (and more expensive) services. The Mills proposal was eventually approved. The Ocean Steam Navigation Company was incorporated in May of 1846 and immediately found it difficult to raise the needed capital. The northern German cities and states of Bremen, Frankfurt, Prussia, Saxony, and Oldenburg, all anxious to establish communication by steamship with the United States, loaned the company 279,100 thalers and, in return, nominated Herman Oelrichs, a Bremen merchant, to the company's Board of Directors. Mills transferred his rights to the company and for a short while became its general agent under Christian H. Sand, its president. Other directors included Horatio Allen, of the Novelty Iron Works, and Mortimer Livingston, one of the proprietors of the Havre Line of sailing packets.[2]

Two steamships were ordered from the Westervelt and Mackay shipyard in New York. The two partners had no steamship experience, but in 1846 there were few who had. The firm was known for the excellent sailing characteristics of the many packets they had produced since the partnership was formed in 1841.

The premier steamship of the line, the *Washington*, had her first frame raised October 15, 1846; was launched on January 30, 1847;[3] and was engined, outfitted, and registered at the New York Custom House on May 29, 1847, a breakneck pace for building any vessel of 1,640 tons. She was twice the size of her only possible prototype, the *Southerner,* completed in September of 1846, yet that smaller steamer had taken much longer to build. She was planned to be larger and faster than the British liners sailing under the Cunard flag from Liverpool to Boston. Like them she was propelled by a pair of side-lever engines. Comparative dimensions were:

Steamer	Date	Length	Beam	Depth	Engine Bore	Engine Stroke
Hibernia	1843	219′	35.8′	24.2′	77 1/2″	7′-6″
Washington	1847	230′-5″	38′-9 1/2″	31′-0″	72″	10′-0″

In size the *Washington* was larger; the speed question, however, will be deferred until later.

Construction was under the supervision of Thomas Brownell, Lieutenant, U.S. Navy. The keel was 16″ square with a 5″ shoe to protect it from grounding. The white oak floor timbers were double 10″ fore and aft and 20″ deep on centerline tapering to 16″ at the bilges, and the frames above were gradually reduced to 6″ thickness at the deck. They were spaced 2′-6″ center to center. A center keelson 3′-0″ x 16″ was fitted over the keel. To support the engines there were built-up side keelsons 3′-0″ deep tapering from 36″ at the bottom to 30″ at the top. The oak outer planking ranged from 4″ to 6″ thick and the ceiling, of yellow pine, varied from 8″ to 6″ thick. There was a flat transom with stern windows looking aft and, at the bow, a cutwater with headrails and trail boards supporting a full-length figurehead of George Washington. The lower portions of the wheel boxes were connected to the sides of the ship by sloping planking, similar to the *Southerner*'s, and their protuberance was emphasized by a painted black-and-white port strake. In later years an all-black hull helped conceal the paddle boxes and improved her appearance. No visible guards extended beyond the housings around the wheels. Their under sides, unfortunately, were very low; a small roll would put them in the water. The general hull shape followed that of a packet.

On deck were extensive houses near the wheels for engineers' quarters, stores, stalls for cattle to provide fresh meat and milk, and zinc-lined ice houses. Over all this a raised deck ran thwartships between the paddle boxes and was reserved for navigating and piloting the ship. At the stern came a house for the steering wheel with the master's and deck officers' quarters adjoining. Four large skylights on the spar deck served the public rooms below. Most of the deckhouses were painted green, a popular sailing-ship color.

One deck down, the Grand Saloon was 85′ long, from the transom to its forward bulkhead, and was lined with staterooms on each side. Nearby were baths, water closets, and the barber shop. The interior was white enamel decorated in gold. The skylights, which provided ventilation as well as light, had the latter element somewhat dimmed by trunks of stained and ground glass ornamented with portraits of American statesmen and the arms of European countries. The Grand Saloon also served as the dining room. Forward were pantries and service spaces. Ahead of the engines and boilers came the smoking room, with further staterooms port and starboard, and, still farther forward, the second-class quarters. The berthing for the seamen was right at the bow. In all, there were three decks; on the lower ones were mail rooms, storerooms, baggage rooms, and cargo. Both cargo and the coal were loaded through ports cut in the ship's sides. The passenger capacity in the first cabin was 140, in the second 44, and 300 measurement (40 cubic foot) tons of space were allotted for cargo.

Three masts were fitted, the two forward ones square-rigged with interchangeable yards. The foremast, however, was 78′ long to the main's 80′, but the topmasts were the same, 50′. Spencers were carried on the fore and main. Masts were short and no topgallants were carried.

Five of the seven boats were of the Francis type, made of galvanized iron. They were the first widely used iron craft in the United States and were recognized as being lighter and tighter than wood boats. Built-in buoyancy tanks kept them afloat even if they capsized.

The Novelty Iron Works supplied the machinery, a pair of 72″ bore by 10′-0″ stroke engines, and two return flue boilers. Although the hull was greatly enlarged over previous steamships, the cylinder dimensions were increased by a moderate amount and were not too difficult to cast, bore, and assemble. The boilers were 12′ in diameter by 36′ long, each with three furnaces. A number of horizontal flues of varying diameters carried the hot gases forward to combustion chamber and six 16″ return flues led them back to the base of the stack.[4] Steam was generated at 12 psi and the large paddle wheels were intended to turn at 11 rpm. Forced draft fans discharged air below the grates. The boiler capacity was badly underestimated and they never supplied enough steam.

All the coal bunkers were enclosed by sheet-iron plating. In the fire rooms, the wood ceiling was lined with cast-iron plates to reduce the fire hazard.

On Monday, May 24, the *Washington* steamed away from the Novelty Works wharf on her trials, to find New York harbor shrouded in fog.[5] Tuesday she went to sea and was tried under steam and under sail while proceeding as far south as Barnegat Light. Although her prospective commander, Frederick Hewitt, was aboard, Captain John Maginnis, president of the New York Pilots Association, was in command. The Novelty Works

Washington, original appearance in 1847. *Courtesy of Peabody Museum of Salem.*

chartered a steamboat, the *John Marshall,* to meet her on her return and had between three and four hundred guests on board, many of them the mechanics who had worked on the *Washington*'s engines. There was still work to be done; she was not to be coppered until the completion of the first voyage. Furthermore, mechanics were still working on her the day she sailed.

With about 127 passengers, the *Washington* left Pier 4 on the North River at 4 P.M. on June 1, 1847, the first fully powered American transatlantic liner to be completed.[6] She was bound for Bremerhaven, on the Weser River, and would stop at Cowes Roads, off the Isle of Wight, to permit passengers bound for England or France to disembark into tenders for nearby Southampton.

In addition, there were about one hundred invited guests who were transshipped to the *John Marshall* off Staten Island. The two vessels proceeded well out to sea, tolling their bells as they went. It should be noted that the steam whistle was not yet a recognized signaling device, and many steamers did not even have one. It was a byproduct of the railroad locomotive's development that was later adopted by the ocean steamship.

Several facts reported in the newspaper accounts of her trials and departure should be particularly emphasized. First, on her sea trials, 400 tons of iron ballast were placed on board, supposedly to simulate the weight of cargo that might be stowed below. In addition, 375 tons of coal had been loaded. All this added up to a large weight, well down in the hull. Second, when she left New York, her drafts were 21′-6″ forward and 20′-0″ aft. Third, there was a visible list to port. The draft figures were excessive for a vessel of this size and neither trim by the bow nor an angle of heel is an accepted operating condition.

What had happened was that both the ship's weights and her stability were out of control. The shipbuilder had not realized how heavy the paddle wheels, shafting, and raised machinery elements were and had not made his ship wide enough to keep her upright. The *Georgia* had too much beam and too much stability. She could operate but was not a comfortable vessel. The *Washington* was the opposite. She had too little stability, a far greater design mistake; furthermore, her machinery weights were too far forward.

All that could be done was to install ballast aft in the bottom of the ship to offset the topside weights and correct the trim. Any coal improved the stability, since it was stowed low down, but its beneficial effect would vanish as it was burned. The additional weight made the draft excessive and this, in turn, buried the paddle wheels too deeply, caused them to operate inefficiently, and brought overhanging wheel enclosures close to the water where they were likely to become swamped and damaged. If stability is low, a small off-center weight causes a large list, a nuisance and an indication of a dangerous loading condition.

When the *Washington* sailed, as the pride of the American merchant marine, the general public expected her to be an instant success, far faster than the Cunard liners that, for seven years, had delivered the British mails. Both the *New York Herald* and *Commercial Advertiser* predicted a ten-day crossing. She was a sad disappointment. Her owners, wisely, had made no special claims in advance of sailing. They must have been aware of the wanting stability and the buried wheels; they would soon receive news of other malfunctions.

Only hours after she got to sea, her engineers found the air pumps taking water from the jet condensers more rapidly than it could be disposed of.[7] The hot wells were too small and the piping discharging what was not returned to the boilers was undersized. On June 4 the engines were stopped for 4 1/2 hours to make repairs and ten days later, just before land was sighted, another stop was necessary. Instead of anchoring off Cowes to unload mail and passengers, the *Washington* headed into Southampton to arrange for machinery alterations and to take on extra coal. She had consumed far more than had been anticipated.

In place of a ten-day passage, she took all of fourteen. Moreover, the seven-year-old Cunard liner *Britannia* had beaten her by two days on a concurrent passage from Boston to Liverpool.

To keep the hot wells from overflowing into the hull, temporary modifications were made before she proceeded to Bremen, and new castings were ordered to replace the existing ones.

From Southampton, Captain Hewitt, acting as his own pilot, took the *Washington* into the North Sea and the Weser River, where she was met on June 18 by two local steamers that escorted her to her moorings. A dock was being constructed especially for her, but was not yet ready. One of her escorts took off the mail bags and the assistant postmaster general, Major Hobbie, who had traveled with them. The second tender transported her eighty-four passengers upriver to Bremen. The ship, the port, and the city were all elaborately decorated for the occasion. The Burgomaster came down to meet the *Washington*, bands played, salutes were fired, and a grand dinner for two to three thousand people was served at the Hunters Club.[9] Next day came a state dinner at which a six-foot model of the *Washington* was presented to the Bremen Senate. Returning, forty-six passengers for New York and four for England were aboard when she arrived at Southampton on June 27. The New York passengers were put up at local hotels while the new hot wells were being installed. She sailed on July 10, but even before she passed the Needles, the coal just loaded proved too much for the grate bars in her furnaces. These were replaced but the second set was destroyed less than twenty four hours out.[10] Having no more spares, she was forced to turn back to Southampton and have new grates made that could stand up to the potent qualities of British coal. It may be remembered that the *Massachusetts* had had identical troubles on her first sailing from Liverpool. A second departure was made on July 15; this time New York was reached in fifteen days, one day longer than the disappointing eastbound passage.

Captain Hewitt must have made strong protests about his ship's lack of stability. He had used sail on a number of occasions when there were favorable winds. Most likely, as coal was burned, the ship heeled more and more as her stability gradually decreased. The *Washington* was taken in hand and drastic steps were taken to cut down her topside weights.[11]

With the exception of the wheel house aft — no forward pilot house was fitted — the weather deck was swept clear of erections. Only a light raised platform was left amidships. New quarters for the officers had to be built in the tween decks. The forward smoking room was removed, as were all the second-class quarters. Some new first-cabin staterooms were installed aft below the saloon and the cargo space was enlarged to seven hundred tons; the paddle wheels were reduced in diameter so that their dip became more reasonable. While the ship was on the sectional dry dock at the foot of Pike Street, the bottom was coppered. Captain Hewitt remained in charge, but the entire staff of engineers was replaced. The new Chief Engineer, Mr. Mars, had five years' experience on the Spanish steamship *Lion*, built in New York in 1841.

The alterations took most of August and September, but the *Washington* was ready to sail on September 23. She arrived at Cowes in fifteen days. Returning, she met bad weather for eight days, which drove her off course. There were anxious hours as a result of spontaneous combustion of coal stowed under the boilers. Three fire pumps were required and nine hours elapsed before the fire was put out.[12] It seems foolhardy to stow fuel in such a location! Eighty three passengers were safely landed at New York on November 9, 1847.

A second ship was laid down at the Westervelt and Mackay yard some time after the *Washington* was launched. Although similar in outward appearance, she was not identical. Had she been delayed further, she might have benefited from the discovery of the *Washington*'s faults. Her register dimensions were 234'-11" by 39'-6" by 31'-0", which made her about five feet longer and a foot wider. Her tonnage came out to 1,734 instead of 1,640, and the extra beam helped her stability.

The intended name was *Lafayette,* and the delay in starting was because of financial problems. The Ocean Steam Navigation Company was advertising for $300,000 in stock subscriptions simultaneously with the *Washington*'s maiden voyage.[13] The name *Lafayette* implies a continued interest in a French branch for the line.

But before completion the name *Hermann* was substituted and a full-length figurehead of that Germanic tribesman who had vanquished the Roman legions under Varus was carved for her cutwater. The *Hermann* was launched on September 30, 1847, a week after the *Washington* sailed on her second voyage.[14]

The sizes of her strength members indicate that she was somewhat more heavily built than the *Washington;* her engine keelsons were deeper and her outer planking was doubled with layers 3 1/2" to 6" thick.[15].

Following the *Washington's* modifications, the *Hermann's* weather deck was kept clear except for small houses near the paddle boxes to serve as entryways to cabins below. Again there was no forward pilot house. Even the windlass for hoisting the anchors was below decks. The Grand Saloon was longer and wider than the *Washington's* and below it were installed "family" rooms for four to six passengers.

After five months at the Novelty Works being fitted with her engines and boilers, parctically duplicates of those on the *Washington*, the *Hermann* was tried on March 7, 1848, on a short run down the harbor to Sandy Hook lightship.[16] A second trial followed four days later; two days were spent at sea and 275 miles covered. This time the line wanted assurance of continued performance. But Neptune does not always favor those who invade his realm.

The *Hermann* left New York March 21 and was greeted by a storm six hundred miles out. For forty hours she lay to, head to the sea, in winds that made it impossible to set a steadying sail. The headrails were smashed, part of the overhanging housing of the wheels was torn away, and the hull deflected so much that both injection pipes carrying water into the condensers were broken. With Halifax 270 miles away, Captain Crabtree put in there on March 28 to repair engine and piping damage.[17] She was supposed to sail from Southampton on April 20, but, owing to the delay, could not leave the Weser until April 19. After further engine repairs at Southampton, she steamed homeward on May 6. Arriving off Sandy Hook on the 21st, she picked up a pilot but had to wait thirty six hours for fog to clear before she could come up to her dock. The passage to Sandy Hook took fourteen days eighteen hours, if a five-hour longitude correction is added to the reported time. Perhaps the most interesting passengers were a dozen camels and two Arab drivers, especially imported from Egypt by the S. B. Howes Circus.[18]

The next sailing date, May 31, had to be put off until June 20 because of further alterations. While the *Hermann* was meeting unforeseen difficulties, the *Washington* continued to encounter adversity.

Winter weather prevailed for her third voyage. Sailing from Southampton on December 19, 1847, she had a "very boisterous passage," severe gales, and heavy seas that washed away her larboard paddle box. For five days she had to lay to, making no progress whatsoever. The engines were misbehaving and had to be stopped on several occasions. Eighteen days out, badly short of coal, she headed for Halifax, where fog kept her outside the harbor for a day. A safe entry was made on January 10. Before she arrived there, her two cylinders had to be braced in position by props cut from the mizzen topmast and wedged in place to keep them from swaying back and forth. Flexing of the hull had fractured the piping leading to the forced draft blowers.[19]

Fifty-six badly needed hours were spent at Halifax patching up the engines so that the *Washington* could limp home. Her arrival on January 16, 1848, was one month out of Bremen and twenty-eight days from Southampton. Another month intervened before she could sail again.

The *Washington's* troubles multiplied. Homeward bound on her sixth voyage, the crosshead for her starboard air pump broke in mid-Atlantic but, despite reduced power for a week, she managed to reach New York in sixteen days from Cowes.

To sum up the situation, let us consider the following quotations from an 1850 letter by Charles H. Haswell, engineer-in-chief of the U.S. Navy.

> Steamers WASHINGTON and HERMANN of the New York and Bremen line.
>
> These vessels cost, when first put in operation, respectively $252,000 and $360,000.
>
> Their frames are not solid below; they are built of white oak and pine. Their frames are only abut bolted, being treenailed at other points. . . .
>
> From the want of stability of these vessels (without even the spars of a naval or cruising steamer), it is necessary, in order to effect an Atlantic passage, that they should be immersed to a draught of water, when leaving port, equal to that of a razeed ship-of-the-line or a first-class frigate; and in order to attain a fair speed it is necessary, from their insufficient proportions and the incapacity of their boilers, to expend the enormous amount of from 55 to 60 tons of coal per diem of 24 hours.
>
> From their deficiency of construction, their want of suitable materials, their insufficient fastenings, and the want of experience in their design and direction, they have involved during the brief period of their existence, the expenditure of sums equal to their first cost.
>
> Many of the engine and boiler pipes of these vessels have been broken in their passages, from the working of their hulls.
>
> During the period elapsing from the first commencement of the service of these vessels, they have, by loss of time incurred in repairs and alterations, been prevented from the performance of one-third of the mail service for which they were designed, and for which their owners contracted.[20]

To summarize the Haswell passages omitted above, the *Washington* completed three voyages in 1847 and five in

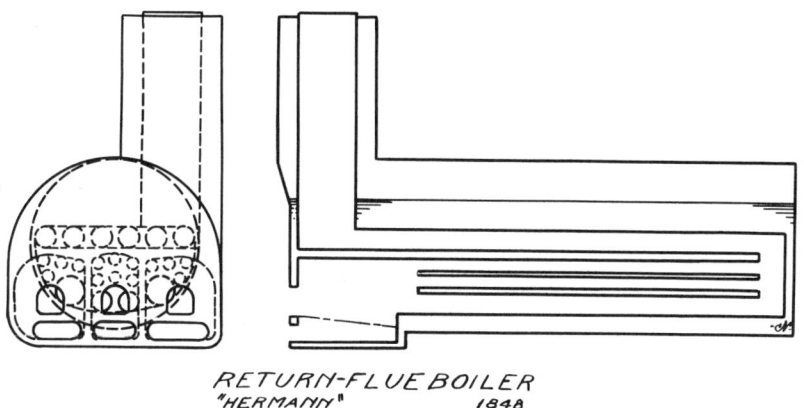

Original return-flue boiler for the *Hermann*.

1848 while the *Hermann,* starting in March of 1848, did only four. In 1849, when both steamers were available, the *Hermann* made five voyages and the *Washington* four. Furthermore, since the original contract was for four steamers and twice-monthly departures, the possible return from two ships at monthly sailings was reduced to $200,000. But, with nine voyages in its third year of operation, the line collected only $150,000.

One has to admire the management of the Ocean Steam Navigation for not abandoning a dismal situation. Neither did they appeal to Congress for more funds. Instead, they set to work to remedy, at their own expense, the defects in their two far-from-satisfactory steamships. It was not possible to install diagonal strapping in their all-too-flexible hulls; these had to be reinforced by additional timber inside the ceiling. The low speed and excessive fuel consumption were also attacked. The engines, after piecemeal repairs and improvements made as various parts failed, were in reasonable order and were definitely large enough. Although they never produced 2,000 horsepower, as announced before they ever turned over, a conservative estimate would show 1,100 IHP as a possible rating. Actual outputs had been 700 to 800. The boilers, however, could not supply sufficient steam, and what they did produce used inordinate amounts of fuel.

Erastus W. Smith, chief engineer of the Bremen Line, decided to try two different boiler modifications.[21] For the *Washington,* the Novelty Works, the original builder, was consulted. The interior furnaces and flues were removed and new ones substituted. Instead of three furnaces side by side, there were now nine in each boiler, located at three different levels. The firing was much more difficult, but there was 60% additional grate area. The flues were arranged in a complex fashion with the combustion gases traveling the length of the boiler three times, instead of two, and the stack was moved forward as a result.[22]

The *Washington* was out of service from November 6, 1849, until March 20, 1850.

For the *Hermann* an even more drastic solution was

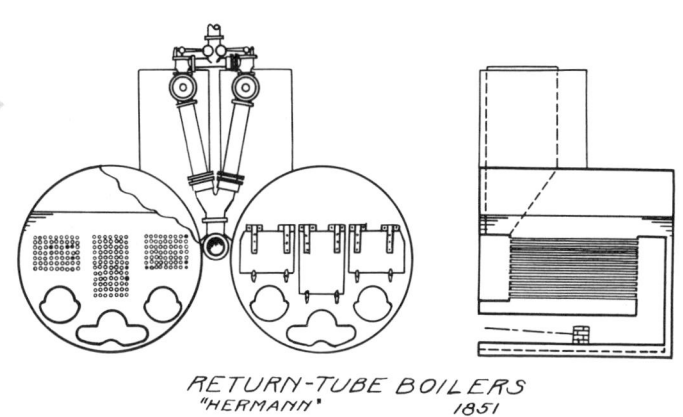

Modified return-tube boilers for the *Hermann*.

133

chosen and this time the firm of Mott and Ayres was employed. The two original boilers, each 36' long, were cut in two and four 14'-0" ones made from their shells, with new heads and interiors of Mr. Smith's design.[23] Three cylindrical furnaces per boiler led the hot gases to a combustion chamber, whence they were brought back to the uptakes by a series of horizontal tubes. As a fire-tube boiler it was somewhat similar to Ericsson's for the packet *Massachusetts,* but with the parts arranged more simply and logically. Two stacks were needed, one for each pair of boilers. The *Hermann* was withdrawn a year after the *Washington* and spent the winter from October 9, 1850, until March 29, 1851, undergoing changes. The forced-draft blowers, which had proved very troublesome, were eliminated from both boats. In addition to the boiler changes, there were a number of alterations to the effective paddle-wheel diameter; it was a simple matter to move the floats in and out without any structural modifications to the wrought-iron wheels. Various diameters were tried at different times. The largest was 39'-0" and the smallest 34'-8" for the *Washington*. The latter was the final one reached by 1851. In the *Hermann's* case, 36'-0" seems to have been the eventual value settled on. Different sizes and arrangements of floats were tried and, finally, single ones 7'-6" wide and 2'-8" deep applied to the *Washington* and 8'-0" by 2'-0" ones to the *Hermann*. There were 28 per wheel.

The appearance of the steamers changed radically. Originally it was impossible to tell them apart, even though the rather short stacks tucked in between the paddle boxes, together with black-and-white painted port strakes, made them distinctive steamers, different from anything else afloat. The *Washington* now had one stack, very far forward, the *Hermann* two, very close together, all much taller than the originals. They were painted red with black tops. Although the bark rig was retained, their masters usually stowed the main yards on deck, making them appear as barkentines.

Some engineering details of the *Hermann* were published after her alterations and are worth recording, since most numerical quantities of this kind are lacking.[24] When loaded she drew 21'-0"; after her fuel was burned she arrived at 17'-9" or an average draft of 19'-4" during a voyage. At the midway point she displaced (weighed) 2,645 tons of 2,240 lbs. For each additional inch of draft, the weight increased 17.4 tons, a number called by the naval architect "tons per inch." The ratio of the area of the waterplane to its surrounding rectangle, or "waterplane coefficient," was .776. The area of the midship section divided by its circumscribing rectangular area, the "midship section coefficient," was .909, a low value because of the eight-degree deadrise of the ship's bottom. The "block coefficient," or volume below the waterline divided by the volume of a rectangular block of the same length, beam, and draft, was .546, a measure of how fine the hull was.

The draft change of 3'-3" during a voyage combined

Hermann, with two stacks after her second set of boilers was installed in 1851. *Courtesy of Peabody Museum of Salem.*

with the tons per inch shows that 575 tons of coal, drinking water, and stores were consumed on an average voyage, New York to Cowes. Roughly 550 tons of this would be coal.

If we turn our attention to the speeds after modification, we find that the *Washington* had a small improvement, and the *Hermann,* with the more extensive changes, became so much faster that, for the rest of her career, she always crossed in a day's less time than the *Washington.* Her extra foot of beam was most helpful, for she carried less ballast to slow her down. Every ton of weight added to a ship increases her horsepower.

The *Washington* was tried out in New York harbor on March 15, 1850, and sailed five days later. She made a disappointingly slow voyage beset by bad weather. The owners and the engineers were nevertheless much gladdened by reports that the boiler modifications had cut her coal consumption almost in half and that 500 tons remained on board when she arrived at Cowes.[25] Returning, she did better, arriving at New York in 13 1/2 days, despite having to change course several times to avoid icebergs. A passenger list of 166 was another encouraging sign. A year later, after further boiler flue improvements and in the best of weather, she left Cowes on May 21, 1851, and reached New York on June 2 with a passage of 11 days, 22 hours, the shortest she had ever made.[26] At the time she had 170 passengers, 180 tons of freight from Bremen, and 70 more from France. If we add five hours for the longitude difference and assume the distance to be 3,120 miles, the average speed was 10.7 kts, a most unusual rate for the *Washington*. In 1852 her fastest crossing was New York to Cowes in 13 days, 4 hrs. 45 minutes, which would correspond to 9.8 knots.[27] The *Hermann's* best crossing was a westward one of 11 days, 10 hours 15 minutes, with a speed of 11.4 knots, another exceptional passage. In general the *Washington* made just under 10 knots and the *Hermann* just over 10 1/2 under the best of conditions. But in bad weather their speeds dropped rapidly. Before the changes, numerous 15- and 16-day crossings were at 8 to 9 knots, no better than the 1840 Cunard steamers.

Strangely enough, none of the many writings on transatlantic steamships have said anything about the freight carried. Speed and regularity of sailing immediately attracted shipments of small but costly items. Steamers could not take heavy weights or bulky products, but had space enough for valuables and charged high rates for transporting them.

Prior to 1850 Cunard had a corner on British goods of this character; furthermore, its home port, Liverpool, was a convenient one for shipping purposes. For Holland, Belgium, the German states, France, or Switzerland, all of which had products to ship to the United States, Liverpool was too far away, and had few steamship connections to the continental ports. The Bremen Line filled a vacuum. As soon as it was established, cargoes from the vicinity of Bremen were offered. In addition, French and Swiss goods, particularly the silks, satins, laces, jewelry, clocks, and watches for the expensive shops throughout the United States, were sent over by special steamer from Havre, only a few hours away from Cowes, and transshipped while English passengers were being loaded.

Even after another American line was opened direct to Havre, the volume of goods was such that the Bremen steamers still carried a considerable share. The two companies cooperated by staggering their sailings and aided rather than competed with each other in delivering a steady flow of continental luxuries.

As early as 1849 there were cargoes of 100 tons of German and 100 tons of French goods (these tons are always 40 cubic foot ones) worth $650,000. Three years later an exceptional shipment of 740 tons valued at $3,000,000 traveled on the *Hermann* when she arrived in January of 1853. Passenger numbers continued to grow with time. When the same steamer left Cowes on August 17, 1853, she had 244 passengers and so much freight that temporary shelters had to be erected on deck for both the animate and inanimate overflow.[28]

The Ocean Steam Navigation Company, beset by costly modifications, could not raise the necessary capital to complete a second pair of mail steamers. Their third hull had been ordered from Westervelt and Mackay, but it and the unfulfilled part of their mail contract were turned over to others, who set up the Havre Line. Since one company was an offshoot of the other, and since both made a stop at Cowes, they worked closely together.

By the start of 1852 the Bremen Line had finally achieved success in all its activities. Both its ships were strengthened and operable, and their machinery was made economical. Passengers and freight had increased so much that at times the westbound shippers had to reserve space a month in advance. That year, for the first time, a sailing was planned every month. Merchants on both sides of the Atlantic were gratified; the German states had increased their exports to this country by 200% since the line started.

Unexpected troubles, however, were still to arise. The *Hermann* left New York on February 28, 1852, in charge of an experienced harbor pilot, John Martineau, but ran aground in a heavy swell, knocking off the forefoot of the stem and starting a leak.[29] Because she was a mail steamer, she was allowed to use the new dock at the Brooklyn Navy Yard for repairs. She sailed a second time, on March 7, but four days out fractured the center shaft adjacent to the port engine, and had to return again. It was replaced and she was ready for the April sailing of the line. Unhappily, the twelve scheduled departures resulted in eleven completed voyages.

The strangest event of the year took place on the *Washington*, under Captain G. W. Floyd. She had left New York on March 27 and all went well until April 13, when she should have been about a day from Southampton. After lunch the captain declared that the noon position was in error; he changed course, and by nightfall was headed north with lights visible to starboard. At 2 A.M. the second mate stopped the ship to avoid running aground.[30] Captain Floyd had refused to alter course. Passing craft were hailed and the ship found to be north of the Cornwall-Devon peninsula in the Bristol, rather than the English Channel. Almost out of coal, she was ordered about and headed for Land's End, with the captain staggering around the deck in a frenzied state and the chief officer unwilling to take command. It took the combined powers of the chief engineer, the doctor, and four mates to persuade Captain Floyd that he had to put in to Milford Haven for coal. Sailing from there on April 16, the *Washington* again headed for Land's End and was narrowly saved from running aground when her helmsman disobeyed the captain's orders. The doctor declared Floyd out of his mind; he was overpowered and forced into his stateroom. From there he escaped, acquired a supply of liquor, tried to go on deck, but was forced below and confined, with waiters and passengers standing guard. At last Elisha M. Fitch, the mate, took over. At Cowes the American consul, warned by a passenger who had left the ship at Milford Haven, was waiting with a group of British physicians, who declared the master temporarily insane and removed him to an asylum ashore. The passage and its many detours had taken twenty-two days, the longest time ever for the *Washington*.

Mr. Fitch was retained in command and achieved fame two years later when his ship came to the aid of a Boston packet, the *Winchester*, in midocean. The latter had sailed from Liverpool April 9, 1854, with almost eight hundred immigrants crammed into every available space. Starting on the 17th, she was gradually dismasted by increasing gales until only the stumps of her fore and mizzenmasts remained above the deck. Her boats were swept away and the able-bodied passengers were put to work pumping. During the next ten days she met a number of small sailing craft, each of which took off as many passengers as she could, whenever the weather permitted such a transfer.

The *Washington* left Cowes on April 26 and sighted the *Winchester* on May 2. A gale was blowing, a high sea was running, and her hulk was rolling heavily, shipping water, trying to make headway under a jury foremast and a sail on the stump of the mizzen. For over two weeks the packet had been in a disabled condition, kept afloat by pumping; there were at least five feet of water in her hold. The weather was so bad that a boat sent by the *Washington* was dashed against the packet's side, but Mr. King, first officer, and four seamen were pulled aboard safely.[31]

Not until the next day could the rescue begin. All the passengers and fifty barrels of provisions were rowed across to the *Washington*. Meanwhile, the pumps having been abandoned, the *Winchester* was filling rapidly and sank an hour after the last boat left her. About 300 passengers had been taken off by the sail that fell in with the wreck before the *Washington*'s providential arrival. Taking on board 445 passengers and thirty two crew, the *Washington* found herself with about 775 persons to house and feed.[32] As a first effort, Captain Fitch had every house on deck cleared out to make room for the women and children. Awnings were spread to shelter the rest. Then all hands were set to work clearing the lower decks, the cargo spaces, and the storerooms. As the cargo and the baggage came up on deck, the rescued immigrants were moved down and, in twenty four hours' time, everyone was below decks except for the sick, whom the ship's doctor preferred to keep topsides. Seven deaths were attributed to exposure and exhaustion, but the remainder reached New York either May 13 or 14.

Upon the outbreak of the Crimean War in 1854, France and England began to charter transports to carry troops, horses, artillery, and supplies to the Black Sea. Cunard Line lost so many steamships to the war service that the New York half of its operations, begun in 1848 just as the Bremen Line was starting, had to be abandoned after December of 1854. The shorter route to Boston was maintained. The three American mail lines to Liverpool, Havre, and Bremen all gained freight and passengers. As a result, 1855 became the most successful year in the Bremen Line's history. On eleven voyages it carried 1,188 passengers eastbound and 2,027 westbound.[33] Cargoes of British origin increased markedly. Even the U.S. Post Office made a handsome profit; it collected $130,663 on letters and $268,623 on newspapers carried by the line, twice the annual subsidy.[34]

Only the European weather was adverse in that banner year. When the *Hermann* arrived at Southampton on February 13, 1855, she received word that the Weser was blocked by ice. The mail bags were sent across the Channel and overland, escorted by a special messenger who collected and brought back the Bremen ones.

The *Washington* did not receive advance word of a similar situation. She left Cowes December 20, 1855, found she could not enter the Weser and, while under a North Sea pilot, grounded off the coast on the 22nd but got off seven hours later. On her way back she encountered a gale in the Straits of Dover and collided with an unidentified sailing ship. She made her way into Southampton with a paddle box stove in. The combination of bad weather and Christmas delayed her getting into dry dock and held up her departure until January 6, 1856. If this voyage is added to the eleven actually completed in the calendar year, 1855 was the first in which the steamers actually sailed once a month, strictly in accordance with their mail contract, and the only time they were ever able to claim the full $200,000 subsidy.

December of 1856 severely tried the *Hermann's* officers and crew. She sailed from Southampton for New York on the 3nd and had covered 900 to 1,000 miles when the pivot bearing for one of her side levers suddenly gave way.[35] A heavy gale was blowing; the steamer was rolling and pitching so heavily that she could make only the barest headway. The sudden unloading of one side lever wrecked many other parts of the engine. Everything was shut down promptly, the cylinder disconnected from its crank, and the vessel started again with the remaining engine driving both wheels. She returned to Southampton on the 14th. The Havre liner *Fulton* took over her passengers and mails. The repairs proved long and costly. She left Southampton on February 12, 1857, to pick up the Bremen mails and returned, via Cowes, to New York with sixty-five passengers.[36]

The year 1857 started inauspiciously for the Bremen Line. The *Hermann* had been repaired at a high price; repairs always cost more abroad. Although the line's first five-year mail contract had been extended to ten, it would expire in June. Congress, disgusted by the general performance of the Collins Line in Liverpool, showed no signs of appropriating money for it or any other European mail contract. Moreover, the end of the Crimean War marked the return of scores of steamships to their original owners.

In June of 1855 it was reported that 123 steamships and 108 sailing ships totaling two hundred thirty thousand tons, a fleet of enormous size, were under British charter.[37] As these came home, interrupted routes were resumed and many new services planned. Cunard Line returned to New York, much to the regret of the American lines. The Hamburg American Line, heretofore a sailing-packet operation, put two iron screw steamers, the *Borussia* and *Hammonia,* into operation from Hamburg to New York in June and July of 1856. Not to be outdone, merchants of Bremen founded the North German Lloyd in December of 1856 and ordered four steamships of over two thousand tons apiece. Like the Hamburg liners, they were iron screw steamers built in England and Scotland.[38] These were to be ready in 1858. For the first time the Bremen Line would face competition. Furthermore, a British steamer, the *Queen of the South,* sailed from Bremen April 25, 1857, via Southampton, for New York.

On the American side of the Atlantic, Cornelius Vanderbilt, whose shipping operations from New York to Nicaragua during the early years of the gold rush had been shut off by revolutions and local wars, turned to the transatlantic trade and, in 1857, sent his steamships to Havre and Bremen. The *Ariel* inaugurated the Bremen operation when she sailed from New York on April 16 of that year.

Vanderbilt, in his usual manner, cut fares. The Bremen Line charged $130 in the upper cabins, $110 in the lower, and $60 in the second cabin (the word *class* was only occasionally used at that time), with the same price charged whether the passenger disembarked at Cowes or Bremerhaven. The Vanderbilt steamers, *North Star* and *Ariel,* quoted $80 for the first, $50 for the second, and $30 for the third cabin, which was actually steerage.[39] Vanderbilt was the first American operator to challenge the sailing packets, which had so far carried the growing immigrant trade.

With no new appropriations, all the postmaster general could do was to pay the Ocean Steam Navigation Company the actual postage collected on the American mail they carried. Revenues, however, declined disastrously as other lines entered the field. Worst of all, the company did not have funds to replace its aging ships. Its only choice, under the circumstances, was to complete the existing contract and go out of business. The expiration date was June 1, 1857, ten years from the day the *Washington* departed on her maiden voyage. The same vessel was sent on one trial voyage afterward, leaving New York on June 13 and Southampton on July 15. She arrived home on the 29th and two days later was offered for sale, together with the *Hermann.*[40] She had completed 54 voyages and the *Hermann* 47, making 101 for the line. The *Hermann* had embarked on one other in February of 1852 but had had to return to New York because of a broken shaft.

The Bremen Line had sent abroad the first pair of fully powered Atlantic steamships built in the United States and had established a new trade route to continental Europe via the English Channel port of Southampton. Both steamers were almost failures: the *Washington* lacked stability and, without diagonal strapping, proved so flexible that machinery fractures were common; the *Hermann* seems to have been a little more staunchly constructed but met similar strength problems. She had more beam, more stability, and, one suspects, better lines. Their machinery was identical but their original boilers produced too little steam for the engines at far too large a fuel consumption.

The Bremen Line rose bravely above their difficulties, strengthened and modified the *Washington* extensively to make her stable enough for safe operation, and rebuilt the boilers for both steamers, all at a great increase in cost and interruption of service. After these and other changes the *Washington* was still slow, the *Hermann* regularly making faster passages. While the latter was the better boat, she suffered greater misfortunes in engine failures later in her career.

For the last five years of the line's existence, the two steamships were taken for granted by the public, running with regularity and carrying excellent cargoes from Germany, France, and Switzerland, and passengers in numbers ranging from a few in the winter up to well over two hundred in the popular summer season, always more westbound than east. Yet, having opened and developed the German route, their last year saw three competitors starting and a fourth in sight. Simultaneously, their mail contract and its fixed payment per voyage ended.

A severe business depression began in the fall of 1857. Consequently, no purchaser was found for the *Washington* or *Hermann* until 1858, when they were sold at a June auction for $46,000.[41]

The newly created California, New York, and European Steamship Company advertised the *Hermann* as sailing on August 21, 1858, for California and continuing northward to Canada's Fraser River, where a new gold rush was developing. The *Washington* would follow on October 16.[42] The intended sailing date found the *Hermann* at her Moore Street pier with repairs not quite completed.[43] Blocking access to her deck was a heavy derrick that had been used to replace the tall stacks, both of which had been removed. Only a single gangway was clear and it was in continuous use, for stores were belatedly being loaded. Some five hundred passengers, many accompanied by friends, arrived and found it impossible to get their baggage on board. Passengers, stores, luggage, and bystanders jammed the pier in a hopeless state of chaos. Not until evening was an announcement made that the sailing would be delayed until the 23d.

Under Captain Edward Cavendy, her former commander and now one of her new owners, she did sail, but had only $300 aboard to finance the voyage. Arriving at Rio de Janeiro on September 18, coal and provisions had to be bought. Further coal was obtained at Lota, Chile, and Valparaiso, reached on October 24. At all ports purchases were made on credit and bonds on the ship left as surety.

At Valparaiso orders from New York dated September 6, 1858, were received, instructing the *Hermann* to call at Panama,[44] to pick up passengers from the *Washington*, which was supposed to stop at Aspinwall on her way to the Pacific.

By this time the orders were obsolete. The *Hermann*'s New York bills were unpaid and the California and European group had been replaced by an organization of even more doubtful probity calling itself The American Atlantic and Ship Canal Company.[45] The *Washington* was to leave New York on November 6 for San Juan del Norte in Nicaragua. The *Hermann* was to pick up passengers at San Juan del Sur (no mention was made of the fact that she already had 500 aboard).

On the 9th of November the *Hermann* reached Panama. Captain Cavendy found no passengers and no news of the *Washington*, which he expected to meet. There was some coal but, because the freight bill was unpaid, he could not take delivery. There was an agent of the American Atlantic and Ship Canal Company, of which he had never heard, with no credentials, no funds, a large, unpaid hotel bill, and a letter ordering the *Hermann* to San Juan del Sur. There were also private letters telling him of his own personal difficulties on account of the ship's debts. After he consulted with local suppliers, who refused credit, and with the American Consul, it was decided to send the vessel directly to San Francisco. What coal and provisions were still on board would be used. She headed north under Mr. Patterson, the first officer, while Captain Cavendy went to New York to try to stave off financial disaster to himself and his family.

The *Washington* did leave New York on November 7 and reached San Juan on the 18th. None of her 300 passengers was permitted to land, for the Nicaraguan authorities knew nothing of the *Hermann*, or any other craft, to take them on their way. For a week the *Washington* lay at anchor until news arrived that the *Hermann* had left Panama long ago, with no intention of making a Nicaraguan landing. Captain Churchill headed for Aspinwall on November 26, arriving there two days later. There he found himself in as awkward a position as Captain Cavendy had been earlier. The passengers held through tickets but there was no steamer to take them. There were no instructions. The line's agent, who had met the *Hermann*, had sailed on her and his hotel bill was still unpaid.

The Pacific Mail Steamship Company agreed to make a special fare reduction for the *Washington*'s exploited passengers. About two-thirds of them had enough funds to proceed on the *Sonora*, while the rest had to return to New York on the *Washington*. They disembarked, many in an almost penniless condition, on December 11, 1858. Completely unabashed, her agent, J. P. Yelverton, advertised a second sailing to Nicaragua and added, "If the HERMANN should fail to connect, we guarantee to take the passengers to California in the WASHINGTON."[46] When she did leave New York on December 26, 1858, it was for Nicaragua and San Francisco. The *Washington*'s passage was extremely long; it must have been marred by breakdowns. She did not reach San Francisco until July of 1859.

The *Hermann* arrived there on November 27, 1858. She was seized by the authorities for her debts in February of 1859, and sold by the U.S. Marshal to Captain George Wright for $40,000. After the *Washington*'s arrival, the Brazilian consul libeled her for unpaid bills at Rio de Janeiro.[47]

By March of 1860 the latter had been purchased by the Pacific Mail Steamship Company, which made essential repairs but used her only as a spare steamer on rare occasions.[48] She was subsequently cut down to a coal hulk and was at anchor in the Bay of Panama as late as May of 1865.[49]

Both steamships entered limbo once they reached the Pacific. They were not adapted for tropical service to Panama and were too costly to send on the northern run to the Columbia River. Most of their remaining careers were spent at anchor. By this time the *Hermann* had lost her bowsprit, clipper bow, and figurehead; the stem was now straight, with a forward rake.[50] She passed through several hands, making one trip to the Pacific Northwest and, later, in the winter of 1862–63, another to Panama for M. O. Roberts's service.[51]

The Pacific Mail Steamship Company planned a transpacific operation starting in 1867 and needed a stores ship at Yokohama. The *Hermann* was bought and overhauled for this purpose.[52] She sailed from San Francisco on March 1, 1867, and was used as a floating stage to which the great new 360' side-wheelers could tie up. Since she was still an operational steamship, a rare thing in Japanese waters, she was soon found more profitable in actual coastwise service and put to work in 1868. Even the feudal Japanese authorities chartered her, from time to time, as a transport. While on her way from Yokohama to the Straits of Sangar, she was wrecked, about seventy miles out, on Point Kawatzu. The date was February 13, 1869, and over half of the 350 Japanese troops aboard were lost.[53]

Despite her initial difficulties, the *Hermann*, the better of the Bremen liners, eventually demonstrated her builder's skill in strengthening a hull that was originally too weak. At the age of nineteen she steamed across the Pacific; at age twenty one she was carrying more passengers than at any time in her Atlantic career when an error in navigation brought her to an untimely end just when she had started a second career. The *Washington*, on the other hand, saw ten years of life for her original owners; she was too slow to find work thereafter. Her end is obscure. The 1865 report putting her at Panama is at odds with another stating that she was sold and scrapped at San Francisco in 1864.[54]

Notes

1. David Budlong Tyler, *Steam Conquers the Atlantic* (New York, 1939), pp. 142-49.
2. *New York Herald,* June 6, 1847.
3. Ibid., May 26, 1847.
4. B. H. Bartol, *A Treatise on the Marine Boilers of the United States* (Philadelphia, 1851), pp. 20-21.
5. *New York Tribune,* May 25, 1847.
6. *New York Herald,* June 2, 1847.
7. Ibid., July 7, 1847.
8. *Illustrated London News,* July 3, 1847.
9. *Merchants Magazine* 17 (July-December 1847): 357-64.
10. *New York Commercial Advertiser,* July 30, 1847.
11. *New York Herald,* September 16, 1847.
12. Ibid., November 10, 1847.
13. Ibid., June 6 and July 26, 1847.
14. Ibid., September 29, 1847.
15. Ibid., March 13, 1848.
16. Ibid., March 8, 1848.
17. Ibid., April 5-6, 1848.
18. Ibid., May 23, 1848.
19. Ibid., January 16-17, 1848.
20. House Executive Document 124, pp. 149-51.
21. Charles B. Stuart, *Naval and Mail Steamers of the United States,* 2d ed. (New York, 1853), pp. 139-40.
22. Bartol, *Marine Boilers,* p. 22.
23. Ibid., p. 36.
24. *Journal of the Franklin Institute* 28, 3d ser. (July-December 1854): 53-57, 121-28 (notes contributed by B. F. Isherwood).
25. *New York Herald,* May 3, 1850.
26. Ibid., June 3, 1851.
27. *Journal of the Franklin Institute* 25, 3d ser. (January-June 1853): 172-75. (notes by B. F. Isherwood).
28. *New York Commercial Advertiser,* September 1, 1853.
29. *New York Herald,* March 1, 1852.
30. *New York Tribune,* May 18, 1852.
31. *New York Commercial Advertiser,* May 15, 1854.
32. H. Parker and Frank C. Bowen, *Mail and Passenger Steamships of the Nineteenth Century* (Philadelphia, ca. 1930), p. 300.
33. *New York Herald,* January 1, 1856.
34. *Merchants Magazine* 34 (January-June 1856): 366.
35. *New York Tribune,* January 1, 1857.
36. *New York Commercial Advertiser,* February 28, 1857.
37. Ibid., June 14, 1855.
38. N. R. P. Bonsor, *North Atlantic Seaway* (Prescott, Lancashire, 1955), pp. 111 and 167.
39. *New York Herald,* advertisements, May 4, 1857.
40. *New York Commercial Advertiser,* July 31, 1857.
41. *New York Times,* June 18, 1858.
42. *New York Commercial Advertiser,* August 5, October 10, and October 14, 1858.
43. *New York Herald,* August 22, 1858.
44. Ibid., December 14-15, 1858.
45. *New York Commercial Advertiser,* October 20, 1858.
46. Ibid., December 16, 1858.
47. *New York Herald,* September 22, 1859.
48. Ibid., March 10 and October 8, 1860.
49. *San Francisco Alta California,* May 25, 1865.
50. John Haskell Kemble, *San Francisco Bay* (Cambridge, Md., 1957), p. 61. Picture shows the *Hermann* in the background.
51. John Haskell Kemble, *The Panama Route, 1848-1869* (Berkeley and Los Angeles, Calif., 1943), p. 230.
52. John Haskell Kemble, "Side-Wheelers across the Pacific," *The American Neptune* (January 1942), pp. 25-26.
53. *New York Times,* March 29 and April 22, 1869.
54. Kemble, *Panama Route,* p. 251.

8
The *United States*

ALTHOUGH the *Washington* and *Hermann* were the first outright steamships designed and built as Atlantic liners, a third, the *United States*, completed her maiden voyage just nine days after the *Hermann* finished hers. Of somewhat larger size, she had a higher horsepower and, in concept, was a more advanced design. Her origin is unclear. There were at least thirty three owners[1] listed when the vessel was completed and a number of statements published that she was designed for the New York to New Orleans trade.[2] Among the owners were Charles H. Marshall, of the Black Ball Line of sailing packets; Henry Chauncey, of the Pacific Mail Steamship Company; Charles Morgan, a long-time owner of steamboats on the East Coast and the Gulf of Mexico; Theodosius F. Secor, head of the iron works bearing his name; William H. Webb, her builder; and Captain William G. Hackstaff, her master.

Marshall was the largest shareholder and the person responsible for operating her.[3] The author does not believe the New Orleans rumors; Marshall had never engaged in that trade. As a master, agent, and owner, his entire career was spent with the Black Ball Line's Liverpool service. One possibility is that the New Orleans designation was to camouflage Marshall's real intent to try a steamship on the Atlantic in advance of his rival, Edward Collins of the Dramatic Line. By the time she was completed, the size alone made the *United States* an ocean steamship, not a coastwise one.

William H. Webb had built thirty vessels prior to the *United States*.[4] Only four were steam powered, three

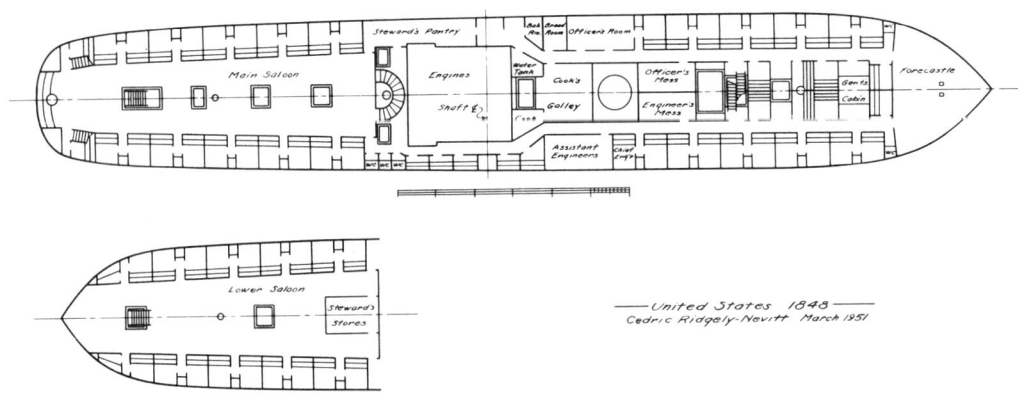

Cabin plans of the *United States*.

ferryboats and a 161' steamboat, the *Genil,* in 1845. Thirteen had been large packets of 1,000 to 1,200 tons, having lengths up to 174'. The *United States,* 244'-7" long by 40'-0" beam by 30'-10" depth, with a tonnage of 1,857, was a great advance in every respect when her keel was laid on January 30, 1847. Less than seven months later she slid down the ways into the East River at the foot of 6th Street on August 20 and was towed to the T. F. Secor wharf at 9th Street. On her spar deck she measured 256' long and, from billet head to taffrail, all of 277'.

She was strongly built and had five rows of 36" deep pitch pine keelsons to reinforce the bottom under the engines and boilers.[5] Three full decks ran from bow to stern and partial, or "orlop," decks were fitted forward and aft of the machinery. Coal was carried on either side of the engines and on the orlop decks; the designed capacity was 860 tons. Nine hundred measurement tons of cargo could be carried on the lower deck; if conditions warranted, part of the space could be converted into staterooms. Aft, on the same deck, was a 70' long cabin for dancing and public entertainment. Directly above it, on the main deck, was the dining saloon, with the best available staterooms port and starboard. At the bow were the sailors' quarters and, forward of the boilers, a cabin with further staterooms on either side. On the spar deck, proceeding from forward to aft, came the anchor windlass, a scuttle leading to the forecastle, and a house covering the stair to the forward cabin. Near the wheels were the mates' rooms, the fiddley (gratings to allow hot air to escape from the boiler room), a skylight, and the crew's galley. A larger house provided lounging space and the companionway to the after saloon. At the stern another house contained a second entrance to the saloon, the steering wheel, and four water closets for the gentlemen passengers. At odds with most of her contemporaries, the cabins and interior were finished in "plain style." Published accounts indicate 160 passengers but her cabin plan shows only 126 berths available.[6]

The *United States* represented a major advance over the *Washington* and *Hermann* in that she had a true steamship hull rather than that of a sailing packet with paddle boxes hung over the sides. The transom stern was replaced by a more seaworthy counter. For the first time on a steamship, the deadrise was reduced almost to zero and the bottom made practically flat. The bow was considerably sharpened, at the load water line, but the full deck line, fine stern, and maximum section well forward elements of a sailing ship were all retained. The hull was sturdy enough to resist the impact of the sea. The flat bottom was particularly important, for it put the engines low in the ship; as a result, there were no stability problems.

The direct successors of the *United States,* the Savannah steamships *Cherokee* and *Tennessee* of 1848-49 and the Pacific Mail's *California* and *Panama* of 1848, have already been described. Although smaller in size, they carried on and refined the design concepts formulated so successfully by William H. Webb for his first steamship.

The *United States'* machinery consisted of a pair of side lever engines of 80" bore by 9'-0" stroke and four return-flue boilers designed and built by T. F. Secor and Company.[7] The intended steam pressure was 12 psi, the engines cut off (stopped admitting steam to the cylinders) at 4'-6" stroke, and the wheels turned at 12 rpm. The latter were 34'-8" in diameter with two paddles 8'-4" by 1'-8" mounted on opposite sides of each of the 28 arms. Split paddles were stylish in 1848. Fuel consumption ran about 48 tons per day.

It is likely that the *Crescent City*'s single engine was a duplicate of the two for the *United States;* all three were built concurrently.

Webb recorded that the ship, with 835 tons of coal aboard, drew 14'-10" at both the bow and stern.[8] Filling the boilers brought the draft up to 15'-0". When fully loaded with stores, fresh water, passengers, and cargo, the maximum draft would be 17'-0", but the average in operation was in the vicinity of 16'. By skillful design Webb had managed to produce a steamship of the same beam as the *Hermann* but with four feet less draft. The *United States* was neither burdened by ballast nor slowed down by excessive weight. There was no iron strapping, but she was so soundly constructed that she did not need it.

A barkentine rig had topsails on the fore and mainmast. There were long poles, however, that would permit the addition of topgallants if her master felt that they were necessary. Gaffs for spencers were fitted on the fore and main. She was a nicely proportioned steamer with two light-colored stripes along the hull. Furthermore, the paddle boxes were painted white above the sheer line. These features tended to reduce the high-sided appearance resulting from her large depth and low draft. Surviving plans show her after Webb had cut her down a deck and converted her into the German warship *Hansa.* The author has redrawn these and has reconstructed the original appearance from the *Hansa* material and an Endicott lithograph showing her as built. The cabin plans are the earliest surviving ones for any American passenger liner. Those for the *Massachusetts* were drawn after she became a naval vessel.

On February 26, 1848, the *United States* steamed down the East River for her first trials.[9] Off Staten Island she met Cunard Line's *Hibernia,* outward bound for Liverpool. She passed her and continued out to sea for seven miles. The engines turned up to 14 1/2 rpm and all worked well. A second trial followed on March 14 but this time 120 guests came along. The start was marred by a loose key somewhere in one engine, which forced the ship to

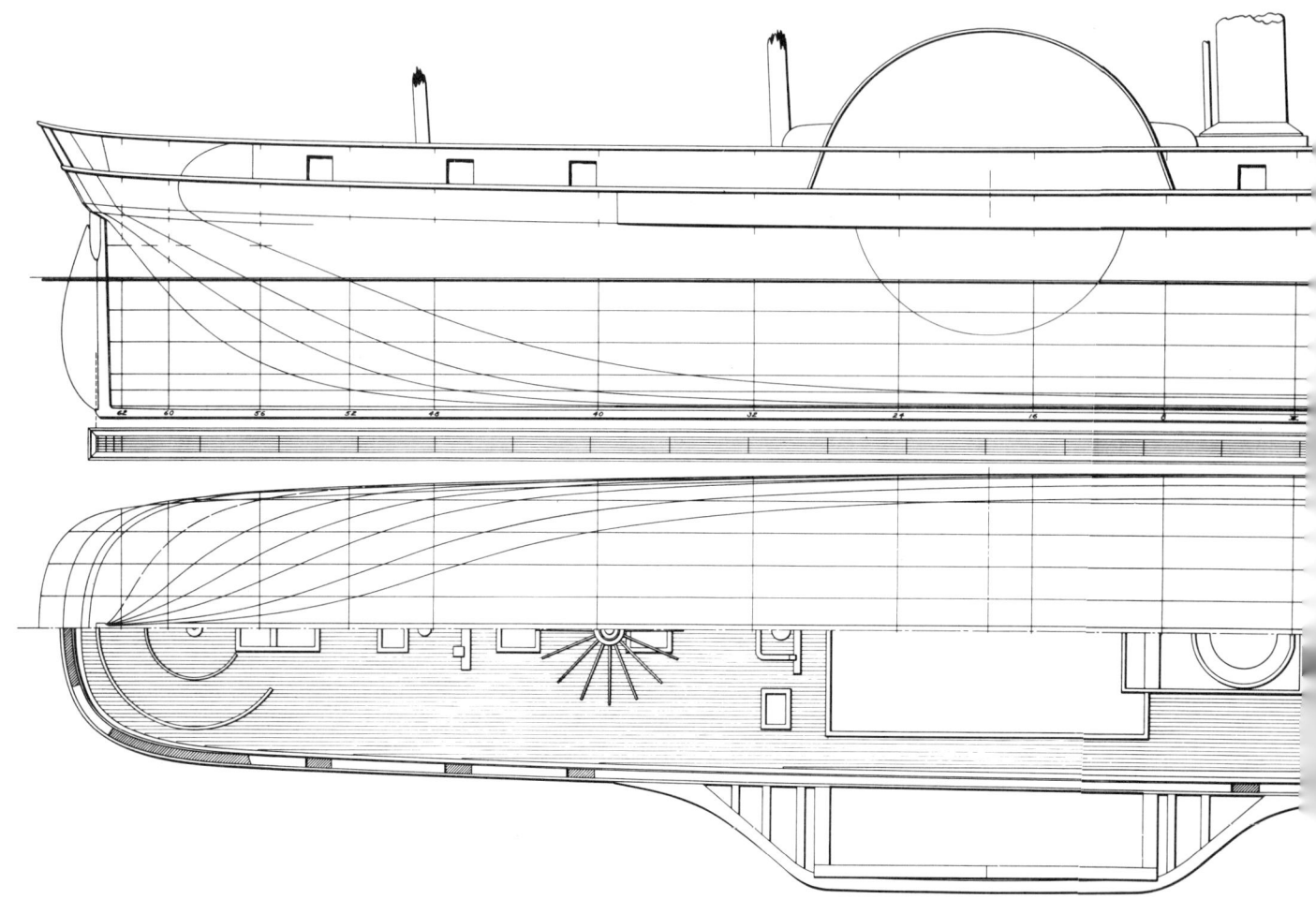

Hansa, formerly *United States*, 1848, lines and deck.

anchor off Hoboken until it could be adjusted.[10] From the Sandy Hook lightship Captain Hackstaff set a southeast course for about seventy miles. Returning, she headed directly into a "double-reefed topsail" breeze and made about 9 knots. The guests were pleased that no spray came over the bow, although the shrouds and bulwarks aft of the paddle wheels were covered with ice where spume fell on the ship. The *United States* docked at Peck Slip on the evening of the 16th just as her passengers were served a gala dinner to celebrate her success.

The first sailing was announced for April 8, at a fare of $120 from New York to Liverpool.[11] Captain Hackstaff, a veteran of twenty years on the Atlantic, commanded and, below decks, Chief Engineer Spencer supervised a staff of six assistant engineers, twelve firemen, and twelve coal passers. The crossing was uneventful for the forty six passengers on board. The hull was so tight that it did not have to be pumped out at any time during the voyage, and the engines worked perfectly. It took thirteen days and twenty hours from Sandy Hook to an anchorage off George's Dock at Liverpool. The speed was low, 9.3 knots, but the bottom of the ship was badly fouled from lying in the East River, uncoppered, for eight months.

While at Liverpool she was docked, the bottom scraped, and copper installed. Her presence was a most unwelcome one to the Cunard officials, who cut rates so that she would receive no freight or would carry it at a loss. Instead of £7 per ton, freight was quoted at £4 for the *Hibernia*, and £2-10 for the *Niagara*'s maiden voyage, to start on May 20, three days after the *United States* would sail. Thereafter, normal rates were reinstated.[12] From Liverpool on May 17, with forty three

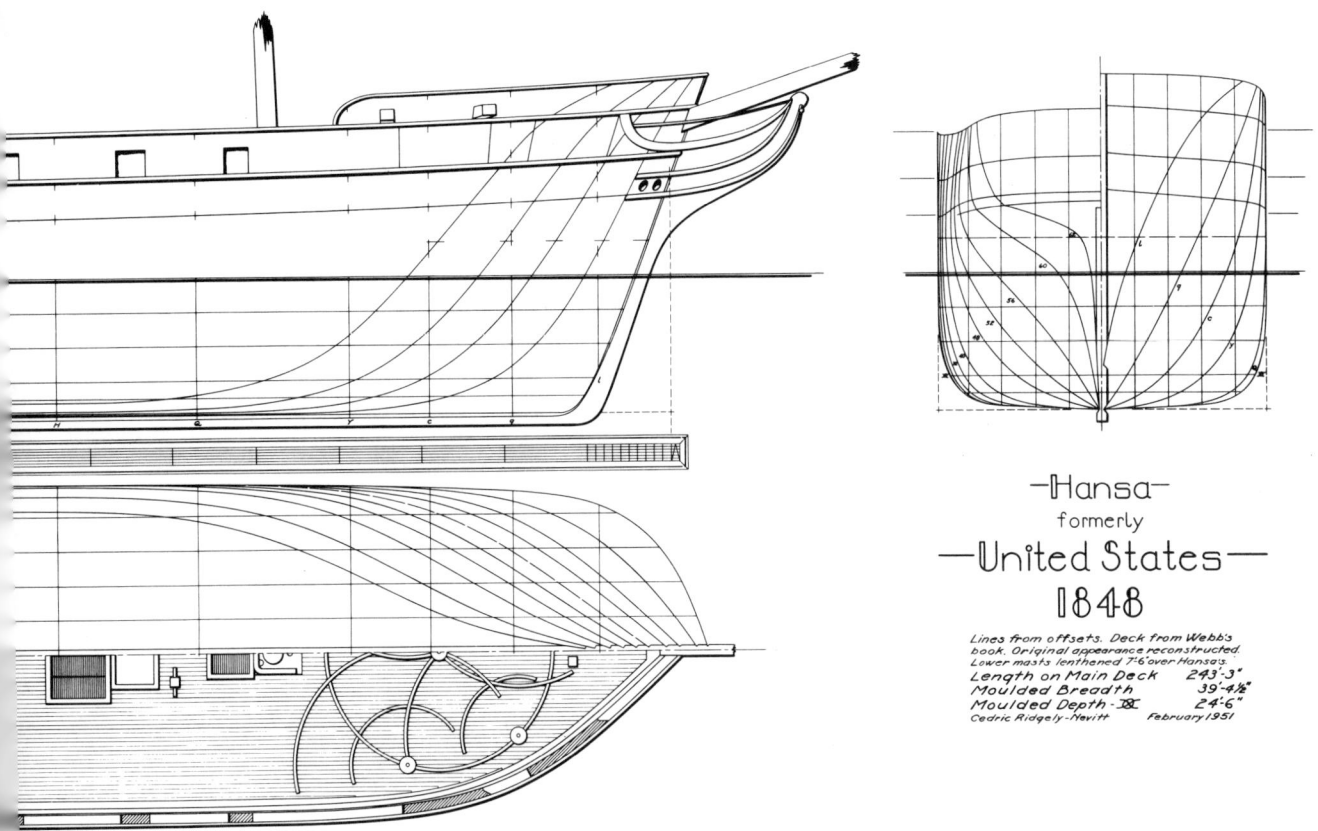

Hansa
formerly
United States
1848

Lines from offsets. Deck from Webb's book. Original appearance reconstructed. Lower masts lengthened 7'6" over Hansa's.
Length on Main Deck 243'-3½"
Moulded Breadth 39'-4½"
Moulded Depth IX 24'-6"
Cedric Ridgely-Nevitt   February 1951

passengers, the *United States* made an excellent first day's run of 274 miles, but then met heavy gales from Cape Clear to the Grand Banks of Newfoundland; she reached Quarantine in exactly 14 days.

In the face of such cutthroat competition, Liverpool was abandoned and Havre substituted for her second voyage. Departing from New York June 10, she made Cowes thirteen days later. At Havre 112 passengers were loaded, together with the largest cargo ever shipped from there for New York. Leaving on July 12, she was off Sandy Hook 12 days 16 hours later. The average speed, 10.4 knots, was a very good one for 1848, especially so since she plowed into head seas for eight days.[13] Citizen G. Poussain, the ambassador from the new French Republic, seized this golden opportunity to travel by steamship.

So far, the *United States* had performed admirably, free of the failures that beset the *Washington* and *Hermann*. But on her third voyage ill fortune caught up with her. She made an outbound passage of 12 days 9 hours to Southampton, then steamed to Havre, where she took aboard eighty passengers and a large cargo. Homeward, she touched at Cowes on September 3 and had gone as far as the Scillies when a defective casting in one of her condensers gave way. It supported the large boss on which a side lever pivoted. The damage was so serious that she put back to Southampton. Part of her passengers were transferred to the *Hermann* and part to the *Acadia*. Smith and Ashby of Southampton made the replacements. On October 12 a second departure from Cowes was made and New York reached on the 26th, after a long delay because of fog.[14]

By this time the owners were finding that a single steamship sailing at intervals of about two months was a

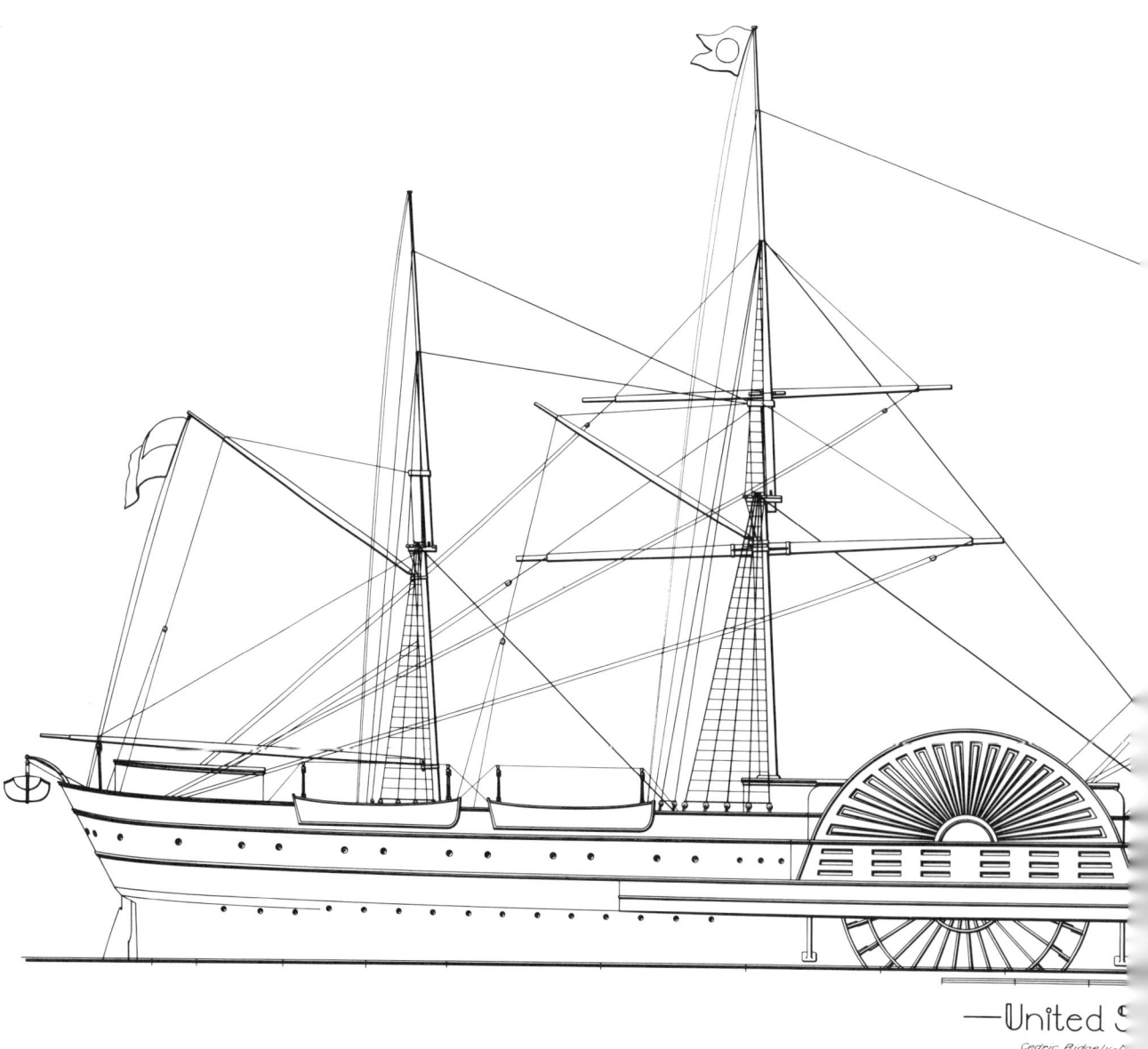

United States, sail plan.

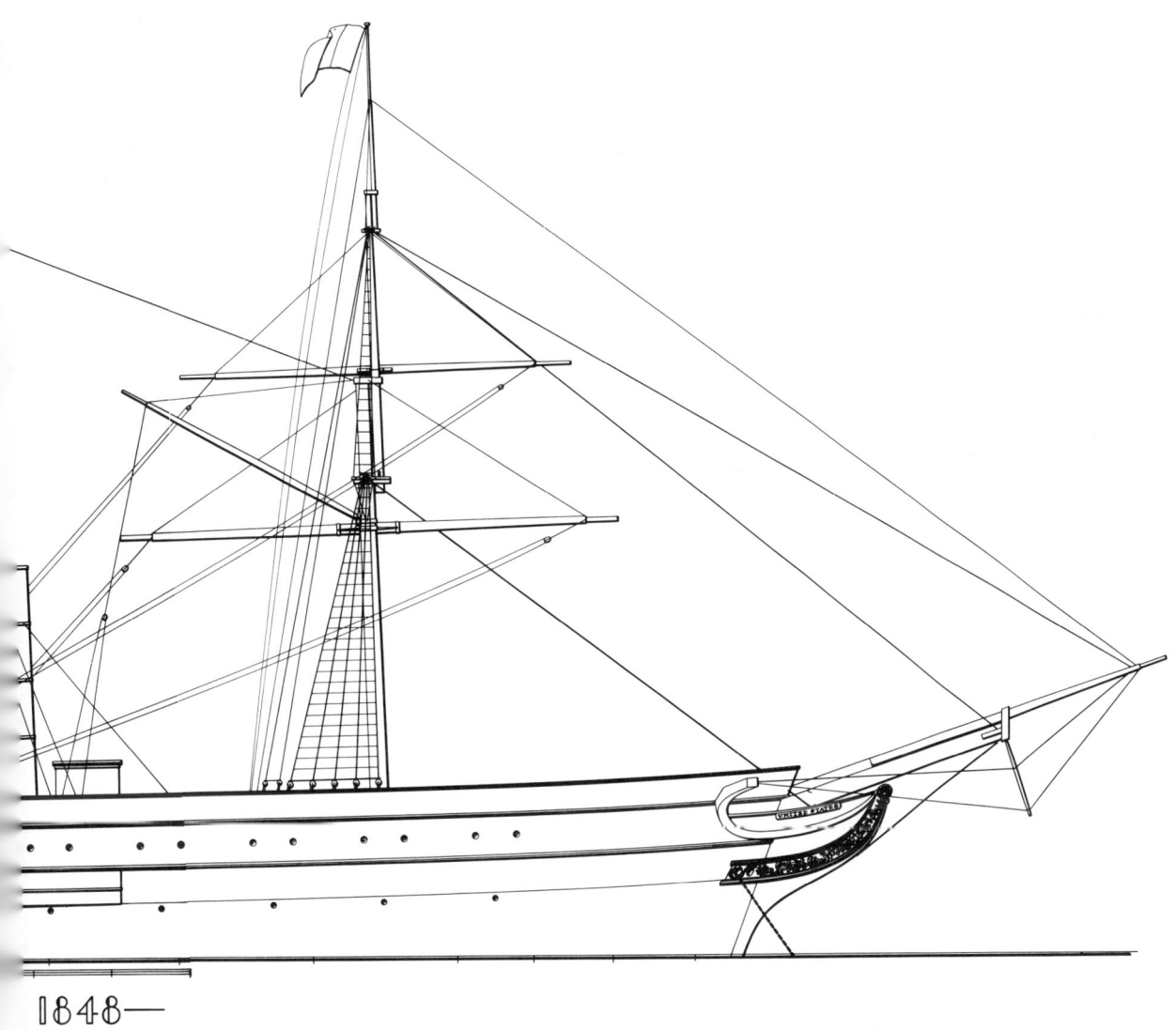

costly venture. Before her, only those British and American lines having mail subsidies had been able to survive. As a further experiment, the *United States* tried a voyage to New Orleans, a route that had already acquired two steamers earlier in the year, the *Falcon* of 891 tons and the *Crescent City* of 1,291, both smaller and less expensive to operate. Marshall advertised fares of $75 first cabin and $20 steerage. The *United States* sailed on November 4, via Havana, and reached New Orleans ten days later. Returning, she docked at New York six days eight hours from the Mississippi's pilot station at Balize.

A fourth Atlantic voyage followed. The *United States* left New York on December 5, 1848, and reached Havre on the 20th. Departing from Southampton on January 9, 1849, she met seventeen successive days of westerly gales. After passing the Grand Banks she ran into an even more severe storm and, simultaneously, floating ice that carried away the buckets on her wheels and damaged a paddle box. She put into Halifax for repairs and badly needed coal. Leaving February 3, she arrived at New York on the 5th. Her seventy passengers were much relieved to debark after twenty seven days battling the elements. As always, there was an excellent cargo and it came through unscathed.

Although further sailings had been advertised for March, May, and July of 1849, her owners received an offer for the steamer at so tempting a price, $265,000, that they could not turn it down. There was no great financial return in sight from her continued operation.

The year 1848 had been a time of political upheaval throughout Europe. One outcome was the formation of the Confederated German States, which were trying to set up an overall form of government to replace many individual, totally independent units that have since been combined with others into the country we now call Germany. A parliament met at Frankfurt and authorized the formation of a North Sea fleet under Admiral Rudolf Brommy. Its headquarters was at Brake, on the Weser River.[15] Most of the vessels were secondhand steamships bought in England and converted for naval use.

The *United States* was purchased in February of 1849 and a contract made for Webb, her builder, to convert her into a steam corvette. The administration of President Polk had cooperated with the German States, had allowed some of their cadets to be trained on board the *Saint Lawrence* of the U.S. Navy, and had ordered the Brooklyn Navy Yard to assist Webb with the conversion.[16] When Zachary Taylor was inaugurated a few weeks later, the official attitude was reversed and the Navy Yard ordered to stop.

Webb removed the uppermost of the hull's three decks, cut gunports in the sides, and reduced the rig. She was armed with three 25 centimeter (10″) pivot guns, and eight 20 centimeter (8″) howitzers.

Although the work was completed in less than three months, delivery was another matter. Because of the war over Schleswig-Holstein, Mr. Bille, Denmark's Minister to Washington, objected to the work done on the *United States* and the possibility of her sailing as an enemy warship. A bond was posted to assure that no hostile purposes were intended, and an American crew assembled under Captain Nathaniel Palmer. On May 31, 1849, she left New York, still under the American flag and still under the name *United States*. After her arrival in England, she was turned over to her German owners, armed, and renamed *Hansa*. She became the flagship of Admiral Brommy's short-lived fleet of two sailing frigates and nine steam-powered vessels.[17]

Neither the officers nor the crew shipped for the crossing proved satisfactory. Less than twenty four hours out the *United States* ran aground on Nantucket Shoals.[18] There she remained until seventy tons of coal were thrown overboard to lighten her. After that she proceeded but, owing to faulty operation, damaged one boiler so badly that it had to be taken out of service. On June 16 she arrived at Liverpool, where the boiler was repaired.

The Confederated States proved such an unstable association that its Brake vessels were soon put up for sale. Six steamers were bought in 1852 by England's General Steam Navigation Company[19] and two more by a pair of Bremen firms, W. A. Fritze and Carl Lemkuhl.[20] The latter were the *Erzherzog Johann*, formerly the Cunarder *Acadia* of 1840, and the *Hansa*, ex-*United States*. Both were rebuilt as passenger liners and the first renamed *Germania*. Bremen had been impressed by the Ocean Steam Navigation's service and decided to enter the field with steamships of her own. The two were ready by the summer of 1853 and their sailings announced with grossly exaggerated sizes—3,000 tons for the *Hansa* and 1,600 for the *Germania*—at fares of $100 for cabin and $40 for steerage.[21] The *Germania* sailed from Bremen August 2; the *Hansa*, under Captain Geerken, followed on the 30th, arriving at New York on September 20 with 469 immigrant passengers. Eastbound, she cleared on October 3, stopped for coal at Southampton, and reached Bremerhaven on October 20, 1853. A second voyage started October 29 and delivered 359 passengers to New York on November 17. Advertisements stated that the two would provide monthly sailings, but the long sea route and the *Germania*'s low speed made such an operation impossible. On November 29 the *Hansa* cleared New York for Bremen and the *Germania* followed four days later. Both were laid up for the winter.

The *Germania* again led the way in 1854, and was followed by the *Hansa* from Bremen on July 21. Although crowded with 659 passengers, the fifteen-day passage was considerably shorter than the ones made by the *Washington* or *Hermann*. One more voyage was made by the *Hansa* from Bremen September 22 and returning there on November 3. In a life of six years, the *Hansa* had made only eight Atlantic voyages in commercial service, four under the American flag and four under that of the City of Bremen. Her career was interrupted by the Crimean War. No doubt Fritze, Lemkuhl, and Company found a charter as a transport financially more rewarding than two voyages a year in an emigrant service to the United States. On March 24, 1855, she left the Thames for Genoa by way of Spithead and Gilbraltar.[22] After she reached Marseilles April 20, the author has found no record of her except that she had returned to England prior to November 18 and was turned back to her Bremen owners.

In 1857, while the *Washington* and *Hermann* were making their last voyages, the *Hansa* reappeared to make one more round trip. In the meantime new boilers had been installed and the engines modified.[23] The machinery changes do not seem to have been successful; she took all of twenty three days from Bremen on April 9 to New York with 505 passengers. She stopped at Boston short of coal and landed an additional seventy eight immigrants there on April 29. Eastward she did better, leaving New York on May 7 and arriving at Bremerhaven in seventeen days with a passenger list of 93.

It was obvious that the *Hansa* could not compete with the iron screw steamships being built for the new Bremen and Hamburg services. But once more military affairs came to her rescue. This time transports were needed to help quell the Indian Mutiny of 1857. On October 10 Captain Geerken and his ship arrived at Cowes for a charter to the Honourable East India Company. After its completion she passed into the hands of a British Member of Parliament, John O. Lever, who renamed her *Indian Empire*.

Mr. Lever was somewhat the counterpart of Edward K. Collins in that he had grandiose dreams of a steamship line carrying the mails at high speeds (supported by an equally high subsidy). Lever proposed sailings from Galway, Ireland, by way of St. John's, Newfoundland, to Portland, Boston, or New York. Galway had, since 1851, been connected by railway to Dublin, which could, in turn, be reached by Irish Channel steamer and rail from London. While no one really wanted to travel to the foggy coasts of Newfoundland, that island had achieved fame since it had been joined by submarine cable and telegraph to New York. In theory, news could travel speedily to Galway, then across the shortest possible stretch of water to St. John's, and, finally, by wire to New York. In practice, the combination of two long train journeys and a channel crossing was avoided by any knowledgeable English traveler, who could deliver himself and his baggage with far greater ease to the water's edge at Liverpool or Southampton. The less-demanding Irish emigrant, however, was pleased to sail from a port near home. Moreover, the rising tide of emigration could fill all the steerage berths made available. Galway harbor had little to offer; any steamship had to anchor well out from shore and have its passengers and freight lightered out. There were no shipyards or dry docks for repairs and no covered piers for loading tenders. The poor would tolerate the inconvenience.

Mr. Lever and his associates, "a company of English and Irish gentlemen, most of whom had little knowledge and, certainly, no experience of the business they were about to undertake," had to organize and operate a line before Parliament would consider a definite mail contract.[24] Mr. Lever's *Indian Empire* was chartered to make the first voyage. She steamed to New York, via Halifax, without a Newfoundland stop. Not to be outdone by its German predecessors, the Atlantic Steam Navigation Company advertised the *Indian Empire* as a steamship of 5,000 tons.[25] Reconditioning had been carried out at Southampton. She left there on June 13, 1858, but the two local pilots taking her into Galway managed to run her onto a submerged rock. She was freed, but her sailing was delayed until June 19, 1858, after it had been ascertained that no serious damage was done.[26] On board were a mere eleven passengers. On her way across she sustained a broken piston rod and, after an eight-hour delay, was able to continue on one engine. As a result, she took seventeen days to New York with a stop at Halifax. Evidently engine repairs were made at the American port, for she returned to Galway, sailing from New York on July 23, in thirteen days, but was then sent back to Southampton for further work. She arrived August 9 and was docked to inspect and repair her bottom.[27] Her second Irish voyage was almost as slow as the first. Leaving Galway on the 28th of September, she took sixteen days to New York but had been held up by the weather at Halifax and, to compound her difficulties, had run short of fuel and was forced to put into Newport, Rhode Island, for coal. Because of fog off Sandy Hook she could not dock until October 14. Two hundred fifty passengers had sailed from Galway, one hundred five of them for New York.[28]

The *Indian Empire* could never boast the luck of the Irish. She cleared New York for her final eastward passage on October 23, encountered engine problems, put into Halifax for "adjustments," and proceeded on the 30th. Then, about 350 miles from Galway, she encountered successive storms that forced her to lay to for seven days with almost no progress. She had completely exhausted her coal, hence cotton from her cargo, woodwork, and spars were burned to reach Broadhaven, Ireland, on November 26.[29] For a line that was promising rapid delivery of passengers and mails, the *Indian Empire*'s showing was less than ideal. Other steamships, both screw and paddle, were substituted. For the next three years nothing further is known of her. In 1862 she was purchased by Z. C. Pearson of Hull, who brought her to the Thames for repairs. When at anchor on July 23, off Deptford, she caught fire and was badly gutted; both decks and masts were gone, but the bowsprit, paddle boxes, and stack remained.[30] The sunken wreck was raised; five years later it was reported that her hull had been lying in Victoria Dock in London for several years and, finally, sprang a leak and sank on May 4, 1866.[31]

The *United States* was a steamship whose performance was adversely affected by her owners, most of whom were firms of limited financial resources and little experience in the management of steamships. Her active career, ten years, saw her under three flags as an Atlantic liner and under one as the flagship of the short-lived North German Navy. During all this time there were four voyages, as originally built, to Liverpool and Havre; five, after her second reconstruction, from Bremerhaven, usually via Cowes, to New York; and, finally, two between Galway and New York. In addition there were an ocean crossing as a naval vessel manned by a civilian crew, a round trip from New York to New Orleans, and two considerable periods of transport duty. Eleven and one half Atlantic voyages are little to show for over a decade of life. Most of her career was at anchor awaiting an opportunity to earn her keep. Only during her earliest months, under the American flag, was she continuously employed. Her most successful work comprised three Havre voyages as a carrier of high-priced cargoes and many first-class passengers. But, lacking a mail subsidy, she could not operate in the winter without a substantial loss. Her endless lay-ups were bad for her machinery, which deteriorated between operations. Her highly innovative hull made her an excellent sea boat, strong, dry, able to survive in any weather. If the Ocean Steamship Company could have traded the *Washington* for the *United States*, it would have had a happier existence and a far superior steamship.

Notes

1. William H. Webb, Certificate Book, vol. 1.
2. *New York Commercial Advertiser*, March 13 and July 6, 1847.
3. *New York Herald*, Marshall Obituary, September 24, 1865.
4. Henry Hall, *Report on the Shipbuilding Industry of the United States* (Washington, D.C., 1882), p. 117, gives a complete list of all the vessels built by Webb.
5. *New York Herald*, March 16, 1848.
6. William H. Webb, *Plans of Wooden Vessels*, (privately printed and distributed by the author, New York, ca. 1895), vol. 1.
7. B. H. Bartol, *A Treatise on the Marine Boilers of the United States* (Philadelphia, 1851), p. 24.
8. Webb, Certificate Book, vol. 1.
9. *New York Herald*, February 27, 1848.
10. Ibid., March 16, 1848.
11. *New York Commercial Advertiser*, March 14, 1848.
12. Harnden & Company Circular, quoted in the *Nautical Gazette*, August 6, 1903, pp. 82-86.
13. *New York Commercial Advertiser*, July 26, 1848.
14. Ibid., September 25 and October 26, 1848.
15. *Schiffahrtsmuseum der Oldenburgischen Weserhafen* (Oldenburg, Germany, ca. 1968), pp. 20-26.
16. *New York Herald*, May 11, May 24, and June 19, 1849.
17. Caspar F. Goodrich, "America's Part in Founding the German Navy," U.S. Naval Institute Proceedings, (1924), pp. 208-13.
18. *New York Herald*, June 29, 1849.
19. L. Cope Sanford, *A Century of Sea Trading, 1824-1924, The General Steam Navigation Company Limited* (London, 1924), p. 53.
20. Charles E. Lee, *The Blue Riband* (London, ca. 1930), p. 61.
21. *New York Commercial Advertiser*, June 25, 1853.
22. *London Times*, March 27 and November 18, 1855.
23. *New York Herald*, May 5, 1857.
24. W.S. Lindsay, *History of Merchant Shipping* (London, 1876), 4:264.
25. *New York Herald*, advertisement, June 26, 1858.
26. N. R. P. Bonsor, *North Atlantic Seaway* (Prescott, Lancashire, 1955), p. 161.
27. B. W. Ginsberg Records, a series of manuscript volumes once owned by the late Frank C. Bowen at Gravesend, England.
28. *New York Herald*, October 11 and 13-15, 1858.
29. *New York Commercial Advertiser*, December 17, 1858.
30. *London Times*, July 25, 1862.
31. *Mitchell's Steam Shipping List*, May 11, 1866.

9

The Collins Line, 1850-1858

EDWARD K. Collins, as early as 1835, was the New York agent for a line of sailing packets in the New Orleans trade.[1] Two years later he founded the Dramatic Line to Liverpool with newly built sailing ships named *Garrick, Siddens,* and *Sheridan,* all of them larger, more luxurious, and, on the average, faster than any previous group to cross the Atlantic. Furthermore, the careful attention paid to the comfort of the passengers and the excellent standard of service on board made these the immediate favorites of all who sailed on them. In 1841 he proposed to the United States postal authorities the construction of four steamships to carry the American mails across the Atlantic.

Collins was a well-known New York businessman respected by his contemporaries, and he had a particular flair for impressing others with the audacity and magnitude of his schemes. He became the most successful press agent in the marine field, as well as a superb lobbyist in the halls of Congress. Together with several financial backers, he submitted a further proposal to the postmaster general on March 6, 1846, for a line of mail steamships sailing from New York to Liverpool twice a month for eight months of the year, and once a month during the winter, for the annual sum of $385,000.[2] A year later T. Butler King, chairman of the House Naval Committee, added the proposal, as a rider, to a bill for constructing four naval vessels, and he managed to steer the package (which also led to the founding of the United States Mail and the Pacific Mail Steamship Companies) through a lame duck session of Congress. It was passed on the last possible day, March 3, 1847, and approved by President Polk.[3]

Five steamships, each of at least 2,000 tons and 1,000 horsepower, were to be built, four at once and one to follow.[4] The first was to be ready by May 1, 1849. Construction was to be under naval supervision and they were to be readily converted to war steamers. The watch officers were to be nominated by the United States Navy, but the masters would be appointed by the owners. The contract was to run for ten years from the start of service at sea. It was signed on November 1, 1847, five months after the Bremen Line's first sailing. J. Y. Mason, Secretary of the Navy, acted for the government while Edward K. Collins, James Brown, and Stewart Brown were the other principals. The Browns were cousins representing the private banking firm of Brown Brothers. That organization became the financial backer of the New York and Liverpool United Mail Steamship Company, the name chosen for incorporation. The title was far too long; no one used it. Even the company's letterhead was abbreviated to "New York and Liverpool U. S. M. S. S. Co." Collins Line was its universal designation and, since its operation was left in Collins's hands, it was, and still is, an appropriate title. Its president, however, was James Brown.

The specified date for beginning service was unrealistic; Collins, through his Congressional friends, obtained a new act establishing June 1, 1850, in its place. He had submitted specifications for the steamships to the Navy Department in March of 1847, well before any contract was signed. Whoever prepared them was not competent to do so. The hulls were to be larger than anything afloat, 300′ long, 45′ beam, 33′ to 35′ depth of hold, and of 3,000 tons. Two engines of 100″ bore by 10′

stroke and 2,000 horsepower were to be installed. At the time, the *Southerner*, 196' long, of 1,100 tons with less than 400 horsepower, was the largest American steamer in actual service, and the *Washington*, 1,640 tons, the largest under construction. The proposed timber sizes were inadequate and the frame spacing, 4'-0", would produce a flimsy hull incapable of holding together. There was no iron strapping.

John Lenthall and Samuel Hartt, both experienced naval constructors, proceeded to straighten out the amateurish proposal.[5] They advised that a length on load water line over 270' was unsafe, as was any frame spacing over 3'-0". The bottom of the ship was to be filled in solid for its entire length, not the 100' to 150' proposed. Diagonal iron strapping 1" x 5" was essential. The sizes of most of the structural members were increased. Many extra knees were added to stiffen the hull. Collins protested against all these changes but, in later years, was to claim credit for insisting upon exceptionally strong, well-built hulls! The fact that naval professionals rescued these steamers from Collins's inferior design has gone unnoticed.

One of the apparently unsolved mysteries about the Collins liners is who did design them. John W. Griffiths claimed that David Brown, of the firm of Brown and Bell, was engaged to undertake the design, that Griffiths assisted, and that a half-model to show the hull's shape was made.[6] He stated that the project fell through because the first steamers were ordered from William H. Brown instead. The two Browns were unrelated. Since David Brown had retired from business in 1846, this is an unlikely story.[7] A second eminent name often quoted as the designer was George Steers, but this is backward thinking; Steers did not become famous until 1851, when he produced the yacht *America*. Three years earlier he was a skilled shipwright who had worked in various shipyards, but his design experience was limited to yachts, pilot boats, and small steamboats. It was most unlikely that he would be entrusted with the largest wood steamships yet contemplated. His real role is best described by a statement published in 1857: "In 1848, Mr. Steers was found in the yard of Wm. H. Brown, in the capacity of foreman. He laid down in the mold loft the lines of the steamship 'Atlantic,' the first of the Collins Line, and assisted in the direction of her building. It afterwards fell to his lot to lay down the ill fated 'Arctic' and assist as foreman in her construction."[8]

When the order for the first two steamships was placed with William H. Brown, that builder automatically became the designer. It was such an obvious matter to the professionals of Brown's day that no one ever bothered to make a specific statement of the fact. A first-rate builder normally designed his own ships. Collins showed remarkably good judgment in contracting with the foremost steamship designer in the United States. Brown, who had designed and built the *Southerner*, *Northerner*, *Crescent City*, and *Empire City*, would prepare a half-model and supervise the design. But, as a busy shipyard manager, he would employ a skilled loftsman, Steers, to expand the lines to full size on the floor of his mold loft. From them molds would be made to show the actual shape and bevel of each frame. Construction details were worked out on the first hull and duplicated in the second. No plans would be made and no record would survive except, perhaps, the half-model. But the thousands of half-models made by New York shipbuilders from 1800 to 1870 were just useless objects once Manhattan's wood shipbuilding era ended; a minute fraction of them still survives.

Since four extremely large hulls would strain the resources of any builder, only two went to William H. Brown. The others were awarded to Jacob Bell, who had previously been a partner of David Brown, adopted son of Noah Brown of the firm of Adam and Noah Brown. The latter firm had built the *Fulton* and *Connecticut* back in 1813 and 1816. Brown and Bell, their successors, had turned out both packets and steamboats with great success. The *Roscius* of Collins's Dramatic Line and the *Hoqua* for the China trade were two of their most famous sailing ships. Their steamship experience consisted of a pair of 687 ton, side-wheel warships built for the Spanish authorities at Havana.[9] Named the *Lion* and the *Eagle*, they were completed in April of 1841. Bell took over the yard when Brown retired.

Bell's two steamships had slightly different lines and slightly different dimensions, but, to the beholder, appeared identical with the William H. Brown pair and followed them in their general arrangement. The Brown steamships were to be powered by the Novelty Iron Works, while Bell's had Allaire engines. Originally, there were to be two engines of the side lever type having 95" bore and 9'-0" stroke but, after Collin learned that the *Asia*, then building for the Cunard Line, was to have 96" by 9'-0" cylinders, he had the second ship from each builder powered with 95" by 10'-0" engines and increased the boiler sizes accordingly. As a result, each of the four sets of machinery constituted an individual design. In reality, the Collins Line started its career with four closely similar steamships, but each one had unique hull and engine characteristics.

Steamship	Completed	Length	Beam	Depth	Tonnage	Builder
Atlantic	4/26/50	284'	45'-11"	22'-11 1/2"	2845	Brown
Pacific	5/24/50	281'	45'-0"	32'-3"	2707	Bell
Arctic	10/25/50	285'	45'-11"	22'-11"	2856	Brown
Baltic	11/16/50	282'-6"	45'-0"	22'-6"	2723	Bell

The register dimensions show that the Brown steamers were several feet longer and a foot wider than the Bell ones. The admeasurer for the *Pacific* chose the upper deck for his depth dimension, but, since depth was not used in the computation of tonnage, no size variation for the *Pacific* followed. One known difference in shape was that the *Atlantic* and *Arctic* had flatter bottoms.[10]

The real break from tradition was the bow shape; the stem raked slightly forward and the bowsprit was completely abandoned. The *California* and her sisters had true steamship bows but still kept spike bowsprits. The Collins ones went all the way. The only vestige of a sailing ship, deadrise, was below water where it could not be seen. The sterns were new in shape: they were short, with a minimum of overhang, and rounded without horizontal knuckles at the wales. Why the change was made is not clear; their counters were difficult to plank and, to the present-day eye, unattractive.

The final scantlings were very heavy and the timbers were put together with multiple fastenings and reinforcement of the highest quality. Each frame was made up of two floors 12″ wide and 20″ to 21″ deep, with the upper portions double 10″ timbers. They were spaced 30″ to 36″, made of oak, live oak, and chestnut. The spaces between the floors were filled solid from bow to stern and out to the bilges. The yellow pine planking ran from 9″ thick at the keel to 5″ minimum thickness and was edge bolted for extra strength.[11] There were 6″ by 1″ diagonal iron straps[12] four feet apart between the frames and the planking. The ceiling was yellow pine, 12″ thick at the bilges and 7″ elsewhere. Again, edge bolting was required. Keelsons supporting the engines were built up to 22″ x 42″ deep and, under the boilers, 22″ x 27″. The enormous beams supporting the overhanging guards were 22″ square. Fastenings below the waterline were copper, and all iron fastenings were galvanized to prevent rust and insure a long life.

The three continuous decks were called spar deck, main deck, and lower deck; orlop decks were added forward and aft of the machinery space. The dining saloon was forward on the main deck, 55′ long and the full width of the ship. Doors could be opened forward to lengthen it another 15′. The galley was overhead, on the spar deck, with dumbwaiters to the pantry below. No live animals were carried, as they had been on the *Washington,* but an ice house of 35-ton capacity provided cold storage space. In the pantry were taps fed by iced coils from overhead water tanks so that ice water was always available.[13] Low-pressure steam from the engine room was piped up for two purposes. First, a part of it was condensed to provide culinary water for the galley. Then, in cold weather, a second part was used to heat the public rooms. Winter passages, for the first time, became livable occasions on shipboard. A number of new systems were tried out. There were bell pulls to an annunciator board registering which stateroom rang. In the dining saloon was another device to summon waiters to the pantry when orders were ready. Since the steering wheel was at the stern, a telegraph was needed to signal orders from the bridge.

By the time these ships were ready, the New York daily papers had already exhausted their superlatives in describing the interiors of earlier steamships. The chief elements on the Collins liners that pleased everyone had been copied from the *Washington* and *Hermann.* These were the stained glass trunks carrying light and air from the spar deck skylights down to the public rooms on the two lower decks.[14] An addition was a brass trellis to protect the glass. The upper cabin on the *Pacific* was done in satinwood with mirrors all around and the lower one had white enameled pillars wreathed with bronze chaplets.

The dining saloon had hanging racks over each table to hold the passengers' wine bottles; between meals these were hoisted up to the carved and ornamented ceiling to keep them out of way (and, presumably, out of reach). Three thousand five hundred bottles were carried in the ship's wine cellar, a number working out to seventeen per passenger!

Accommodations for about two hundred first-cabin passengers were fitted, mostly in double staterooms, but connecting doors could be opened for family groups. Five especially large cabins were designated bridal suites. As a means of reducing the fire hazard from oil lamps, these were placed so that the passengers could not meddle with them. For the public rooms they were hung inside the skylight trunks. Each pair of staterooms was supplied with one in an overhead box having ground-glass panes to light the rooms, but access was reserved to the steward, who could reach it from an alcove between them. The spittoons had a properly nautical motive with tops shaped like sea-shells. When new they were sea green or sky blue.[15]

In spite of her straight stem, each steamship had a larger-than-life figurehead, usually listed as a "sea monster" in the ship's papers. The *Atlantic*'s was a Triton with a horn in his hand, ascending from the sea and supported on either side by a golden mermaid.[16] When first completed, all were three-masted barks of moderate sail area, having nothing above topsails on the fore and main. In 1853 or 1854 the mizzens were removed. The black hulls had a bright red decorative stripe from bow to stern at the level of the guards. The stacks were black with red tops. The overall appearance was original, all parts were well proportioned, and four remarkably handsome steamships were created.

It has already been noted that the machinery consisted of two side lever engines driving paddle wheels with steam generated by four boilers. Collins employed two Chief

Atlantic, original appearance. *Courtesy of The Mariners Museum, Newport News, Va.*

Engineers of the U.S. Navy, William Sewell and John Faron, Jr., to prepare engine specifications. Faron, in 1848, left the Navy to become the Collins engineering superintendent. He also prepared detail designs for all the boilers. The final engine designs, however, were worked out by the Novelty Works for the Brown steamers and by the Allaire Works for the Bell pair. For the latter, Charles W. Copeland, who had designed the first American side-lever engines for the U.S.S. *Mississippi*, was responsible. The *Atlantic*'s and *Pacific*'s cylinders were 95″ bore by 9′-0″ stroke and were 5″ larger in bore and 1′-0″ greater in stroke than the very large ones made for the *Ohio* and *Georgia* in 1849.

Details of the *Pacific*'s engines indicate them to have been conventional in their arrangement, with extensive Gothic ornamentation on the condenser, hot well, and steam chest castings.[17] The bedplates were 32′ long, 9′ wide, and weighed 40 tons.[18] Each served as a foundation, formed the bottom cylinder head, provided steam and water passages, acted as the lower part of the jet condenser, and had sockets for rugged wrought iron pillars supporting the shaft bearings. The condenser, in addition, supported large pivots carrying the side levers.

Contemporary sources state that surface condensers were contemplated,[19] but it is probable that the engine builders rejected the idea as impractical.

The main steam pipe was all of 24″ in diameter, even larger than the paddle shafts, which were 19 1/2″ and had a total length of 71′ port to starboard. There were three separate sections, each with its own bearings, since it was impossible for a flexible wood hull to keep a single shaft perfectly straight. The two outer shafts had bearings on the guards and on the outboard side of the engine. Connecting these was a center section with bearings on the inboard side of each engine and crank pins fixed in place. The latter were, however, slightly too small in comparison with the holes in the crank webs of the outer parts into which they were fitted. Such planned flexibility was essential on any side-wheel steamship.

The chief innovation was the use of Martin type water-tube boilers; these were essential to assure an adequate steam supply. But, because jet condensers were actually used, the boilers had to operate at low pressures to cope with the salt water continually pumped into them from the sea. The four boilers were rectangular in plan view, arranged close together so that they could be served by a single stack. Each had a steam chimney rising from it around one quarter of the stack's circumference. The forward pair were a little narrower than the after ones in order to fit into the tapering hull. Each had eight furnaces, four on a low level and four above, with the after fireroom at the forward end of the engine room and the

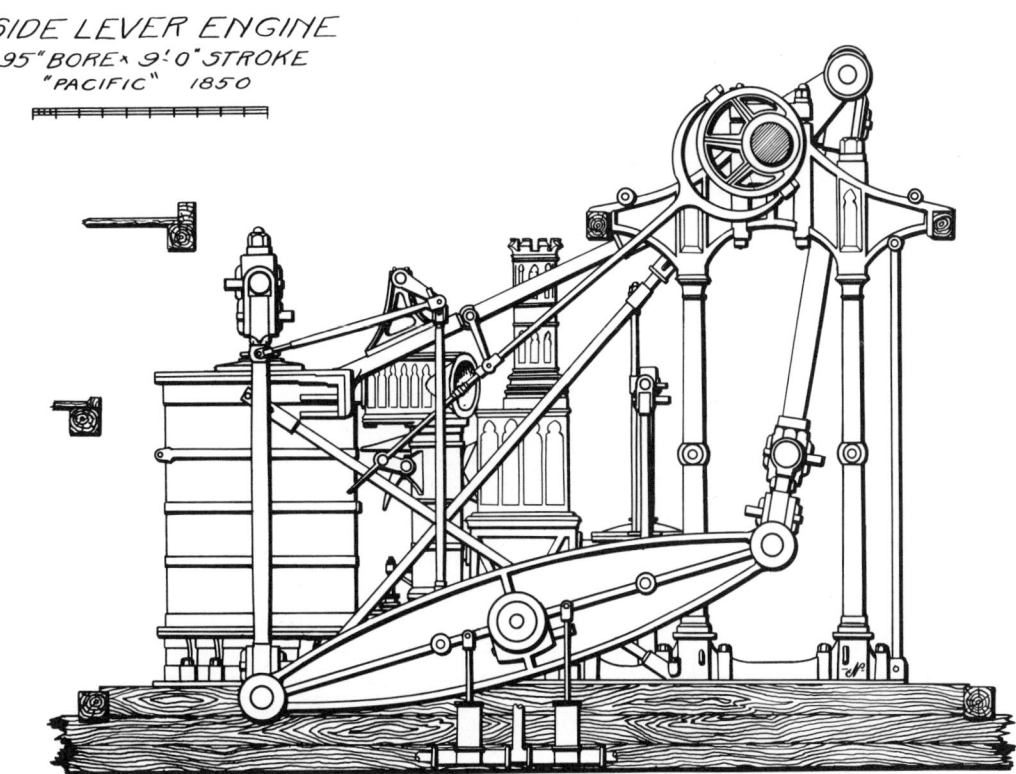

Side lever engine 95″ bore by 9′-0″ stroke for the *Pacific*.

forward one ahead of the boilers. The hot gases from the fires were carried through four large rectangular flues for the length of the boiler before going up the stack, and in each flue were hundreds of 2″ diameter vertical brass tubes with water inside.[20] Thus a very large square footage of surface was attained to transfer the heat to the water. Although designed for 14-psi steam pressure, a somewhat higher value could be attained. Anthracite coal was burned eastbound and bituminous westbound.

Two other novelties were introduced. Steam whistles finally made their appearance, although the ship's bell (every sailing ship had one) was still the standard signaling device on the ocean. In fact, one correspondent of the *New York Herald* stated that he had never heard the whistles used during several crossings that he had made.[21]

A more useful innovation was the engine-room telegraph, constructed by Charles Howland of New York, fitted with a gong to attract the engineer's attention and transmitting the following signals: "head," "back," "slow," or "hook on." The first three were used when starting and maneuvering. During these periods the automatic valve gear was disconnected and the steam and exhaust valves worked by the engineers with long hand levers called "starting bars" in order to provide absolute control and prevent the engines from stopping on dead center where they could not be started again. Once the vessel was well clear of the dock, "hook on" was the order for the engineer to connect the valve gear to the paddle shaft. After this was done it became possible to attain full speed ahead. Under hand operation only five to six revolutions per minute could be made in either direction.[22]

If the machinery differences between the steamers are listed, the major one was the increase in stroke from 9′-0″ to 10′-0″ for the last two, the *Arctic* and the *Baltic*. Such a change would produce an 11% increase in horsepower but would add only 4 percent to the speed. The boilers had to be widened and equipped with more tubes. Naturally, they were heavier. Their dimensions were 22′-0″ long, 14′-3″ high, 15′-0″ wide, for the after pair, and 14′-0″ for the forward ones, with a total of 5,624 tubes 2″ diameter by 5′-0″ and 5′-6″ long.[23] The paddle-wheel shafts were increased to 20 1/2″ in diameter; in way of the bearings they were 22 1/2″. The enormous side levers became 23′ from center to center, 5′-6″ deep, with a minimum thickness of 3 1/2″.

The complex system for operating the steam and exhaust valves was one designed by Horatio Allen for the

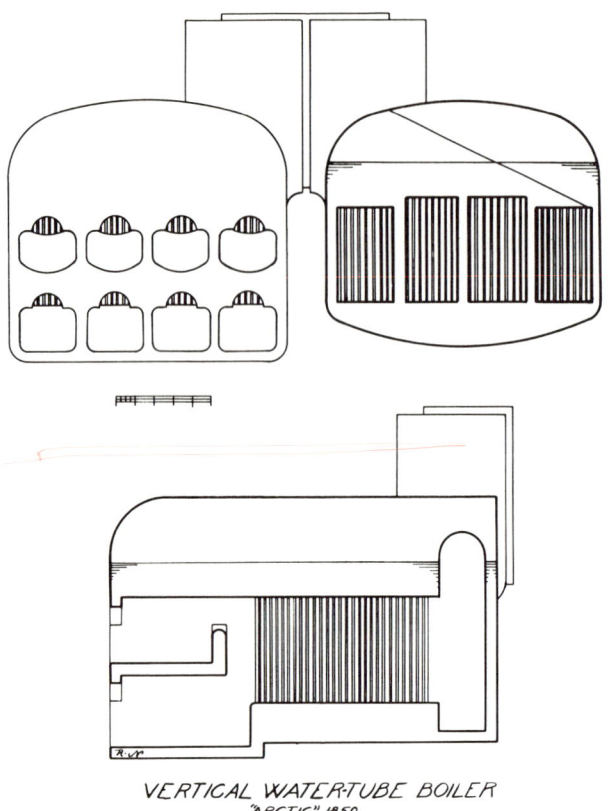

Vertical water-tube boilers for the *Arctic*.

	Maximum	Minimum
Average speed, knots	13.08	10.85
Steam pressure, psi	16.8	13.2
Revolutions per minute	18.3	14.5
Indicated Horse Power	2122	1450
Coal consumption, tons per day	79	68

For comparison, the *Arctic,* with a longer stroke engine and larger boilers, on a fast passage to Liverpool made 13.13 knots and burned eighty seven tons of coal per day. Unfortunately, no horsepower measurements were made but, on another occasion, 2,300 IHP was attained.

In review, the Collins liners were the finest product that American ship and engine builders could produce. Edward K. Collins had no restrictions imposed by his financial backers and forged ahead with lavish plans for a spectacular steamship line. The steamships exceeded the stipulations of the mail contract; instead of 2,000 tons, they were 2,700 to 2,850. While 1,000 horsepower (without a clear definition as to whether it was Nominal or Indicated) was required, the value installed was over 2,000 and may have been as high as 2,500 IHP under optimum conditions. The costs were enormous and difficult to ascertain with accuracy. Published values ran from $550,000 to $790,000 per ship. The author prefers an intermediate one, $675,000.[27]

Collins and the Browns were carried away by dreams of speed and grandeur. No one ever examined the competitive situation or stopped to consider whether a $2,700,000 investment, enormous for its day, was justified. Since 1840 the Cunard Line had sent passenger and mail steamships from Liverpool to Halifax and Boston. In 1848 they had added a New York service that was rapidly becoming the more important of the two. With an orginal fleet of five steamships, experienced crews who stayed with the steamers—American practice was to ship a new one for each voyage—and an established group of satisfied merchants wo found that their goods were delivered regularly without losses, plus, best of all, an excellent mail contract from the British government that helped offset the large voyage losses encountered in the winter when there were few passengers, the Cunard organization was in a very strong position. It had already survived the incursions of other steamships that had entered the Atlantic trade with glowing prospects and then abandoned it as unprofitable. The *Great Britain* was a well-known example. She had an interesting history but her original owner could not afford to keep her running. Spectacular ships are not necessarily profitable ones.

The Congress of the United States became infected with Collins's enormous optimism and passed special legislation allowing him extra time to build the steamers

Altantic and *Arctic.* The Allaire engines chose a Stevens cut-off (developed by Francis Bowes Stevens, nephew of Robert L. Stevens) for the *Pacific,* but replaced it by a Sickels one on the *Baltic.*

There was considerable experimenting with the shape and size of floats on the paddle wheels.[24] The *Atlantic* and *Arctic* had thirty six in number, the *Baltic* thirty two, and the *Pacific* twenty eight. The *Atlantic*'s wheels were 35'-0" in diameter and the others varied from 35'-0" to 36'-0" with different sources disagreeing as to what value each ship had. In all cases, the wheels had half as many arms running all the way to the shaft as the number of paddles. The extra floats were carried on short spokes supported by the outer rims on the wheels.

To operate the engines and to dig out of the bunkers and shovel eighty tons of coal per day into thirty two hungry furnaces, ten engineers, twenty four firemen, and twenty four coal passers were necessary.[25]

The *Journal of the Franklin Institute,* the best engineering magazine available, gives the following summary of five voyages by the *Pacific,* from January through August of 1851:[26]

and permitting the Secretary of the Navy to make advance payments of $25,000 per month per ship once their hulls were afloat.[28] These payments were to be deducted, later, from the $385,000 yearly subsidy. In effect, an interest-free loan by the United States to a firm of bankers, Brown Brothers, was authorized!

Excellent progress was made throughout the building period. The first pair, *Atlantic* and *Pacific,* were started in early 1848. Despite the fact that Brown, with the *Atlantic,* received the first contract, Bell progressed even faster on the *Pacific,* so that both were launched on February 1, 1849. The hulls were towed to their respective engine builders' wharves where all further work was accomplished. By October of 1849 the four boilers and the two enormous bedplates, the largest and most awkward items, had all been installed in the *Atlantic.*[29] The engines were being erected, piece by piece, on the bedplates; the cylinders and condensers were in place. Since, in New York, no commercial dry dock yet existed capable of handling such large vessels, the Brooklyn Navy Yard made its new graving dock available for coppering their bottoms.

First to be completed was the *Atlantic.* Although a sailing for Liverpool was announced as early as April 6, 1850, there were inevitable delays. After a final trial on April 22, the steamer docked at the line's Canal Street pier for the first time and was thrown open for public inspection the next day.[30] Even then, women were still sewing carpets in the dining saloon and painters were at work on the cabins.

On Saturday, April 27, the newest, largest, and fastest American steamship afloat, the *Atlantic,* left New York for Liverpool. Saturdays were chosen for sailings to avoid conflict with the Wednesdays pre-empted by the Cunard steamers. Collins selected experienced packet captains, each of whom was required to buy at least one share of the line's stock in order to hold his position. James West, formerly of the *Shenandoah,* commanded the *Atlantic.* His mates were naval officers, as stipulated in the mail contract. J. W. Rogers, the Chief Engineer, and his assistants came from the merchant service.

The first crossing by the *Atlantic* was somewhat reminiscent of the *Washington*'s three years earlier. The popular view was that this time Uncle Sam would really show John Bull how to build and operate a steamship. She sailed at noon with 137 passengers but, finding the Lower Bay shrouded in fog, anchored off Quarantine until 4:00 P.M.[31] Then, getting under way, she passed the *Ohio,* outward bound, and left her a mile astern by the time Sandy Hook was reached. From then on the passage became a frustrating one. There was continuous trouble with paddle-wheel floats breaking loose. Driftwood off Sandy Hook started the damage and further failures continued throughout the trip. Six days out a valve in the air pump of the port engine ceased functioning, forcing the engineers to shut down for thirty hours while they rigged up temporary exhaust piping so that one engine could operate without a vacuum. This reduced its horsepower and lowered its efficiency. Liverpool was reached in 13 days 3 hours from New York.[32] Since none of the wet docks was large enough, the *Atlantic* had to anchor in the stream and send her passengers ashore by tender. The Cunard liner *America,* sailing four days later, did much better with an 11-day 12-hour crossing. Repairs at Liverpool delayed the *Atlantic*'s return until May 29. There were seventy three passengers and two bags of mail. The British Post Office used only the Cunard liners. She made an improved time of 11 days, 10 hours, 15 minutes, fifteen hours less than the *America*'s on a four-days-earlier sailing. The *Atlantic* left New York again on June 15, making an excellent passage of 10 days 11 hours at a speed of 12 1/4 knots.[33] Her best day's run was 319 miles at over 13 knots. Coming back, against the prevailing wind and current, she reached New York on July 21 in ten days twenty hours, an equally satisfactory performance.

The second steamer, the *Pacific,* was ready for her maiden voyage on May 25, 1850, four weeks after the *Atlantic*'s first sailing. Under Captain Ezra Nye with sixty three passengers aboard, she had an uneventful twelve-and-a-half-day passage to Liverpool. While in the Mersey she was opened to visitors and their admission fees were donated to a local school for the blind. With two steamships available, sailings were scheduled at two-or three-week intervals from May until October.

For the *Arctic,* William H. Brown arranged the most spectacular series of launchings New York had ever seen. Some 30,000 spectators gathered in and around his yard at 12th Street and the East River on January 28, 1850.[34] First the steamboat *New World,* 216' long, was launched with steam up. As soon as she became waterborne her wheels were turned to take her out on a trial run. The *Bangor,* a 225' steamboat, followed her. She was towed away to clear the river for the crowning event of the day, the *Arctic*'s launch. After the cheering ended, she was towed to the Stillman and Allen wharf, adjacent to the Brown yard. As the second Collins ship engined there, the time required was reduced to nine months instead of the sixteen needed for the *Atlantic*'s completion.

The *Arctic* was tried on October 17, 1850, and then set out for Liverpool on the 26th.[35] She got no farther than Staten Island, where fog forced her to anchor overnight. She reached Liverpool November 7, making the crossing in 10 days 12 hours, if we start with her delayed departure. Coming home, however, she met extremely heavy weather and took 14 days 13 hours, reaching New York on December 4 from Liverpool on November 20.[36] To operate her were Captain James C. Luce, four deck officers, thirty six crew, thirty eight stewards, ten engineers, and forty eight firemen and coal passers, or a total of 137 on a vessel built to carry 200 passengers.[37]

Jacob Bell held a double launch on February 5, 1850, the New Orleans packet *Saint Louis*, 1,050 tons, and the *Baltic*, last of the quartet.[38] On November 16, 1850, The *Baltic* sailed for Liverpool under Joseph J. Comstock, who later became the most respected steamship master of his day. In later years he assisted on the first trial of the *Great Eastern* and was chosen by William H. Webb to deliver the American-built steam frigate *General Admiral* to Russia and the enormous ironclad *Dunderberg* to Cherbourg after her purchase by Napoleon III.

The *Baltic* reached Liverpool in just under twelve days, with bad weather all the way. Coming back, conditions were even worse; cabin lights were stove in and she had to put into Provincetown on December 28 short of coal. Some passengers and the mails were sent to Boston on the pilot boat *Friend*. Fortunately, the brig *Palm*, with 250 tons of coal, was found there. Captain Comstock purchased her cargo, transshipped it, and sailed December 30, arriving at New York on New Year's day of 1851, eighteen days after her Liverpool departure.[39]

Collins and the Browns had done remarkably well to complete four splendid steamships and get them off to sea in the seven-month period from April through November of 1850. The line, the builders, and the engine builders had worked well together under Collins direction. Thirteen voyages were begun in 1850, with the *Atlantic* completing six of them. Now that all were in service, the intended twenty voyages per year could easily be achieved. Collins, instead, scheduled twenty four for 1851. The contract's requirement for a fifth steamship was given a token consideration by listing a steamship called the *Adriatic* in the company's newspaper announcements, but, for the next four years, she was no more than a paper promise. Collins assured Congress and the general public that, since his four steamers added up to over 10,000 tons, the contract requirement for five 2,000 tonners was, to all intents and purposes, met. As long as the line performed well and made news with its speed records, the fifth ship was tacitly overlooked by the Post Office and the Navy.

The completed steamships, however, were undergoing teething difficulties. Collins, the master of public relations, kept their problems out of the public press and we do not know what all the defects were. There were nevertheless rumors of gangs of men from the engine builders working day and night between voyages. Moreover, if one lists their arrivals and departures, one finds that three of the four were withdrawn for periods of two to four months.

The *Pacific*, after four voyages, was missing from October of 1850 until January of 1851. The *Arctic* followed, making no trips from August to December of 1851. The *Baltic*'s problems must have been especially serious. She managed to complete only five voyages in 1852, being out of service for January and February and again from July to October.

Returning from her sixth voyage the *Atlantic* left the Mersey on December 28, 1850, meeting a storm so severe that the pilot could not be dropped. After nine days of driving into continuing gales, the center paddle shaft broke, wrecking one of its journal bearings and making both wheels unusable. High seas delayed for three days the removal of floats and the lashing of the wheels in place before she could proceed under sail. At first she headed for Halifax, then for Bermuda, but, the winds becoming unfavorable, she turned about and sailed 1,400 miles back to the Irish port of Cork, where she anchored on January 22.[40] Her mails and passengers were sent on via the Cunard steamer *Africa* while a second Cunard liner, the *Cambria*, was persuaded, for £3,000 plus freight, to make an extra stop at Queenstown and take aboard the cargo. All this was most mortifying; it was too spectacular an event to keep out of the papers.

The *Atlantic* was towed to Liverpool and had her damaged parts replaced by the Soho Works at Birmingham, which, some forty six years earlier, had built the *North River*'s engine. Other needed modifications were made during a long, six months' detention at Liverpool. One marked improvement was the construction of a new dining saloon on deck instead of below. The other steamers had been so modified, or would be as the occasion arose.

The best report of the mysterious ills is by an American marine engineer who, many years later, stated:

> With using salt water (in the boilers), the scale in the tubes was troublesome, and the unequal expansion of the front and back tubes caused them to leak, so that the expense for cleaning and repairs was heavy. The unequal expansion of the parts of the engine, due to the design of the bracing, caused breaks and consequent expense, yet the engines did good service, the ships made good voyages and with reasonable economy of fuel.[41]

The out-of-service intervals just listed did not solve all their troubles. The *Atlantic* had two other overhauls, one in the spring of 1852 and one in the winter of 1852-53, both of about seven weeks' duration, and the *Arctic* was withdrawn again for three months in the winter of 1853-54, ostensibly because of bottom damage off Tuskar rock, at the southeast tip of Ireland.

Known difficulties were reported for both the *Pacific* and the *Baltic*. The former left Liverpool on August 10, 1853, and, when five days out, fractured her center shaft, just as the *Atlantic* had earlier.[42] Normally there was little load on it in smooth water, but it was severely stressed when the ship rolled and put all the power on the buried wheel while the high one spun in the air. The *Pacific* was more fortunate than the *Atlantic*, for she was able to continue on one engine and one wheel. Despite eight days under these circumstances, she made New York in thirteen days, an extraordinary performance. Two years earlier she had completed her eighth homeward passage with one cylinder disconnected after a failure of a crosstail and a broken connecting rod, but with both wheels operating. Her log showed, on that occasion, 11.4 knots with two cylinders and 8.9 with one.[43] But with one wheel lashed, even though its floats were removed, she must have been difficult to steer. The *Pacific* was withdrawn from August 23 until November 26 while a new shaft was forged and installed.

Paddle shafts presented severe problems throughout the wood shipbuilding era. The only suitable material was wrought iron, and such large members had to be made up of many small bars heated to a very high temperature and pounded together by blows of a steam hammer. Temperature control was difficult and the hammer's impact not always enough to completely weld all the pieces together on the inside of a shaft. There was no way to find interior defects until, in service, cracks formed and reached the surface where they could be seen. Often a complete break was the first sign that a flaw existed. Statistically, the author has found at least one shaft fracture per year in large American steamships from 1850 to 1870. The *Atlantic*'s occurred in 1851; the Bremen liner *Hermann* and the Havre liner *Franklin* both had breaks in 1852. The *Pacific* followed in 1853. The Charleston liner *Isabel* suffered in 1854 and then the *Union*, chartered to the Havre Line, in 1855. Others followed.

The *Baltic* had no shaft problems but needed both cylinders replaced. At the same time, all the boiler tubes had to be renewed.[44] This work was done from April until June of 1854, when the ship was four years old.

The Collins line, in effect, was a three-ship line with a fourth out of service for routine overhaul—the author has not attempted to list these short absences—or withdrawn for two or three months for major engine repairs, and, if a serious accident were met, even longer. Not until 1853 were all four running at the same time. Then, in August, the *Pacific*'s shaft fracture disrupted the temporarily happy situation.

Collins, throughout the years, was assuring everyone of the speed and luxury of his steamships, and periodicals were publishing tables comparing the superior performance of the Collins liners with the Cunard ones. Whenever we discuss Atlantic speed records, it must be emphasized that reliable data are difficult to find and even more difficult to check. There are numerous problems. One is the matter of dates, which may represent sea time, in which the day began at noon; the afternoon of September 30, for example, would be reported in the ship's log as October 1. The other is that no standard time zones existed and no regular corrections for longitude differences. Sailing times could be quoted from the dock, from Sandy Hook, from the bell buoy off the mouth of the Mersey River, or any other arbitrary point. Steamers were often fogbound outside New York or had to await the tide at Liverpool. As a result, the author has frequently found two sailing dates for the same voyage (one land, one sea) and sometimes three different times (affected by longitude corrections, starting, and ending points). Reasonably trustworthy sources are the reports and year-end summaries in the *New York Herald*, which, in the early 1850s, are from dock to dock and include all delays but omit any longitude correction. Five hours must be added to the westward times and subtracted from the eastward ones. A second, even better, source consists of quotations from the engineering logs of the Collins liners collected and published by B. F. Isherwood, a meticulous engineer, in the *Journal of the Franklin Institute*. The author considers the latter more accurate. However, in a number of cases, the two agree within an hour or two. Unfortunately, he does not have any British log books for comparison and he suspects that the Collins crossings were more carefully collected than were those of their competitors. Furthermore, the data on the westward voyages were more fully reported than on the eastward ones. The New York papers frequently quoted record homeward passages while blissfully neglecting the outbound ones, which must have been faster because of favorable wind and current conditions.

The first record of importance was reported when the *Atlantic* completed her second voyage.[45] The corrected time was 10 days 11 hours, in comparison with the *Europa*'s 10 days 16 hours made ten days earlier. If we round off the distance from New York to Liverpool to 3,080 nautical miles, the *Atlantic*'s speed was 12.3 knots and the *Europa*'s 12.0. Unfortunately, the new *Asia* had just completed a passage from Boston to Liverpool, via Halifax, about 2,890 miles, in 9 days 15 hours at 12.5 knots. From this point on the Collins liners, as they appeared, were hard driven, and every effort was made to outdo their Cunard rivals. The older *Europa*, *Canada*, *Niagara*, and *America* of 1848 were easily surpassed, but the 1850 pair, the *Asia* and *Africa*, proved almost as fast as the Collins vessels.

In the spring of 1851 the *Pacific* became the queen of the fleet by making her seventh eastbound passage in 9 days 20 hours at 13.1 knots, just after a westward one of 10 days 1 hour at 12.7 knots.[46] These were records in both directions. She had left Liverpool on April 9 and returned there in triumph on May 20.

The *Arctic* and *Baltic*, having larger engines, should have been faster, but it was not until February of 1852 that the *Arctic* was able to make a better eastbound time than the *Pacific*.[47] Nine days 16 hours 41 minutes gives 13.14 knots, and she must have been helped by a following wind, for she burned only 866 tons of coal.

The *Baltic* was the only one of the four to achieve voyages from Liverpool to New York, against the prevailing wind, in less than ten days. She sailed from Liverpool August 6, 1851, and reached New York in 9 days 18 hours 45 minutes. Three years later, just after receiving new boiler tubes and new cylinders, she did even better, again in summer,[48] crossing in smooth weather from Liverpool on June 28, 1854, to New York in 9 days 16 hours 53 minutes at 13.23 knots. This is the fastest westward passage ever made by the Collins Line.

A crossing in fewer than ten days was a great rarity. Perhaps once a year, at the most twice, when weather conditions were at their best, the boilers newly scaled, and the engines at the peak of their condition, speeds over 13 knots could be attained. After 1855 there were no more crossings in less than ten days. For more normal conditions, the author has taken all the crossing times from 1852 through 1854 and finds that they average 10 days 22 hours at 11.76 knots eastward and 11 days 19 hours at 10.87 knots westward; the year round average, then, is 11.31 knots. The best possible speed, on the other hand, was for a single day's run of 338 miles at 14.1 knots by the *Pacific*; all four, however, made or exceeded 330 miles per day at some time in their careers.

Despite the ships' machinery problems and despite their extravagant coal consumption, the line operated on schedule and delivered the mails. There had been thirteen eastbound sailings in 1850. In 1851 there were twenty four, four more than the contract required, and in 1852 plans were made for twenty six. The added sailings were to meet Cunard's number; Collins publicly laid claim to purely patriotic motives for providing such splendid steamships, such high speeds, and a fortnightly service regardless of cost.

His efforts were remarkably successful in attracting both passengers and freight. A comparison with Cunard's New York operations for 1851[49] shows that the Collins steamers carried 4,156 passengers to their rival's 4,112. In terms of westbound cargo, however, most of which was manufactured goods from England to the United States, Cunard carried 40% more than Collins, but had to make a serious cut in its freight rate to do so.

The American steamships were, without doubt, superior from the passengers' point of view simply because they were larger:

Steamship	Year	Length	Beam	Depth to Upper Deck
Atlantic	1850	284'	45'-11"	32'-0"
Asia	1850	266'	40'-8"	27'-6"
America	1848	251'	38'	25'-3"

The breadth was particularly important; it allowed larger staterooms against the sides, and wider public saloons between the rows of cabins. The public spaces were, however, almost empty during winter sailings. Carrying thirty to fifty passengers, their fares did not even pay for the coal burned. On March 23, 1851, the *Arctic* completed a homeward passage with fifteen on board. The summer record that year was achieved by the *Pacific*, with 238 tourists headed for London's Great Exhibition at the new Crystal Palace. Despite the crowd, she made a record passage. The 1851 average was eighty seven passengers per crossing on a year-round basis. Trans-Atlantic travel, first class, at $130 outbound and $167 returning on a fast mail liner, was a luxury for the wealthy few. They could take their dogs for $24 additional. No steerage passengers were carried and only a small number of second-class passengers. The sailing packets still transported an overwhelming majority of the traveling public.

Perhaps the best-known passage of the line occurred during its first year of operation, when P. T. Barnum, the impresario, brought the renowned Swedish singer Jenny Lind to America. All the arrangements were announced well in advance by Collins and Barnum, both of whom reveled in publicity. The *Atlantic*'s sailing, from Liverpool on August 21, 1850, was the first occasion when every stateroom was engaged. A special tender brought Miss Lind; Signor Giovanni Belleti, basso; Jules Benedict, composer and conductor; and their party out to the *Atlantic* as she lay at anchor in the Mersey. There was a dense crowd on shore to see her off and Captain West saw to it that the "Swedish Nightingale" was on top of the paddle box waving her hankerchief as his ship got under way.[50] At sea there was very bad weather; three days out a wave swept the forecastle, breaking an arm off the figurehead and injuring three sailors, but on the fourth day the sea moderated and Miss Lind sang in the evening. A benefit concert was arranged; £64 was collected and distributed to the sailors and the firemen.

At New York, on September 1, there were crowds to welcome Miss Lind and once more she was on the starboard paddle box, this time with Messrs. Belletti and Benedict, to see the sights and to be seen. Edward Collins saw to it that he boarded the ship in advance of P. T. Barnum and presented the singer with a bouquet some three times the size that her manager produced. The crowds cheered and the visitors were flattered; all of New York was aware of the *Atlantic* and the famous diva.

By the end of 1851, despite universally favorable notices, the Collins Line was on the verge of bankruptcy. The expensive steamships required large interest, depreciation, and insurance payments. The engine and boiler repairs were very costly. The coal consumption had been higher than anticipated. True, they had attracted a satisfactory number of passengers and a reasonable amount of freight. Yet the cold facts were that mail payments of $385,000 per year were not enough. The four extra trips that had been made—and they were added in the winter when fuel consumption was very high and passenger lists very low—simply increased the losses. They made a superb impression on the public and the postmaster general, but they were financially unsound.

In 1852 sailings every two weeks, obviously at further losses, were continued and, simultaneously, Congress was petitioned for a larger subsidy. Both houses debated the subject extensively and, although the request for aid was made on January 10, 1852, the matter was not settled until July 12.[51] The *Baltic* had been out of service during the winter. To speed the lawmakers along, she was sent to the Potomac and a grand banquet was arranged for the President, Cabinet Members, Senators, Representatives, and foreign diplomats. Two steamboats carried guests from Washington to her anchorage at Alexandria. There were no speeches, but food and wine were liberally provided. "Many were the salutes fired by inoffensive corks in all directions," reported one newspaper. Some Senators questioned whether it was appropriate to accept Mr. Collins's entertainment, but they were outvoted. The *Baltic* was back at New York on March 5, 1852, and sailed for Liverpool the next day.[52]

The petition to Congress proved better in its rhetoric than its accounting; Collins never released enough figures to reveal the line's true financial position. He claimed a loss of almost $17,000 per voyage despite his $19,250 subsidy. The Havre Line, meanwhile, was profitable, with mail payments of $12,500 per voyage, and the Bremen Line broke even on $16,667. Eventually, a new rate of $33,000 was adopted for twenty six round trips per year, but, instead of the problem's being settled by Congress for the contract's eight remaining years, the new rate was effective through the end of 1854, now seventeen months off. After that date the higher payments would continue, but they could be terminated by Congress with six month's notice. Should this occur, the original terms, $385,000 for twenty voyages, would be re-established. The new subsidy was voted by narrow margins in both houses and would not have passed at all without the clause permitting termination of the extra payment.

Nevertheless, $858,000 per year was a magnanimous answer to the Collins Line's request. It would cover the present service and, without doubt, was liberal enough so that past losses could, eventually, be paid off. But any business that counted on 50 percent of its income from governmental appropriations was a hazardous venture.

During the long six months of Congressional review, the line continued to function steadily. The steamships were kept in the best of condition and no passenger was allowed to suspect that the service would collapse the day that adverse action was taken in Washington.

The masters of the steamships operated them at high speeds and adopted a cavalier attitude toward anything that got in the way. Captain Nye, early in the *Pacific*'s career, steaming up the Hudson on October 27, 1850, managed to smash in the port bow of the ship *Esmerelda* with his starboard guard.[53] The latter had just tied up at the end of Pier 4. One eminent passenger who witnessed the collision was Commodore Vanderbilt.

Some years later, coming up the Mersey at night on January 9, 1855, the *Pacific* ran into a brig anchored there, the *Corinthian*, sinking her but rescuing the twelve crew members.[54] The Admiralty court ruled the steamship at fault; the loss was estimated to be £6,000. Captain Nye was replaced before the end of the year—luckily, as we shall see, for him.

The *Baltic*, under Captain Comstock, after leaving Liverpool on July 28, 1855, ran down a schooner, the *Sarah Ann*, off Holyhead.[55] Again, everyone seems to have been rescued. The passengers collected $200, which was presented to the captain's small daughter, and $300 for the crew.

At first the Collins steamships suffered little from the perils of the sea. It is generally recognized that the larger the vessel, the less likely it is to meet waves of its own length; these are the waves that cause ships to pitch badly, take water over their decks, and lift their bows clear of the sea. As the largest craft on the Atlantic, the Collins ships performed remarkably well and suffered little damage. The vertical stems and the location of their guards, high above the waterline, were helpful qualities. The *Washington* and *Hermann* suffered from low guards and damaged head rails when they pitched their bows under. The *Atlantic* did carry away some bulwarks and had parts of the wheel houses stove on an early return from England, but she was able to make repairs and sail again on October 12, 1850, only three days after reaching New York. Just four years later, homeward bound, she met such heavy seas that, for twenty four hours, she headed into them with her wheels turning at 6 rpm but making only fifty five miles from noon of September 12 to noon of September 13, 1854. The starboard paddle box was carried away and her cutwater badly damaged. This may be the occasion when her figurehead was removed; it shows up clearly in early prints and paintings but is missing (as also is the mizzenmast) from the S. Walters lithograph dated June 25, 1855. None of the four ships carried figureheads in their later years and all had their mizzens removed.

Captain Luce took the *Arctic* too close to Tuskar rock on May 18, 1854, damaging his bilge planking; the steamer heeled over and slid free, but had to return to Liverpool.[56] Her pumps controlled the leakage, even though there was a wait until May 23 before a large grav-

Baltic, in later years after her mizzenmast was removed.
Courtesy of The Mariners Museum, Newport News, Va.

ing dock became empty. Thirty of her 126 passengers chose to proceed on the *Asia*, which took over her mails, but the remainder stayed on board until the *Arctic* could sail again on May 28. Once she was home she was withdrawn for an extensive overhaul.

The years 1853 and 1854, now that the higher subsidy was in effect, were splendid ones for the line. More and more passengers traveled on its steamers. Merchants, impressed by speed and regularity, sent more cargoes on them. Millions of dollars in California gold regularly went on eastward trips. The *Baltic* sailed from New York on May 28, 1853, with a sold-out passenger list of 229. On her next voyage, from Liverpool on July 27, she had as much cargo as she could carry—its weight always had to be limited to keep from burying the paddle wheels too deeply—and left almost as much behind her. Prudent English shippers began to reserve space well in advance and could no longer rely on American steamers to take any overflow from the Cunard liners.

The first tragic event in the line's history took place in the fall of 1854. The *Arctic*, following her bottom damage off Tuskar, underwent three months of overhaul. On September 2 she returned to service and headed East on her twenty fourth voyage. Westward from Liverpool, on September 20, she had a full load of passengers, many of them intimately connected with the ship and the line.[57] Mr. Collins's wife and two children, Henry, 15, and Mary Ann, 19, were coming home from Europe.

Captain Luce had his son, William, a crippled boy of 11, with him. A whole party of Browns, from the banking firm that had financed the line, were aboard, together with a representative from the Novelty Iron Works, George F. Allen, whose wife was one of the Brown group.

Seven days out from Liverpool, on the 27th, the *Arctic* was about forty miles off Cape Race, Newfoundland, and, despite fog, was running at full speed.[58] Just afternoon a small, bark-rigged, iron-propeller with all sails set, appeared out of the fog and ran into the *Arctic*'s bow. The little steamer, later identified as the French *Vesta*, 250 tons, was carrying a large number of French fishermen from the islands of Saint Pierre and Miquelon, just south of Newfoundland.[59] Her bowsprit and a portion of her bow were missing after the collision, her foremast had fallen, and she was visibly in trouble. Captain Luce, unwisely assuming that his own ship had little damage, circled the wreck twice and launched a boat. Some of the *Vesta*'s terrified passengers also put a boat over.

But the *Vesta* was iron and had watertight bulkheads. The *Arctic* was wood and had none. Through holes in the starboard bow some sixty feet aft of the stem, the sea poured in. It was the engineers on watch who warned the captain of rising water. An attempt to get a sail over the holes was rendered impossible because of twisted iron from the *Vesta*'s bow. The ship's carpenter could not reach the damage from inside or outside his ship. Once the *Arctic*'s critical condition became evident, Captain

Luce headed for the land at full speed, abandoning both the *Vesta* and his own boat, and ruthlessly ran down the *Vesta*'s boat and crushed it under one of his wheels. One lone fisherman was pulled aboard the *Arctic*.

Water put out the fires in thirty minutes and reached the second cabins in forty five. The remaining boats were launched without discipline and with little seamanship; there was considerable loss of life as they were dropped, upended, and commandeered by crewmen who leaped from the deck into them. Because they were of the Francis type, with buoyancy tanks, they survived the ordeal but had neither provisions nor water in them. A raft was started by lashing spars and gratings together and it, too, was filled with terror-stricken people long before it could be made seaworthy. When the *Arctic* sank, a piece of one paddle box broke clear and provided a second raft, which Captain Luce, George Allen, and others managed to climb aboard from the ice-cold waters in which they were floundering.

Two boats reached Broad Cove, Newfoundland, two days later. A third boat, and one lone survivor from the many people who had crowded the uncompleted raft, were picked up on September 28 by a Canadian bark, the *Huron*, bound for Quebec. A day later the ship *Lebanon* was met and about half the *Huron*'s passengers were transferred to her.[60] When the *Lebanon* reached New York, October 11, 1854, she brought the first news of the *Arctic*'s fate. The *Huron* landed her group at Quebec on the 13th.

The ship *Cambria*, from Glasgow, encountered and searched through the *Arctic*'s floating wreckage. From it she picked up ten survivors, among them Captain Luce, George Allen, and one passenger, the three still alive on the paddle box out of twelve who had reached it. Of the 282 passengers, twenty three were saved. The fishermen rescued from the *Vesta*'s boat was doubly fortunate and reached shore safely. Of about 150 in the crew, six officers and fifty five men survived. There was little discipline and less heroism. In many cases there was very little common sense. The seamanship displayed had nothing to recommend it. The loss of the *Arctic* was a disgrace as well as a tragedy.

The real seaman on the spot was Captain Alphonse Duchesne, of the *Vesta*, who cut away the wreck of his foremast and jettisoned everything he could to raise his shattered bow. After shoring up his collision bulkhead and dragging sails over the bow, he brought the *Vesta* slowly into St. John's, Newfoundland, on September 30.

While the loss of the *Arctic* created an enormous stir, it did not disrupt the line's operations. The insurance, $540,000, was a financial windfall. A further $300,000 on the cargo reimbursed the shippers. Moreover, sailings had been scheduled on a three-boat basis while the *Arctic* was being overhauled. Furthermore, the expansion of the Crimean War had created a great demand for transports to carry the British and the French armies to the Black Sea and to supply them after their arrival. So many of the largest and fastest Cunarders were chartered that their New York to Liverpool branch had to be abandoned at the end of 1854. The Collins steamships took over their rival's passengers and freight and even annexed the Cunard sailing day, Wednesday instead of Saturday eastbound, and Saturday for Liverpool departures.

The year 1855 was the apogee of the Collins Line. The long process of withdrawing ships and modifying machinery had ended. The three remaining liners sailed on schedule every two weeks throughout the year. Their average speeds were well maintained, although records were avoided and only one eastward passage was below ten days and two westward below eleven.[61] Each steamship sailed strictly in turn, except for the *Pacific*, which missed one passage for a quick overhaul. In her place a Charleston steamer, the *Nashville*, sailed from New York on March 21, 1855, and arrived Liverpool on April 3, in 12 days 16 hours. The *Nashville*, owned by Spofford and Tileston, was less than half the size of a Collins steamship, 215′-6″ long with a tonnage of 1,220. She was about a year old but had already made two Atlantic voyages on charter to the Havre Line. As a much smaller steamer, she was somewhat slowed by heavy seas. Westbound, she arrived in New York on April 22 with fifty one passengers, having taken fifteen days for the crossing. It was not the slowest eastward passage of the year, for the *Atlantic* had taken seventeen days when beset by continuous March gales. The only other interruption was due to a minor engine repair that delayed the *Atlantic*'s April 21st departure from Liverpool for two days.

A record of 7,176 passengers was carried in 1855, 138 per trip, a great improvement over the eighty seven average for 1851. These, together with full cargoes and the magnificent new subsidy, justified the ordering of a new steamship, the long-delayed *Adriatic*. The 1855 appropriation bill had contained a reminder: "*Provided*, That Edward K. Collins, and his associates, shall proceed, with all due diligence, to build another steamship in accordance with the terms of their contract, and have the same ready for the mail service in two years from and after the passage of this act."[62]

All Atlantic crossings prior to the Civil War were notably different from those in recent years. First, the average temperature of the sea has been slowly increasing with time. As a result, pack ice and drifting icebergs were found much farther south than they are today. Second, there were no tracks established for east and westbound steamships in order to avoid both ice and collision. Most masters chose the shortest, and most northerly, course between New York and St. George's or the English Channel. Westbound, the ships came close to Newfoundland's

Cape Race and crossed the Grand Banks as a means of determining their position. If that cape were shrouded in fog, soundings were relied upon. The steamships ran in the general direction of the land, passing close to Cape Cod and following the south shore of Long Island. In the event of a shortage of coal, they could easily reach Halifax or, as the *Baltic* did on her maiden voyage, even Provincetown.

The colder climate and the northern track increased the perils from ice, whether in the form of individual bergs or floating fields of pack ice, and many steamships reported encounters.

The spring of 1854 was a particularly difficult one. The *Hermann,* leaving Southampton on March 29, used sails alone for eighty miles through immense quantities of ice in order to avoid damaging her side wheels.[63] The *Baltic,* eastbound, entered an opening in an ice field on March 9 and was unable to clear the hazards until three days later. During that time she had to run over 300 miles south and west between bergs and pack ice. The *Pacific,* headed westward at the same time, arrived in New York on March 23, 1854, fifteen days from Liverpool, having met gales, vast amounts of ice, and numerous individual icebergs.

The spring of 1856 was a repetition of 1854. There was a new mail steamer on the Atlantic, the magnificent *Persia,* constructed by Cunard to outdo the Collins steamers. The British Admiralty, who had technical control, had finally consented to the use of iron, instead of wood, for a mail steamer, but still insisted on side-wheel propulsion. Hostilities in the Crimea having ended, Cunard's New York service was resumed in January of 1856; the *Persia*'s maiden voyage was the second on that route. Because she had departed Liverpool on January 26, she was listed as overdue in New York on February 8, but limped in a day later. She had met heavy weather and, on February 2, several large ice fields. She had run into a small iceberg, splitting it and badly damaging her starboard wheel. Her immensely strong iron hull survived the impact without damage.[64]

Three days before the *Persia,* the *Pacific,* with about 150 pasengers, left the same port never to be heard from again. She may have run into a berg or may have been crushed by ice. Since no wreckage was ever recovered, despite prolonged searches by several vessels, no evidence exists.

There is ample evidence as to ice; when the Havre Line's *Arago* reached New York on March 1, Captain Lines advised all masters to go south of the Grand Banks. He had found ice along their eastern edge, had steered through drifting bergs from February 22 to 24, had been forced to stop on numerous occasions, and had, eventually, turned due south for 200 miles before he could find a path clear of danger.[65]

Although the *Arctic*'s loss did not disturb the company's operations—in fact, it may have been helpful in making it schedule its sailings more tightly to keep the ships at sea where profits could be made—the loss of the *Pacific* was more than a disaster. Two steamships could not keep up twenty six sailings per year and the new *Adriatic* was nowhere near completion.

Public opinion reversed itself; instead of congratulating the masters for their speed records, it now castigated them for operating at unsafe speeds. In Washington, the unfavorable publicity reduced the already shaky Congressional support for the line. Now that Cunard had resumed its New York run with two large steamers, the *Arabia* and the *Persia,* as well as the older *Asia* and *Africa,* 1856 became a year of mounting losses for the American line. Cunard's safety record of never having lost a passenger was a memorable achievement.

The *Pacific* had been scheduled to leave New York, on what would have been her 41st voyage, on February 16, 1856. As an immediate replacement Collins chartered a coastwise steamship, the *Quaker City.* She was a handsome, two-stacked craft built in 1854 to operate from Philadelphia to Charleston. Later she was used on the Mobile route. Of 1,426 tons, she was somewhat larger than the *Nashville,* the only previous charter.

She left New York on the 16th, arrived at Liverpool March 2, departed from there March 5, and arrived New York on the 21st.[66] The return passage took fifteen and one half days and the weather was bad in both directions. She was advertised to sail again on April 1, but instead returned to her Mobile owners.

As a replacement, Collins obtained one of the most unsuccessful transatlantic liners ever to remain afloat, the *Ericsson.* Built in 1853, her hot air engines had proven to be John Ericsson's greatest failure. Two inclined steam engines were later installed but they did not have enough horsepower. Because of her many machinery modifications, she did not get to sea until 1855. From June of that year to March of 1856 she had completed four unprofitable transatlantic voyages for John B. Kitching before Collins chartered her.

Her size, 1,902 tons, and her length, 253'-6", were reasonable; the passenger capacity, 130 berthed in sixty four staterooms, was quite suitable, but the speed was woefully inadequate. No doubt her charter rate was cheap; no one really wanted her, even if her coal consumption was low. Perhaps an advantageous rate is sufficient explanation for such an unsuitable choice. For Collins did choose her and ran her from March of 1856 until August of 1857 for eleven voyages. The traveling public would have nothing to do with her; the largest passenger list in the *New York Commercial Advertiser* is thirty nine, arriving at New York on June 13, 1856. The smallest, for a winter sailing from Liverpool in February of 1857, is four.

Her first voyage was one of her best, for she met smooth seas in both directions. Leaving New York on March 29, 1856, she took fourteen days to Liverpool; sailing from there with ten passengers she completed the return trip in the same time, reaching New York April 30. But in July she took nineteen days for a homeward passage.[67] One cylinder broke down on the third day out and the other did all the work for the rest of the trip.

Passengers not only boycotted the *Ericsson*, but they also steered clear of the *Atlantic* and *Baltic*. Naturally enough, the new record-beater *Persia* was a great attraction. In April of 1856, two months after the *Pacific*'s nonarrival, the *Persia* sailed from New York on the 2d with 227 passengers;[68] the *Atlantic* followed ten days later with 105. Even the slow, elderly *Washington*, of the Bremen Line, had 172 when she left on the 19th.

Twenty six voyages were made in 1856, just as the mail contract specified, and it is a miracle that Collins was able to keep a rigid schedule with two fast steamers and a slow one, but the number of passengers carried in both directions dropped to 3,685, roughly half the 1855 total. Cunard delivered over 10,000 — 5,378 on the New York route and 4,767 on the Boston one. A further problem was that the postal and naval authorities were dissatisfied with the *Ericsson*. When Collins did not take any action on the suggestion that she be replaced, the Secretary of the Navy cut her mail payments. During the latter part of 1856 $90,000 was paid, whereas Collins claimed $198,000.[69] He insisted that the *Ericsson* met all the essentials of his contract. He claimed she was over 2,000 tons, which was untrue.[70] Further, he quoted a power "over 1900 horses," a gross exaggeration, and boasted that her strength was unequaled by any other steamer in the mail service, a statement impossible to confirm or deny.

Attorney General Cushing, writing to Secretary of the Navy Dobbin, summed up the matter by saying that the contractors ". . . instead of making complaint in these premises should congratulate themselves for the liberality with which they have been treated by the government." and refers to their operating two inspected and approved steamships when their contract called for five.[71]

The government was about to end its liberality. In August of 1856 Congress voted to terminate the $858,000 subsidy and to revert to the original value of $385,000. The reasonable thing for Collins to do, at this point, was to continue running for the remaining six months at the higher rate and abandon the service. But Edward Collins and James Brown did not act like reasonable men.

They advertised the first sailing of the new *Adriatic* for November 22, 1856. That was entirely impossible; she was suffering from engine defects that would delay her for more than a year. The *Atlantic* and *Baltic* had completed two years of hard service with such tight schedules that no time could be spared for major repairs. Both badly needed new boilers, at a cost of $110,000 per vessel.[72] Undaunted by the new expenditures, Collins went ahead with the reboilering and, simultaneously, authorized the installation of watertight bulkheads and all the general rebuilding that had to be done when the decks and cabins above the boilers were torn out. A further million dollars was spent on the *Adriatic* during the 1855-57 period. That supposedly hard-headed bankers would advance so much money to a declining business seems unbelievable; Edward K. Collins was indeed a persuasive person.

January of 1857 began in the best of style, with the *Baltic*, the *Ericsson*, and the *Atlantic* sailing at their ordained fortnightly intervals. But in February the *Baltic* and *Atlantic* both had to be withdrawn, the former for six months and the latter for four. For six weeks there were no sailings at all. The *Ericsson*, the only steamship left, made one trip out on March 14 and, eight weeks later, another on May 9. To effect some improvement in service, a Spofford, Tileston, and Company's liner, the *Columbia*, was chartered for two voyages. She was a new steamer, a later and improved *Nashville*, whose engine was just getting properly run in after two trips to Charleston. With her came that veteran of ocean steamships, Michael Berry, who in 1846 had brought out the *Southerner*. On June 6, 1857, the *Columbia*, flying the Collins flag and with Berry in command, steamed past Sandy Hook for Liverpool. Although a small ship of 1,347 tons, she did far better than the *Ericsson* with east and westbound crossings of less than twelve days. On her second trip, leaving New York July 18, she improved her own record, making the trip in 11 days 4 hours.[73]

The *Atlantic* returned in June, the *Baltic* in August; the *Ericsson* and *Columbia* completed their last voyages during the latter month. From then on only the two original Collins liners were retained. Irregular sailings at two or four week intervals were possible until the end of November 1857, when the *Adriatic* set out on her single voyage for the line.

Up through the time of the *Adriatic*'s trip, the regular steamships and their chartered assistants had transported 1,516 passengers in 1857, compared with 3,685 in 1856 and 7,176 in 1855. There were seventeen sailings, three less than the contract required, and only two departures from New York between January and June.

The construction of the *Adriatic* spanned the line's descent from its greatest success to its ultimate collapse. In effect, her beginning was a contract awarded in 1853 by the Imperial Russian Navy to William H. Webb for a very large steam-powered frigate.[74] One of the major problems, when constructing a large, heavily built steamship, was that of assembling the numerous oak knees, breasthooks, and all the curved timbers of the proper sizes and shapes to form the floors and frames. The builder could not bend heavy timbers; he carefully selected curved limbs and trunks of the right size and shape. Every knee came from a naturally grown crook where a branch or root joined the trunk of a gnarled oak so that the wood's grain, growing around the intersection, gave it strength. The bigger the vessel, the greater the problem became. Webb had acquired timber for the vessel, but the outbreak of the Crimean War in 1854 led to the suspension of his contract. Collins ordered his new, very large steamship from James and George Steers, who set up a new shipyard to build her on the East River next door to the Webb yard. By purchasing Webb's carefully assembled timber, the brothers could start immediately, rather than spending months finding the material. Despite the fact that she was the largest wood vessel extant, the hull advanced rapidly and was ready for launching on April 7, 1856. The shipyard was filled with so many spectators that the crowd overflowed onto the bark *Grapeshot* and the steamship *St. Louis*, which were tied up at adjacent piers.[75] Men filled the decks and climbed the rigging, and some athletic souls even entered a paddle box of the steamer and sat on the paddle wheel looking out through the open slots in its housing.

Webb took advantage of the occasion to stage an unannounced launch, the screw steamer *Cuba*.[76] After she was towed away, tugs arrived and removed the *Grapeshot* and the *St. Louis*, together with their protesting passengers. When the *Adriatic* slid down the ways, she gained so much momentum, despite dropping her anchors, that she ran all the way across the river and cut into a pier on the Williamsburg side. A tug, the *Titan*, extricated her and took her to the Balance Dock, where her bottom would be coppered. The damage amounted to a few scratched planks on the *Adriatic*. The pier was less fortunate.

Although the largest wood vessel afloat, she was somewhat smaller than the four-masted clipper ship *Great Republic*, built by Donald McKay in East Boston in 1853. The latter had burned while loading in New York, and, after raising and rebuilding, ended up a considerably smaller vessel. It is also interesting to see how the *Adriatic* compared with the iron paddle steamer *Persia*. Because British and American tonnage laws differed, the author has recomputed the latter's tonnage by American rules.

	Length	Beam	Depth of Hold	Tonnage
Original *Great Republic*	335'	53'	38'	4555
Rebuilt *Great Republic*	302'	48.4'	29.2'	3356
Adriatic	345'	50'	33'-2"	4145
Persia	376'	45.3'	29.8'	3580

The *Adriatic*, in typical Collins fashion, was to be larger, faster, grander than anything afloat, regardless of cost. Her length on the load waterline was 343'-10"; her actual, as opposed to official, length on deck was 351'-8".

Her floor timbers were 22" deep and 13" to 16" thick, spaced from 33" to 36" apart. The diagonal strapping was 5" by 7/8" iron. Her light draft was 17'-1 1/2" and her designed draft 20'-0", but, with the maximum amount of coal, stores, and cargo aboard, it increased to about 21'. At the 20' value, she displaced 5,233 tons.

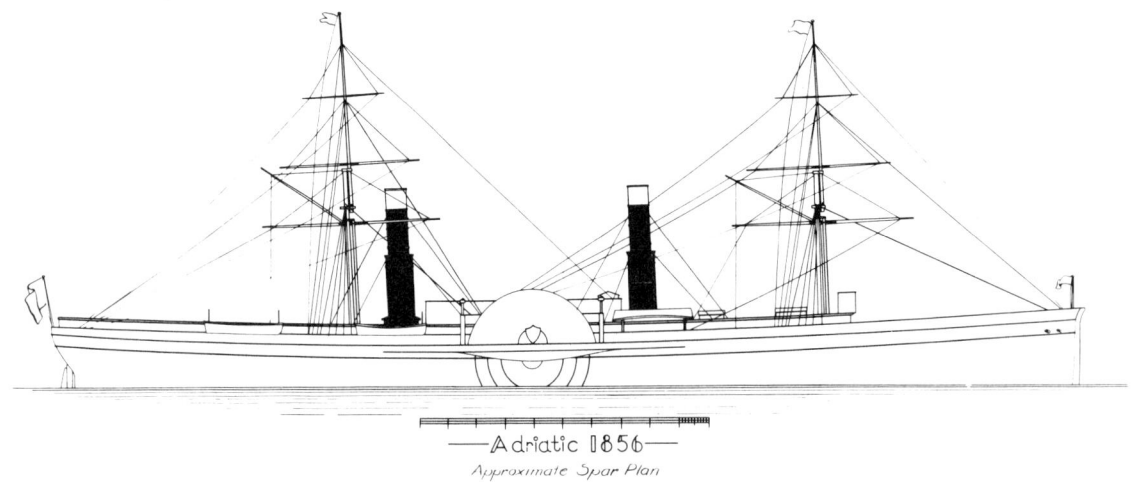

Adriatic, approximate spar plan.

Dining saloon on the *Adriatic*. Author's collection.

Dead light, with water in the boilers but no stores, cargo, or fuel, she displaced about 4,240 tons. For each additional inch of draft, 28 3/4 tons had to be added to the ship.[77] This made the *Adriatic* about 30 percent larger than the *Arctic* in terms of actual weight.

At the bow the only decorations were three small scrolls with a barred shield. At the stern was a gilded eagle above two scrolls bearing *Adriatic* and *New York* in ornamental letters; otherwise the simple exterior gave no hint of the vessel's size. A light hurricane deck, really the top of the deckhouse, extended for three-quarters of her length and had two pilot houses, a prominent one forward, steamship fashion, and a low one aft, as in a sailing ship, with bell signals between them.[78] Officers' staterooms and a twenty-foot-square skylight over the engines were situated between two towering stacks. Adjacent, to take advantage of hot air rising from below, were the laundry's drying rooms.

In the long deckhouse, on the spar deck, was the after dining saloon, 75' long by 28' wide, with the usual stained-glass trunks descending from skylights overhead. A bird decorated every pane. Not only were there side windows for the space itself, but there were further ports admitting light beneath the velvet-cushioned seats along the sides. Through glass in the spar deck they helped illuminate the saloon below. Forward of the saloon came the pantry, the galley, and then the machinery space. Cold rooms holding 100 tons of ice were installed and a dishwasher made its first shipboard appearance. Aft of the dining saloon was the main stairway down to the first-class quarters. Right aft were store rooms and the smoking room.

The first-class saloon, aft on the main deck, had ornamentation wherever one looked. The ceiling was covered with arabesques. There were long mirrors for the ladies and 120 paintings in wax colors with an explanatory text under each. An example, for one entitled "Water," read:

A female figure crowned with reeds and almost naked, emblematic of her purity. Her position elevated, signi-

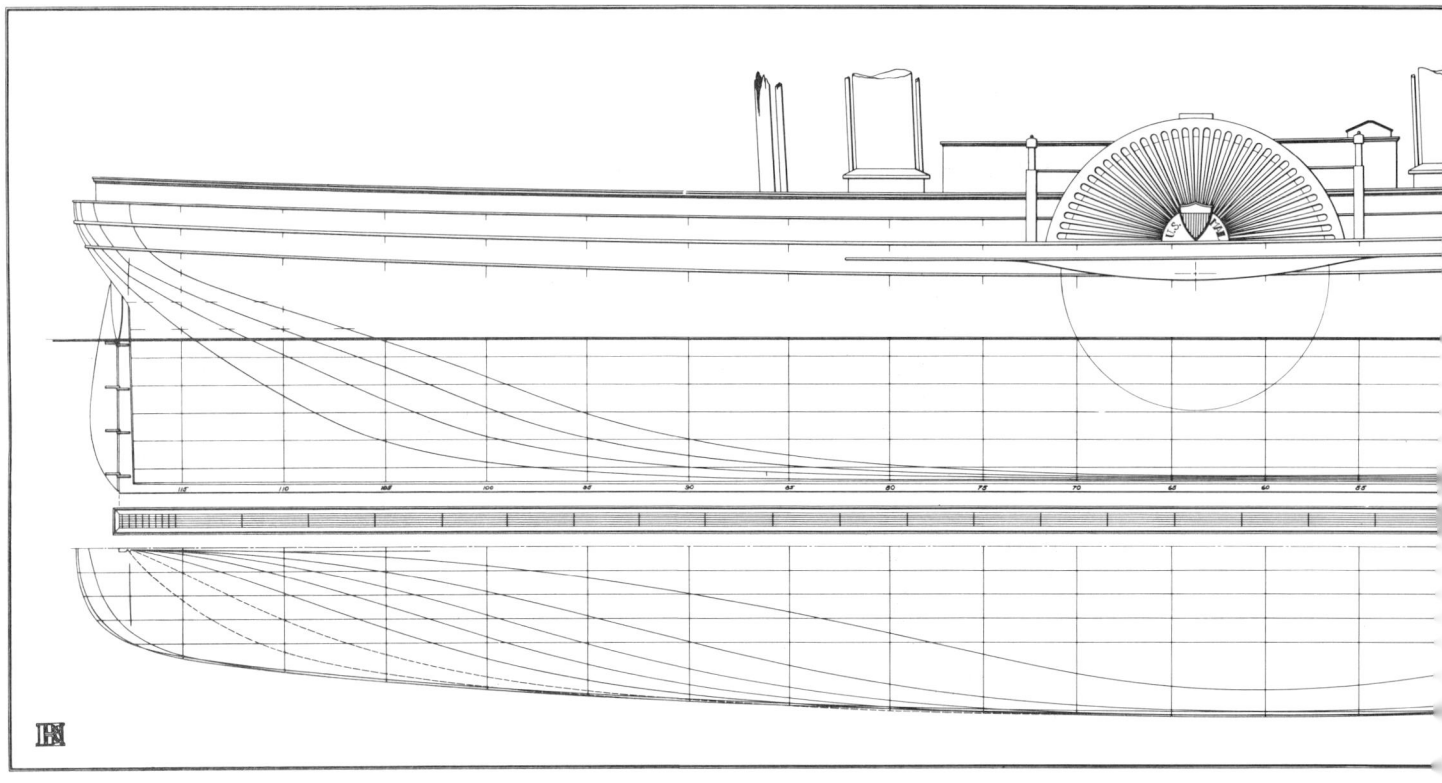

Adriatic, 1856, lines.

fying that from the mountains descend the rivers, indicated by an urn of running waters. The trident of the God of the Seas and the nets are characteristic of the element; the multitude and variety of the shells, the richness of her production.

Thirty six alcoves along the sides led to staterooms. In the hot and unpleasant regions along either side of the uptakes and the engines were berths and mess rooms for the firemen and coal passers, and, walking forward, one entred the second cabin and a stair up to a forward dining saloon, also in the deckhouse. That was paneled in black walnut and upholstered in green morocco leather. The second cabins omitted the carved flowers, vegetables, fruits, animal heads, and other decorations considered so tasteful in the first-class spaces.

All passengers were berthed on a single deck, the main, with stewards and servants below. Quarters for about 300 first class and 100 second were provided.

As a consequence of the *Arctic*'s loss, watertight bulkheads were installed, although it was never possible to attain real tightness in a wooden ship. They had two layers of oak, each three inches thick, with pitch and felt in between. Eight metallic boats were fitted, but, as always, their total capacity fell well below the number of persons on board. That discrepancy would not be corrected until sixty years later, when the sinking of the *Titanic* brought the matter to the attention of a horrified public.

A two-masted rig had square sails on both masts, but, for much of the *Adriatic*'s life, her main yards were stowed on deck, giving her the appearance of a half brig. She was far too big to sail, but canvas could be set to diminish rolling, a particularly important precaution on a paddle steamer. Her anchors weighed 3 1/2 tons each, and their 600' chains had links welded up from 2 1/2" diameter iron bars.[79]

The designer of the *Adriatic*, George Steers, has invariably been credited for her success. His half-model survives at the Mariner's Museum. A take-off from it is also extant in the Naval Plan files at the National Archives. The author has used both in reconstructing the lines of the steamship. The former has inlays locating the guard beams and the shaft centerline; the latter shows the stem, the rudder, and the bow decoration. Additional details are from photographs and lithographs. In one respect the author has departed from the model take-off, which shows a molded beam of 48'-0". He has expanded it to 50'-0" to correspond to the dimension written on the Steers' drawing. It was quite common to modify wooden ships during their construction.

The hull, without question, was rugged, well designed, and a great success. James R. Steers, the older brother of George, must be given much of the credit for its rapid

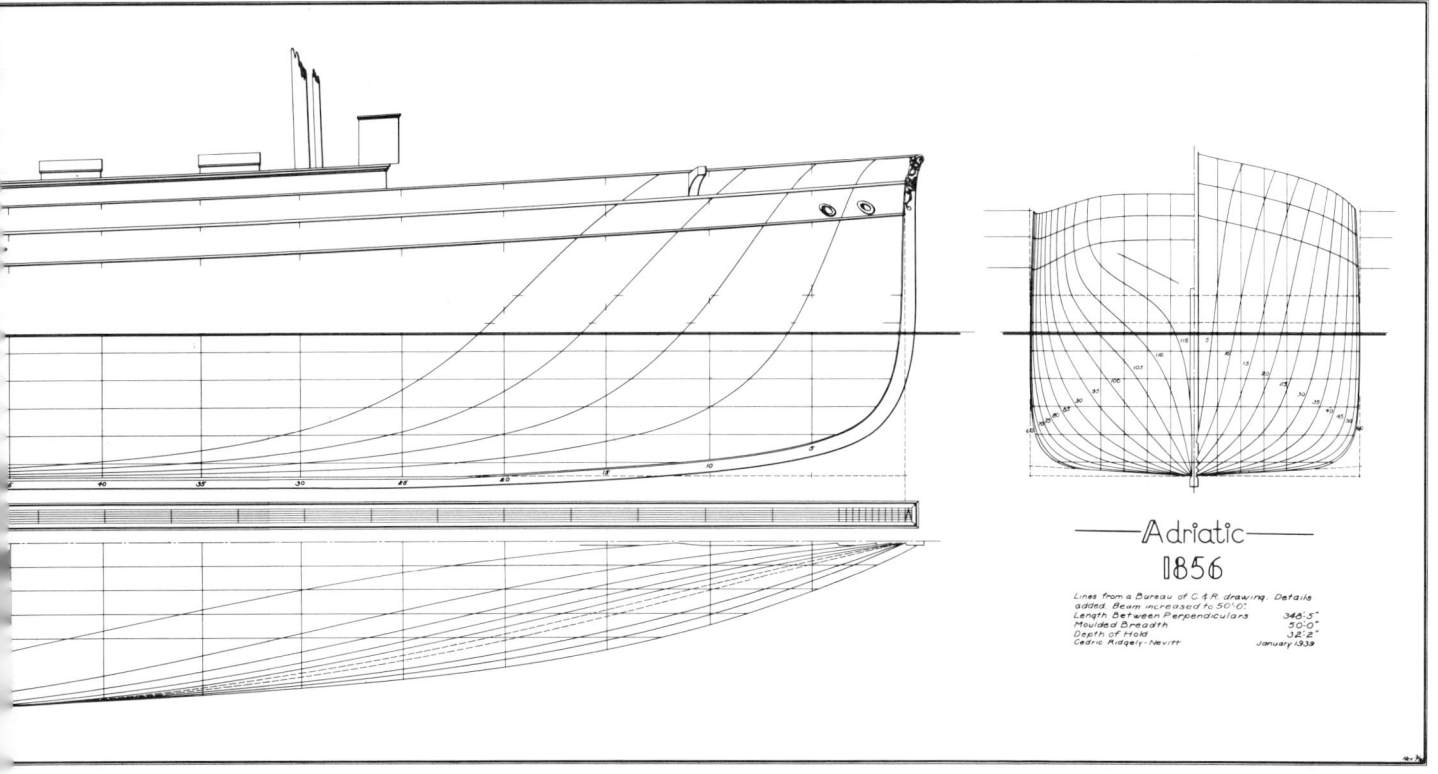

Adriatic 1856
Lines from a Bureau of C & R drawing. Details added. Beam increased to 50'-0".
Length Between Perpendiculars 346'-5"
Moulded Breadth 50'-0"
Depth of Hold 32'-2"
Cedric Ridgely-Nevitt January 1939

construction and great strength.[80] He was the businessman in the family, who ran the shipyard, was the general organizer, and was the person who carried out his brother's plans and ideas. Just after the hull was completed, George Steers was killed when the horses he was driving ran away. At the time, September of 1856, he was only thirty six years old.

When the *Adriatic*'s machinery is considered, it was far less successful than the hull. The exclusive use of side lever engines on all the major ocean steamships so far considered has been noted. Other types existed, however, and were tried out. Cornelius Vanderbilt, in 1850, chose walking beam engines for his steamships on the New York-Nicaragua route. These had been used for decades on rivers and sound steamers. A newer type that had proved itself in England and Scotland was the oscillating engine. Devoid of levers, beams, or linkages, it connected directly to the crank pin. The major complication was that each cylinder had to be mounted on a pair of trunnions, large projections supported by bearings, so that it could rock back and forth as the shaft revolved. The trunnions were hollow; steam was piped to the cylinder through one and exhausted to a condenser through the other.

The elimination of all moving parts between the piston and the crank saved weight, space, and cost. The disadvantage was that the piston rod had to turn the cylinder and the resultant wear on alternate sides of its stuffing box made the latter difficult to keep tight; steam leaked out and air leaked in. The packing for the trunnion joints was sometimes troublesome. But the advantages for the most part outweighed the disadvantages and the lightness was essential for the fast paddlers on the Clyde, the Irish Sea, and the English Channel.

The Baltimore firm of Murray and Hazelhurst produced the first set of oscillating engines in the United States for the 852-ton *Republic* in 1849.[81] Two cylinders 54" bore by 6'-0" stroke turned 25'-6" diameter paddle wheels at 14 rpm. The *Republic* was built for a short-lived Baltimore to Charleston line.

After its first three steamships, *California*, *Panama*, and *Oregon*, the Pacific Mail Steamship Company ordered two larger vessels, the *Golden Gate* and an almost-sister that was transferred to United States Mail ownership and completed as the *Illinois*. Both were highly successful and both had two oscillating engines of 85" bore by 9'-0" stroke. Allaire engined the *Illinois* and Novelty the *Golden Gate*.

Their success led to the adoption of oscillating engines by the Havre Line for the *Arago* of 1855, which had two 65" bore by 10'-0" engines built by the Novelty Works. Thus, when the order for the *Adriatic*'s engines was placed with that firm sometime in 1855, the type had been built for a period of five years and had reached large sizes, with indicated horsepowers up to 1,500. In spite of previous successes, several items went radically wrong

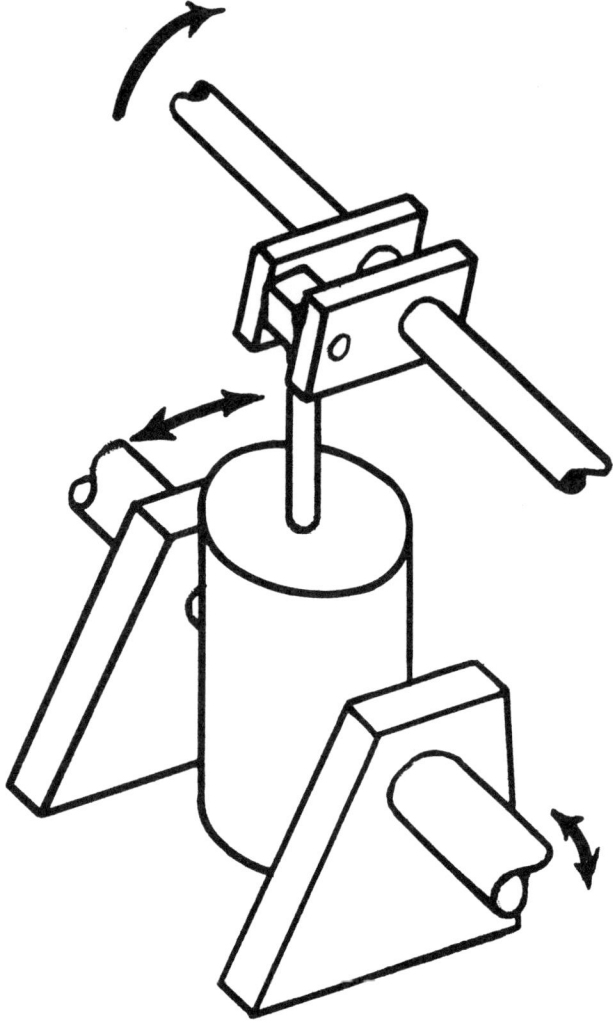

Diagrammatic arrangement of an oscillating engine.

made up of thousands of small-diameter tubes cooled by sea water. It will be remembered that most earlier paddle steamers had jet condensers and, consequently, salt water in their boilers. There had, however, been a number of surface condensers tried out, usually in small screw steamships. With fresh water there were no scale problems and a steam pressure of 25 psi, ten pounds above the designed value for the *Atlantic,* could be used. Steam was applied by eight boilers of the vertical water-tube type, but with shorter tubes and fewer furnaces than the previous Collins ships had. In particular, the 2″ tubes were located so that no flames could play directly on them. Six boilers were 20′-3″ long, the other two 17′-0″, with widths of 13′-0″ and heights of 14′-0″. The boilers were arranged althwartships on either side of a center firing aisle. The engines were intended to be an improvement over those of the *Arago,* which had just been completed. With surface condensers, fresh water in her boilers, and a steam pressure of 30 psi, she became a remarkably economical liner.

The major untried feature on the *Adriatic* was the use of unorthodox steam and exhaust valves and special gear to operate them. They were of Horatio Allen's design and consisted of tapered plugs, 2′-6″ in diameter, turning back and forth. Although these had performed well on a small scale, they proved recalcitrant when enlarged. Instead of turning, the heat expanded them so much that they stuck solidly in place. After months of modifications, they were replaced by conventional poppet valves that had worked successfully ever since James Watt introduced them.[84]

Meanwhile, a second problem arose; the Sewell condensers did not provide for enough tube expansion when heated by the exhaust steam.[85] Furthermore, the 12,000 tubes had rubber packing, which deteriorated rapidly. After various attempts to make them tight, new Pirsson condensers were installed instead.

Next, the operating gear for the new valves proved unsatisfactory. Mr. Collins, in disgust, ordered the Novelty Works to stop. He contracted with Dickerson and Sickels to design and install new gear. It was a poor choice. E. N. Dickerson was a lawyer enamored by steam, and F. E. Sickels was an inventor and engineer. At that time they were designing engines for the U.S. Navy that would, when completed, prove incredibly complex, excessively heavy, and totally unsatisfactory.[86] For Collins they designed a new cut-off and valve gear and had the parts made by other iron works, because the Novelty engineers predicted failure and refused to have anything to do with the work. After a one-hour trial under steam, Collins canceled their contract and ordered them off the ship. The Novelty Works took over and installed a third set of valve gear.

During this time the original air pumps had been

during the design and construction of the two 100″ bore by 12′-0″ stroke engines of 3,300 IHP for the *Adriatic.* The condensers, valves, and valve-operating gear all had major defects. As a consequence, the *Adriatic,* first announced for service in September of 1856, could not make a voyage until November of 1857.

The two enormous cylinders sloped at 45 degrees to the vertical, one forward and one aft of the shaft. Their axes were about one foot off centerline, one to port, the other to starboard, and each drove an individual crank with an 18° angle between the two and a connecting link, from one crank pin to the other, exactly on the hull's centerline.[82] The link was loosely fitted on the pins to provide flexibility between the port and starboard shafts. The piston rods were 14″ in diameter and the shafting 26 1/2″.[83] The wheels were 40′-0″ diameter with sixteen sets of main arms and thirty two floats 12′-0″ wide by 36″ deep. In order to increase the steam pressure and thereby improve performance, two Sewell surface condensers were installed, with their condensing surfaces

168

found too small and these, too, were replaced. The final pair were 42" bore by 5'-0" stroke. The series of failures, modifications, attempts to improve, and eventual replacements held up the *Adriatic* for a year, just when the line needed her most.

All this added greatly to her cost. At the time of launching, a value of $850,000 was quoted.[87] When completed, it had risen to $1,000,000 and, in later years, another $90,000 would be spent on further engine modifications.

Final trials started on Friday, November 13, 1857, when the steamship headed out to sea and ran back and forth fifty miles off the Sandy Hook light ship.[88] At night she anchored. Saturday she proceeded farther out to sea while her engineers gradually increased her speed and adjusted her bearings. The usual practice was to start with large clearances and gradually tighten up the bearings, continually checking to see that they did not overheat. A new engine was noisy, but, as the bearings wore in and adjustments were made, it became quieter and smoother in its operation. On Sunday came a long run south of Long Island, then around Block Island, and back to New York, a distance of 275 miles. The trials started with 900 tons of coal on board — 1,200 were planned for ocean service — at a draft of 20'-0". The steam pressure did not exceed 16 psi and there were no accurate speed measurements.[89] Coming up New York Harbor, a distance of 18 miles was made in slightly over an hour.

Everything worked well. The steamship took her place at the line's Canal Street Pier, where thousands visited her in the last few days before her departure, announced for noon, November 21, 1857. In command was James West, senior master of the line, who had taken the *Atlantic* on her maiden voyage. The crew list included four mates, surgeon, purser, four quartermasters, two carpenters, boatswain, thirty six seamen, chief engineer, three assistant engineers, six fire and boiler superintendents, four oilers, twenty four firemen, thirty six coal passers, steward, three assistant stewards, three stewardesses, thirty six waiters, two cooks, two pastry cooks, baker, barkeeper, barber, two engineer's messmen, six storekeepers of various kinds, and a housekeeper. The total was 183 for 400 passengers.[90]

Sailing day began in the best of fashion, although only thirty eight passengers appeared. There was $380,000 in specie bound for England. The mails arrived in great style in a special wagon drawn by twelve "gaily caparisoned horses." From then on, the *Adriatic* continued to show her reluctance to go to sea, and reporters published conflicting stories. The *Tribune* said that there was a low tide, that the vessel was hard aground, and that, in spite of tugs, she could not be freed.[91]

The *Courier and Enquirer*, which seems to have had better sources of information, reported that a sleeve inside the pipe connecting two exhaust valves on one cylinder was broken and that the pieces could be heard rattling around inside when the engine was turned over.[92] Its replacement delayed the sailing for two days until Monday, November 23.

The largest tug available, the *William H. Webb*, had been chartered and filled with guests for the abortive departure. She was back again on the 23d for the great event. There was a strong wind blowing upstream, plus current in the same direction.[93] The *Adriatic* proved so large that three tugs could not control her. As she came out of the slip her paddle box grazed the nearby *Atlantic*, and she collided with the *Webb*. One of the *Adriatic*'s boats was crushed and some slats on her paddle box were broken, but no serious damage was done. The *Webb* had a dent in her bow. By the time the various craft were sorted out, the *Adriatic* was headed toward Albany and was carried a mile upstream before she could be turned around and started toward the ocean. Her competitors reacted in different ways: Cunard's *Arabia* fired a nine-gun salute, while Commodore Vanderbilt's flagship, the *Vanderbilt*, ignored her.

Sandy Hook light ship was passed at 4 P.M. after a delay to adjust the piston-rod packing. On December 3, at 9 P.M., she was off Port Lynas, but low tides prevented her coming up the Mersey until the afternoon of December 4.[94] The elapsed time of 10 days 4 hours from dock to Port Lynas gave a speed of 12.5 knots despite several stops to cool shaft bearings. The *Persia*, however, had made a 9-day 17-hour crossing twelve days earlier.[95]

The *Adriatic* left Liverpool at 3:50 P.M. December 9 and arrived at her dock on the 21st at 10:00 A.M. Heavy weather for most of the trip and extensive delays because of overheated bearings resulted in an eleven-day twenty three-hour crossing. Eighty passengers returned with her, a good showing for a winter voyage. This time the *Persia* sailed three days later and took even longer, 12 days 22 hours.

The Collins Line, dying for the last year, was on the verge of extinction. Two weeks after the *Adriatic*'s sailing, the *Atlantic* left New York on December 5. Then came a six-week hiatus. On January 16, 1858, the *Baltic* followed, reaching Liverpool on the 28th; she sailed from there on February 3, with only thirty passengers, into a series of cold, bitter, westerly gales. She reached Quarantine on the 18th, only to be delayed by a case of smallpox. When she docked, February 19, 1858, the Collins Line was at an end. Its mail contract had not expired, but its financial resources had. A further sailing advertised for the *Atlantic* on February 13 was never made.

April 1, 1858, found the *Atlantic* and *Baltic* moored on one side the Collins pier; on the other was the *Adriatic* — all dead, crewless steamships.[96] About one hundred spectators had gathered to attend a sheriff's sale. The United States presented liens of $115,500, plus interest, against the first two steamers. Charles N. Potter and Stewart H. Brown presented one for $500,000 on all three steamships as a result of a loan made in May of 1855, with its interest unpaid since November of 1857. The City of New York claimed taxes of $139,000. Finally, a sum of $3,000 was due from the *Atlantic* for unpaid wages to her crew. With over $750,000 already assessed against the three steamships, no one offered a bid except Mr. Potter, on behalf of Dudley Fuller, who was awarded all three for the token sum of $50,000. Fuller was acting for James Brown and Brown Brothers.

The Collins Line, which had started operations in 1850, aided by a substantial subsidy, rose to immediate fame with superb steamships making record crossings. Passengers were delighted with the service. Businessmen applauded the speed and regularity with which the American mails reached England. Shippers were anxious to see Cunard Line's freight monopoly broken and welcomed the reduction of almost 50 percent in rates that followed the appearance of competition. In two years' time, despite these public successes, the line was on the verge of financial failure. Only an annual increase in mail payments from $385,000 to $858,000 saved it. By 1855 the Crimean War's demand for steam transports eliminated all its direct competitors and made it the major transatlantic mail and passenger line until 1856. But three disasters struck in close succession. The *Arctic* was lost in collision with the *Vesta* late in 1854, the *Pacific* vanished without a trace in February of 1856, and Congress canceled the increased subsidy in August of the latter year. Without enough steamships and, simultaneously, squandering large sums on the *Adriatic* and her unfortunate engines, the operation almost ended. After February of 1857 service was irregular, mail contracts were not fulfilled, and passages on chartered steamers became long and slow. No longer a first-class operation, the line's remarkable standard of performance ceased. For its final year it was a travesty of its former self and ended, unmourned, in February of 1858. At no time during its career did its stockholders ever receive a dividend.

In eight years of operation, 185 voyages were started from New York and all but two completed. The line's first steamship, the *Atlantic*, was its most successful; she completed fifty five voyages. After her came the *Baltic* with fifty. The *Pacific* was lost on her 40th; the *Arctic* on her 24th. The last steamship, the *Adriatic*, completed only one voyage. In addition, the line chartered the lumbering *Ericsson*, which made eleven voyages; the *Columbia*, for two, the *Nashville* for one voyage, and the *Quaker City* for one.

Two names are still remembered. E. K. Collins, who ran the line but lost his family when the *Arctic* sank, and James C. Luce, Captain of the *Arctic*, who was forced to end his seagoing career but undertook a new one as a marine surveyor. Collins, on the other hand, abandoned all marine activities.

The contribution, or lack of it, by Brown Brothers has never reached the printed page. They do not appear to have restrained Collins when he particularly needed restraint. In later years they were to invest in the Pacific Mail Line, the fortunes and policies of which they seem to have affected adversely. They did not understand steamship operation and in that field did not use their funds wisely.

The five steamships constructed by the line were technical triumphs even though they were economically unsound. They were very large, fast, extremely seaworthy, and most comfortable. All had machinery problems in their early years but, eventually, overcame their difficulties. They carried the American flag on the high seas and were magnificent examples of the New York shipbuilders' act.

Notes

1. Carl C. Cutler, *Queens of the Western Ocean* (Annapolis, Md., 1961), pp. 208, 210 and 211.
2. House Executive Document 124, pp. 75-78.
3. David B. Tyler, *Steam Conquers the Atlantic* (New York, 1939), pp. 146-47.
4. *Merchants Magazine and Commercial Review* 16 (January-June 1847): 419-20.
5. S. Hartt and John Lenthall, Manuscript Specifications dated April 12, 1847. Naval Record Group 45, File AD, 1812-1859, Design and General Construction of U.S. Ships, U.S. National Archives.
6. *U.S. Nautical Magazine and Naval Journal* 6, (September 1857): 441.
7. "The Old Shipbuilders of New York," *Harpers New Monthly Magazine*, (July 1882), p. 231.
8. *U.S. Nautical Magazine and Naval Journal* 5 (July 1857): 243.
9. *New York Herald*, January 30 and April 21, 1841.
10. Ibid., February 1, 1849.
11. Charles B. Stuart, *Naval and Mail Steamers of the United States*, 2d ed. (New York, 1853), pp. 199-200.
12. *New York Tribune*, April 18, 1850.
13. *New York Herald*, July 1, 1850.
14. *New York Commercial Advertiser*, April 23, 1850.
15. *Chambers Edinburgh Journal* (January-June 1850), pp. 408-10.
16. *New York Tribune*, April 22, 1850.
17. *Tredgold on the Steam Engine, Marine Engines and Boilers* (London, ca. 1851), Division B, "Ocean Steam Navigation in the United States," 1: 26-32 and Plates 1-4.
18. *New York Herald*, February 1, 1849.
19. Stuart, *Naval and Mail Steamers, p. 126.*
20. B. H. Bartol, *A Treatise on the Marine Boilers of the United States* (Philadelphia, 1851), pp. 42-46.
21. *New York Herald*, October 14, 1854.
22. Ibid., October 15, 1854.
23. Stuart, *Naval and Mail Steamers*, pp. 202-5.
24. Ibid., p. 127; Bartol, *Marine Boilers*, pp. 42-50.
25. *Merchants Magazine* 24 (January-June 1851): 127-28.
26. *Journal of the Franklin Institute* 26, 3d ser. (July-December 1853): 391-401, and 24 (July-December 1852): 55.
27. John Austin Stevens, Jr., Memorial of the Chamber of Commerce of the State of New York to the Senate and House of Representatives, (New York, 1864), p. 11.
28. House Executive Document 124, pp. 40-41, 81-82.
29. *New York Commercial Advertiser*, October 19, 1849.
30. *New York Herald*, April 24, 1850.

31. *New York Tribune,* April 29, 1850.
32. *New York Herald,* May 22-24, 1850.
33. Ibid., July 12, 1850.
34. Ibid., January 29, 1850.
35. *New York Tribune,* October 17, 1850.
36. *New York Herald,* December 5, 1850.
37. *Merchants Magazine,* 24 (January–June 1851); 127-28.
38. *New York Herald,* July 1, 1850.
39. Ibid., December 31, 1850.
40. Ibid., February 17, 1851.
41. T. Main, *The Progress of Marine Engineering* (New York, 1893), p. 31.
42. *New York Commercial Advertiser,* August 24, 1853.
43. *Journal of the Franklin Institute,* 26, 3d ser. (July-December 1853): 391-401.
44. Ibid., 28 (July-December 1854): 201-2.
45. *New York Herald,* July 12, 1850.
46. *Journal of the Franklin Institute* 26, 3d ser. (July-December 1853): 391-401.
47. Ibid., 25 (January-June 1853): 32-42.
48. Ibid., 22 (July-December 1851): 195 and 28 (July-December 1854): 201-2.
49. *Merchants Magazine* 27 (July-December 1852): 242-49.
50. *New York Herald,* September 2, 1850; *New York Tribune,* September 2, 1850; *New York Commercial Advertiser,* September 2, 1850.
51. Tyler, *Steam Conquers the Atlantic,* pp. 201-13.
52. *New York Times,* February 28 and March 1-4, 1852.
53. *New York Herald,* October 28, 1850.
54. *New York Commercial Advertiser,* January 25, 1855.
55. Ibid., August 8, 1855.
56. Ibid., June 2 and 7, 1854.
57. Alexander Crosby Brown, *Women and Children Last* (New York, 1961), pp. 213-18, lists 282 passengers, a larger number than any previously reported. The usual newspaper practice, however, was to list only first-cabin passengers.
58. *New York Commercial Advertiser,* October 11-14, 1854.
59. Alexander Crosby Brown, "The Steamer VESTA," *The American Neptune,* (1960), pp. 177-84.
60. Brown, *Women and Children Last,* p. 120.
61. *New York Herald,* January 1, 1856.
62. Edson B. Olds, Speech on the Appropriation for the Collins Steamers, Delivered in the House of Representatives, February 15, 1855 (pamphlet), p. 1.
63. *New York Commercial Advertiser,* April 14, 1854.
64. Ibid., February 8-11, 1856.
65. Ibid., March 1, 1856.
66. Ibid., March 21, 1856.
67. *New York Tribune,* July 29, 1856.
68. *New York Commercial Advertiser,* April 2, 12, and 19, 1856.
69. E. K. Collins appeal received by Isaac Toucey, April 13, 1857, National Archives, NRG 45.
70. E. K. Collins letter, April 21, 1857, National Archives, NRG 45.
71. Letter, February 28, 1857. National Archives, NRG 45.
72. *Merchants Magazine* 39 (July-December 1858): 391-92.
73. *New York Herald,* January 1, 1858.
74. Ibid., December 29, 1853.
75. Ibid., April 8, 1856.
76. *New York Times,* April 8, 1856.
77. *Journal of the Franklin Institute* 31, 3d ser. (January-June 1856): 391.
78. *New York Herald,* October 7, 1856.
79. *Illustrated London News,* December 19, 1857, p. 606.
80. *The U.S. Nautical Magazine and Naval Journal* 6 (July 1857), p. 247.
81. *Journal of the Franklin Institute* 18, 3d ser. (July-December 1849): 400-401.
82. *New York Tribune,* August 27, 1857.
83. *Journal of the Franklin Institute* 38, 3d (July-December 1859): 338-39.
84. *New York Times,* October 22 and 31, 1857, and April 14, 1860.
85. *New York Herald,* December 18, 1856.
86. Frank M. Bennett, *The Steam Navy of the United States* (Pittsburgh, Pa., 1896), pp. 161-66.
87. *New York Herald,* April 8, 1856, and August 27, 1857.
88. *New York Tribune,* November 16 and 21, 1857.
89. *New York Herald,* November 17, 1857.
90. *Illustrated London News,* December 19, 1857, p. 606.
91. *New York Tribune,* November 21, 1857.
92. *Morning Courier and New York Enquirer,* November 23 and 24, 1857.
93. *New York Times,* November 24, 1857.
94. *New York Herald,* December 19, 1857.
95. W. S. Lindsay, *History of Merchant Shipping* (London, 1876), 4:603.
96. *New York Herald,* April 2, 1858.

10
The Havre Line
Part 1:
1850-1861

IN May of 1846, when the Ocean Steam Navigation Company was incorporated, its mail contract offered two alternatives. It could build four steamships and send them to Bremen, with two sailings per month from New York, for an annual payment of $400,000. If preferred, it could dispatch one steamship per month to Bremen and one to Havre for $350,000. Sufficient capital proved difficult to obtain, even for the first vessels ordered, the *Washington* and *Hermann.* After the disappointments arising from these far-from-ideal steamers, the line was unable to raise further funds to complete the two others required by their contract. Its financial resources were strained to the utmost in rebuilding, altering, and improving its existing craft.

Nevertheless, a third order had been placed with Westervelt and Mackay for a steamship larger and more powerful than the *Hermann.* She was launched on August 31, 1848,[1] as the *Lafayette,* and sent to the Novelty Iron Works for her machinery. Soon, however, all work was stopped for lack of money.[2]

One of the Bremen Line directors was Mortimer Livingston, a partner in the firm of Fox and Livingston, which had since 1830 operated a most successful line of sailing packets between New York and Havre.[3] Livingston set up a new organization to take over the alternate part of the Ocean Steam Navigation Company's contract covering two steamships sailing once a month to Havre, via Cowes or Southampton, and the remaining mail subsidy of $150,000 per year. Called at first the New York and Havre Steam Navigation Company, the name was regularly shortened to Havre Line. Its organization was completed sometime in the summer of 1849.

The Havre Line, unlike the Collins one, was operated by a group of experienced businessmen who intended to achieve profit, rather than glory, from their shipping enterprise. Wisely, they did not complete their arrangements until the five-year Bremen contract had been extended to ten. Then the *Lafayette* was taken over and a second steamship ordered from Westervelt and Mackay. Intended to be the *Havre,* she was, before completion, renamed *Humboldt* after an eminent scientist. The *Lafayette* became the *Franklin.* Because they were contracted for at different times and their designs developed to suit different operators, they bore no resemblance to each other, even though they shared the same shipbuilder and the same engine works. Only in tonnage did the agree.

The *Franklin* was 264' long, 41'-8" beam, and 25'-9" deep, with a tonnage of 2,183. When leaving port she drew 19'-5"; when arriving, after the coal was burned, 16'-3".[4] She had the bow of a sailing ship with a full-length figurehead and a three-masted bark rig. The bowsprit was short and bore no jibboom. Square sails as high as topgallants were set on the fore and main.

According to Captain Skiddy, who supervised her construction for the Navy, she was "well proportioned for a

172

mail steamer." She was designed to carry 124 passengers in the first cabin, 597 measurement tons of cargo, 918 tons of coal, and 24 tons of fresh water in iron tanks on the orlop decks.[5] Captain Skiddy listed the following weights for the ship:

Hull	1,100 tons
Engines and boilers	850
Coal	918
Stores and Cargo	375
Masts and Anchors	60
Crew and Passengers	30
Load Displacement	3,333 tons at a draft of 18'-6"

It is probable that the actual weights were higher; the author prefers the 19'-5" figure for the loaded draft.

The Novelty Works supplied two 93" bore by 8'-0" stroke side lever engines, intermediate in size between those of the *Hermann* and the Collins liners. There were four return-flue boilers[6] designed for 15 psi, with three furnaces, three successive passages of smoke through flues, and the uptakes at the opposite end from the furnaces. A single stack served all four boilers. Two fire rooms were needed, one forward of them and the other aft. The Novelty Iron Works seems to have hit just the right combination of engine size, grate surface, and heating surface in this steamship; she became the faster of the two. The paddle wheel's diameter was a little smaller than usual, 32'-2", with twenty eight floats 11'-8" by 2'-0" dipping 6'-9" into the water at mid-voyage.

Although ordered considerably later, the second vessel progressed far more rapidly. She was launched October 5, 1850, two years after the *Franklin*, but was ready by May of 1851, seven months behind her predecessor. The *Humboldt* was patterned after the Collins liners with a plumb stem, a round stern, and side lever engines by the Novelty Works, identical in size to those installed in the *Atlantic*. With dimensions of 283' by 40'-0" by 27'-2" and a tonnage of 2,181, she was longer and narrower than the *Franklin* and had fully 5' less in beam than a Collins liner.

The figurehead was a bronze bust, rather than a wood carving, and was set into a recess in the vertical bow. A barkentine rig with only two square sails, course and topsail, spread considerably less canvas than the *Franklin*. The two side lever engines, 95" bore by 9'-0" stroke, should have been more powerful than the *Franklin*'s. The boilers were larger, each having eight furnaces instead of three, and a very complex gas flow pattern through large flues divided into three sectors by partitions. They seem to have been more reliable than the Collins ones but were less successful than the *Franklin*'s. Everything about the *Humboldt*—the longer narrow hull, the bigger engines, the larger boilers—should have made her the faster vessel. The paddle wheels were made larger, 34'-2" in diameter with thirty six paddles 12'-3" by 2'-2" to suit the increased power.[7] Yet operating experience belied theory; she was regularly beaten by the *Franklin*. A comparison of four westward crossings shows speeds of 10.75 knots for the former to 11.28 for the latter.[8]

Both were well built, iron strapped, and free from the structural failures encountered in the *Washington* and *Hermann*. Westervelt and Mackay had learned from past mistakes how to construct a large steamship's hull. The *Humboldt*, complete, cost $560,000.[9]

Under the command of James A. Wotton, late of the Havre packet *Admiral*, the *Franklin* completed her trials on September 25, 1850, a full two years after her launch. Saturday, October 5, was a festive day for the Havre Line. Not only did the *Franklin* sail for France, but the *Humboldt* was launched. Yet the whole organization had been started just fifteen months earlier. Livingston must have been an excellent executive who had no difficulty in raising the needed capital and converting it quickly into ships.

The *Franklin* took with her thirty five passengers and a considerable cargo. She anchored at Cowes on October 18 to discharge passengers and mail for England and then proceeded on to Havre. Returning, she left Havre on November 1st, Cowes the 2d, and docked at New York on the 16th with a corrected time of 13 days 22 hours for 3,156 miles.[10] The Cunard liner *Cambria*, sailing from Liverpool on the 2d, arrived at Boston in 14 days 15 hours over the much shorter distance of 2,849 miles. The *Franklin*'s speed was 9.45 knots to the *Cambria*'s 8.12; obviously the weather conditions were very bad. Perhaps the most interesting passengers on the *Franklin* were a troupe of circus performers from Franconia and their eight trained horses.

On her second voyage she was able to fill all her 24,000 cubic feet of cargo space with French and Swiss jewelry, laces, silks, gloves, cottons, and luxury goods worth over $2,000,000.[11] This brought in $18,000 in freight, her forty passengers contributed about $6,000 in fares, and the mail subsidy added $6,250 for the one-way crossing. Sums of this size for a mid-winter voyage showed that the Havre Line would succeed. But the mail, paying 20 percent of the income, was the critical feature. Without this margin, the earlier *United States*, on the same route, had not been profitable.

The *Humboldt* was to depart from New York on March 8, 1851, but delays held her up until May 6, when she sailed under Captain David Lines. When overheated engine bearings forced her to slow down on the 12th, her passengers glumly watched Cunard's *Asia* steam past.[12] The former took 11 days 15 hours to Cowes, while the *Asia* made Liverpool in thirty seven hours less time. Returning, the *Humboldt* left Havre on the 4th of June,

173

Cowes the next day, and took somewhat under thirteen days from Cowes to New York.

Sailings at four-week intervals followed. The dates were always two weeks after the Bremen Line's departures, so that the two organizations complemented each other. After April of 1851, the Havre, Bremen, and Collins Lines gave the American traveler a choice of a sailing every fortnight from New York to Cowes or, at the same interval, to Liverpool. The Havre pair were extremely regular, making twelve sailings per year, strictly in accordance with their mail contract. Once a year came a longer interval, in December, when traffic was at a minimum; the extra time was used for docking and overhaul. The *Franklin* and *Humboldt* did not have to be withdrawn for major changes as did Collins and the Bremen liners. They arrived and departed on schedule, delivered their passengers, mails, and cargoes, and only rarely got into trouble.

At no time did they try to break speed records, and, as a result, their coal bills were never excessive. The *Franklin*'s chief engineer provided data covering three consecutive voyages in the summer of 1852.[13] The average speed eastward was 11.55 knots and westward 10.55. The mean steam pressure was 15.75 psi, the rpm 14.35, and the measured horsepower 1,466. In the Collins line chapter it was noted that the *Pacific* made her best crossing at 13.08 knots with a horsepower of 2,122. The *Franklin*, then, operated at 70% of the Collins liners' power. The coal for a crossing might be 750 tons all the way to Havre against 900 on the shorter route to Liverpool. Speed was expensive.

The first interruption to a long list of sailings exactly on schedule came after a stormy passage when the *Humboldt* left Havre on January 13, 1852, and Cowes a day later, heading into a series of heavy gales. Her bulwarks were wrecked by seas over the bow, paddle boxes were stove in, a boat was swept overboard, and her rudder was so damaged that she steered only with the greatest difficulty. She put into Halifax on February 1 for coal and temporary repairs, sailed the next day, and delivered thirty five passengers to New York on February 5.[14] The time taken for her repairs forced the Havre Line to cancel a sailing.

The *Franklin*, later in the year, had her first severe accident. Eastward on October 26, 1852, three days out of New York, her main shaft broke in way of the starboard engine. Because both the first and second assistant engineers were in the engine room, everything was speedily secured. One wheel was lashed in place, sail was set, and the vessel continued on one engine and one paddle wheel, making 150 to 200 miles per day and reaching Southampton four days later.[15] Day and Ballock, at that port, made a new intermediate shaft and completed its installation in a month's time. She had arrived November 8, 1852, and sailed again, from Cowes, after a quick trip to Havre, on December 14. The weather was so bad that she had to stop at Halifax on the 29th for extra coal.

Just as the merchants of Bremen had been delighted to have a steamship service to New York, so their Havre counterparts were equally pleased with a direct means of sending their luxury items abroad. Small steamers had regularly carried French and Swiss goods across the English Channel to Cowes, or Southampton, for transshipment to the *Washington* and *Hermann*. Now that double handling could be eliminated, the Havre liners regularly carried full cargoes. The demand for space increased so rapidly, however, that the Bremen liners, sailing alternately with the Havre ones, could still carry almost as much continental goods as they had before.

Some rearrangement of the cabin spaces must have been made late in 1852, for the *Franklin* and *Humboldt* advertised second cabins at $70, in addition to first, starting in September. Also, by 1853, extra space for cargoes up to 800 tons had been found. Record amounts worth $2,500,000 and passenger lists as high as 320 were reached. With the 320 passengers, in April of 1853, came 111 "superior" merino sheep. Other blooded livestock was carried from time to time.

On October 22, 1853, the *Humboldt* left New York on her sixteenth voyage, still under Captain Lines. Bad weather and the fact that she had to heave to for two days while unspecified engine repairs were made, resulted in a slow fourteen-day crossing to Cowes. With ninety passengers she departed from Havre on November 23 and was slowed down so much by continued storms that she had to head for Halifax for coal. On December 6 she took on a pilot, but, because of poor visibility, struck a rock off Sambro light below the harbor's entrance. She backed off but was holed so badly that she had to be beached to prevent her sinking.[16] She grounded at Portuguese Cove, ten miles from Halifax, with four to five feet of water in her holds. There was no loss of life and about half her cargo was taken out before the ship broke up. By December 15 breakers had completed her destruction. A Spofford and Tileston steamer, the *Marion*, was sent to pick up the salvaged cargo, much of it damaged by water.

A second Charleston liner, the *Nashville*, was hired as the *Humboldt*'s replacement. She had just returned to New York, on January 13, 1854, from her maiden voyage. Although a highly successful coastwise steamer, a larger and faster successor to the *Northerner*, the *Nashville* was small for the Havre Line. When filled with enough coal for the trip across the Atlantic, there was neither space nor buoyancy left for westbound cargo, the financial mainstay of the company. She was chosen because she could cope with the small winter passenger load and carry the scheduled mails. Her great advantage was that she was available. The only large steamships that might have been chartered were the two great failures of the U.S. Mail Line, the *Ohio* and the *Georgia*. Livingston knew better than to take a chance on them.

Three days after reaching New York from Charleston, the *Nashville* cleared for Havre on January 16, 1854, commanded by Captain Berry, senior master of the Spofford and Tileston fleet. On February 15 she sailed from Havre and, a day later, Cowes. After passing the Grand Banks, heavy gales seriously delayed her, swept away a boat, and damaged a paddle box. At 1 A.M. on March 5 a light was sighted and thought to be Fire Island. In fact, the *Nashville*, without a recent observation to fix her position, was off Barnegat Light on the New Jersey coast, and, as a result, ran aground at Little Egg Harbor. No damage was done; she backed off without assistance and reached New York the same day.[17] Her second round trip was less eventful and took thirteen days to Cowes in March and sixteen days from Havre to New York in April. She then returned to the Charleston route.

Another coastwise liner was chartered from the same owners, this time the *Union*. She was almost identical in size to the *Nashville* — 215' by 34' by 22' with a tonnage of 1,200, but had several more attractive features. Instead of one, she had a pair of side lever engines of 60" bore and 7'-0" stroke; in addition there was a three-masted rig rather than two. Noted for speed, she had spent her first year in the New Orleans trade, her second to Charleston, the early part of 1853 from New York to the Isthmus of Panama, and, finally, had returned to Charleston. For almost two years the *Union* would serve as the *Humboldt*'s replacement.

Her first sailing from New York was scheduled for May 6, 1854; she passed Eddystone light thirteen days later en route to Cowes. Returning, she sailed for Havre on June 7 and reached New York on the 23d. Although slower than the *Franklin*, she met her scheduled departure dates, every eight weeks, without difficulty. Coming home from her seventh voyage, in May of 1855, she ran onto Sable Island in a dense fog and remained there fifteen hours until fifty tons of coal had been hoisted up and thrown over the side.[16]

On October 20, 1855, she left for Havre and was steaming through a calm sea in pleasant weather when, two days out, a shaft gave way.[19] She returned to New York on the 25th.

Some seven weeks later, she sailed on December 17 for her tenth and last call at Havre. She left there January 19, 1856 and did not touch at Cowes, but was delayed so much by winter storms that she had to call at Halifax on February 2, and Newport on the 6th, for extra coal, and did not reach New York until February 7. At nineteen days, it was her longest Atlantic passage.

Like the *Humboldt*, the *Franklin*, through no fault of her own, fell victim to the hazards of the sea. Eastward from New York on June 3, 1854, she made an excellent twelve-day crossing to Cowes. Coming home she left Cowes on July 6 with 160 passengers and 700 tons of cargo worth almost $1,000,000. On the morning of July 17, in a dense fog, she grounded at full speed near Moriches Inlet, about fifty miles west of Long Island's Montauk Point. For four days the skies had been so overcast that no observation could be made. She swung broadside to the sandy beach but was not holed. The passengers were landed and sent on to New York, via the Long Island Railroad. Tugs were dispatched to her assistance, but during the night a rising sea broke against her with such force that she parted her anchor chains and was forced over the outer bar, on which she had been resting, and ended up about fifty feet from shore. The surf remained so high that the three craft intended for her rescue could do nothing and the schooners sent for her cargo did not dare come close. Nevertheless, a line was rigged from the main masthead to the shore and cargo slid down it one case at a time. The passenger's baggage and the valuable items were put into surf boats and rowed out to a steam tug, the *Leviathan*, while the more prosaic items were carried across the sand and loaded into boats in Moriches Bay. Three days after the stranding the hull was badly hogged and bilged, the remaining cargo water soaked, and the wreck abandoned to the underwriters.[20] Further attempts to get her off were futile. By September 12 the bedplates of the engines were broken and the hull so battered that it could be seen to bend and twist when struck by the waves.[21] Finally, the wreckage was sold at auction for $1,625 to a Mr. G. S. Lewis.[22]

Within less than a year's time the line had lost the two fine steamships completed for it four years earlier. Both had been seaworthy, realiable, economical craft and both had run aground in a dense fog. Comparatively unknown vessels, they were the most successful Atlantic liners yet constructed in the United States. They ran regularly, pleased their passengers, collected their mail payments, attracted large amounts of extremely valuable cargo, and regularly made such satisfactory profits that their owners were eager to continue the Havre service and could easily find the capital to replace them.

The line's first action was to charter vessels to continue the regular mail service and to relieve the cargo glut that had been building up at Havre, even before the *Franklin*'s loss, largely because of the *Union*'s limited capacity. As much as possible was transferred to Cowes and placed on the *Washington* and *Hermann*, but they, too, did not always have enough space.

A British auxiliary screw steamer, the *Indiana*, had been pressed into service, just before the *Franklin*'s loss, and dispatched as an extra steamer by the line's European agent, Croskey and Company. She left Havre July

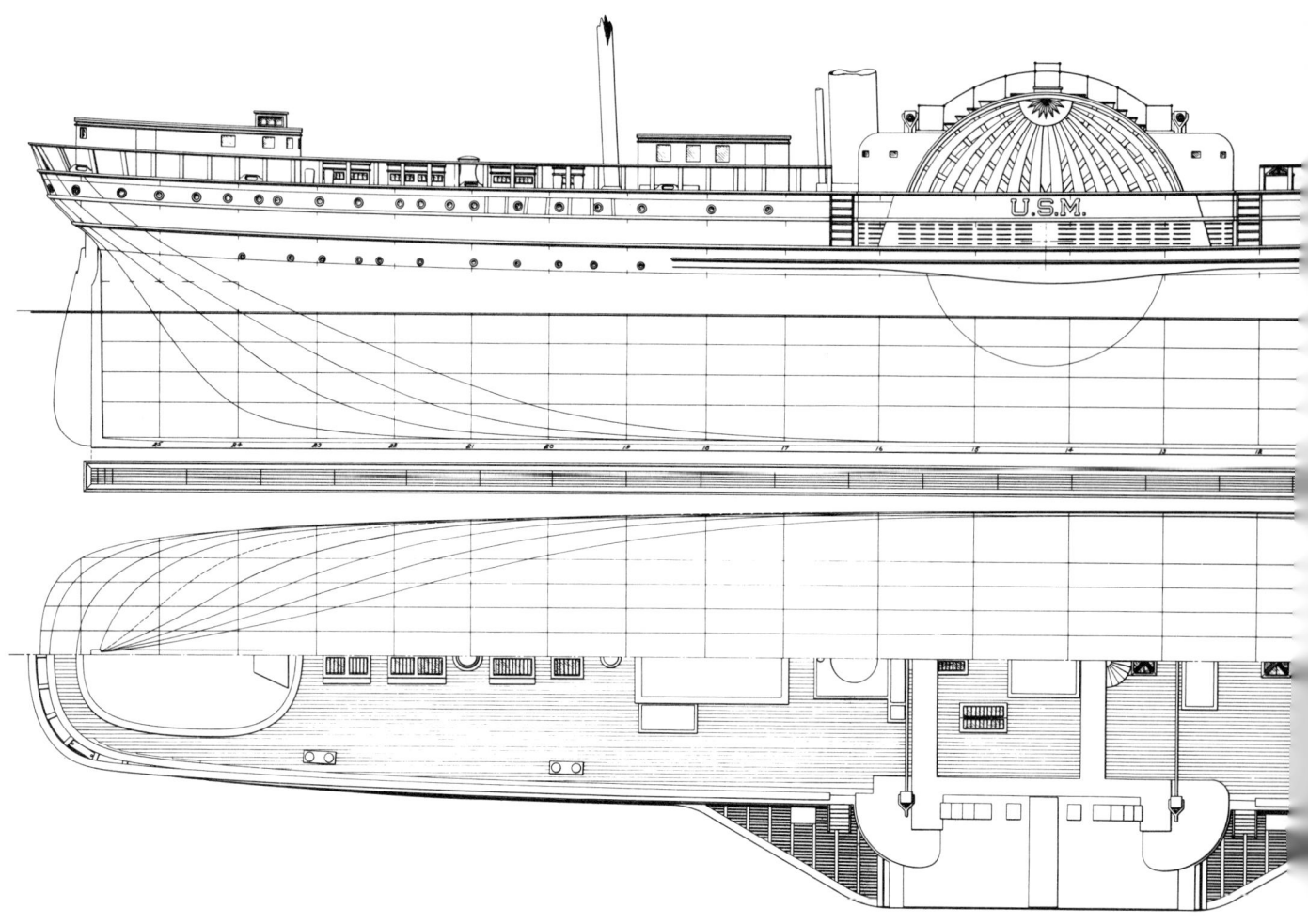

Arago, 1855, lines and deck.

16, 1854, the day before the *Franklin* grounded, laden with over a million dollars worth of goods, forty prize merino sheep, and eighteen Spanish jackasses for breeding purposes.[23] Her owner, the General Screw Steam Shipping Company,[24] was considering an Atlantic venture; the *Indiana*'s sailing was a trial of the route. Although she made another voyage, the Crimean War intervened and British competition did not develop.[25]

As a direct replacement for the *Franklin*, a new side-wheel steamer, the *St. Louis*, was chartered from the Pacific Mail Steamship Company.[26] She was a vessel of large size, 270'-2" by 35'-2" by 25'-7", with a tonnage of 1,621. Originally intended for tropical service on the Pacific Coast, her upper works were modified and the overhanging guards removed to make her more suitable for the stormy Atlantic. Jacob Westervelt and Company, successors to Westervelt and Mackay, were her builders. She was a sister to the *Sonora*, from the same yard. The pair were most unusual in having their boilers and a single stack located aft of the paddle wheels. All the Cunard liners, from the *Britannia* of 1840 through the *Africa* of 1851, were so arranged. American steamships, on the other hand, put their boilers forward. For machinery, the Morgan Iron Works, formerly T. F. Secor and Company, provided two 50" bore by 10'-0" stroke walking beam engines turning wheels of 30' diameter with twenty eight floats 9' wide by 16" deep.[27] An unusual feature was the installation of a Pirsson surface condenser for the port engine and an older-style jet condenser for the starboard one.

Two return-flue boilers, 30' long by 13' wide, with six furnaces each, supplied steam at 22 psi. She had two masts and a fore topsail schooner rig.

The *St. Louis* ran trials on July 21, four days after the *Franklin*'s stranding.[28] On board were Captain Skiddy, who had supervised her construction for the Pacific Mail Line; Mr. Aspinwall, the latter's President; Mr. Fox of the Havre Line; George Quintard of the Morgan Iron Works; and John McGimm, veteran New York pilot. All seems to have gone well enough to justify the inevitable banquet on board. Conversion, other than the removal of

176

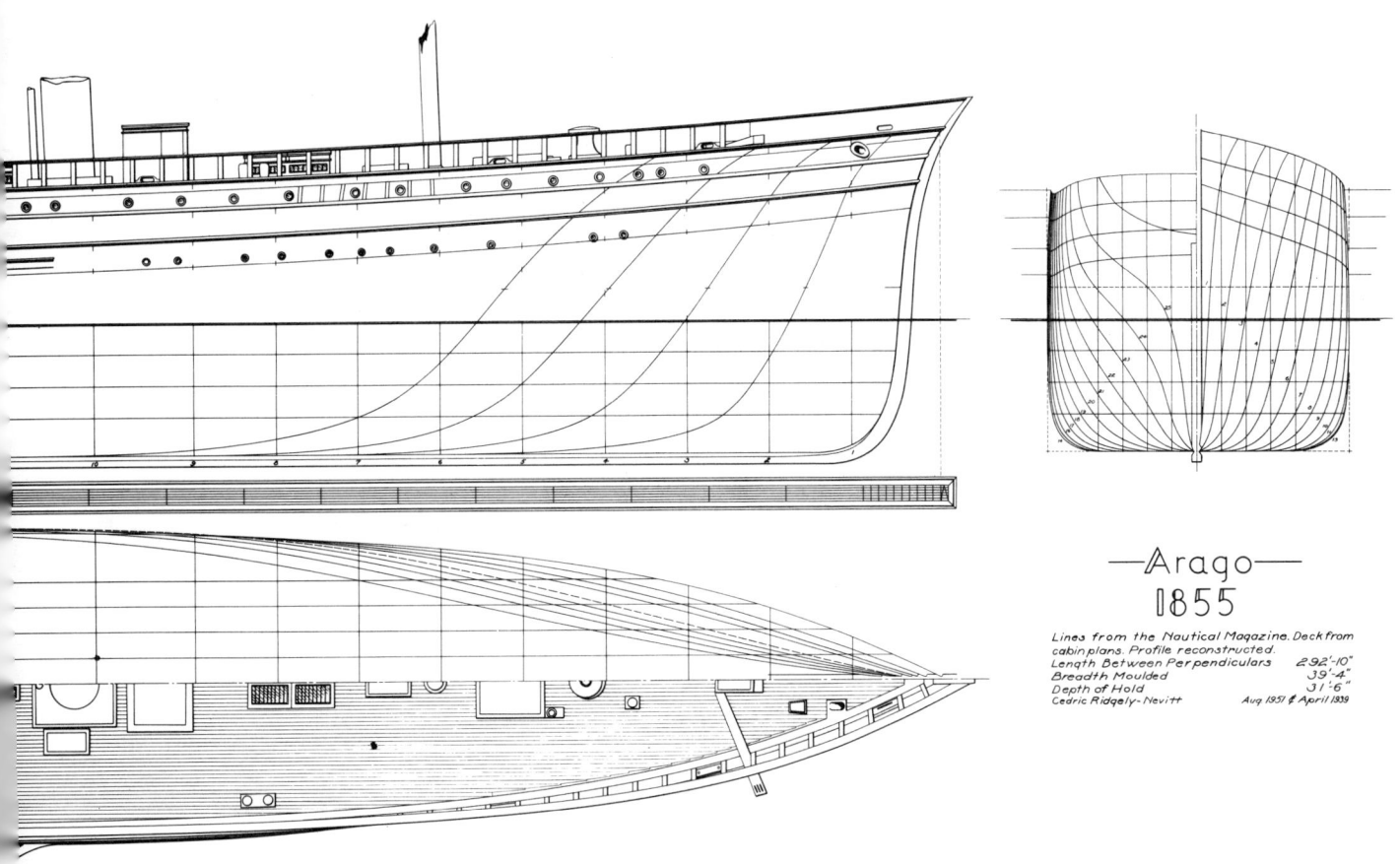

the overhanging guards and any cabins on them, must have been minor, for it was completed on August 1, the day she sailed for Havre under Captain Asa Eldredge, formerly of the *Pioneer*.

The *St. Louis* was a remarkably satisfactory steamship for the line to find and put to work in two week's time. She was larger than the *Union*, was designed for the long sea passage from Panama to San Francisco, and was therefore provided with both ample coal bunkers and cargo space; moreover, she had the advantage of following her sister, the *Sonora*, so that any minor hull and machinery problems had been found and corrected before her completion. She had room for 160 cabin passengers and boasted a dining saloon on deck, like the Collins liners, complete with polished oak, white panels, and gold ornaments. The space for 600 steerage passengers was, obviously, not needed, but could be used for cargo. In addition, the *St. Louis* had a nice turn of speed. She reached Cowes on August 13 in 11 days 17 hours, at 11 knots, and carried on to Havre the same day, arriving there in 12 days 4 1/2 hours. Returning she did equally well, 11 days 15 1/2 hours Cowes to New York. To gladden New York's bon vivants, she brought along 150 baskets of champagne. Her increased steam pressure, 17-24 psi, and less blowing down of the boiler with the surface condenser, gave an extremely low coal consumption of twenty five to thirty eight tons per day.

The *St. Louis* alternated with the *Union* in making two voyages in 1854 and three in 1855, completing her intended last trip on June 19, 1855. When the *Union* suffered a broken shaft, she was chartered once more and sailed from New York October 29, 1855, on her sixth and final transatlantic voyage. When she reached New York on December 6, she was turned back to the Pacific Mail.

The Havre Line reacted quickly, when disaster struck it, to keep up its service, to deliver the mails, and to satisfy its shippers. Once the temporary measures were running smoothly, new steamships were ordered. Funds for the *Humboldt*'s replacement were subscribed by the Ocean Steam Navigation Company; Mortimer Livingston; William Iselin, the line's agent in Havre; Captain Lines; and Jacob Westervelt.[29] At the time, the cost was estimated to be $400,000 but, before completion, grew to $500,000. J. A. Westervelt and Company, sometimes referred to as Jacob Westervelt and Son, was to build a steamship of about 2,200 tons with all the latest improvements. Captain William Skiddy was to oversee the work. Erastus W. Smith, former chief engineer of the Bremen Line, had moved to the same post with the Havre Line and supervised the design and construction of a pair of oscillating engines supplied by the Novelty Iron Works.

The keel of the new steamship, eventually named *Arago*, was laid during the third week of June 1854, less than seven months after the *Humboldt*'s loss.[30] On January 27, 1855, she was launched and by June of the same year was ready to sail. A year's construction time for a major liner was a fine example of what could be accomplished when a noted shipbuilder, an experienced engine works, and an owner with clear ideas of what he wanted collaborated. The end result was the first of a pair that were to become the most successful side-wheel liners ever built of wood for Atlantic service.

With the *Arago*, the American liner started developing in a new direction. She was not an outgrowth of the Bremen steamships, the Collins liners, or of the previous Havre vessels. Instead, her ancestors were two highly successful coastwise vessels constructed for the California trade. The Pacific Mail Steamship Company had in 1850 ordered two large steamers of original design to improve their West Coast service. At the time, the line had fleets on both the Atlantic and Pacific, as did their rival, the United States Mail Steamship Company. In early 1851 the two came to terms, the former agreeing to retain the Pacific and the latter the Atlantic part of the California service. The Pacific Mail completed the *Golden Gate*, then being built by William H. Webb, and turned the almost-duplicate, *Illinois*, at the Smith and Dimon yard, over to the U.S. Mail. Both were of about 2,100 tons in size and just under 270′ long, and both had large passenger capacities. Each was driven by a pair of oscillating engines of 85″ bore by 9′-0″ stroke. In appearance they had short bowsprits built into the hull, one stack forward of the wheels and another aft, and a wide expanse of flush deck unencumbered by houses. Instead of high bulwarks around the spar deck, there was an open rail with galvanized netting stretched between wood stanchions.

The use of oscillating engines cut the engine-room length, and the rearrangement of boilers, two forward and two aft, put both fire rooms adjacent to the engine room; the engineer on watch could see what was happening and act accordingly. In addition to compactness, the

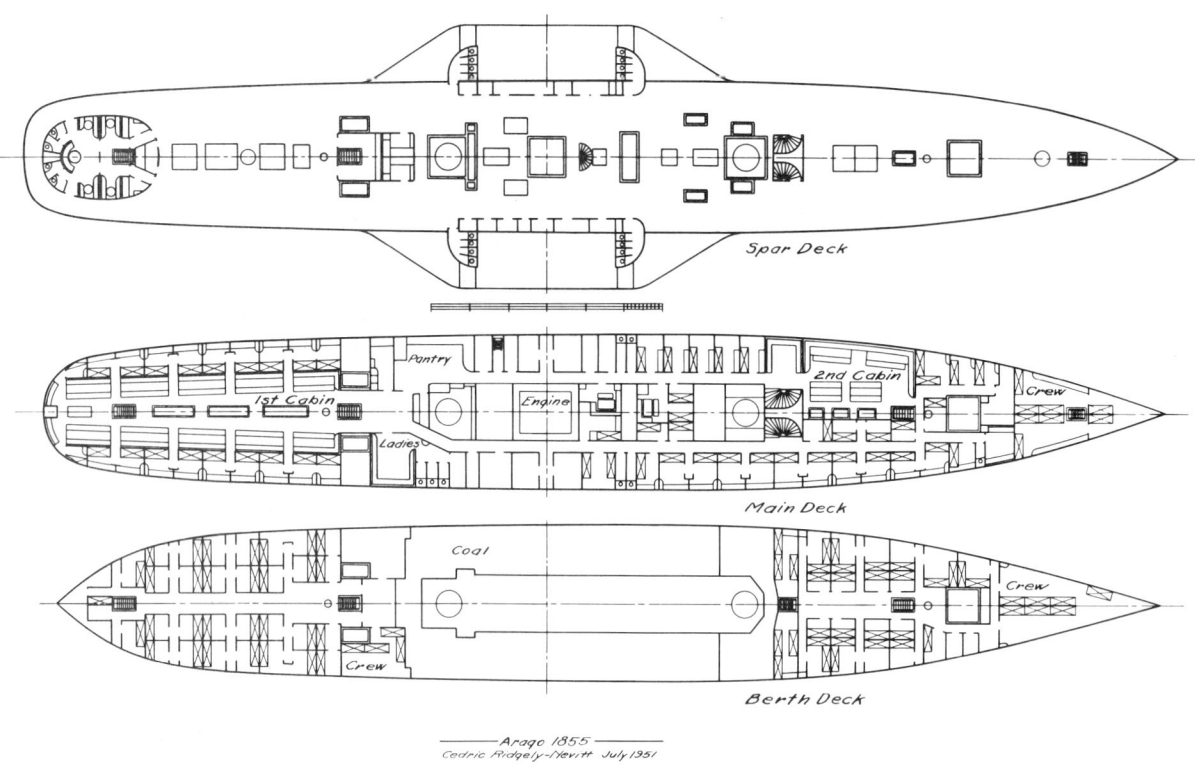

Cabin plans of the *Arago*.

engines were lighter than the side lever type. Both vessels were completed in August of 1851. After some initial engine problems, the *Illinois* began to make record runs between Aspinwall and New York. George Law, never a man to decry his achievements, advertised her as the "fastest steamer in the world!"[31] At the same time, the *Golden Gate* was making a name for herself as the speed queen of the Pacific with a 290-mile day's run at 12.1 knots.[32]

The new Havre liner was an *Illinois,* or a *Golden Gate,* lengthened 20' deepened 2' with an unchanged beam. Her dimensions were 290' by 40' by 31'-6", and her tonnage was 2,240, about the same size as the *Humboldt*'s. John W. Griffiths, editor of the *Nautical Magazine,* computed her technical data at the 15'-0" molded water line (distance above the top of the keel) and found her displacement to be 2,954 tons; her block coefficient was .578, giving her an even sharper hull than a Collins liner, her midship section coefficient .948, and her waterplane coefficient .70. These would produce a fine hull very easy to drive. She could carry 900 (measurement) tons of cargo and 900 (weight) tons of coal.[33]

The *Arago*'s keel was 16" square, the center keelson 16" x 18", and enormous built-up side keelsons 36" x 60" supported the engines. Frames and floors were spaced 30" apart, tapering in depth from 21" at the keel to 6 1/2" at spar deck. The floors, 15" wide, were filled in solid. The frames gradually narrowed to 11" width at the deck. Diagonal iron straps 5" x 7/8" spaced 4' apart extended from the heads of the floors to the clamps below the spar deck. The yellow-pine ceiling strakes tapered from 14" thick at the center keelson to 5" under the spar deck and were edge-bolted throughout. The oak bottom planking ran from 10" thick at the garboards, next to the keel, to 4 1/2" at the bilges. Only the deck planking was light, 3" pine at the uppermost deck, 3 1/2" at the main. The engines and boilers were enclosed by a rectangular watertight compartment with its boundaries made of two thicknesses of yellow pine. Thus, if the hull were flooded, the machinery could, in theory, be kept dry and in operation. The arrangement was a direct consequence of the *Arctic*'s rapid sinking. In practice, it is doubtful that either real tightness or safety was achieved. Because all the coal had to be stowed outside the longitudinal bulkheads, they had to have holes cut in them for it to reach the fires. And, while the openings had iron doors, it is unlikely that they could ever be made tight. Moreover, since they were always open at sea, it is questionable whether they could be closed in an emergency. No bulkhead in a wood ship could achieve complete tightness; it stopped at the ceiling and water could find its way between the ceiling and the planking, gradually leaking past any such barrier. All it could do was slow down the flooding process, not stop it.

The *Arago*'s hull had three full decks from stem to stern. On the spar deck was a Brown capstan forward to raise the anchor, an improvement over the pump-brake windlass shown on the *Tennessee*'s plans. Aft of it came companionways for access and skylights to light and ventilate the spaces below. A large forward hatch was fitted for loading cargo. A house near the mainmast provided cushioned seats and the grand staircase down to the first cabin. Aft was a rounded house containing the steering wheel, staterooms for the master and mate, the smoking room, four water closets, and an after stairway. Storage spaces lined the inboard side of the paddle boxes, a number of them leading to the ice and vegetable rooms on the deck below. Oddly enough, there was no forward pilot house. Perhaps Captain Lines, who was overseeing her, was still enough of a packet ship sailor to insist on steering his ship from the stern. Traditions were difficult to change.

On the main deck, the dining saloon was aft and extended for a full one hundred feet, with two rows of tables and a series of open galleries on centerline to let light down to the berth deck. Suspended over the tables were "French fancy lamps" and shelves with cut glass bottles, sugar bowls, and other items needed when the tables were set. On either side of the ship were short passages giving access to double staterooms. Large glass ports were let in the outboard side of each room. Forward, on the starboard side, was the ladies' boudoir, bath, and water closets. To port were the captain's main cabin and the pantry, complete with a bar and steam tables. In way of the wheels, where minor seepage was common, came the vegetable rooms and icehouses.

To starboard, still proceeding forward, were more first-cabin staterooms. On the port side, amidships and forward, were officers' rooms, stewards' rooms, galley, messrooms, and the second-class dining saloon. Going down to the berth deck, the second-cabin passengers were forward and the less luxurious first-cabin staterooms aft. On either side of the engines and boilers were coal bunkers. The total passenger capacity came to about 250.

Two masts, both square rigged, carried courses, topsails, topmast studding sails (of doubtful utility), spencers, and jibs. Yards were interchangeable between masts, the fore and main ones measuring sixty feet long, a very moderate size when compared to those found on a clipper or a packet. The after boiler extended under the mainmast, which had to be stepped on the berth deck instead of on the keelson. The same boiler negated an after hatch on centerline. Instead, port and starboard hatches were used to serve the cargo spaces outboard of the longitudinal bulkheads.

The Havre Line benefited greatly from the services of Erastus W. Smith, their resident engineer. In his earlier work with the Bremen Line he had been instrumental in

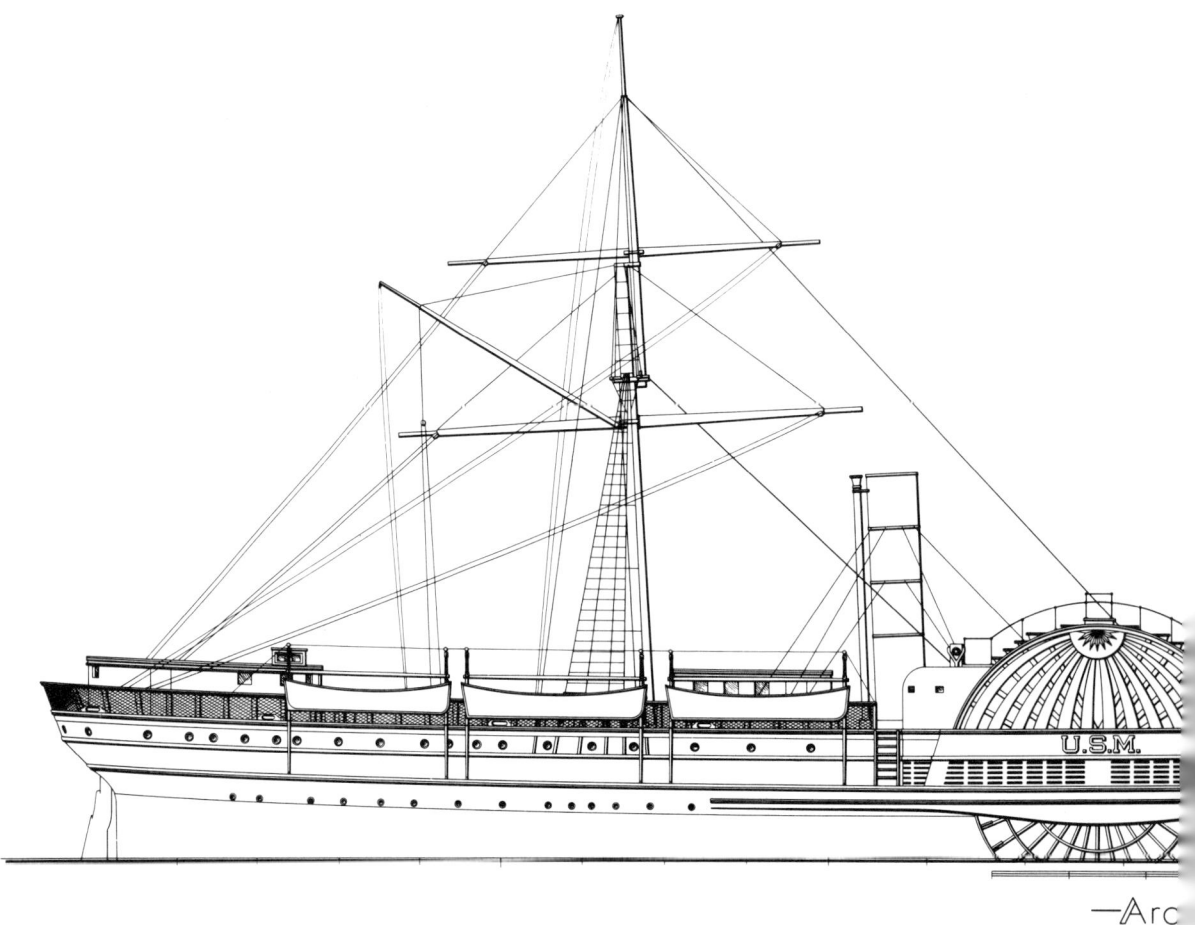

Arago, sail plan.

solving their engine and boiler problems. He saw to it that the *Arago*'s machinery was designed to be the most economical possible rather than the most powerful. To achieve this end, many modifications were made to the engine and boiler designs of the *Golden Gate,* the largest oscillating installation yet produced by the Novelty Iron Works. The steam pressure was increased from 12 psi to 30. As pressures increase, fuel consumption goes down. To keep the boilers from scaling up, a Pirsson surface condenser was supplied so that fresh water would be pumped back into the boilers. The pros and cons of the surface condenser have been discussed before. It had been tried out as early as 1845 in the *Massachusetts* and the experimental installation in the *St. Louis* for one engine and a jet condenser for the other has just been noted. The *Arago* was the largest American seagoing ship to try it and, after a change in tube materials, became the first major liner in which it performed successfully.[34]

As a result of the higher pressure, it was possible to power the *Arago* with a pair of 65″ bore by 10′-0″ stroke engines, which produced about 1,450 Indicated Horsepower and drove the steamer at a speed of eleven knots. The *Golden Gate*'s were larger, 85″ by 9′-0″, but delivered less horsepower. Despite the increased power, the *Arago* consumed forty five to fifty tons of coal per day to the *Golden Gate*'s sixty.[35] The Collins liners devoured ninety tons per day.

The detail designs of the machinery were, of course, by the staff of the Novelty Works. One cylinder was forward of the shaft, one aft, each at an angle of about 24° to the vertical. Each piston rod turned an individual crank just off the ship's centerline, and, exactly on center, a wrought iron "union link" joined the two crank pins and provided flexibility between the port and starboard shafts. Valves to admit and exhaust the steam were of the

180

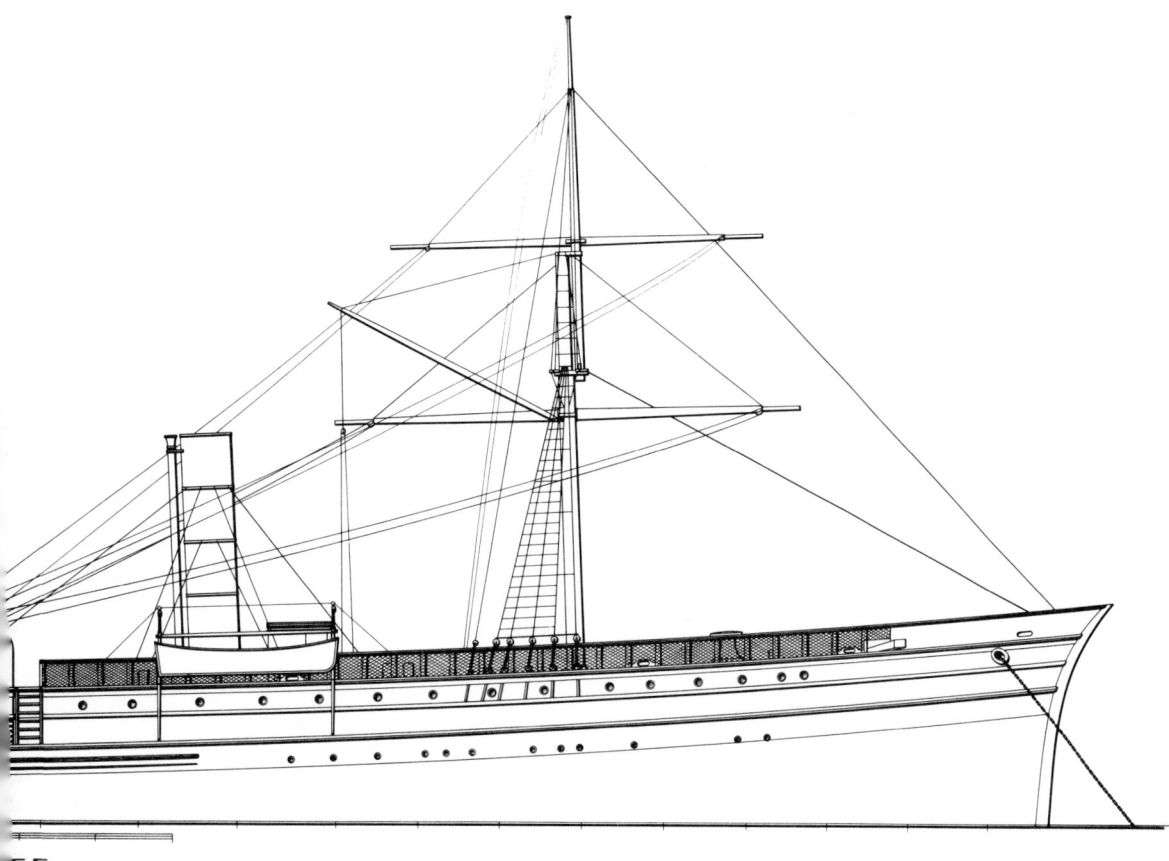

double balanced poppet type, for ease in lifting by hand, and were controlled by an Allen and Well's cut-off, which could be adjusted while the engine was running. It had the further advantage that the valve gear functioned going astern as well as ahead. The Collins engines had always required hand operation when backing. The 33′-0″ diameter paddle wheels had buckets 10′-0″ wide by 20″ deep.

Only two boilers were installed, both being extremely large, of the conservative, return-flue type that had given trouble-free service in many vessels. The after one was 28′-5″ long and had a single set of return flues so that both the stack and the furnaces were at its forward end. The forward one was 29′-3″ in length with double-return flues so that the combustion products traveled its length three times from the furnaces located aft to the stack forward. Both boilers were 14′-6″ in diameter made of 3/8″ thick iron and had two tiers of furnaces and grates.

Around the stacks were steam chimneys 7′-6″ in diameter and 12′ high.

A small donkey boiler furnished steam for two independent pumps, each of 700 gallons per minute capacity; it could be used in port when the main boilers were shut down. The pumps could also be served by the regular boilers, and each could pump water into the ship as a fire pump or, out of it, as a bilge pump.

Just after the *Arago*'s keel was laid, the stranding of the *Franklin* made a second new steamship necessary. The normal procedure would have been to order a sister to the *Arago*. It happened, however, that the Smith and Dimon yard had a partially completed hull on their hands that, for some obscure reason, had had no work done on it for a year or more. It had been ordered by the Pacific Mail Steamship Company as a successor to their *John L. Stephens*, which had been delivered in 1852. It was to be 280′ long, 40′ in beam, and 26′ deep. When about two-thirds complete, work was stopped.[36]

The Havre Line, by taking over the unfinished hull and modifying it to suit their requirements, could obtain an early delivery. The length was increased slightly, the structure reinforced by adding a second layer of 3 1/2" oak planking from the keel up to four feet above the waterline. Further additions were eleven deep keelsons and an extra set of iron straps on the longitudinal bulkheads either side of the engine and boiler rooms. Considering all these changes, a launching date of December 1855 was anticipated, but the yard outdid itself; she slid into the East River on September 4.

Named the *Fulton*, her dimensions were almost identical to those of the *Arago*.

Steamship	Length	Beam	Depth	Tonnage
Arago	290'-0"	40'-0"	31'-6"	2240
Fulton	287'-6"	40'-10"	32'-0"	2307

She was somewhat heavier and could carry only 800 tons of coal at a draft of 17'-6" to the *Arago*'s 900. Externally, the *Fulton* could be identified by her short bowsprit and topgallant sails. The *Arago* had neither. On deck was a forward pilot house and a midship structure in way of the paddle wheels containing the engineers', doctor's, and purser's cabins; galleys; companionways; and skylights over the engine. Aft, the spar deck arrangement was close to the *Arago*'s, but had a large, single, cargo hatch on the starboard side rather than a pair of small ones. Loading the port hold must have been most inconvenient.

Below decks there was little difference between the two. On the orlop decks were a mail room forward, lined with iron, and a vault aft for the stowage of 6,000 bottles of wine—a more than ample supply for 250 passengers.

The engines were quite different from the *Arago*'s. They came from the Morgan Iron Works from designs by Miers Coryell, its chief engineer. The two oscillating cylinders were still 65" bore by 10'-0" stroke but were inclined at an angle of 45° to the vertical, both turning a single crank. The engine was modeled on that of the ill-fated *San Francisco*, the Pacific Mail steamer that sank in

Fulton, with coal barges alongside. *The Nautical Photo Agency*.

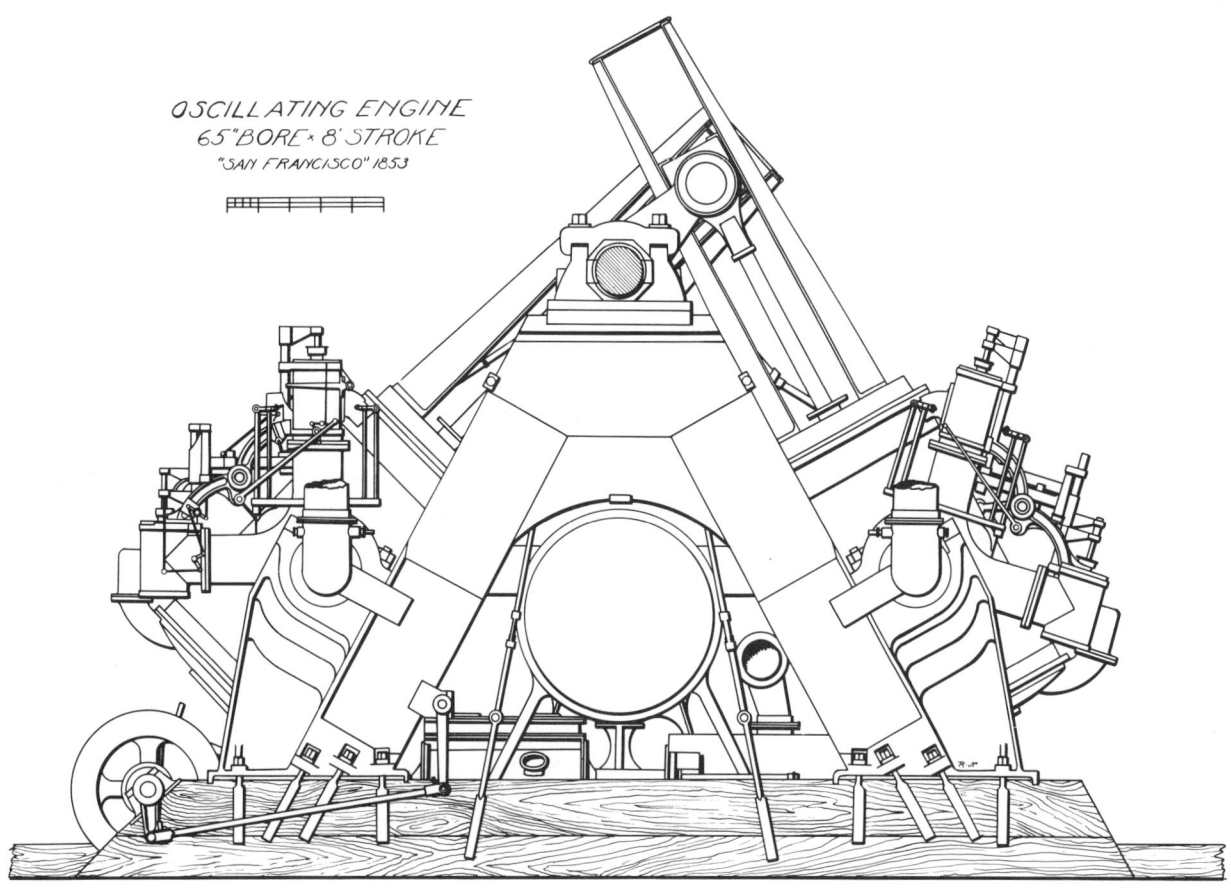

Oscillating engine 65″ bore by 8′-0″ stroke for the *San Francisco*.

January 1854 when disabled by overpowering gales.[37] One unusual feature consisted of combined cast and wrought iron, A-shaped supports for the 24 1/2″ diameter trunnions. A second was the fitting of guides for the piston rods so that the cylinders were rocked back and forth without developing side forces on the piston rods or their stuffing boxes. Wear and leakage were greatly reduced. A cut-off and reversing gear developed by Hermann Winters, of the builder's staff, permitted stopping and reversing without the use of hand gear.[38]

The boilers were of the Martin type, with vertical brass tubes to increase their heating surface, an improved version of the ones in the Collins liner. Two of the, 30′ long by 12′ by 14′, each with seven furnaces, had somewhat different internal arrangements, the forward one incorporating 2,500 2″ diameter tubes and the after one 1,900 2-1/2″ ones. The two stacks extended 41′ above the grates and provided enough natural draft so that no blowers were needed. The paddle wheels' diameter was 31′ with 28 single buckets 9′-0″ by 18″. The maximum beam over the guards, outside the wheels, came to 65′-6″.[39] The complete machinery installation was 105′ long by 18′ wide and was, as in the *Arago*, surrounded by so-called watertight bulkheads. It is difficult to understand why the *Fulton*, which had more advanced boilers, was fitted with jet condensers and why her rated steam pressure, 25 psi, was lower than the *Arago*'s. Actually, she started out at only 20 pounds to avoid scaling and plugging of the tubes. Second thoughts must have arisen for, by the time she was nearing completion, it was announced that Pirsson condensers would be put in after her second voyage and that a considerable fuel saving would be expected. The change was made, and although there were early tube failures, after they were replaced with a different grade of copper they performed admirably.

The *Arago* had a more attractive profile. Her masts and spars were in just the right places, while the *Fulton*'s were not. We are fortunate that enough data has survived to make an extremely accurate reconstruction of the former possible. The *Nautical Magazine* material includes lines, mast, and spar dimensions, and a detailed description. Deck plans were published in a French treatise on Atlantic steamships.[40] An excellent oil painting by Philip Tanneur survives,[41] as do several lithographs. Best of all, there are a number of actual photographs showing details of the spar deck arrangement that were taken when she was a Civil War transport.[42] The author has combined this evidence to produce the accompanying drawings of the *Arago*.

The Havre Line's action in retaining Captains Lines and Wotton is worthy of note. Both their vessels went aground while attempting to make a landfall, the *Humboldt* after a pilot had boarded her, and the *Franklin* after four days of navigation by dead reckoning. Both losses wre considered normal hazards of the sea. Many other craft, before and since, were wrecked approaching New York or Halifax. As respected masters of long experience, these two were assigned to oversee the construction of the new steamships they were to command.

On June 2, 1855, the *Arago*, under David Lines, steamed past Sandy Hook on her first transatlantic crossing. Thirteen days later she reached Cowes with 240 passengers and $2,400,000 in specie.[43] An abstract of her log shows daily runs varying from 215 miles to 294. The measures taken for fuel economy proved successful; she burned forty seven tons of coal per day. Coming home she passed the Needles at 1 A.M. on July 5 and reached Sandy Hook at noon on the 16th for an 11-day 16-hour passage at 11.1 knots. Her last day's run, 304 miles at 12.7 knots, was exceptionally good. The 126 passengers passed the usual resolution praising the master and his ship. In part it stated:

> We take great pleasure in expressing to you, as the unanimous sentiment of the passengers in the steamship ARAGO, of whom we are a committee, the satisfaction they have felt in their passage from Havre and Southampton with regard to the excellent qualities of the ship as a sea-boat, the efficiency of the officers, and your vigilance and skill as a commander. Her accommodations for passengers, the cleanliness and good order which are maintained in her internal arrangement, and her steadiness and ease in rough weather, are all that could be desired by families or individuals crossing the Atlantic.[44]

For once, the statement was more than a routine gesture; it was true in every respect.

The chartered *St. Louis* was retired and the *Arago* sailed alternately with the *Union* until October of 1855; when the latter suffered a broken shaft and the *St. Louis* was acquired again.

The *Fulton*'s first sailing was announced for December 15, 1855; instead, the repaired *Union* took her place. The *Arago* left New York January 12, 1856, and the *Fulton* was ready, four weeks later, to follow her. Thus the Havre Line started 1856 with two new steamships, the finest pair of wood transatlantic liners ever constructed. They were large, comfortable, economical ships that ran regularly and made money. Unlike the Collins liners, they garnered no speed records; neither did they end disastrously. Over the years little was reported of their activities and the excellence of their design was appreciated only by those in the shipping and shipbuilding business—now long dead and, like the ships, forgotten.

The *Arago* was a final vindication for Jacob A. Westervelt after the poor performance of his first steamship, the *Washington*. The *Fulton*, the last major vessel built by Smith and Dimon in their three decades of partnership, was a fitting conclusion for a firm that started so successfully, in the steamboat field, with the *James Kent*. The engine builders performed just as admirably. Their casting methods had improved, their forgings were superior products, and their earlier failures had been solved by the time-tested method of trial and experience.

When the *Fulton* left New York, under James A. Wotton, on February 9, 1856, she met bad weather, ice on the Grand Banks, and severe gales both eastward and westward.[45] She arrived at Cowes on the 25th. Returning she left Havre on March 12, Cowes the next day, and arrived at New York on the 29th, making the passages to and from Cowes in sixteen days. Despite the long crossings, her best day's run, 288 miles, was a very good one.

Until 1856 there had been no real competition on the Havre route. That year Cunard announced that a fleet of iron screw steamers, which included the *Etna, Jura, Emu, Alps*, and *Lebanon*, just released from their Crimean War duties, would operate between New York and the French port, charging a first-class fare of $100 to the Havre Line's $130.[46] A second competitor was the French Compagnie Franco-Américaine, which put four more iron screw steamers, *Barcelone, Alma, Lyonnais*, and *Vigo*, into operation that year.

Neither became a serious rival; the Havre Line's regular sailings, excellent ships, and long record of fine service, exending clear back to the days when only sailing packets were available, garnered the discriminating passengers and the cream of the freight. The British ships moved on to more profitable routes and the French line failed after the *Lyonnais* was lost in November of 1856. Earlier in the year the *Arago* had reached New York on June 17, 1856, with forty passengers taken over from the *Alma* after the latter had damaged her screw and had to abandon a voyage.

To outward appearances, 1857 should have been as disastrous a year for the Havre Line as it had been for the Bremen and Collins organizations. The Bremen Line's mail contract expired on June 1 and, because the Havre Line had taken over half of the contract, the latter's $150,000 yearly subsidy ended. In its place the postmaster-general offered to pay the total postage collected on the American mails carried by the line, roughly $88,000 per year.[47]

A second blow was the sudden death of Mr. Livingston at the age of 49.[48] His associates, members of the Fox family, took over the management of the line.

Perhaps the gravest problem was the advent of Cornelius Vanderbilt's magnificent new steamship, not-too-modestly called the *Vanderbilt*, which left New York on May 5, 1857, for Cowes and Havre. In his usual way, the Commodore tried to eliminate competition by cutting fares and selecting the most awkward possible sailing dates. The *Vanderbilt*'s return from France was timed to coincide with the *Fulton*'s regular sailing. The new ship beat the *Fulton* a full two days. But, rather than risk anything but the most profitable season, her owner had the *Vanderbilt* make only five voyages before being laid up for the winter. She regularly burned twice as much coal per day and required a crew of 140 to the *Fulton*'s 97.[49] Furthermore, travelers found her quarters spartan and her service equally so. Unwary merchants could always send their freight to the pier, but they could never be sure that a Vanderbilt steamer would sail when announced.

While the Vanderbilt Line continued and had two steamers running to Havre in later years, it never provided a year-round service; moreover, steamers other than the *Vanderbilt* were often small and too slow. The vessels were a nuisance to the Havre Line, from 1857 through 1860; they skimmed off the passengers looking for bargain rates, provided a new steerage service, and took over some of the mail and freight, but never provided serious enough competition to force the Havre liners to cut their fares. For once Vanderbilt did not emerge as the victor.

Even the appearance of the *Adriatic* in 1860, as a single ship operated by the North Atlantic Mail Steamship Company, did not interfere with the regular sailings of the *Arago* and *Fulton*. They were such a satisfactory combination that they survived both American and foreign competition, continued to carry most of the American mails to Southampton and Havre, and, eventually, provided the only American sailings across the Atlantic after the outbreak of the Civil War. The only change was that Captain Henry A. Gadsden took over the *Arago* in December of 1860, when Captain Lines went ashore to act as one of the company's agents.

A year later the *Arago* made the final voyage of the line, leaving New York on November 9, 1861, with General Winfield Scott and his family among the 140 passengers on board. Homeward bound, she left Havre on December 11 and docked in New York the day after Christmas. This was her eighty fourth crossing of the Atlantic.

In eleven years, despite hazards of the sea, loss of a mail contract, and the development of competition, the line had averaged 12 voyages per year by means of four steamships built to their order and three others chartered as necessary. Their first two vessels, both successful, were wrecked—the *Humboldt* in 1853 on her sixteenth voyage, and the *Franklin* in 1854 on her twenty second. These were replaced by the Charleston Line's *Nashville* for two voyages, the *Union* for ten, and the Pacific Mail's newly completed *St. Louis* for six. Two exceptional liners were built to replace the ones lost. Starting in 1855 the *Arago* completed forty two round voyages and the *Fulton* thirty six. The grand total of all voyages is 134.

The Havre Line, as the last of the three formed to take over the European mail contracts authorized by Congress, had the least engineering and technical problems to solve. Furthermore, it was well and conservatively managed and took over an established packet ship route that had always been profitable. The men who ran it were professionals. They did not make promises or put on a show. They ran their ships on schedule, carried luxury goods from Havre to New York, and acquired a loyal traveling public that appreciated their ships and their services. The Collins Line made speed records and failed. The Havre Line ran quietly, paid dividends, and succeeded. Its subsequent story will be considered in chapter 18.

Notes

1. *New York Herald,* September 1, 1848.
2. *New York Commercial Advertiser,* June 28, 1849.
3. Carl C. Cutler, *Queens of the Western Ocean* (Annapolis, Md., 1961), pp. 394-95.
4. *Journal of the Franklin Institute* 27, 3d ser. (January-July 1854): 37-44.
5. William Skiddy letter, April 27, 1850, to Commander Charles W. Skinner. U.S. National Archives, NRG 45, File AD 1812-1859.
6. B. H. Bartol, *A Treastise on the Marine Boilers of the United States* (Philadelphia, 1851), pp. 52-53.
7. Ibid., p. 54-55.
8. Frederick Mone, *Treatise on American Engineering,* (New York, 1854), part B, "Marine Engines," pp. 7 and 26-28, Plates I-XIII.
9. *New York Commercial Advertiser,* December 7, 1853.
10. *New York Herald,* October 7, 29, & November 19, 1850.
11. Ibid., January 17, 1851.
12. Ibid., June 3, 1851.
13. *Journal of the Franklin Institute* 27, 3d ser. (January-June 1854): 37-44.
14. *New York Commercial Advertiser,* February 6, 1852.
15. Ibid., November 21, 1852.
16. Ibid., December 6, 7, 13, and 16, 1853.
17. *New York Herald,* March 6, 1854.
18. Ibid., May 27, 1855.

19. *New York Commercial Advertiser,* October 25, 1855.
20. Ibid., July 18, 20, 22, & 25, 1854.
21. *New York Herald,* September 12, 1854.
22. *New York Commercial Advertiser,* October 20, 1854.
23. *New York Herald,* August 3, 1854.
24. Marischal Murray, *Ships and South Africa* (London, 1933), pp. 16-20.
25. *New York Commercial Advertiser,* October 23, 31, and November 27, 1854; N. R. P. Bonsor, *North Atlantic Seaway,* (Newton Abbott, London and Vancouver, 1975), 2d ed., 1: 273-74.
26. *Journal of the Franklin Institute* 28, 3d ser. (July-December 1854): 339-41.
27. Ibid., 27 (January-June 1854): 353.
28. *New York Herald,* July 22, 1854.
29. Ibid., June 2, 1854.
30. Ibid., June 18, 1854.
31. *New York Commercial Advertiser,* May 13, 1852.
32. *Journal of the Franklin Institute* 24, 3d ser. (July-December 1852): 112-13.
33. *Monthly Nautical Magazine and Quarterly Review* 2 (September 1855): 486-94, and *U.S. Nautical Magazine,* 3 (November 1855): 90-92.
34. *Journal of the Franklin Institute* 36, 3d ser. (July-December 1858): 234-36.
35. Charles B. Stuart, *Naval and Mail Steamers of the United States,* 2d ed. (New York, 1853), p.213.
36. *New York Herald,* July 4, 1853 and March 27, 1855; *New York Evening Post,* April 26, 1855.
37. Edouard A. Stackpole, *The Wreck of the Steamer SAN FRANCISCO* (Mystic, Conn., 1954).
38. *U.S. Nautical Magazine and Naval Journal,* May 1856, pp. 110-13.
39. *Journal of the Franklin Institute* 31, 3d ser. (January-June 1856): 36, 339-42. The wheel diameters are given as both 34' and 31'. The author considers the latter more likely.
40. Eugène Flachat, *Navigation à Vapeur Transocéanienne,* (Paris, 1866), Atlas, Plate 16.
41. Catalogue of the Marine Collection at India House New York, 1935, p. 3.
42. U.S. National Archives, Army Signal Corps photographs.
43. Ginsberg records.
44. *Monthly Nautical Magazine* (September 1855), p. 494.
45. *New York Herald,* April 7, 1856.
46. Bonsor, *North Atlantic Seaway,* 2d ed., 1: 84, 342-43.
47. Thomas Rainey, *Ocean Steam Navigation and the Ocean Post* (New York, 1858), p. 106.
48. *New York Tribune,* August 26, 1857.
49. New York Board of Underwriters, Committee Report, "Causes of the Loss, USM Steamer CENTRAL AMERICA" (New York, 1858) 2d part.

11
Screw-Propelled Steamships, 1850-1855

Paddle Wheels and Propellers

BY the end of the 1850 three subsidized mail lines were providing year-round sailings to Liverpool, Southampton, Bremen, and Havre. It was only natural that other shipping interests would turn their attention to the Atlantic and would construct additional steamships to attract some part of the increasing business that the new lines had generated. It was clear to informed shipping men that the mail liners had been expensive craft to build. Because their horsepower and speed requirements were high, their engines were large, heavy, and very costly. The coal consumption was prodigal, and this, combined with heavy machinery, demanded very large hulls in order to provide the necessary buoyancy. We have seen that early steamships cost anywhere from $270,000 for the *United States,* to $675,000 for the *Atlantic.*

For the period, the paddle wheel was the only possible solution for a high-powered steamship having a hull constructed of wood. Large wooden hulls are inherently very flexible. They bend, distort, and change shape under stress. Engines, on the other hand, must be kept in shape if they are to run properly; otherwise the bearings bind and overheat. In a paddle steamship the cylinders, paddle shafts, and shaft bearings were located close together so that rigid iron foundations and braces could keep the moving parts in proper alignment, even though the ends of the vessel might be working up and down in a seaway. The provision of joints in the paddle shaft allowed some misalignment between port and starboard engines and shafts, and permitted some flexibility in the transverse direction.

An important advantage of side-wheel machinery was that it turned so slowly—one revolution every four or five seconds—that there were no vibration problems because of unbalanced weights. A second virtue was that floats or paddles if damaged by ice or driftwood, could be replaced, even at sea, far from any dry dock.

The chief disadvantages were weight and cost. Another drawback was that only limited changes in weights aboard the ship were possible. If too much coal and cargo were loaded, the wheels were deeply buried, became inefficient, and the ship slowed down. Equal problems arose when there was too little weight aboard and the wheels, barely immersed and rolling into and out of the water, became ineffective.

A standard way of reducing the weight, space, and cost of machinery was to make it rotate faster; the screw propeller was the one available means of achieving this. It was in this direction that those who wanted to build lighter and cheaper liners tried to go. They were, in addition, far too willing to accept low horsepowers and speeds, both of which further reduced the engine weight and the amount of coal carried. The combination of all these features led to considerably smaller steamships for the Atlantic. Furthermore, since the propeller was completely under water, a greater variation in draft was possible without impairing its efficiency.

The disadvantages, however, were to prove serious. Propeller damage could not be repaired at sea and early propellers were remarkably fragile. When the shaft revolutions were increased, at first to 35, later to 40, or even 60 rpm, unbalanced weights caused severe vibration at the engine. Propellers were mounted behind massive timber propeller posts and ahead of equally heavy rudder posts, neither of which was streamlined. As the blades turned through the irregular eddies in the aperture left for the screw, they set up further vibration at the extreme stern. This was always accompanied by objectionable noise. Wooden hulls, being flexible, were most unsuited for severe vibration, whether generated by the engine or the propeller. Moreover, the shafting between the two had to be kept in line, and, as the hull flexed, the shaft tended to rub in its bearings and this led to wear, damage, and, ultimately, failure.

Provided powers were kept low and the hulls reasonably short, the wood screw steamers did not have too severe vibration or shafting problems. The application of large powers and long shafts had to await the introduction of iron hulls rigid enough to keep the shafting in line and strong enough to prevent the engine from tearing out its holding down bolts. Rivets would hold, even under severe vibratory stresses, when the treenails and spikes fastening a wood hull together would not.

As long as wood was employed for American steamships, the largest ones were propelled by paddle wheels. Not until the early 1870s, when new shipyards were established to build iron hulls, could the propeller take over. Prior to this, only the smaller, second-class steamers of moderate power could have screws.

In practice, paddle wheels and propellers developed simultaneously, but for different categories of steamships. The Stevens family first experimented with propellers on small launches, but then wisely adopted side wheels for their large commercial steamboats. The first American liner to carry paying passengers on the Atlantic was a low-powered, auxiliary-screw steamship, the *Massachusetts,* in 1845. But two years later, when the first fully powered liner, the *Washington,* was completed, no one ever seriously considered that she should be anything but a side-wheel steamship. That the screw propeller suddenly appeared and immediately proved its superiority over paddles is a notion fostered by popular story tellers, not by informed engineers. Both methods of propulsion had their places and both were built at the same time. With a few important exceptions, each was used where it was most suitable.

The year 1851 was to see the appearance of five American screw steamships for the Atlantic. Three years later the demand for transports for the Crimean War would attract several more from coastwise service and transfer them to European waters. These, together with an Australian gold-rush steamer, will be considered in the order in which they crossed the North or South Atlantic.

The *Lafayette,*

1851

Launched from the Greenpoint shipyard of Perrine, Patterson, and Stack on January 18, 1851, the *Lafayette* was begun as a Havre packet but sailed to Liverpool instead.[1] She had a number of owners, including John G. Williams; Charles Stoddard, her master; Samuel Engle; David B. Moses; the three partners who built her, William Perrine, Ariel Patterson, and Thomas Stack; the two partners who were to engine her, Peter Hogg and C. H. Delamater; and Mary B. Dale. She was 210' long, 32'-6" beam, 19'-0" deep, and had a tonnage of 1,059. There were two complete decks, a flying dragon as a figurehead, and three masts. Two cylinders of 50" bore by 3'-8" stroke were mounted at a 90° angle to each other over the crank shaft,[2] and turned a four-blade, 14'-0" diameter cast iron propeller weighing 6 tons. Steam was supplied by two Montgomery type boilers, with vertical water tubes and a forced-draft blower discharging air under the grates of their furnaces.

On her upper deck was a short forecastle and a long deckhouse containing the main saloon and cabins for 90 passengers. Her lower deck had second-class quarters for 150 and space for 700 measurement tons of cargo.[3] A clutch in the shaft allowed her propeller to spin freely when under sail.

The *Lafayette,* with one hundred guests aboard, steamed away from the Hogg and Delamater wharf at the foot of New York's West 13th Street, on April 21, 1851, without sufficient coal or ballast to completely submerge the propeller. There was a grand banquet the next day as she was coming up the Delaware, but Captain Stoddard considered it wise to absent himself, once the toasts were proposed, and returned to the wheel house. For Philadelphia's first American-built Atlantic liner arriving only four months after the British *City of Glasgow* a public reception, complete with salutes, welcoming steamboats, and crowds along the waterfront, was to be expected.

Originally scheduled for May 7, the *Lafayette* left three days later, carrying eighty seven passengers and a "fair" amount of freight.[4] She anchored off New Castle, Delaware, for engine adjustments, and reached the open sea at noon on May 11. When three days out, her air pump failed, making the condenser useless. Her resourceful engineers rigged an exhaust pipe to the quarter deck and operated her as a high-pressure steamboat exhausting to the air (like a steam locomotive) but with a considerable increase in fuel consumption.[5] As a result, she had to put into Cork, Ireland, for coal and reached Liverpool on June 2 after a 23-day passage. On June 17 she sailed, but on the 22d, was stopped for engine repairs when she was sighted and passed by the Bremen liner *Hermann*. Seven days later, in a dense fog, she collided with the English brig *Jane,* which was carrying a timber cargo from Quebec. The *Lafayette* lost her bowsprit, jibboom, and cutwater, and sank the *Jane*. Somehow, in the collision, three of the propeller's four blades were broken off, perhaps by entanglement with wreckage. Afterward, the *Lafayette* rescued the brig's crew and found herself still tight. She put into New York on July 7 for repairs.[6]

One such eventful Atlantic voyage was enough for her owners. After repairs she was advertised for a sailing to San Juan del Norte, Nicaragua, and Chagres, still under Captain Stoddard. She left New York on August 28 and was at anchor at Chagres on September 11 when a fire broke out. A few passengers for the return had come aboard. The steamboat *Gorgona* tried to tow her into shallow water but the fire engulfed her so rapidly that she sank before reaching it.[7] With the exception of one coal passer, everyone managed to escape.

The *S. S. Lewis,*

1851-1853

While New York was making its reputation as the city whose ship and engine builders created the large sidewheel steamship for ocean service, Philadelphia, concurrently, was developing a successful series of smaller screw-propelled vessels for coastwise, shallow-draft, and towing purposes. It had a shipbuilding heritage going back to colonial days, long before the Revolutionary War, and a number of iron works capable of constructing engines. Merrick and Towne, for example, had engined the first screw propelled warship, the U.S.S. *Princeton,* as early as 1843, from designs by John Ericsson. By 1851, single screw steamers from 100 to 160 ft. in length were building for Maine waters, for Florida ones, and for a regular line from Philadelphia to Charleston. Even a steam yacht had been started.

In many cases these were the direct result of efforts by Captain Richard F. Loper who had, like Ericsson, obtained a number of patents covering the design of propellers and the engines to turn them. Loper was a much more practical, if considerably less flamboyant, engineer; as a result, his vessels were remarkably successful. The original Loper propeller had sheet-metal blades riveted to a large cylindrical hub.[8] It was considerably simpler than the Ericsson design with its cast inner section, a strengthening ring, and sheet-metal outer blades. Because thin blades could not stand large forces, Loper soon changed to a cast propeller, which turned out to be a stronger and more efficient design.[9] Not only did Loper design the engines and propellers for the craft in which he was interested; he would usually undertake the entire construction contract, then sublet the hull to a shipbuilder and the machinery to an iron works. He advised the shipbuilder on the hull details to insure that it was structurally and hydrodynamically suitable for the mode of propulsion. The first registration (or enrollment) at the Philadelphia Custom House, as a result, often shows Loper either as a whole or part owner, sometimes for a short period, sometimes for a long one.

The largest and most powerful propeller-driven steamship yet constructed in Philadelphia was begun in early 1851. Ordered by Richard Loper, E. Lincoln, and Samuel Reynolds, she was intended for the California trade. Before launching, she was purchased by Boston's Harnden Express Company for the Atlantic. Then, to operate her, a subsidiary called the New England Ocean Steamship Company was incorporated.

The hull was built at Kensington, now a part of Philadelphia, by Theodore Birely and Son, the machinery by J. T. Sutton and Company's Franklin Iron Works, both under the personal supervision of Loper.[10]

The *S.S. Lewis,* 216.0' by 32.6' by 16.3', about equal in size to the *Lafayette,* measured 1,103 tons. She is the only early screw steamer for which detailed scantlings are available, and these show her to have been strongly built and cleverly reinforced in critical areas.[11] All the frame and most of her planking were white oak. The keel was 16" deep at the stem, 20" at the stern, and was sided 15". At the bow there were four timbers: a cutwater; the stem; an apron behind it; and, finally, a stemson making over 8' of solid wood 15" thick. Aft were a vertical sternport 16" x 15", a 7'-0" space for the propeller, and a rudder post. Heavy composition (probably brass) reinforcements were bolted on either side to tie the keel and the two vertical posts together. Furthermore, iron and composition plates were added to reinforce the propeller post where it was bored out for the stern tube and shaft. There were many wood knees inside the hull and metal ones aft connecting the propeller post to a 20" square transom timber.

Frames were 26" apart but, unlike the paddle steamers, the bottom spaces between the floors were not filled in except in way of the engine. In this highly stressed region, a special set of diagonal iron straps 60' long by 5" wide by 1" thick were notched into the floor timbers before the bottom was planked. Such bottom reinforcement had not been necessary in side wheelers. The usual diagonal straps, on the sides of the hull, were smaller, 4" x 5/8" spaced 5'-0" apart. On centerline was a deep keelson made of two 15" square timbers; further outboard came 15" x 15" bilge keelsons. The bottom ceiling was light, 3 1/2" thick, but on the sides it was increased to 6" with three strakes 7" x 12" as clamps supporting the deck beams. The bottom planking started with 8" x 12" garboards against the keel, then became 4" thick up to the wales where thirteen strakes of side planking were 5 1/2" x 7" wide. Above water the planking was hard pine 4" thick. Every plank had two locust treenails and two composition spikes 1/2" square by 8" long wherever it crossed a frame.

In all, not counting the strapping, there were 125,000 pounds of iron fastenings, 10,000 of copper, and 5,000 of composition. No wood ships was ever any stronger than her fastenings and the *Lewis* had an extremely well-fastened hull. The combination of carefully planned reinforcement at the stern, special strapping beneath the engine, and thorough fastening throughout shows that Loper and Birely were acutely aware of the vibration problems found in propeller-driven steamships.

The *Lewis* had a sailing-ship hull with fine, slightly hollow waterlines forward; 6" of deadrise, giving her bottom a slight vee; a billet head supported by carved trail boards; and an old-fashioned transom stern. This stern permitted a gilded eagle to spread its carved wings across an ornamental stern board. Tasteful carving always enhanced a ship.

Two full decks were fitted, partial orlop decks, and, on the spar deck, a 190' long house whose sides followed the curve of the bulwark to form a gangway 7'-6" wide on either side except dead aft, where a short house extended out to the sides. Here were located the wheel house, a smoking room, and three water closets. In the deckhouse were officers' staterooms at its forward and after ends. Amidships were the engine-room entrance, galley, mess rooms, an ice house, companionways, and a saloon 20' long. Right at the bow was a small topgallant forecastle, but the crew were berthed below on the main deck.

On the main deck aft was the great cabin for passengers, with thirty staterooms opening from it. Their inmates could not escape the propeller noise. The wide transome had four stern windows to light the saloon. The rest of the deck, abaft the crew's quarters, was used for second-cabin passengers. The orlop decks could be fitted up for steerage passengers or reserved for cargo. They also had coal bunkers near the fire room. In the after hold were ten iron tanks holding 12,000 gallons of water. Cargo was carried in the forward hold. The after engine-room bulkhead was 60' forward of the sternpost, and the propeller shaft, 67' long, was made in three sections, each 11" in diameter. There were two cylinders at 45° to the vertical driving a 17"-diameter, overhead crankshaft 14' long. At one end was an 8'-diameter gear with a face width of 3'-6", which drove a smaller pinion on the propeller shaft below it. The gear's teeth were of wood to reduce noise and to allow replacement by the ship's engineers. The propeller shaft turned 1 3/4 revolutions for every one of the engine's.[12] By keeping the engine rpm low, unbalanced forces were reduced to a minimum. To support all this complication, a cast-iron bedplate 14' long by 16' wide and at least 4" thick was bolted to the wood bottom structure. Like those for paddle engines, it was a complex casting incorporating exhaust passages from the cylinders to the condensers and part of the air pumps. The cylinders were 60" bore by 3' 1" stroke. The propeller was 13'-0" in diameter and had four blades.

Two boilers 18' long by 11' in diameter were connected to a separate steam drum 8' in diameter by 6' long, which was fitted in lieu of steam chimneys. The boilers were forward of the engine and extended above the orlop deck.

The *S.S. Lewis* was given a barkentine rig, three smartly raking masts, and four square sails on the fore. Lower masts were varnished and, from the doublings up, everything was painted black. The fore yard was all of 60' long. Shrouds were set up through rollers on the rail and the conventional channels, chains, and deadeyes of a sailing ship eliminated.

Four Francis lifeboats 30' long were carried. The forward capstan was operated by hand brakes from the topgallant forecastle. This kept the men powering it a deck away from the cable being hauled in, a most convenient feature.

The hull was black with a gilded stripe along the planksheer, the decks were bright pine, the inside of the bulwarks were fawn colored, and the deckhouse was paneled and painted with imitation graining to make it look like varnished oak. The bottom was to be coppered to the 18' waterline forward and to the 19' aft.

The *S.S. Lewis* was scheduled for launching on June 10, 1851, and quickly completed.[13] Since the engine parts were smaller and lighter than those for a paddle steamer, they could be put in while the ship was on the ways. Hence the decks could be completed, cabins installed, and the steamer practically finished in advance of launching. She was tried out on September 4 with steam at 18 psi and the engine making 22 rpm. After overcoming some bearing problems, the *Lewis* sailed September 13 for Boston.[14]

On board were a hundred invited guests. Going down the Delaware, a distance of 106 miles was covered in nine and a half hours, not counting a stop made to adjust the bearings. She reached the Delaware Breakwater to find a storm raging outside; Loper decided to stop rather than expose his guests to high seas. On Monday, September 15, she remained at anchor, but proceeded that night, for she was expected in Boston for a grand jubilee to celebrate the completion of the first railroad from there to Canada. The *Lewis* was to take President Fillmore and other dignitaries on a tour around Boston Harbor. Against a northeaster, she steamed along at an average speed of 7.5 knots and arrived just in time for the celebrations.[15]

She was docked, coppered, and handed over to a new master, Captain Cole, formerly of the steamship *Tennessee*. She left Boston October 4, 1851, heading for Liverpool and carrying 45 first class and 60 other passengers.[16] After 10 1/2 days, during a violent gale, the propeller was carried away, forcing her to sail the rest of the way.[17] Liverpool was reached in seventeen and one half days on October 21. Being pulled up the Mersey by two tugs was a sad experience for the Bostonians, who had hoped to compete with the Cunard paddlers and had even named their vessel after Boston's Cunard Line agent. An English propeller with wider blades and a lower pitch was installed; the *Lewis* started home on December 8 only to run out of coal and have to put in to Halifax twenty days later. Boston saw her on January 3, 1852. Within a month she was sold to Cornelius Vanderbilt for $150,000 and, on March 5, 1852, left New York for San Francisco. It was a long, slow trip; she did not arrive until July 7. She was loaded down with 653 passengers taken aboard at Panama and San Juan del Sur. Most of them had traveled from New York to Nicaragua on the *Northern Light*, had crossed the Isthmus, and had been marooned at San Juan, without adequate shelter or supplies, for three weeks during the rainy season awaiting the arrival of the *Lewis*.[18] Seventeen had already perished and 19 more were to die en route to San Francisco. Asiatic cholera was adding to the normal yellow fever toll.

For almost a year the *Lewis* served on the Pacific, carrying enormous numbers of passengers northbound and a smaller number of gold miners going home. There were almost no regulations covering the safety of passengers at that time, but the Vanderbilt boats disregarded what there were. In the fall of 1852 she was libeled for overloading.[19] On January 4, 1853, a breakdown just off San Francisco disabled her.[20] A steam tug, the *Goliah*, towed her in and she was out of service for almost two months. Then, northbound from San Juan, she ran past San Francisco in a dense fog and went aground about 3 A.M. on April 9, 1853, some eight miles north of cove where the *Tennessee* had been wrecked, in exactly similar circumstances, one month earlier. The *S.S. Lewis* was impaled on sharp rocks. All 385 passengers and the crew were ashore by noon. The combination of rocks, surf, and an advancing tide reduced the hull to wreckage in a

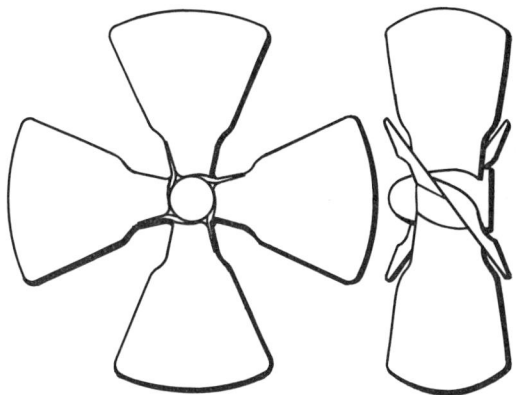

LOPER PROPELLER

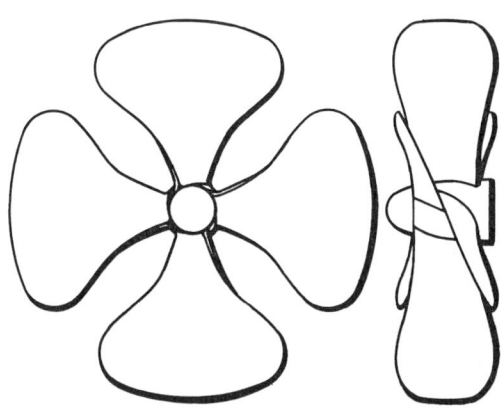

ENGLISH REPLACEMENT
"S.S. LEWIS"   1851

Two propellers installed on the *S. S. Lewis*.

day's time. A contemporary comment was: "As far as we can ascertain, no charge of negligence or incapacity has been brought against her commander; but the loss of steamers on this coast has been extraordinary, the Vanderbilt Line coming in for a full share."[21] Three of the line's steamships, the *Pioneer*, *Independence*, and *S.S. Lewis* had been lost in an eight-month period.

The *Pioneer*,

1851-1852

A more ambitious steamship than the *S.S. Lewis* was the *Pioneer*, named in honor of Daniel Boone. She was built by Jacob Bell at New York and was the only screw steam-

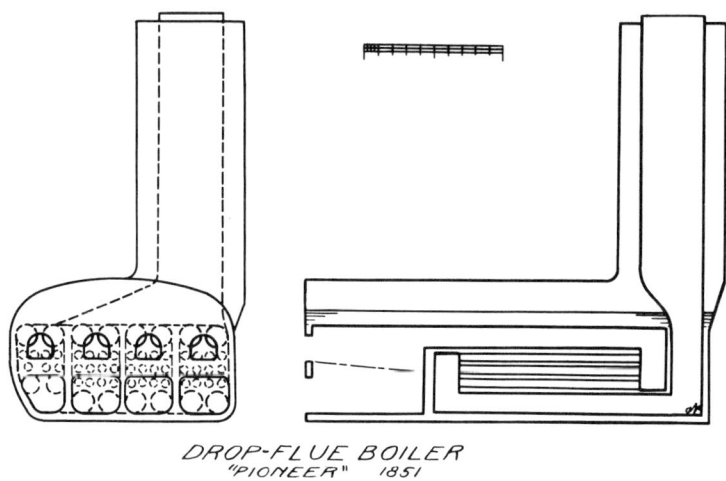

Drop-flue boiler for the *Pioneer*.

ship to come from his yard after the Collins liner *Baltic*. Asa Eldridge, late master of the sailing packet *Roscius;* Jacob Bell; and Edward Gillilan were the principal owners, although many others held small shares. She was large, 1,833 tons, 218'-4" long by 42'-6" beam with a depth of hold of 29'-0". Her hull and rig were basically those of a sailing vessel, with a long overhanging bow carrying a full-length figurehead of Boone, a bark rig having very tall masts with topgallants on the fore and main, and even studding sails on the fore. Only the stack between the fore and main, together with a mainmast slightly forward of its normal position, testified to her steam propulsion. Why Bell, after building purely steamship hulls for the Collins Line, reverted to a sailing ship model for their successor is not known.

There were two return-flue boilers, 29' long, 16' wide, and 10' high, generating steam at 15 psi, each with an excessively tall steam chimney extending up for 18' above the boiler shell.[22] The two uptakes could not be joined into a single stack until they were above the spar deck. The arrangement was a safe one, for no wood adjoined the hot inner stack until the deckhouse was reached, and ample ventilation to cool the breechings was available there.

The engines were derived from a type that had been developed by John Penn and Sons in England and consisted of pistons annular in shape and an inner cylinder, or trunk, attached to each piston and extending out through both cylinder heads. No piston rod was required, just a connecting rod from a heavy pin at the center of the cylindrical trunk to the crank. It was a cumbersome arrangement but it kept down the engine's height. The diameter of the cylinders, two in number, was 85 1/2", large enough for a paddle wheel steamer; the coaxial inner trunk, sliding up and down with the piston, was 39"in diameter, and the stroke was 4'-3". Charles W. Copeland designed the machinery for the West Point Foundry. Simultaneously, an identical set of engines was made for another steamer, the *City of Pittsburgh*.[23]

The cylinders were directly over the crank shaft, one ahead of the other, and were designed to operate at 36 rpm turning a 16' diameter iron propeller with three blades weighing 11 tons and having an extremely large pitch, 34' at the blade tips. The engine fitted into a space no more than 21' long. The whole arrangement appears, to modern eyes, to have invited troubles. With a mean draft of 19', a 16' diameter propeller would never stay immersed when the *Pioneer* was pitching in a seaway. The moving up and down of pistons in any engine results in unbalance and vibration. The lighter they are, the less the difficulty. The added weight of the trunks, moving with the pistons, made the whole problem worse. Only because the revolutions were very low was it possible to use such a seemingly unsatisfactory design. One practical matter springs to mind. How did the engineers get to the wrist pin, inside the trunk, to oil and grease it while the engine turned?

The *Pioneer* was launched April 5, 1851, had her machinery installed, and steamed toward Sandy Hook on August 11 for trials but "broke some of her steam pipes."

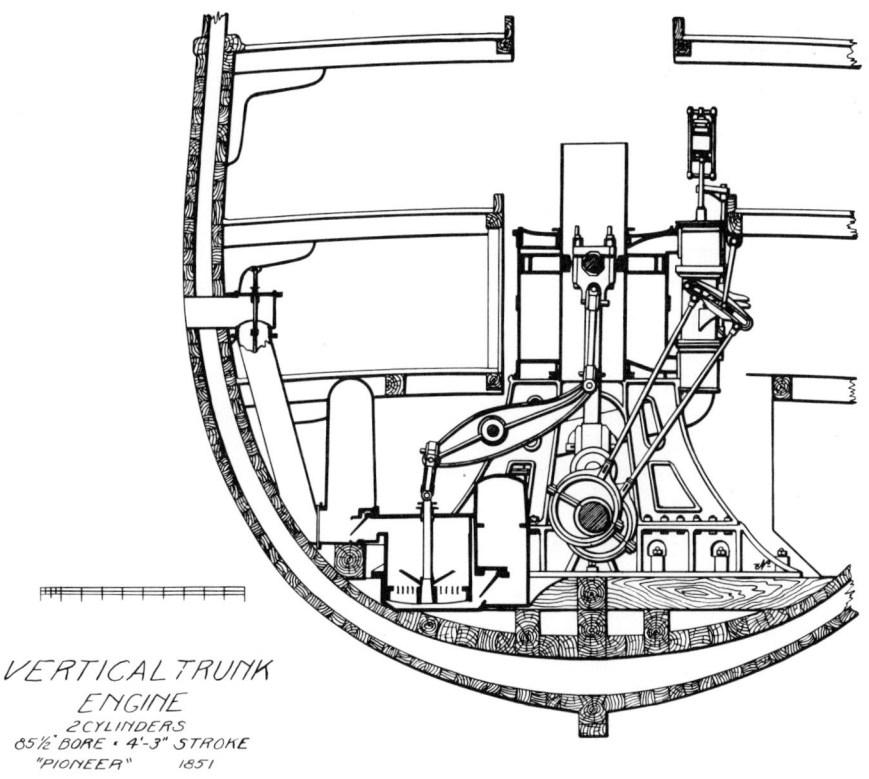

Vertical trunk engine 85 1/2″ bore by 4′-3″ stroke for the *Pioneer*.

She had to be towed back by steamboats having the appropriate names of *Hercules* and *Samson*.[24] There must have been extensive changes, because two months intervened before she was ready for sea again. Spofford, Tileston, and Company were agents when she sailed from New York, on October 18, 1851, for Liverpool, with Captain Eldridge in command. She carried at least 25 first-cabin passengers, as well as 137 in the second cabin and steerage. As the first American steamer designed to carry steerage passengers, she was a true pioneer. With a passage time of almost 19 days at 6.75 knots, the passengers might as well have sailed across. The return took her from Liverpool on November 22 to New York on December 14 in 22 days. Eastbound, she was no faster than the auxiliary packet *Massachusetts*. A grounding on the New Jersey coast about 12 miles north of Barnegat Light on the 12th slowed her return.[25] On board were 145 passengers.

Her owners, disappointed by such slow passages, disposed of her to Spofford and Tileston, who fitted her up for the California trade where anything that floated could find a cargo and, no matter how slow, passengers. She left New York January 24, 1852, for Chagres, on the Isthmus of Panama, and San Juan del Norte, the corresponding Nicaraguan port. Only forty passengers were aboard when she left Chagres February 6 for San Juan and New York. On March 18, under Captain G. W. Kitteridge, she cleared at New York for San Francisco. En route, a transfer to Cornelius Vanderbilt made her part of the Commodore's newly organized West Coast service from Nicaragua to San Francisco. The passage was a very long one. After reaching the Pacific she put into Talcahuano Bay, about 250 miles south of Valparaiso, Chile, for coal and while there was caught in a gale, dragged her anchors, and struck a reef.[26] Leaking from the damage, she continued on her way even though constant pumping was required.

Passengers had left New York on Vanderbilt's *Daniel Webster* in June of 1852, had crossed to the Pacific, and had spent weeks impatiently awaiting the *Pioneer* at San

Juan del Sur.[27] History repeated itself and both cholera and yellow fever developed. Despite the *Pioneer*'s leaky condition, the whole lot was loaded aboard and the voyage continued. After touching at Acapulco, matters worsened. Two of her three propeller blades broke off and, simultaneously, her coal was running out. In addition to steam pumps, fifty men were working day and night to keep her afloat. When heavy seas worsened her chance of survival, the *Pioneer* took refuge in St. Simon's Bay, some 200 miles short of San Francisco. At 3 A.M. on August 17 she was beached to prevent her sinking; there was no loss of life aside from the deaths that had already occurred from sickness contracted in Nicaragua. Although Captain Kitteridge and Vanderbilt's Nicaraguan agents had acted disgracefully in taking passengers on a damaged ship, the demand for transportation to the fabled shores of California would continue unabated and other unsuspecting travelers would rush to fill the steamships of the line, regardless of condition, overcrowding, or disease.

Captain Robert H. Waterman, of packet and clipper ship fame, purchased the *Pioneer*'s wreck, salvaged her, and brought her to San Francisco. Some of her spars and fittings were transferred to the clipper ship *Gazelle*, which reached port on October 1, 1852, badly mauled by storms off Cape Horn.[28] The *Pioneer*'s engine was taken out and her damaged hull rebuilt. About a year and a half later she was ready to sail for the Chincha Islands on February 17, 1854. She was incorrectly called a "clipper ship" although she still retained her bark rig.[29] A cargo of guano was brought around Cape Horn.

New York showed no interest in the *Pioneer* as a sailing bark; neither an arrival nor a departure was ever reported. Nevertheless, she had reached Brooklyn by September of 1854, there to spend the next eleven months without employment.[30] The *Post* commented that she had been "unlucky and appears friendless on the other side of the East River."[31] Further work on her hull was necessary, for she was at the E. F. Williams shipyard in May of 1855 and on the Sectional Dock for bottom work for two weeks in August. Sometime in the middle of the latter month she slipped quietly out of the harbor.

It is hoped that the *Pioneer*'s later, unrecorded career improved upon her earlier one. As a fine-hulled vessel built by a respected shipbuilder of long experience, she should have been successful under sail.

The *City of Pittsburg*,

1851-1852

There seems to have been an extensive effort, on both sides of the Atlantic, to establish a line of screw steamships between Philadelphia and Liverpool. On the eastern side, the Liverpool firm of Richardson Brothers, whose most interested partner was William Inman, set up the Philadelphia Steam Ship Company and bought the 1,609 ton, iron screw steamer *City of Glasgow*.[32] Completed in the spring of 1850, she had made four trips to New York before her purchase, and had already proved herself to be the first successful, or, better still, the first profitable Atlantic screw steamer. Her first voyage for Inman Line, as the company was popularly termed, came in December of 1850.

Three months later the American press announced that a second steamer was under construction in Scotland, a third in New York, and that a fourth would be built in Philadelphia.[33] The name of the second was given as the *City of Philadelphia* but, in fact, *City of Manchester* was substituted.[34] The third was named *City of Pittsburg* and, because of her poor performance, no fourth vessel was built in the United States and no joint British-American operation ever developed.

The *City of Pittsburg*, built at Williamsburg by Perrine, Patterson, and Stack, was the largest early American propeller and was of the same size as her British running mates. They, however, had iron hulls; she had wood.

Steamship	Length	Beam	Depth	Tonnage
City of Glasgow	227'	34'	25'	1609
City of Pittsburg	246'-8"	40'	31'-4"	1875
City of Manchester	261'-8"	36'-2"	25'-3"	2109

A two-cylinder trunk engine identical to the *Pioneer*'s was placed in the *Pittsburg*'s short 21' engine room and turned a 16'-2" diameter, three-bladed screw with a pitch varying from 30' to 36'.[35] The boilers were different, however, three in number, with much lower steam chimneys.

Three classes of passengers were carried. The upper deck had a short forecastle and a long deckhouse containing the first-class saloon. Located aft, it was paneled in satinwood and decorated in white, green, and gold.[36] Amidships were the galleys, pantries, service spaces, and firemen's quarters, and, going forward, a second-cabin saloon. On the main deck aft were staterooms for sixty first-cabin passengers. The 160 second-class berths were located forward, away from the propeller's thrashing. Three hundred steerage were relegated to the lower deck along with store rooms, water tanks, and coal bunkers. One thousand tons of cargo could be stowed.

The *City of Pittsburg* was completed in October 1851 and, under Captain C. Stotesbury, who had supervised her construction, she left New York for Philadelphia on October 21. After a delay in order to load the very large amount of cargo awaiting her, she left Philadelphia on October 26 and reached Cork in 17 days 4 hours under steam at the very low pressure of 10 psi. The speed was 8.5 knots, the propeller rpm 32, and 519 tons of coal were burned. She set out at the extremely deep draft of 22' forward and 23' aft. In light of her horsepower, 777, a third of that on a Collins liner, the fuel consumption was excessive.[37] Although there was fine weather, she ran short and had to call at Cork. Her total time to Liverpool became a disappointing nineteen and one half days, no better than the *Pioneer*'s. Both the latter and the *S.S. Lewis* were still in port when the *City of Pittsburg* arrived.

The British-built *City of Manchester* had just completed a 14 day 7 hour-eastbound passage, wharf to wharf, and had averaged 260 nautical miles per day, well above 10 knots, for five successive days.[38]

Heading down the Mersey on November 29, 1851, the *City of Pittsburg* was able to use power for just three days before losing her propeller. After consulting with his passengers, Captain Stotesbury decided to proceed under sail but encountered winter gales all the way. Her bark rig proved its worth, but, not until January 9, 1852, did the *Pittsburg* pick up a pilot some twenty miles south of Cape Henlopen, at the mouth of Delaware Bay. The next day she anchored at the Breakwater and sent her purser ashore to telegraph the news of her long overdue arrival. Her Philadelphia agents dispatched the *City Ice Boat* and a new screw towboat, the *America,* to her assistance, but floating ice, gales, and low tides detained her for several days. The towboats were lashed alongside to save her from grounding when her anchor chains gave way. By January 15 she had been towed as far as New Castle, Delaware, and was met by the steamer *Kennebec,* which brought new anchors and chains from New York, took off her passengers, and ferried them up to Philadelphia. The *City of Pittsburg* did not reach her home port until January 18, nine days after making her landfall.[39]

Permission was granted by the Philadelphia Navy Yard for the use of its sectional dock in replacing the propeller and coppering her bottom. The naval authorities took the opportunity to try out their new ship-transfer installation on someone else's vessel rather than their own. Using a hydraulic ram and greased skids, the *City of Pittsburg* was slid from the dock onto shore in one day's time and, later, pushed back again.[40] Moving, 2,800 tons of ship a distance of 300 feet was a major engineering achievement.

Advertisements by the Philadelphia and Liverpool Steamship Company's Philadelphia agent, Thomas Richardson, listed all three *Cities*, the *Manchester, Glasgow,* and *Pittsburg,* but only the first two continued to operate. The last was just another unsuccessful propeller built at a cost of about $300,000. When she was withdrawn, the scheme for a joint American service with Inman collapsed. Like the *Pioneer,* she was sent to the Pacific. Many extra steerage berths were added prior to her departure.

She was to sail in early June of 1852, but not until June 23 could she leave Philadelphia for New York to load freight. On July 17 she departed, under the auspices of her usual agent, Thomas Richardson, but was intended for a revived Empire City Line's Panama-San Francisco service. She coaled at Coronel, Chile, and arrived at Valparaiso four months out of New York on November 22. At 2 A.M. on the 24th, as she lay at anchor with banked furnaces, a fire broke out just forward of the stack. Her chief engineer tried to raise steam for his fire pumps but the flames spread rapidly.[41] One hundred fifty nine passengers and crew were rescued by boats from other vessels in the harbor. The British flagship, H.M.S. *Portland,* was particularly active in the rescue.[42] The *City of Pittsburg,* insured for $279,000 (roughly her original cost), was probably more profitable as a loss than as an operating steamer. Not only was her one Atlantic crossing an excessively long one, but also her trip from New York to Valparaiso. She was both underpowered and uneconomical to operate.

The *South Carolina*, 1851-1887

Nothing is known about the appearance of the *South Carolina* beyond the limited data given on her official document issued at New York on December 15, 1851. With dimensions of 200'-6" by 37'-3" by 30'-0", she measured 1,301 tons, had three masts, a round stern, and a billet head. Jabez Williams constructed her at Williamsburg; the author has been unable to identify her engine builder. She was ordered by the South Carolina Atlantic Steam Navigation Company, which had Robert Caldwell as president. On December 31, 1851, she cleared at the New York Custom House for her home port, Charleston, but did not sail until January 26.[43] Engine problems seem the most likely reason for the delay. The trip took four days, a long time in comparison with the many sixty-hour passages of the *Southerner.* Back in New York on February 23, she brought in a cargo of cotton and rice, turned around at a leisurely rate, and headed South again about March 12. On April 17, 1852, she left Charleston bound for Liverpool with a full load of cotton but, while crossing the harbor's treacherous bar, grounded and damaged her propeller. Changing course, she reached New York April 22 for repairs.[44] A month later she was on the sectional dry dock for several days.

Her owners must have given up the idea of a Charleston line to Europe, for she was sold at auction for $105,000 on June 11.[45] A second translatic attempt was made from New York on August 5, 1852, again for Liverpool, but under a new master, Shirley C. Turner, and a new agent, John Ogden. The "driving wheel" was damaged and further repairs were needed before she finally went to sea on August 11.[46] She reached Liverpool on the 29th, her eighteen-day passage was slightly shorter than those of the *Pioneer* and the *City of Pittsburg*,[47] but still too slow to compete with the sailing packets. Westbound she carried 72 passengers when she arrived at New York on November 23, 1852, after a long, twenty-three-day crossing. Her consignee was now R. Caldwell, which indicates that her original owners may still have been involved in her operation. In January of 1853 there was another auction sale.[48] Her engine was removed and she was redocumented as a sailing ship on March 30, 1853, owned by George A. Trenholm of Charleston. She was later sold foreign and began to trade out of Liverpool. For a time she carried Australian emigrants.[49] In 1863 she was owned by Nicholas and Whittle, in 1874 by James Alexander. By the late 70s her rig had been reduced from a ship to a bark as a means of cutting down the crew size. *Lloyd's Register* lists her for the last time in 1887, with Liverpool as her home port and Miss. E. S. Stanton as her owner. Thirty-four years under sail demonstrated that Charleston's attempt at an Atlantic liner produced a superior vessel but a not-very-satisfactory steamship.

The Failure of the Screw Steamer on the Atlantic

The year of 1851 had begun brightly. The Great Exhibition and its Victorian marvel, The Crystal Palace, were to open in London. More travelers than ever before would cross the Atlantic for business or pleasure. During that year five wood screw-propelled steamers were completed for five American organizations eager to profit from the expanding trade. They sailed from Philadelphia, Boston, New York, and Charleston. As we have just seen, only the *Pioneer* managed to get across the ocean and back again

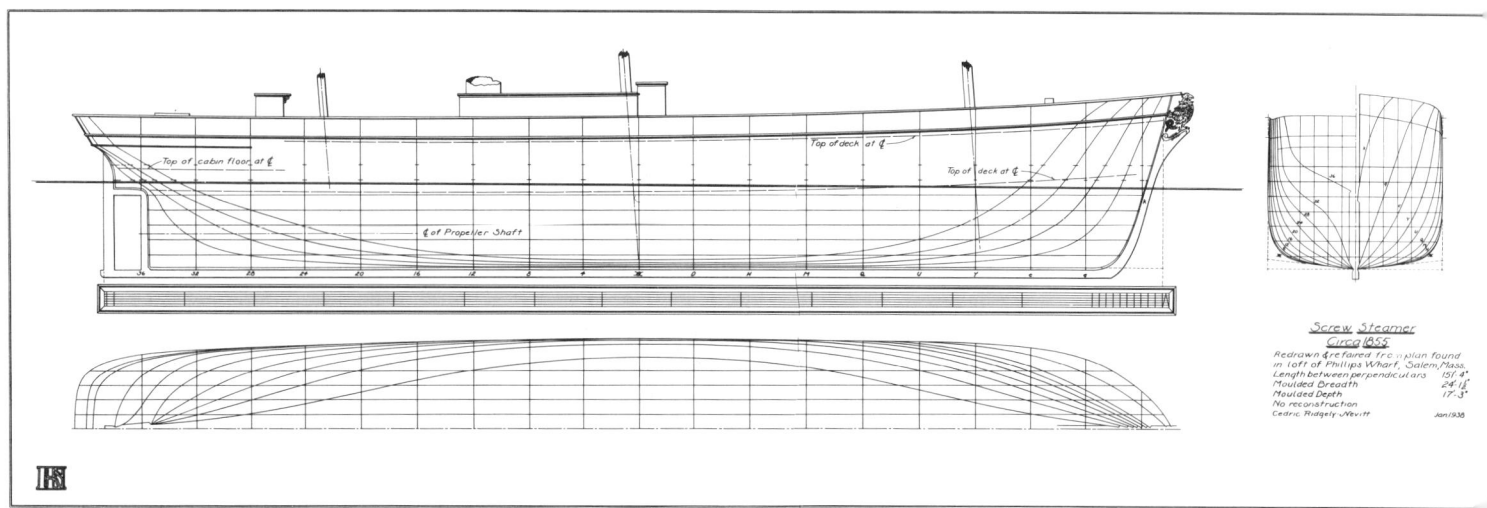

Screw steamer, circa 1855, lines.

Screw steamer, inboard profile.

without suffering an engine casualty, breaking off one or more blades of her propeller, or running out of coal. But with a crossing time of nineteen days eastward and twenty two westward, the *Pioneer* offered little advantage over a sailing ship despite the extra money poured into building and operating her. She could not even equal the speeds that had been achieved by the first Cunard liners in the 1840s.

The debacle was an expensive one for all who had invested in these ventures, for the cost of the steamships ran from $200,000 to $300,000 apiece and each had to be sold at a lower figure. Worst of all, the screw steamer itself suffered a damaging loss of prestige. The merchants of New York, the center of the nation's maritime industry, began to look askance at the type. As a result, all major ocean and coastwise steamers for the next decade would be designed and built as side-wheel steamships. Abroad, however, where an iron hull could provide a rigid foundation for the shaft and engine, the screw steamship advanced rapidly.

At least five other American screw steamers crossed the Atlantic prior to the Civil War; all but one were attracted by the high charter rates paid for steam transports during the Crimean War, which broke out in 1854. The exception was lured to Australia by tales of gold discoveries. They were all small vessels that had originally been constructed for coastwise operation. Three of them performed far better than their larger predecessors.

Although the screw steamers of over 1,000 tons had received a serious setback in their development, smaller ones continued to be built and to operate successfully. A wood hull under 200' long was not too flexible a foundation, and an engine below 500 horsepower did not create too serious a vibration problem. These modest craft, starting with Ericsson engines and propellers in the 1840s, continued to be built in increasing numbers. Loper engines and propellers replaced the Ericsson ones in the 1850s. Their plans, if any were drawn, have vanished along with the many wood shipyards that closed their gates after the Civil War.

The only detailed drawing of one of these small steamships that the author has ever seen was found decades ago in the loft of Phillips Wharf in Salem, Massachusetts, and is now in the Peabody Museum's collection. It has neither date nor identification. The vessel is an 1850 to 1855 propeller. It has been redrawn and shows a three-masted coastwise steamship around 150' long and of 470 tons. Despite her small size, she is iron strapped for strength. The hull has some steamship characteristics, such as no tumblehome, flat sides, no bowsprit, and a maximum section amidships, but still reflects her sailing-ship ancestry in deadrise, a raking stem, and a figurehead. The pilot house, galley, and officers' quarters would be in the midship house, while the after cabin would contain state rooms for a limited number of passengers. The main engine is of the inverted (cylinders over crankshaft), direct-acting type that eventually replaced all the more complex geared, trunk, or vee-types tried out in earlier craft. The two cylinders are about 24" bore by 2'-0" stroke turning a propeller 9'-0" in diameter.

In every respect she appears to be a well-designed little steamship, but just because a plan exists does not give definite assurance that she was ever built. Her drawings are included here as the one available set that shows what a small, mid-nineteenth century, propeller-driven steamship was like.

The *West Wind*, 1851-1854

The Thatcher shipyard in Wilmington, Delaware, undertook a small steamer for George W. Aspinwall of Philadelphia. Intended for the California trade, she was 144' by 26.1' by 13.1', measured 460 tons, had two decks and a transom stern. She was taken to Philadelphia for engining by Reaney and Neafie's Penn Works, where a two-cylinder, direct-acting engine of 30" bore and 2'-2" stroke driving an 8'-2" diameter propeller was installed.[50] Steam was supplied by a single return-flue boiler 9 feet in diameter by 20 feet long and condensed by a Pirrson surface condenser.

Nothing ever went well for the *West Wind*. First she was sold at a U.S. Marshal's sale because of debts to the Penn Works.[51] At that time, September 17, 1851, she was almost completed. Bender, Wright, and Company acquired her. Her trials, on October 4, gave a speed of ten miles per hour at 15 psi steam pressure, while her screw turned at 48 rpm. She had space for sixty five passengers and a large cargo. Under R. H. Leese, a part-owner, she cleared at Philadelphia on October 24 for Cape May, New Jersey. Her subsequent carer is very rarely reported. She arrived at New York in early January with a new master, Captain William Savage, and left about January 31, 1852, for Chagres. In April she was supposed to sail from New Orleans for Chagres but never arrived at New Orleans. About May 16 she reached New York with only thirteen passengers from San Juan del Norte, Nicaragua. It was a wretched trip, with many of the passengers, crew, and even the master very sick.[52] Heavy gales compounded their misery and she had to put into Norfolk for extra coal on May 12. Thereafter she started for New Orleans on the 31st but stopped at Mobile instead. From July until September she was laid up in Brooklyn.[53]

On September 19, under a Captain Smith, she left, overloaded with passengers, for the gold fields of Australia.[54] On this trip she had a three-masted bark rig. Whether it was original, or whether extra sails were added for the trip, we do not know. Certainly it was necessary, for there was a continuous series of engine breakdowns starting one day out of New York. After she was repaired at Porto Praya, in the Cape Verde Islands, the casualties continued. She made Ascension Island and lay there for sixty-three days while her engineers did what they could. Then, ten days out, there were more troubles; one cylinder had to be disconnected and she carried on under the other. When she reached Africa and Table Bay, her long-suffering passengers protested to the American consul that the *West Wind* was unseaworthy and her provisions inferior. An official survey was held and it reported that the vessel was seaworthy but that the bolts holding the main shaft bearings were broken and a few "trifling" repairs were needed to her engine. The ship was mortgaged to raise money,[55] repairs were made, and she was supposed to continue on to Australia March 7. On April 28, 1853, she arrived off Port Phillip Heads, Victoria, with 153 long-suffering passengers, over seven months from New York.[56] Entering the harbor she fouled a vessel, the *Prince Albert*, but continued up to Melbourne.

After a coastwise trip to Sydney starting May 19, and a second collision, this time with the pilot boat *Anonyma*, when she returned, the *West Wind* was offered for sale, but no one wanted her. Two unprofitable trips to Port Adelaide followed. Her Australian papers show W. M. Tennent and Thomas Smith as owners. Thereafter, she was anchored in Hobson's Bay.

Two weeks later, February 27, 1854, a fire ended her career. An owner was later convicted of arson. At what is now Port Melbourne, her remains were broken up and her boiler left on the beach, where it was referred to in sailing directions for the next forty years.

Under the supervision of Captain Loper, Reaney and Neafie had produced many successful engines for screw steamers. Without his restraining hand, they seem to have cut corners on the *West Wind*'s machinery. She was no testimonial for the screw steamship. Nevertheless, her rusting boiler became a useful landmark.

William Penn. Courtesy of The Mariners Museum, Newport News, Va.

The *William Penn*,

1851-1856

Exact contemporaries of the transatlantic propellers just discussed, a pair of small steamers were being built in Philadelphia to inaugurate a passenger and freight service between that port and Boston. These were completed in 1851 and appropriately named *William Penn* and *Benjamin Franklin*.

In general, the *William Penn* was an enlarged version of the unidentified steamship described in an earlier section, similar in every respect except for the addition of a cut-water, trail boards, and a bowsprit. The deck houses, passenger cabins, engines, and boilers were all located in exactly the same places but the size was increased to 183' by 26.4' by 13.2' depth of hold to the second deck. The depth to the upper was 19.5'. Like the *S.S. Lewis*, the *William Penn* was built by Theodore Birely and Son and had the transom stern favored by Birely and a billet head, and she measured 613 tons. She was diagonally strapped and had her bottom, under the engines, filled in solid for 40' of her length. A barkentine rig of moderate size was fitted with three square sails on the foremast and a very short bowsprit and jibboom. She was unusual in having projecting channels to increase the spread of the shrouds. The *S.S. Lewis,* from the same builder, had her shrouds set up on the bulwarks and most steamers led them to the deck inside the bulwarks. R. F. Loper was again responsible for much of her design. The engine of the *William Penn* came from Reaney, Neafie, and Company. It was a great improvement over those used in most of the screw ships so far described in that it had two inverted, direct-acting cylinders of 34" bore and 2'-10" stroke. The propeller was 4 bladed, 10'-0" diameter by 20'-8" pitch.[57] Steam came from two return-flue boilers 21 long and 8' in diameter with forced draft fans. Although the cylinders had the simplest (and, therefore, the best) possible arrangement, the necessary auxiliaries were still awkward and complex. Each cylinder had its own condenser and air pump, the latter being driven from an auxiliary shaft geared down to turn at one-half the speed of the crank shaft.

The *Benjamin Franklin* had an identical hull with a two-cylinder Loper engine supplied by the I. P. Morris Works. It also had separate air pumps, but jet condensers were built into the engine frame and the *Penn*'s geared shaft was eliminated.[58] Although the keels for both vessels were laid the last week of January in 1851, their

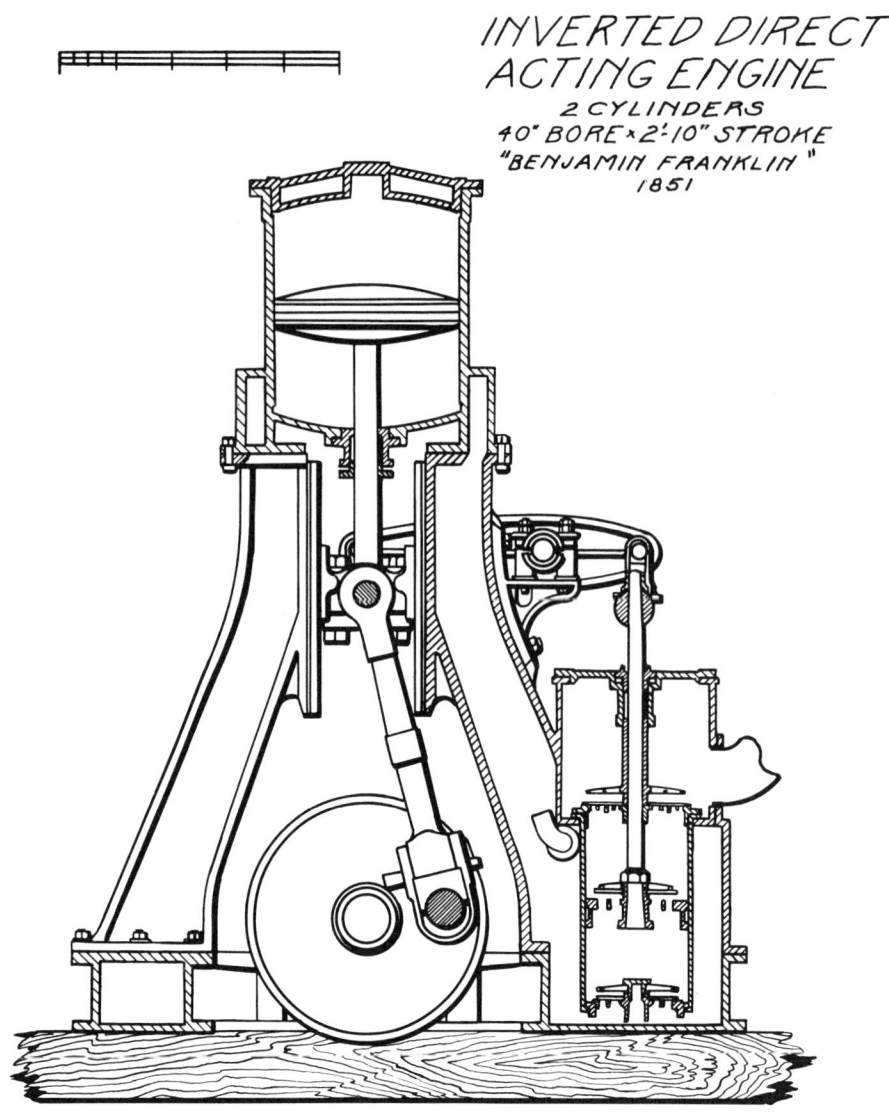

Inverted direct acting engine 40″ bore by 2′-10″ stroke for the *Benjamin Franklin*.

engines had been under construction for some time and were ready to go aboard.[59]

The *Benjamin Franklin* was launched July 14 and the *William Penn* on the 26th. The latter had all her machinery installed, her masts stepped, and had been rigged so that she was almost complete; as a result, she was the first finished. Her owners were Samuel Reynolds, E. Lincoln of Philadelphia, and J. B. Lincoln of Boston; the first two were also owners of the *S.S. Lewis*, which was fitting out at the time the *Franklin* and *Penn* were launched.

Trials were run September 1, 1851, with a large party aboard, and a distance of 40.5 miles covered in 3 hours and 16 minutes. When steaming along at 28 psi steam pressure, her propeller turned at 65 rpm, a very high value for a direct-connected engine. As completed, she could take forty passengers in her after cabin and 760 measurement tons of cargo in her holds.[60] In view of her short sea route, only 1,500 gallons of fresh water were carried in a small tank on the main deck aft of the engine. On September 6 she sailed for Boston under Captain Zimri Wheldon. One week later the *Benjamin Franklin* followed and, for the remainder of 1851, a weekly service between Philadelphia and Boston was maintained but, alas, did not prosper and was terminated early in 1852.[61]

On February 25, 1852, the *William Penn* arrived at New York for the first time in her career, and, on March 2, left for Chagres. She was back again March 30, having sailed from Chagres on the 20th and coaled at Kingston, Jamaica, on the 24th. Although both vessels were very small for so long a route, the *Franklin* tried a similar trip from Philadelphia. In April they were advertised for New

York to New Orleans operation at fares of $60 first class and $25 steerage.[62] The *Penn* led the way from New York on April 22 and continued until March of 1853. Both were then laid up in Brooklyn.

The *William Penn* is reported, seven months later, as leaving New York on October 7 for Charleston; five round trips were made to that city with the *Franklin*, again, as a running mate. The *William Penn*'s usual time from Charleston to New York, seventy two hours, corresponds to 9 knots, while the paddle steamer *Southerner* regularly made 10 1/2. Certainly these small steamers could not complete with Spofford, Tileston, and Company's four large paddlers on the route. The Lincolns had to face up to owning a pair of ships of a size and speed ill fitted for any of the services tried so far. On March 2, 1854, both were sold at auction, the *Penn* bringing $37,750 and the *Franklin* $34,750.[63] McCready, Mott, and Company of New York were the purchasers. From this point on, our knowledge of the *William Penn* comes from her new master, John Codman, who published an entertaining account of her ensuing operations some forty years after the actual events took place.[64]

Captain Codman stated that N. L. McCready and Company intended her as the first steamship for a new line from New York, via the Strait of Gibraltar, to Marseilles, with an announced sailing date of May 20, 1854. Although neither passengers nor cargo appeared, she sailed. There was no reason to send her across unless her owners counted on a Crimean War charter. Although there was a minor collision in the fog with the Havre liner *Franklin*, westbound into New York, neither vessel was damaged. Whatever the intended plans, Marseilles, when she arrived, was suffering from a cholera epidemic and no ship from that port could be admitted elsewhere without a long period of quarantine. Captain Codman and his wife and child, as did everyone who could afford it, abandoned the port for the country and settled down to wait until the heat of summer passed and the disease died down.

In the fall the *William Penn* was chartered to the French government for 50,000 francs per month, was loaded with stores, ammunition, and troops, and sailed November 1 for the Crimea by way of Constantinople. In the Dardanelles she ran into fog, but, in accordance with orders, followed a French gunboat. Both ended up aground on Nagara Point. The local Turkish pasha unloaded the ship to help get her off, and accepted a token payment in champagne. Captain Codman had taken aboard an ample supply, which would smooth his ship's way on more than one occasion. After three days' delay the *Penn* proceeded on to the Ottoman capital. Thence she headed for Kamiesh, as the improvised French harbor near Sebastopol was then called. The *William Penn* continued between Marseilles and there, usually loaded down with stores and ammunition in the hold, 400 troops unmercifully crammed into the tween decks, and, often, the open upper deck encumbered with stalls and horses.

Later in the winter there came engine and boiler repairs. Once they were completed and she was at sea again, a following gale was encountered in the Gulf of Lyon. Waves broke over the stern, demolished the horse stalls, and washed two unfortunate animals down the engine room skylight onto the cylinder heads. The ship had to stop until their carcases could be hoisted out and thrown overboard. Seas swept the deck and the fires were nearly extinguished, but after the troops were put to work bailing, control was regained and the *William Penn* continued on her way.

As the war progressed, sick and wounded were brought back from Kamiesh to Gallipoli, in the Dardanelles. Further trips were made to Algiers to pick up Zouaves stationed there, but the relationship between the Americans and the French authorities was never happy. The latter insisted on getting the most out of the high price they were paying and, whenever an engine or boiler broke down, reduced their payments. A new contract for 2,000 English pounds per month was negotiated with the Turkish authorities, who operated in a far more informal manner. From then on, the ship and her officers enjoyed a leisurely existence, usually running from Constantinople to nearby Black Sea ports such as Varna, now in Bulgaria, or Eupatoria.

One time, while at anchor in the latter port, the *William Penn*'s propeller fouled the anchor cable of a nearby vessel and was hopelessly damaged. Remembering his French experiences, her master continued on his regular route under sail, but he always had a fire built so that smoke billowed from the stack when leaving or entering a harbor. Eventually, he was able to purchase a spare propeller from another steamer at Constantinople that could be fitted to his propeller shaft. The *Penn* was trimmed by the bow, a wooden cofferdam built around the stern, and the new screw installed by his chief engineer without having to use a dry dock. All this may have been a slick Yankee trick to Captain Codman, but it was a remarkable piece of engineering work by the chief engineer, whom he never bothered to name.

The last, and one of the very few, dates quoted by Codman is April 2, 1856, when the news of peace reached him. At that time he was still under charter to the Turks. Regrettably, he neglected to tell us what happened next. Certainly the *William Penn* was sold abroad and never returned to American waters.

The *William Penn* should be remembered as the first American screw steamer to operate successfully for a period of at least five years in a wide variety of waters, often a long distance from the iron works needed to replace any engine part that failed. She may have lacked financial success in the coastwise trade, but as an available steamer, she earned large sums, first as a French and, later, as a Turkish transport.

The *Granite State*,

1852-1855

There have been a number of obscure steamers that crossed the Atlantic at least once in their lives. The *Granite State* is such a vessel. All we know of her origin was that she was Philadelphia-built for R. F. Loper. She had two decks, three masts, a scroll head, a square stern, and was screw propelled. Her dimensions of 174.6' by 24.4' by 13.2' resulted in a tonnage of 582.

After completion, in mid-July of 1852, Loper sent her to New England under Captain David L. Wilcox, who had both supervised the construction of, and commanded, several Philadelphia steamers. She visited Boston and Bangor, Maine, and, while carrying a group called the "Cape Cod Association" on an excursion from the former city to Providence, damaged her propeller.[65] The casualty was minor, for she made the excellent time of forty nine hours from Boston back to Philadelphia, arriving there the weekend of August 14, 1852. The Penn Works was remarkably efficient; a new propeller was ready and installed, while she was dry-docked, on the 19th.

Certain Philadelphia entrepreneurs were considering a steamship service to Charleston at that time, and there was some talk of purchasing the *Granite State*. Instead, Loper decided to try the route himself, sending her out on September 11 carrying twenty cabin passengers and a full load of freight.[66] The nine-day round trip was followed by another, leaving September 22 with nineteen cabin and six steerage passengers as well as a "satisfactory" cargo.

The steamer than vanished from view for two years. The next reports found by the author cover a series of voyages between Boston and Baltimore, starting at the former port on June 3, 1854, and ending in October of the same year.[67] Again she disappeared for the winter of 1854-55 but, on March 17, undertook a single round trip from Boston to Philadelphia. On March 30 she cleared Boston for Philadelphia, arriving there on April 3 to be completely refitted and given new spars—probably her rig was increased for her prospective ocean voyage—and to have another propeller installed on May 24, since one blade of the old one had broken off.[68] Five days later, May 29, 1855, under David L. Wilcox, who had returned as her master, the *Granite State* steamed down the Delaware bound for Liverpool and arrived without any untoward incidents on June 16.[69] The eighteen-day passage from Philadelphia corresponds to a speed of 7.5 knots, a good performance for so small a vessel. The Lytle List shows that the *Granite State* was sold abroad; it is our last report on her.

The *Star of the South*,

1853-1868

By 1853 the Loper steamers had become so successful that he obtained an order for ten identical vessels for the Parker Vein Coal Company. These 500-ton colliers were built by a number of Philadelphia shipyards and engined by several different iron works. The "model," or lines of the entire group, was developed by Theodore Birely, the mold loft work done at his shipyard, and molds supplied by him to all the other builders to insure their uniformity. The continued Birely-Loper association had, over a period of four years, developed the finest screw steamers in the United States. John Ericsson may have introduced the propeller, but Richard Loper made a success of it. Ericsson is remembered, Loper forgotten.

While these colliers were under way, Loper obtained an order from Messrs. J. W. and T. P. Stanton for a steamer to run from New York to New Orleans.[70] When completed, she became a particularly attractive, schooner-rigged craft of 960 tons, a little smaller than the *S.S. Lewis* but 50 percent larger than the *William Penn*. Birely and Son, as usual, received the hull contract, while the machinery went to J. T. Sutton and Company's Franklin Works. All had collaborated on the *S.S. Lewis*. With dimensions of 206.1' long by 31.2' beam by 15.6' depth of hold to the lower of her two decks, the *Star of the South* had a billet head and a round stern replacing the flat transoms of her predecessors. Loper was still experimenting with different engine types and accepted an unorthodox, two-cylinder, geared trunk engine of 52" bore by 3'-0" stroke turning at one-half the propeller's speed. The screw was 10'-6" in diameter by 17'-0" pitch. An equally unusual boiler carried the combustion products up and down through 5" diameter flues. On trials, the *Star of the South* made 13 1/4 miles per hour with 36 psi steam pressure and 90 rpm at the propeller. With such a high pressure there must have been a surface condenser, although none is mentioned in her published description.

A cabin was provided with adjoining staterooms for forty passengers. The dining saloon was in a deckhouse and featured sofas mounted on transverse rails. At meal times the port and starboard ones were shoved together, forming the bases for the dining-room tables.[71] Hold space was rated at 7,000 barrels capacity. There were four Francis boats.

On May 18, 1853, Captain William R. Gardner headed the *Star of the South* down the Delaware River bound for New Orleans; when she returned, however, it was to New York, and all subsequent sailings were from there.[72] James Marks took over as master for her second trip; thereafter, she spent several days in dry dock at New York, probably being coppered. For the next eighteen months she continued on the New Orleans route and northbound usually made the trip in eight days; her record occurred in February of 1854 when she took only six, a remarkably good time for a 960-ton steamship in the dead of winter.[73] She regularly carried both first and steerage passengers, as well as freight.

There were no sailings in 1855 until March 15, when she was chartered to take freight to Aspinwall for the Panama Railroad. On her return, bringing a cargo of coffee and pimientos, she reached New York on April 19.

In June, Roche Brothers & Coffey advertised her for Liverpool with fares of $100 in the first cabin, $35 in the second, and $25 in the third.[74] She left New York on June 20, reaching Liverpool July 7. The seventeen-day passage was the shortest yet for any American propeller. A charter from one of the Crimean War participants must have been arranged, for she continued on to the Black Sea. Her transport duties are not known. The next news heard from her comes two years later when, still under Captain Marks, she reached New York on July 8, 1857, sixty four days out of Marseilles. The extreme length of the passage resulted from two separate encounters with gales so severe that she might well have perished on the way. She was first partly dismasted near the Azores, with her rudder sprung and considerable hull damage. After the wreckage had been cleaned up and temporary repairs made, she continued on her way only to meet a southeast gale on July 1, which tore all the canvas off her yards; with the rudder already in a hazardous condition, she was completely out of control for twelve hours; her mainmast went overboard and its wreckage had to be cut away to save the ship.[75] But the indefatigable James Marks again patched up his ship and brought her into New York under some kind of a jury rig. Although not explicitly stated, the whole voyage must have been under sail. It is likely that, after extensive wartime operations, her single boiler had given out.

The next six months were spent installing new boilers and making alterations for a new owner, S. L. Mitchill and Sons, who had three paddle steamers operating between New York and Savannah.[76] These were the *Augusta*, *Alabama*, and *Florida*, all developments of, and successors to, the *Cherokee* and *Tennessee*. Under Captain Thomas Lyon, formerly of the *Cherokee*, the *Star of the South* sailed on January 9, 1858, for Savannah. She did well and was kept running until the outbreak of the Civil War.

After a charter at $600 per day to the U.S. Quartermaster Corps, starting May 21, 1861, she was kept in continuous operation throughout the conflict. In November of 1861 she was part of Admiral S. F. DuPont's invading fleet that captured Port Royal, South Carolina. On the way she ran into and sank a cattle transport named the *Peerless*.

In April of 1862 she recovered the ship of the line *Vermont* that had been missing for weeks after having broken loose from a steamer towing her to Port Royal.[77] When she was delivered, under tow instead of sail, the *Vermont* was a sad example of how archaic the sailing battleship had suddenly become. She was outmoded, an enormous unarmed hulk useful as a barracks or a floating warehouse, but no longer a great offensive weapon. Yet the far smaller *Star of the West*, hauling her in, was always in demand.

The steamer spent much of her time between New York and the Port Royal naval base but, after the fall of New Orleans, made many voyages there, often with way stops. In September of 1863 she towed the U.S.S. *Alabama* from New York to Portland, Maine. The latter, now a gunboat, was one of her Savannah running mates, still under her original name, and not the Confederate raider, C.S.S. *Alabama*. Later the same year, she was ferrying troops and supplies to General Grant's armies via Fortress Monroe, Alexandria, the Potomac, and City Point on the James River. Her last transport duty was from New Orleans July 1, 1865, to New York on the 9th.

Less than a month later, far too short a time to eliminate the scars of arduous war duty, she cleared New York August 7, 1865, for New Orleans, under charter to Livingston, Fox, and Company, who used her for only two voyages. Two wars and too little maintenance had taken their toll. From October of 1865 to June of 1866 she was at New York. During this lay-up she was sold to Pendergast Brothers and Company who planned a Baltimore cargo service starting on June 9.[78] Her next appearance was in the disaster column when she was towed into Baltimore on November 9. On her way from Charleston to New York she had suffered a broken propeller shaft and had to sail into Hampton Roads. Repairs were made; afterward the Pendergasts offered her up for sale by means of advertisements appearing irregularly from May of 1867 to January of 1868.

On January 25, 1868, the *Star of the South* left New York on her last voyage. She was headed for Santiago, Cuba, under the auspices of J. and S. Badell. The sailing date had been postponed from January 18, whether because of lack of cargo or her deteriorating condition, we do not know. Her return voyage was suspiciously long. She left Santiago on February 16, put into Savannah for coal twelve days later. She did not arrive at New York until March 9.[79] All this sounds as if her engine and boilers

203

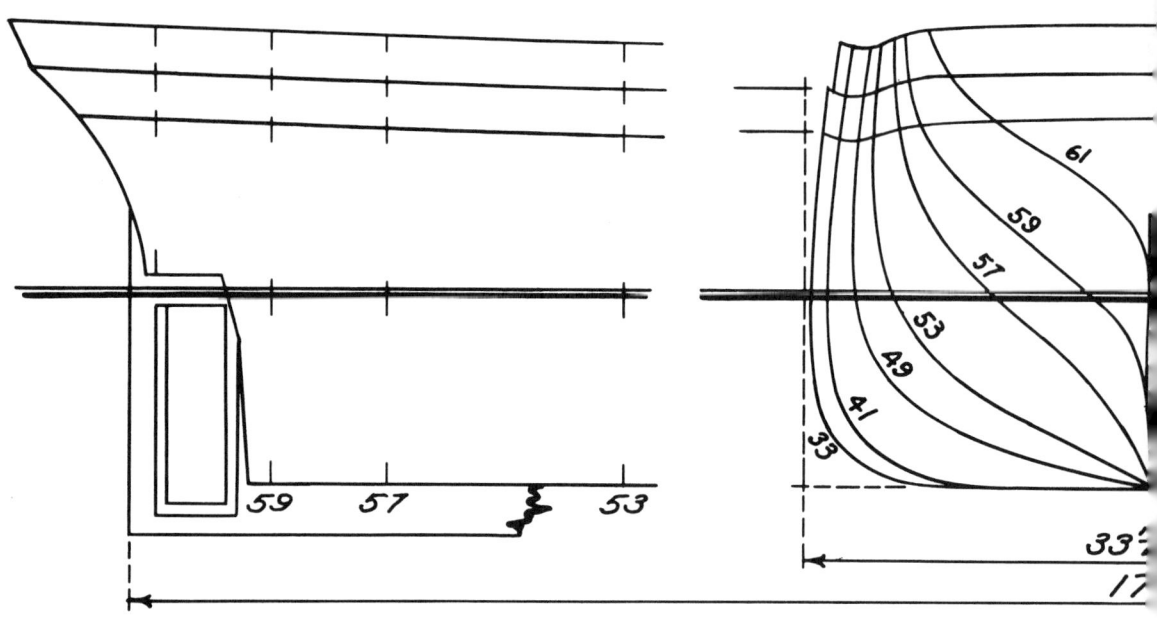

North Carolina, 1854, lines.

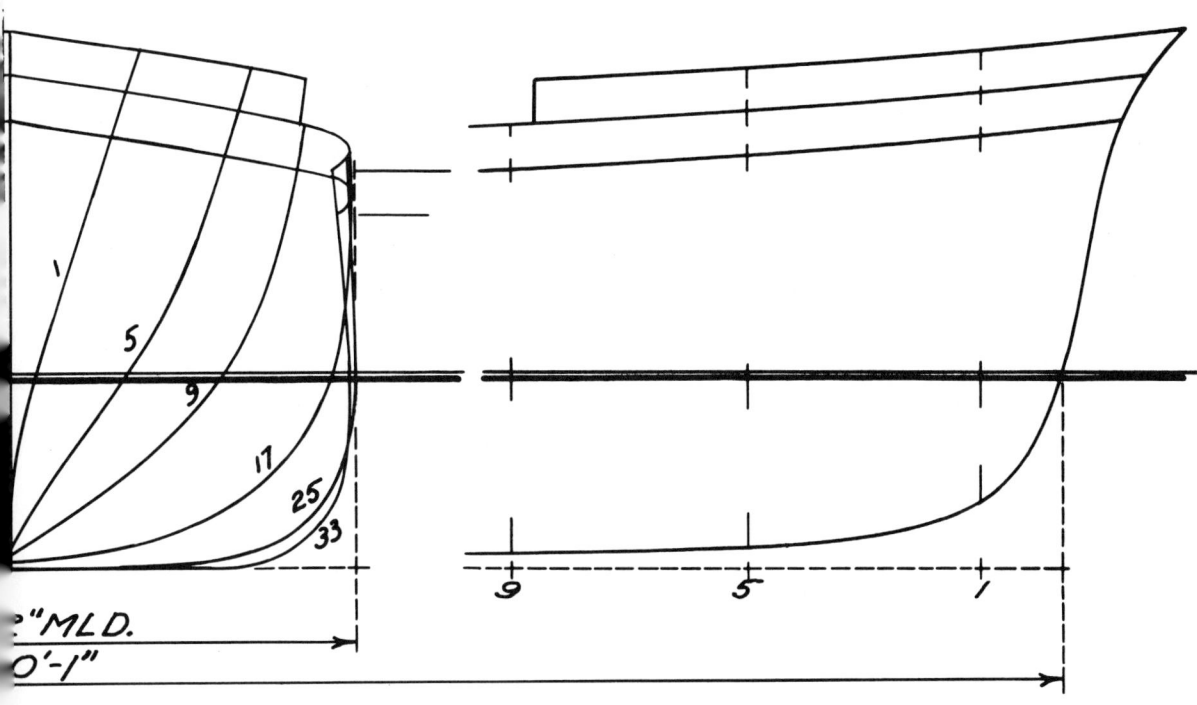

were in a very bad state. The *Shipping and Commercial List* shows that she vanished, without a Custom House clearance, from New York in April. No doubt she was towed away for scrap.

The ending is anticlimactic for the first entirely successful wood screw steamship encountered in these pages. A profitable year in the New Orleans trade, in direct competition with the U.S. Mail Line's large side-wheel liners, was followed by two years as a Crimean transport, and, best of all, by over three years of steady operation to Savannah as the only propeller in a fleet of four steamships. The next four years saw her as a transport in her second war. A good example had been set and many more screw steamships would succeed the *Star of the South*.

The *North Carolina*,

1854-1855

The earliest steamship built to lines drawn by John W. Griffiths, the *Georgia,* has been discussed in an earlier chapter. She suffered from excess beam and directional instability. Since that time Griffiths had produced a number of successful designs for side-wheel steamers that included the *John L. Stephens,* for the West Coast, and the *Keystone State,* for the East. His first screw steamer was the *North Carolina,* built by Vaughan and Fisher in Philadelphia in 1854. She was intended to be a rule-beater with a very large cargo space and a low tonnage.[80] To achieve this, Griffiths gave her a single deck with a short forecastle and a long poop; in terms of the low tonnage, 672, she was somewhat larger than the *William Penn*. She was 173.2′ by 31.1′ by 13.3′, barkentine rigged, and intended to carry freight between Philadelphia and Wilmington, North Carolina. Alexander Heron, Jr., was her owner. For this service Griffiths restricted her draft to 10′-0″. Her engine, by Merrick and Sons of Philadelphia, had a single cylinder 56″ bore by 4′-0″ stroke. It was similar to the *Little Juliana*'s of 1805. Above the cylinder was a crosshead from either end of which descended a connecting rod driving a large gear 6′-4″ in diameter. The gears turned a smaller one, 2′-8 1/2″ diameter, on a propeller shaft located underneath the cylinder. All gears had 10″ wide iron teeth.[81] Another unusual feature was a conical roller bearing to take the propeller's thrust. A clutch allowed the propeller to windmill when sailing. The 9′-3″ screw had three blades and a pitch of 16′-6″; it turned over twice as fast as the engine.

Griffiths, still obsessed with the idea of a fine bow and a full stern, put his propeller into a very small aperture behind blunt water lines. The aperture, if scaled from the published plans, is too small and must have been enlarged when the hull was built. Both the proximity to the propeller post and the full stern lines created bad flow conditions. Despite these disadvantages, the *North Carolina* is said to have made ten miles per hour on trials in a light condition drawing 7′-7″ forward and 8′-2″ aft, with the propeller blades sticking 1′-7″ out of the water. Steam was supplied by two boilers 13′-3″ long by 9′-7″ wide by 10′-0″ high with four furnaces and 3″ diameter return tubes. It was condensed by a surface condenser and, to make the unduly complex engine even more interesting, a small transverse walking beam was driven by the crosshead to operate the air pump. The use of a single cylinder for a screw steamship was unusual, in 1854, although it became popular after the Civil War. The two large gears acted as flywheels to carry the cranks across dead center. Nevertheless, hand-operated valve gear must have been necessary for stopping, starting, and reversing.

In short, the *North Carolina* did not profit from the considerable experience of previous Philadelphia screw steamers. Griffiths's hull and Merrick's machinery were poor examples of what could be achieved. After two trips between Philadelphia and Wilmington, in November and December of 1854, it was decided to send her abroad for sale or charter. For the ocean trip, the waist, between the poop and forecastle, was closed in, making her a two-decked steamer.[82] Passenger quarters were added and advertisements announced that she would sail from Philadelphia for Liverpool on February 3 carrying saloon passengers for $90 and third-class passengers for $25.[83] Captain Washington Symmes was listed as master and Thomas Richardson and Company as agents. She was badly overloaded, for she drew 14′ when she sailed. The fat stern and the deep draft combined to make her steer badly. At sea, in bad weather, she was unmanageable under sail but able to steer herself under steam. This, combined with strained upper works, led her captain, when five days out, to turn back. About one hundred miles from the Delaware Breakwater, her cast-iron propeller lost two of its blades, but Philadelphia was reached under the remaining one. Drastic measures were taken to correct her steering. The keel was deepened for its whole length, the forefoot filled out, and a larger rudder installed.[84] The damaged upperworks were repaired, the leaking hull recaulked, and a new propeller of 15′-0″ pitch installed.

When she left the Delaware Breakwater for the second time, March 23, 1855, she was less heavily loaded, drawing 12′-3″ forward and 13′-6″ aft and displacing 1,195 tons. This time her steering proved satisfactory. On the trip, the propeller turned about 69 rpm under steam at 23 psi, cutting off at one-half stroke. Coal was consumed at the low rate of 12 tons per day while she was making 8.5 knots.

In Saint George's Channel, southeast of Waterford, Ireland, the *North Carolina* was struck on the port bow by the ship *Robert*, outbound from Liverpool to New Orleans. The collision, which took place about 1 A.M. on April 8, sank the steamer in ten minutes; all on board were rescued.[85] The *Robert* took the captain and twenty one crew members to Liverpool. Eleven others spent the night in boats and were rescued at daylight.

The *North Carolina* is difficult to explain. She came from a city that had Theodore Birely and Richard F. Loper continuously perfecting the screw-propelled steamer. Yet her builders went to a New York naval architect of extremely strong prejudices for her lines. And Griffiths repeated the mistakes he had made six years earlier in the design of the *Georgia*. With excessive beam for her draft came excessive rolling; as a result, she strained her upper works. With too full a stern she steered badly. The author feels that it was just as well she did not reach Liverpool and add to the woes of anyone who might charter or purchase her.

Notes

1. *New York Herald*, January 17, 1851.
2. *Journal of the Franklin Institute* 21, 3d ser. (January-June 1851): 420.
3. *Philadelphia Public Ledger*, April 23, 1851.
4. Ibid., May 5 and 12, 1851.
5. *New York Herald*, June 18, 1851.
6. *New York Commercial Advertiser*, July 7, 1851.
7. *New York Herald*, October 5 and 7, 1851.
8. Robert Macfarlane, *History of Propellers and Steam Navigatio.* (New York, 1851), pp. 119-20.
9. *Journal of the Franklin Institute* 23, 3d ser. (January-June 1852): 128-30.
10. *Philadelphia Public Ledger*, May 31, 1851.
11. *Boston Atlas*, July 31, 1851.
12. *Journal of the Franklin Institute* 22, 3d ser. (July-December 1851): 215.
13. *Philadelphia Public Ledger*, May 31 and July 17, 1851.
14. Ibid., September 9 and 17, 1851.
15. Ibid., September 17 and 22, 1851.
16. N.R.P. Bonsor, *North Atlantic Seaway*, 2d ed. (Newton Abbott, London and Vancouver, 1975), 1: 225.
17. *New York Herald*, November 3, 1851.
18. Ernest A. Wiltsee, *Gold Rush Steamers* (San Francisco, Calif., 1938), p. 82.
19. *New York Herald*, October 12, 1852.
20. Wiltsee, *Gold Rush Steamers*, p. 91.
21. *San Francisco Alta*, April 10 and 11, 1853.
22. *Journal of the Franklin Institute*, 22, 3d ser. (July-December 1851): 50-51.
23. *Appleton's Mechanics' Magazine*, January 1852, pp. 1-2.
24. *New York Commercial Advertiser*, August 12, 1851.
25. Ibid., December 15, 1851.
26. *San Francisco Alta*, August 21, 1852.
27. *New York Times*, October 4, 1852; *New York Commercial Advertiser*, July 16, 1852.
28. *San Francisco Alta*, November 1, 1852.
29. *New York Commercial Advertiser*, March 15, 1854.
30. *New York Shipping and Commercial List* (September 1854-August 1855).
31. *New York Evening Post*, May 5, 1855.
32. Bonsor, *North Atlantic Seaway*, 2d ed., 1: 199, 200, 218, and 220.
33. *Journal of the Franklin Institute* 21, (January-June 1851): 139-40.
34. *Philadelphia Public Ledger*, October 22, 1851.
35. *Journal of the Franklin Institute* 23, 3d ser. (January-June 1852): 178-82.
36. *Philadelphia Public Ledger*, October 3, 1851.
37. *Journal of the Franklin Institute* 23, (January-June 1853): 178-82.
38. *Philadelphia Public Ledger*, November 11, 1851.
39. Ibid., January 12, 13, 16, and 19.
40. Ibid., February 12, April 24, April 27, and May 1, 1852.
41. *New York Times*, December 13, 1852.
42. *Illustrated London News*, January 1, 1853.
43. *New York Shipping and Commercial List*, January and February 1852.
44. *New York Herald*, April 23, 1852.
45. *New York Commercial Advertiser*, June 11, 1852.
46. *New York Herald*, August 7 and 11, 1852.
47. *Willmer & Smith's European Times*, (Liverpool) September 4, 1852.
48. *New York Herald*, January 1, 1853.
49. Carl C. Cutler, *Queens of the Western Ocean* (Annapolis, Md., 1961), p. 294.
50. *Journal of the Franklin Institute* 22, 3d ser. (July-December 1851): 213.
51. *Philadelphia Public Ledger*, July 16, September 18, and October 6, 1851.
52. *New Orleans Picayune*, February 28, March 17, and May 22, 1852.
53. *New York Shipping and Commercial List*, May-September 1852.
54. *New York Commercial Advertiser*, May 18, 1853.
55. *Philadelphia Public Ledger*, May 21, 1853.
56. Ronald H. Parsons of Lobethal, South Australia, has kindly supplied me with the material on the *West Wind* in Australian waters.
57. *Journal of the Franklin Institute* 22, (July-December 1851): 213-15.
58. Oliver Byrne, *The American Engineer, Draftsman, and Machinist's Assistant* (Philadelphia, 1853), p. 77 and Plates XI, XII.
59. *Philadelphia Public Ledger*, February 2, 1851.
60. Ibid., September 2, 11, and 15, 1851.
61. Ibid., February 2, 1852.
62. *New York Herald*, April 14, 1852.
63. *New York Commercial Advertiser*, March 3, 1854.
64. John Codman, *An American Transport in the Crimean War* (New York, 1896).
65. *Philadelphia Public Ledger*, August 20, 1852.
66. Ibid., September 3, 9, 13, 22, and 23, 1852.
67. *New York Shipping and Commercial List*, 1854-1855.
68. *Philadelphia Public Ledger*, September 25, 1855.
69. *New York Commercial Advertiser*, June 28, 1855.
70. *Philadelphia Public Ledger*, April 18 and May 2, 1853.
71. Ibid., May 16, 1853.
72. *New Orleans Picayune*, May 29 and June 4, 1853.
73. *New York Shipping and Commercial List*, July 23 and 27, 1853.
74. *New York Herald*, June 20, 1855.
75. *New York Commercial Advertiser*, July 8 and 9, 1857.
76. *American Lloyds*, 1863.
77. *New York Commercial Advertiser*, March 4 and April 14, 1862.
78. *New York Tribune*, June 5, 1866.
79. *Journal of Commerce*, January 17, 27; February 29; March 10, 1868.
80. *Monthly Nautical Magazine*, (January 1856), pp. 294-95.
81. *Journal of the Franklin Institute* 28, 3d ser. (July-December 1854): 411-12.
82. *Monthly Nautical Magazine*, (February 1855), pp. 337-40; (July 1855), pp. 336-38.
83. *New York Herald*, January 28, 1855.
84. *Journal of the Franklin Institute* 29, 3d ser. (January-June 1855): 395-96.
85. *New York Herald*, April 27, 1855.

12
Experimental Craft

The Hot-Air Ship *Ericsson*, 1853-1892

PERHAPS the most successful invention to spring from the fertile mind of John Ericsson was the hot-air engine. Today it is forgotten, but, when it first appeared in 1833, it was particularly celebrated because no one could satisfactorily explain why it worked.[1] Ericsson explained at great length and demonstrated that he was just as ignorant of its thermodynamic performance as his contemporaries. It was perfected during the 1850s and became the first successful small source of power manufactured in the United States. It could operate on almost any fuel and could be installed in every home, store, or factory that was piped for illuminating gas. As a self-contained unit it avoided the problems of boilers, piping, firemen, condensing water, and all the complexities of a steam engine. In effect, it was the electric motor of its day and was used to drive pumps, printing presses, elevators, cranes, and even sewing machines.

Between 1840 and 1850 Ericsson built eight "caloric" engines, as he called them, gradually improving details and increasing sizes as he went along.[2] By 1851 a large engine of 30" cylinder diameter was running successfully. Not until 1857, however, were engines manufactured as a standard item in a series of sizes going down to cylinders as small as 6" bore, but by the outbreak of the Civil War thousands had been built. Their manufacture and sale were carried on, not by Ericsson, but by business associates who had purchased a part interest in the device.

The inventor, never content with small experiments, decided that the way to demonstrate the tremendous economy of his new prime mover was to install it in an ocean-going ship. His contagious enthusiasm enlisted further financial aid from a wealthy young New Yorker, John B. Kitching, who owned a part of the "caloric" engine, and a number of others. The hull came from the Williamsburg yard of Perrine, Patterson, and Stack, builders of the *Lafayette* and *City of Pittsburg,* who held some of the shares of the new vessel that was to be named *Ericsson*. After the keel laying in April 1852, the work advanced rapidly; the hull was launched on September 15, the machinery was installed, and everything was made ready for her first trial on January 4, 1853. She was a little bigger than the *City of Pittsburg,* the largest propeller to date, and a little smaller than the Havre liners *Franklin* and *Humboldt*. The tonnage was 1,902 and the dimensions 253'-6" by 39'-8" by 26'-6". At the time of her trials, with 200 tons of ballast aboard, she drew 17'-0". There was no figurehead, but the stern carried a figure of Ericsson being crowned with a laurel wreath by two allegorical figures representing the United States and Great Britain.[3] The inventor's friends demonstrated artistic, as well as financial, faith in their hero.

On the spar deck were the sailor's and firemen's quarters in the forecastle; a center house over the machinery; and, in other deckhouses, the dining saloon, pantry, library, smoking room, and officers' rooms.

Ericsson, original appearance. *Courtesy of The New-York Historical Society, New York City.*

Below decks were 64 staterooms for 130 passengers, all finished in Gothic style with white woodwork and "chaste" gilding. Space for 12 passengers' servants was provided in the hold, forward of the engine room, while cargo was stowed on the whole length of the deck extending above the poor menials' heads. "By this arrangement accommodations are provided for a class of persons who have formerly had no particular space allotted to them." There were two masts, a brig rig, and four small stacks, two for smoke, two for air exhausts, all painted white with gold tops. No one could ever mistake her for any other craft.

The machinery was simply enormous. On centerline were four working cylinders, each 168", or 14'-0", in diameter, with a 6'-0" stroke.[4] They were single-acting, the pistons moving up under air pressure and down from sheer weight. The top of the piston was open to the engine room. Extending up from it were eight iron rods connecting the lower piston to an overhead one operating in an inverted supply cylinder of the same stroke but 137" in diameter. That was also single-acting and had the underside of its piston open to the engine room. When the two descended, outside air was pulled into the supply cylinder. On the up stroke, it was compressed and discharged into a large tank, called a "receiver," on one side of the cylinder, where it remained under pressure during a down stroke. A valve from the receiver to the lower cylinder then opened, permitting the compressed air to enter the working cylinder. As it entered, it was heated by passing through a chamber filled with wire mesh called a "regenerator." Then, for the upward stroke, the air in the lower cylinder was further heated by a furnace burning anthracite coal directly below it. As the hot air expanded, driving up the working piston and, simultaneously, compressing the next charge of air in the supply cylinder, useful work was done. Four cylinders were necessary to turn the paddle shaft continuously, since only the upstroke generated power. Finally, on the downstroke, the hot air from the working cylinder escaped to the atmosphere after passing through the regenerator and heating up the wire mesh to make it ready, in turn, to warm the next supply of air from the receiver to the cylinder.

It was the regenerator that Ericsson contemplated with particular pride. He had the mystical idea that it enabled the engine to use the same heat over and over again with only minor losses. In the half century before the *Ericsson* was built, a mysterious fluid called "caloric" had been postulated to explain certain heat phenomena. He claimed that the caloric from the cylinder was completely captured by the regenerator and returned to the incoming air to drive the engine once again. His ideas were remarkably close to the theory of perpetual motion, but thermodynamics as a science was just beginning to be understood and we cannot blame Ericsson for his lack of understanding.

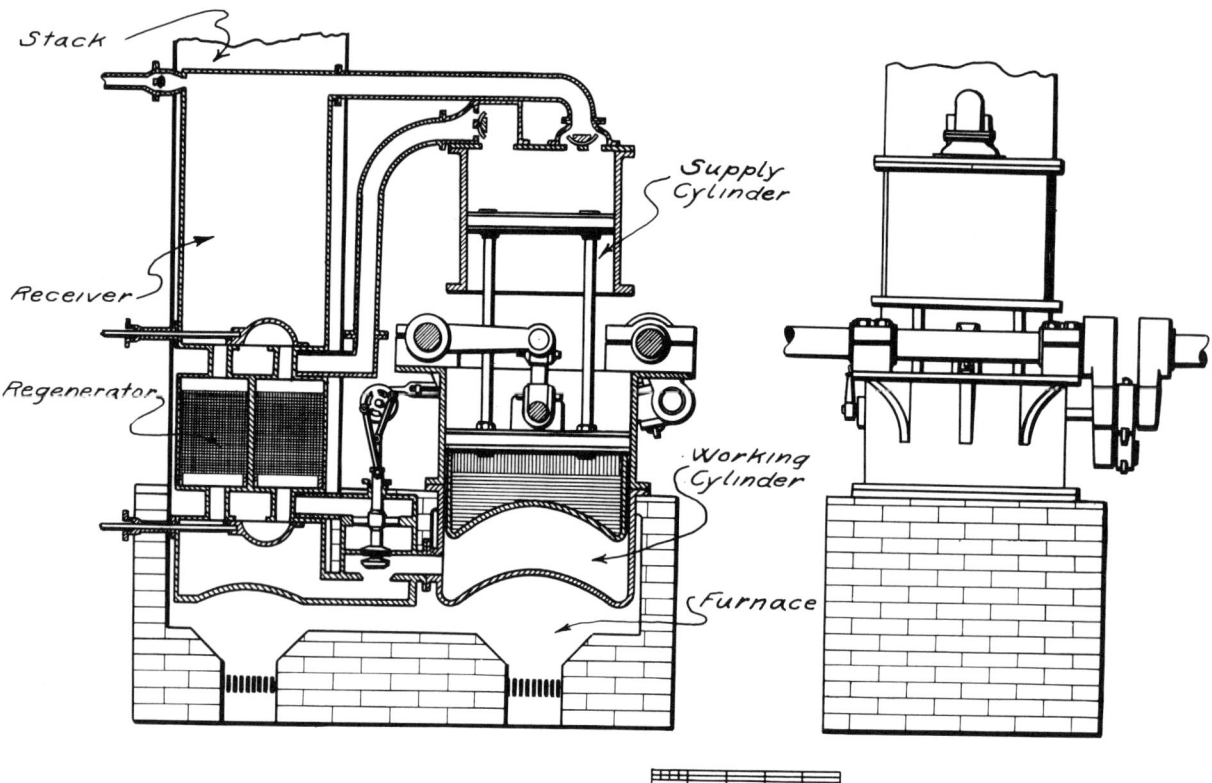

Hot air engine 60″ bore by 3′-0″ stroke.

The regenerators were made of endless lengths of 1/16″ diameter wire woven into sheets 6′ long and 4′ wide. A 24″ high stack of sheets was piled together to form a heat exchanger for each cylinder and, as we have seen, it was first heated by exhaust air passing through the pile in one direction and then cooled as the supply air traveled in the other. There must have been a considerable loss in energy just forcing air through such a maze twice per revolution.

This massive collection of cylinders, receivers, regenerators, pipes, and valves were arranged along the ship's centerline, with two vertical cylinders forward and two aft of the paddle shaft. Each pair was connected by levers to one connecting rod leading up at a 45° angle to a single crank on the paddle shaft. The forward and after rods were therefore at 90° to each other and the cylinders in each pair were at the opposite ends of their stroke; it was impossible to stop the engine on dead center. But how all this was stopped, started, or reversed is unknown. No satisfactory drawings of the engine have been found and no printed description mentions such practical matters. The smallest things in the engine room were the furnaces, located directly under the large cylinders and heating their 1 1/2″ thick, dome-shaped head casting.[5]

Hogg and Delameter, who were always willing to construct whatever Ericsson demanded, built the machinery. It turned a pair of conventional paddle wheels 32′-0″ diameter with 28 buckets 10′-0″ by 20″. When first completed, the total cost was listed as $320,000; the engine accounted for $130,000 of it.[6]

According to its inventor, it was impossible for the *Ericsson* to consume more than seven tons of coal per day. A mere 140 tons (two days' supply for a Collins liner) would take her across the Atlantic. Ericsson scoffed at the idea that the cylinder heads, in direct contact with the flames, might overheat and deteriorate rapidly. To show how cool things were, visitors were invited to climb into the cylinders and ride up and down on top of the pistons! Because the latter were hollow and 6′ thick, and their interiors were filled with insulation, the top actually was at a reasonable temperature. The engines turned so slowly,

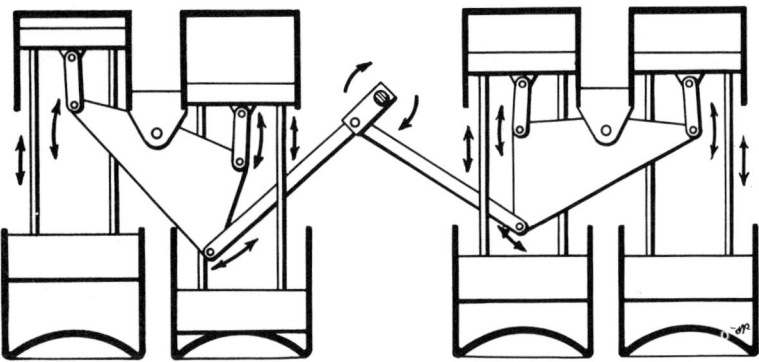

FOUR CYLINDER SINGLE-ACTING
HOT AIR ENGINE
DIAGRAMMATIC ARRANGEMENT
"ERICSSON"   1853

Diagrammatic arrangement of the four-cylinder hot air engine for the *Ericsson*.

a mere eight to nine revolutions per minute, that everyone enjoyed the ride.

One item never mentioned by Ericsson was a small steam engine needed to compress air into the receivers before the engine could start.[7] Would not a small "caloric" engine have been more suitable?

The inventor was remarkably proficient at producing accurate working drawings and Hogg and Delamater were equally adept at creating the machinery shown by him. He and Cornelius Delamater were firm friends. The caloric ship and her machinery, as a result, evolved with unusual speed, five months from keel laying to launching and three from launching to trials. The manufacture and installation of the four enormous supply cylinders, the four even-larger working ones, the corresponding receivers, and the regenerators, complete with valves, valve gear, and the complex mechanical leverage system of beams, bearings, and connecting rods driving a single crank on the paddle shaft were superb achievements.

By December 20, the engine was operating at 6 1/2 rpm, although the ship was still moored to a pier.[8] A builder's trial was held on January 4 and 5 without publicity,[9] and was successful enough for Ericsson to schedule a public trial on January 11, 1853, to which representatives from all the New York newspapers were invited, as were others from as far afield as Boston, Buffalo, and Cincinnati. The trip was carefully planned to keep the journalists continously instructed, entertained, fed, and fortified. At least one nonbeliever, Orson Munn of the *Scientific American*, sneaked aboard without benefit of an invitation.

Under the command of Alfred B. Lowber, the *Ericsson* headed down the Upper Bay at 9:30 A.M. Ericsson used a model of his engine to instruct his guests on how it worked. Extensive figures were given out predicting the performance, the economy of the installation, and its great superiority to steam. Tours of the ship followed; the engine room was remarkably cool and everyone was impressed by the fact that only one fireman and one engineer were on duty. The guests climbed onto the pistons, had an exhilarating ride, and held a formal meeting to pass resolutions praising the inventor, his ship, and his backers. The trip took about two hours; lunch was served and no one seems to have been bothered by the *Ericsson*'s very slow speed.

Her designer had predicted that the air pressure would be 12 psi, that the wheels would turn 12 rpm, and that the engine would deliver 600 indicated horsepower and drive the ship at 8 to 10 miles per hour.[10] In fact, only 8 psi was developed, the wheels turned 9 rpm, the speed was hopelessly slow, about 6 1/2 knots, and a careful modern computation of the horsepower makes it a mere 250.[11] It comes as a shock to realize that 450 tons of machinery with four 14' diameter cylinders produced the same horsepower as its inventor's steam plant for the auxiliary packet *Massachusetts* in 1845. The latter had two cylinders 24" in diameter with a stroke of 3'-0".

Most journalists have very short memories (often remarkably inaccurate) and little grasp of engineering. Charles A. Dana of the *Tribune* outdid all the others: "The demonstration is perfect. The age of Steam is closed; the age of Caloric opens. Fulton and Watt belong to the Past; Ericsson is the great mechanical genius of the Present and the Future." A more rational report came from a Philadelphia paper, after the initial intoxication wore off: "Captain Ericsson intends in July next, to take his ship to England. The experiments of hot air engines for ocean steamships have proved a dead failure. The only economy obtained appears to be an economy of speed, and this can be effected in any steamer, if a saving of a few tons of coal is more desirable than a quick passage."[12]

Messrs. Kitching and Ericsson decided that a visit to Washington to enlist the aid of the Navy Department and Congress was appropriate. The *Ericsson* left New York for Norfolk on February 15, 1853, the only guest for the trip being Captain Sands of the United States Navy.[13] She was visited by President Fillmore, President-elect Pierce, and delegations from both houses of Congress.[14] On the last day of Fillmore's administration, an amendment to the Naval appropriation bill adding $500,000 for a screw frigate with a caloric engine was defeated.[15] Since the average speed during the *Ericsson*'s return was 4 1/2 knots,[16] the wisdom of the Senate was most commendable.

The original cylinders were replaced by two new, double-acting, inclined ones of 72" bore and 8'-0" stroke. The air pressure was increased. After seveal unsatisfactory trials,[17] Ericsson claimed a speed of 11 miles per hour on April 27, 1854, with the paddles turning 12 to 13 rpm.[18] That trial was free of invited guests or unprejudiced observers. On the way back from Sandy Hook, while cargo ports on the lower deck were open, a sudden squall struck, the ship heeled over, water poured in, and she sank to the bottom of the Hudson off the Cunard Line docks at Jersey City. Because the water was relatively shallow, it was possible to close up the openings and pump her out, and by May 12, 1854, she was afloat and had been towed to an East River pier. Ericsson's veracity with regard to speed and success, as well as Captain Lowber's seamanship in opening ports so close to the waterline, are questionable.

Mr. Kitching, who owned thirty of the fifty shares in the ship, must have decided that he could no longer underwrite the great inventor's new propulsion schemes. Ericsson himself held no part of the ship. The second set of engines was removed and a steam plant designed and installed, but since the four original stacks were retained, her peculair appearance remained unaltered. Two of them were real; the others now acted as engine-room ventilators. They were so small that she always looked incomplete.

The new machinery, still of Ericsson design, had two inclined cylinders of 62" bore by 7'-8" stroke. The type, because it required a very long engine room, had not proved popular outside of naval vessels, but since it had fewer moving parts, it was cheaper than the usual side lever engine. It also fitted the ship's long, narrow engine room. The United States Mail steamer *Falcon* was an earlier example powered by inclined engines taken from Ericsson's equally unsuccessful Hudson River steamboat, the *Iron Witch*. Two vertical tubular boilers supplied steam, and surface condensers were used to insure fresh feed water. No changes were made in the wheels.

Kitching took the precaution of employing an eminent engineer, Charles H. Haswell, to supervise the trials and make an unbiased report on his steamship's performance.[19]

On May 8, 1855, one year, one month, and a day after she sank, the *Ericsson* put to sea. With the engines cutting off at 45 percent of stroke and turning 13.4 rpm, a speed of 11 knots was logged. The average pressure was 22.6 psi and the condenser vacuum 27.5" of mercury. Coal consumption, under these ideal conditions, ran to 21 tons per day, a very low figure. Haswell and Kitching were pleased by the economy, but time would show it to be a poor solution; the fuel consumption was, for the second time, low simply because the horsepower was so low. The 11-knot speed could never be achieved in ocean service against wind and waves. Kitching had a large, underpowered steamship on his hands and Ericsson had, once more, failed to produce satisfactory engines. If we look into the future, we find that similar situations would continually reappear. His monitor engines during the Civil War were so weak that the vessels were almost unmanageable. The large, supposedly fast, cruiser *Madawaska*, in 1867, was never able to steam for the distance specified for her trials because of her enormous fuel consumption and, even over a reduced distance, made less than 13 knots instead of a contractual 15.[20] Ericsson never designed a creditable power plant for any large steamship, merchant or naval. He did design a very successful hot-air engine that could produce up to 25 horsepower, and a considerable number of small steam engines for screw steamers when no more than 250 horsepower was required. But he was far too impatient to increase sizes gradually and correct his past mistakes as he went along. And it was always someone else who footed the bill when his designs did not live up to his promises.

The only successful parts of the *Ericsson*'s machinery were her two surface condensers and their separate, steam-driven, propeller-type pumps to circulater sea water through them. After two years of service (many of the early ones failed in a month or two), her chief engineer wrote an enthusiastic letter extolling them.[21] He was particularly pleased with the elimination of the deafening noise created when salt water boilers were blown down.

The year 1855 should have been a good one for a new steamship to start her career. A grand Exhibition was being held in Paris and the continued absence of foreign liners from the Atlantic because of Crimean War service created a brisk demand for passenger space. Mr. Kitching advertised his steamship as sailing from New York directly for Havre and taking only first-cabin passengers at $130, the fare quoted by the Havre Line.[22] She sailed June 16, made a crossing in 14 days 4 hours, arrived at Havre on the 30th and did just over 10 knots on her best day.[23] Rubber air-pump valves failed and forced her to stop for eight hours while they were replaced. The old *Washington*, the slowest American liner on the Atlantic, left the same day and arrived at Cowes 12 hours earlier. Returning, the *Ericsson* reached New York August 22 after an even slower 17-day passage. One voyage showed that a single steamer could not compete with the monthly sailings of the faster Havre Line or the cheap fares of Vanderbilt's European Line.

The *Ericsson* was shifted to Bremen, where it was hoped that the *Washington* and *Hermann* would provide less competition. The fare was reduced to $120, despite the longer trip, with a second cabin at $80. Sailing from New York on September 15, her next crossing took fourteen days to Cowes. The sea speed came to little more than 9 knots. Once she was home again, in early November, Kitching had second thoughts about the North Sea part of the voyage and announced that she would run to Havre during the winter. This she did, sailing from New York on November 21 and January 25 of 1856. The second of these eastward passages took twenty days to Southampton, but for the last 350 miles she was towing a Dutch brig, the *Anna Maria*, which she had found in distress.

It is unlikely that the *Ericsson* earned her keep on her single Bremen voyage or the three to Le Havre. After her return to New York, March 13, 1856, she was chartered to the Collins Line whose *Pacific* had vanished without trace. Collins had to find another steamship in order to maintain his fortnightly mail sailings. The *Ericsson* flew the Collins flag for the next year and a half. During that time she made eleven voyages to Liverpool, all at her usual leisurely rate; she was completely outmatched by her fleet running-mates, the *Atlantic* and *Baltic*.

No doubt Kitching was delighted to find employment for his white elephant and was willing to accept a moderate return in order to get something back from the half million dollars he and his associates had so unwisely invested. On March 29, 1856, the *Ericsson* sailed for Liverpool for the first time. The general public had by this time been frightened by the *Arctic* and *Pacific* disasters and was well aware of the *Ericsson*'s lack of speed. She returned on May 1, to New York with only thirty-eight passengers, fifteen days from Liverpool. On other trips she carried even smaller numbers. Her performance has already been dealt with in the Collins Line chapter; on her later voyages mail payments were reduced because of her inadequate speed. Aside from an engine break down on July 12, 1856, three days out of Liverpool, her crossings were uneventful.[24] She completed the trip under the other engine. After the final voyage, which brought her and twelve passengers back to New York on August 6, 1857, she was returned to her owners.

In the fall of 1857 Bremen was served by the Vanderbilt Line's *Ariel* and *North Star*, while Havre had three outstanding American steamships, the *Arago* and *Fulton* of the Havre Line, and the magnificent new *Vanderbilt*. The *Ericsson*'s agents chose the former port as the lesser of two evils, and offered to take three classes of passengers at fares of $80, $50, and $30 when she departed from New York on September 16. An unusual feature was that no stop was scheduled at Cowes or Southampton. The low horsepower and moderate fuel consumption enabled her to steam all the way without coaling. For this trip, C. H. Sand, former head of the Bremen Line, acted as a joint agent with Dunham and Company. Although a second trip was advertised, she was laid up for the winter. In the spring Dunham made a new effort to utilize her. For $750 one could start a "pleasure voyage" from New York on May 1 to Gibraltar, Malta, Alexandria, Jaffa, Constantinople, Athens, Naples, and home again. No great interest having arisen, the *Ericsson* was rescheduled for direct Bremen service, leaving May 8.[25] Upon the completion of the voyage, June 20, 1858, her owners gave up.

The *Ericsson* made a special trip to take the New York Seventh Regiment to City Point, on the James River, came home empty, and was laid up for the next three years.[26]

Chartered by the U.S. Army's Quartermaster Corps on October 7, 1861, for $1,200 per day, the *Ericsson* entered wartime service as one of twenty three steam transports sailing with Flag Officer Samuel F. DuPont's invasion fleet headed for Port Royal, South Carolina. Other major steamships assembled for the occasion were the *Atlantic* and *Baltic*, the Vanderbilt liners *Ariel* and *Ocean Queen*, and the *Empire City*. The fleet of 84 vessels of all kinds departed from Hampton Roads, Virginia, on October 29.[27]

After being out of service from January through July of 1863, the *Ericsson* was rechartered on August 24 at a more realistic figure of $700 per day for one ten-day trip. Her use, however, was sporadic. Even at the height of the Civil War's demand for steamships, she was employed by the Army for just twenty-one months out of the forty-six in which she was available.[28] During the war years, Edward Dunham and Company were listed as the owners; they had taken over the management of the ship starting with her first voyage to Bremen in 1855.

In January of 1862 the *Ericsson* was part of General Burnside's expedition to capture Roanoke Island, but her deep draft prevented her from entering Hatteras Inlet. Turning back on January 26, she was driven out to sea by a storm and then consumed five more days before reaching Fortress Monroe, Virginia. There she discharged her load of gaily uniformed but miserably seasick Zouaves.

Frequent trips to Port Royal and occasional ones to Key West, Dry Tortugas, and Ship Island, in the Gulf of Mexico, followed. As monitors were completed, a fleet of ironclads was being asembled to attack Charleston. As an afterthought, John Ericsson designed rafts to be pushed ahead of them that would blow up any mines or obstructions. It seems particularly fitting that, in February 1863 under a naval charter, the *Ericsson* started South loaded with explosives for these Ericsson contrivances that were to be attached to DuPont's monitors.[29] Four of the rafts, too heavy to be loaded on board, were towed astern, but proved manageable only at speeds below four knots. Not finding the fleet off Charleston, the *Ericsson* turned back to Port Royal and delivered her cargo and one remaining raft, the other three having broken loose in a gale.[30] It was lucky for the underpowered monitors that they did not have to push these ungainly objects.

A second naval affair was a brief cruise out of New York from June 28 to July 9, 1863, searching for privateers. In her final passage as a transport, she arrived at New York March 13, 1865, two days from Beaufort, North Carolina.

Marshall O. Roberts, starting in October of 1862, had organized a service between New York and San Francisco to compete with the two lines that shared the California mail contract, Vanderbilt on the Atlantic and the Pacific Mail on the Pacific. Usually advertised as "M. O. Roberts' Line," it later became the Central American Transit Company, and, finally, the North American Steamship Company. In 1864 and 1865 there were monthly sailings to Nicaragua by the *Golden Rule*, on the eastern part of the route, connecting with the *Moses Taylor* or the *America* at San Juan del Sur on the Pacific. May of 1865 saw the *Golden Rule* wrecked on Roncador Reef.[31] As a temporary measure the *Ericsson* was chartered. She cleared New York on July 20, 1865, for Greytown (formerly San Juan del Notre) and made four voyages at monthly intervals. On any one of these she transported more passengers than she had during her entire career to Havre, Bremen, or Liverpool. Her October 19th arrival brought 849 into New York.[32] The passage was a very slow one, thirteen days, and made her late departing on her last voyage. She left the 23d of October instead of the 20th, her regular sailing date. The charter ended on November 22, 1865, and she was sent to the Sectional Dock in New York for overhaul. Roberts had purchased a much faster steamship, the *Santiago de Cuba*.

The North American Lloyd Steamship Company will be covered in more detail in a later chapter. It had purchased three steamships to establish a steerage service from Bremen. Two of them, the old *Atlantic* and *Baltic*, proved good investments, but the third, the *Western Metropolis*, suffered a long series of breakdowns. To replace her, the *Ericsson* was chartered and made the line's second sailing from New York on March 15, 1866, via Cowes. She reached New York again on May 2, twenty days from Bremen, sailing north of Scotland, via the Pentland Firth, loaded down with 557 immigrants.[33] On her second voyage, departing May 24, she made Cowes in both directions and took seventeen and one half days from there to New York, arriving on July 27. Her speed was diminishing as her age increased.

Immediately thereafter she was chartered by the Continental Mail Steamship Company, William Salem and Company, Agents, for their brief operation to Antwerp. The *Ericsson* made one voyage, and an equally slow steamship, the screw steamer *Circassian*, another before the operation was abandoned. Fares of $90, $62.50, and $37.50 were charged and Havre used as a Channel port rather than Cowes. The *Ericsson*'s New York sailing was August 23, 1866. Home again on October 16, she disembarked 350 passengers after a twenty day passage from Havre. Other trips were advertised but never made. Thus, at the age of thirteen years, the *Ericsson*'s life as a steamship came to an end. She had made four voyages to Bremen or Havre for John B. Kitching, who underwrote much of her cost; eleven to Liverpool for Collins; two to Bremen for Dunham and Company; two to the same port for North American Lloyds'; one to Antwerp for Salem; and four to Greytown. A total of twenty Atlantic voyages and four to Nicaragua is not impressive. To these should be added about two years' service to the Quartermaster Corps as a transport and to the U.S. Navy for special wartime services. During her entire career she had one master, Alfred B. Lowber, who owned three out of her fifty original shares. From February until June of 1867 she was offered for sale and then put up for auction on August 14.[34]

Her real career, however, was about to begin. The

engines were taken out in the fall of 1867 and she was rebuilt and rerigged as a three-masted sailing ship for W. W. Sherman. The deck was modified and a half-poop fitted. The conversion was completed early in 1868; she loaded cargo and cleared New York for Liverpool on February 17.[35] From there, on May 6, she sailed to San Francisco, arriving August 23 and departing for Liverpool October 15.[36] She was not only a very large sailing ship, but a very speedy one as well. Her first three passages produced sailing times of seventeen days to Liverpool, 109 days to San Francisco, 103 days back to Liverpool, each excellent, but the sum of the three consecutive runs resulted in an outstanding, and possibly unbeaten, record.[37]

In March and April of 1874, when there were four departures from San Francisco of noteworthy ships loaded with grain for Liverpool,[38] the two best passages were by the *Ericsson* and a celebrated Webb clipper, the *Young America*, each taking 103 days. Donald McKay's last ship, the *Glory of the Seas*, sailed six days after the *Ericsson* and required 118 days for the trip. The fourth, the English ship *Wasdale*, was a day slower than the *Glory of the Seas*. The *Ericsson*, freed of the encumbrance of machinery, was at last in her own element. Her splendid hull, with the right wind, could sail faster than it had ever steamed. For practically a quarter of a century more she continued on her smoke-free way.

From 1868 through 1875 the *Ericsson* was engaged in the grain trade between San Francisco and Liverpool, with a voyage via Callao, Peru, to Mejillones and another to Newcastle, New South Wales, in Australia. An 1876 trip made Manila and Iloilo, in the Philippines, with a return to New York. The years 1878 and 1879 saw her in the coastwise trade from San Francisco to Puget Sound ports. In 1880 came a second Australian visit and one to Shanghai the following year. The years 1882 and 1883 were spent coastwise, and 1884 saw her last call at Liverpool. Further Australian trips, including Sydney and Melbourne, as well as Newcastle, were made in 1886 and 1888. In 1889 she went South to Valparaiso. Most of her sailings were coastwise from 1889 to 1892. There were several changes of ownership and successive home ports were New York, Boston, and San Francisco. Late in the 1880s, George Plummer, who had long been her master, acquired her.

At long last the fates caught up with her; on November 19, 1892, she was wrecked on Entrance Island, near Barclay Sound, British Columbia. She was en route from San Francisco to Nanaimo when she went ashore in a gale. No lives were lost.[39]

The *Ericsson*'s second career of twenty-five years under sail was a great success; her first, after the installation of her third set of engines, was a sorry story of intermittent operations, mostly at a loss. Her steam voyages have already been summarized. Under sail she made at least eight trips to Liverpool; four to Australia; others to Shanghai, the Philippines, Callao, Valparaiso, Mejillones, and New York, plus innumerable voyages from San Francisco up and down the West Coast of the United States. As a steamship she was laid up many times, as a sailing ship a few, but these were necessary delays awaiting the grain harvest and an assured full load from San Francisco to Liverpool. She was a splendid testimonial to her builders, Perrine, Patterson, and Stack, if not to her namesake.

The *Ocean Bird*, 1855-1873?

The *Ocean Bird* was the exact opposite of the *Ericsson*. Instead of an unusual machinery installation in a conventional hull planned by a marine engineer, she had a conventional engine installed in an unusual hull conceived by a naval architect. The one point in common was that John Ericsson reveled in the production of untried machinery just as much as John W. Griffiths did in revolutionary hulls. Each had some successes amidst a larger number of failures. Both are remembered for their successes and never for the magnitude and frequency of their failures. Neither was able to finance his own experiments; both were persuasive enough to convince others that large investments in their brainchildren would return rich rewards.

The project that eventually produced the side-wheel steamship *Ocean Bird* got under way sometime in 1853. Her financial backer was a civil, mechanical, and locomotive engineer by the name of William Norris, a resident of Philadelphia. She was intended to make a seven-day passage from St. John's, Newfoundland, to Galway, Ireland, the nearest European port. Since she was to be unhampered by past prejudices, Griffiths set up a shipyard at Greenpoint, on what is now the Brooklyn side of the East River; he would build, as well as design, this radical new steamship.

The plans envisioned two decks and dimensions of 225' long by 37' beam with a depth of hold of 16'.[40] Various drafts were quoted, ranging from 6'-0", with a displacement of 600 tons, to 6'-6" and 750 tons.[41] The full-load value was also listed as 1,000 tons, with 300 of it in coal. Other reports gave a depth of 21'-0", and measurement tonnages of 800, 1,000, and 1,267 tons. Griffiths seems to have been designing and building simultaneously and not to have known, at any given interview, what he was really going to do. There is no worse method of ship design; if decisions are not made in ad-

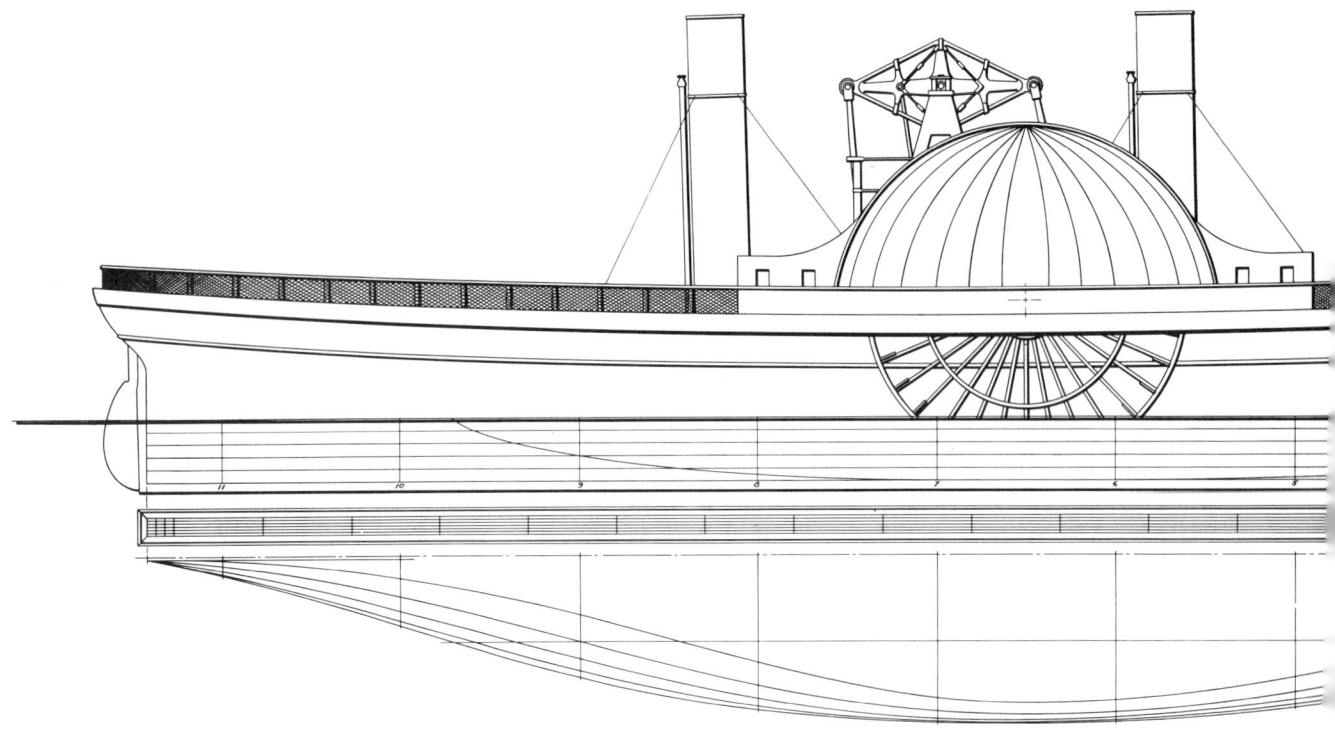

Ocean Bird, 1854, lines.

vance of construction, there will be endless costly changes.

Speed, however, always grew with the telling. First the steamer would make 20 knots in all weathers; second, the seven days on the short run to Galway became six days; and after that, six days from New York to Liverpool. A lithograph was issued showing a "Six Day Steamer." No freight would be carried, just eighty passengers. The engine was to be a single walking beam one of 72″ bore by 12′-0″ stroke, [42] driving enormous, 36′-0″ diameter paddle wheels and having an output of 750 horsepower.

Writing after the event, Griffiths claimed that he had guaranteed a speed of 16 1/4 knots, or a seven-day passage to Galway in winter weather, and had agreed to build the hull for $120,000.

It is clear that what he really planned was a seagoing steamboat of incredibly light draft, having almost no superstructure and no overhanging guards. She was to be very lightly built, to save weight, but her wooden hull was to be reinforced by a series of iron bulkheads and keelsons in order to provide strength and rigidity. There were four lengthwise bulkheads—two single ones outboard of the engine spaces, and a double centerline one, at the bow and stern but eliminated amidships. The two inner bulkheads were only 15″ apart and served to carry the ship's fresh water between them. Transverse iron bulkheads at either end of the machinery spaces were attached to both the centerline and the side ones, the latter forming the inner boundaries for the coal bunkers. All of these extended only to the second deck. The *Ocean Bird* anticipated the *Arago* and *Fulton* in having a watertight compartment around the engine and the boilers. As in the Havre liners, there were doors in the wing bulkheads so that coal could be passed through. If the hull were holed, it seems unlikely that anyone would stay below long enough to close the openings, clogged as they were with coal and dust.

The keel and stem were only 12″ by 13″, of oak; moreover, they were tapered down to fair smoothly into

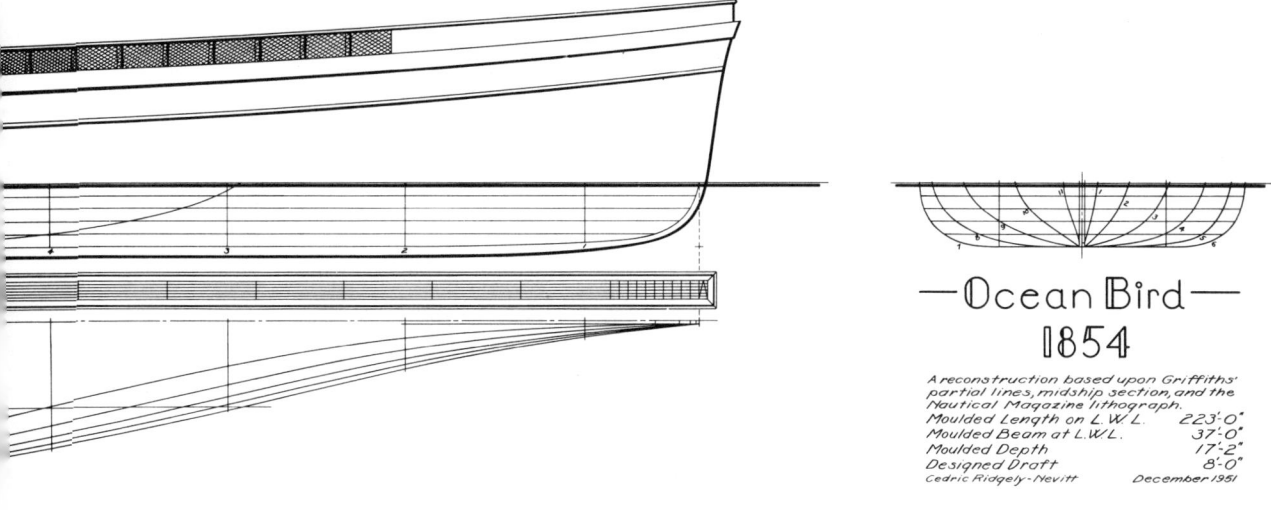

the planking; the actual width at the bow was 1/2". Frames were 36" apart amidships, increasing to 42" at the ends. Even in way of the engine, the oak floor timbers were a mere 11" wide by 12 1/2" deep and the frames tapered to 4 3/4" at the gunwale. Yellow pine planking varied from 4" on the bottom to 2 1/2", with heavier wales and sheer strakes.

To stiffen this slender structure, there were the usual diagonal iron straps and, in a new departure, further strapping fitted horizontally underneath the upper deck beams. Another unique feature, forward of the engine room, was a series of sloping braces from the centerline bulkheads, at the second deck, out to the frames to support the bow against the impact of waves. Masts were planned, but they would be hinged and folded down when not in use.

Two Griffiths drawings survive at the Smithsonian Institution in Washington. One is the lines to the outside of the planking—a most unusual feature—up to the designed waterline, and the other a midship section showing the iron tie rods and turnbuckles that were required in all side-wheel steamships to support the guard beams outside the hull. These, in turn, carried the outer paddle-shaft bearings. Although no ceiling is shown, it must have been fitted inside the frames.

The author has redrawn these and added a conjectural profile above the waterline, taken from the crude lithograph published by Griffiths in volume 1 of his *Nautical Magazine*.

According to her designer's later calculations, the steamship would displace 1,137 tons when provisioned and ready for sea with 100 tons of coal on board. His hull would weigh a mere 397 1/2 tons. The draft would be 8'-0" to 8'-6". It was planned to launch the hull in December of 1853 and to name her *William Norris*. The *New York Herald* took a more skeptical view of this odd ocean steamboat than it had of the *Ericsson*. She was compared, accurately, to the "flat bottomed Mississippi and Ohio boats." It was noted that her frames were much lighter than those incorporated into a Long Island Sound steamer being built nearby. The promises of speed and seaworthiness were questioned and the conclusion reached that "If, therefore, the WILLIAM NORRIS achieves what her constructor undertakes, she will be the greatest naval wonder of the age."[43]

Just before the intended launch, Norris declared himself bankrupt and Griffiths had to stop work. What there was of the steamboat was sold at auction.[44] She was purchased by Captain John Graham who asked Griffiths to alter her so that she could be used in the coastwise trade. Griffiths spurned the idea and refused to change his design in any way. The unfinished vessel was turned

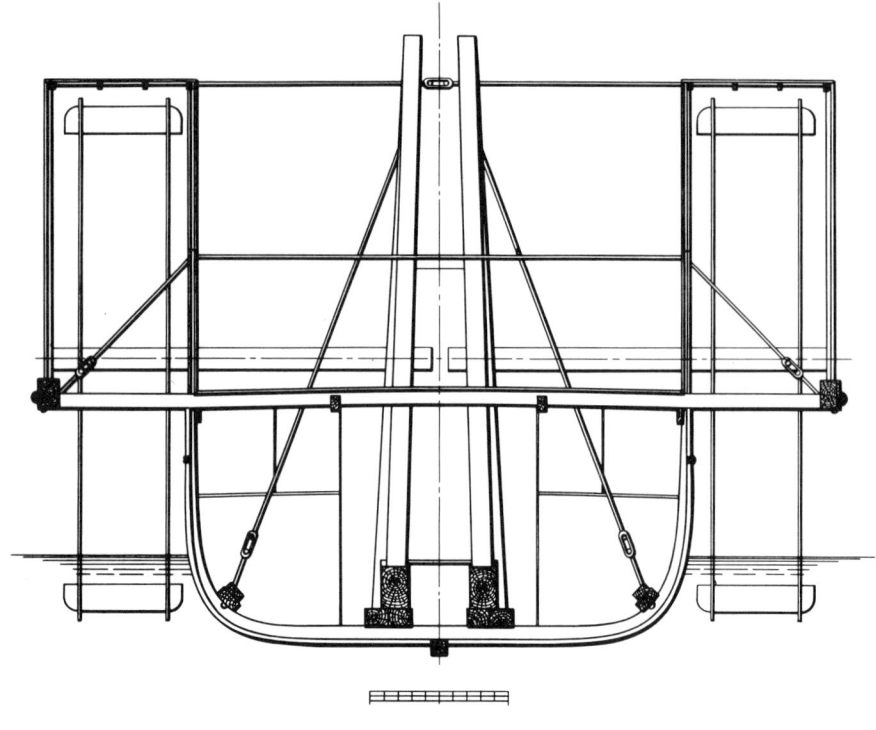

Ocean Bird, midship section.

over to a more amenable firm of shipbuilders, Cornelius and Richard Poillon, but over a year elapsed before a much-changed steamer was completed. She was deepened; another whole deck added, and extensive deckhouses installed; eventually, she was assigned a tonnage of 1,467 with register dimensions of 229′-4″ by 36′-8″ by 25′-0″. A second, and probably more accurate, source gives her depth of hold as 16′-0″ to the main deck and 23′-0″ spar deck.[45] Quarters for two hundred cabin and five hundred steerage passengers were installed, the fuel capacity was increased to seven hundred tons, and a brigantine rig was fitted. She was christened *Ocean Bird*. A flying-bird figurehead was mounted on her stem and the design repeated on her paddle-box decorations. The changes made her into a handsome two-funneled steamship.

The installed power was reduced but a beam engine retained. Supplied by Messrs. Guion and Boardman of the Neptune Works, it was 65″ bore by 12′-0″ stroke, with valve gear cutting off the steam at one-half stroke in the down direction but five-eighths when moving up. The wheels had 28 paddles 7′-9″ wide by 1′-10″ deep, were 33′-0″ in diameter, and were to have a 4′-10″ dip at a draft of 10′-0″. Four small return-flue boilers generated steam at 25 psi. All were 9′-6″ wide and 10′-2″ high, but the forward ones were 20′ and the after ones 22′ long, each having two furnaces.[45]

In October of 1854 it was announced that she was fitting out at the foot of 8th Street, on the East River, for Mobile service. Six days later the possibility of a trip to the Mediterranean and a Crimean war charter was mentioned.[46] A third proposal was to send her to Portland, Maine.

Work progressed so slowly that June of 1855 arrived before she went on trial. On the 21st her wheels turned at 17 rpm under a steam pressure of 23 psi at a draft of 9′-3″. According to Griffiths, she made seventeen miles per hour "against wind and tide."[47] After completion and documentation on July 7, 1855, the *Ocean Bird* saw no immediate service.[48] It was fall before a sailing from New York to New Orleans, via Havana, was scheduled. With Graham as owner, master, and agent, she sailed November 2, reaching New Orleans on the 10th. Her time to Havana was 4 days 5 hours, and from there to Balize, the Mississippi River pilot station, forty hours. The *New Orleans Picayune* commented on her being "an exceedingly fast boat."[49] From New Orleans on the 15th she touched at Havana and arrived home on November 25 with twenty-nine passengers. The cargo consisted of cotton bales, barrels of flour, and sacks of wheat. Then a blank period ensued. Captain Graham left her tied up to Pier 36 on the North River while he went back to sea in command of the steamer *St. Lawrence* and sailed for Havana and Nassau on December 21, 1855.

Pajaro del Oceano, ex-*Ocean Bird*. *Courtesy of The Mariners Museum, Newport News, Va.*

In the spring of 1856 the *Ocean Bird* was sold to Ramon de Herrera of Havana, who retained her name after translating it into Spanish. On June 4, as the *Pajaro del Oceano,* she left New York for Havana under Captain de Villar. When she returned on June 23, she made a fast, but not a record, passage of ninety-six hours at an average speed of 12.7 knots. There were eighty-nine passengers, thirteen for New York and seventy-six for Cadiz.

She probably sailed on the 25th, the same day she cleared at the New York Custom House.[50] If so, it took her eighteen days to reach Vigo, in Spain's northwest corner. From there she headed South for Cadiz and through the Strait of Gibraltar to Barcelona. She left that Mediterranean port on July 31 for Havana.[51] A second voyage brought her from Havana to Cadiz on November 24, 1856.[52] According to Liverpool reports she did not leave until May 8, 1857, but in light of the time interval, it is possible that another, unrecorded, Havana voyage may have intervened. Or, on the other hand, she may have suffered damage and needed six months for repairs to her hull or engine. After these two confirmed voyages we lose track of her.

It seems unlikely that a small, light-draft steamship, still only a little beyond a steamboat in her basic characteristics, could possibly continue operating on the high seas. Her large beam and low draft would result in excess stability and a quick, exceedingly uncomfortable roll. A walking beam, way up in the air, under these circumstances, seems a most unsuitable way to propel an ocean steamship. It is more likely that she was used then, as we find she was later, as a local mail steamer between Havana, Puerto Rico, and Saint Thomas in the Virgin Islands.[53]

From time to time the *Pajaro del Oceano* came to New York for overhaul. In 1858 she arrived on March 13, six days from Havana, after a far slower passage than her first one. Extensive work was obviously needed, for she did not depart until May 1. The Spanish firm of J. M. Ceballos acted as her regular New York agent. The light hull may have been a continuing problem, because a year later further major repairs were required. On this occasion over four months were consumed between her arrival August 26, 1859, and her departure January 14, 1860. Aside from a regular listing in *American Lloyds,* all track of her vanished during the Civil War. On January 10, 1867, she left Havana for New York, where four new boilers were awaiting her at the Fulton Iron Works. She ran into a storm off Cape Hatteras that strained her so badly she had to put into Norfolk on the 19th with twelve feet of water in her hold and all her pumps struggling to keep her afloat.[54] Because the dry dock at the Navy Yard was already occupied by a sailing packet, the *Enoch Train*, she had to wait a day before she could get in. There the hull was completely recaulked and new copper installed before she could continue.[55] Leaving Norfolk the last day of February, she reached New York on March 2. The final date quoted for her was May 7, 1867, when she was still listed at New York.[56] Although no clearance for Havana was published, she must have departed for her home port. Nothing is known about the remainder of her career, but because an American steamship, the *Niagara*, of the Old Dominion Line, was purchased late in 1873,[57] rebuilt, and renamed *Nuevo Pajaro del Oceano*, we can assume that the original of that name had ceased to operate.[58]

The *Ocean Bird* is a fascinating and frustrating steamer. In design she was a throwback to the seagoing steamboat that had proved such a dangerous experiment in the 1830s. Yet John W. Griffiths was aware of the dangers and took measures to overcome them. He eliminated guards and sponsons, cut superstructure to a minimum, and strengthened a frail wooden hull by means of well-thought-out iron bulkheads, keelsons, and strapping. Her great defect was the excessive stability, always present in a shallow-draft vessel, that must have racked and strained the wood gallows frame supporting her walking beam. A failure of the engine in heavy weather was an ever-present threat. Griffiths, despite his published texts on naval architecture, never seems to have mastered stability.

The fact that Graham and the Poillon brothers deepened the hull and added to its weight and draft may have made the difference between success and failure. They brought her nearer to a seagoing steamship. Although we are uninformed as to her actual service, she did survive and apparently operated for at least eleven, and perhaps seventeen, years under the Spanish flag. But the available information is so limited that it is impossible to tell whether she was an asset or a constant nuisance to her owners, tolerated because they could not afford anything better. Her replacement, the *Niagara*, was small and inexpensive. Worst of all, we have no evidence as to the *Ocean Bird*'s performance on the two Atlantic voyages we know she made. There are frequent hints that she was fast. A thirdhand quotation of Joseph Carter, chief engineer of the Norwich Line's *City of Worcester* in 1891, who served on her from Spain to the West Indies, reads: "She is the fastest steamer to go I ever saw."[59]

Notes

1. William Conant Church, *The Life of John Ericsson* (New York, 1890), 1: 73-75.
2. Ibid., 1: 185-89.
3. Articles descriptive of the Caloric Ship *Ericsson* and of her Trial Excursion. . . (Washington, 1853), pp. 4, 5, quoting the *New York Courier and Enquirer*, January 12, 1853.
4. *New York Evening Post,*, January 12 and 13, 1853.
5. *New York Tribune*, January 12, 1852.
6. Articles descriptive, p. 19, quoting the *New York National Democrat*.
7. *Scientific American*, January 22, 1853, p. 149.
8. *New York Herald*, December 18 and 21, 1852.
9. *New York Shipping and Commercial List*, January 8, 1853.
10. *New York Tribune*, January 12, 1853.
11. Eugene S. Ferguson, "John Ericsson and the Age of Caloric," Contributions from the Museum of History and Technology: Paper 20. (Washington, D.C.: Government Printing Office, 1960), pp. 59-60.
12. *Philadelphia Public Ledger*, May 3, 1853.
13. *New York Herald*, February 14 and 18, 1853.
14. Ferguson, "John Ericsson," p. 47.
15. David B. Tyler, *Steam Conquers the Atlantic* (New York, 1939), p. 190.
16. *Scientific American*, March 26, 1853, p. 221.
17. *Journal of the Franklin Institute*, 26, 3d ser. (July-December 1853): 64; 27, (January-June 1854): 135 and 211-12.
18. *New York Herald*, May 6 and 12, 1854.
19. *Monthly Nautical Magazine*, (July 1855), pp. 332-35.
20. Frank M. Bennett, *The Steam Navy of the United States* (Pittsburgh, Pa., 1896), pp. 532-33.
21. John Mars letter, September 4, 1857, to John B. Kitching, Ericsson papers. New York Historical Society.
22. *New York Commercial Advertiser*, May 29, 1855.
23. *Monthly Nautical Magazine*, (August 1855), pp. 445-46.
24. *New York Tribune*, July 29, 1856.
25. *New York Commercial Advertiser*, advertisement March 27 and April 15, 1858.
26. Church, *John Ericsson*, 1: 197.
27. *New York Herald*, November 6, 1861.
28. House Executive Document 337, 40th Cong., 2d sess. Vessels bought, sold, and chartered by The Quartermaster Corps since April 1861; hereafter called House Executive Document 337.
29. *New York Herald*, April 10, 1863.
30. Official Records, Ser. 1, 13; 519, 520, 669, and 670.
31. John Haskell Kemble, *Panama Route, 1848-1869* (Berkeley and Los Angeles, Calif., 1943), p. 228.
32. *New York Herald*, October 20, 1865.
33. Ibid., May 3, 1866.
34. *New York Journal of Commerce*, February 14 through June 4, and August 10, 1867.
35. American Lloyds Register, 1868.
36. *New York Shipping and Commercial List*, 1868.
37. *New York Herald*, February 26, 1869.
38. Basil Lubbock, *The Down Easters* (Boston, 1929), p. 64.
39. E. W. Wright, *Lewis & Dryden's Marine History of the Pacific Northwest*, (1895; reprint ed., New York, 1961), p. 405.
40. *New York Herald*, November 6 and December 29, 1853.
41. *New York Commercial Advertiser*, September 12, 1853.
42. *Monthly Nautical Magazine*, (August 1855), pp. 411-14.
43. *New York Herald*, November 6, 1853.
44. *Monthly Nautical Magazine*, (March 1855), pp. 459-60.
45. *Journal of the Franklin Institute*, 29, 3d ser. (January-June 1855): 274.
46. *New York Herald*, October 12 and 18, 1854.
47. *Monthly Nautical Magazine*, (July 1855), pp. 382-84.
48. *New York Shipping and Commercial List*, April throgh December 1855.
49. *New Orleans Picayune*, November 11 and 17, 1855.
50. *New York Commercial Advertiser*, June 24 and 25, 1856.
51. *Willmer and Smith's European Times* (Liverpool), July 26 and

August 16, 1856, and May 30, 1857.
52. Ibid.
53. *New York Herald*, November 22, 1858; American Lloyds Register, 1861 and 1863.
54. *New York Herald*, January 23, 24; March 3 and 6, 1867.
55. *New York Journal of Commerce*, January 22; February, 4, 8, and 23, 1867.
56. *New York Shipping and Commercial List*, May 8, 1857.
57. *Nautical Gazette*, October 25, 1879.
58. American Shipmasters' Association, Record of American and Foreign Shiping, 1874-75.
59. A 1938 typescript by John W. Griffiths's granddaughter, Miss Marian Virnelson, summarizing information about the *Ocean Bird*. Author's collection.

13
The Vanderbilt European Line, 1855-1860

The Nicaragua Route

CORNELIUS Vanderbilt began his career as a steamboat captain on the *Mouse of the Mountain* in 1818. After eleven years in the employ of the Gibbons family, in charge of their New Jersey to New York steamboat operations, he had assembled enough capital to proceed independently. By 1831, in cooperation with his brother, Jacob Vanderbilt, another steamboat captain, he was operating his own boats on the Hudson River and, by sharp practice and extreme fare cutting, made such a nuisance of himself that the established lines bribed him, for $100,000 plus a payment of $5,000 per year, to give up the Hudson. In 1835 he transferred to Long Island Sound and, as we have already seen, invested in the *Chancellor Livingston* in Maine waters. With the continued expansion of his interests to include other steamboats scattered anywhere from Maine to South Carolina, he was promoted from "Captain" to "Commodore" in the newspaper accounts of his exploits. By 1845 he owned more steam vessels than anyone else and had accumulated a fortune of at least $1,200,000.[1] It is ironic that today's general public thinks of him as a railroad entrepreneur despite his forty five years of continuous involvement with ferries, steamboats, and steamships.

Throughout this period, Vanderbilt, who captained many of his own vessels, played an active role in their design. He was intensely interested in their arrangement, speed, and performance, and, as a keen businessman, in any aspect affecting the costs of construction or operation.

He regularly used the firm of Bishop and Simonson to build his steamboats. Joseph Bishop, however, withdrew at the end of 1849 and Jeremiah Simonson, a nephew of the Commodore, carried on the business alone; both versions of the firm worked amicably with their most frequent customer, who did not haggle over prices when he had a trusted relative to carry out his verbal orders.

Following the discovery of gold in California, Vanderbilt saw a glittering opportunity for anyone who could deliver passengers and freight to that distant shore. The United States Mail Line had been slowed by the late deliveries of their *Ohio* and *Georgia*, and the Pacific Mail had started with three small steamers, none adequate for the sudden and unanticipated demand for space aboard.

Vanderbilt investigated a new route via Nicaragua, over five hundred miles shorter than the Panama one, sent down an emissary armed with ample funds, and obtained the right to build a canal there. Until this could be completed, he received the additional right to establish other means of transit and this he did.

From the Wilmington, Delaware, firm of Harlan and Hollingsworth, one of the few American shipyards building in iron, he ordered two small stern-wheel steamboats, the *John M. Clayton* and *Sir Henry Bulwer*, to ascend the San Juan River from the Gulf of Mexico. The iron hulls were necessary because of rapids, shallows, rocks, and frequent groundings. From Simonson he obtained a steam scow 110' long, named *Director*, to cross Lake Nicaragua, and, in addition, the *Prometheus*, the first of a long line of ocean steamships that he would own.[2] Although not an Atlantic liner, the *Prometheus* is important because all other Vanderbilt-Simonson steamers descended directly from her. Moreover, Vanderbilt never believed that any distinction existed between a coastwise steamship and a transatlantic one. He would regularly shift his vessels back and forth as the need arose. Because Bishop and Simonson had built but one steamship before the *Prometheus*, the ill-conceived *Ohio* that had followed whatever designs the U.S. Mail had supplied, and Vanderbilt, as always, had clear ideas as to what he wanted, the new vessel bore little resemblance to any earlier craft.

The most important change was a completely different type of engine for an oceangoing steamer. All the major examples from 1847 to 1851 had side lever engines placed low in their hulls. Because every mail liner had to be approved by the Navy Department, the choice had been a military one. Nothing protruded above the deck to be shot away. It appealed to the operating engineer, for everything was concentrated in one place, where it could be inspected, greased, and touched by hand to see if its bearings were overheating. Furthermore, ship motions were a minimum below decks, even when the ship was rolling badly with first one wheel and then the other out of the water.

Vanderbilt chose to disregard the advantages. He had captained many beam-engined steamboats on the sheltered water of the Hudson River and Long Island Sound. He knew the beam engine to be the simplest and lightest that could be built. Moreover, it could be constructed at the lowest possible cost.[3] As far as he was concerned, the cheapest engine was the best and that was all that mattered. His one concession to the rigors of ocean service was that he sometimes installed two engines rather than one, so that some possibility of returning to port could be anticipated if one broke down.

There was one essential feature of the beam engine that permitted it to perform, even when its beam was gyrating madly through the air as the hull beneath it rolled and pitched. Its moving parts were so far apart that it could withstand considerable misalignment as the flexible hull yielded. Otherwise, it never could have operated, as it had for years, on the lightly built steamboats that bent, twisted, and changed shape over the years. The heavy wood gallows frame supporting the beam was stiff longitudinally but flexible enough transversely to bend as needed. In addition, a very artfully designed set of iron tie rods and turnbuckles absorbed tension loads and anchored the trunnion bearings for the beam and the shaft bearings in place under a wide range of deflections. The side-lever engine was the antithesis of this; its bedplate, pillars, and rigid struts held all machinery parts firmly in alignment, with only a loose joint in the shaft to compensate for the transverse deflections of the hull.

When launched on August 3, 1850,[4] the *Prometheus* had a conventional hull 230'-6" long by 33'-0" beam by 20'-2" depth of hold, with a sea horse as a figurehead, three masts, and a tonnage of 1,207. The depth quoted was to the second deck. The two engines, of 42" bore by 10'-0" stroke, came from T. F. Secor, who had been building beam engines since 1838.[5]

Vanderbilt built ships for the Atlantic portion of the long trip to California. These could be seen and inspected by prospective passengers at New York. He bought whatever existing craft he could find, whether suitable or not, for the longer part of the journey on the Pacific. The traveler had to take what there was, once he had crossed Nicaragua, regardless of whether she was a large screw steamer like the *Pioneer*, leaking badly, or a well-designed but small side-wheeler like the *Pacific* (not the Collins liner), of 1,003 tons.

When the *Prometheus* left New York on December 26, 1850, she was bound for Havana; San Juan del Norte, the hamlet at the mouth of the San Juan River; and on to Chagres, the usual Panama transshipment point, since arrangements for the Pacific servie had not been completed. When she returned on January 21, 1851, she carried 244 passengers and a half-million in gold! It was an auspicious start.[6]

A second steamship was completed in 1851, the *Daniel Webster*, a smaller craft of 1,035 tons and, therefore, equipped with a single beam engine. She was the exception to the rule in that she was built by William H. Brown, but set a new example with her 56" by 10'-0" stroke engine from the Allaire Iron Works. Vanderbilt had become a part-owner of that establishment and, after the *Prometheus*, used it exclusively to engine his steamers.

By the fall of 1851, when the Nicaragua route was in operation and served by steamships running more or less regularly on both the Atlantic and the Pacific, fares were drastically cut and passengers swarmed aboard the Vanderbilt vessels.

A second and far more impressive Simonson steamship was launched on October 25, 1851, and christened *Northern Light*. She was a remarkably successful paddler with a long career ahead under several owners. Fifty percent larger than the *Prometheus,* she had a long, low, flush-decked hull and a shallow draft of 14′ to provide the necessary stability to hold up two heavy walking beams far above the deck. She measured 1,767 tons, with dimensions of 253′-6″ by 38′-2″ by 22′-6″,[7] had three decks and a vertical stem, and did away with unnecessary frills such as stern decorations or a figurehead. There was just a vestige of gilt scrollwork on either side of the stem and two white and red stripes around the hull at the level of the guards. Her sides were painted dark green, a Vanderbilt innovation. Black stacks with red tops were copied from the Collins liners. A two-masted rig had square sails forward. Immediately abaft the foremast was the pilot house; aft of that came a series of deckhouses. Below, on the main deck, was the first-class dining saloon, 40′ long by 36′ wide, extending the full width of the hull.

Berths for 250 in the first cabin, 152 in the second, and 400 to 500 in the steerage were provided, and the total would prove none too large in actual service. There were no gaudy interiors; Vanderbilt chose nothing more than paneling painted to look like oak. The two Allaire engines of 60″ bore and 10′-0″ stroke driving 33′-0″ diameter wheels were supplied steam by four flue boilers.

Solid bulwarks extended from the stem almost to the foremast. Aft of them came open rails, which helped make the hull look long and sleek. The two masts and two stacks, one forward of the walking beams and the other aft, gave her a most attractive profile, one that would be repeated again in later steamships. Her cost was listed at $290,000.[9]

Because the West Coast service was temporarily disrupted, the *Northern Light*'s maiden voyage from New York, starting April 5, 1852, headed for Panama and returned by way of San Juan del Norte with a stop at Havana. Coming home she took three and one half days from San Juan to Havana and exactly ninety six hours, at 12.8 knots, from Havana to Sandy Hook. On her second trip, departing May 5, she was supposed to connect with the *S.S. Lewis,* then on her way to the Pacific. Her hapless passengers had to wait three weeks at San Juan del Sur without adequate shelter before the latter vessel arrived; thirty-six died before reaching California. Yellow fever and cholera were normal hazards.

Following the *Northern Light* came the *Star of the West,* in September of 1852, almost a duplicate of the *Prometheus,* and the *North Star* in the spring of 1853, a repeat of the *Northern Light.* But by the time the last steamer was nearing completion, Vanderbilt's position had undergone one of the many abrupt changes that punctuated his business ventures. Unable to raise money for a canal, he and his partners had set up the Accessory Transit Company to handle the passengers crossing Nicaragua. In September of 1852 Vanderbilt resigned from its presidency. Three months later he offered his seven steamships to the Accessory Transit for $1,350,000, of which $1,200,000 was to be paid in cash and the remainder in a year's time.[10] The vessels were the *Pro-*

North Star, original appearance. *Author's collection.*

metheus, Daniel Webster, Northern Light, and *Star of the West,* all designed to his order and operating on the Atlantic. The other three, on the Pacific, were purchased steamers, the *Brother Jonathan, Pacific,* and *S.S. Lewis.* The offer was accepted and Vanderbilt's only remaining tie was the New York agency for the line, a most profitable office, for it collected a 2 1/2 percent fee on every ticket sold.

The Cruise of the *North Star*

Commodore Vanderbilt was left as the owner of one uncompleted steamship intended for Nicaraguan service but no longer needed. He announced that he would outfit her as a yacht for a leisurely European tour.

The so-called yacht, christened the *North Star,* was launched on March 10, 1853, with most of her machinery aboard. She measured 262'-6" by 38'-6" by 28'-0". Her tonnage of 1,867 made her, statistically at least, 100 tons larger than the *Northern Light.* Her two 60" bore by 10'-0" beam engines were identical in size, if not in every detail, to her predecessor's. Including the wheels, they weighed about 230 tons. Four drop-flue boilers 23'-11" long by 11'-0" wide by 10'-6" were mounted in pairs forward and aft of the engines. Each pair had a 5'-0" diameter by 36' high stack and steamed at 20 psi. The length of the machinery spaces came out to 104'-6".[11] The normal bunker capacity was 600 tons of coal. Paddle wheels 33'-0" in diameter had 28 floats 8'-0" by 1'-6", dipping 7'-9" in the water at the designed draft of 14'-0". The hull scantlings were light in comparison with a Collins or a Havre liner. The oak keel was 15" x 14" and the deadwood at the bow and stern 13" thick. Floor timbers were 12" wide by 13" deep and the bottom was not filled solid, for no extensive bedplate was required for the beam engines; instead, built up keelsons 26" wide by 60" deep formed the engine foundations and others, 15" x 32", supported the boilers. The bottom planking was 3 1/2" to 5" thick oak and the ceiling 6" pine. There were two continuous decks and a partial deck at either end. Simonson became one of those gifted New York shipwrights who understood that well-arranged fastenings were even more important than timber sizes. His unfortunate experience with the *Ohio* seems to have spurred him on toward superior structural achievements for his later hulls.

The after cabin was fitted up in the normal, passenger-saloon manner, but was far more luxurious than usual with maple and rosewood paneling, expensive furniture, and a ceiling covered with "chaste and beautiful" designs.[12] The deck below was left without its usual cabins so that it could serve as a ballroom. In short, the Commodore did not complete the steerage and second-class spaces on the lower decks and installed exceptional first-class quarters. Her profile, with its two stacks and brigantine rig, was the same as the *Northern Light*'s except for shorter deckhouses. It made her a lean and handsome steamer.

To command his yacht the Commodore chose Captain Asa Eldridge, formerly of the packet *Roscius* and, more recently, of the *Prometheus.* In 1856 he would vanish with the ill-fated Collins liner *Pacific.* After a trial to Sandy Hook and back, a party of twenty-five assembled for the cruise. Included were Mr. and Mrs. Vanderbilt, one son, eight daughters, a grandson, seven sons-in-law, one daughter-in-law, a doctor, a minister, and the wives of the professionals.[13] To see them off, several hundred well-wishers were to travel as far as Sandy Hook, where they would transship to the famous, four-stacked steamboat *Francis Skiddy.* Everyone boarded at the Allaire Iron Works wharf, at the foot of Corlears Street on the East River, during the morning of May 9, 1853. Three minutes after sailing, on a swiftly ebbing tide, the *North Star* was hard aground and the excursion brought to an embarrassing end. Fortunately for him, Captain Eldridge was not in charge, but a pilot was.

After the yacht was pulled off, the dock in the Brooklyn Navy Yard was made available by special permission of the Secretary of the Navy. The damage to the shoe of the keel was slight—a few copper plates were scored; everything was repaired and the *North Star* steamed down the harbor at 8 P.M. on May 20, firing her saluting cannon and sending up rockets. Because she was the first beam-engined steamer to head across the Atlantic, some special celebration was in order.

The passage was a smooth one with runs of 260, 285, and 300 miles per day reported by Mr. Choules, the invited minister, who published an account of the voyage. The best, on June 29th, was said to be 337 miles at 14 1/2 rpm.[14] The report is badly in error; the distance is equal to the paddle wheels rolling over a solid surface, such as a road. In a yielding medium, water, only 270 miles would be possible at that rpm. One wonders whether the clergyman was willfully exaggerating, or was it the ship's master? Southampton was reached on June 1 with a widely reported passage time of 10 days 8 hours 40 minutes. Again, someone gave out false information. The actual time works out to eleven and one half days at 11 knots.[15] To duplicate her New York experience, the Solent pilot ran the *North Star* on a mud bank; she was soon freed.

On the 13th of June the mayor of Southampton and some of its leading merchants invited the Vanderbilts to a banquet; the Commodore responded by making an excursion around the Isle of Wight on the 14th. About four hundred guests boarded the *North Star* for the occasion. Previously, thousands had inspected and admired her during her two weeks in English waters. After putting the

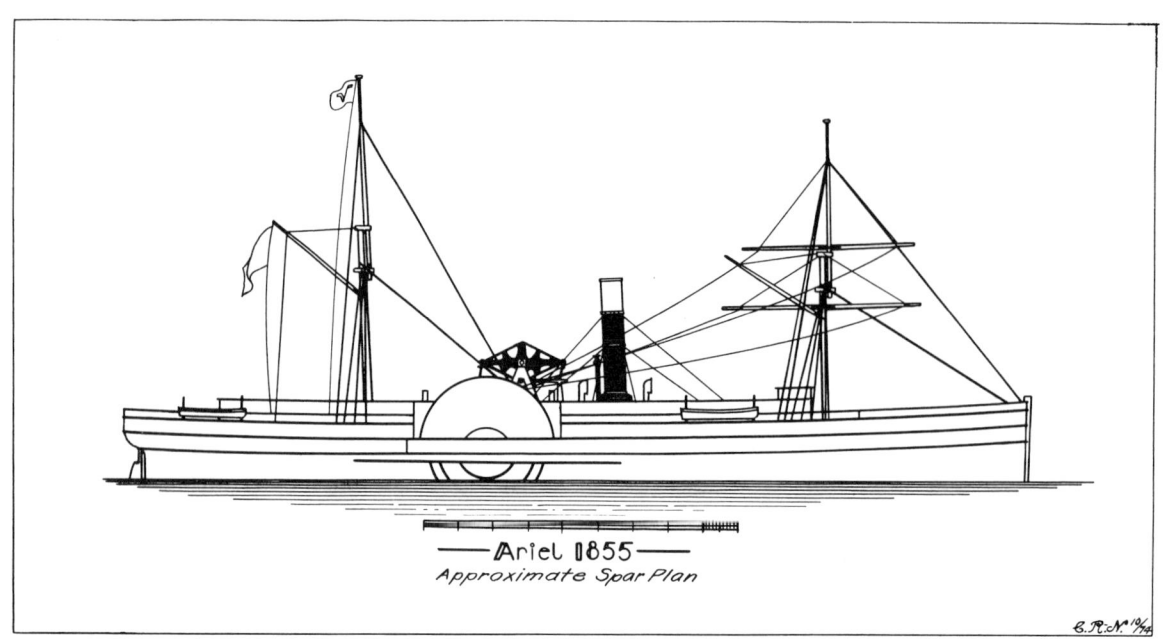

Ariel, approximate spar plan.

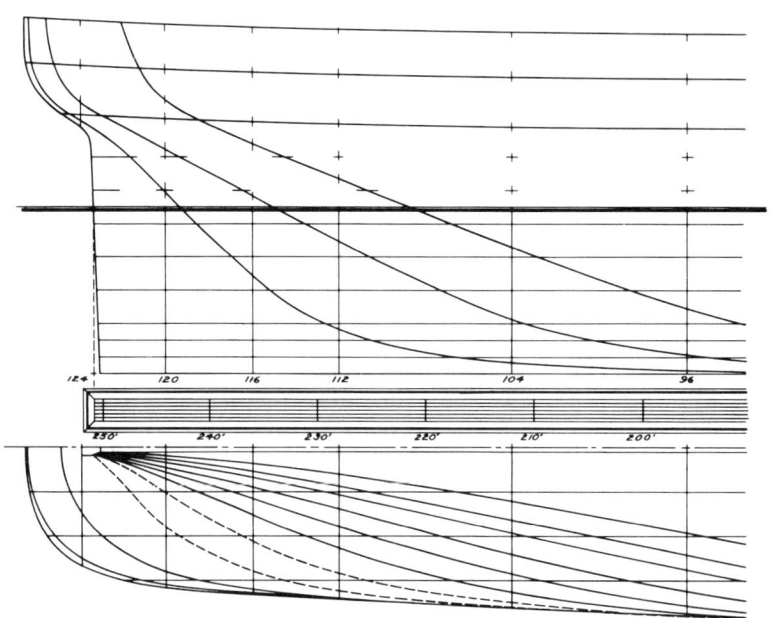

Ariel, 1854, lines.

visitors ashore, the *North Star* sailed the same evening for the Baltic and the Russian fortress of Kronstadt, the first American steamship to call there since the *Savannah* in 1819. She left Kronstadt June 29, stopped at Copenhagen, and put into Havre for a stay from July 6 to the 25th while everyone piled ashore to visit Paris and see the sights.

There was much publicity. The Commodore was pleased with everything; everyone was pleased with the Commodore. The whole trip went smoothly except in Italy, where various authorities could not understand her peaceful intent and suspected that she was part of a conspiracy against the Papal States.[16] As a result, she was unable to stop at Civitavecchia or Naples. Instead, she went on to Malta, then headed for Constantinople. From the Ottoman capital on August 26, via Gibraltar and Madeira, she reached New York on September 23, 1853. The last leg of the trip took 10 days 20 hours at the slow speed of 10.6 knots, based on a minimum distance of 2,765 miles between these ports. The Reverend Mr. Choules made things sound better by quoting 2,930 miles and a fast run of 306 miles on September 20.[17] His alleged average daily travel throughout the trip was 259 miles and his coal consumption of twenty-eight tons per day were both too good to be true. As an author he is neither entertaining nor accurate. He is said to have done the praying and Vanderbilt the swearing aboard the *North Star*.

Much had changed while Vanderbilt was away. The Accessory Transit Company had been taken over by Charles Morgan, one of his associates in that venture, and Cornelius K. Garrison, former mayor of San Francisco and Pacific agent for the line. Morgan assumed the New York agency and took over Vanderbilt's commission on its fares. Furthermore, various sums due the Commodore for items he had sold the company were unpaid. A claim of no profit from the land transits had nicely eliminated other promised payments. Vanderbilt was furious, it was just the sort of thing he would do, but had never believed that it could be done to him.

At once the *North Star* was fitted out with staterooms and steerage berths and sent from New York, on February 20, 1854, to Aspinwall, in competition with both the U.S. Mail Line, serving that port, and Vanderbilt's former liners, now under Morgan and Garrison, to Nicaragua.[18] The latter service, oddly enough, was still thought of as the "Vanderbilt Line," while the *North Star*, connecting with the *Yankee Blade* and *Uncle Sam*, operated on the Pacific by Edward Mills, was now advertised as the "Independent Line." Within a year the latter had cut fares so drastically that both the Nicaraguan operators and the U.S. Mail, in desperation, bought up its ships to keep them out of service. By a joint agreement, the U.S. Mail purchased the *North Star* for $300,000 and paid Vanderbilt an additional $100,000 to give up the California trade. The *Uncle Sam* and *Yankee Blade* were bought by Morgan and Garrison.

Meanwhile, the *North Star* had made six voyages from New York to Aspinwall before the sale. Having arrived in New York on August 25, 1854, as an actual Vanderbilt

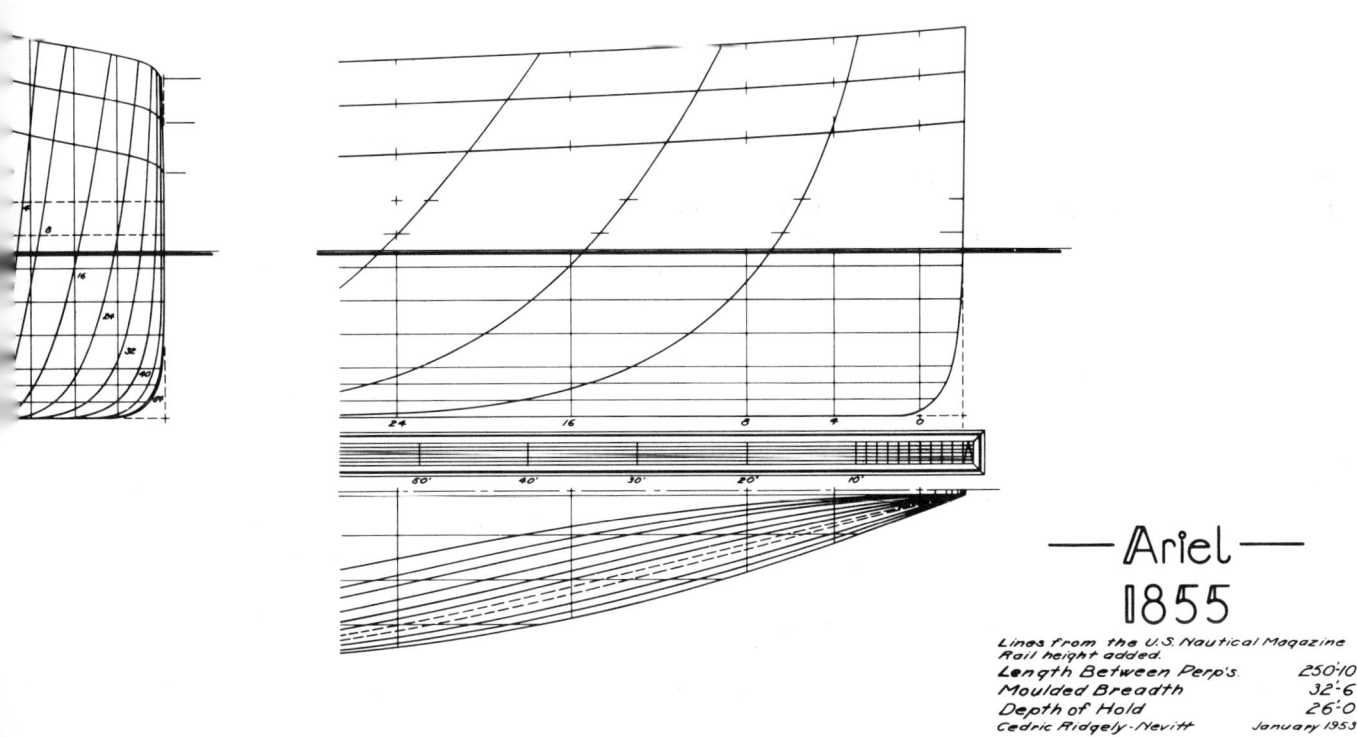

—Ariel—
1855

Lines from the U.S. Nautical Magazine
Rail height added.
Length Between Perp's.   250'-10"
Moulded Breadth   32'-6"
Depth of Hold   26'-0"
Cedric Ridgely-Nevitt   January 1953

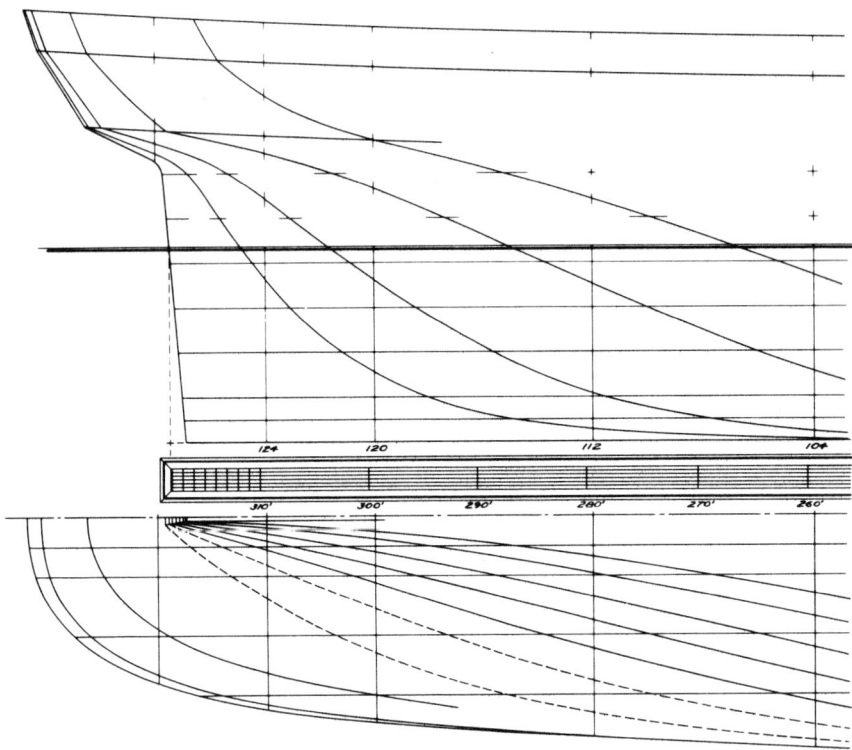
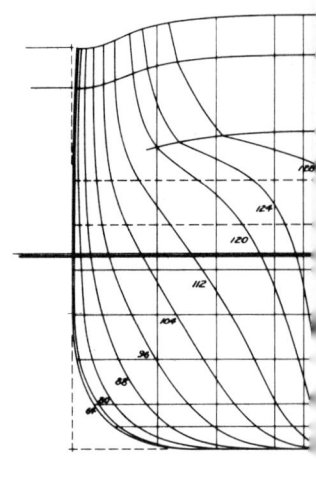

Vanderbilt 1856

Lines from the U.S. Nautical Magazine.
Stern profile altered.
Length Between Perp's. 318'-2"
Moulded Breadth 47'-6"
Depth of Hold 31'-9"
Cedric Ridgely-Nevitt March 1952

Drawn at 30" Frame Spacing
31" Spacing Makes L.B.P. 328'-10"

Vanderbilt, 1856, lines.

liner, she sailed on September 20 for the U.S. Mail Line. She made five more calls at Aspinwall but, after she docked at New York on February 7, 1855, Vanderbilt reacquired her for use on the Atlantic.

The European Route

Steamship building was in Vanderbilt's blood by now; Simonson was already well along with a successor to the *Star of the West*, a moderate-sized vessel of 1,295 tons with a brigantine rig, a single stack, and a single-beam engine. Named the *Ariel*, the register dimensions were 252'-6" by 32'-6" by 16'-3". The *Journal of the Franklin Institute* gave other depths of hold: 19' to the main deck and 26' to the spar deck.[19] John W. Griffiths computed the displacement, at a 16' draft, to be 2,360 long tons.[20] The block coefficient of .63 and the waterplane coefficient of .76 show her to have a considerably fuller hull than the Havre Line's *Arago*, which had .578 and .70 respectively.

Because timber sizes were often larger than those on the *North Star*, even though the *Ariel* was smaller, it is believed that she was intended for transatlantic service and that Simonson had decided that extra strength was necessary. The ceiling, for example, was 8" instead of 6" and the floor timbers 18" deep in lieu of 13". Frames were very closely spaced, 24". The usual diagonal strapping, 4 1/2" x 3/4" iron bars, was fitted. There were two complete decks plus short ones in the hold at either end. The straight stem bore a small amount of scrollwork; on the stern was a gilt eagle supporting a United States shield,[21] and the paddle boxes bore flying eagle decorations. Unlike the green *Northern Light*, the hull was black with a red stripe. Berths for 284 cabin passengers

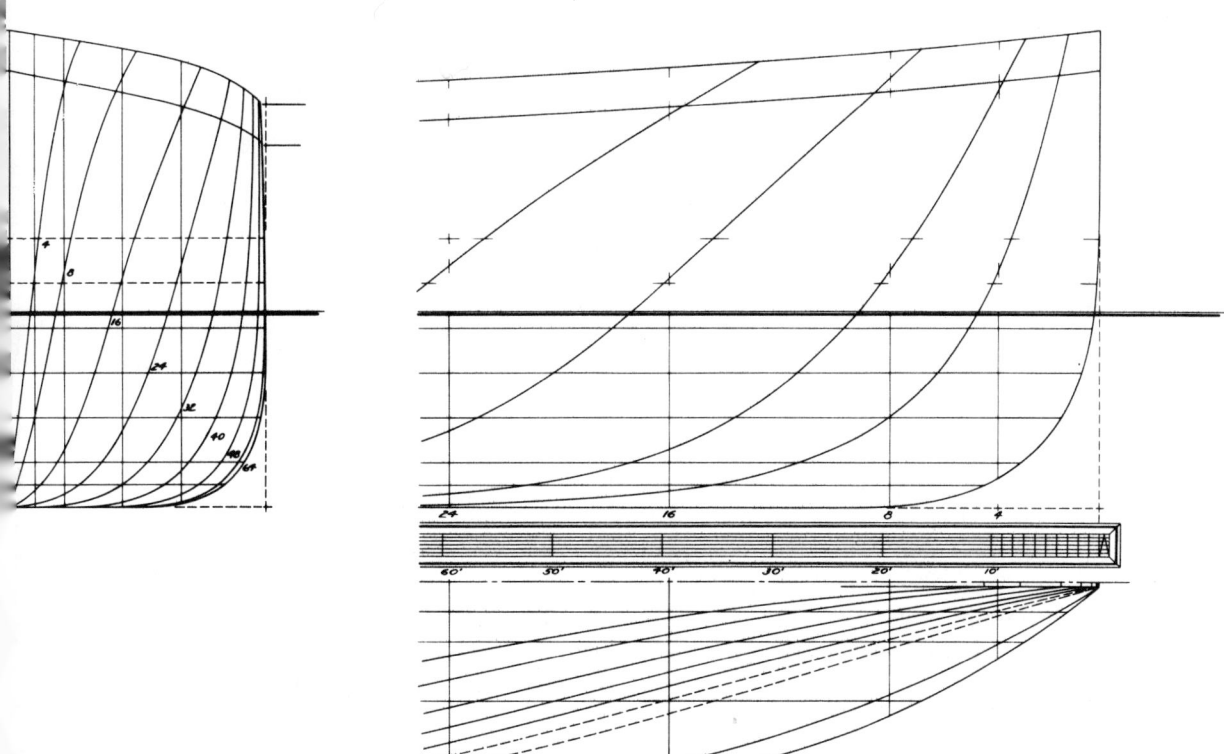

were supplied. On the main deck the dining saloon was forward, the main cabin amidships, and the ladies' cabin aft.

The Allaire engine was 75" bore and 11'-0" stroke, and turned 33'-0" diameter wheels, each with 28 buckets 8'-0" by 18" deep. Two return-flue boilers had cylindrical iron shells 11'-2" in diameter by 24' long, each with three furnaces and a tall steam chimney. The 6'-2" diameter stack was so high, 48', that no forced-draft blowers were necessary. In short, the simplest possible engine and the most rugged type of boilers were installed. Six hundred tons of coal could be stowed.

Like previous steamships in this series, the *Ariel* was nearly complete and had all her machinery on board when she was launched on March 3, 1855.[22] At first it was planned to fire the boilers and raise steam, but the idea was dropped because of shallow water; the rudder had to be left off to keep it from plowing through the mud.[23] Trials were deferred until April. Vanderbilt opened her to the public, claiming that he wanted everyone to see that she was, in every respect, "equal to any other ship." He also claimed that her tonnage was 2,300, a gross exaggeration.

When the spring of 1855 arrived, only two steamships were available to the irrepressible Commodore. Nevertheless, he had written the postmaster general offering to carry the mails from New York to Liverpool twice a month at $15,000 per voyage, alternating with the Collins liners, which were being paid $33,000.[24] When Congress did not act on his proposal, Vanderbilt selected the Havre route, rather than the Liverpool one, for his first Atlantic venture. The Havre Line, currently operating with chartered steamships, appeared a more vulnerable competitor. The *North Star* inaugurated Vanderbilt's European line on April 21, 1855, direct from New York to Havre without a stop at Cowes. Fares of $130 first cabin and $75 second were advertised; the Havre Line's were identical. Immediately after the *North Star* departed, with 149 passengers and $164,000 in specie, fares were cut to $110 and $60 for all subsequent crossings.[25] Daniel

Torrance, one of the Commodore's sons-in-law, was appointed agent for the line.

The *North Star* made the trip in about fourteen days at a speed below 10 knots. She left Havre May 19 and arrived home on June 1 in thirteen days. Outbound she carried American displays for the Paris Exhibition that was held that year, and homeward 100 baskets of champagne.

The *Ariel* attracted 191 passengers and $186,000 specie for her maiden voyage. She left May 19, under Captain Peter E. Lefevre, and reached Havre on the 31st, making a passage faster than the *North Star*'s, a good performance when one considers that she was shorter and had less power. On her return she docked on June 22 after a thirteen-day passage fighting westerly gales.[26] Thereafter, sailings were made every three weeks, but after two voyages by each steamship, it was decided to touch at Cowes in both directions, in the hope of attracting additional passengers. The *North Star* departed eastbound on July 21 and the *Ariel* on August 11, and these crossings were the first to touch at Cowes.

For once, a Vanderbilt operation was not an immediate success. The Collins Line, at the height of its fame, carried more travelers in 1855 than it ever had before. The Havre Line had an enviable reputation and put its new *Arago* into service in June. She surpassed the two beam-engined interlopers in every particular. Furthermore, the first-cabin Atlantic traveler was a far more sophisticated individual than the illiterate adventurer who fought to board any California-bound steamship, regardless of her owner's reputation, as long as the fare was low. But as a result of the Crimean War's demands for steamships, there was considerable cargo left over for the *Ariel* and *North Star*. The latter reached New York, on her second voyage, with 655 tons of freight but only 100 passengers. In the fall, the *North Star* had a record number, 212, when she sailed from Cowes on September 22. Each steamship made five voyages in 1855; the last, by the *Ariel*, left New York November 3 and arrived home December 9.

Vanderbilt in 1856 was much occupied with complex financial and political maneuvers connected with the Nicaragua route. He was too busy to bother with his Atlantic venture and left the *North Star* and *Ariel* idle for most of the year.

When the U.S. Mail's *Crescent City* ran aground and became a total loss in December of 1855, the *North Star* had been chartered as her replacement for a trip to Havana and New Orleans. After reaching New York on January 28, she saw no further service until May of 1856, when she sailed on the 10th for Cowes, Havre, and Bremen. Westbound she left Bremen June 7, made Havre the 10th, and Southampton the 11th, entering the harbor of New York June 24 with eighty passengers -- a poor showing.[27] It was the only European trip that year. After it the *North Star* never raised steam again until the spring of 1857.

Her smaller companion, the *Ariel*, completed a single voyage, also under charter to the U.S. Mail Line, from New York on July 21, to Aspinwall; she landed about $1,500,000 in gold on August 13, 1856.[28]

The major event of the year, at least for the future of the line, was the completion of a magnificent new steamship larger than any American vessel afloat, if we except the uncompleted *Adriatic*. Originally called the *C. Vanderbilt* but shortened to *Vanderbilt* after launching, this splendid steamship was laid down in a new Greenpoint shipyard in early 1855. Simonson had moved away from Manhattan to find sufficient space for building larger craft.

A great unknown is just how long she was. Her Custom House register is erroneous; it gives her length as 311'-6" and her tonnage as 3,360, both values being too low. The builder's half model was loaned to John W. Griffiths, who published a set of lines taken off it.[29] That drawing has been enlarged by the author, faired, and the stern shape corrected to agree with photographs of the ship. It shows a molded length from the rabbet at the stem to that at the stern, measured on a waterline 17'-6" above the top of the keel, of 318'-2". Griffiths, however, says she was 331' on the spar deck and 328' on the waterline. The *New York Times* and *Herald* give 335' on deck, with the *Times* agreeing on 328' at the waterline. Because 340' is quoted by several sources, it is probably the length overall. Jeremiah Simonson's notebook lists 335' by 48' by 32'. Finally, the Statistical Records of the U.S. Navy are wildly wrong with a 250' length.[30]

What the author believes happened was that the half model was built for a frame spacing of 30". This corresponds to the *Nautical Magazine*'s drawing and the accompanying enlargement by the author. During construction, the spacing was increased to 31" (not the 32" value in Griffith's text). Having 128 frames, the change would add 10'-8" to the ship's length and this would make the molded length on the waterline 328'-10". If another 2'-2" are allowed for the stem and sternpost, the result would agree with Griffiths' 331' on deck. This is still 4'-2" shorter than her builder's figure. For comparison, the registered dimensions of the *Northern Light* were 3'-6" longer and the *Star of the West* 1'-8" shorter than their builder's. If we adopt 331' for the official length and revise the tonnage accordingly, the beam and depth remaining unchanged, the probable dimensions for the *Vanderbilt* would be 331' by 47'-6" by 31'-9", and the tonnage would increase to 3,600. These figures would produce a 342'-0" length overall, within two feet of the longest published value. In comparison with the *Adriatic*, the *Vanderbilt* was 14' shorter, 2'-6" nar-

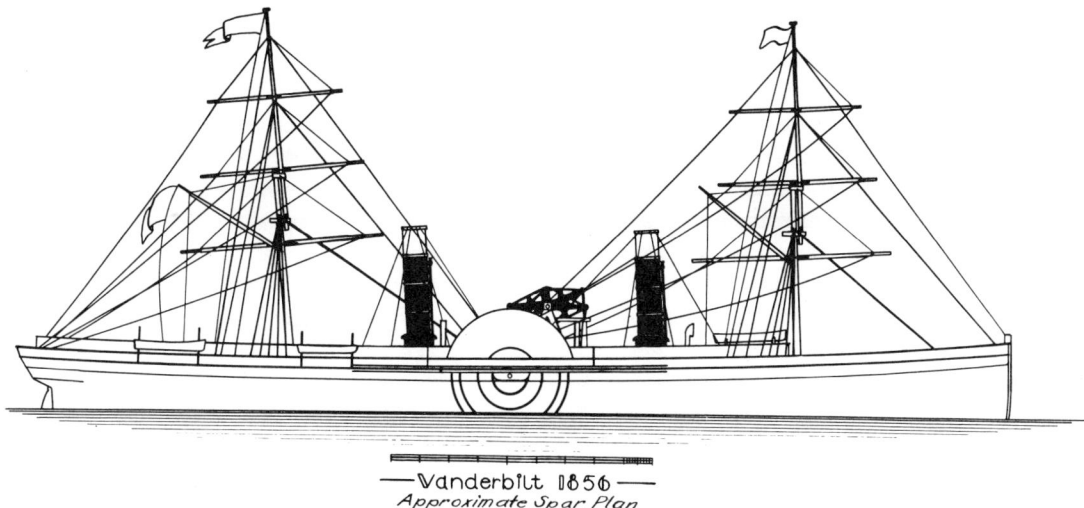

—Vanderbilt 1856—
Approximate Spar Plan

Vanderbilt, approximate spar plan.

rower, and 1'-5" shallower. In consideration of tonnage, she was a 13% smaller steamship. Without question, she was intended as a rival; Vanderbilt had decided to show Congress what an unsubsidized liner could do. The ultimate goal, of course, was a mail contract.

Whereas the *Adriatic* represented the largest, most luxurious, most expensive possible liner that incorporated the latest style of marine engine, the *Vanderbilt* was to have the strongest and lightest hull possible, equipped with adequate, but unostentatious, quarters, and was to use the tried and true beam engine that had performed reliably for several decades. The *Adriatic* had endless delays, and her cost, over $1,000,000, helped bankrupt the Collins Line. The *Vanderbilt* was built for $600,000 to $800,000, $220,000 of it for her machinery, and was ready a year in advance of her owner's decision to put her to work. When she did go to sea, it was her size and speed that made the Vanderbilt Line a paying proposition.

She was launched on a windy, wintry day, December 10, 1855. About 5,000 spectators braved the weather, and the ladies of the party watched from the windows of a workshop in order to stay warm. Immediately after the event she was towed away to the new balance dry dock where she was lifted out to have her bottom coppered.[31] The strength of the hull is indicated by massive floors of white oak 15" x 21", with the whole bottom filled in solid. This mass of timber had 1 1/4" diameter iron bolts 6' to 8' long tying it together in the longitudinal direction. The frames were mixed white oak, live oak, and locust. The 350 diagonal straps were 5" x 7/8" iron 40' long and weighed almost 100 tons. The outer planking was white oak 6" thick. There were five decks, the upper three running the full length of the ship, and a series of watertight bulkheads. The combined engine and boiler room was 114' long and had bunkers located outboard of it holding 1,200 to 1,400 tons of coal.

Aft, on the main deck, was a dining room 108' long and 25' wide, with an unusually high ceiling 8' above the deck. Off it opened twenty staterooms having two fixed and one folding berth in each. They were small in size, only 8'-10" by 6'-0". Fourteen other staterooms were uncomfortably located in the vicinity of the pantry and passageways. On the same deck forward was a dining saloon with forty five adjacent staterooms containing 138 berths. Farther forward was a "cabin" with 33 berths, obviously not for the demanding passenger. Above, on the upper deck, was a long saloon 258' in length, 30' wide, with forty-one staterooms. A total of five hundred passengers was contemplated. A new departure was the fitting of interior windows so that passengers could watc the mighty engines. Anyone who has watched a beam engine at work knows what a hypnotic sight it is.

Comments were not favorable. One passenger stated: "I am told that Mr. Vanderbilt is his own ship's architect. If so, I think he would be effectively cured of his vanity, or his parsimony, by taking a first class passage in the fore saloon of his own ship, which he has taken pride in naming after himself."[32] Another unenthusiastic report said, "... if she is not furnished in the same luxurious styles as others, the great strength which is claimed for her will be better appreciated by her passengers," and continued by speaking of the "great economy" of space.[33] It was a nice way of saying *cramped*.

The engine installation was an example of extreme conservatism. No surface condensers were fitted and no forced-draft blowers. The two beam engines, 90″ bore by 12′-0″ stroke, had a cylinder volume of 81% of that being supplied for the *Adriatic*. If the latter had 3,300 IHP, then the *Vanderbilt*'s would be about 2,600 at the same steam pressure; ridiculous figures as high as 4,000 had been printed. James P. Allaire, while still alive, was no longer associated with the iron works bearing his name. Messrs. Secor and Braisted were the proprietors directly responsible for the *Vanderbilt*'s engines. The shafting was 25″ in diameter and the paddle wheels were built up with three sets of flanges and wheel arms. Their diameter was 41′-0″ and 36 blades 10′-0″ by 2′-0″ dipped 8′ into the water at a 20′ draft. The four boilers were the only break with tradition; they had 3″ diameter fire tubes returning the flue gas from the back to the firing end of the boiler and were, therefore, of a design anticipating the "Scotch" boiler of the 1870s. These boilers, half as many as the *Adriatic* had, were 28′-6″ long, 13′-11″ wide, and 13′-6″ high, exclusive of their steam chimneys, and weighed about 60 tons apiece. Eight furnaces per boiler were installed. The smoke pipes were 8′-8″ in diameter and 40′ tall.[34] The fuel was bituminous coal, with an anticipated consumption of 100 tons per day for 2,500 indicated horsepower.[35] Had surface condensers been supplied, she would have been less prodigal a coal consumer.

On July 19, 1856, under Captain Peter E. Lefevre, the first master of the *Ariel*, the *Vanderbilt* made a short trip down the habor as far as Fort Hamilton carrying several hundred Allaire workmen. Fifteen miles per hour and a maximum of sixteen revolutions per minute were reported.[36] After this her sea trials took her to Washington, D.C., where President Pierce, his cabinet, and members of both houses of Congress saw her on July 26.[37] The visitors "were treated rather cavalierly by Commodore Vanderbilt. The anticipated feast was not forthcoming, nor was there a single bottle of champagne cracked to enliven the occasion." Edward K. Collins and John Ericsson's friends had shown themselves far better hosts when their respective steam and hot-air ships had been sent to impress the country's lawmakers. Because the same lawmakers were not impressed enough to vote the Commodore the mail contract he wanted, the *Vanderbilt* reached New York on July 29, 1856, only to be laid up along with the *Ariel* and *North Star*.

The spring of 1857, however, saw the advent of two Vanderbilt European routes. The new *Vanderbilt* would provide an express service from New York to Havre starting on May 5 and touching at Cowes. Other sailings would follow at about six-week intervals. First-class fares would be $100 to $130, depending on the accommodations, and second ones $75. The size of the ship was advertised as 5,268 to 5,408 tons, with a typical disregard for truth. In July the fares were reduced to $100 and $50, again in accordance with Vanderbilt's standard policy of undercutting his rivals.

The second service was a two-ship one to Bremen, via Cowes and Havre, performed by the *North Star*, at an inflated size of 2,500 tons, and the *Ariel*, magnified to 2,000. Fares were low, $80 and $50. The *Ariel* was to sail first, on April 16, the *North Star* following on May 16; from then on, there would be four-week intervals between sailings.[38]

If we consider the *Vanderbilt* first, a new set of engineering trials were run on April 18, 1857. Loaded with 650 tons of coal, she drew 17′-9″. The engines turned 17 rpm under a steam pressure of 18 psi. Coming up the harbor she was reported as making the seventeen miles from Southeast Spit to Castle Garden in fifty five minutes against an ebb tide. On May 5 she went to sea sped by the cheers of thousands who had gathered along the waterfront.[39] There were 212 passengers and $445,000 in specie aboard; a fast passage was predicted. She ran through fog all the way to the Grand Banks and then into bad weather. Captain D. L. Wilcox slowed down for neither; she reached Cowes in 9 days 21 hours 15 minutes at an average speed of 13.15 knots.[40] It was almost as good as the *Baltic*'s 1854 record of 13.23 knots, but not up to the *Persia*'s 9 days 4 hours 35 minutes to Liverpool at 13.95.[41]

Coming back, she left Havre on June 2 and Cowes on the 3d, with 101 passengers and 500 tons of freight, and made New York on June 14 after an 11 day 1 hour crossing against heavy seas and a head wind. Her sailing coincided with the Havre Line's *Fulton*, which took two days longer.

For her second trip, the *Vanderbilt* was to leave New York on June 20, in order to race the *Atlantic*. The former had 190 passengers and $850,000 gold to the *Atlantic*'s 127 and $1,128,581. The *Vanderbilt* reached Cowes on June 30, a full day ahead. Westbound, she won again by fourteen hours. She had 120 passengers to the *Atlantic*'s forty-five. It must be remembered, however, that the Collins Line by 1857 was on the verge of collapse and its former passengers, frightened by the losses of the *Arctic* and *Pacific*, had transferred to what they hoped were safer steamships.

On her third voyage, the *Vanderbilt*, under Captain Edward Higgins, formerly of the *Hermann*, created a record for the Cowes-Southampton route. She left her pier at New York on August 1 at 12:20 P.M. Once she was at sea, every single day's run was at least 300 miles, and a maximum of 350 was achieved on August 9. That corresponds to 14.5 knots.[42] From Sandy Hook to the Needles, off the Isle of Wight, the sea time came to 9 days 8 hours. Unfortunately, the *Persia* had just lowered her Liverpool time to 9 days 3 hours in July, so she could boast 14.05 knots to the Vanderbilt's 13.75.

Just to show that the third eastward crossing was not a fluke, the fourth, starting September 12, proved a mere two hours longer. The third westward passage, the best made that year, was nowhere near the *Baltic*'s record of less than ten days; the *Vanderbilt* left Cowes on August 23 and reached New York in 10 days 15 hours 25 minutes with an average of 12.2 knots.[43] The fifth and final 1857 voyage, leaving New York October 24, had her largest passenger list, nearly 600, but was a slow one, taking all of eleven and one half days to Cowes. Bad luck and bad weather had caught up with the *Vanderbilt*. Owing to fog, she ran onto a sandbank before reaching Cowes Roads. Her speed was low, a large tug was passing, and, on a rising tide, she was soon afloat.[44]

Homeward, she left Cowes at midnight November 14 with 305 passengers and $250,000. Off Cape Race, Newfoundland, she should have met the newly stationed Associated Press steam yacht, which would rush the news ashore and send it by cable and telegraph to New York, but wintry and tempestuous weather intervened so that no meeting took place. A fishing schooner was hailed, chartered, and sent into Saint John's with the dispatches. Upon her arrival at New York on November 26, she was laid up for the winter.

During her five voyages she had averaged 13.05 knots eastward, 11.75 westward, and 12.40 overall, between New York and Cowes, despite delays of every kind, waiting for pilots, for fog to clear, or for missing connections off Newfoundland. The average of six of the *Arctic*'s voyages in 1852 was 12.19 east, 11.14 west, 11.70 mean.[45] Thus, as a longer ship with only slightly more horsepower, she easily outdid the Collins liners. But the British *Persia,* because she was longer still and had 3,600 indicated horsepower, made higher speeds in both directions. In addition, as a larger ship, she held her speed better in heavy weather. The Atlantic record was securely held by this remarkable Cunard liner.

The other two Vanderbilt steamships ran to Bremen against the Ocean Steam Navigation Company's *Washington* and *Hermann,* both old and slow. When their mail contract ended, the *Hermann* ceased running in June of 1857 and the *Washington* in July. Thereafter the Vanderbilt pair carried the American mails, not for any fixed subsidy, but for the total amount of land and sea postage paid on them; the same offer had been declined by the Bremen Line as inadequate to keep it operating. For the first time Vanderbilt could apply the coveted description "U.S. Mail Steamer" to the *Ariel* and *North Star.*

The *Ariel* seems to have been unsuccessful in attracting German immigrants at the second-class fare of $50. She made five sailings from New York from April 16 through November 28, but never reported more than 184 passengers when she returned. On July 29, 1857, at the height of the summer season, she landed just 84.

On December 31 she started home from Cowes carrying 100 passengers, a large freight, and the mail. She had reached Longitude 29°31′ West and was driving into strong winds and heavy seas when, on January 6, her starboard shaft fractured just outside of the hull.[46] The wheel was chained and roped in place; had it dropped off, the steamship would have capsized. The weather, already bad, was made worse by severe squalls. The foreyard was broken and the sails blown away, and the ship lay helpless without sail or power for twenty-four hours. Smashed glass ports in the stern flooded the after cabin; companionway doors and a skylight gave way and filled the forward one half-full of water. A drag on a nine-inch cable was put over the bow to head her into the sea, but the line chafed and parted. Eventually a storm trysail was set aft, the cabins were bailed out and openings nailed up, and the fore yard was fished and rehoisted. Thirty-six hours after the fracture, the *Ariel* was making headway at four to five knots under her port wheel. Queenstown was reached on January 15, 1858.[47] The *Ariel* had almost proved herself too small a steamship for the stormy North Atlantic.

The Vanderbilt Line took immediate action. First it agreed to pay the fares of the passengers on the next Cunard steamer to Boston or New York. Second, it canceled all winter sailings that had been planned and advertised. After the *Ariel*'s lucky escape, no further effort was made to establish a year-round service and the delivery of the American mails had to depend on foreign vessels. Somehow the *Ariel*'s difficulties proved an advantage to the line. The passengers and the press praised the steamship, her master, and her owner for their performance under stress.

The *North Star* found 1857 less strenuous than the *Ariel* had. On her first sailing, May 16, she was pitted against the *Hermann*—Vanderbilt always liked to win his races—and reached Cowes a day ahead of her. She seems to have been more acceptable to immigrant passengers than the *Ariel,* for on October 22, when completing her third westward trip, she brought 273 into the country. Two stacks and two engines made her a more impressive sight. Her last voyage that year left New York on December 26 and returned there February 13, 1858. A March sailing was canceled as soon as word arrived of the *Ariel*'s damage.

In 1857 each of the three steamships made five sailings from New York. Their cut-rate fares attracted 4,863 passengers, in both directions, during seven months of operation. The Bremen Line, in its final six months, took 1,008 and the Havre Line, operating throughout the year, 3,252; neither had steerage space. The German-owned Hamburg Line, primarily for immigrants, completed its first year of service with iron screw steamers, the *Borussia* and *Hammonia,* and carried 5,142 passengers, 70% of them in the westward direction.[48]

The year 1858 started like 1857. The *Vanderbilt* opened the season on April 10 by sailing for Havre, with a brief stop at Cowes, carrying two classes of passengers. First cabin at $100 to $120 was slightly below the Havre Line's fares. She was well patronized and transported up to 346 passengers per trip eastbound and 355 westbound. Sailings were at six-week intervals, except for the last, which started on October 2 and, after reaching Havre, continued on to Bremen. Her deep draft resulted in her grounding while descending the Weser. As a result, she had to be docked at Southampton on November 15, 1858.[49] A sixth voyage was advertised but, wisely, canceled.

The rest of the Vanderbilt service was somewhat irregular, with changes of route and steamships. A new, low steerage fare of $30 was established to attract more passengers. The *Ariel,* after reaching Queenstown in January of 1858, had temporary repairs and, still under one wheel, headed for Liverpool where she could have a new shaft made. After its installation, she steamed to Bremen in April of 1858, loaded her passengers, touched at Southampton on April 21, and docked at New York on May 4. Thereafter she completed three uneventful voyages to Cowes, Havre, and Bremen. On the fourth she sailed from New York on October 30, 1858, and met strong November gales that slowed her badly. On November 12 a sea struck the ship, damaging some joiner work, starting the caulking around the guard beams, and knocking Captain Ludlow to the deck, severely injuring one knee. Because the Weser River was covered with ice, Bremen had to be omitted. She sailed from Cowes December 1 and met a gale on the 9th that forced her to head North, running before its mountainous seas. Even this became dangerous; a drag was prepared, and an attempt made to head her into the sea. At once waves broke over the ship, houses on the guards were swept away, the main hatch was broken in, and the Captain and the two mates were severely injured, the master so badly that he died after being carried below.[50] It was the First Officer, Mr. Brown, who, despite injuries, contrived to bring the ship into Halifax on December 20 and to New York on December 26, 1858, 25 days out of Cowes. Captain Ludlow's body had been packed in ice to avoid burial at sea.

For the second time, Neptune demonstrated that the *Ariel,* a rather small steamer having a limited draft, a low freeboard, and a beam engine, straining at its tie rods, was not a safe ship to brave his realm. Her two narrow escapes, however, did not deter Vanderbilt from using her again on the high seas.

The *North Star* made one trip to Bremen with the usual way stops, departing on April 17, 1858. She was then diverted to the Havre operation as a running mate for the *Vanderbilt,* making sailings in June, July, and September, and finishing her last westward crossing October 14.

Since 1853, when the Commodore was eased out of the Nicaragua route by his business partners, he had been conducting a series of attacks on them and their operations. Affairs in Nicaragua were in a constant state of flux.[51] There were revolutions, invasions by Costa Rica, and American-financed expeditions by "General" William Walker, who seized power and became president of the country.

A former Vanderbilt steamship, the *Northern Light,* had been particularly active in connection with Walker's affairs. On December 24, 1855, she was scheduled to leave New York for Punta Arenas, Nicaragua, under Captain E. L. Tinklepaugh.[52] Of 400 to 500 passengers on board, 350 were said to be "colonists" intending to settle in the country; in reality, they were troops for Walker's army. Despite protests by the United States District Attorney, she sailed but was stopped by the revenue cutter *Washington* and the U.S.S. *Vixen.* Only bona-fide passengers with through tickets to California were permitted to continue. Shortly thereafter Vanderbilt regained control of the Accessory Transit Company and the *Northern Light.*[53] But because the crossing concession had been canceled by Walker, she spent the next year and a half in idleness.

The United States Mail Steamship Company chartered her for three trips to Aspinwall, starting on September 21, 1857, and ending December 27. She was out of service during January and February, but suddenly put to sea again for Aspinwall on March 10, 1858. The operator this time was a Vanderbilt creation called the New York and San Francisco Steamship Line. After a single voyage, the *Northern Light* was laid up again until June, when she was shifted to the European route to replace the *North Star* when she was withdrawn from Bremen and promoted to the Havre branch of the line. Because the *Northern Light* was almost a duplicate, she could substitute for the *North Star* with ease. Her maiden voyage across the Atlantic began on June 12, 1858, taking thirteen days to Cowes. She continued on to Havre and Bremen; she was six years old and had spent almost two years of her life tied to a pier while Vanderbilt collected $40,000 (later $56,000) per month for not engaging in the California trade. She delivered a moderate passenger list, 200, to New York, 93 of them in the steerage, when she docked on July 27. In contrast, she had brought 615 from Aspinwall in March. A second Bremen voyage was completed on September 21.

To summarize the 1858 operations, the *Vanderbilt* and *North Star* made seven voyages to Havre; The *Ariel,*

North Star, and *Northern Light* seven to Bremen; and the *Vanderbilt* a single trip to Havre and on to Bremen for a total of fifteen voyages. Business had improved; 6,819 passengers were transported instead of 4,863 the year before. The new steerage fare helped. The Hamburg Line did even better with 9,254, mostly steerage, and the Havre Line, despite the severe competition, had its best year in history, 3,704 passengers, most of them first class.

The year 1859 was a very busy one for Mr. Vanderbilt and his steamships. The event of greatest interest to him was the expiration of the postal contracts with the United States Mail, on the Atlantic, and with the Pacific Mail, on the Pacific. Next to collecting blackmail for not operating his steamships, the Commodore most desired a large mail contract on an essential and busy route. He had always claimed that the lines that possessed such financial bounty had never exploited their advantages properly. In January, therefore, he announced that three of his Atlantic liners, the *Ariel, North Star,* and *Northern Light,* would sail to the Isthmus of Panama. He actually used only the latter two. The service started with the *Northern Light* from New York on March 10, 1859, bound for Aspinwall on the first of ten monthly voyages. The *North Star,* after being strengthened by an extra layer of planking and receiving new boilers, started later, June 20, and was able to complete six that year. On her fifth she left New York on October 20. Five days later, in a thick fog, she ran hard aground off La Plana Island, one of the French Keys. On board were 878 passengers and crew, including over 200 officers, sailors, and marines sent out for the U.S. Navy's screw steamer *Saranac.* The passengers were landed, but the naval draft assisted in throwing overboard 420 tons of coal, putting out anchors, and hauling her off. After six days of backbreaking work, Captain Jones had his ship afloat again; he reloaded the passengers from the tents (made of sails) in which they lived ashore, then headed for Fortune Island for water, Jamaica for coal, and Aspinwall.[54]

The *Ariel*'s near loss, in 1857, had led to favorable publicity. The *North Star*'s grounding had the opposite result. After the event it was found that Captain Jones was below, eating dinner; the mate, supposedly in charge, had left the pilot house; and only a helmsman was on duty when she grounded at full speed. The boats put overboard were missing masts and sails, leaked badly, and often had neither rudders nor thole pins. The steamer was undermanned and carried only six sailors to cope with any emergency; it was the *Saranac*'s crew that labored to get her off. They found that the *North Star* had no spare canvas, hawsers, blocks, or rope. They had to unrig the ship to create tackles and make do with worn-out mooring lines. Since the ropes were rotten and the blocks in execrable condition, both gave way in the attempt to haul the vessel off.[55] Finally, it was the ship's engines that finished the job, once she had been freed of the weight of most of her coal. Lieutenant M. Carrington Watkins, U.S.N., Executive Officer of the *Saranac,* wrote: "Mr. Vanderbilt should be prosecuted for endangering the lives of passengers and the safety of the mails under his charge." One hundred nineteen petty officers and seamen signed a supporting letter.[56]

On October 5, 1859, Vanderbilt's newly incorporated Atlantic and Pacific Steamship Company received a provisional contract to transport the United States mails twice monthly from New York and New Orleans to California for $187,500 per year, via the Isthmus of Panama.[57] After the completion of the Panama Railroad in 1855, connecting Aspinwall on the Atlantic side, with Panama on the Pacific, its four-hour crossing was a tremendous improvement over the rivers, rapids, portages, lakes, and mules encountered in Nicaragua. As a result, Vanderbilt wound up his affairs there, never to return. During 1858 he had come to terms with Charles Morgan whereby the latter bought a pair of steamships that Vanderbilt had built and operated in the Gulf of Mexico purely to plague him. Vanderbilt, in turn, took over a large steamship ordered by Morgan that was intended to connect Nicaragua and San Francisco, and acquired a half interest in the *Sierra Nevada* and *Orizaba,* which had previously served that route.[58] The latter pair formed the West Coast branch of Vanderbilt's Atlantic and Pacific Steamship Company, the name of the new operation.

The uncompleted steamship was to be the *Queen of the Pacific* and, as a name, it was entirely appropriate. She would have been larger than any other vessel operating to San Francisco. The Commodore called her *Ocean Queen.* She was built by Stephen G. Bogert, who had leased a part of the Westervelt shipyard for the occasion.[59] Her official dimensions were 327' by 42'-0" by 22'-6" to the second deck, with a tonnage of 2,801. The depth to the spar deck, 30', and her length on keel, 315', made her about the length and depth of the *Vanderbilt,* but, because the beam was narrower, her tonnage was much lower. She was strongly built and, in addition to diagonal strapping, had a lengthwise iron strake outside the frames at the height of the deck clamps, to which the upper ends of the straps were riveted. As originally designed, she would have had overhanging decks extending out to the guards, like the later Pacific Mail steamers, and large ports for ventilation in tropical waters. There were three decks, a brig rig, and an eagle figurehead. The machinery space was 141"-9" long and contained a single walking beam engine 90" bore by 12'-0" stroke from the Morgan Iron Works. Three return-flue boilers were to supply steam at 20 psi under natural draft.[60] Because of the odd number, one stack was 85" and the other 77" in diameter; the author does not know which was the fatter of the two. At a guess he would place the large one and its two boilers forward of the engine. The paddle wheels were 38'-0" in diameter with thirty-two floats and were to turn at 15 rpm. The single cylinder of the *Ocean Queen* was identical in size to the two driving the *Vanderbilt*. With 50% of the horsepower she should, in theory, make 80% of the *Vanderbilt*'s speed, but, as a narrower and lighter steamship, she could do a little better than that. Her bunker capacity, 800 tons, gave her a longer cruising range. Speed was (and is) an expensive luxury; horsepower and fuel consumption increase enormously when an extra knot is demanded.

On April 8, 1857, the *Ocean Queen* started down the launching ways but came to an abrupt halt when her stern plowed into the river bottom's mud. Her hull was severely tried and only her heavy scantlings and well-placed fastenings prevented a disaster. Almost four days of arduous toil with jacks, derricks, towboats, excavations, and renewing of the launching ways went by before she was pulled off on April 11.[61] Two years passed before she was put to work.

Once she had been taken over by Vanderbilt, she was put into Jeremiah Simonson's hands for completion in accordance with his uncle's ideas. The overhanging guards were cut away, the tropical ports closed up, and she ended up looking very much like the *Vanderbilt*. Her cost, around $450,000, was much lower. She was a splendid running mate for the *Vanderbilt* when she was put to work on the European service in 1859. With the *North Star* and *Northern Light* on the Panama route, the *Vanderbilt* and *Ocean Queen*, nicely matched in size if not in speed, were joined by the totally unsuitable *Ariel* to make fortnightly sailings to Havre from April until November. Bremen was abandoned; even Vanderbilt could not compete with that city's North German Lloyd Line.

The *Vanderbilt* made the year's first sailing from New York on April 24, 1859, under Captain Lefevre. Westbound, she arrived off Sandy Hook light ship during the night of May 20, 9 days 8 hours from the Needles,[62] her fastest passage to the United States. Fog held her outside the harbor for 12 hours. To the light ship her speed averaged 13.75 knots and, on her best day, she did 14.9.

Two weeks after the *Vanderbilt*, the *Ariel* put to sea with a good passenger list of 218. After another two-week inerval, the *Ocean Queen*'s master, Charles P. Seabury, took her out on May 21. She had 249 passengers and $1,500,000 in specie. A leisurely 11 1/2 day maiden crossing was a day better than the *Ariel*'s. Leaving Cowes on June 8 she took over thirteen days coming home and was held up by fog for six hours before she could enter the Lower Bay. The remainder of the year was unexceptional. The *Vanderbilt* completed six voyages, the *Ocean Queen* five, the *Ariel* four, all at two-week intervals with the exception of an October 22d sailing that should have been by the *Ariel* but was canceled. Instead, she was transferred to the Aspinwall route, when the *North Star*'s grounding had laid her up for repairs. On November 21, 1859, the *Ariel* steamed toward Aspinwall and remained on that service, together with the repaired *North Star* and the *Northern Light*, until the outbreak of the Civil War.

On the Havre route, fifteen voyages were completed in 1859 and a total of 5,844 passengers delivered, about 200 less than 1858, the peak year of Vanderbilt's transatlantic venture. Fares were delicately shaded to suit the drawing power of the steamships. The *Ariel* charged $80 first cabin, the *Ocean Queen* $80 to $100, and the *Vanderbilt* $100 to $120. The *Ariel* carried third class, at $30; the *Ocean Queen* had second, at $50; and the *Vanderbilt* the same, at $60.[63] One suspects that second cabin on the largest was no better than third on the smallest steamer. The year's termination date was December 18, the day the *Vanderbilt* reached home. Winter sailings were advertised but never made;[64] in view of the line's previous practice, it is unlikely that they were ever intended. The advertisements were to entice the postal authorities into awarding an Atlantic mail contract that would make winter sailing profitable.

Even before the nine-month period, starting in October of 1859, when Vanderbilt operated under a provisional contract transporting the United States mails to California, he began to add to his steamship fleet. His optimism was justified when the temporary arrangement was made permanent. The most important was the *Champion*, 235' by 35' by 18',[65] built of iron by Harlan and Hollingsworth in Wilmington, Delaware. This was the yard that had produced the iron steamboats needed for river work in Nicaragua. The Allaire Iron Works supplied a pair of walking-beam engines of 42" bore and 10'-0" stroke to power her. A single-stacked steamship, she looked much like the *Ariel*, but had a shallower draft because an iron hull was lighter than a wooden one. Although conservative in insisting on a copy of both the hull arrangement and the machinery type that had worked so well on his other steamships, Vanderbilt was the first American owner willing to accept an iron hull for a large ocean steamship. The *Champion* was intended to bolster the West Coast service and left New York for San Francisco on October 23, 1859. Concurrently, steamships of the United States Mail Steamship Company were coming up for sale as M. O. Roberts liquidated that company's assets. Vanderbilt, in February of 1860, bought the *Illinois* and the almost-new *Moses Taylor* at auction for the small sum of $25,000 each.[66] The latter was never put to work and eventually was sold back to Roberts, but the former was chosen to run with the *Vanderbilt* to Havre. In her day the *Illinois* had been a fine, fast steamship. Now, sadly, she had been poorly maintained and her boilers were in bad condition. But

Vanderbilt put her to work and, simultaneously, laid up his new *Ocean Queen*, the one really suitable vessel he had to partner the *Vanderbilt*. It all seems illogical. The early disposal of the *Moses Taylor*, which had his favorite means of propulsion, a pair of beam engines, seems even stranger! No one could anticipate the Commodore's actions.

The year 1860 saw six voyages by the *Vanderbilt*, starting on March 24, and five by the *Illinois*, which sailed on April 7 under Captain Seabury, transferred from the *Ocean Queen*. The *Illinois*'s first Atlantic passage to Cowes consumed almost 16 days, but her 11 day 17 hour return was more creditable. Eastward, her passenger list was fifty and the slow trip left her dangerously short of fuel. On her fifth and final westward crossing, leaving Havre on October 10 and Cowes the 11th, she ran into similar difficulties and had to put into Halifax for coal. She reached New York on October 30 in a total of twenty days. November 18, when the *Vanderbilt* docked, ended the service.

Evidently Vanderbilt still liked this one steamship without a beam engine, for he overhauled her during the winter, replaced her boilers, and installed new iron bulkheads around the machinery spaces.[67]

Although Vanderbilt intended operating the *Illinois* and the *Vanderbilt* in 1861, the advent of the Civil War changed his plans. Thus 1860 became his final year on the Atlantic, and a not particularly inspiring one. The number of passengers, 3,765, was no better than the Havre Line's total. In the latter case, it was their best year and, in the former, a decline of 900 from 1859 and 1,900 below the record year of 1858.[68] The appearance of a former Collins liner, the *Adriatic*, on the Havre route had reduced the *Vanderbilt*'s glamor. She now had a larger, faster rival that was well-run and appealed to the discriminating traveler.

In six years of operation, 1855 through 1860, a total of sixty-seven voyages had been made to Bremen and Havre. The *Vanderbilt* led the way with twenty-two and, surprisingly, the far smaller *Ariel* was second with eighteen. The *North Star* completed fifteen, the *Ocean Queen* and *Illinois* five each, and the *Northern Light* two. By curtailing operations in the most severe months of the year, Vanderbilt had demonstrated that he could send steamships powered by beam engines across the Atlantic. He may have been more lucky than wise; on two occasions the *Ariel* came very close to foundering. No one else ever chose to build a beam-engined steamship specifically for the stormy North Atlantic and no one ever did operate one on that ocean under a mail contract requiring regular sailings throughout the year.

Yet the European Line was unique. It was the only large-scale American operation that ever succeeded on the Atlantic without a mail contract.

The Panama Route

About the time that the European line was starting its last year, Vanderbilt's Atlantic and Pacific Steamship Company was negotiating with the Pacific Mail Steamship Company in order to resolve the rivalry and existing duplication of services on both the Atlantic and the Pacific parts of the Panama route to California. The result was that the latter purchased all Vanderbilt steamships on the Pacific side, except the *Champion*, for $250,000 plus 5,000 shares of stock.[69] Vanderbilt would continue the eastern part of the operation while the Pacific Mail would take over the western one, and, after payments to the Panama Railroad, the receipts from passengers and freight for the next five years would be split, with 30 percent going to Vanderbilt and 70 percent to the Pacific Mail.[70] From this point on, Daniel B. Allen, Secretary of the Atlantic and Pacific Steamship Company, the son-in-law and trusted agent of Commodore Vanderbilt, became both the operating head of the Vanderbilt Line and a director of the Pacific Mail Company to represent Vanderbilt's stock interest. The Commodore was bringing many of his steamship affairs to a close and becoming increasingly involved with railroads.

From March of 1860 on, every remaining Vanderbilt steamship became a part of the major, New York to Aspinwall, or the minor, New Orleans to Aspinwall, service. The *Champion* was brought back from the West Coast and the *North Star* and *Northern Light* put to work, as a nicely matched pair, to Aspinwall. The *Ariel* had been summarily recalled from the Havre route in October of 1859. The *North Star*, which began as a yacht, and the *Ariel*, an Atlantic steamship, became coastwise ones. All that was necessary was the addition of more steerage quarters; tween decks now had hundreds of berths installed with access ladders placed in what were once cargo hatches; a conversion took only a few days.

The *Vanderbilt* was too big and too expensive for the Aspinwall trade, but the *Ocean Queen*, laid up for 1860, was added in 1861. The *Illinois*, however, was sold back to Marshall O. Roberts. Vanderbilt was remarkably partial to his own creations, whether suitable or not. From 1860 to 1865, most were ill-suited because they were too small. Overloading and overcrowding were the rule.

The *Ariel*

If we follow the careers of the former Atlantic liners, let us start with the *Ariel*. She steamed to the Isthmus of Panama from November of 1859 through February of 1861, then served twice as a Civil War transport with another Aspinwall voyage, in June of 1861, sandwiched in between. She was part of Flag Officer DuPont's fleet to Port Royal, South Carolina, in October. Back on the Aspinwall run for a year, she was southward bound with 850 passengers on December 7, 1862, when she encountered the Confederate raider *Alabama*. Captained by Raphael Semmes, the raider was lying off Cape Maysi, the eastward tip of Cuba, hoping to capture the northbound *Champion* and her cargo of California gold. Because the *Alabama* had her telescopic stack down and sails sets, she did not appear dangerous until she clued up her sails, raised her stack, and pursued the *Ariel*. She flew the American flag until within gunshot, then changed colors and fired a blank. This was ignored, but when a real shot cut a piece out of his foremast, Captain A. G. Jones stopped his ship.[71] Semmes proposed landing the passengers on a nearby island and burning the steamer, but Jones expostulated and brought up the hardships for the women and children aboard. Eventually, the ship's money, $5,500, was taken, together with the arms and ammunition of the 126 marines on board. Jones signed bonds for $125,000 for the steamship and $135,000 for the cargo, to be paid thirty days after the independence of the Confederate States was achieved.

Both the mails and the passengers' baggage were undisturbed. After several days' detention, the *Ariel* was allowed to go on to Aspinwall but, as a safeguard, she took no treasure northbound. She left Aspinwall on the 19th, touched the Key West for coal, and arrived New York December 28.

A naval convoy system was immediately organized, with a cruiser to escort the mail steamers on their thrice-monthly sailings through dangerous waters. There were, however, no further incidents. The *Ariel* continued on her Isthmus sailings until the summer of 1864. She and the *Champion,* the smallest and least satisfactory steamships, were then relegated to the New York-New Orleans route; the *Ariel*'s last arrival from Aspinwall was on June 26, 1864. After four New Orleans voyages she was chartered, on December 31, 1864, by the Army's Quartermaster Corps for further transport work.

When the *North Star,* on the Aspinwall run, suffered a shaft failure early in 1865, the *Ariel* was called back for two trips in April and May. Then, through July, she brought Federal troops home from a number of ports, including Alexandria, Savannah, and New Orleans. After that came her final trip to Aspinwall in August of 1865. The *Ariel* completed over forty Aspinwall voyages for Vanderbilt, in addition to eighteen transatlantic ones. One suspects that her small size and low operating cost endeared her to him, if not to the mass of passengers packed aboard her.

In September of 1865 the Pacific Mail Steamship Company, after a new set of negotiations, agreed to take over the entire mail service from New York to San Francisco and to purchase most of the steamships of the Atlantic Mail Steamship Company, the latest name for the Vanderbilt Line. The latter would be paid $2,000,000 in Pacific Mail stock and would be allowed to retain its New York-via-New Orleans-to-Aspinwall branch.[72] Although the *Ariel* was included in the ship transfer, her new owners, who took over in November of 1865, recognized her unsuitability and made no plans to use her.

A new and successful line, the New York Mail Steamship Company, had set up a service to New Orleans once Federal forces had secured that major port. Starting in 1863, it ran from New York via Havana, and by 1865 had sailings twice a week, often with chartered vessels. The Pacific Mail supplied steamers on several occasions, one being the *Ariel* for a single trip from New York on December 20, 1865, and from New Orleans January 6, 1866. After that she lay idle for two and one-half years.

Ruger Brothers, whose operations will be discussed in detail in a later chapter, sent out a series of chartered steamers during the spring and summer seasons of 1868 through 1870, mainly from New York to Bremen. The *Ariel* was reactivated for two of their voyages in 1868. On the first she left New York, under Captain Jones, who had commanded her at the time of her capture by the *Alabama*. She sailed June 13, 1868, and reached New York on July 30, nineteen days from Bremen. Her second return brought in 650 passengers on September 23. Next year she was the first Ruger steamer out, leaving New York on March 13, 1869, for Copenhagen, via Cowes and Bremen. She sailed from Copenhagen on April 16, bringing in 750 passengers to New York on May 6, the date marking the end of her Atlantic career. If we add these three voyages to the ones for Vanderbilt, she completed a total of twenty-one, second only to the *Vanderbilt*, which had a record of twenty-two, the greatest number for any beam-engined steamship on the Atlantic.

But at the age of fourteen the *Ariel* was not yet ready for retirement. The Pacific Mail Steamship Company needed a small, economical steamer for its new feeder line out of the Japanese port of Yokohama, connecting with its large new transpacific steamships running from San Francisco. The *Ariel* was overhauled and sent halfway around the world, via the Cape of Good Hope, to Hong Kong. From New York on October 14, 1869, she made Saint Vincent November 2, Table Bay on the 26th, Mauritius December 14, Singapore January 5, and eventually Yokohama on February 15, 1870, after a four-month trip. At first she ran between Yokohama and Shanghai, but later was shifted to the Japanese port of Hakodate on the Island of Hokkaido.[73] Three years after her Pacific debut, she left Yokohama on October 26, 1873, northbound for Hakodate, and the next evening struck an uncharted reef. Although the passengers, crew, and mails were landed safely, the damaged hull slipped off its supporting rocks into deep water and sank until only a topmast was visible.[74] It was a sad end for a well-built steamship that, finally, during the last three years of her life, had found the proper niche for herself. She was operating coastwise in smooth water under a sound management that did not continually overload her.

The *North Star*

As for the *North Star*, she was last reported when she ran aground in October of 1859. The damage proved slight and she was soon returned to her Aspinwall run, where she completed thirty five voyages between June of 1859 and July of 1862. She, the *Ariel*, and the *Northern Light* were providing three sailings per month.

In August of 1862 she made a single voyage to New Orleans, then a mail trip to Aspinwall and back, and was chartered, at $1,200 per day, by the Quartermaster Corps, as one of the transports for General Banks's expedition to Ship Island and New Orleans.[75] The *Northern Light*, *Illinois*, and *Moses Taylor* were all part of the fleet leaving New York on December 4, 1862, and joining many more transports out of Hampton Roads on the same day. The *North Star* had aboard the 41st Massachusetts Regiment, General N. P. Banks, and his staff. Ship Island was reached on December 13 and New Orleans two days later.

Northbound, from New Orleans on January 30, the *North Star* was off the New Jersey coast the night of February 9, 1863, near the Atlantic Highlands, when she sighted another steamship almost dead ahead. Both changed course but their maneuvers resulted in the *North Star*'s cutting into the other's hull just forward of the wheels. She turned out to be the *Ella Warley*, a captured blockade runner that had started life as the Charleston-to-Havana liner *Isabel*. The impact knocked one of the *Warley*'s boilers off its foundation and escaping steam scalded everyone in the engine room. The *North Star*'s bow was badly damaged, but, though leaking badly, she lowered her boats and picked up the survivors from the stricken *Warley*. Soldiers aboard the *North Star* helped jettison cargo from the forward hold and rolled hogsheads of sugar aft to raise the bow. Holes were stuffed with mattresses and pillows. The *North Star* was kept running back and forth outside New York all night long in order that the large pumps attached to her main engine could be kept at work. Once it was daylight she proceeded to her dock where, in smooth waters, it was found that the donkey pumps could cope with the inflow and it was safe to stop the ship.[76]

After a refit, she returned to her usual Aspinwall routine with interruptions for transport work in April of 1864 and occasional periods out of service. Her last Aspinwall voyage, believed to be her 49th, began on February 3, 1865, and proceeded in uneventful fashion until she was within twenty minutes of her destination. First, one shaft snapped, then, after reaching port on the other wheel, an inspection showed the remaining shaft seriously damaged and about to let go.[77] The *Costa Rica* towed her north, leaving Navy Bay on March 26 and delivering her to New York on April 5. It was a remarkably fast trip; the *Costa Rica* took along her regular passengers, treasure, and mails and treated it as a routine sailing. This was the *North Star*'s last call at Aspinwall.

New shafts were quickly fitted; she was under charter, at $641 per day, from May 28, 1865, through July 23 as a transport. From the Army she went to the New York Mail Steamship, which company used her for four trips to New Orleans. She sailed at four-week intervals, starting from New York on July 29, 1865. October 21 saw the start of her fourth voyage but off Cape Hatteras she ran into a hurricane that swept away one stack, smashed both paddle boxes, and broke the steering chains. In uncontrollable condition she fell into the trough of the sea, rolled violently, and started leaking so badly that her pumps could not keep the water down. On board were four hundred passengers and three companies of troops. Some manned hand pumps while others broke out the cargo and threw it overboard. The firemen had water up to their knees as they threw lumps of coal into the furnaces by hand to maintain some steam and keep the pumps running! On October 25 the level was within four inches of drowning out the furnaces, but the wind died down on the 26th and she reached Norfolk on October 27, 1865.[78] The *North Star* was just one of many vessels caught by the storm; the beam-engined *Republic*, formerly the *Tennessee*, foundered off Savannah and a number of sailing craft were dismasted.

Once she was patched up enough to reach her home port, the Pacific Mail decided that the *North Star* was not worth keeping.[79] She was sold to the Novelty Iron Works, which removed her machinery and, in June of 1866, offered the hull for sale.[81] It was broken up at New London, Connecticut.[82]

During twelve years of active service the *North Star* had made at least sixty voyages to Aspinwall under various auspices. She had cruised extensively in European waters that included the Baltic, the North Sea, the English Channel, the Mediterranean, and the Bosporus. In 1855 she made the first sailing of the Vanderbilt Line to Havre and followed it with fourteen others. Counting the yachting trip, she completed sixteen transatlantic voyages. In addition to these were five to New Orleans in commercial service and others as an army transport. She was a moderate-draft steamer with a particularly pleasing profile, whose design was expanded a few years later to create the *Vanderbilt*.

The *Northern Light*

After her two voyages on the European route in 1858, the *Northern Light* lay idle for five months. But on March 30, 1859, she left New York for Aspinwall on the first of a long series of monthly sailings that kept her at work until October of 1862. Her performance was truly remarkable. In forty-three months she made forty-two voyages; only March of 1860 did not see her at sea. The fact that all dry dockings and repairs could be made without interrupting her schedule testifies to the superb reliability of her Allaire beam engines and return-flue boilers. The *North Star*, on the same route, had missed three monthly sailings out of 36, a more normal state of affairs.

In November of 1862 the *Northern Light* joined the *North Star* as a transport on General Banks's expedition to Ship Island and New Orleans. She left New York December 3 or 4 and departed from New Orleans on January 17, 1863. She resumed her monthly Aspinwall sailings on February 11, 1863, but after her July voyage was out of service for eight months.

She returned to sea as a transport, chartered for $800 per day, in April of 1864. A month later she was back on the Isthmus mail run, starting May 23. Monthly sailings through August were completed and then came a long period as a transport. Her last military voyage ended July 7, 1865, when she brought troops from Port Royal to New York.

Her career, as we have seen, practically parallels that of her almost-sister, the *North Star*. In August of 1865 both were chartered to the New York Mail Steamship Company to convey passengers and freight to New Orleans. The *Northern Light* made three round trips, her first starting on August 9.

Homeward from New Orleans on October 18, she ran into the hurricane that ended the *North Star*'s career. Gale winds rose on the 23rd and increased steadily. Although her paddle boxes were wrecked and the rudder carried away, she was, somehow, able to make her way to the naval base at Port Royal, South Carolina. There she was not welcomed; governmental red tape prevented repairs from starting immediately.[83] After the essentials were completed, she left October 28 and reached New York on the 31st. Her arrival coincided with the end of Vanderbilt's and the start of the Pacific Mail's management of the Atlantic portion of the California mail service.

Unlike the *North Star*, she was taken in hand, refitted, and repaired by her new owner. It used her just twice—one voyage to Aspinwall in June and a second in August of 1866—then laid her up.

The New York and Bremen Steamship Company had a single year of existence, 1867, at a time it could charter old wooden paddlers cheaply, send them to Bremen, and fill them with steerage passengers anxious to reach the reportedly golden shores of America. The *Northern Light* was chartered to sail on May 18, 1867, for Bremen via Southampton. She had a large passenger list, 944, but no cargo when she reached New York on June 27.[84] Her second voyage was less crowded; she arrived home on August 29, 19 days from Bremen and 16 from Cowes, with 410 passengers.

In 1868 she returned to the Atlantic with one voyage to Bremen for Ruger Brothers from New York on August 22, docking there October 8 after a 19-day return from Bremen. On March 27, 1869, she headed for Copenhagen with way stops at Cowes and Bremen, also for the Rugers, sailing from Copenhagen on May 1 and arriving at New York on May 18, 1869. This was her sixth and last transatlantic round trip.[85]

Late in 1870 the Pacific Mail sold her to Henry F. Hammill for $25,000.[86] The *Northern Light*, under charter to Livingston, Fox and Company, undertook three coastwise trips. On October 1, 1870, she left New York for New Orleans. Coming home, despite her age, she made a creditable seven-day passage from New Orleans on the 27th. In November she traveled to Savannah for a load of cotton and, northbound, stopped at Norfolk for coal. On December 17 she made her final voyage from New York to Havana and New Orleans; it was completed on January 19, 1871.

Although not broken up until March of 1872,[87] her effective career ended when she was nineteen years old. During that time she had made six Atlantic voyages, two for Vanderbilt, two for the New York and Bremen Steamship Company, and two for Ruger Brothers. All except the last were to Bremen. At least 52 round trips to Aspinwall were completed for Vanderbilt plus three more for the U.S. Mail Line and two for the Pacific Mail. The author has not attempted to keep track of her early Nicaraguan voyages under Vanderbilt, Morgan, and Garrison, but estimates that at least forty were completed. New Orleans saw her arrive and depart three times for the New York Mail Steamship Company and twice for Livingston, Fox, and Company. To this must be added a voyage to Savannah for the latter and many trips to many ports during thirteen months as a Civil War transport. She was a busy ship and a staunch one; her two beam engines ran for long periods without overhaul or accident. She was luckier than her semisister, the *North Star*, for she survived the hurricane that badly damaged the latter.

The *Ocean Queen*

The *Ocean Queen*, after the *Vanderbilt* the largest of the fleet, had been tied up from December of 1859 until October of 1861. An exorbitant charter rate of $2,000 per day[88] brought her out of lay-up and put her to work, along with the *Ariel*, and *Illinois*, and the *Vanderbilt*, transporting troops to capture Port Royal, in October of 1861.[89]

May of 1862 found her acting as a hospital ship for a voluntary organization known as the Sanitary Commission, which provided doctors, nurses, supplies, and hospitals to augment the primitive medical arrangements established by the Army.[90] The *Ocean Queen*, on loan, made a single trip from Yorktown, Virginia, to New York laden with sick and wounded soldiers. The Quartermaster Corps decided that she was too costly for such nonmilitary duty and put her to work transporting healthy troops instead.

When returned to Vanderbilt, she was at long last assigned to the Panama route, starting on October 11, 1862. She became the last of is European liners to carry the California mails; as the largest, she should have been the first. Once there, however, she continued for many years. She made the final voyage for the Vanderbilt organization, sailing from New York on October 16, 1865 and returning there on November 10. The Pacific Mail Steamship Company withdrew her for a year, and presumably remodeled her quarters to bring them up to their much higher standards. She was the only Vanderbilt steamship big enough to warrant continued use. In fact, she proved to be a most satisfactory running mate for the *Henry Chauncey* and *Arizona*, both especially constructed for Aspinwall service and, like her powered by single walking beam engines.

On October 1, 1866, the *Ocean Queen* steamed out of New York for the Pacific Mail. On the 8th she sighted the disabled brig *John Hastings*, which had lost her foremast, main topmast, jibboom, and rudder in a gale. She was leaking and had her only boat damaged. Everyone was rescued except the steward, who had been killed by falling tophamper.[91] The *Ocean Queen* completed three round trips in 1866, six in 1867, eight in 1868.

After a single voyage in 1869, she was chartered to Ruger Brothers for a transatlantic voyage to Bremen by way of Cowes eastbound, and Copenhagen westbound. She left New York April 17, three weeks after the *Northern Light* and five after the *Ariel* had departed for the same ports. The westward crossing was completed on May 31. She was then laid up until chartered again by the Rugers for a longer route that included Havre; Bremen; Brouwershaven, in Holland; Swinemunde, the port for Stettin; and Christiansand, Norway. It must have been a commercial success, for 1,137 passengers were crammed into her. Outbound, she had taken from March 3, 1870, to April 1 to reach Swinemunde. Homeward, she left there on the 7th and reached New York on the 1st of May. In June of 1870 she returned to the Aspinwall route and carried on regularly until April 11, 1872, when she ran aground off San Salvador Island while headed south.[92] By throwing overboard some of her cargo, she was able to kedge off. She completed the voyage but had to be docked in New York to repair her damaged keel.

The *Ocean Queen* continued with regular sailings until the fall of 1873. Her final voyage started from New York on August 20 and ended there on September 16. That year saw the arrival of the *Colon* and *Acapulco*, two new iron screw steamships built to replace the wooden paddlers that had completely dominated the Isthmus of Panama trade for a quarter of a century. The *Ocean Queen* lingered on as a spare steamer in New York, or its environs, until 1876, when she was broken up.[93] It is believed that she was turned over to John Roach, the eminent shipbuilder of Chester, Pennsylvania, who was building iron steamships for the Pacific Mail, and that her scrap value was credited toward their cost.

The *Ocean Queen* was, when she was built, one of the largest steamships in the United States, but she saw little use prior to the Civil War. She made seven Atlantic voyages, five for Vanderbilt in 1859 and two for Ruger Brothers in 1869 and 1870. From 1862 on, however, she spent most of her time on the New York-Aspinwall route for Vanderbilt and the Pacific Mail Steamship Company. Although there were periods of inactivity, she made seventy-nine voyages in this service. As the only Vanderbilt steamship that the Pacific Mail considered worth retaining, she became a most useful addition to their fleet of magnificent side-wheelers.

The *Costa Rica* and the *New York*

The post-Civil War careers of all the Vanderbilt European liners that were put on the California mail service have been reviewed. The queen of the fleet, the *Vanderbilt*, will be considered later.

There remain, however, two other steamships, the *Costa Rica* and *New York*, each of which made an Atlantic crossing for the Pacific Mail Steamship Company. These represented the culmination of the Vanderbilt-Simonson-Allaire collaboration and were built especially for the Aspinwall service of the Atlantic Mail Steamship Company, successor to the Atlantic and Pacific Steamship Company, which officially held the California mail contracts from October of 1859 until November of 1865. As we have seen, the two companies, both Vanderbilt creations, actually provided the Atlantic portion of the service, and their steamships, *Ariel*, *North Star*, *North-*

ern Light, and *Champion*, were too small to do the work properly. Aside from the addition of the *Ocean Queen*, nothing was done to alleviate this unfortunate situation prior to the completion of these new steamships in 1864 and 1865.

The *Costa Rica* and *New York* were not identical, despite the fact that they looked almost alike and were completed within seven months of each other.

Steamship	Length	Beam	Depth	Tonnage	Date
Costa Rica	269'	38'-10"	27'-0"	1950	1864
New York	292.6'	41.7'	26.5'	2217	1865

The engines were of 81" bore by 12'-0" stroke for the *Costa Rica* and 90" x 12'-0" for the *New York*. It should be noted that a new tonnage law made all values, from 1865 on, more realistic measurements of the actual volume of the hull and houses. As a result, the numbers quoted above are not comparable.

Both were single-stacked steamships with a brigantine rig, a single-beam engine, and a light hurricane deck from stem to stern. The Vanderbilt steamship had become almost identical to a Pacific Mail one, much to the advantage of its passengers. Forward of the paddle boxes the superstructure was entirely enclosed and painted white above the bulwark line. Aft there were open promenades outboard of a centerline deck house.

For the first time a Vanderbilt steamship was given luxuries: the *New York* had two saloons aft, one above the other, each with twenty staterooms "fitted up in superb style." Adjoining the upper saloon were two bridal suites and bronze statues of "Night" and "Morning" guarded the grand staircase. The upper saloon had magenta red upholstery and the lower green plush; both had Brussels carpets. A smoking room with card tables was provided for the gentlemen. Quarters for 200 first cabin, 170 second, and 500 steerage were installed.[94]

The *Costa Rica* took almost a year to complete after her 1863 launch; she was not ready until July of 1864. On her first trip, she left New York on the 5th under Captain Alfred T. Jones but, coming back, made a very long passage, twelve days, from Aspinwall to New York on the 29th. Whatever the difficulty, it was quickly overcome; she was back in service by August 13 and made sailings once a month through August of 1865. She replaced the *Ariel*, which was shifted to the New Orleans route.

After the whole fleet was taken over by the Pacific Mail in 1865, she was chartered to the New York Mail Steamship Company, who sent her to New Orleans in late November and again at the end of December 1865. After reaching New York on January 22, 1866, she was put back on the Aspinwall route for three voyages in February, April, and May of 1866.

The *New York* was launched on June 16, 1864, just as the *Costa Rica* was nearing completion. The Civil War slowed her progress and over a year elapsed before she was advertised to sail from New York on September 1 of 1865. Her maiden voyage was brilliantly completed the 23d with a new record of 6 days 11 hours 30 minutes from Aspinwall to Sandy Hook, nine and one half hours faster than the *Baltic*'s time for the same run.[95] The average speed works out to be 12.7 knots.

The *New York*'s late completion resulted in only two voyages under the Vanderbilt flag before she was transferred to the Pacific Mail. For the latter, leaving New York on December 21, she continued to Aspinwall at monthly intervals through September of 1866. Further trips were made in December of 1866 and April of 1867.

In 1867 the Pacific Mail Steamship Company needed steamships in the Orient for the extension of their recently organized mail service from San Francisco to Japan and China. It has already been reported that the Bremen liner *Hermann* was sent out first as a stores ship, and then pressed into actual operation. Other craft from both their East and West Coast fleets would follow. Included were the two newest Vanderbilt steamships.

The *Costa Rica* left the Panama route in June of 1866 to be fitted out for Pacific waters. She was advertised for Hong Kong and would take a limited number of first-class passengers.[96] She sailed on October 11, but when 120 miles out ran into such tempestuous weather that the guards around her port wheel were damaged by high seas and the ensuing leaks forced her to return to New York on the 17th.[97] No doubt a heavy load of coal had brought the guards close to the sea. Although she was docked and repaired, nothing further was attempted until spring. She cleared again, April 1, 1867, under Captain Furber for Japan via St. Vincent, Mauritius, Singapore, and Hong Kong. All went well; she reached Singapore on June 11 and Nagasaki on July 12; continuing on to Yokohama, she became the first steamship available for the Pacific Mail's branch line from Yokohama to Shanghai.[98]

The *New York*, after thirteen Aspinwall voyages, arrived home on April 22, 1867, and was withdrawn to prepare her for the long trip around the Cape of Good Hope. She sailed on August 3, 1867, for Hong Kong via Capetown, St. Vincent, Mauritius, and Singapore.[99] The *New York* reached Hong Kong's magnificent harbor on October 16 and arrived at Yokohama November 3 to join the *Costa Rica*.

The Pacific Mail was lucky to have such a large, economical steamship available. Their *Great Republic* made a delayed arrival at Yokohama on April 7, 1868, having made the last 2,000 miles from San Francisco on one wheel with a broken shaft.[100] The *New York* not only

Costa Rica on the launching ways. She was christened *Commodore* but renamed before completion. *Courtesy of Peabody Museum of Salem.*

replaced her to Hong Kong and back, but also carried on in her place for a round trip to San Francisco. Further vessels were sent out in later years, the *Ariel* from the East Coast, and the *Golden Age* and *Oregonian* from the West, as trade between Japan and China increased. In 1871 there were four sailings a month between Yokohama and Shanghai.

Both private and public Japanese agencies became increasingly alarmed by such a foreign operation in their waters; eventually the four remaining steamships, the *Ariel* having been lost in 1873, were taken over between 1874 and 1876, along with piers and shore installations, and the branch line became a part of the Mitsubishi Steamship Company. The *New York* had been sold to the Japanese Government for $250,000 in 1874, to serve as a transport to Formosa. She was renamed *Tokio Maru* and turned over to Mitsubishi the next year.[101] The other three were priced at $160,000 apiece and sold directly to Mitsubishi in December of 1875. The *Costa Rica* became the *Genkai Maru*, the *Golden Age* the *Hiroshima Maru*, and the *Oregonian* the *Nagoya Maru*. Despite their new Japanese nationality, all sailed with American or British masters and, very probably, engineers.

The *Tokio Maru* continued in active service until 1884 and the *Genkai Maru* to 1885.[102] Several more years passed before they were broken up. Like their Simonson-built predecessors, the *Costa Rica* and *New York*, the last ocean steamships ever constructed to the order of Cornelius Vanderbilt, were noteworthy examples of the shipwright's art. They were built to last, and they survived long after their contemporaries were forgotten. Each had made one trip across the South Atlantic, by way of the Cape Verde Islands and Capetown, South Africa, on her way to the Orient.

The U.S.S. *Vanderbilt* and the *Three Brothers*

The *Vanderbilt*, the flagship of the line and the longest-lived steamship of her era, remains for consideration. At the end of her 1860 European season she had, as usual, been overhauled during the winter lay-off. She was advertised to sail for Havre on May 4, 1861,[103] but Vanderbilt, after the attack on Fort Sumter, canceled all plans for an Atlantic service. By June she had been chartered, at $2,000 per day, and was loading stores, ammunition, shells, cannon, and mules. Although supposed to sail under secret orders, the New York newspapers gave, correctly, her destination as Fort Pickens, adjacent to Pensacola, Florida.[104] A combined army and navy force had been assembled to relieve it. She took aboard a large wood scow, which would be needed to ferry men and supplies to the besieged garrison. After a successful mission, she called at Fort Jefferson, in the Gulf, and reached New York on July 25, 1861.

Her next important task came in October. Many of the ocean steamships of the United States were assembled at Hampton Roads, Virginia, as the first major expedition into Confederate territory. The *Vanderbilt*, *Atlantic*, and *Baltic*, the largest and most powerful of twenty three chartered steamships, were assigned to lead the three columns of a transport fleet.[105] An even larger number of sailing vessels were filled with troops and supplies. The *Great Republic*, the largest sailing ship afloat, went to sea on the end of the *Vanderbilt*'s tow line, as the eighty-four-vessel armada stood out to sea on October 29, 1861, headed for Port Royal, South Carolina. Flag Officer S. F. DuPont commanded the naval forces and Major General Thomas W. Sherman the 16,000 troops on board the transports. DuPont established a major Union base for the blockading fleet, which would, with time, almost close off the Confederate seaports south of the Virginia Capes.[106] A second *Vanderbilt* voyage to Port Royal was completed at New York on December 6.

In March of 1862 the country was electrified by news that the Confederate ironclad *Virginia* had destroyed two major warships, the *Congress* and the *Cumberland*, at Hampton Roads. Her sloping, armored sides proved impervious to shot from the federal fleet of unarmored vessels. The War Department telegraphed a request to Vanderbilt that he sink or destroy the *Virginia*.[107] In consequence of such a flattering approach, Vanderbilt offered the *Vanderbilt* as a ram to run down and sink the ironclad. In his enthusiasm, his intent to lend her to the United States was interpreted, by both President Lincoln and Secretary of War Stanton, as an outright gift. To this the Commodore made no objection. Stanton accepted her in a letter dated March 20, 1862.[108] Vanderbilt supplied the crew and sent her to Hampton Roads to join the *Illinois*, *Arago*, and *Ericsson*, all chartered by the Navy and already there with steam up ready to run down the slow and clumsy enemy when and if she came out of Norfolk. Instead, the *Virginia*, ex-U.S.S. *Merrimack*, was burned by her own crew on May 11, when the Confederates abandoned that city.

The War Department turned the *Vanderbilt* over to the Navy; she was sent to the Brooklyn Navy Yard to be converted into a cruiser. By November of 1862 she had two large 100- pound Parrott rifles fitted, one forward and the other at the stern.[109] Deckhouses were removed to make room for these pivot guns. Ports were cut in her sides for a dozen nine-inch smooth bores.

Although a naval officer, Charles H. Baldwin, was appointed as her master, John Germain, who had been her Chief Engineer, was retained in that most essential position. He had held the same post on the *North Star* when she took the Vanderbilts abroad. After two short cruises on the Atlantic, the *Vanderbilt* set out from New York on January 10 for Hampton Roads. She was then ordered to head for the West Indies and on to Brazil and Africa, if necessary, seeking the Confederate raider *Alabama*. Her plans were rudely changed in February, when off Saint Thomas, in the Danish Virgin Islands, she came within the sphere of Acting Rear Admiral Charles Wilkes. Because he greatly outranked Acting Lieutenant Baldwin, Wilkes took her over for his own use as a flagship. Under his orders, despite her being in Danish waters, she seized the blockade runner *Peterhoff* on February 25 as the latter left St. Thomas.[110]

The *Vanderbilt* captured another, the *Gertrude*, off Eluthera on April 16. Wilkes, who had previously caused a grave international crisis by forcibly removing Confederate officials from the British mail steamer *Trent*, was relieved from his command by the irate Navy Secretary, Gideon Welles, and the *Vanderbilt* was again ordered after the long-vanished *Alabama*.

At this point a long struggle began between the *Vanderbilt*'s engineers and her boilers. She had been sent out as the fastest American steamship afloat to pursue and capture the elusive *Alabama*. In her years as a transatlantic liner there had never been any problem in maintaining her speed. Because she had jet condensers, her boilers used sea water; this, at a pressure of 20 pounds per square inch, deposited insoluble scale on the interior tubes and heated surfaces around the boiler furnaces. At the end of each trip, however, the boilers were drained and men crawled inside to chip off the scale and keep it from insulating the metal and causing it to overheat. This kept the heat transfer rate up and the coal consumption down. In the ten days available in New York or the week at Havre, there was ample time for this absolutely essential operation. In addition, if tubes did overheat or corrode, the Allaire Works would send skilled artisans to the pier to replace them.

In naval service her engineers had to raise steam at a moment's notice to pursue any enemy sighted. A cold boiler would take hours to heat up. The *Vanderbilt* regularly cruised on two of the four boilers; the others were kept hot with banked fires. Her Chief Engineer had neither the tools nor the mechanics that had kept the *Vanderbilt*—and all the other liners, for that matter—in superb condition. After long periods at sea her boilers deteriorated, the scale built up, the tubes corroded, and eventually they became so thin that the crew did not dare hammer them to knock off the scale.

The *Vanderbilt*, once freed of Admiral Wilkes, headed for Brazil and, at Pernambuco on July 6, 1863, hearing news of the C.S.S. *Georgia*, reversed course for Trinidad to search for her in vain. At Rio de Janeiro, from July 14 to August 2, work was done on the boilers; plates below the furnaces had become badly corroded.[111] But by this time all tube scaling had to be stopped. As soon as 1,200 tons of coal were loaded and the boilers patched, she crossed the South Atlantic to the island of St. Helena, where a coal price of $31.50 per ton led Captain Baldwin to put to sea again. After leaving on August 17, bad weather slowed the ship so much that he considered it prudent to turn back and, despite the price, buy 400 tons. Against head winds, 46 tons per day had to be burned to make 7 1/2 knots. St. Simon's Bay, near Capetown, was reached on August 30, ten days from St. Helena. The slow passage was disastrous. The C.S.S. *Georgia* had left that same bay the previous night. The *Vanderbilt* steamed for Cape Agulhas in pursuit but sighted a distressed Dutch bark, the *Johanna Elizabeth*, partially dismasted and with a damaged rudder. Captain Baldwin decided that mercy for her passengers took precedence over war and towed her a hundred miles back to the safety of St. Simon's Bay. There his coal was low and his boilers had to be shut down for further work. The forward stack was so badly rusted that it could not withstand a gale; the aft one was little better.[112] Eight days were spent repairing and loading 904 tons of coal. On September 11 she steamed away for Mauritius, off the East Coast of Africa, a likely port for the *Georgia* to choose for docking and bottom cleaning.

Once again the *Vanderbilt* was at the wrong place; the *Alabama* arrived at Capetown on September 16. During the former's useless search one boiler had had to be emptied and patched up under way. Further work was needed at Port Louis, Mauritius. There, a crack was discovered in the starboard paddle shaft about two feet inboard of the ship's side. The boiler work took two weeks. Because there were no news of either the *Georgia* or the *Alabama*, the *Vanderbilt* departed on October 10 and reached Table Bay on the 22d. There the English authorities refused to supply coal, so she sailed from Capetown October 25 for Angra Pequena, where she found a British ship, the *Saxon*, loading American cargo that had been seized by the *Alabama*. The *Vanderbilt* put a prize crew on the *Saxon*, sent her and her twice-captured cargo to New York, and then took on 293 tons of coal that had been shipped out from Capetown for the *Alabama*. She turned her bow toward Brazil and reached Bahia November 19, via St. Helena on the 10th. On December 21, 1863, she was back at St. Thomas, where further boiler work idled her until New Year's Day.[113]

On January 10, while chasing a blockade runner, the front of one boiler gave way, even though a reduced steam pressure of 14 psi was being carried. It is difficult to understand why no one was killed. Baldwin continued on three boilers, all leaking. Several tubes had given way; their ends had to be plugged in order to keep on going. The shaft crack, after a long period without growth, was also opening up. It was more than time for the *Vanderbilt*'s cruise to end; a week later she was at New York.

Extensive repairs, at a cost of $200,000, kept her out of

service until September of 1864.[114]

She left New York to patrol off Nova Scotia. Summer had brought a yellow fever epidemic to Bermuda and Nassau; as a result, healthier Halifax had become a temporary center for the transshipment of goods to the blockade runners bound for Southern ports. When the *Vanderbilt* returned to Boston a month later, she was ordered to Hampton Roads to join the fleet being assembled under Admiral David D. Porter. By November 1 she was off Wilmington, North Carolina, temporarily a part of the blockading fleet, awaiting the planned attack on Fort Fisher, which guarded the approaches to that city.

On December 24 and 25 she steamed into a position about a mile from the fort. Ahead of her was a long line of wooden warships with broadside batteries. Astern and closer in were ironclads and monitors. She fired about 800 shells and received no hits. After the first attack failed, a second was made, with the *Vanderbilt* assigned practically the same position with three handsome but archaic steam figures, the *Wabash, Minnesota,* and *Colorado* nearby.[115] During the actual attack, from January 13 to 15, the *Vanderbilt* performed a variety of tasks. On the 13th her boats ferried troops ashore from the transports. She fired a few rounds on the 14th and 15th but was kept busy delivering shell to other ships. Finally, she sent out a large landing party, made up of her own men, to help storm the fort. Once it surrendered, during the night of the 15th, Captain Pickering, who had replaced Baldwin, was called to the Admiral's flagship, the *Malvern,* and given dispatches to take to Fortress Monroe announcing the great victory. About half of the *Vanderbilt*'s crew were still ashore and could not be rounded up until dawn. Porter was furious that his fastest ship had not departed instantly, and even more furious when an Army transport, the *Atlantic,* reached Hampton Roads ahead of the *Vanderbilt* and sent unofficial reports ahead by telegraph. It seems poetic justice that a Collins liner beat the very steamship that had been designed to outdo her. The *Vanderbilt* went to New York in January of 1865 for docking and replacing her copper.

As the Civil War was ending, she was assigned to routine chores. In May of 1865 she went to Charleston to tow the captured Confederate ironclad *Columbia* to Norfolk. The pair arrived off Cape Henry on May 25, 1865. Having delivered the *Columbia,* she picked up a Federal monitor, the *Onandaga,* on May 29, and towed her to New York; they arrived safely on the 31st.[116] Briefly she was tied up as a receiving ship at the Portsmouth Navy Yard and then, in August, sent to Philadelphia to prepare for a long trip to the West Coast.

Secretary Welles, having decided that a modern warship should be stationed at San Francisco, selected a double-turreted, twin-screw monitor, the *Monadnock*. A special squadron, under Commodore John Rogers, was organized to accompany her, with the *Vanderbilt* as its flagship. The side-wheel frigate *Powhatan* and the screw sloop *Tuscarora* were added to make a fleet of four steamships widely varying in type, size, and speed. These set out from Hampton Roads on November 2, 1865. Rather to the surprise of the naval world, the *Monadnock* managed to steam all the way and actually survived several storms.[117] It might be noted that she was not an Ericsson design. The original *Monitor* had foundered in a storm.

A considerable delay developed at Valparaiso, Chile, where Rogers's fleet found a state of war existing between Spain and three South American countries, Chile, Peru, and Equador. Rogers interrupted his voyage, from March 1 until April 17 1866, while trying to bring about a settlement between the warring parties and to prevent the Spanish fleet, under Admiral Mendez Nunez, from bombarding Valparaiso and Callao. In both cases the Americans had to put to sea to get out of the way of a Spanish attack. The *Powhatan* and *Tuscarora* joined the Pacific Squadron and the two remaining vessels reached San Francisco June 22. The voyage was a long, slow one with many stops for coal. The monitor, at her best, could do no better than six knots.

Special alterations were made at the Mare Island Navy Yard to prepare the *Vanderbilt* for a diplomatic mission. Queen Emma of Hawaii had been touring the United States and needed a suitable means of returning home. What could be better than the finest steamship under the American flag? On September 5, 1866, the broad pennant of Admiral H. K. Thatcher, Commander of the North Pacific Squadron, was hoisted. Joseph P. Sanford, Commander, U.S.N., who had brought her west, remained as her master. Queen Emma came aboard on the 26th on an incognito visit to inspect her quarters. She arrived, officially, October 12, together with her large retinue, the Admiral's wife, and her daughter. Next day the *Vanderbilt* steamed out to the Pacific and reached Honolulu nine days later. Homeward she entered the Golden Gate December 3, bringing with her E. M. McCook, American Minister to the Hawaiian Isles, and his wife. This was her last voyage under steam. Although nominally the flagship of the North Pacific Squadron, she swung at anchor until March 13, 1867, when that duty ended.[118] She was much too expensive a ship for the economy-minded years following the Civil War.

On March 27, 1873, six years later, the *Vanderbilt* was sold for $42,000 in gold to George Howes, who represented himself and his brothers. They renamed her, appropriately, *Three Brothers,* and found they had a real bargain. The engines were sold for scrap at $20,000; $10,000 worth of pig iron ballast was discovered aboard and, when other salable items were disposed of, their net cost was $7,000.[119] Not only that, but the seventeen-year-old hull was in remarkable condition.

Coombs and Taylor, at Hunters Point, under the supervision of Captain Robert W. Waterman, converted her into a three-masted sailing ship. The upper works were rebuilt so that the plumb stem and low forward flare vanished. Her sheer was increased and a new upper deck installed. Once the work had been completed, at a cost of $175,000—again in gold and not inflated paper dollars—nothing above the waterline showed that she was anything but a sailing ship. From there down, however, it was Jeremiah Simonson's hull that was to make her just as fast a windjammer as she had been a steamer. She was now the largest American sailing ship afloat. The mainmast was a single stick 98'-6" long and 44" in diameter, while both fore and main yards were 100' by 20" at the slings.[120]

Her second career began on October 24, 1873, when two steam tugs, the *Neptune* and the *Rescue*, towed her out to sea.[121] These carried visiting dignitaries and a band for the occasion, and were joined by an excursion steamer, the *Princess*, loaded with well-wishers. Eight miles outside the Golden Gate the *Three Brothers* proceeded under an eminent master, George Cumming, formerly of the clipper *Young America*. The wind was fair and even skysails were set. Her 112-day passage around Cape Horn to Havre was an excellent one and her best day's run, 314 miles, gave her a speed of 13 knots under sail. She carried the largest wheat cargo ever loaded in California. From Havre she sailed to New York and there took on an equally large general cargo, 4,800 measurement tons, but head winds and calms resulted in a long, 132-day passage to San Francisco. She arrived on November 8, 1874.[122]

Five more trips were made, usually to Liverpool in the grain trade, the best being her fourth under sail, 103 days out in the spring of 1877 and 112 days home with 3,473 tons of coal delivered to San Francisco on January 8, 1878. After reaching Liverpool in April of 1880 she was sold for $48,000 to Mr. John Williams and transferred to the British flag.[123] She continued in the San Francisco trade, but from Antwerp, until 1885, when she was sold to the Anchor Line and sent to Gibraltar as a hulk for coaling steamships. On her last trip she sailed from Cardiff December 23, 1885, loaded down with coal, and reached Gibraltar in fourteen days.[124]

The *Three Brothers'* third career as an anchored coaling depot lasted 43 years, until 1929 when she was sold for £450 to Spanish shipbreakers.[125] A useful life of seventy three years for a wood hull over 300 feet long is unprecedented. During that time she spent the first year of her life unused, then made twenty-two Atlantic voyages as the crack steamship of the Vanderbilt Line. Without her that organization would never have prospered. After the outbreak of the Civil War she was first a transport, then a cruiser ranging from the West Indies to Brazil, the Cape of Good Hope, Mauritius, and even off Madagascar, in a vain search for the Confederate cruisers *Alabama* and *Georgia*. She chased blockade runners and took part in the greatest naval bombardments of that war, the two attacks on Fort Fisher. Afterward she reached the Pacific via the Strait of Magellan, became involved with South American wars, albeit as a peacemaker rather than a belligerent, and carried a queen to Hawaii. As a steamship her life was highly varied from 1856 to 1867. As a sailing ship, from 1873 to 1885, she was unusual in terms of her enormous size, but followed a prosaic career, under both American and British registries, carrying record cargoes of grain from California to Europe and coal or general cargo back again. San Francisco cherished her under the nickname of *Big Brother*. But finally, her hull was so well built of such exceptional materials that she could continue on as a floating coal hulk at Gibraltar from 1886 to 1929. She outlasted every wooden oceangoing steamship of her era and every important commercial sailing ship built before the Civil War. Jeremiah Simonson, her builder, is practically unknown—just someone who constructed steamers for Cornelius Vanderbilt. It is time that we honored the artisan as well as the capitalist. Nevertheless, Vanderbilt, once a steamboat captain, collaborated with Simonson in their design but was wise enough not to overrule his master shipwright in technical matters.

Vanderbilt and His Steamships

The Vanderbilt Line, first to Nicaragua in 1851, then to Havre and Bremen from 1855 to 1860, and, finally, to the Isthmus of Panama from 1859 to 1865, was original in many respects. It introduced the walking beam engine as a means of powering oceangoing craft. It was the first American steamship line to attract large numbers of steerage passengers. Always aggressive, it went out of its way to seize upon the same sailing day as any weaker rival and regularly undercut fares.

Although it was a tremendous financial success carrying gold-seekers to California, its European operations were less spectacular. Other lines had better service, well-trained crews, satisfied passengers, and, since they operated under mail contracts, sailed their ships on schedule throughout the year. The Atlantic line was the least profitable of Vanderbilt's many maritime ventures. In the end, he was fortunate in transferring his activities to the Panama route and, eventually, disposing of his badly run, often overcrowded vessels to the Pacific Mail Steamship Company.

One striking fact emerges when the original Custom House Registers of the Vanderbilt vessels are inspected. In all but two cases, the *Costa Rica* and the *New York,* Cornelius Vanderbilt was the sole owner. No one else had the money to build ships as he wanted them, lay them up or sell them when he chose, and operate them in the way he desired. The Vanderbilt Line, whether to Nicaragua, Europe, or Panama, was the creation of a single individual in the best Horatio Alger tradition.

Notes

1. Wheaton J. Lane, *Commodore Vanderbilt* (New York, 1942), pp. 52-77.
2. Ibid., pp. 93-95.
3. *Journal of the Franklin Institute* 21, 3d ser. (January-June 1851): 419-20.
4. MS notebook, Launches B. & C. S. (Jeremiah Simonson's list of hulls built 1838-66), n.d. Steamship Historical Society of America.
5. James P. Baughman, *Charles Morgan* (Nashville, Tenn., 1968), p. 242.
6. *New York Herald*, December 27, 1850, and January 21, 1851.
7. The depth dimension on her register appears to be erroneous; to agree with that of her successor, the *North Star*, it should be 28'-0".
8. *New York Herald*, March 14, 1852.
9. Ibid., March 10, 1860.
10. Lane, *Commodore Vanderbilt*, pp. 102-4.
11. *Journal of the Franklin Institute* 25, 3d ser. (January-June 1853): 275.
12. *New York Herald*, April 30, 1853.
13. John Overton Choules, *The Cruise of the Steam Yacht NORTH STAR* (Boston and New York, 1854), p. 25.
14. Ibid., p. 31.
15. *New York Herald*, June 13 and 16, 1853.
16. Ibid., September 23, 1853.
17. Choules, *Cruise*, pp. 348-51.
18. John Haskell Kemble, *The Panama Route, 1848-1869* (Berkeley and Los Angeles, Calif., 1943), pp. 69-70.
19. *Journal of the Franklin Institute*, 29, 3d ser. (January-June 1855): 285.
20. *Monthly Nautical Magazine*, (August 1855), pp. 417-21.
21. Kemble, *Panama Route*, p. 215.
22. *New York Herald*, March 27 and April 23, 1855.
23. *New York Evening Post*, February 1, 1855.
24. Lane, *Commodore Vanderbilt*, p. 144.
25. *New York Herald*, April 23 and 24, 1855.
26. Ibid., May 20 and June 23, 1855.
27. *New York Commercial Advertiser*, May 10 and June 25, 1856.
28. Ibid., July 22 and August 14, 1856.
29. *U.S. Nautical Magazine and Naval Journal*, (June 1857), pp. 180-82 and folding plate.
30. Official Records, ser. 2, 1: 230.
31. *New York Tribune*, December 11, 1855.
32. Lane, *Commodore Vanderbilt*, p. 152.
33. *New York Herald*, July 21, 1856.
34. *Journal of the Franklin Institute*, 31, 3d ser. (January-June 1856): 280.
35. *New York Tribune*, April 27, 1857.
36. *New York Herald*, July 21, 1856.
37. Ibid., July 24 and 27, 1856.
38. *New York Commercial Advertiser*, April 20 and July 22, 1857.
39. *New York Tribune*, April 20 and May 6, 1857.
40. *New York Herald*, January 1, 1858.
41. Lindsay, *Merchant Shipping*, 4: 603.
42. *New York Herald*, August 24, 25, and 27, 1857.
43. Ibid., January 1, 1858.
44. *New York Tribune*, November 21, 1857.
45. *Journal of the Franklin Institute*, 25, 3d ser. (January-June 1853): 32-42.
46. *New York Commercial Advertiser*, February 2, 1858.
47. *New York Herald*, January 31 and February 1, 1858.
48. Ibid., January 1, 1858.
49. *Willmer and Smith's European Times*, (Liverpool), November 6, 1858.
50. *New York Commercial Advertiser*, December 27, 1858.
51. Ernest A. Wiltsee, *Gold Rush Steamers* (San Francisco, Calif., 1938), pp. 186-94.
52. *New York Commercial Advertiser*, December 26, 1855.
53. Lane, *Commodore Vanderbilt*, pp. 117-26.
54. *New York Herald*, November 16, 18, and 22, 1859.
55. *New York Times*, November 22 and December 21, 1859.
56. Ibid., February 13, 1860.
57. Kemble, *Panama Route*, pp. 85-86.
58. Lane, *Commodore Vanderbilt*, p. 133.
59. *U.S. Nautical Magazine*, (May 1857), p. 150.
60. *Journal of the Franklin Institute*, 36, 3d ser. (July-December 1858): 203.
61. *New York Tribune*, April 9, 10, and 13, 1857.
62. *New York Herald*, May 22, 1859.
63. Ibid., May 5, 1859.
64. *New York Commercial Advertiser*, December 6, 1859.
65. The Custom House Registers for many of the Vanderbilt steamships give depths that are obviously incorrect. The author has quoted more reasonable values in the Appendix. All are from contemporary sources.
66. *New York Herald*, February 28, 1860.
67. Ibid., February 11, 1861.
68. Ibid., January 2, 1860, and January 1, 1861.
69. Kemble, *Panama Route*, pp. 93-94.
70. Lane, *Commodore Vanderbilt*, pp. 171-72.
71. *New York Commercial Advertiser*, December 29, 1862.
72. Kemble, *Panama Route*, p. 100.
73. Edward Kenneth Haviland, "American Steam Navigation in China," *The American Neptune*, (1957), pp. 311-12.
74. *New York Times*, December 2, 1873.
75. *New York Herald*, December 21, 1862.
76. Ibid., February 11, 1863.
77. Ibid., March 18 and April 6, 1865.
78. Ibid., October 28, 29, and 30, 1865.
79. *New York Journal of Commerce*, November 21, 1865.
80. *New York Shipping and Commercial List*, May 1, 1866.
81. *New York Journal of Commerce*, June 22, 1866.
82. Kemble, *Panama Route*, p. 237.
83. *New York Herald*, November 1, 1865.
84. Ibid., June 28, 1867.
85. Ibid., May 19, 1869.
86. Kemble, *Panama Route*, p. 238.
87. *New York Shipping and Commercial List*, March 6, 1872.
88. *New York Commercial Advertiser*, April 8, 1862.
89. *New York Herald*, November 6, 1861.
90. (Frederick Law Olmsted), *Hospital Transports*, (Boston, 1863), pp. 29-42.
91. *New York Commercial Advertiser*, October 22, 1866.
92. *New York Journal of Commerce*, May 4, 1872.
93. *New York Maritime Register*, 1874-1876; Kemble, *Panama Route*, p. 239.
94. *New York Herald*, August 31, 1865.
95. Ibid., September 24, 1865.
96. *New York Journal of Commerce*, September 7, 1866.
97. *New York Shipping and Commercial List*, October 24, 1866.
98. Haviland, "American Steam Navigation," pp. 311-13.
99. Ibid., pp. 311 and 313.
100. Kemble, *American Neptune*, (1942), p. 35.
101. *Golden Jubilee History of Nippon Yusen Kaisha*, (1935), p. 5.
102. Haviland, "American Steam Navigation," p. 314.
103. *New York Herald*, February 11, 1861.
104. Ibid., June 11, 1861.
105. Ibid., November 6, 1861.
106. Official Records, ser. 1, 12: 225, 259, 260, 261.
107. Lane, *Commodore Vanderbilt*, p. 176.
108. Official Records, ser. 1, 7: 148-49.
109. Ibid., ser. 2, 1: 230.
110. Ibid., ser. 1, 2: 60, 97-104, 159.
111. Ibid., 2: 407-9.
112. Ibid., 2: 426-27, 445-47, 466-69, 475-77, 480-82, 573.
113. Ibid., 2: 504, 576, 577, 586-88.
114. *New York Herald*, December 12, 1864.
115. Official Records, ser. 1, 11: 245, 308-9, 425, 518-22, 613.
116. Ibid., 12: 137, 141, 152.

117. Robert Erwin Johnson, *Rear Admiral John Rogers, 1812-1882* (Annapolis, 1967), pp.282-97.
118. Typescript summary of dates from *Vanderbilt* logs, no author, n.d. Navy Department, Ship History Division.
119. *New York Journal of Commerce,* April 28 and May 10, 1873.
120. F. C. Mathews, "The Clipper Ship THREE BROTHERS," *Pacific Marine Review,* (October 1921), pp. 581-83.
121. *New York Times,* October 25, 1873.
122. Ibid., November 17, 1874.
123. *New York Maritime Register,* October 27, 1880.
124. Ibid., January 6 and February 3, 1886.
125. Basil Lubbock, *The Down Easters,* (Boston, 1929), p. 84. Most of Lubbock's account is taken directly from Mathews.

14

The North Atlantic Steamship Company, 1859-1860

ON September 30, 1859, both California mail contracts expired. The United States Mail Steamship Company, which had served New York, New Orleans, Havana, and Aspinwall, was content to wind up its affairs. The Pacific Mail Steamship Company, which ran from Panama to San Francisco, was a far stronger company, better organized, better run, and highly profitable. It decided to set up an Atlantic service and hoped to receive a new contract to carry the mails all the way from New York to San Francisco. Cornelius Vanderbilt plunged into the void, as a new, vigorous contender seeking the same prize. He had anticipated the event and had already placed three of his Atlantic liners, the *North Star, Northern Light,* and *Ariel,* on the New York to Aspinwall route.

The Pacific Mail joined forces with the Panama Railroad to create an Atlantic service under the name of North Atlantic Steamship Company. The former contributed $400,000 and the latter $500,000 to its capital. For around $780,000 they purchased, on July 9, 1859, the three remaining Collins liners, *Atlantic, Baltic,* and *Adriatic,* all owned by Brown Brothers, who had bought them at the auction held in April 1858.[1] For the moment the *Adriatic* was neglected, but the *Atlantic* and *Baltic* were refitted for the Isthmus trade. The lower staterooms were removed in order to create steerage quarters and new, free-standing berths were installed.[2] Forward of the paddle boxes, the deckhouses and hurricane deck were extended out to the sides, thereby providing new first-and second-class quarters. Two sailings a month, on the 5th and 20th, were to start in October. These had been the United States Mail's regular departure dates. Vanderbilt adopted them too.

On October 5, 1859, the *Baltic,* under Captain Alfred G. Gray, former master of the U.S. Mail's *Star of the West,* departed from New York with 900 passengers, 1,200 tons of coal for the round trip, and 250 tons of freight. Missing were the United States mails, which at the last moment had been given under a provisional contract to Vanderbilt. They traveled on his *Northern Light,* a smaller and slower steamer than the *Baltic.* There was a great deal of last-minute fare cutting, and both lines had runners out waylaying prospective passengers with promises of a $50 steerage ticket all the way from New York to California.[3]

The *Atlantic* and the *North Star* sailed October 20.

When the former left Aspinwall, homeward bound on November 2, carrying $1,568,000 in treasure, the latter had not even arrived.[4]

The North Atlantic's two splendid steamships were an immediate success, as far as their passengers were concerned. New speed records were made between New York and Aspinwall, the *Baltic* cutting the time, southbound, to 6 days 22 hours[5] and the *Atlantic*, northbound, to 7 days 1 hour.[6] The North Atlantic steamers regularly carried many more passengers and far more treasure, but always at a high cost; the Collins liners had never been built with economy in mind. Vanderbilt's steamships were smaller, slower, cheaper to run; moreover, his enormous financial resources made him a formidable rival. The Pacific Mail, over the years, had many chastening clashes with Vanderbilt's fare cutting. It chose to negotiate rather than continue operating while its deficits mounted. The outcome, discussed in the preceding chapter, was that Vanderbilt acquired the coveted five-year mail contract but came to an agreement with the Pacific Mail whereby it provided the West Coast service and he the East Coast one. Consequently, the North Atlantic Steamship ended its Aspinwall operations in March of 1860 after eleven voyages. The *Atlantic* made five and the *Baltic* six, the latter leaving New York March 5 on the line's last trip. She arrived home on March 25, bringing in 420 passengers and $895,000 specie.[7]

Once the older steamships had been modified and put to work, the *Adriatic* was considered. Her new owners were convinced that her engines' much-modified valve gear was still faulty, even though it had served for her single Atlantic voyage in 1857. A committee of eminent marine engineers, H. B. Renwick, Miers Coryell of the Morgan Iron Works, and W. C. Everett of the Novelty Works, advised a completely new design, which was built and installed by the Novelty Iron Works under Everett's supervision; it proved a complete success.[8] The original paddle-shaft bearings were too small, having suffered damage on her maiden voyage. Larger replacements were made. On trials, February 21, 1860, the *Adriatic* made 6 knots at 10 rpm, 14 1/2 at 14, and reached 16 1/2 at 17 in rough weather with a gale blowing. Captain Joseph J. Comstock, the Collins Line's master of the *Baltic*, was put in charge and brought with him a Mr. Robinson, his former Chief Engineer.[9]

The *Adriatic* was to operate as a one-ship line to Havre sailing at about six-week intervals. The Havre Line was making monthly departures and the Vanderbilt European Line provided irregular sailings at intervals ranging from two to four weeks. Because Vanderbilt now held considerable Pacific Mail stock and the Pacific Mail was actually operating the North Atlantic service, the two collaborated in scheduling the *Adriatic*, *Vanderbilt*, and *Illinois* so that there were no conflicts of sailing day and no possibility of a race between the two largest and fastest steamships under the American flag.

When she left New York at noon on April 14, 1860, the *Adriatic* carried 197 passengers and $605,000 in specie. After dropping the pilot at 3:00 P.M., she steamed slowly along with the wheels turning at 11 rpm in order to run in her new journal bearings. The revolutions were gradually increased to 15 over the next several days. On April 20 and 22 there were stops of about thirty minutes each to make adjustments. The Isle of Wight was reached at 11 P.M. on the 24th; she docked at Southampton the morning of the 25th. After the first day she averaged better than 300 miles per day at about 12.7 knots, her best run being 335 at 14.2 if we take the day as twenty three and one half hours long.[10] The time of 10 days 6 hours to Cowes was a respectable crossing time but a full day longer than the *Vanderbilt*'s 1857 record passages.

Coming home her bearings no longer overheated; she was able to make a sea distance of 3,067 miles in 9 days 17 hours 30 minutes for an average speed of 13.14 knots. The coal consumption was large, 110 tons per day, and the engines operated at 21 psi steam pressure and a 25-inch vacuum in the condenser while the wheels turned 12.6 rpm. The highest pressure was 24 psi, the corresponding revolutions 15.5, and the best day's run 360 miles on May 18.[11] She had started from Havre on May 8, sailed from Cowes on the 9th with ninety passengers, and reached Sandy Hook lightship on May 19 only to be delayed three hours by dense fog. After this relatively slow start, the four subsequent voyages were much faster. All eastbound crossings, as well as two westbound, were completed in less than ten days.

Her best was her second trip, sailing from New York on June 2 and reaching Cowes in 9 days 13 hours 30 minutes at 13.6 knots. Her slowest was westbound from Cowes on September 13 in 10 days 21 hours 30 minutes at 11.9 knots.[12]

The real measure of a ship's speed is not the maximum obtained in the best of weather under the most favorable conditions but rather the average over a long-term basis, facing all wind and weather states, when the engines and boilers are not just out of the machinists' hands, and when the ship's bottom is no longer smooth and new. The year 1860 provided the single occasion when three magnificent side-wheelers, the *Vanderbilt*, *Adriatic*, and *Persia*, were operating simultaneously. The last, an exact contemporary, was built in 1856 as the first iron mail steamer of the Cunard Line fleet. Reliable passage times are available for all three. The summaries in the *New York Herald*'s January 1, 1861, issue have been checked against detailed arrival and departures in the marine columns of the *Commercial Advertiser*, *Tribune*, and

Herald itself. These show minor differences in arrival or departure times but support the total passage length. The *Persia*'s times are also given in Lindsay, vol. 4, appendix 9, and check to the nearest hour in every case. The *Herald* values, which by this time include the necessary five-hour time correction, were therefore accepted. An average of 3,120 miles from New York to Cowes was used for the *Adriatic* and *Vanderbilt* and 3,080 miles to Liverpool for the *Persia*.

The *Vanderbilt* made six voyages and the *Adriatic* five between March and November. Within exactly the same period, the *Persia* completed five. In the latter case, Queenstown, renamed Cobh after Irish independence, had been added as a stop on her westbound voyages, but because no records for the delays there are available, only the eastbound passages are considered. The average speeds work out to be:

Steamship	No. of Passages	Speed in Knots Eastbound	Westbound
Persia	5	13.40	
Adriatic	5	13.27	12.85
Vanderbilt	6	12.55	12.43

These averages are conclusive evidence that the *Persia* was the fleetest, but only slightly faster than the *Adriatic*, while the *Vanderbilt*, despite the fact that she had averaged 13.75 knots in August of 1857, was not in the same category as the others.

Without doubt the *Adriatic* was, at long last, a superb steamship in every respect. Best of all, she was run in high style by the Pacific Mail Steamship Company, the largest single operator under the American flag. Passenger lists, after her first trip, were extremely large, especially so if it is remembered that there was no steerage space. On the *Adriatic*'s July sailing from New York there were 315 aboard, and on her last trip from Cowes, leaving October 26, she bore 312 passengers and 420 tons of freight.

The arrival at New York on November 5, 1860, marks the end of the *Adriatic*'s all too brief career as an American liner. For the final voyage of the year, the North Atlantic Steamship Company substituted the older *Atlantic*. She departed from New York on November 17 and returned home again with a small number of passengers, sixty-three, braving so late a crossing of the North Atlantic. There was an excellent cargo of 800 tons that included watches, jewelry, and $850,000 in coin.

Following the Vanderbilt Line's example, sailings were suspended for the winter months. In 1861 a two-steamer service was planned, starting with the *Adriatic* on April 6, and the *Baltic* a fortnight later.[13] National and international developments intervened, however, and all the line's steamships were diverted to other uses.

The Atlantic Royal Mail Steam Navigation Company, better known as the Galway Line, had a checkered history, beginning with the *Indian Empire*, ex-*United States*, in 1858. A year later it received a contract from the British postal authorities to carry the mails from Galway, Ireland, via St. John's, Newfoundland, to Boston and New York, with fortnightly sailings. Four iron paddle steamships ordered for the service proved both structurally weak and deficient in speed; worst of all, they were not completed by mid-1860, when the mail service was to begin.[14] Various steamships were chartered as a stopgap measure, and in February 1861 the *Adriatic* was bought for $437,500. She was manned by an American crew for her delivery voyage, originally scheduled for March 9, 1861, and put under the command of Jefferson Maury, her former First Officer.[15] At about that time, the Cunard liner *Australasian* was overdue at New York, having turned back after breaking off two propeller blades shortly after leaving Liverpool. The *Adriatic*'s departure was rescheduled for March 13 in order to take the *Australasian*'s passengers and mails. These were put ashore at Queenstown on the 23d, after a fast 9 day 4 hour 30 minute crossing; the *Adriatic* continued on to Cowes and Havre.[16]

Back at Southampton, still operated by her American complement, the *Adriatic* ran her official trials on the 27th at a very light draft, 17'-2" forward, 18'-10" aft, and with only 113 tons of coal aboard.[17] The steam pressure was 25 psi and, with her paddles turning 17 to 18 rpm, she averaged 15.91 knots on four runs over a measured mile. This made her the only steamship of the Galway Line fast enough to meet their mail contract's requirements. She was docked and coppered, had extensive steerage quarters installed, and left Southampton for Galway on April 17, this time with a British crew; the Americans were still aboard and would return home as passengers. A marked change in appearance resulted from the removal of the small but conspicuous forward pilot house and changes in the funnel heights. At one time they were reduced, at another increased in comparison with the original values.

On her first British flag voyage, carrying 572 passengers with only 16 in first class, she left Galway on April 23, 1861, reached Saint John's on the 29th, and New York on May 2. The northern route had serious drawbacks; she was delayed twenty-nine hours by ice and fog off St. John's.[18] Despite the delay, the total time of 9 days 12 hours was extremely good. The line's new paddler, the *Columbia*, crossing earlier in the month, had taken eighteen and one half days, Galway to Boston. Eastbound, the *Adriatic* had a remarkable passenger list for that direction, 463. Then as now, there must have been a great many loyal Irish wanting to visit their homeland. As many as 337 of them took advantage of a bargain fare of $30 from New York to any town on a railroad in Ireland.

Her second voyage was extremely fast, 8 days 22 hours to New York and two hours less returning.[19] Thus, the *Adriatic* was the first liner to make Atlantic crossings in less than nine days. To go from Galway on June 5 to St. John's had taken 5 days 12 hours.[20]

All service was suspended after the *Adriatic*'s second voyage; the Galway Line had failed to meet its contract requirements. Two years later, however, it was revived and a new agreement reached covering sailings from Liverpool alternately to New York and Boston, with calls at Galway and St. John's in both directions.[21]

During the suspension period, the Civil War in the United States and President Lincoln's blockade of all Southern ports had strained diplomatic relations between Washington and London. The *Trent* affair in November of 1861, when Captain Charles Wilkes, in the U.S.S. *San Jacinto*, stopped the Royal Mail Steamer *Trent* and forcibly removed two Confederate commissioners traveling on her, further aggravated matters.[22] One direct result was the British decision to send troops to Canada. The *Adriatic* was chartered as a transport. She was loaded at Southampton with 820 men of the Grenadier Guards[23] and sailed on December 20, 1861, for Halifax, Canada's chief ice-free port. She arrived December 31 and eight days later sailed for St. John, New Brunswick,[24] where the troops landed. She went back to Halifax and thence went to Sydney, at the northern tip of Cape Breton Island, where she was trapped by ice. Attempts to free her damaged the stem and wheel floats. After spending the winter frozen in, she returned to Halifax in April of 1862. Diplomatic tension having subsided, troops of the 96th Regiment were brought back to England. In May of 1862 she was at Spithead.

When the Galway Line resumed in August of 1863, an iron side-wheeler, the *Hibernia*, made the first trip to Boston and was followed, two weeks later, by the *Adriatic* from Liverpool on August 28, Galway on September 1, touching at St. John's on the 8th, and arriving at New York on the 11th. The trip was marred by a burst steam pipe her first day out,[25] but the 686 steerage and 50 cabin passengers were not seriously delayed. Two more voyages were completed with sailings from Liverpool on October 24 and December 19, 1863. On the latter she reached New York January 3, 1864. Returning, she left on the 11th and, because of damage by ice, was unable to make her St. John's call. Galway was reached on the 23d and Liverpool on January 29. Within a month the Galway Line had reached its end. It had promised much and produced nothing but debts. Of the many steamers tried, only the *Adriatic* delivered the mails on time. The city of Galway was alone in mourning the line's demise.

Thus ended the *Adriatic*'s career as an Atlantic liner. Although eight years had passed since she was launched, she had completed just twelve and one half Atlantic voyages, one for the Collins Line, five for the North Atlantic Steamship Company, one as a British transport, five as an Irish emigrant carrier for the Galway Line, and one eastward crossing, on which, although flying the American flag, she transported passengers and the English mails that had been left stranded by Cunard's *Australasian*. The last and greatest of the Collins liners suffered from two unsatisfactory owners and saw but one summer of operation in a style that befitted the largest and fastest wooden side-wheeler afloat. But even the North Atlantic Steamship Company found a single vessel of this caliber a costly proposition and was pleased to dispose of her when a buyer appeared.

For five years the *Adriatic* lay near Southampton before she was purchased by the Liverpool firm of Bates and Company.[26] The tug *Cruiser* towed her away on September 21, 1869, and at Liverpool her engines were removed and she was given a temporary rig as a three-masted ship. The project must have proceeded very slowly, for two years elapsed before its completion. Loaded with 3,800 tons of coal, she left Birkenhead on September 30, 1871. Various destinations, including Bombay[27] and Rio de Janeiro[28] were quoted. She actually became a floating-stores hulk at one of the several mouths of the Niger River, at Bonny, on the southwestern coast of Africa in Nigeria. Lloyds Register for 1874 lists the African Steamship Company as her owner. In 1885 she had to be beached to prevent her sinking and was abandoned.

There are sources that state she was actually used as a sailing ship after 1869, and even mention the San Francisco trade for her.[29] These confuse her with two smaller sailing ships of British registry bearing the name of *Adriatic*, which called at American ports between 1874 and 1878.[30] The ex-steamer never did so.

The *Atlantic*'s and *Baltic*'s Final Years

If we go back to 1861, we find that the remaining North Atlantic steamships, the *Atlantic* and *Baltic*, were in continuous use, throughout the Civil War, as transports, usually under charter to the Army's Quartermaster Corps. Starting in April they earned high rates of $2,000 per day. Later in the year these were reduced to $1,500, in 1862 to $1,200, and, finally, to $1,000.[31] Officers and crew were invariably supplied by the owners; the vessels went wherever directed by the military authorities. On April 7, 1861, the *Atlantic* left New York to deliver reinforcements to the garrison at Fort Pickens, off Pensacola, Florida. Two days later, in a heavy gale, she had to heave to in order to prevent injury to the horses she was carrying. Further troops and guns were loaded at both Key West and Fort Jefferson, on Dry Tortugas Island.[32] On the night of the 16th the *Atlantic* towed twenty boats from various vessels in the squadron close to Fort Pickens to ferry six hundred men and their supplies ashore. Thus reinforced, the fort survived the Confederate menace and Pensacola became an important federal naval base.

The *Baltic* left New York a day later than the *Atlantic* on a secret mission to Charleston harbor in a vain attempt to reinforce Fort Sumter. Both she and the accompanying expedition were under the direct command of Gustavus V. Fox, an ex-naval officer, now Assistant Secretary of the Navy. But on April 12, 1861, as the *Baltic* and the revenue cutter *Harriet Lane* steamed toward the Charlestown bar, they found the fort being bombarded by South Carolina's shore batteries. The next day its commander, Major Anderson, surrendered. He and his men were sent out to the *Baltic* by the Havana steamer *Isabel*. The *Baltic* was somewhat battered by a collision with the U.S.S. *Pawnee* on April 14, but when she departed the next day she flew at her main truck a flag that had flown over Fort Sumter during this, the opening engagement of the Civil War.[33] Both Anderson and his flag would return in 1865.

Less than a month later, she helped evacuate the Naval Academy from Annapolis, Maryland, to Fort Adams, at Newport, Rhode Island.[34]

In October of 1861 both steamships became part of the extensive fleet assembled at Hampton Roads under Flag Officer DuPont. Their former rival, the *Vanderbilt,* joined them to lead out the three columns of transports that sailed on October 29. The *Atlantic* acted as the flagship of the transport fleet and the *Baltic* towed out a clipper ship, the *Golden Eagle.* There was bad weather and the many ships carrying 16,000 troops were badly scattered. The *Baltic* grounded at Hatteras Shoals but was lucky enough to escape those treacherous waters.[35] In the end Port Royal, South Carolina, was reached and a landing made with little opposition. Both steamers would make many more trips between New York and the major naval base established there, often touching at Fortress Monroe, Old Point Comfort, Virginia, on the way, carrying troops, reinforcements, supplies, wounded, and essential passengers. Although extremely valuable as large, fast steamships, their loaded draft of twenty feet or more kept them out of the many shallow regions along the Atlantic and Gulf coastal waters.

The *Atlantic* left the transport service late in 1863 for a commercial voyage. The Panama Railroad, a co-owner, needed her for a single trip to carry freight but no passengers from New York on November 25 to Aspinwall. She arrived back at New York on December 26 and resumed her transport work. Near the end of the war, both steamers were operating mainly between New York and Fortress Monroe, bringing down men and material to General U. S. Grant's armies in Virginia.

The *Atlantic* was one of many transports at the second, successful attack on Fort Fisher, at the mouth of the Cape Fear River, near Wilmington, North Carolina. As an army transport she got away earlier than the navy's *Vanderbilt* to deliver the first news of that great victory to Fortress Monroe on January 17, 1865.

The *Atlantic* docked at New York on March 26, 1865, to complete her last trip as a transport; the *Baltic* followed on April.

The fall of 1865 found the Pacific Mail Steamship Company completing arrangements to take over Vanderbilt's California mail service from New York to Aspinwall. The line was well aware of the small size and cramped quarters previously available on the *Ariel, Northern Light,* and *North Star* and intended replacing them as soon as possible. As a first step toward improving the service, the *Atlantic* and *Baltic* were for a second time sent to Aspinwall, together with the *Henry Chauncey,* just completed. The last named vessel made the first Pacific Mail sailing from New York on November 1, 1865; the *Atlantic* followed on the 11th; the *Baltic* on the 21st. As expensive ships to run, not very well suited for tropical climes, they were a stopgap measure; the *Atlantic* saw three voyages and the *Baltic* one. But with the arrival of the former at Quarantine, off Staten Island, on February 1, 1866, the Collins pair came to the end of fifteen years of hard service as first-class liners and exceptional transports. From now on they would play less glamorous roles as immigrant carriers.

The firm of Ruger Brothers had strong ties with various North German States and set up a steamship service, the North American Lloyd Steamship Company, from New York to Bremen via Cowes or Southampton. Some of their craft were chartered, but, as a basic fleet, the *Arctic, Baltic,* and a large beam-engined coaster, the *Western Metropolis,* were purchased.

The *Atlantic* opened the service from New York on February 22, 1866, and from Bremen on March 22. The return was slow, 18 days from Bremen and 14 from Cowes, but financially a very successful trip, for 900 westbound passengers took advantage of the new line's low fares. Her second voyage, from New York on April 13 and from Bremen on May 10, had even more, 1,158 immigrants, three of whom were born on the way over. Further Bremen sailings in May, July, and September were equally well patronized; the *Atlantic*'s four homeward trips brought over 4,000 passengers, a greater number than the entire Collins Line had carried from Europe in all of 1855.

The *Baltic* made two voyages from New York, on April 27 and June 14. Other chartered steamships provided eastward sailings at irregular intervals of one to three weeks from February to July. The *Atlantic* closed out the 1866 operation when she arrived home at New York on September 25. Then, like Vanderbilt, the Rugers ceased their operations for the winter.

Despite this highly successful summer, the line's three vessels were sold to a new firm, the New York and Bremen Steamship Company. The *Atlantic, Baltic,* and *Western Metropolis* continued on the Bremen via Cowes run, with the chartered *Northern Light* joining them. As she had for the Rugers, the *Atlantic* opened the service, this time even earlier in the 1867 season, with a New York departure on February 7, followed by the *Baltic* two weks later. There seem to have been even more German immigrants available, for the *Atlantic* reported passenger lists as high as 1,727 and the *Baltic*, 1,754.

The *Atlantic* was twice damaged. On her first westward trip she left Southampton on March 9 with 500 passengers. A week later she ran into such a severe gale that waves stove in her bulwarks, washed everything movable off the deck, damaged her forecastle and companionways, and swept the forward boats overboard.[36] Six of the crew, working on deck, and one passenger were injured.

The same situation recurred on her sixth voyage. Leaving Bremen on November 24 and Southampton on the 28th, she ran into an even worse storm on December 6, which demolished the houses on the guards, flooded the forecastle and the captain's quarters, stove in the forward hatch, and even damaged the after houses on the spar deck. Furled sails were blown out of their gaskets and whipped to shreds. The hull was so severely punished that nine deck beams were broken and the rudder head sprung; by good fortune it did not give way completely and leave the steamer uncontrollable. Captain Charles Hoyer brought his damaged ship to New York only to be further delayed by gales, snow storms, and fog that kept him off Sandy Hook for thirty-six hours.[37] The *Atlantic* finally docked on December 13, 1867, nineteen days from Bremen. This disastrous crossing was her last under steam, as well as the last by the New York and Bremen Steamship Company.

In all, the *Baltic* completed five voyages and the *Atlantic* six for this latest owner. But these aging steamships consumed far too much coal and required too large an engineering staff to keep their four boilers fired and two side-lever engines running. Isaac Taylor considered it prudent to end the line and, on June 30, 1868, the *Atlantic* was sold at auction for $41,000.[38] She lay in Brooklyn's Erie Basin while William D. Andrews and Brother advertised her for sale from September of 1868 through April of 1869. In 1870 she was scheduled to be converted to a sailing ship,[39] but the scheme came to naught and she was towed to Cold Spring Harbor, Long Island, for dismantling in 1871.[40]

Of all the American side-wheel liners, the *Atlantic* is the undisputed doyenne in terms of number of Atlantic voyages. There were fifty-five for Collins, a single one for the North Atlantic Steamship Company, four for North American Lloyd, and six for the New York and Bremen Steamship Company, making a grand total of sixty-six. If we add her five years of continuous transport service, it becomes obvious that she was a remarkable steamship. William H. Brown, her designer and builder, produced not only the first Collins liner but also the most successful one of the lot. True, there were early machinery difficulties, but these were overcome by the Novelty Iron Works and soon forgotten during the rest of her seventeen years of useful service.

The *Baltic*, a similar but not identical vessel from the Jacob Bell yard, never did quite so well. Each successive owner preferred the former and used her longer. As a result, the *Baltic* completed fifty-seven transatlantic voyages, nine less than the *Atlantic*. Similarly, she made seven to Aspinwall to the *Atlantic*'s nine. Nevertheless, she proved to be longer lived.

The *Baltic*'s engines were removed early in 1868 and her hull was put up for sale.[41] It must have been in good condition, for she was converted into a sailing ship under the ownership of Isaac Taylor and Charles Luling. The underwater work was done at the Erie Basin dry dock in February of 1869 and the rerigging completed in March. She was an extremely large sailing ship, measuring 2,552 tons, carrying 10,000 yards of canvas, and having fore and main yards 90 feet long. The *Three Brothers*, ex-*Vanderbilt*, would set 15,000 yards and have a 100-foot main yard when she was converted four years later.[42] Because of the *Baltic*'s unusual size, a donkey boiler was installed and a steam winch to help load cargo. George D. Sutton, her New York agent, advertised her as "the magnificent A-1 extreme clipper ship *Baltic*," and the *Herald* printed, "... her owners expect she will be the fastest clipper ship ever put under sail." She was very heavily loaded when she passed down the North River on April 2, 1869, drawing 23'-6" of water. She arrived at San Francisco August 19, loaded grain, and made a fast passage to Liverpool in 106 days. Speeds as high as 16 knots were achieved and it was decided that she could stand an increase in sail area.[43] Leaving Liverpool on April 15, 1870, the *Baltic* reached San Francisco in August. She continued on a yearly voyage basis, carrying grain eastbound to England, and manufactured goods or coal westbound, until 1872, when she picked up a guano cargo at Mejillones, Chile, but, having been strained by gales off Cape Horn, had to put into Rio de Janeiro on May 25, 1872. Over a thousand tons had to be unloaded and transshipped to another vessel, the bark *Victoria*, before the leak could be reached and repaired.[44] She left for Liverpool August 7. That year, 1872, she became the largest American-built sailing ship afloat when the *Denmark*, formerly the clipper *Great Republic*, sank at sea.[45] A year later the *Three Brothers* took over the title. Under the 1865 tonnage rules the *Baltic* rated 2,552 tons and the *Three Brothers* 2,972.

March 1, 1873, saw the *Baltic* entering New York harbor from Liverpool on her first visit since 1869. She was docked, stripped of her metal, recaulked, and had new sheathing installed before continuing on to San Francisco. She carried on in the grain trade from California to Europe, but return cargoes were becoming harder to find. Her last San Francisco sailing took place in October of 1876 with 3,248 tons of wheat for Queenstown, Ireland.[46] In July of 1877 she went on to Bremen, where she was laid up for three years. Early in 1880 a cargo of salt and iron was loaded and she sailed on March 28 for either New York or Boston, still owned by Isaac Taylor and still flying the American flag. It was a heavy load and extremely bad weather strained the hull so much that, leaking badly, she had to put into Fayal, the Azores, on April 27. The iron was unloaded, but it was impossible to repair the *Baltic* properly. She needed many new fastenings and much caulking; no dry dock was available.

A steam pump was sent out from Boston and put aboard. It could be driven from the winch boiler. Temporary stanchions were installed in the hold and old knees put in to keep the hull from working.[47] With the leakage reduced to only one inch of water per hour, which the pump could handle, she sailed July 14 in ballast for Boston. But after she arrived there on August 12, even the skilled New England shipwrights could not repair her without prohibitive costs. The *Baltic*'s name was dropped from the active list of American ships in November of 1880. In eleven years as a sailing ship she had made six trips from San Francisco to the British Isles, usually to Liverpool. Returning, she sometimes went directly and sometimes by way of New York.

Notes

1. John Haskell Kemble, *The Panama Route, 1848-1869* (Berkeley and Los Angeles, Calif., 1943), p. 84.
2. *New York Herald*, September 5, 1859.
3. Ibid., October 6, 1859.
4. Ibid., November 10, 1859.
5. Ibid., December 25, 1859.
6. *New York Commercial Advertiser*, February 10, 1860.
7. *New York Herald*, March 26, 1860.
8. *New York Times*, April 14, 1860.
9. Unidentified clipping dated April 16, 1860, in William H. Webb's scrapbook, Webb Institute of Naval Architecture.
10. Undated clipping, Webb scrapbook.
11. *New York Herald*, May 12, 1860.
12. Ibid., January 1, 1861.
13. *New York Commercial Advertiser*, February 8, 1861.
14. N.R.P. Bonsor, *North Atlantic Seaway* (Prescott, Lancashire, 1955); p. 162.
15. *New York Commercial Advertiser*, February 25 and March 1, 1861.
16. Ginsberg records.
17. *Journal of the Franklin Institute*, 42, 3d ser. (July December 1861): 53.
18. *New York Commercial Advertiser*, May 3, 1861.
19. *New York Herald*, January 4, 1862.
20. Ginsberg records and *New York Herald*, June 12, 1861.
21. Bonsor, *North Atlantic Seaway*, p. 164.
22. Naval History Division, Navy Department, *Civil War Chronology*, 1 (Washington, 1861), p. 54.
23. *Illustrated London News*, March 15, 1862.
24. Ginsberg records.
25. *New York Herald*, September 9, 1863.
26. *London Times*, September 22, 1869.
27. *Engineering*, October 6, 1871.
28. *Nautical Gazette*, September 16, 1871, p. 95.
29. H. Philip Spratt, *Transatlantic Paddle Steamers*, 2 ed., (Glasgow, 1967), p. 59; J. H. Isherwood, "Collins Liner Adriatic," *Sea Breezes*, (February 1857), p. 107.
30. *New York Maritime Register*, 1870-78.
31. House Executive Document No. 337. 40th Cong., 2d sess.
32. *New York Herald*, May 2, 1861.
33. Official Records, ser. 1, 4: 254-55.
34. Ibid., 4: 399.
35. *New York Herald*, November 14, 1861.
36. *New York Journal of Commerce*, March 27, 1867.
37. *New York Herald*, December 14, 1867.
38. *New York Journal of Commerce*, July 3, 1868.
39. *New York Shipping and Commercial List*, January 5, 1870.
40. *New York Times*, September 16, 1871.
41. *New York Journal of Commerce*, August 17, 1868.
42. *New York Herald*, February 25; March 7, 26, 29, and 30, 1869.
43. Ibid., April 4, 1869 and January 24, 1870.
44. *New York Maritime Register*, July 14 and September 2, 1872; *New York Herald*, June 22, 1872.
45. Ibid., March 14, 1872.
46. *New York Shipping and Commercial List*, October 21, 1876.
47. *New York Maritime Register*, May 19 through November 17, 1880.

15

Occasional Atlantic Liners Part 1

THE screw steamers that made Atlantic voyages before the Civil War have been considered in an earlier chapter. In this one it is planned to trace the careers of the side-wheel steamships, most of them designed for coastwise operation, that, at some time during their lives, were diverted to the Atlantic. All were completed before 1861.

The *Union*
1851-1856

After establishing the first totally successful steamship service from New York to Charleston with the *Northerner* and *Southerner*, Spofford, Tileston, and Company joined forces in 1848 with the Charleston firm of Mordecai and

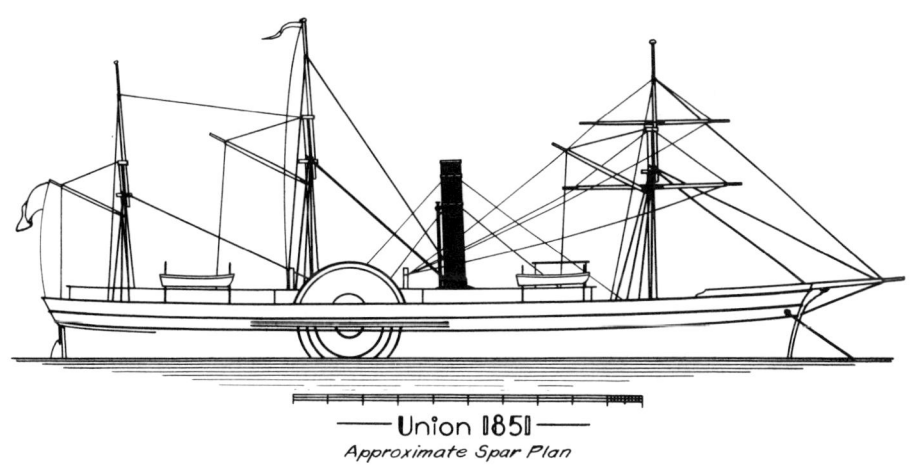

Union, approximate spar plan.

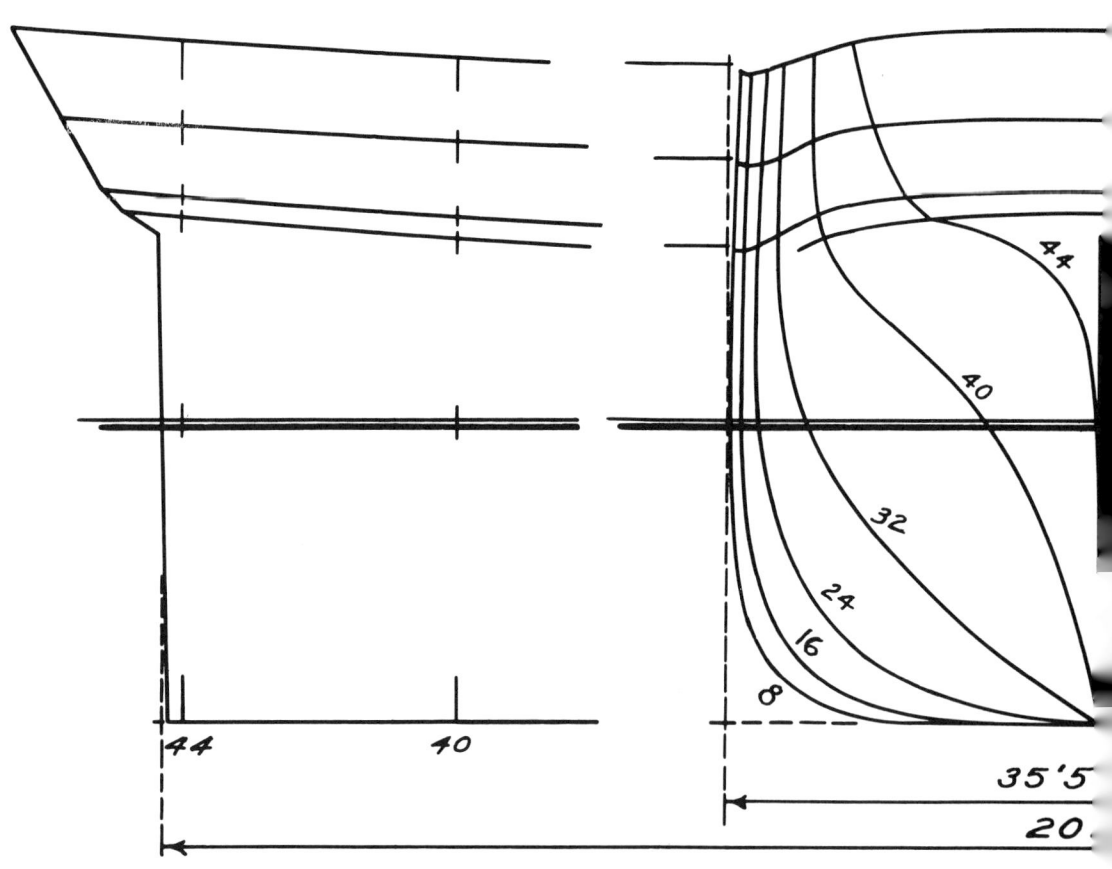

Union, 1851, lines.

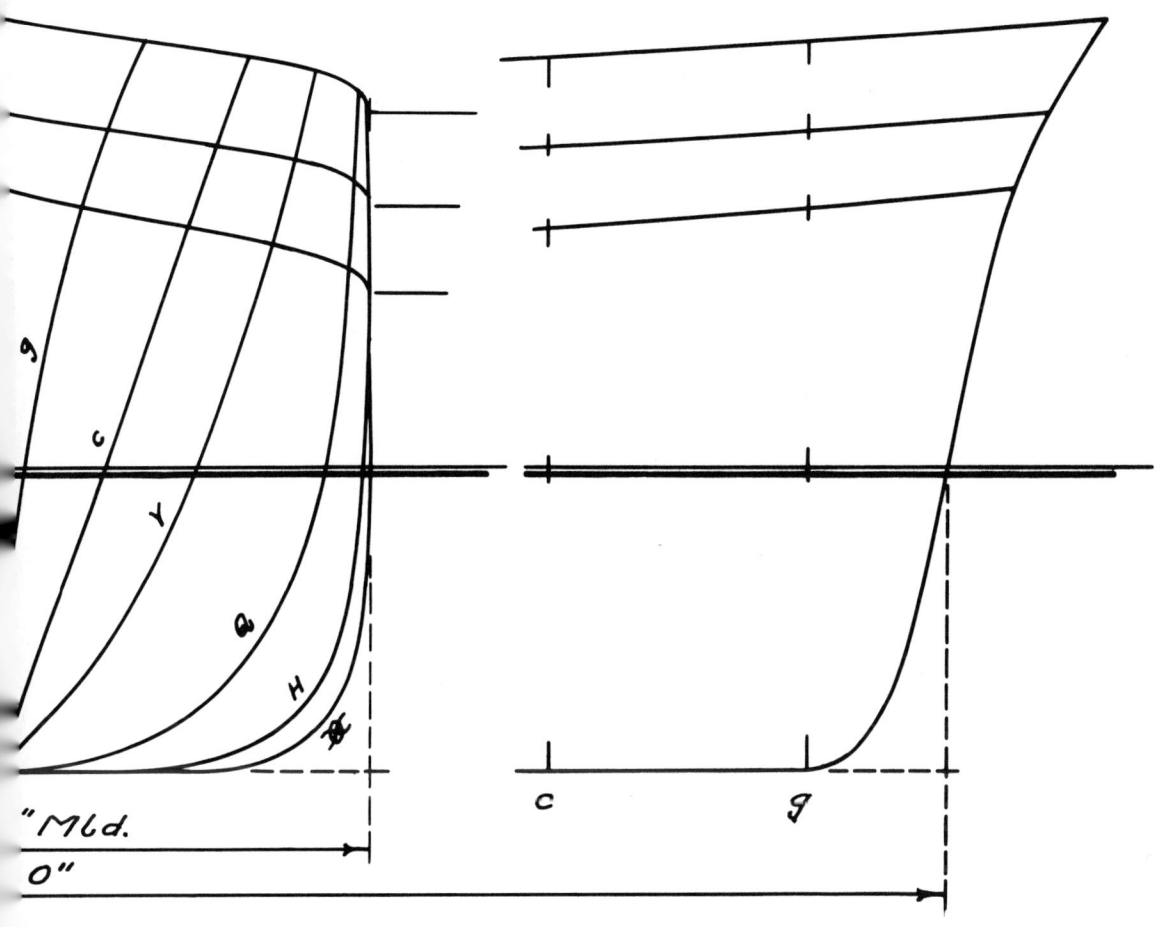

Frame Spacing
X-Q   2'-6"
Q-Y   2'-8"
Y-g   2'-10"
I omitted

Company to extend their service and carry the mails on from Charleston to Key West and Havana with the *Isabel*. They recognized the enormous importance of regular sailings to attract passengers and freight. With this in mind, they continually built new steamships somewhat in advance of their actual needs, and usually had a spare ship available to take over in the event of delay or accident. Other operators, aware of this policy, often approached them to charter, or even purchase, one of their steamships, and from time to time they were able to do so — at a considerable price but on very short notice. The *Northerner*, for example, was sold early in her career for use on the Pacific. Three later steamships, the *Union*, *Nashville*, and *Columbia*, were chartered when urgently needed by the Havre and the Collins Lines.

The *Union* had been ordered from William H. Webb, who laid her keel on April 19, 1850, and launched her, less than five months later, on September 9.[1] As a replacement for the *Northerner*, she was slightly larger, 1,200 tons instead of 1,102. Her most important difference was the installation of two side-lever engines instead of one. A straight-stemmed steamship with two full decks, a spike bowsprit, and a large eagle as a figurehead, she had a moderate three-masted rig with two square sails on the foremast. There was a full-height forecastle forward and a deckhouse aft for three-quarters of the ship's length, which contained officers' rooms, two passenger saloons, and eighteen double staterooms at the stern.[2] A dining saloon forward, main saloon aft, twenty-four triple staterooms, and four family rooms with six to nine berths were on the second, or main, deck. Below that came a second-class dormitory space and the steerage. The hull was 215' long, 34' in beam, and 22' in depth of hold, and was designed for a draft of 13'. The Allaire side-lever engines, 60" bore by 7'-0" stroke, turned 30' paddle wheels.

When completed and commanded by Captain Thomas S. Budd, late of the *Northerner*, she opened a new service from New York to New Orleans. It was a joint venture between Spofford and Tileston and another operator, Davis Brooks, who supplied the *Winfield Scott*. Sailings twice a month were planned. The *Union* sailed from New York on March 8, 1851, and arrived home exactly one month later bringing in passengers, cotton, and California gold that had been transshipped at New Orleans. The voyage was most successful; the new engines worked beautifully; no bearings overheated; no stops were necessary to tighten up their brasses. Outward the trip took less than eight days dock to dock, and homeward 6 days and 6 hours, Balize to Sandy Hook.[3] Her fourth trip made a new speed record, 6 days 7 hours dock to dock, and included an astonishing run on July 7 of 357 miles at almost 15 knots.[4]

Most of 1851 continued with monthly sailings to New Orleans, but when her consort, the *Winfield Scott*, was sent to the Pacific, the *Union*, in December, went on from New Orleans to Chagres. She then was shifted to her owners' usual run. On January 31, 1852, she cleared New York for her first trip to Charleston. In the late spring, when a quite similar Webb steamer, the *James Adger*, was delivered, Spofford and Tileston had four liners available and could increase their Charleston sailings to twice a week. The others in their fleet were the *Southerner* and *Marion*. For a year the *Union* steamed regularly in and out of Charleston harbor. She was then chartered by Davis Brooks and his New York and San Francisco Steamship Line, which had her former running mate, the *Winfield Scott*, on the Pacific. The *Union* made three voyages to Aspinwall, always returning by way of Kingston for coal. Her first departure was March 5, 1853, with others in May and June. The first of these was delayed by a broken crank on her port engine. Leaving Aspinwall on March 18, she ran into a gale and a heavy cross sea about 180 miles from Kingston. Possibly the heavy rolling led to the damage. She put back to Porto Bello, arriving there on the 24th, and then turned back to Aspinwall. Although a damaged crank would seem a major problem, temporary measures were completed in a week; she sailed again, on the 31st of March, but had to be nursed along at a slow speed; it took her six days to Jamaica and thirteen to New York.[5]

After her last arrival from Aspinwall, June 28, the *Union* returned to the Charleston route and continued there through April of 1854. In January of that year, however, she was chartered briefly by the Pacific Mail Steamship Company to search for their missing steamship, the *San Francisco*.

In May she was used by the Havre Line to replace the wrecked *Humboldt*. No alterations were made; she arrived at New York from Charleston on April 30, 1854, filled her holds with coal, and sailed for Havre, via Cowes, on May 6. Her regular master, Captain Richard Adams, took her across. As a rather small steamship on a long and often rough ocean passage, the *Union* could not be pushed; the extra weight of coal buried her wheels far too deeply. Eastbound she took thirteen days from New York until she passed Eddystone light, near Plymouth.[6] Westward she sailed from Havre on June 7 and reached New York on the 23rd, sixteen days later. Many drifting icebergs along the route slowed her down. A total of ten transatlantic voyages were made for the Havre Line.

On her sixth homeward passage she met extremely heavy weather and had to be put into Halifax on April 1, 1855, seventeen days out of Cowes, short of coal.[7] Her next westward trip was also extended, this time by running aground on Sable Island during a thick fog. She remained there fifteen hours until fifty tons of coal had been thrown overboard and most of her fresh water pumped out; she then came off and proceeded on to New York, arriving May 26, sixteen days from Cowes.[8]

On October 22, 1855, just two days out of New York, the *Union* was steaming to the northeast in fair weather and a smooth sea when suddenly one of her paddle shafts fractured. She put back into New York on the 25th.[9] She was not the first vessel on the Atlantic to suffer this type of casualty; four liners had preceded her, the *Atlantic* in 1851, the *Hermann* and the *Franklin* in 1852, and the *Pacific* in 1853. A new shaft was forged, machined, and installed in about six weeks. She left New York on December 17 for her tenth and last Havre voyage reaching Cowes December 31. From Havre on January 19 she skipped the usual Cowes stop and was delayed so much by winter gales that two extra stops for fuel were necessary. The first was Halifax on February 2, and the second Newport on the 6th. She docked at New York February 7, 1856.

In the meantime, she had been purchased by Austrian interests who planned an operation from Trieste to Malta.[10] Still under Captain Adams, she sailed from New York on March 15 for Trieste, making Gibraltar on April 1 and Malta eight days later.[11] Nothing is known about her subsequent career.

The *Nashville*, 1854-1863

The most famous Spofford and Tileston steamer, the *Nashville*, was a development of the *Union*, with her rig reduced to make her a two-masted, fore-topsail schooner instead of a three-masted bark. The standard hull arrangement of a topgallant forecastle and a long deckhouse was retained, but a graceful sailing-ship bow with a billet head, cutwater, and trail boards was reintroduced in place of a straight, utilitarian stem. Probably as an economy measure, a single engine was chosen instead of two. The *Nashville* was built by William Collyer at the foot of 19th Street on the East River under the supervision of Captain Michael Berry. She was launched September 22, 1853, and ready for trials on December 28. Her dimensions were 215'-6" by 34'-6" by 21'-9", with a tonnage of 1,220. The Novelty Iron Works provided a side lever engine of 86" bore by 8'-0" stroke, turning wheels 32'-0" diameter, each with 28 floats 10'-0" by 20". The two boilers were 24'-0" long, 12'-3" wide, and 12'-3" high, of the return-flue type, with five furnaces apiece burning anthracite coal. Steam-driven blowers supplied cool air to the fire room.[12] The maximum steam pressure was 28 psi and the highest rpm 19. The original bunker capacity was 185 tons of coal, but, if needed, more could be loaded in her cargo holds.

Under Captain Berry the *Nashville* sailed for Charleston on January 4, 1854. She had no sooner returned to New York, on January 13, than she was impressed to replace the stranded Havre liner *Humboldt*; three days later, January 16, she was outward bound for Havre. As in the *Union*'s case, a small steamship with deeply buried wheels lost much of her usual speed. The *Nashville*, further slowed by storms, took seventeen days to Cowes. Westbound from Havre on the 15th and from Cowes a day later, she found herself, in poor visibility, off Barnegat Light and south of New York.[13] Mistaking it for Fire Island Light, she grounded at Little Egg Harbor, but got off without damage and reached New York on March 5.

Her second trip was shorter: thirteen days to Cowes and sixteen from there to New York. After arriving on April 28, she was turned back to her owners. The *Union* replaced her on the Atlantic and continued on the Havre Line's service until a new steamship could be completed. The *Nashville* ran to Charleston until the spring of 1855.

The Collins liner *Pacific* was being overhauled and, as replacement, the *Nashville* was obtained for one voyage to Liverpool. Departing New York on March 21, 1855, she required thirteen days eastward. Westbound she arrived on April 22, fifteen days from Liverpool. Less than two weeks later she was back in coastwise service, where she spent the next six years shuttling between New York and Charleston and making an excellent reputation for herself as a fast liner of the highest class.

Just before the hostilities marking the start of the Civil War broke out in Charleston Harbor in the spring of 1861, the *Nashville* left New York April 6 on one of her regular voyages. Because L. M. Murray, her commander, was a secessionist, he kept his ship at Charleston.[14] The *Nashville*, as a result, became the first cruiser commissioned by the Confederate Navy. She was armed with two brass twelve-pound guns and, under Lieutenant Robert B. Pegram, late of the U.S. Navy, ran the Federal blockade on October 26, 1861, coaled at Bermuda four days later, and headed for British waters. Storms damaged the bulwarks, paddle boxes, and hurricane deck. Approaching the English Channel on November 19, she met and captured an American ship, the *Harvey Birch*.[15] As a prize she was a disappointment; her hold carried nothing but ballast. The crew were taken off and put in irons and the ship was burned.

Two days later the *Nashville* steamed into Southampton. She had to dock there to repair bottom damage received when she grounded while crossing the Charleston bar and the storm damage on the way over. She was closely watched by the U.S.S. *Tuscarora* until the British authorities ordered both to leave their waters. Should the *Nashville* leave first, the *Tuscarora* would be detained twenty-four hours before she could follow. Thus, on February 3, 1862, the *Tuscarora* watched the *Nashville*

Nashville as a Confederate cruiser. She captured and burned the *Harvey Birch* on November 19, 1861. *Courtesy of The Mariners Museum, Newport News, Va.*

sail openly and safely.[16] Slowed by gales, Lieutenant Pegram made Bermuda on February 20, picked up a pilot, and headed for Beaufort, North Carolina, capturing a schooner, the *Robert Gilfillan*, on the way. That, too was set afire. Approaching the Federal blockading fleet, the *Nashville* raised the American flag and passed through with ease.

February 28 found her alongside a wharf at Morehead City, North Carolina. On board was an essential cargo of newly printed Confederate currency, stamps, and the endless printed forms needed by the new government.

In order to escape capture, as General Burnside's expedition to North Carolina approached, the *Nashville* fled on March 17, could not get past the Charleston blockade, and put into Georgetown, South Carolina, instead. There she was renamed *Thomas L. Wragg* after a transfer to supposedly British owners, Fraser, Trenholm and Company, one of the most successful firms sending steamers and their cargoes through the Union blockade.[17] Although she spent only four months as a blockade runner, she is said to have completed eight successful trips into Carolina ports.[18] Despite the relatively short distance to Nassau, it is difficult to see how she could have penetrated the blockade so many times. One suspects Southern propaganda.

In July of 1862 the *Wragg* delivered a cargo of Enfield rifles, ordnance stores, and drugs to Savannah and loaded cotton to take out.[19] At the direct order of Gideon Welles, Secretary of the Navy, Admiral DuPont took special pains to see that his South Atlantic Blockading Squadron was on the alert for her emergence. As a result, she was bottled up for eight months.

During her enforced idleness she was sold to Captain T. Harrison Baker who, on November 5, 1862, obtained a letter of "marque and reprisal" making her a privateer with the right to capture Union vessels for private profit.[20] Despite her new designation and a new name, *Rattlesnake*, she was still unable to leave the Ogeechee River, where she was protected by the guns of Fort McAllister.

On February 27, 1863, the *Rattlesnake* planned to break out but ran hard aground near the fort. The monitor *Montauk* and three small gunboats, the *Dawn*, *Wissahickon* and *Seneca*, ventured up the river and, at sunrise of the 28th, fired on the steamship and the fort. The ironclad came within 1,200 yards of the *Rattlesnake*, while the others stopped at a safer distance. Shells from the *Montauk*'s 11- and 15-inch guns set the helpless *Rattlesnake* on fire. When the flames reached her magazine at about 9:30 in the morning, she blew up. The *Montauk* survived the fire of the fort's guns but had her bottom damaged by a mine while going down river. She had to be beached until her crew could install a temporary patch and pump her out.[21]

The *Columbia*, 1857-1875

The third Spofford and Tileston steamship to make a transatlantic voyage was the *Columbia*, which had been built at Greenpoint, Long Island, now a part of Brooklyn, by William Collyer in 1857. As usual, she was similar to the line's preceding vessels but a little larger and a little faster. The *Columbia* was 234' by 35' by 23', with a tonnage of 1,347, making her about ten percent bigger than the *Nashville*. The hull arrangement, single stack, and two-masted rig were almost identical. For once, there was no bowsprit. The single side lever engine of 85" bore by 9'-0" stroke, built by the Novelty Works, produced about 12% more horsepower than the *Nashville*'s.[22] Under the best of conditions she could go from Charleston to New York in forty-seven hours at 13.4 knots. As usual, too, she was built under the supervision of Captain Berry. After a November 27, 1856, launch it took five months to install the machinery and complete her.

May 4 of 1857 saw her maiden voyage, under Captain Berry, to Charleston; she arrived home on the 12th in fifty-two hours. Her second trip was completed on the 25th with an improved time, forty-nine hours. Concurrently, the Collins Line had just ended the *Ericcson*'s charter because the naval authorities refused to pay a full mail subsidy on such a slow steamer. The *Columbia* was substituted; as a new, fast coaster, although smaller than the *Ericcson*, she made possible a definite improvement on the deteriorating service of the line. She was quite large enough for the limited numbers still traveling on Collins steamships.

Still under Captain Berry, she left New York on June 6, 1857, with forty-nine passengers, and reached Liverpool in 11 days 8 hours, a time only thirteen hours longer than the *Atlantic*'s two weeks later.[23] From Liverpool July 6, she took 11 days 17 hours westward. She did even better from New York on the 18th in 10 days 18 hours, but met strong winds returning so that a full twelve days elapsed before she docked on August 17, 1857.

Immediately thereafter, on her way toward Charleston, she was caught in the September hurricane that destroyed the U.S. Mail Line's *Central America*. The *Columbia* was more fortunate; she had only a paddle box wrecked and her deck cargo, including two thoroughbred horses, swept overboard.[24]

After September of 1857 she ran to Charleston in company with the *Nashville*, *James Adger*, and *Marion*. In January of 1861, when the clouds of war were gathering, South Carolina's authorities blocked the regular Charleston ship channels in order to keep any Federal reinforcements from reaching Fort Sumter. The *Columbia*, leaving by one of the shallower, less-used exits, ran hard aground near Sullivan's Island. Her cargo had to be unloaded and her passengers transferred to the *James Adger*.[25] On January 29 three steamboats were able to pull her off but the hull was strained and leaking. She was undergoing repairs in New York when Fort Sumter was attacked; thus, unlike the *Nashville*, she remained under her owners' control. On May 4, 1861, she sailed as a transport loaded with New York Militia bound for Washington.[26]

Spofford, Tileston, and Company abandoned Charleston and shifted their terminus to Havana; late in May the *Columbia* began her first Cuban voyage. After New Orleans was captured by Northern forces, the line was extended on from Havana. The *Columbia* made nineteen voyages to that delightful Creole city between January 6, 1863, and October 1864; thereafter she started turning around at Havana and continued on the shorter route for the rest of her useful life.

At the end of May in 1866, when she was running with the steamships *Eagle* and *Morro Castle*, Spofford and Tileston sold out to the firm of Garrison and Allen; Garrison had at one time been a partner with Vanderbilt and Morgan during their Nicaraguan adventures. Allen was Vanderbilt's son-in-law, who adopted for the service "Atlantic Mail Steamship Company," the name that had been used by Vanderbilt's mail line to Aspinwall. A year later George B. Hartson took over, and, in later years, S. G. Wheeler replaced Hartson as principal agent.

The *Columbia* steamed serenely along, making regular sailings regardless of all the changes in ownership, but succumbing to the weather upon occasion. She survived her second hurricane in late 1865. It was an exceptionally severe one that inundated Key West, sank the *Republic*, ex-*Tennessee*, and severely damaged both the *North Star* and *Northern Light*. Southbound on October 22, three days out of New York, the *Columbia* had her decks swept by the seas. Three of four horses on board were lost, the purser's room was flooded, and all the ship's papers were washed overboard. The steamer rolled so violently that the swinging cabin lamps fell from their hooks, leaving the terrified passengers in darkness. The second engineer guarded the fire-room ladder with a drawn pistol as the only way to keep the firemen below and at work. The passengers, however, chose to honor the Captain, D. B. Barton, with a gold medal, and the Chief Engineer with a gold watch.[27] The man who saw that steam was kept up did not even get his name in the papers!

The *Columbia* was replaced by an iron screw steamer, the *Crescent City*, early in 1872. After being hauled out for repairs, she was kept as an extra steamship.[28] But about four months later the *Columbia* was back again on the Havana route, with occasional stops to deliver the mails to Nassau. She sailed for Havana on June 13, 1872, and continued regularly until her arrival at New York on January 11, 1873. After that the failing Atlantic Mail was liquidated. The *Columbia* was sold at auction for $21,000 to W. H. Starbuck on March 11.[29] Both she and the *Morro Castle* were later acquired by William P. Clyde and Company, who assigned them to the New York-Havana trade once more. The *Columbia* made seven voyages between June and November of 1873. She was laid up for a time but reactivated, in July of 1874, as part of the "Great Southern Freight Line" to Charleston. Thus she ended her career, after seventeen years, just as she began it. Her last New York arrival was September 29, 1874. In 1875 she made a trip from New York to Boston, arriving on May 1; there she was surveyed, found unseaworthy, and broken up.[30]

The three Charleston steamships built for Spofford and Tileston between 1851 and 1857 were chartered by either the Havre Line or the Collins Line as replacements for larger Atlantic liners that had been lost through perils of the sea or withdrawn for repairs. Together they completed a total of fifteen voyages to Havre or Liverpool. The *Union* made ten of these for the Havre Line; the *Nashville* one for Collins, two for Havre; and the *Columbia* two for the Collins Line. The *Union* completed an additional crossing to Trieste when sold to Austrian owners, and the *Nashville* one ocean voyage from Charleston to Southampton as a Confederate cruiser. All were somewhat small for the stormy North Atlantic; they must have been badly overloaded when coaled for such a long voyage. As a consequence, they were considerably slower on the high seas than they had been while running coastwise. Nevertheless, they were staunch and seaworthy, and the *Columbia*, the last and largest, made quite respectable crossings as long as the weather was good and the sea smooth. The steamers were invaluable to the parties who had to charter them, and to Spofford and Tileston they were extremely profitable, whether on their designated routes or under charter to others.

The *Illinois*, 1851-1888

The fastest steamship ever owned by the United States Mail Steamship Company, the *Illinois*, has already been discussed as the progenitor of the Havre Line's *Arago* and *Fulton*. In 1851 she represented a major advance in the design of wooden steamships. Ordered by the Pacific Mail Steamship Company to supplement their original *California-Panama-Oregon* trio, the *Illinois* was 266'-6" by 40'-10" by 22'-0" depth of hold to the main deck and 29'-6" to the spar deck. She was coppered to 17'-0", her anticipated draft, and, to prevent corrosion, had all copper fastenings up to the 20' waterline.[31] Her tonnage of 2,123 made her about three-quarters the size of the Collins liner *Atlantic* and one and three-quarters times as large as the Charleston steamer *Union*. During her construction, the Pacific Mail abandoned its Atlantic route and, after her completion, sold her to the U.S. Mail Steamship Company.

The hull had two complete decks and was very strongly built, with the whole stern and the major parts of her frame of live oak, the longest-lasting timber available. Floor timbers were 20" deep, sided 17", the keel and center keelson 18" square, both of white oak, while ten side keelsons were each made of two 14" x 14" yellow pine timbers. For further longitudinal strength, ten of the ceiling strakes, at the bilges, were 12" x 14" pine. A 7" thickness was used from there up. The yellow pine bottom planking was 4" thick and the wales 6". There was, of course, iron strapping. Her builders, Smith and Dimon, gave her beautiful lines to go with her excellent structure.

The use of powerful oscillating engines was one of the *Illinois*'s innovations. Hers came from the Allaire Works and had two cylinders 85" bore by 9'-0" stroke that delivered about 1,150 indicated horsepower in normal operation. They swung back and forth on trunnion bearings 30 1/2" diameter to which were led 28" diameter copper steam and exhaust pipes. The two air pumps, 51" diameter by 4'-6" stroke, were large enough to be steam engines themselves. The side wheels, 33'-6" in diameter, each with 28 double paddles, turned on 19 1/2" shafts. The 56 buckets were 10'-6" long, 15" wide, and 3 1/2" thick, and the arms supporting them 5" x 1 1/4" iron.

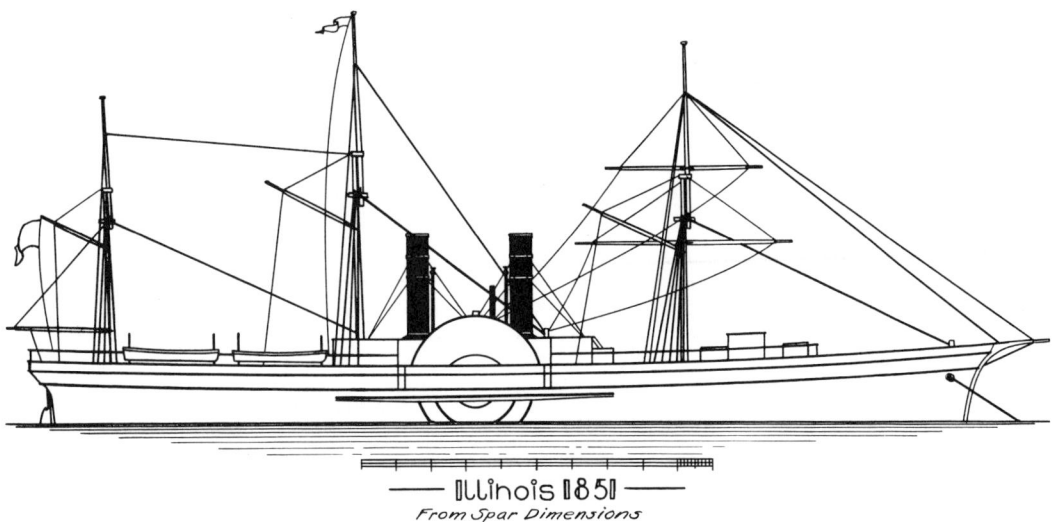

Illinois, spar plan.

Two boilers were placed forward of the cylinders and two aft; while only 12' long, they were 16'-3" wide 13'-0" high. Each had five furnaces, 315 return tubes 3" in diameter, and each pair was served by a stack 6'-6" in diameter extending 42' above the main deck. Adding together 242 tons of boilers, 120 tons of water, and 600 tons of machinery makes 962 tons to propel the ship.

The *Illinois* had a particularly attractive silhouette with her very short bowsprit, barkentine rig, clean sheer line, and two stacks nicely located with regard to the wheels and the masts. She was launched on February 7, 1851,[32] and was ready to sail from New York to Chagres on August 28, 1851.

Southbound the *Illinois* stopped at Havana on her way to Chagres, then discharged her passengers at the latter port for them to cross the Isthmus and board the much smaller *Tennessee* at Panama. Leaving Chagres on September 9, she coaled at Kingston on the 12th but ran into bad weather and had to put into Norfolk for additional fuel. Early in the morning of September 20 she arrived at New York, having on board 419 passengers and $1,350,000 in gold.[33] Two more voyages were made at monthly intervals; thereafter she had to be withdrawn, probably for engine modifications.

Whatever the problem was, it kept her out of service for five months. But when she resumed she immediately began to show her speed. She sailed from New York on April 26, 1852, for Aspinwall which, as the starting point for the Panama Railroad, had replaced Chagres as the line's terminus; she left there May 8, touched Havana May 13, and docked at New York on the 19th. Southbound the time was 7 days 10 hours 40 minutes. Havana was reached in 3 days 15 hours northbound and New York 3 days 19 hours from Havana; all these were new speed records. From Havana she averaged 13 1/2 knots, but, best of all, she made 347 miles or 14 1/2 knots during a single day.[34] She was faster than the Collins liners! Sails had been set and a following sea helped her along. The steam pressure was 18 psi, her wheels turned 11 1/2 rpm, and she burned sixty-three tons of coal in twenty-four hours.[35]

The *Illinois* made monthly sailings throughout 1852. On January 11, 1853, approaching New York at the end of her twelfth voyage, she met dense fog and anchored off Sandy Hook. The next morning, as she was getting under way, the starboard cylinder burst, apparently without injury to the engineers near it.[36] Replacement kept her idle until March 21, 1853, when she resumed monthly departures.

The major problem for the *Illinois* was her coal consumption, sixty tons per day, the highest of any of the U.S. Mail's steamships after the *Georgia* and *Ohio*, both of which had been quietly withdrawn and left to rot at out-of-the-way piers. Therefore, whenever her owners could manage it, other steamships were substituted. The *Illinois* was never used at her full capacity of twelve voyages per year; instead she made eight to ten and, occasionally, only two to four.

Northbound from Aspinwall, in August of 1857, she grounded on Colorado reef, about seventy miles from Havana, and could not get off for three days. Two hundred tons of coal were thrown over the side before the combined pull of the *Empire City* and a Spanish naval steamer, the *Blanco de Gray*, could get her off.[37]

During the final year of the U.S. Mail's career the

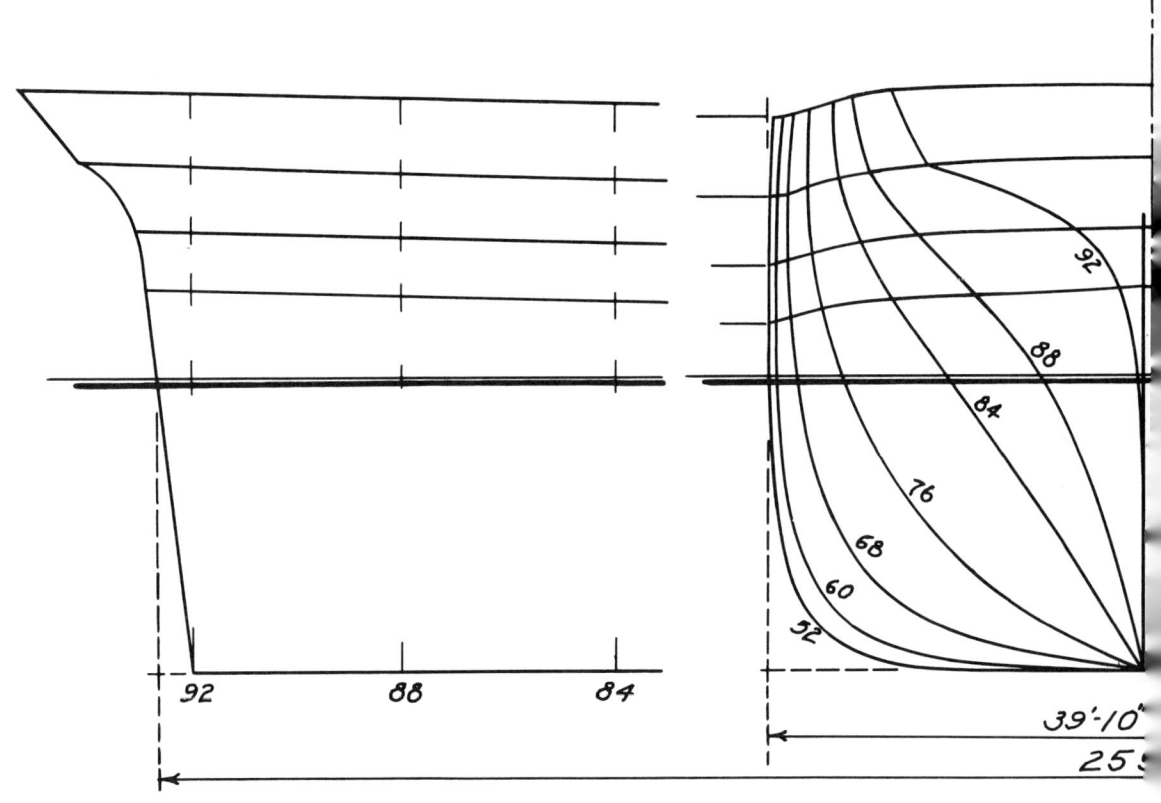

Illinois, 1851, lines.

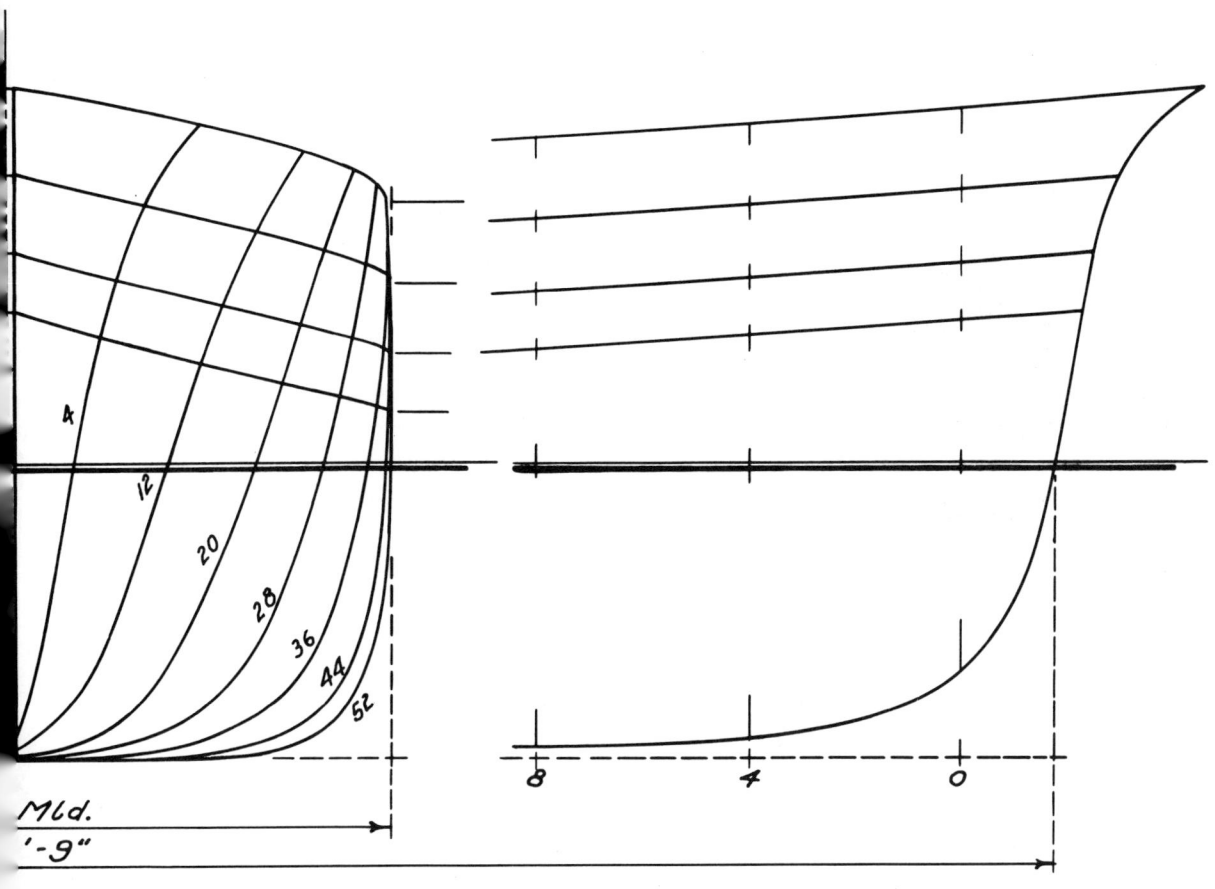

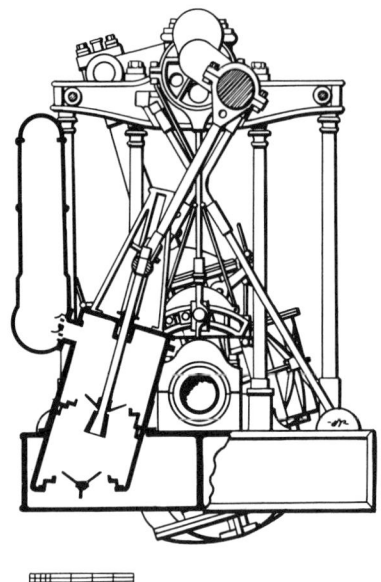
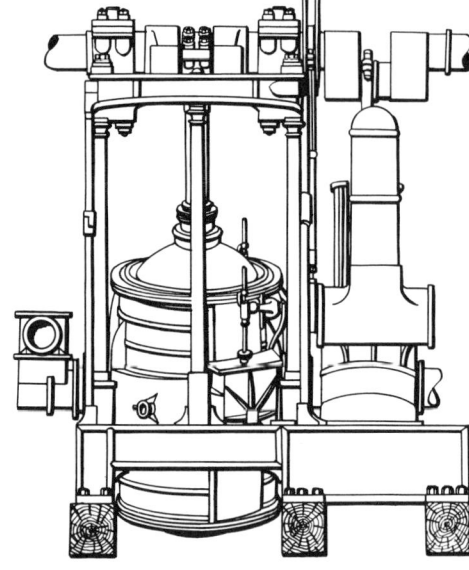

OSCILLATING ENGINE
85" BORE × 9'-0" STROKE
"ILLINOIS" 1851

Oscillating engine 85" bore by 9'-0" stroke for the *Illinois*.

Illinois sailed from New York in January, February, and March of 1859. On her last voyage, leaving March 7, she had to put into Havana, northbound, on March 23 under her starboard cylinder with her port one damaged. Three hundred and forty passengers and $1,000,000 in gold were transshipped to the *Granada*; a further 100 passengers and 105 mail bags went to the *Empire City* to be taken to New York.[38] The *Illinois* stayed behind to coal and make temporary repairs; her delayed arrival date at New York was March 30. She made no more sailings during the remaining life of the line. Its final trip, on September 30, 1859, was by the *Star of The West*.

Although the damaged machinery of the *Illinois* had been put in order, her boilers were in poor condition; the U.S. Mail neglected to keep up its fleet as the end approached. The *Illinois* was sold at auction to Commodore Vanderbilt at a bargain price of $25,000.[39] The Vanderbilt Line could not have done much to improve her run-down state; she was purchased in late February of 1860 and sent to sea two months later.

Under Captain C. P. Seabury the *Illinois* began her first Atlantic crossing on April 7, 1860, bound for Cowes and Havre. She would spend the spring and summer alternating with the larger and faster *Vanderbilt* at irregular intervals; the *Illinois* always followed two weeks after the *Vanderbilt*, while the latter, in turn, sailed four weeks after the *Illinois*. As a pair they were hardly matched, the *Illinois* averaging three days longer on the eastbound passage and almost four more westbound. She was in no condition to duplicate her earlier speed records. On her first crossing she did not reach Cowes until April 23, taking 15 days 16 hours, but homeward she made the same distance in 11 days 17 hours, her fastest crossing of the year, arriving New York on May 13.[40] Her fifth and last westward trip was from Havre on October 10, 1860, and from Cowes the next day, but bad weather delayed her so much that she had to put into Halifax on the 28th for coal and did not reach New York until the 30th, making it an extended nineteen-day passage.

Vanderbilt considered the *Illinois* worth retaining on his European route; during the winter of 1860-61 she was thoroughly overhauled, with new boilers put in and new iron bulkheads erected around her machinery spaces. Cargo was actually loaded and a Vanderbilt Line sailing to Havre scheduled for April 6, 1861; even a clearance had been filed at the New York Custom House.[41] Simultaneously, she was being prepared for a secret expedition to Fort Pickens, just off Pensacola, Florida. A second clearance was obtained April 8 for Indianola, Texas; it was to mislead Southern sympathizers. In fact, she reached Fort Pickens on the 19th after touching Key West on the 17th. In addition, the *Atlantic, Baltic,* and *Vanderbilt* brought troops and supplies to the beleaguered fort.

Because Vanderbilt was unable to use the *Illinois* on his European route, he disposed of her to her former owner, M. O. Roberts. The Army's Quartermaster Corps chartered her, at $1,500 per day, for the DuPont expedition to Port Royal. She steamed out of Hampton Roads on October 29, 1861. On her way, in adverse weather and poor visibility, she grounded off Hatteras shoals, losing one of her stacks, but got off without serious injury.[42] Continuing as an Army transport, she traveled up and down the East Coast throughout 1862 until she joined General Banks's expedition to Ship Island and New Orleans in December. She then operated as a transport from New York to New Orleans through August of 1863.

In October of 1862 Marshall O. Roberts reentered the California trade via the Nicaragua route that had been opened by Vanderbilt and closed by war, revolution, and the unsavory acts of his colleagues. Because the preparations in Nicaragua took longer than expected, Roberts shifted his first transits to Aspinwall and the Panama Railroad. Thus the *Illinois* went back on the route for which she had served for years. She sailed from New York on October 3, 1863, and had completed eight Aspinwall voyages when she arrived home on July 17, 1864.

After that she was on continuous trooping service for the Quartermaster Corps for the next year. On January 27, 1865, she was sold outright to that agency for the enormous sum of $400,000, roughly her original cost.[43] Roberts must have been overcome with delight. Her last passage under steam was from New Orleans on July 1, 1865, via Key West and Fortress Monroe, to New York on the 12th. By October she was advertised for sale but found no buyers.[44]

Somewhat later, when a hospital ship was needed at the New York Quarantine station off Staten Island, the War Department agreed to lend the *Illinois* for that purpose. Her engines were sold for scrap value and removed early in 1866. The space vacated was to be equipped with additional staterooms and washrooms in order to increase her capacity. When a British steamship, the *Virginia* of the National Line arrived at New York in April of 1866, crowded with 1,043 passengers, she reported twenty-one sick and the deaths of thirty-six passengers and two crew members from Asiatic cholera. The existing hospital ship, the old *Falcon*, although rotten and in poor condition, was used to take off the sick and the *Illinois* to receive the steerage passengers who had no symptoms.[45] Another infected steamship, the *England*, of the same line, soon arrived and helped fill both vessels. Unfortunately, the precautions did not prevent the dread disease from reaching shore.

In 1870 the *Illinois* was put on the balance drydock at New York for an extensive rebuilding. Jeremiah Simonson undertook the work that removed the masts, cut ports in her sides, added new deckhouses, installed kitchens that could feed 1,000 patients, and put in a small boiler and steam pump. Water for cooking and washing was distilled.[46] As the *Herald* described her, "When all the repairs and alterations designed shall have been perfected a more comfortable and commodious home than this ship for those sick with infectious and contagious diseases will not exist."[47] Five years later the British liner *Canada* ran into her and cut her hull almost to the waterline, but she remained afloat and was patched up and put back in service.[48] Despite occasional repairs, she gradually deteriorated and, in 1887, was found unserviceable, towed to the Brooklyn Navy Yard, and sold for scrap in January of 1888.[49]

So passed a vessel that thirty-seven years before had helped create a new pattern of design for the American steamship. In her day she had demonstrated that oscillating engines could be lighter, less expensive, and take up a smaller space than the side lever ones that preceeded them. Hers, unfortunately, were not so reliable as they should have been, but the ones that followed did perform with great regularity; they incorporated improvements that would not have been achieved without actual construction and testing at sea. Her career as an Atlantic liner was confined to the 1860 season, during which she made five voyages to Cowes and Havre for Vanderbilt's European line. She was its only steamship without a walking beam engine. A handsome steamer and a fast one, she reestablished the reputation of the Smith and Dimon yard after it had been besmirched by the *Georgia*.

The *Golden Age*, 1853-1890

The last ocean steamship to come from the shipyard of William H. Brown was a superb example of his original design and painstaking workmanship. Several myths, with no basis of fact, have been reprinted far too many times. Well-known sources say she was the fifth steamer for the Collins Line and even give her the erroneous name of *Adriatic*.[50] In other instances, William H. Webb is credited instead of William H. Brown.[51] But the Navy Department, which had to approve the design of any mail steamer, would never have approved the use of a walking beam engine exposed to shot and damage. Furthermore, Collins, a superb press agent, would have gleefully announced the start of a fifth ship for his line. The *Golden Age*, to use the name eventually chosen for the steamship, had very little published about her until she was finished. Brown seems to have been a retiring, although

Golden Age. Courtesy of The New-York Historical Society, New York City.

remarkably talented, person. He let his steamers' reputations speak for him.

George Steers, foreman in Brown's yard in 1852, was assigned the task of producing lines for the new vessel. He had just achieved international fame for the design of the yacht *America*. The *Golden Age* was large for her day, 272′-10″ by 41′-10″ by 25′-1″; her tonnage, 2,281, made her about three-quarters of the size of a Collins liner. She had been ordered by the Empire City Line for Panama to San Francisco service and was to be named after the latter city. Before her launching, the hull was sold to the New York and Australian Steam Navigation Company.

She was a handsome ship equipped with guards extending a long way forward and aft of the paddle boxes that were smoothly faired into the hull rather than forming local protuberances. These overhangs were supported by a series of sloping braces ahead and abaft the wheels. A three-masted bark rig had three square sails on the fore and the mainmasts, and an extensive spread of canvas. She was planked up to the spar deck forward but was open aft to provide sorely needed ventilation in the tropics. Extensive deckhouses containing highly ornamented saloons were fitted on the spar deck from the pilot house to the mizzenmast.

The machinery, supplied by Quintard and Merritt of the Morgan Iron Works, was unusual in two respects. First, the single cylinder beam engine of 83″ bore and 12′-0″ stroke was the largest yet installed in an ocean steamship.[52] Second, there were two enormous boilers 40′ long equipped with eight furnaces at each end and flues leading to uptakes and a single stack at the center of their length.[53] Her 27 psi steam pressure was higher than usual and her 34′ diameter wheels very large. Her final cost came to $400,000.[54] She was to operate from Australia, which like California had suddenly become important when gold was discovered there, to Panama. It would become possible to cross by the Panama Railroad, then partially complete, and make connections for England on the Royal Mail Steam Packet Company's steamships. She would coal at Tahiti on her way from Sydney to Panama.

During her trials on September 17, 1853, the *Golden Age* went out to the Sandy Hook Lightship, where she awaited the Collins liner scheduled to sail at noon. When the *Atlantic* appeared, the *Golden Age* followed her, spurted past, and crossed her bow to show off a speed of

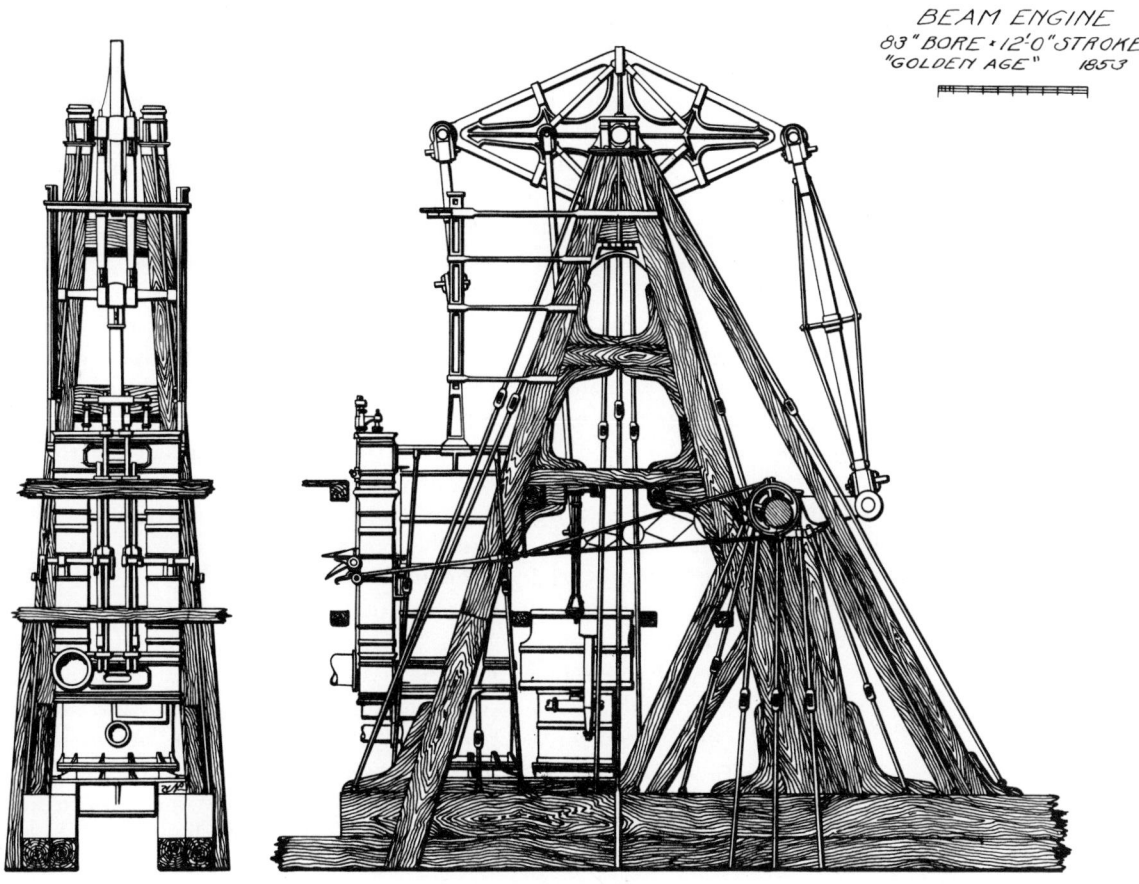

Walking Beam engine 83" bore by 12'-0" stroke for the *Golden Age*.

almost 15 knots. Like any such demonstration, this one was heavily weighted in the *Golden Age*'s favor, for she was light and the *Atlantic*, at the start of a voyage, was so loaded down with fuel that her wheels were deeply buried.

Lieutenant David D. Porter, U.S.N., formerly of the *Georgia*, was selected to deliver her. She had been advertised to sail as early as May 15, 1853, for Melbourne and Sydney, at fares of $350 and $375 for first-class cabins, $275 second, and $200 third. The date was put off until September 28, when the *Golden Age* left her pier. She got no farther than Staten Island and was forced to anchor off Quarantine.[55] Her 161 passengers had to wait two days while a mysterious hole that a disgruntled workman had drilled in one of her boilers was repaired. On September 30 she headed out to sea as the second beam-engined liner to cross the Atlantic. The *North Star* had preceded her in May. The *Golden Age* made Liverpool in 11 days 18 hours.[56] Passengers and cargo were taken on for the gold diggings in Victoria but Porter refused to carry the British mails because he considered the payment offered too low.[57] In a grand gesture, however, he would take private letters free. The *Golden Age*'s departure, on November 28, was marred by her running into a wharf on the Birkenhead side of the Mersey, which caused a week's delay to repair her cutwater and stem.[58] Setting out again on December 5, she coaled at St. Vincent and sailed from Capetown on January 16, 1854. Her first Australian call was at King George's Sound in forty-seven days of actual steaming time from England; fourteen more days were spent in port and coaling. Melbourne was reached February 20 and Sydney the 24th.

Coal was the great problem. No transpacific lines and no coaling stations existed. To feed her hungry furnaces, 1,200 tons had to be sent out by a fleet of sailing ships to Tahiti and then unloaded by hand. While the fuel problem was being tackled, the steamship spent three months in coastwise operation between Melbourne and Sydney.

The *Golden Age* left Melbourne on May 5, 1854,

carrying a number of officials as guests who spent a night aboard when the steamer anchored. They left the next morning as she proceeded on to Sydney.[59] On the 11th, having taken on additional passengers, she headed for Tahiti and made a 13 day 12 hour passsage at 10 1/2 knots. Her best days' run was 272 miles.[60] From Tahiti, on May 30, another nineteen days were required to reach Panama. About eighty of her 170 passengers and $200,000 in gold crossed the Isthmus and were loaded aboard the Royal Mail Line's *Magdalena*.[61] They reached England in sixty-seven days from Australia. Because the cost of transporting coal from England was excessive, the owners abandoned their grandiose plans for a large fleet and monthly transpacific sailings. Like many a pioneer, the *Golden Age* had operated at a loss.

The Pacific Mail Steamship Company, in need of larger steamships for their California service, bought the *Golden Age* and found in her exactly what they wanted. Her large passenger capacity, 1,200 in three classes, and the quarters arrangement incorporating overhanging guards and excellent ventilation, were ideal for tropical waters. Subsequent sailings demonstrated that a single walking beam engine was remarkably reliable. As a result, later Pacific Mail steamships were based on a combination of the best characteristics of the *John L. Stephens,* which had arrived on the West Coast in 1853, and the *Golden Age*. These two, together with the *Golden Gate* and the *Sonora*, became the line's Panama-San Francisco fleet for the next decade.

On October 1, 1854, the *Golden Age* took aboard the passengers and mails brought down from New York by the *North Star* and sailed from Panama for the first time.[62] Further voyages were routine until she departed San Francisco on April 17, 1855, with 750 passengers and $1,300,000 in gold. On the night of April 28, about 200 miles from Panama, she struck a rock and had to be beached near Quicara Island to prevent her from sinking. The *John L. Stephens*, northbound, took off the passengers and treasure and turned back to Panama, where both were landed on May 3.[63] The leak was controlled by a sail pulled under the *Golden Age*'s bow, the ship pumped out, refloated, repaired without difficulty, and put back into service by the middle of June.

In 1858 she underwent other repairs and it may have been at this time that her mizzenmast was removed. It also seems that some modifications must have been made to her engine or her boilers, for when she went back into service she was credited with a new speed record of 11 days 7 1/2 hours from Panama to San Francisco. Subtracting the stops at Acapulco and Manzanillo, the running time became 10 days 23 1/2 hours, corresponding to 12.3 knots.[64] The *Golden Age* became the Pacific Mail's largest and fastest liner. Furthermore, she must have been one of the most economical, for she was kept on for fifteen years of hard, continuous work.

In 1869, however, she was relieved of her coastwise duties and sent across the Pacific to join the *Costa Rica*, *New York*, and *Oregonian* on the line's new Yokohama-Shanghai route. On the way, three days before reaching Yokohama, she ran into a typhoon on October 15 that stove a paddle box and, somehow, damaged one boiler. Perhaps it broke free from its foundation.[65]

After almost five years on the Shanghai branch, she was sold for $160,000 to the Mitsubishi Mail Steamship Company, which took over the entire Shanghai operation from the Pacific Mail. The transaction was completed in December of 1874 and the *Golden Age* renamed *Hiroshima Maru*. She was still active on the Japanese coast between Yokohama and Yokkaichi early in 1890, but in August was at Kanagawa for breaking up.[66]

To the best of the author's knowledge, the *Golden Age* had the longest career as an operating steamship of any wooden side-wheeler that ever carried passengers across the Atlantic. Thirty-seven years is unique. A delivery voyage almost around the world from New York to Panama was an auspicious start, even though it was interrupted by three months on the Australian coast. Fifteen years on the San Francisco to Panama route followed by five from Japan to China made her one of Pacific Mail Steamship Company's most successful steamships. Then another fifteen in Japanese hands brought her career to a close. But even then she outlived the exceptionally well-built *Costa Rica* and *New York* which, although a decade younger, were scrapped ahead of her. She was William H. Brown's last ocean steamship and truly his masterpiece.

The *Tennessee*, 1854-1865

Almost all of the side-wheel steamships discussed to date had been built either on the East River side of Manhattan Island or directly across that body of water in what is now called Brooklyn but in the 1850s was subdivided into the individual towns of Greenpoint, Williamsburg, or a much smaller Brooklyn. Most of the screw steamships came from Philadelphia's Delaware River shipyards. There was still a third maritime city with a long history of wood shipbuilding and a more recent one of iron works capable of producing engines and boilers. Baltimore had already built many steamboats plus one very successful coastwise steamship, the *Isabel*, in 1848. The keel for its second oceangoing steamship was laid in 1853 at the John A. Robb shipyard. She had been ordered by the Baltimore and Southern Packet Company, James Hooper, President, to run to Charleston.

In most respects she was almost identical to the *Nashville* and measured 210′ by 33′-11″ by 16′-11 1/2″, with a tonnage of 1,149. Her topsail schooner rig was similar, but her bowsprit and topmasts were shortened. The carved trailboards and ornamental billet head were omitted. The chief difference was a walking beam engine of 72″ bore and 9′-0″ stroke instead of a side lever one, turning 28′ wheels. Although not nearly so handsome, she was the *Nashville*'s equal in other respects. She was named *Tennessee* and, just like the Webb-built *Tennessee* of 1849, had the state's arms emblazoned on her paddle boxes. She was designed for one-hundred passengers and 5,000 barrels of cargo on a draft of 13′-6″.[67]

At her launching, the tallow used to grease the sliding ways squeezed out and she stuck. Several days were required to block her up, dismantle the ways, and regrease them; the second attempt, August 31, 1853, proved so successful that she started ahead of time.[68]

Her engine and a pair of return-flue boilers were built and installed by Charles Reeder, Jr. On trial February 7, 1854, she sailed from Pell's Point with Captain Lewis Parish in command. The steam pressure varied from 8 to 25 psi (18-20 was considered normal), and covered a 13 1/2 mile distance in 53 minutes.[69] She departed from Baltimore on March 14 with a full load of freight and with both Messrs. Hooper and Reeder aboard to see how she performed.[70]

The *Tennessee* came home in fifty-four hours from Charleston, arriving March 27 and bearing a large cargo of rice and cotton, fifty-nine passengers in the first class, and twenty-nine in steerage. It was the first of twenty-eight voyages, usually at two-week intervals, between Baltimore and Charleston. One way fares were $17.50 first and $8.00 second cabin and included meals. The northbound passage times ranged from forty-six to fifty-six hours.

While she was considered an excellent ship, properly run, there just was not enough year-round traffic to make the *Tennessee* profitable. After arriving home on February 26, 1855, she was put up for auction on March 22.[71] Apparently no satisfactory bid was received, for on June 16 the *Tennessee* cleared Baltimore for Southampton, still with James Hooper and Company as her agent.[72] There were no announcements and no advertisements; she just departed.

Westbound, Mortimer Livingston of the Havre Line was listed as her consignee. It is likely that he had chartered her as an extra steamer to carry freight that had piled up at Havre awaiting delivery to the United States. The *Tennessee* left Southampton August 17, 1855, and Havre on the 19th; after eleven days of punishing weather she put into Halifax, in damaged condition, on September 4 for coal. Three days later she reached New York.[73]

Sometime before the end of the year she was acquired by S. de Agreda Jove and Company for the first American steamship service to South America.[74] The *Tennessee* left New York on January 8, 1856, for La Guaira and Porto Cabello, Venezuela, with way stops at San Juan, Puerto Rico, and St. Thomas, Virgin Islands. This pioneering voyage was completed, on February 9, as the first of five trips to Venezuela, the last starting June 21, 1856, and ending July 25. Thereafter, a shorter one to San Juan and St. Thomas (but not to Venezuela) ensued.[75] We can infer an unprofitable operation, for the line ceased and the *Tennessee* changed hands. Her last Venezuela trip had only fifty-six passengers and a moderate freight.[76]

Charles Morgan and his associates, who had annexed Cornelius Vanderbilt's Nicaraguan operations, bought the *Tennessee*, jammed her with extra berths, and sent her out from New York on October 6, 1856, to San Juan del Norte. From there she made New Orleans on October 22. Further trips from both New Orleans and New York followed. "General" William Walker was leading a group of American soldiers of fortune at that time, and he attempted to seize Nicaragua for the benefit of himself and, supposedly, for the owners of the *Tennessee*. Their steamships frequently carried recruits to Walker's army but called them "immigrants" about to settle in Nicaragua. On December 24, 1856, the *Tennessee* had three-hundred aboard. All legal niceties had been settled between Morgan and the New York District Attorney, John McKeon, to make sure they could leave without interference.

On Christmas Day the elements intervened; a gale was met, the paddle shaft broken, and the steamship forced to put into Norfolk.[77] The recruits proceeded on a chartered Charleston liner, the *James Adger*. The *Tennessee* was quickly repaired and sailed again on January 29, 1857, with 300 paying passengers and 250 immigrants—an uncomfortable total for so small a vessel. After this sailing, military matters closed the Nicaragua crossing and the *Tennessee* transferred to Aspinwall, sometimes by way of San Juan. Late in the summer she came home, via San Juan, with 275 "deserters" (as the immigrants were now called) from Walker's defunct army. After they had disembarked at New York on the 18th, all Nicaraguan steamship service came to a halt.

Morgan sent the *Tennessee* to the Gulf of Mexico, where he had been continuously engaged in steamship operation ever since 1837. She left New York on September 30, 1857, for New Orleans and was put to work from there to Vera Cruz, Mexico, twice a month. Aside from an occasional trip to New York for docking, repairs, and machinery overhaul, she found herself continuously and gainfully employed until the outbreak of the Civil War. Together with several other Morgan vessels, she was seized by the Confederate authorities but lay unused at her New Orleans wharf. She was in first-rate condition, having had a new set of boilers installed during an 1860 call at New York.

When New Orleans surrendered to the Federal forces, the *Tennessee* was seized, on April 29, 1862, by a contingent of Admiral David Farragut's marines. Within a week's time she was manned with a crew from the U.S.S. *Varuna*, given a token armament of two 20-pound Parrott rifles, and began to tow sailing colliers up the Mississippi to fuel Farragut's steamships.[78] Her moderate size, relatively light draft, and high speed made her an extremely useful addition to the Union fleet as it worked its way laboriously up the Mississippi toward Vicksburg.

The *Tennessee* carried coal, fresh provisions, powder, and shell upriver and dispatches down. Farragut found that her passenger cabins could take his entire staff and that she could get into many waters that excluded his large, lumbering flagship, the *Hartford*. Starting in the spring of 1863, he employed the *Tennessee* as his flagship whenever mobility became essential. He was aboard her and attacking Port Hudson when Vicksburg surrendered, July 4, 1863. His ships had split the Confederacy in two.[79]

Sometimes a flagship, sometimes a dispatch vessel, she was kept busy until after the Battle of Mobile Bay. As a result of that victory, the captured Confederate ironclad *Tennessee* became a Union warship. To avoid confusion, the Secretary of the Navy ordered, on September 1, 1864, that Farragut's *Tennessee* be renamed *Mobile*.[80] She was sent to the Western Gulf of Mexico, where she chased blockade runners and captured the *Friendship, Fox, Montgomery, Emily*, and *Louisa*, most of them schooners.[81] Another runner, the *Jane*, was destroyed to avoid their fate.

Off the Rio Grande, in October of 1864, the *Mobile* was caught at anchor by a sudden storm and was swept by waves that carried part of her battery overboard and strained her hull. Because her boilers were now in sad condition as a result of long, hard use, Farragut regretfully sent her to New York for overhaul. She saw no further active service. On March 30, 1865, she was sold to Russell Sturgis for a mere $25,000.[82]

The next two months were spent installing completely new passenger quarters.[83] Once more her name was changed, and the *Mobile*, ex-*Tennessee*, became the *Republic* before being chartered to William H. Robson's New York-New Orleans line. Her first postwar departure took place on May 13, 1865. Further sailings were made on June 15 and July 19 but, while the third voyage was under way, her charter was taken over by H. B. Cromwell and Company, who retained her on the same route.

Her fifth New Orleans voyage started on October 18, 1865. The *Republic* anchored overnight off Staten Island and proceeded the next day. Near Savannah she ran into the hurricane that almost sank the *North Star* and disabled the *Northern Light*. On the 24th, heading into an overwhelming wind and sea, her slowly turning wheels stalled and would not carry the single engine past dead center; she was left powerless, drifting, and at the mercy of the elements. Paddle boxes and the bulwarks were washed away. Steam was raised on the donkey boiler to start all the independent pumps, but the next day, as cargo was being thrown overboard to lighten the laboring ship, that boiler broke down and water poured into her holds. A raft was constructed from three spars and four boats were launched before the wreck broke in two during the late afternoon of October 25. Those in the boats were rescued but the eighteen last seen on the wave-swept raft perished.[84]

The actual cause of loss is unique; the author knows of no other case where an engine stuck on dead center in a gale. The catastrophe ended the career of a very useful vessel that had demonstrated that Baltimore's builders could produce just as satisfactory a steamship as New York's. Admiral Farragut wrote of her ". . .she is so fast and so fine a vessel that I was unwilling to let her go. . . ."[85]

The *St. Louis*, 1854-1879

In 1853 the Pacific Mail Steamship Company awarded contracts to Jacob Westervelt and Company for two steamships. Although intended to be exactly alike, the fact that they were built consecutively, rather than simultaneously, may have led to minor differences.[86] The first, completed in March of 1854, was named *Sonora*, and the second, which followed four months later, became the *St. Louis*. The latter was 270'-2" by 35'-2" by 25'-7", with a tonnage of 1,621 to the *Sonora*'s 1,616. They were unusual in that the boilers were aft of their engines, British style. No other American steamship of the 1850 decade was so arranged. The *St. Louis*, like so many other mail steamers, was built under the supervision of Captain William Skiddy. Because she was intended for tropical operation, she had overhanging guards extending well beyond the wheels in the lengthwise direction, and a dining saloon on deck where the windows could be opened to keep it cool.[87] Its interior paneling consisted of polished oak with white panels and gold decorations. The luxurious quarters for first-cabin passengers included a nursery to keep the children out of the way. There was ample space for a baker, doctor, barber, children's attendant, and the others needed to keep 160 first-class passengers in the best of moods. Steerage quarters took care of 600 to 700 less-fortunate travelers. A two-masted topsail schooner rig had a small sail area and the stem curved slightly forward to support a short bowsprit.

The machinery came from the Morgan Iron Works and consisted of a pair of walking beam engines of 50″ bore and 10′-0″ stroke. Two return-flue boilers 30′ by 13′ by 12′ were fired by six furnaces apiece and supplied steam at 22 psi.[88] There was bunker space for 550 tons of bituminous coal (no anthracite was shipped to the West Coast). Each 30′ side wheel had twenty-eight buckets 9′-0″ x 16″. Her most unusual feature was the presence of two completely different condensers, the older jet type for the starboard engine and one boiler and a Pirsson surface condenser to port. As a result, there were two independent power plants whose coal consumption and maintenance problems could be compared. The draft was low, 14′-4″ loaded, to provide the stability necessitated by two heavy walking beams placed high above the deck.

A gala trial, complete with invited guests and a banquet, was held: William Aspinwall, of the Pacific Mail, was host; Mr. Fox, of the Havre Line, a guest; Captain Skiddy, in general charge; John McGimm, a veteran New York pilot, in command; and Messrs. Quintard, of the Morgan Works, and Pirsson were present to see that the engines and the condenser functioned properly.[89] The date was July 21, 1854.

Instead of heading for the Pacific, the *St. Louis* was chartered by the Havre Line to replace the *Franklin*, which grounded on Long Island on July 17, 1854, and conversion work was begun to make her fit for the stormy North Atlantic. All the projecting guards were removed to avoid damage by waves.

The modifications held her up until August 1, when she departed under the command of Asa Eldridge. She arrived at Cowes on August 13 in 11 days 17 hours and continued on to Havre the same day. Her return took 11 days 15 hours 30 minutes from Cowes and 13 days 17 hours from Havre. Eastbound she carried steam at 17 to 22 psi and made daily runs of 255 to 295 miles with a low fuel consumption of 25 to 32 tons per day. One hundred and sixty-two passengers and 350 tons of German and French goods were delivered to New York on September 12.

Between August of 1854 and June of 1855 she made five voyages between New York and Havre, with the smaller *Union* as her running mate. She was then laid up, but when the *Union* had to turn back with a broken shaft, she was coaled and loaded with the *Union*'s passengers and cargo, all within two days, and sailed for Havre on October 29, 1855. This became her sixth and last transatlantic voyage. After arriving home on December 6, she was turned over to the U.S. Mail Steamship Company, which had purchased her from the Pacific Mail for $250,000 in August.[90]

Her first trip to the Isthmus of Panama was a special one to transport troops westward. There were Indian problems in the Pacific Northwest. She left New York on December 10 for Old Point Comfort, Virginia, where she took on 860 officers and men of the 9th Regiment, U.S. Infantry, for Aspinwall.[91] Heavy weather and seasick soldiers were the rule, and all concerned were pleased when the troops boarded the Panama Railroad the day before Christmas and headed for the Pacific.

Her next departure, however, was a normal U.S. Mail sailing on January 21, 1856, as a temporary substitute for the *George Law*. Thereafter she was kept as a spare boat and not used again until October of 1857, when she replaced the *Central America*, formerly *George Law*, which had foundered in a hurricane. This was followed by a chartered trip to New Orleans for William Nelson and Sons in December of the same year.

The year 1858 saw three Aspinwall voyages, from New York in March, July and November, as well as one to New Orleans in September. Her final trip for the U.S. Mail Line started on April 5, 1859, as a substitute for the *Illinois*, incapacitated by an engine breakdown. It seems wasteful that the *St. Louis* was purchased at practically her original price and then used for only seven Aspinwall and two New Orleans voyages in the next four years. In 1859, when the U.S. Mail ceased operation, she was reacquired by the Pacific Mail, which sent her to New Orleans in February and May of 1860. Then, presumably after modifications to suit Pacific conditions, she steamed South from New York on November 22 bound for San Francisco by way of Rio de Janeiro, the Strait of Magellan, Valparaiso, and Panama. Early in February of 1861 she reached San Francisco and departed for Panama on the 21st. Some seven years late, she finally undertook the service for which she had been designed. Despite the interval, she became so successful that she was kept on it until 1866 when newer and larger steamships took over. In 1873 she was listed on the line's local Panama to Acapulco operation.[92] In 1874 she was not. Thereafter came extensive periods of idleness, but in 1877 she was sent to Panama Bay as a storehouse and workshop for the Pacific Mail.[93] Two years later she was still there but had become a dormitory for the line's stevedores.[94]

The *Quaker City*, 1854-1871

The year 1854 was a banner year that produced many outstanding steamships. The *Nashville* and *St. Louis* were completed in New York and the second *Tennessee* in Baltimore. A fourth, and equally successful coaster was the *Quaker City* from Philadelphia, a place famous for its propellers. She, however, was a side-wheeler.

The American Steamship Company, M. V. Baker, President, wanted a fast liner to operate between Philadelphia and Charleston; she was built by Vaughan and Lynn and engined by Merrick and Sons. Her dimensions of 234.4′ × 35.7′ × 19.8′ (the Philadelphia Custom House used decimal dimensions while the New York one quoted feet and inches) and a tonnage of 1,426 put her size midway between the *Tennessee* and the *St. Louis*. Her machinery shared the characteristics of a coastal and an Atlantic liner. It had a single side lever engine of 85″ bore and 8′-0″ stroke, as in the *Nashville*. But instead of the coaster's usual two boilers and single stack, she had the four boilers and two stacks of a liner.

A handsome steamer, she looked somewhat like the *Illinois*, with a two-masted rig, instead of three, and a wider stack spacing. The paddle shaft was very far aft, only 91′ forward of the stern post.[95] Two complete decks were fitted and the power plant took up 76′-4″, or almost a third of her length.[96] The usual Charleston steamer arrangement of a forecastle and a long after deckhouse was adopted.

The machinery design was progressive. The small boilers, each 14′ long by 10′-4″ by 11′-1″, were of the water-tube type, with two hundred individual tubes of 3″ diameter. To keep them from scaling up, a Pirsson surface condenser was installed. The wheels were 31′-0″ in diameter and had 10′-0″ × 1′-8″ floats dipping 5′-3″ into the water at a designed draft of 12′-0″.

The *Quaker City* was launched on May 2, 1854, and completed so rapidly that she ran trials on the Delaware River September 6 and 7. At the time she had 205 tons of coal aboard (her total capacity was 270) and displaced 1,475 tons at a draft of 9′-9″ forward and 11′-6″ aft. With steam at 21 psi and wheels turning 14.15 rpm, she covered seventy-seven miles upstream in five hours, aided by a two-knot tidal current. The speed through the water was 13.4 knots and indicator cards showed that her engine produced 750 horsepower.

Her agents, Heron and Martin, advertised fares of $20 first cabin and $6 steerage, Philadelphia to Charleston.[97] Under Captain J. H. Hodgden, she steamed on this route from the fall of 1854 to the summer of 1855. The service does not seem to have paid; Philadelphia's trade to Charleston was far more limited than New York's. Mortimer Livingston, of the Havre Line, had planned to charter her and even advertised that she would leave New York for Havre on June 23, 1855;[98] the *Tennessee* was used instead and, in August of 1855, the *Quaker City* was offered for sale.[99]

She was bought by a group of Mobile, Alabama, merchants who employed her between that city and New York.[100] She made her first southbound trip on December 5, 1855, via Havana. A second voyage followed in January. Off Sandy Hook on February 13, ice delayed her overnight, but she was able to dock next day. There she found another organization anxiously awaiting her.

The *Pacific* was overdue and the *Quaker City* had been obtained as a replacement. In forty-eight hours of hectic activity she was unloaded, coaled, and provisioned, and she sailed on February 16, 1856, with the passengers and mails intended for that ill-fated Collins liner. She arrived at Liverpool on March 2, sailed on the 5th, and was home the 21st. Bad weather slowed her in both directions; times were fourteen-and-a-half days eastbound and fifteen-and-a-half returning. After this precipitant introduction to the turbulent North Atlantic, she resumed her calmer Havana and Mobile sailings, eventually completing nineteen voyages; the last began at New York on November 7, 1857, and ended there on the 30th. Her Mobile owners' expectations of continued success were dashed by the severe financial panic of 1857.

She lay idle at New York for almost a year before acquisition by Hargous Brothers for their Louisiana and Tehuantepec Company, which had been set up to develop a new route to California via Mexico's Isthmus of Tehuantepec. The land journey was longer than those across Panama or Nicaragua, but, as the northernmost possible crossing, the sea part would be much shorter. Moreover, the Post Office was interested in the route. It had awarded a mail contract of $250,000 for a year's trial, with departure required from New Orleans twice a month.[101] The mail contractors provided a steamship across the Gulf of Mexico to the mouth of the Coatzacoalcos River and paid the Pacific Mail Company to take on the passengers and mails at Ventosa, on the Pacific side.[102]

Under the command of R. W. Shufeldt, the *Quaker City* left New York for New Orleans on October 9, 1858, to begin sailings to that hostile Mexican shore. She had to cross a dangerous bar at the mouth of the Coatzacoalcos River in order to transfer her passengers to a small steamboat, the *Suchil*, which would take them from the town of Minatitlan up to the village of Suchil, where they would board stage coaches for 120 miles of comfortless travel to Ventosa.[103] From New Orleans on October 27, the *Quaker City* met good weather and found no problems at her Mexican port. The mails reached San Francisco on November 14 in record time.

On other occasions during her five-month service, seas were often so high that she could not cross the bar at the river's mouth; the mails and the loudly protesting passengers missed their connections on the Pacific side. Returning from her last trip on the route, the *Quaker City* came up the Mississippi on April 8, 1859. So far the new service had not been popular and she had never reported more than 123 passengers, eastbound. She was replaced by a smaller vessel, the *Habana*, and returned to New York.

Starting on April 27, 1859, still owned by Hargous and Company, she began regular and, for the first time in her career, continuously profitable sailings from New York to

New Orleans via Havana. As a fast steamship with excellent accommodations, she attracted the passengers and freight that would keep her operating until the outbreak of the Civil War. In July of 1859, with a time of 3 days 14 hours Havana to New York, she made a remarkably fast passage,[104] at the phenomenal speed of 14.2 knots, well above the *Illinois*'s record of 13.5. Later that year she suffered a most unusual accident.[105] On October 7, southbound and about fifty miles past Cape Hatteras, one of her engine's side levers broke and the crosstail between the levers tore itself free; nevertheless, it still remained attached to the connecting rod and the latter was, in turn, left on the crank pin. The momentum of the ship kept the wheels turning and the attached rod and heavy forging were carried around through a number of revolutions battering everything within reach. Engine columns were shattered, knocked out of their sockets, and the deck overhead was smashed into splinters. By the time the ship drifted to a stop, the terrific pounding had destroyed much of the engine except for the cylinder and air pump, both out of reach of the flailing weights. Luckily, the main steam pipe, although bent, did not give way and no one was scalded. Water flowed in through broken pipes and the ship was thought to be sinking. The donkey pumps were started and, after a time, kept her clear of water. About one hundred of her passengers were transferred to a bark, the *Dunbarton*, of Boston, together with the necessary provisions for feeding them. The *State of Georgia* towed the *Quaker City* to Norfolk on October 12, where the engine was rebuilt, presumably with new parts sent down from Philadelphia. The ability of marine engineers to rise to an emergency and quickly produce large, heavy pieces of machinery is amazing. In two months' time the *Quaker City* was under steam; she arrived at New York on December 16, 1859, and cleared for New Orleans on Christmas Eve.

Her last prewar sailing to New Orleans left April 5, 1861, and brought her home on the 19th. She was loaded and ready for another but, President Lincoln having proclaimed a blockade of all Southern ports, was taken over on April 25 by the Union Defense Committee of New York. This private patriotic group, having ample funds, chartered the *Quaker City* for thirty days at $1,000 per day; afterward they could either purchase her for $140,000 or renew the charter.[106] Furthermore, the Committee had her armed with two 32-pounders and two 6-pound rifles and fitted out as a cruiser. Although the work was done at the Brooklyn Navy Yard, the charterers paid a grossly inflated bill, hired the crew, and, on April 29, 1861, sent her out under the command of Samuel W. Mather. As a privately armed cruiser without legal papers of any kind, she was unique. She reported to the Flag Officer at Hampton Roads, who was delighted to have her and put her to work at the mouth of Chesapeake Bay examining vessels coming in from the sea and detaining those belonging to Southern owners. Captain Mather was given a Naval Commission on June 1, 1861, but the crew remained civilian and all expenses were still the responsibility of the Union Defense Committee. The Norfolk-bound ship *North Carolina* became her first capture on May 14, 1861.[107]

After further charters by the Defense Committe, then by the Navy Department, the *Quaker City* was purchased by the latter in August for $117,500. She had been under steam for seventy-nine days and had stopped eight Southern craft, when she left Fortress Monroe on September 7, 1861, for the Brooklyn Navy Yard to be refitted for naval service.[108] Her armament was revised and her civilian crew replaced. In December she was sent to the West Indies to search for the Confederate raider *Sumter*. The trip out to St. Thomas proved a great disappointment to her new commander, James M. Frailey. He arrived with three separate leaks in the starboard boiler and had made less than 10 knots.[109] The overworked Navy Yard had given his ship no more than a lick and a promise.

Most of 1862 was spent with the East Gulf Blockading Squadron, cruising off Nassau, where the blockade runners loaded and waited for dark nights and bad weather to dash across to Southern ports. In spite of aggravating problems with her eight-year-old boilers, a number of prizes were captured, including the iron steamer *Adela* of 300 tons.[110] Eventually the boiler situation became so acute that in September she had to be sent to Philadelphia for corrective work. Southbound in October, she grounded off North Edisto but was got off without much damage.

She then joined the Charleston blockading fleet. There she had the misfortune, on the foggy morning of January 31, 1863, to meet the Confederate ironclads *Chicora* and *Palmetto State*. Other Union vessels involved were the *Augusta*, *Housatonic*, *Keystone State*, and *Mercedita*. The *Mercedita* was rammed, the *Keystone State* set afire, and both had their boilers pierced by gunfire. The *Quaker City* was hit by four shells, one of which burst in her engine room and carried away a major brace between the cylinder and bedplate. One of the rods operating her air pump was broken and the top of the pump's dome blown off. Chief Engineer George W. Farrer proved his worth by keeping his ship going throughout the action and making temporary parts so that she could continue on the blockade.[111] Over a month later, northbound for permanent repairs, she fell in with the steamer *Douro* running out of Wilmington with a cargo of 420 bales of cotton and captured her. The *Quaker City* steamed proudly into New York on March 12, 1863, followed by her prize.

No sooner had she returned to duty, this time off

Wilmington, North Carolina, than the *Quaker City* sustained new damage when she collided with the U.S.S. *Connecticut* and had to go North again to have her stem replaced.[112]

In the meantime, the *Douro* had been condemned and sold for a high price of $150,538.[113] In what must have been a remarkably well-planned shuffling of ownership and papers, she steamed to Nassau and resumed her blockade running. On October 11, 1863, when she came out of Wilmington again, the U.S.S. *Nansemond* ran her ashore and burned the wreck.[114]

The *Quaker City* returned to the waters off Wilmington in December of 1863 and remained there throughout the next year. In December of 1864 and January of 1865 she became part of Admiral Porter's fleet attacking Fort Fisher and, after the second successful attack, transported wounded to Norfolk. She was then sent to the Gulf of Mexico as a blockader off the Texas coast.[115] There she captured several sailing craft. In April, while at New Orleans, she pursued the Confederate ram *William H. Webb* until the latter was beached and burned by her own crew. Afterward the *Quaker City* headed for Philadelphia for overhaul, arriving there May 10, 1865.[116] She was auctioned off on June 20 for $35,000, a bargain price, to Arthur Leary.

Like the *Vanderbilt*, she had endless boiler problems throughout the war. In theory, her surface condenser should have prevented them. In practice, a ship stationed off a Southern port had no means of procuring fresh water to replace that which escaped through the safety valves, around the piston rod, and through leaky valves. Her engineers had to pump sea water into the boilers. Although it took longer, salt accumulated and the scale problems of a salt-water boiler eventually developed. Should the condenser tubes leak, as they often did, the scale accumulated rapidly.

The Leary brothers spent almost five months reconditioning their newly purchased steamship before sending her out, in August of 1865, on the first of many voyages from New York to Charleston. Despite the overhaul, she reached the latter port with a damaged engine on August 27, 1865. Her exhausted engineers had to work the valves by hand for the last twenty-four hours of the trip.[117] After corrective work she was joined by two other steamers, the *Andalusia* and *Alhambra*, to provide twice-weekly sailings. In effect, the Learys took over Spofford and Tileston's prewar service.

Before the year was out, the seaworthiness of the *Quaker City* was severely tested by the hurricane that sank the *Tennessee*, drove many sailing vessels ashore, and damaged the *North Star* and *Northern Light*. She left New York October 21 and met rising winds and seas that stove in her paddle boxes.[118] Simultaneously, she plunged into the waves so violently that her bowsprit was torn off. At Charleston she found one boatload of survivors from the *Republic*, ex-*Tennessee*, and returned them to New York. The year 1867 marked her last Charleston trip; she docked at New York March 26.

The *Quaker City* had been advertised for a Mediterranean cruise, the first ever actually to sail from an American port. It was implied that General W. T. Sherman and other dignitaries would travel on her.[119] The General did not materialize, but a journalist by the pseudonym of Mark Twain did. A series of newspaper articles and a volume (most libraries catalogue it under *Fiction*) named *The Innocents Abroad* recorded his impressions of the trip and of his fellow passengers. The five-month tour, covering 14,000 miles at a fare of $1,250, began on June 8, 1867. Dozens of ports were made, including Gibraltar, Marseilles, Leghorn, Athens, Constantinople, Sebastopol, Beirut, and Alexandria. The *Quaker City*'s passengers toured Paris, Rome, the Holy Land, Egypt, and other inland sites. The exhausted sightseers breathed more freely on October 24, when their ship finally turned westward through the Pillars of Hercules. Madeira was left astern on the 28th, Bermuda on November 11; the passengers disembarked at New York on November 19, 1867. For the *Quaker City* there was now no employment in sight, but only a dismal lay-up.

Next year she was chartered by Ruger Brothers for a single voyage to Bremen, sailing from New York on May 30, 1868, and from Bremen June 27 with 497 immigrants.[120] They reached New York in eighteen days.

In 1869 the *Quaker City* came out of lay-up on a charter to Livingston and Fox, for a voyage to Havanna and New Orleans. She cleared at New York on February 27. Northbound she suffered from a machinery "disarrangement" while leaving Havana.[121] Ruger Brothers had planned to send her to Europe, but as a result of the delay they acquired the *Ocean Queen* instead.

In April of 1869 the *Quaker City* was repaired, docked, caulked, and given a new suit of metal sheathing, a sure sign of further use. Then, on June 12, she cleared New York as the British steamer *Columbia*, bound for Jamaica. In fact, she had been bought by Haitian revolutionists as part of a newly organized naval force. Her departure was delayed until June 17 because of suspected arms shipments, but she was eventually allowed to go to sea.[122]

After arriving at Jamaica with a perfectly innocuous cargo of flour and provisions, she unloaded part of it and made for St. Marc, Haiti, where guns had been assembled for her and another American steamer, the *Florida*. The revolutionists armed both and in August they bombarded the town of Gonaives. In September they met two of the Haitian navy's vessels, the *Salnave*, ex-U.S.S. *Maratanza*, and *Alexandre Petion*, ex-*Galatea*

Quaker City at Naples in 1867. *Courtesy of The Mariners Museum, Newport News, Va.*

and a sister of the *Carroll, Worcester, Somerset* trio. When their gunfire disabled the *Salnave*,[123] the *Petion* towed her into Cape Haitien. Later, the *Quaker City* and the *Florida* ventured into the harbor, sank the *Salnave*, and captured her consort.[124]

On December 18, 1869, the rebel fleet arrived off Port au Prince, shelled the presidential palace, and landed troops. The town surrendered, President Salnave was executed, and the revolutionists, now that they were successful, became patriots.

The (unnamed) American captain of the *Quaker City* left Haiti in January; the vessel was sold at auction for $8,000 in June.[125]

By February of 1871 she had become the property of Haiti's Admiral Dejois, who loaded her with five-hundred tons of logwood and started for New York to sell the ship, now called the *Republic*, as well as the cargo. On the way a boiler exploded and left her in sinking condition. The Admiral and his far-from-competent crew were rescued and taken to Bermuda.[126]

Such was the catastrophic end of a highly varied career. She had made three Atlantic voyages, one for the Collins Line in 1856, one as the first American cruising liner in 1867, and one as a steerage steamer to Bremen in 1868. As a coastwise steamship, the *Quaker City* had served Philadelphia, New York, Charleston, Mobile, Havana, and New Orleans. In 1858 she opened the first mail route from New Orleans to California via Minatitlan and Mexico's Isthmus of Tehuantepec. During the Civil War she proved invaluable to the U.S. Navy in maintaining blockades off Florida, Charleston, Nassau, Wilmington, and the Texas coast. Her light draft and high speed enabled her to get into narrow waters and to chase and catch the fast but fragile steamships that continuously tried to break through the Federal squadrons. She was a well-designed, splendidly constructed steamship that upheld Philadelphia's century-old shipbuilding reputation.

Notes

1. William H. Webb, Certificate Book, vol. 1.
2. *New Orleans Picayune*, March 12, 13, and 18, 1851.
3. *New York Herald*, April 9, 1851.
4. *New Orleans Picayune*, July 17, 1851.
5. *New York Herald*, April 14, 1853.
6. *New York Commercial Advertiser*, June 2, 1854.
7. Ibid., April 5, 1855.
8. Ibid., May 28, 1855.
9. Ibid., October 25, 1855.
10. *New York Tribune*, March 12, 1856; *Morning Courier and New York Enquirer*, March 5, 1856.
11. Ginsberg records.

12. *Journal of the Franklin Institute* 27, (January-June 1854): 200.
13. *New York Herald*, March 6, 1854.
14. Ibid., March 6, 1863, and November 26, 1865.
15. Ibid., November 26 and December 2, 1861.
16. Official Records, ser. 1, 1: 745-49.
17. J. Thomas Scharf, *History of the Confederate States Navy* (New York, 1887), p. 797.
18. Marcus W. Price, "Ships That Tested the Blockade of the Carolina Ports, 1861-1865," *The American Neptune* (1948), p. 226.
19. *New York Herald*, September 4, 1862.
20. William Morrison Robinson, Jr., *The Confederate Privateers* (New Haven, Conn., 1928), pp. 129-30.
21. Official Records, ser. 1, 13: 697-707.
22. Undated newspaper clipping, William H. Webb scrapbook. Webb Institute.
23. *New York Herald*, January 1, 1858.
24. Ibid., September 18, 1857.
25. *New York Commercial Advertiser*, January 29, 1861; *Frank Leslie's Illustrated Newspaper*, February 16, 1861.
26. *New York Herald*, May 13, 1861.
27. Ibid., November 3, 1865.
28. Ibid., January 31, 1872.
29. *New York Times*, January 30 and March 12, 1873.
30. *New York Journal of Commerce*, May 3, 1875.
31. House Executive Document 124, pp. 178-80.
32. *New York Herald*, January 1, 1852.
33. Ibid., September, 19 and 20, 1851.
34. *New York Commercial Advertiser*, May 17, 1852.
35. Charles B. Stuart, *Naval and Mail Steamers of the United States*, 2d ed. (New York, 1853), p. 212.
36. *New York Commercial Advertiser*, January 13, 1853.
37. *New York Herald*, August 31, 1857.
38. *New York Commercial Advertiser*, March 28-31, 1859.
39. *New York Herald*, February 28, 1860.
40. Ibid., January 1, 1861.
41. *New York Commercial Advertiser*, April 6-8, 1861.
42. *New York Herald*, November 29, 1861.
43. House Executive Document 337.
44. *New York Commercial Advertiser*, October 24, 1865.
45. *New York Herald*, April 19-22, 1866.
46. *New York Times*, May 15, 1871.
47. *New York Herald*, November 20, 1870.
48. *Journal of Commerce*, September 9, 1875.
49. Heyl, 1: 218.
50. John M. Maber, *North Star to Southern Cross* (Prescot, Lancashire, 1967), pp. 56, 58; Will Lawson, *Pacific Steamers* (Glasgow, 1927), p. 29.
51. Ernest A. Wiltsee, *Gold Rush Steamers* (San Francisco, Calif., 1938), p. 135; Heyl, 1: 183.
52. *Journal of Commerce*, September 17, 1853.
53. *Illustrated London News*, October 22, 1853.
54. John Haskell Kemble, *The Panama Route, 1848-1869* (Berkeley and Los Angeles, Calif., 1943), p. 227.
55. *New York Herald*, March 31, August 15 and September 26-30, 1853.
56. Ginsberg Records.
57. *New York Commercial Advertiser*, December 9, 1853.
58. Maber, *North Star*, pp. 56-57.
59. Will Lawson, *Pacific Steamers*, pp. 31-36.
60. *San Francisco Alta*, July 16, 1854.
61. *New Orleans Picayune*, April 7, 1854.
62. *New York Commercial Advertiser*, October 11, 1854.
63. Ibid., May 12 and 14, 1855.
64. *New York Herald*, April 14 and 26, 1859.
65. *New York Journal of Commerce*, December 30, 1869.
66. Edward Kenneth Haviland, "American Steam Navigation in China," *The American Neptune*, 1957, pp. 313-14.
67. *Journal of the Franklin Institute* 27, 3d ser. (January-June 1854): 199.
68. *Baltimore Sun*, August 29-30 and September 1, 1853.
69. Ibid., February 8 and March 6, 1854.
70. *Baltimore American and Commercial Advertiser*, March 15, 1854.
71. Ibid., March 16, 1855.
72. Ibid., June 18, 1855.
73. *New York Herald*, September 8, 1855.
74. *The Artizan*, February 1856, p. 35.
75. *New York Shipping and Commercial List*, January-August 1856.
76. *New York Times*, July 26, 1856.
77. *New York Herald*, December 24, 25, and 30, 1856.
78. Official Records, ser. 1, 18: 699, 755.
79. Ibid., 20: 252-53.
80. Ibid., 21: 621.
81. Ibid., 21: 698-99, 741; B. S. Osbon, *Handbook of the United States Navy* (New York, 1864), pp. 221, 255.
82. Official Records, ser. 2, 1: 221.
83. *New York Herald*, May 6, 1865.
84. Ibid., October 30 and November 4, 1865.
85. Official Records, ser. 1, 21: 728.
86. Kemble, *Panama Route*, p. 247.
87. *New York Herald*, July 22, 1854.
88. *Journal of the Franklin Institute* 27, 3d ser. (January-June 1854): p. 353; 28 (July-December 1854): 339-41.
89. *New York Herald*, July 22, 1854.
90. Kemble, *Panama Route*, p. 245.
91. *New York Commercial Advertiser*, January 7, 1856.
92. *Nautical Gazette*, February 22, 1873.
93. *New York Maritime Register*, November 14, 1877.
94. *Nautical Gazette*, July 5, 1879.
95. *Monthly Nautical Magazine* (November, 1854), pp. 103-4.
96. *Journal of the Franklin Institute* 27, 3d ser. (January-June 1854): 72; 28 (July-December 1854): 269-73.
97. *Philadelphia Public Ledger*, April 4, 1855.
98. *New York Herald*, June 9, 1855.
99. Ibid., August 18, 1855.
100. Ibid., January 1, 1856.
101. Kemble, *Panama Route*, pp. 79-80.
102. Wiltsee, *Gold Rush Steamers*, pp. 255-56.
103. *New York Herald*, November 6 and 24, 1858.
104. *New York Commercial Advertiser*, July 5, 1859.
105. *New York Herald*, October 10 and 13, 1859.
106. William Morrison Robinson, Jr. *The Confederate Privateers* (New Haven, Conn., 1928), pp. 303-7.
107. The Union Defense Committee of the State of New York, Minutes, Reports, and Correspondence, New York, 1885.
108. Official Records, ser. 2, 1: 186; ser. 1, 6: 186.
109. Ibid., ser. 1, 1: 242, 267.
110. Ibid., 17: 270, 273, 312.
111. Ibid., 13: 593-94, 747-48.
112. Ibid., 9: 175-76, 199, 226.
113. Osbon, *Navy Handbook*, p. 215.
114. Official Records, ser. 1, 9: 233-34.
115. Ibid., 11: 603, 718-19.
116. Ibid., 21: 141-42, 172, 184.
117. *New York Herald*, September 1 and 2, 1865.
118. *New York Times*, November 9, 1865.
119. *New York Herald*, June 1 and 2, 1867.
120. Ibid., May 31 and July 16, 1868.
121. Ibid., April 7, 1869.
122. *New York Journal of Commerce*, June 17, 1869.
123. *New York Tribune*, July 3 and 22, September 4 and 6, 1869.
124. *New York Times*, August 27 and November 30, 1869.
125. Ibid., February 1 and June 24, 1870.
126. Ibid., February 25, 1871.

16

Iron Screw Steamships

THE United States was blessed by nature with magnificent forests from Maine to Florida that provided prodigal supplies of all the varieties of wood needed by her shipbuilders. Great Britain, on the other hand, faced by a timber shortage, was forced to turn to iron early in the steamship era. The techniques for manufacturing and working it into the complex shapes needed for ships had to be developed and as a result strong, light, metal hulls were produced. The *Great Britain* of 1843 was, simultaneously, the world's largest ship and the first iron transatlantic liner.[1] Her hull was so sturdy that it still survives at Bristol in the Great Western Dock where she was built.

On the American side of the Atlantic, prior to 1850, a few experimental iron vessels appeared, most of them barges or river steamboats. In addition, a rather unsatisfactory group of iron revenue cutters were built in 1843 and 1844 and, simultaneously, one highly successful side-wheel steamship, the *Michigan*, for the U.S. Navy.[2] The latter was fabricated at Pittsburgh and erected at Erie, Pennsylvania, for Great Lakes service. Although cut up for scrap in 1949, her bow can be seen as a shore-based monument at Erie.[3]

Rolling mills for iron plate were few and the materials produced, prior to 1830, was of extremely poor quality. The chief reason for the early use of copper boilers was that they were more trustworthy. Iron boilers required quality plates imported from England. Even more difficult to obtain were rolled shapes, the angles, channels, tees, or bulb sections needed for frames, deck beams, clips, and bulkhead stiffeners. Not only were new materials required for the hulls, but radically different modes of construction mandated machine-powered rolls, shears, and punches for working the iron. Unlike timber, it could not be sawed, shaped by hand, or steamed and bent into shape. The sawyers, carpenters, and caulkers from wood shipyards could not cope with this hard, obstinate material, nor did they want to. Riveters with sledge hammers had to be trained to fasten the pieces together. Frames had to be heated red-hot and bent by skilled men who fastened them down to heavy iron bending slabs so they would hold their curved shapes while they cooled.

Iron ships required new means of design, as well as construction, and wood shipbuilders did not understand their problems. As a result, very few existing yards ever made the transition to the new material. Instead, it was the organizations skilled in working iron that expanded their activities to include shipbuilding. The machine shops, iron works, boiler works, and even the railroad locomotive builders, could and did move into the shipbuilding field. They had the tools and the trained workmen needed, but lacked the basic design knowledge to solve naval architectural problems such as buoyancy, stability, seaworthiness, propulsion, and an appropriate hull shape. Someone with the eye and training developed through an apprenticeship with a first-rate master builder in a wood shipyard had to be found to assist those who understood the iron work.

The first successful iron steamboat in the United States, a 60' vessel named the *Codorus*, was built in 1825 by a nail manufacturer named John Elgar, who had worked for the iron works of Davis, Gartner, and Webb in York, Pennsylvania. She was engined by the latter organization. The chief reason for a thin, sheet metal hull was the shallow draft needed to ascend the Susquehanna River.[4] Thus, an individual skilled in working with iron created a hull far lighter and stronger than wood. It was remarkable in having a draft of only six inches!

Iron was an essential prerequisite for a large, screw-propelled ocean steamship. Its strength and rigidity were needed to resist the vibration produced by rapidly rotating engines and propellers.

The first shipyard to regularly construct such steamers was the Harlan and Hollingsworth Company of Wilmington, Delaware. In 1843 the firm of Betts, Harlan, and Hollingsworth installed a slipway near their plant. Their partnership was originally formed to build railroad cars and then branched out to include steam engines, boilers, marine repairs, and, eventually, its first iron vessel, the *Ashland*, 1844. Although only 97'-6" long, she had twin screws geared together and driven by a single cylinder. She was followed by a sister, the *Ocean*, and then the *Bangor*, the first iron oceangoing steamship in the United States. All three served as either transports or naval auxiliaries in the Mexican War. The *Bangor* was 118'-6" long by 24'-2" beam and 8'-8 1/2" depth of hold, had a three-masted schooner rig, and was given a passenger cabin on the after deck. Two propellers 8'-6" in diameter, designed by R. F. Loper of Philadelphia, were turned by independent engines of 22" bore and 2'-0" stroke and were supplied steam by a drop-flue boiler 20'-0" long. Her method of construction was unusual, by later standards, in that the shell plates overlapped each other in clinker fashion (like shingles) and, because angle iron shapes were not available, the frames were bent from rectangular bars with wrought-iron clamps to fasten them to the plating.[5] The *Bangor* was neither handsome nor lucky. Her launching ways broke and she fell over onto the mud bank of the Christiana River. Nevertheless, momentum carried her along until she floated freely. There was no injury. Her woodwork was installed by William A. Thatcher of Wilmington. He was more than a subcontractor; as an experienced ship carpenter trained in the art of wood shipbuilding, he was the actual designer who supplied the *Bangor*'s lines and whatever other drawings were needed to build her.[6]

This 231-ton steamer was begun in October of 1843, launched in May of 1844, but did not start work for the Bangor Steam Navigation Company until the summer of 1845. On her second trip from Boston to Bangor, she sailed August 31 with thirty-four passengers and freight.[7] Next day, while in Penobscot Bay, a fire broke out near her stack. She was run ashore at Islesborough and completely burned out; the flames spread so rapidly that there was time to launch just one of her two boats; alert townspeople rowed out to rescue the passengers and crew.[8] Decks, houses, masts, and most of her fittings had been of wood. Although she was declared a total loss, the *Bangor*'s hull and her engines survived; the hulk was taken to Bucksport, Maine, rebuilt, and put back into service until the Mexican War's demand for steamships led to her sale, for $28,975, to the U.S. Navy in December of 1846. She was rechristened *Scourge* and sent to the Gulf of Mexico. Two years later she was sold to John F. Jeter, a resident of Lafayette, Louisiana, for $2,300.

The shipyard that built her continued on for a long and active career under the name of Harlan and Hollingsworth Company. As late as World War I it was still active as a subsidiary of Bethlehem Steel Company. Up to the time of the Civil War it had completed over seventy iron vessels but only four of them exceeded 1,000 tons. Of these four, just one, the beam-engined *Champion*, was a major ocean steamship. She has already been described as part of the Vanderbilt Line; when she came out in 1859, she was, at 1,419 tons, the largest iron vessel under the American flag.

A second builder of iron craft was Reaney, Neafie, and Company, later Neafie and Levy, of Philadelphia. After an 1844 start, their first oceangoing steamship of any size was the *Oriental* of 1,202 tons, completed in 1861 for the New York, Nuevitas, and Cuba Steamship Company.[9] She saw a year's service before going aground near Cape Hatteras. To replace her, a larger iron steamer, the *Havana*, was ordered from the same firm. Both were propeller-driven. Starting in 1865 under Garrison and Allen, she made occasional voyage to Brazil and was sold to Peru in 1869.

New York, the center of the country's steamship activity, produced a few iron steamers, all by-products of the city's engine works. Hogg and Delamater launched several that were designed by John Ericsson. The Hudson River steamboat *Iron Witch*, of 1846, was one of his major failures; her rebuilt engines ended up in the United States Mail Line's *Falcon* two years later. The successor firm, C. H. Delamater and Company, did, however, build a very satisfactory iron screw steamer named the *Matanzas* in 1860. She was 202' by 29'-10" by 20'-6", with a draft of 12'-6" and a tonnage of 862. Owned by the New York and Matanzas Steamship Company, she was used in the Cuban trade.[10] Fire destroyed her in 1868.

Another iron propeller from New York was the *North Carolina*, a product of the Novelty Iron Works. Of 618 tons, she was delivered to H. B. Cromwell and Company in 1861, seized by Confederates later that year, and outfitted as a blockade runner. Because she crossed the Atlantic, her career will be dealt with shortly.

These occasional experiments by New York engine

builders were not continued and no iron shipyards were established on Manhattan Island.

If one counts the American-built screw steamers above 1,000 tons in the 1863 "American Lloyds' Register of American and Foreign Shipping," one finds:

No.	Material	Type
15	Wood	Passenger and Cargo
4	Wood and Iron	Passenger and Cargo
3	Iron	Passenger and Cargo
1	Iron	Egyptian Yacht

The table is clear evidence of the slow progress in the adoption of the screw propeller and the almost total unawareness that the iron screw steamship would be the accepted design of the very near future.

The four combination wood and iron hulls are of particular interest. All were products of a forward-looking shipyard and engine works established in South Boston by Harrison Loring in 1857.[11] Loring, as a young man, had worked on the construction of one of the 1844 iron revenue cutters. His first major steamships were the *South Carolina* and *Massachusetts* for the Boston and Southern Steamship Company. These 217' screw steamers of 1,165 tons had iron hulls but wood upper works.[12] Completed in 1860, they were among the first steamships purchased by the U.S. Navy for blockade duty.[13] After the war they had long and useful careers as, respectively, the *Juniata* and *Crescent City*.

The *Mississippi* and the *Merrimack*

The *South Carolina-Massachusetts* pair impressed Boston's Union Steamship Company so favorably that it ordered two larger and more powerful steamships for a projected service from Boston to New Orleans.[14] The eminent New England designer, later a naval constructor, Samuel H. Pook, supplied the lines. Pook had, at one time, worked for Samuel Hall, builder of the steam packet *Massachusetts*. He was widely known for his fast clipper ships, the *Red Jacket* being the most famous. Harrison Loring and Company built the engines as well as the hulls.

Because Massachusetts had no suitable rolling mills, the 3/4" and 1" thick iron plating, together with the special shapes needed, were made in Baltimore and shipped to Boston. To avoid delay they came by coastwise steamer, not by the cheaper, but less predictable sailing packet. On March 1, 1861, the first hull was nearly plated and the second had its iron frames set up. The bedplates for the engines, weighing twenty-five tons

Mississippi, original appearance. Photographed as a Civil War transport June 14, 1864. *Photo by U.S. Army Signal Corps.*

apiece, had been cast and hauled from the foundry to the machine shop by thirty-five straining horses. They were the largest castings yet produced in New England.

First to be completed, the *Mississippi* was listed at 274.3' long with a tonnage of 2,008, while the *Merrimack* measured 272' and 1,991 tons. They were, no doubt, identical, for a beam of 39' and a depth of hold of 28' were quoted for both. The designed draft was about 16' and the iron plating extended well above that waterline to the second deck. About 8' of the topsides were planked with wood, probably over iron frames. Deck planking and houses were wood. These were handsome ships, having a sweeping sheer and an open rail from the foremast aft to the counter. The stem curved forward to support a figurehead and a short bowsprit. No trail boards or fancy rails were installed. Because the combination of depth and sheer raised the bow well above the sea, no forecastle was needed; the weather deck continued unbroken from knightheads to taffrail. Three masts provided for a bark rig. A rather short, squat stack with just the right rake was installed forward of the main. Between it and the foremast was a deckhouse and a slightly raised pilot house. In her early years the *Mississippi* had a white band painted around her black stack and the *Merrimack* an all-black one.

The engines were two cylinder, direct acting, of 62" bore by 4'-0" stroke, taking steam from four boilers and turning a 15' diameter propeller. In later years it was reduced to 14'.[15] The designed speed was 12 knots in order to assure a ten-day trip from Boston to Havana and New Orleans. In reality these were iron ships, just as their contemporaries described them. The wood topsides were limited in extent; that part of the hull could and would be altered to suit changing requirements without disturbing the strong, rigid structure below the second deck. Four iron bulkheads divided it into watertight compartments that, in at least one instance, proved their worth.

The *Mississippi* was completed in February of 1862 and was followed by the *Merrimack*, which seems to have been called *Merrimac* when she first came out.[16] Immediately upon completion, J. H. Foster and Company, who owned the Union Steamship Company, chartered both to the U.S. Quartermaster Corps. On her very first trip, the *Mississippi* weighed anchor at Boston on February 21, 1862, carrying a load of 1,400 troops bound for Ship Island and, ultimately, New Orleans.[17] Benjamin F. Butler, Major General, and his staff came aboard at Hampton Roads. First she met a gale that swept her decks, smashed skylights, and poured water into the hull. The troops had to spend the night of the 25th bailing. Three days later, having been driven off course, she ran aground on Frying Pan Shoals, near Cape Fear, and gouged a hole in her bow.[18] When pulled off by the U.S.S. *Mount Vernon*, her forward bulkhead limited the flooding and enabled her to continue to Port Royal.[19] A wood ship would not have survived. She brought the General up the Mississippi and landed him at New Orleans on May 1. Exactly one week later the *Mississippi* headed upriver crowded with 2,000 federal troops.[20]

The *Merrimack* made a less spectacular entry into trooping service. Like her sister, she was initially chartered at $1,350 per day, but by August of 1863 the rate had declined to $850. Both made frequent East Coast runs; New Orleans, Key West, Port Royal, Fortress Monroe, Alexandria, New York, and Boston saw them as men were sent South and casualties brought back. Both met the regular difficulty that plagued screw steamers, broken propeller blades. The *Merrimack* put into New York on May 9, 1863, with two missing. She had sailed from Boston three days earlier. In January of 1864 the *Mississippi* was delayed at New Orleans for a less specific "disarrangement" of her propeller, doubtless a similar casualty.

Under more favorable circumstances, the *Merrimack* was able to take 259 sick and wounded soldiers in 5 days 15 hours from New Orleans to New York in October of 1864.[21] The speed attained, 12.75 knots, made her the fastest, as well as the largest, screw-propelled merchant steamship in the United States. A year later the paddler *Guiding Star* improved on her time by five hours. But, for the first time, the screw steamship had become competitive in every respect with the big side-wheelers that had reigned supreme ever since the start of the steamship era. At the end of the war, the two sisters were purchased by the New York Mail Steamship Company. Their careers will be continued in the chapter dealing with that line.

The *North Carolina*, 1861-1903

Like the four Harrison Loring vessels, the New York-built *North Carolina* had an iron hull with wood upperworks. The arrangement, an entirely American one, never found favor elsewhere. She was a small steamer but had, like the other iron craft of her day, extremely sturdy construction. The shell plating was 1/2" thick, her frames were 4" x 3" angle iron, and her rivets 3/4" in diameter. James Baird, her designer, produced a hull of such strength that it would see forty years of active service. It should be noted that wrought iron, while not so strong as steel, was and is more resistant to corrosion. Many iron steamships, as a result, had extremely long careers, while more modern designs with lighter steel scantlings rusted away and had to be scrapped.

The Novelty Iron Works built and engined this 169'-6" by 29'-2" by 13'-3" steamship of 618 tons with the simplest possible machinery, a one-cylinder engine of 42" bore and 3'-6" stroke, and a single boiler producing steam at 30 psi. The latter was 15'-3" long and had three furnaces and 168 return tubes 3 1/2 inches in diameter. The coal consumption was low, 9 tons per day. On trials a four-bladed propeller of 12' diameter and 18' pitch turned at 60 rpm. A separate steam pump was installed for pumping out any of the ship's five compartments and for supplying sea water to the fire hoses aboard.[22]

The use of a single cylinder needs explanation; this again seems to be a peculiarly American idea. One thinks at once of its getting stuck on dead center. The problem was overcome by carrying on the paddle-wheel tradition in which the engineer avoided closing a throttle valve in the steam line. Instead, when maneuvering, the engine's valve gear was operated by a hand lever so that instant response and absolute control could stop it in mid-stroke. Only when the ship was clear of her dock and heading out to sea was the automatic valve gear hooked on. The hand operation of a screw engine working at 60 revolutions per minute must have been more difficult than a side-wheel one turning at fifteen, but the unsung heroes sweating away in their overheated engine rooms developed the necessary technique.

A second item worthy of remark is the single-boiler installation. By 1860 it was common practice on small steamers and indicated that the gradual improvement in surface condensers had resulted in a fairly reliable supply of fresh feed water. The 1860 boiler had become far more casualty-free than an 1850 one. The major reason was that skilled engineers and firemen were available to operate steamships. A decade earlier finding even a partially trained man was difficult.

On January 27, 1861, H. B. Cromwell and Company sent the *North Carolina*, their first iron steamer, to sea from New York bound for Wilmington, North Carolina. At first she carried freight but, after a few voyages at fortnightly intervals, the agents began to solicit passengers at a one-way fare of $15.[23] She was a single-decked craft, so that all of them must have been berthed in a deckhouse, just as the *Bangor*'s were back in 1845.

The *North Carolina*, on her sixth trip, was interned at Wilmington on the outbreak of the Civil War. Thereafter she saw a series of owners and took on a new name with each. In succession she became the *Annie Childs, Julie Usher, T. D. Wagner,* and *Victory.* She managed to sneak through the Federal blockading fleet at least once in 1862 and a half dozen times in 1863.[24]

Her first reported trip began at Wilmington on February 5, 1862.[25] James D. Bulloch, one of the Confederacy's principal naval representatives abroad, had personally delivered the British steamer *Fingal* to Savannah but could not get a passage back to England. He found the *Annie Childs*, ex-*North Carolina*, loading cotton, rosin, and tobacco, and arranged for her to cross the Atlantic rather than head for the usual blockade runner's ports of Bermuda or Nassau. This made her the first American iron screw steamship to cross to a British port. So small a steamer could not carry enough coal and had to stop at both Fayal in the Azores, and Madeira on the way. Eventually, she had to saw up her spars and burn part of the rosin cargo to reach Queenstown, Ireland, on March 8, 1862. Bulloch described her as a good sea boat but, writing twenty years after the event, confused her with the Boston-built *South Carolina*.[26]

As a blockade runner, under the name *Victory*, she was chased and captured by the *Santiago de Cuba* near Eleuthera Island on June 21, 1863. She had left Wilmington for Nassau four days earlier, loaded down with cotton, tobacco, and naval stores. Once again she burned the rosin in her cargo, as well as throwing overboard her cotton deck load, in a vain attempt to escape.[27] She was sent to a prize court at Boston. As a new and desirable steamship, the U.S. Navy bought her in for $65,000, renamed her *Queen*, armed her, and fitted her out as an ordnance supply ship to carry guns, ammunition, stores, and powder, from Boston and New York to the Federal ships and depots scattered from Hampton Roads all the way to Galveston, Texas.[28]

In February of 1864 she spent a few weeks as a blockader on the Texas coast and captured a schooner, the *Louis*, with a cargo of Enfield rifles, cigars, and whiskey, all essential items for an army. After this martial interlude, she spent the rest of the war in further ordnance voyages.[29]

In the fall of 1865 she was sold for $51,300 to the firm of Smith and Dunning.[30] They provided a seventh name, the *Gulf Stream.* It proved a lasting one; after five adventurous years she would settle down, as a small coaster of no great distinction, to four decades of unglamorous but useful work in Atlantic and Gulf of Mexico waters.

From September of 1865 to June of 1866, the *Gulf Stream* made leisurely trips between New York and Mobile. She was then shifted to her owners' New York and Venezuela Line, the second American attempt at a regular service to Venezuela. In this she joined two small wood propellers, the *Vicksburg* and the *Mercedita*. In April of 1866 the *Vicksburg* opened this new route from New York to Porto Cabello and Laguaira, with stops at Puerto Rico and St. Thomas. She was replaced on August 4 by the *Gulf Stream*, which took over the line's third sailing.[31] In 1867 the *Mercedita* joined the latter to make monthly departures possible. The whole operation was abandoned in April of 1867. The *Gulf Stream* had arrived at New York on March 13 at the conclusion of her fourth voyage. Back in 1856 the second *Tennessee* had made five equally unprofitable Venezuelan trips.

In the fall of 1867 she started six years of trading to Key West, Galveston, New Orleans, Savannah, Charleston, Haiti, Jamaica, and Santo Domingo. Although owned by C. H. Mallory and Company, she was frequently under charter to others. Eventually, Mallory sold her to W. P. Clyde of Philadelphia. Because her original boiler had reached the end of its useful life, Clyde replaced it in 1874, and at the same time cut the ship in two and lengthened her by inserting a new section 50'-10" long.[32] Ten years later the second boiler was replaced by a pair of new ones, but the one-cylinder engine remained throughout her life.[33] No doubt the steam pressure was increased. A phenomenon often observed when a ship is lengthened is that her speed goes up, even though no more horsepower is installed. It is likely, therefore, that the changes made her a faster vessel after 1874, and that some further increase in speed resulted from the 1884 boilers. There is no real evidence, but the fact the Clyde Line could run or charter her for so many years indicates that she could still compete with the newer craft with their higher horsepowers and speeds.

Wilmington, for W. P. Clyde, and Charleston, chartered to James W. Quintard, were her usual ports of call well into the 1880s; after that it was Wilmington alone, with occasional lay-ups in the winter.

October 13, 1887, found her in collision with an equally elderly propeller, the *E. C. Knight*, off the New Jersey coast.[34] The wooden *Knight* sank in twenty minutes; the iron *Gulf Stream* rescued her crew and continued undamaged to Wilmington. Her first approach to disaster came in November of 1888 when she was caught by seas that made her roll so badly that her cargo shifted and heeled her over. As the waves battered in her deckhouse doors, water poured below.[35] Despite four feet of it sloshing in her hold and her pumps choked by debris, she made Charleston safely. A year later, while coming up the Delaware River in a thick fog, she was struck by the *Josephine Thompson* and had to be grounded to prevent her sinking.[36]

The twentieth century opened with the *Gulf Stream* still gamely at work for the Clyde Line, still plodding along under her ancient one-cylinder engine, still stopping and backing under the hand control of some overworked engineer. She left New York on January 29, 1903, carrying a crew of twenty-one and a cargo of cotton for Philadelphia. As night fell a winter gale arose and, in heavy fog, she grounded on the New Jersey coast near Avalon. The local lifesaving station had great difficulty in removing the crew but, despite pounding seas, saved all aboard. The battering was more than the *Gulf Stream*'s forty-two-year old hull could endure; on the 30th it broke in two.[37]

The only unusual feature of the diminutive *North Carolina* was her iron hull. It proved so rugged that, under a succession of owners and names, she was able to make almost every important port south of New York during forty-two years of coastwise service. Not until her 1874 enlargement did she exceed 200' in length and she never quite made a thousand tons. Yet her light draft, her hardiness, and her simple, reliable power plant served her well. She made hundreds of trips to Charleston and Wilmington after her pioneering trial in the Venezuela trade. No one today remembers that the *North Carolina* was the first American iron-hulled screw steamer to cross the Atlantic. Yet she was a far more successful steamship than the British-built *Circassian*, also an iron propeller, that followed her, as an American passenger liner. Their three points in common were screw propulsion, iron hulls, and capture by the U.S. Navy during the Civil War.

The *Circassian*, 1857-1876

At the end of 1861, when the Havre Line ceased its sailings, the American passenger liner vanished from the North Atlantic. A number of British lines, together with two German ones, were only too happy to take over all European sailings to and from the United States. Four years would elapse before any steamship flying the American flag would set out for a European port carrying passengers and the U.S. mails.

The first to do so was the *Circassian*, anything but a typical American steamer. She was an auxiliary screw steamship of low power, built in Ireland and engined in Scotland, that had been captured during the Civil War. Her design was in some respects archaic; in others, progressive. Her history had been, and would prove to be, eventful; she was always in trouble. Not only was she the first American transatlantic liner to sail after the end of the war, but she was also the first iron steamer of American registry to make such a crossing. The *North Carolina* had worn the stars and bars of the Confederacy during her wartime trip to Ireland.

The *Circassian* had been built as an improved version of an earlier steamship, the *Khersonese*, which had been completed in 1856.[38] The hulls were almost identical, but the *Circassian*, in 1857, had been given a higher horsepower engine and a lower sail area. Both had been built by Robert Hickson and Company of Belfast, Ireland. Hickson's yard, in later years, was to become famous as Harland and Wolff. The *Circassian*'s dimensions were 248' by 39' by 23'-6" depth of hold.[39] Under British regulations her gross tonnage was 1,537 and, under American ones, 1,457. After a July 1856 launching, the hull was towed to Glasgow where the machinery was installed by Randolph, Elder, and Company, and the passenger cabins were fitted out.[40] She was then sent to Birkenhead, across the Mersey from Liverpool, where final touches were completed in the dry dock at the Laird shipyard.[41] She had a three-masted ship rig that included topgallants but no royals and was badly underpowered by a two-cylinder, geared, beam engine of 60″ bore by 4′-0″ stroke.[42] The use of transverse walking beams between the vertical cylinders and the cranks was a British arrangement that had few American counterparts. The propeller shaft lay between the cylinders and the crank shaft and was geared up from 33 rpm at the engine to 83 at the propeller. Steam at 17 psi was supplied by two water-tube boilers. A speed of 11 knots and a low coal consumption of one ton per hour were promised, but both were wrong. She was very slow and consumed so much coal that she frequently ran out. The chief item installed at Laird's was disconnecting gear so that, under sail, the propeller could spin freely. Her appearance was that of a sailing ship with a long, overhanging bow, a larger than life-size figurehead of a Circassian warrior, and a small stack abaft the mainmast.

On deck was a narrow house 230′ long. In it her first cabin passengers were berthed amidships to keep them away from the vibration and noise of the propeller. The location was a break from the tradition of putting the best cabins aft, an excellent arrangement for sailing packets and side-wheel steamships. Another new feature was the installation of steam winches to unload cargo.

The hull material and arrangement, the cargo handling, and her water-tube boilers were progressive elements. The fact that she had lots of sail, little horsepower, and disconnecting gear made her a backward craft, a throwback to the *Massachusetts* or the *Savannah*. Her maiden voyage demonstrated the defects that were to bedevil her ever afterward.

The *Circassian* opened the North Atlantic Steam Navigation Company's service from Liverpool to Portland, Maine, with stops at Newfoundland and Nova Scotia. She departed on March 7, 1857. Exactly one week later she crept back into the Mersey; severe gales and lack of power had reduced her headway to zero. Although sails had been set, they were blown to ribbons and her bowsprit was damaged. A second departure was made on March 19 with forty-six cabin and 160 steerage passengers.

The *Khersonese* joined her and the two alternated their sailings until the fall of 1857, when both were chartered to take troops to India. By that time the *Circassian* had three Atlantic voyages to her credit. After her Indian Mutiny work, she was chartered by the Galway Line.

Amidst the motley collection employed by this ill-planned venture, the *Circassian* was a better-than-average choice. She left Galway on October 27, 1858, reached St. John's in nine days, and docked at New York on November 12. The sixteen-day crossing was a good one for her. Her second voyage took eighteen days from Galway to St. John's, arriving on January 29, 1859, short of coal.[43] She did not reach New York until February 11, a month out of Galway, with all coal gone, her spare spars burned, and her screw missing.[44]

During sixteen months with the Galway Line, she completed eight voyages.[45] Her last crossing was from New York on March 27, 1860, to Galway on April 14. Her ownership was continuously changing and led to Richard Taylor in 1861 and Z. C. Pearson in 1862.[46]

After the Galway Line's suspension, she was transferred to the Mediterranean and, from the American point of view, became noteworthy on December 20, 1862, when she sailed from Constantinople with a cargo of saltpeter (used for making gunpowder). Thomas Dudley, the U.S. Consul at Liverpool, sent word that she was bound from Trebizond to a Confederate port.[47] His report was accurate but her westward progress was excruciatingly slow. Late in February she coaled at Cardiff; not until April 27 did she leave St. Thomas, ostensibly for Havana.[48]

On May 4, 1862, a converted New York ferryboat, the U.S.S. *Somerset*, sighted the *Circassian* off the Cuban coast between Havana and Matanzas. She neither changed course nor tried to run away, but just blundered along until a shot through her fore rigging halted her.[49]

One thinks of a blockade runner as a long, low, fast, lightly built steamship evading the Federal fleet on a dark night. The *Circassian* was large, slow, fully rigged, and was stopped in broad daylight. The *Somerset* could not spare the force to man so large a prize and had to tow her into Key West. After purchase by the U.S. Navy for $107,000, its Brooklyn Yard converted her into a bark-rigged supply ship.

One 100-pound Parrott rifle, four 9-inch Dahlgren guns, and a 12-pound rifle were installed. Starting in December of 1862, she made regular trips as a supply steamer from Boston to the Gulf Blockading Squadron. She delivered food, stores, ice, crew, mails, traveling officers—everything except coal—to the many ships stationed from Cape Canaveral on the Atlantic coast of Florida, to the mouth of the Rio Grande in the Gulf of Mexico, and regularly touched at Port Royal, South Carolina.

Like other steamships kept at sea for extensive periods, she had endless boiler troubles; her tubes were constantly becoming plugged with scale and her speed suffered. Despite these problems she managed, in December of 1863, to capture a valuable screw steamer, the *Minna*, from Nassau for blockaded Charleston with a cargo of iron, powder, dry goods, hardware, and the parts for a marine engine, possibly for a Confederate ironclad.[50]

One expects an iron hull to stay tight; the *Circassian*'s was a disappointment. Just before the *Minna*'s seizure she was caught in a storm and found that water was rising in her holds at the rate of twelve inches per hour, a most worrisome matter. As a result, repairs had to be undertaken at Boston in the spring of 1864. Thereafter she continued supplying the Gulf squadrons.

Immediately after the war ended, she was bought at an auction in Boston by Arthur Leary for $71,000 on June 22, 1865, just two days after he had acquired the *Quaker City* for $35,000 in Philadelphia.

Two months later the *Circassian* went to sea under Edward Cavendy, a previous commander of the Bremen Line's *Hermann*. She left New York on August 19 bound for Bremen by way of Southampton as the first American-owned passenger liner on the Atlantic since 1861. The venture augured well when 650 passengers and 450 tons of freight were loaded for the return trip.[51] The iron hull for the second time proved defective; it leaked so badly off the coast of Nova Scotia that water had almost reached the fires in her furnaces before Captain Cavendy could run her aground on the beach at Rocky Bay, near Arichat, Cape Breton Island, on October 20.

Although the *Circassian* was at first thought to be a complete loss, the Columbian Wrecking Company was able to get her off. She came into New York on November 20, 1865, under her own steam, with the leakage controlled by steam pumps. But Arthur Leary found his wood side-wheeler, the *Quaker City*, a far superior investment, and he never tried to use the *Circassian* again.

Repairs were completed during the spring of 1866 and a new line, the Continental Mail Steamship Company, took her over. It was the first American attempt to operate to Belgium's ancient port of Antwerp. The two chartered steamships were both large and slow. One was the underpowered *Ericsson* and the other the equally underpowered *Circassian*. William Salem and Company acted as their New York consignee.

First to sail was the *Circassian*, on July 18, 1866, under Captain Thomas S. Ellis, bound for Antwerp via Havre. She left Antwerp with 180 passengers on August 18, departed from Havre two days later, only to run into gales lasting throughout the crossing.[52] After engine damage on the 27th, she continued under sail for eight days while her engineers made temporary repairs. She then steamed slowly along at a reduced steam pressure but ran short of coal and had to put into Halifax on September 14. Finally, on the 19th, thirty-two days out of Antwerp, she reached New York.

The *Ericcson* did a little better; she left New York on August 23 and returned October 16 in twenty days from Havre with a passenger list of 350. Three of them were born on the way.

After repairs that postponed her intended sailing date, the *Circassian* cleared on October 11 for the line's third voyage. She was still tied up, however, on October 19, at the same wharf where she had been on the 8th.[53] It is not certain whether she sailed and returned or did not sail at all. The Continental Mail had reached its inglorious end.

In 1867 she was chartered to Ruger Brothers for a single trip to Bremen in the steerage trade. Under Captain Ellis she cleared at New York August 10 and was sighted by the *Atlantic* on the 26th, a day away from Bremen. Homeward she had to put into the channel port of Falmouth for further engine repairs on October 9, four days out of Bremen. She left on the 11th with 470 passengers, taking twenty-two more days to New York.[54]

Three weeks later she was heading out to sea, not for Europe, but toward New Orleans under the management of the New York Mail Steamship Company, which seems to have acted as agent for the ship without any attempt to use her on its own passenger service.

The *Circassian* reached New Orleans on December 6, 1867, and sailed January 13 for Liverpool, probably with a cotton cargo. She came into New York on March 26, 1868, after a thirty-day eastward passage and a twenty-two-day return.[55] By that time she had a new owner, Ernest Fiedler, who sent her out again to Bremen. Because she was advertised by both Fiedler and Ruger Brothers, it is believed that she was under Fiedler's control but that he used the Rugers, who had many German contacts, as his agent abroad. She left New York on April 25 and returned from Bremen on May 17, via Falmouth, to New York on June 9, 1868, with 480 immigrants. One birth and one death had taken place.

The Rugers liked her well enough to charter her for a departure to Bremen on July 25. Instead of returning by way of the English Channel, she made a direct passage north of Scotland to New York. It still took twenty days for her 632 passengers to cross. The date she docked, September 14, 1868, marked her end as an Atlantic liner.

In January of 1869 she was advertised for sale but could not find a buyer.[56] After a ten-month lay-up, N. H. Brigham sent her to New Orleans on October 1. Captain Thomas Ellis, her master since 1866, headed her South again on November 11 but, as she had so many times before, she ran out of coal and had to make an unscheduled northbound stop at Savannah on December 10. Eight days later—it was an extremely slow trip—she ran hard aground on Squan Beach, New Jersey.[57] Her iron hull remained tight and permitted her undamaged cargo to be unloaded into lighters. The Coast Wrecking Company sent steamers, the *Relief* and *A. Winants*, to assist her. Anchors and cables were set out, and, as the hull was lightened, she was gradually pulled off.

This latest disaster wrote finis to her career as a steamship. Fiedler laid her up in Brooklyn and, from January of 1870 until early in 1873, tried to sell her.[58] He finally succeeded, for she was moved to a North River pier and, on May 21, 1873, left for Newark, Quebec, and Liverpool.⁵⁹ Her engines had been removed and mizzen yards added to make her a full-rigged sailing ship. She belonged briefly to James G. Ross of Quebec but was given a Liverpool registry in August of 1874 with John S. and James R. De Wolf as owners. She spent the last two years of her life trading from the United Kingdom to New Orleans and, later, to India.

On October 6, 1876, the *Circassian* reached Liverpool from Bombay. She loaded a general cargo and set out, on November 6, for her first trip to New York under sail alone. About December 1 she encountered a small bark, the *Heath Park*, in a sinking condition.⁶⁰ The thirteen-man crew were taken off; their trials, however, were not over for, approaching New York on December 11, the *Circassian* overran her reckoning and went ashore opposite Bridgehampton, Long Island.⁶¹ A gale was blowing, the surf ran high, and not until daylight could the local lifeboat take forty-seven men ashore. The officers remained aboard to assist in her salvage. Once again the Coast Wrecking Company came to her aid. While the cargo was being unloaded, the weather worsened and the hull began to leak. By December 23 there was 16' of water in her; nevertheless, anchors and cables had been laid to pull her off. A crew of thirty-three were put aboard on December 29 in anticipation of pumping out and hauling off at high tide. At this point the weather suddenly worsened and the breakers became so violent that no boat dared approach or leave her. Waves broke over her deck; the mainmast fell; the hull broke in two; and, finally, the mizzen gave way, carrying with it all who had clustered on its shrouds to escape the waves sweeping below. Only four were washed ashore alive.⁶²

An unlucky steamship, the *Circassian* met an unlucky end. Prior to that her iron hull had survived two groundings that would have finished a wooden one. Nevertheless, the sudden appearance of extensive leaks had always demonstrated that poor structural design and defective riveting were present. Worst of all were inadequate engines and boilers whose tubes clogged with scale so that her low initial speed dropped further with time. It is doubtful that any of the six organizations who operated her on the Atlantic ever found her a satisfactory steamship. Only her eight Galway Line voyages, from 1858 to 1860, were ever performed on a regular basis. The North Atlantic Steam Navigation Company used her for three, Ruger Brothers and Ernest Fiedler for two apiece, the Continental Mail Steamship Company for one, and Arthur Leary for the one that ended prematurely at Rocky Bay. These add up to seventeen voyages in twelve years. As the first iron screw steamship to make Atlantic crossings with an American registry, as the first such vessel to sail with passengers after the end of the Civil War, as a particularly handsome vessel, the *Circassian*, if not a complete failure, must regrettably be adjudged as far from being a successful steamship. In this respect she emulated her generic predecessors, the *Savannah* of 1819 and the *Massachusetts* of 1845. But, like the latter, she was extremely useful as a naval auxiliary.

Notes

1. Capt. Claxton, R.N., *Steamship GREAT BRITAIN*, (New York, 1845).
2. John H. Morrison, "Iron and Steel Steam Vessels of the United States, 1825-1905," *Scientific American Supplement*, October 21-November 25, 1905, and Reprint no. 3 of the Steamship Historical Society of America (1945), pp. 4-6.
3. Naval History Division, *Dictionary of American Fighting Ships* (Washington, 1969), 4: 350-51.
4. Alexander Crosby Brown, "The Sheet Iron Steamboat 'Codorus'," *The American Neptune* (1950), pp. 163-90.
5. *Semi-Centennial Memoir of the Harlan & Hollingsworth Company* (Wilmington, Del., 1886), pp. 212-32.
6. Letter by E. C. H. to the *Bath Daily Times*, quoted in the *Nautical Gazette*, March 1, 1883.
7. Francis B. C. Bradlee, *Some Account of Steam Navigation in New England* (Salem, Mass.), pp. 90-93.
8. *New York Commercial Advertiser*, September 3, 1845.
9. Geo. Henry Preble, *A Chronological History of the Origin and Development of Steam Navigation, 1543-1882* (Philadelphia, 1883), pp. 471-72.
10. *Journal of the Franklin Institute* 39, 3d ser. (January-June 1860): 185.
11. Henry Hall, *Report on the Shipbuilding Industry of the United States* (Washington, D.C., 1882), p. 201.
12. J. Leander Bishop, *A History of American Manufacturers from 1608 to 1860*, 3d ed. (Philadelphia, 1868), 3: 282-83.
13. *Boston Transcript*, May 8, 1861.
14. *New York Commercial Advertiser*, March 5, 1861.
15. *New York Herald*, March 27, 1874.
16. Enrollment, Boston, July 7, 1862, NRG41, U.S. National Archives.
17. *Boston Transcript*, February 21, 1862.
18. *New York Journal of Commerce*, March 10, 1862.
19. Official Records, Series I, 6: 674-75.
20. Ibid., 18: 723-24, 744.
21. *New York Journal of Commerce*, October 7, 1864.
22. *Journal of the Franklin Institute* 40, 3d ser. (July-December 1860): 296; 42 (July-December 1861): 202.
23. *New York Herald*, January 28 and April 11, 1861.
24. Marcus W. Price, "Ships That Tested the Blockade of the Carolina Ports, 1861-1865," *The American Neptune* (1948), pp. 196-241.
25. Francis B. C. Bradlee, *Blockade Running during the Civil War* (Salem, 1925), p. 136.
26. James D. Bulloch, *The Secret Service of the Confederate States in Europe* (New York, 1884), 1: 149-51.
27. Official Records; ser. 1, 2: 361-62.
28. Ibid., ser. 2, 1: 187.
29. Ibid., ser. 1, 21: 53, 79.
30. Ibid., ser. 2, 1: 187.
31. *New York Shipping and Commercial List*, August 8, 1866.
32. Original American Lloyd's Register, 1879.
33. *New York Maritime Register*, July 30, 1884.
34. Ibid., October 19, 1887.
35. Ibid., December 5, 1888.
36. Ibid., December 13, 1890.
37. *New York Times*, January 31, 1903.
38. N.R.P. Bonsor, *North Atlantic Seaway* (Prescott, Lancashire, 1955), p. 136.
39. Official Records, ser. 2, 1: 59.
40. *Belfast* (Ireland) *Daily Mercury*, May 24, June 19-20 and July 19, 1856.
41. *Mitchell's Maritime Register*, February 21, 1857.
42. *Willmer and Smith's European Times* (Liverpool), February 21, 1857.
43. *New York Herald*, February 1, 1859.
44. *The Artizan*, April 1859, p. 104.
45. Bonsor, *North Atlantic Seaway*, pp. 160-66.
46. Papers of the House of Commons, 1859, vol. 27, paper 493;

1861, vol. 58, paper 275; 1862, vol. 54, paper 783.
47. Official Records, ser. 1, 6: 530-32.
48. Ibid., 1: 384-85.
49. Ibid., 17: 231-32.
50. Ibid., 9: 341.
51. *New York Commercial Advertiser,* October 21, 1865.
52. *New York Journal of Commerce,* September 15, 1866.
53. *New York Shipping and Commercial List,* October 9 and 20, 1866.
54. *New York Herald,* October 11 and November 3, 1867.
55. *New York Shipping and Commercial List,* December 14, 1867; January 22 and March 4, 1868.
56. *New York Journal of Commerce,* January 5, 1869.
57. Ibid., December 11 and 20-23, 1869.
58. *New York Shipping and Commercial List,* 1870-73.
59. *New York Maritime Register,* May 28, 1873.
60. Ibid., November 11 and December 13, 1876.
61. *New York Times,* December 15 and 31, 1876.
62. *New York Journal of Commerce,* December 13-15, 20-23, 1876, and January 1, 1877.

17

The Baltimore and Liverpool Steamship Company, 1865-1868

IN the fall of 1862 William P. Williams made plans for the Neptune Line to run from Boston to New York with a fleet of steamships sailing around Cape Cod and carrying some passengers but mainly freight. The ocean route would eliminate the transshipment that for many years had been necessary at Providence, Stonington, or Fall River, where the Boston railroads met the Long Island Sound steamers. Because a considerable part of the new route was exposed to the broad Atlantic's wind and waves, oceangoing steamships were essential. Propellers were chosen to achieve lower weight and first cost, when compared to side wheels; they could be just as efficient when there were no draft restrictions. Five "large" steamers were ordered from the New York shipyard of J. B. and J. D. Van Dusen, located at the foot of East 16th Street on the East River.[1]

All were to have a 205' length on deck and a 35'-8" beam with a 20'-0" depth of hold and a draft of 13'. As completed, slightly different dimensions would be quoted, but a tonnage of 1,244 was assigned in every case. The chief variation was an increase in length to 209'. The machinery order was placed with Brooklyn's Henry Esler and Company for two-cylinder inverted (cylinders over the crankshaft) engines of 44" bore and 3'-0" stroke,[2] driving 12' diameter cast iron propellers.

The stem curved forward to support a diminutive bowsprit and a half-height forecastle made for easy anchor handling. The square-rigged foremast was stepped just aft of a slightly raised pilot house. Then came the cargo hatch and a long deckhouse. Rigged as half brigs, all had moderate passenger quarters but large holds. No attempt at speed was made; ten knots could be achieved with less than 500 horsepower. In short, these were excellent examples of the type of steamship to which screw propulsion lent itself; they were moderate in size, moderate in speed, and intended for short coastwise trips so that they did not have to be weighted down with fuel. Their raking masts and a stack well aft gave them an attractive profile.

Named, in the order of completion, *Neptune*, *Glaucus*, *Galatea*, *Proteus*, and *Nereus*, they were immediately purchased by the U.S. Navy for $160,000 apiece, with the exception of the *Galatea*, whose more interesting price tag was $158,523.07. The *Neptune* was delivered to the Brooklyn Navy Yard on September 3, 1863, and the others followed, two in October, and two in February of 1864.[3] Each was armed with one large 100-pound pivot gun and eight to ten smaller rifles or smooth bores. Commissioning dates ranged from December 19, 1863, for the *Neptune* to April 19, 1864, for the *Nereus*.

The *Neptune* and *Galatea* were put to work in the

Caribbean Sea as armed escorts for the Panama mail steamers.[4] The *Nereus* had a minor role at Fort Fisher and the *Glaucus* seems to have had no feats worth recording. The *Proteus*, under Commander R. W. Shufeldt, who had been master of the *Quaker City* before the war, led an adventurous career pursuing and capturing blockade runners, first off the Bahama Banks and later in the Gulf of Mexico. Her first major prize was the steamer *Jupiter* on June 27, 1864, which had thrown her cargo overboard to increase speed but still could not escape; she had been headed for Wilmington.[5] A second important capture was the *Ruby* on February 27, 1865, on her way from Havana to St. Marks, Florida.[6] She was carrying lead for Southern bullets, a metal in short supply.

When the war came to an end, the *Nereus* and *Proteus* were sold at auction to James Hooper, of Baltimore, for $73,000 and $75,500 respectively. On the same day, July 12, 1865, the *Glaucus* and *Neptune* went to John Henderson for $62,000 and $67,000. Both gentlemen were acting for John W. Garrett, President of the Baltimore and Ohio Railroad. The fifth steamer *Galatea*, was sold in September to the Haitian navy and renamed *Alexander Petion*.

The railroads originating at Boston, Philadelphia, New York, and Portland had all profited from the passengers and merchandise landed by various steamship lines. The Baltimore and Ohio sought to improve its position by starting a transatlantic service from Liverpool to Baltimore with a stop at Queenstown to pick up the Irish emigrants who had, ever since the famine of 1848, been deserting their beautiful but impoverished homeland. The railroad hoped to generate revenue for its inland routes; if the new line could pay its own way, the maritime venture would be considered a success.

The four steamships were renamed after Maryland counties: the *Glaucus* became *Worcester; Neptune, Allegany; Nereus, Somerset;* and *Proteus, Carroll.* All were brought to Baltimore for overhaul. First to be fitted out was the *Somerset*, at the shipyard of John J. Abrahams & Sons.[7] Cabins were built by Louis Waggner and furnished by Michael Brothers. James Clark and Company worked on the machinery and installed a new shaft. On trials the engine turned 35 rpm under 15 psi steam pressure;[8] it could deliver 600 IHP. Despite a strike by the caulkers, who had not finished their work even though the *Somerset* was already loading cargo, she was made ready to sail for Liverpool on September 30, 1865, ten weeks after her purchase. Despite her small size, she was advertised to carry three classes of passengers at $90 for first, $65 second, and $32.50 steerage. Just five passengers for Ireland and two for Germany bought tickets. They were outnumbered by a crew of forty-one, which included John L. Sanford, Master, and David S. Frazier, Chief Engineer. A cargo of cotton, tobacco, corn, oil cake, bark, and dye stuff was much more satisfactory.

Saturday the 30th was a day of celebration for Baltimore. Some 5,000 people gathered at Henderson's wharf to see the first steamer of their city's first transatlantic steamship line depart. Since the recent war, only the *Circassian* had preceded the *Somerset* as an American passenger liner bound for a European port. At noon she set out accompanied by her sister, the *Carroll*, which was headed for New York. The revenue cutter *Nemeha*, carrying officials of the Baltimore and Ohio Railroad and other dignitaries; the propeller *Elizabeth*; and the steamboat *Trumpeter* all escorted her out to Chesapeake Bay. Liverpool was reached on October 16, which made the ocean part of the voyage from Cape Henry fifteen days long. For the return, English dry goods, ale, whiskey, salt, pig iron, potash, and soda were loaded, but the passenger list was still meager, only twenty.[9] Her homeward passage equaled her outbound one, sixteen days port to port. There was no mention of a Queenstown stop, although all subsequent westbound sailings of the Baltimore and Liverpool Steamship Company would call there.

A month after the *Somerset*, the *Worcester* sailed from Baltimore on November 29 but suffered an engine breakdown that brought her back into port on December 11.[10] She passed Cape Henry again on the 24th, running into winter gales so boisterous that she lost both her foremast and stack before reaching Liverpool on January 9, 1866. Westbound she met more bad weather and had to put into Halifax for coal. It was a bitterly cold passage; quantities of field ice were encountered and a mess boy died of pneumonia and had to be buried at sea. Her thirty passengers embarked at Liverpool on February 7 and landed at Baltimore twenty-two days later on March 1. These 200' steamships were not ideal craft to send across the North Atlantic in the winter. They were even smaller than Vanderbilt's *Ariel*, a steamship that had been in trouble more than once.

The other ships, the *Allegany* and *Carroll*, spent the latter months of 1865 ferrying freight between Baltimore and New York. The former ran aground on Long Island December 4 in a heavy fog and became a total loss when a storm arose before she could be pulled off.[11]

The *Carroll*, after returning to Baltimore in late December, spent the winter being refitted for the Atlantic. The *Somerset* made the line's first 1866 sailing in January, followed by the *Worcester* in March and then the *Carroll* on April 25; she reached Liverpool in fourteen days. Now that a monthly service could be assured, reasonable passenger lists were attained. The *Carroll* left the Mersey behind on May 6, sailed from Queenstown next day, and reached Baltimore on the 21st with ten cabin and 200 steerage passengers. Coming up the Chesa-

Somerset sailing from Baltimore on September 30, 1865. *Courtesy of The New-York Historical Society, New York City.*

peake, she saluted the *Somerset* on the way down. In all, there were a dozen sailings in 1866, each of the steamers taking four of them. In view of the 1865 experience, there was no December departure.

Each steamer found herself in difficulties at some time during the year.[12] Coming up the Chesapeake, the *Worcester* ran aground on Winter Quarter Shoal on May 10, but was hauled off with the aid of the *Rapidan* and delivered 230 passengers to Baltimore the next day. The *Somerset* was less fortunate. Scheduled to sail for Liverpool on September 28, she was delayed by a high wind and low water. After getting away on the 29th she ran into a sunken pile and was forced to return. Her cargo had to be unloaded to find and repair the leakage; she could not leave again until October 13. The *Carroll*, not to be outdone, collided with and sank the schooner *Doras* in the Mersey. Undamaged herself, she put to sea on October 4.

As a further complication, the *Worcester*'s last passage of the year, from Baltimore on October 30, 1866, was delayed so much that she did not reach Liverpool in time to meet her advertised sailing date, November 28. To replace her, a British screw steamer of 1,800 tons, the *Mexican*, was chartered to sail on December 5.[13]

January of 1867 saw the *Carroll* chartered to Mordecai and Company for coastwise trips to Charleston. These were followed by several to New York before she resumed her Liverpool operations on May 29. On that date she replaced the *Somerset*, which was withdrawn and saw no service of any kind from mid-April until October, when she was taken up by Mordecai for Charleston, Savannah, and New York activities. It is probable that alterations were being made. The Baltimore and Liverpool schedules were rather ragged in 1867. Nine eastbound sailings were made at irregular intervals, four by the *Worcester*, two by the *Somerset*, two by the *Carroll*, plus the *Mexican*'s return to Liverpool. From September to February of 1868 the line ceased operating on the

Atlantic. The *Worcester* had met with difficulties after leaving Queenstown for Baltimore on May 30, 1867. She had crossed as far as 15° West longitude when she lost her propeller; she sailed back to Queenstown and was towed to Liverpool for docking and propeller replacement.

During the fall and winter of 1867-68 the *Carroll* and *Somerset* were chartered by Mordecai to take cotton from Charleston and Savannah to New York. Both then joined the *Worcester* for the Baltimore and Ohio's freight run to New York.

By this time the parent company had perfected plans to substitute new and much larger craft for their small steamships. In January of 1867 an agreement had been made with the North German Lloyd to form a jointly owned company to connect Bremen and Baltimore. Two iron screw steamships of 2,300 tons with large freight and steerage capacities were to be built.[14] The Bremen operators would handle all marine affairs and the railroad would supply the piers, warehouses, and terminal installations at Baltimore. Everything proceeded with dispatch and the first vessel, the *Baltimore*, left Bremen on March 1, 1868. The second, the *Berlin*, followed a month later.

For some unknown reason, the Baltimore and Ohio continued their Liverpool and Queenstown sailings. The *Worcester* set out on February 26, 1868, four days before the *Baltimore*'s maiden departure from Bremerhaven. An innovation for the American craft was a stop at Norfolk before leaving the Chesapeake. Thereafter, approximately monthly sailings continued until September 15, when the *Worcester* left Baltimore on the line's last voyage. From Liverpool on October 14, she took eighteen days to return, including her usual Queenstown stop, and arrived at Baltimore on November 1, 1868. Of the seven sailings that year, the *Somerset* and *Worcester* took three apiece and the *Carroll* one.

These three steamers made twenty-nine voyages across the Atlantic and back during the three-year life of the Baltimore and Liverpool Steamship Company. In addition, the British steamer *Mexican* made crossings both ways under charter. As the only fleet of wooden propellers ever to try the route, they must have been exceptionally seaworthy vessels manned by exceptional seamen. Their original tonnage of 1,244 made them just three-quarters of the size of the *Washington*, the first regular American liner in 1847. Yet they ran with surprising regularity, often making the long distance between Baltimore and Liverpool in fifteen days. The *Carroll* did especially well in October of 1866 with twelve and one-half days Queenstown to Baltimore.

The operation proved to be a marginal one; cargo space was limited and passenger lists were never large. The highest number reported was 250 landed by the *Carroll* on October 23, 1866. In contrast, the German liner *Baltimore*, on her second trip, carried 756 steerage passengers in May of 1868.

Starting in September of 1868, the *Carroll* was used again by Mordecai for three voyages from Baltimore to Charleston, but in December, all three were steaming to New York as part of the railroad's freight service. About March of 1869 that venture, too, ended and they were laid up until the spring of 1873, when all three were purchased for $100,000 by F. W. Nickerson and Company of Boston.[15]

The *Somerset* was being prepared for delivery when, on April 13, she started leaking so badly that a large crew had to man the pumps until she could be put on a marine railway and hauled out. Nickerson at that time had two services out of Boston, one to Savannah and the other to Halifax, Nova Scotia, and Charlottetown, Prince Edward Island. All started on the latter route, but after the first year the *Somerset* spent most of her time on the Savannah one. Further interchanges took place as the need arose.

During the winter of 1873-74 the *Carroll* was sent to the Harrison Loring yard for new boilers and extensive alterations.[16] She came out with a new deck at what had been the original deckhouse top level, with the side planking built up to it. The forecastle had been raised to make her a flush-decker from stem to stern and new deckhouses had been built on top. The bowsprit vanished and the shallow, attractive, fantail stern became so heavy and so deep that it looked like a modern cruiser one. All the extra weight (ballast must have been added to maintain stability) buried her deep in the water.

The pilot house was well forward; aft of it came a house for the officers. A midship one was fitted between the masts and an after one enclosed a dining saloon seating 50 and the best staterooms. A lower saloon had further cabins for intermediate and second-class passengers. Dormitory space for 150 steerage was divided into three sections, one for men, one for women, and a third that was interchangeable. The *Carroll* was almost a new ship when she left Boston on May 2, 1874, for Charlottetown via Halifax, Port Hawkesbury, and Pictou. Encountering spring gales, she rolled and pitched so violently that the chain plates anchoring the fore topmast backstays gave way.[17] Her unsupported tophamper snapped off the foremast at its head and all the entangled spars and rigging crashed to the deck. No injuries resulted, but it was a sorry-looking steamer that put into Halifax on the 4th.

In 1875 she suffered a shaft fracture 135 miles out of Boston but was picked up by the *Worcester* the same day, October 10, and towed back to Boston.[18] Although propeller shafts were smaller and easier to forge than paddle shafts, they still had unsuspected flaws and from time to time gave way. The *Carroll* was plagued by bad luck; she had a second shaft break in 1880, repaired it at Pictou,

Nova Scotia, and then, heading for Prince Edward Island, ran hard aground before she could take on the passengers and cargo awaiting at Charlottetown.[19]

Like the *Carroll*, the *Worcester* and *Somerset* were rebuilt in a similar, but not identical, manner to increase their passenger capacity. The *Somerset* spent most of her time on the Savannah route. She had to put into Hampton Roads with a leaking boiler in the fall of 1875.[20] Then, in August of 1876, a crank pin gave way shortly after she left Savannah and she had to be towed, first to Charleston to unload and later to Boston for repairs.[21] The *Worcester*, like the *Carroll*, broke her shaft twice. The first time was on October 15, 1876, but despite the drag of her propeller she managed to sail into Halifax a day later.[22] In June of 1882 she duplicated the event and had to be towed from Cape Sable to Boston.[23]

This trio of steamers, once they were employed in the coastwise trade, proved to be the right size and the right type for Halifax and Savannah. As a consequence, they spent many long and prosperous years doing exactly what they had been designed to do some ten years earlier. When Nickerson improved his Savannah fleet in 1882 by adding larger, more modern, iron steamers, the *Somerset* became redundant but was kept as a spare boat. Four years later her owner got into financial difficulties as a result of an unfortunate speculation in the guano trade.[24] Harrison Loring, who always took a special interest in the vessels he had built or remodeled, took over the Halifax-Charlottetown route in 1887 and continued to offer weekly sailings for the next six years. Thus the *Carroll* and *Worcester* steamed serenely on while the *Somerset* was scrapped in 1886 or 1887. The former pair ceased running in 1893 and were dismantled on Nut Island in Boston Harbor in 1894.[25]

The three Van Dusen steamships deserve to be remembered as the only wooden screw propelled ones ever to make regular voyages across the Atlantic. The *Worcester* completed twelve, the *Somerset* ten, and the *Carroll* seven, all to Liverpool. Although completely outclassed by both foreign steamships and the much larger paddlers flying the American flag, they still operated with considerable regularity and some degree of financial success from 1865 through 1868. Moreover, in a quarter of a century of coastwise service they showed that the wooden screw steamer had become a commercial success as long as it was used in the proper trade. From Boston they made hundreds of trips to Savannah, Halifax, and Charlottetown.

Notes

1. *New York Herald*, October 6, 1862.
2. *Journal of the Franklin Institute* 50, 3d ser. (July-December 1865): 52, 53.
3. Official Records, ser. 2, 1: 89, 96, 157, 185.
4. Naval History Division, *Civil War Chronology* (pt. 4, 1864) (Washington, D.C., n.d.), p. 33.
5. Official Records, ser. 1, 17: 725, 728, 822.
6. *Civil War Chronology* (pt. 5, 1865), p. 49.
7. *Baltimore American and Commercial Advertiser*, September 29, 30 and October 2, 1865.
8. Edward Hungerford, *Story of the Baltimore and Ohio Railroad* (New York, 1928), 2: 76-79.
9. *Baltimore American and Commercial Advertiser*, November 27, 1865.
10. *New York Journal of Commerce*, December 14, 1865.
11. Ibid., December 6, 1865.
12. Cedric Ridgely-Nevitt, "The Baltimore and Liverpool Steamship Line," *Steamboat Bill* (1965), pp. 83-86.
13. *Liverpool Journal of Commerce*, December 3, 1866.
14. N.R.P. Bonsor, *North Atlantic Seaway* (Prescott, Lancashire, 1955); p. 168.
15. *New York Maritime Register*, April 16, 1873.
16. *Nautical Gazette*, May 16, 1874.
17. *New York Journal of Commerce*, May 6, 1874.
18. Ibid., October 13, 1875.
19. *New York Maritime Register*, August 25 and September 8, 1880.
20. Ibid., December 1, 1875.
21. Ibid., August 16 and 23, 1876.
22. *New York Herald*, October 17, 1876.
23. *New York Maritime Register*, June 14, 1882.
24. *Nautical Gazette*, March 6, 1886.
25. Francis B. C. Bradlee, *Some Account of Steam Navigation in New England* (Salem, Mass., 1920), p. 127.

18
The Havre Line Part 2: 1865-1867

ALTHOUGH the Civil War had disrupted the transatlantic sailings of every other American steamship service, the Havre Line not only carried on its regular sailings throughout 1861, but even advertised that the *Arago* would leave for Havre in March of 1862 and that the *Fulton* would follow in April.[1] Instead, the *Fulton* was chartered as a transport as early as February 15, 1862, by the Army's Quartermaster Corps, at $1,500 per day, and she became so useful that she was kept in almost continuous operation until the summer of 1865. Only from August of 1862 to March of 1863 was she not employed.[2]

The *Arago* was taken over by the Navy Department in March of 1862 to join the *Vanderbilt, Illinois,* and *Ericsson* at Hampton Roads, where Commodore Goldsborough was to keep them ready, with steam up, waiting for the Confederate ironclad *Virginia* to venture out.[3] They were to pursue, ram, and sink her. The *Arago*'s civilian officers and crew, however, refused to serve; they had not signed on for so hazardous a venture. Only Captain Henry A. Gadsden and his Chief Engineers were willing. We do not know what happened next, but the problem must have been solved. Official orders were issued on April 12 to all steamships to be ready to ram. A month later, after Norfolk had surrendered and the *Virginia* had been destroyed by her own men, Secretary Welles discharged his would-be rams. Starting May 29, the *Arago* joined her former running mate in the transport fleet of the Quartermaster Corps, but for unknown reasons was paid at a lower rate, $1,100 per day. Governments have always practiced unjust discrimination.

The *Fulton* made her first wartime trip to Ship Island, in the Gulf of Mexico, and then to Hatteras Inlet, off the North Carolina coast, as part of General Burnside's expedition, in March of 1862. Thereafter, both she and the *Arago* had fairly tranquil careers ferrying officers, troops, dispatches, and mails up and down the East Coast, mostly between New York and the naval base at Port Royal, South Carolina. Occasionally they went as far as New Orleans. There were a few interludes of excitement, all connected with the blockade of the Confederacy's coast.

On November 5, 1863, off Wilmington, North Carolina, the *Fulton* sighted an 800-ton iron sidewheeler, the *Margaret and Jessie*, a celebrated blockade runner, hotly pursued by three naval vessels, the *Nansemond, Keystone State,* and *Howquah*.[4] As the fastest steamship present, the *Fulton* outran all the others, firing her 20-pound Parrott gun as she went. She forced the *Margaret and Jessie* to surrender and put a prize crew aboard before the *Nansemond* caught up and tried to take over.[5] G. M. Walker, the *Fulton*'s First Officer, temporarily acting as her master, informed the naval contingent that "having captured the prize he was able to take her to New York." She was said to be worth $125,000 and her cargo $300,000 although a quarter of it had been jettisoned during the chase.

On her next northbound voyage from Port Royal, the

Fulton repeated herself. This time her regular master, James Wotton, was back aboard. On November 21, another blockade runner, the *Banshee*, was sighted; the *Fulton* gave chase and was joined by a gunboat, the *Grand Gulf*, and a transport, the *Delaware*. Off Cape Lookout the *Fulton* took possession and put aboard Walker and a crew that included a cavalry officer and twelve troops from her passenger list. The latter were to repel boarders from the *Grand Gulf* in case she claimed the prize.[6] The *Banshee*, a 220' steel steamship built at Liverpool, was a splendid example of the type especially designed for blockade running; she reached New York, towed by her captor, on November 23, 1863.

Without aid or hindrance of any naval vessel, the *Arago* captured the steamer *Emma* on July 24, 1863, running out of Wilmington with a full load of cotton and naval stores.[7]

On more than one occasion the *Arago* assisted craft in distress. The disabled U.S.S. *Augusta*, for example, was towed from Port Royal up the coast. Although the weather was very bad and the towline broke several times, Captain Gadsden delivered her to Hampton Roads on December 12, 1864. There General B. F. Butler ordered the *Arago* to abandon that tow[8] and take over a much larger steamer, the *Guiding Star*, which had been disabled by a broken paddle shaft; the two reached New York on December 16, thirty-seven hours from the Roads.[9]

The first news that the Confederates had abandoned Charleston was brought North by the *Arago* when she arrived at New York on February 21, 1865. The end of the war was imminent and her military days were numbered; her last trip as a transport left Port Royal on July 21 and arrived at New York three days later. The *Fulton* had completed her trooping service on July 15. Their final charter rates were $1,000 per day for the *Arago*, $1,200 for the *Fulton*. The continued difference still remains a wartime mystery. Both were returned to the Havre Line and, by September 1, were at 9th and 10th Streets and the East River undergoing repairs and reconstruction.

Four years of military charters at high rates meant that ample funds were available to put these fine steamships back into the very best of order. The *Arago* went to the Morgan Iron Works for a machinery overhaul and had two new boilers, weighing 140 tons apiece, installed. Her joiner work was replaced and new cabin furnishings were installed.[10] Upon completion she could carry exactly the same number of passengers, 250, that she had when she was new. The *Fulton* had similar work done, but as the second to be finished she attracted no attention from the press. About $400,000 was spent refitting the two. Henry Gadsden of the *Arago* and James Wotton of the *Fulton* continued as their commanders. The New York and Havre Steamship Company was now managed by Isaac Bell, its president. Despite severe postwar inflation, fares remained at $132.50 first and $85 second cabin, the prewar prices.[11]

The *Arago* left New York on November 25, 1865, bound for Havre, but, instead of Cowes, touched at Falmouth December 8 to land the English passengers and mails. Beset by terrific gales, she made a slow return passage from there on the 20th to New York on January 6, 1866. The *Fulton* followed, eastbound, on December 23, 1865, and from Falmouth January 25 with a thirteen-day passage home. Thereafter, there were regular sailings every four weeks. Conditions, however, were far less favorable than they had been in the 1850s.

The government of Napoleon III had been instrumental in setting up the Compagnie Générale Transatlantique, better known as the French Line, to provide service from France to Mexico and the French West Indies and, starting in 1864, added sailings to New York. Three iron paddlers, the *Washington, Lafayette,* and *Europe*, were completed in Scotland for the Havre, Brest, and New York route in 1864-65 and were soon followed by the *Napoleon III* and two screw steamships, the *Péreire* and *Ville de Paris*.[12] Late in 1866 the French Line was providing fortnightly sailings to and from New York and pampering passengers with their admirable wines included in the fare. Despite high rates, $180 first and $100 second, these large and very comfortable ships, together with excellent service, attracted a high class of passengers. Needless to say, French and Swiss luxury goods, a major contribution to the Havre Line's earlier success, now traveled in French bottoms. American cotton, at a much lower freight rate, went eastward on the American steamers.

The Havre Line, in the fall of 1866, cut its fares to $100 and $80 first cabin and to $50 for second, but to no avail.[13] Its management decided to withdraw and put its two steamships up for sale on December 22, 1866. Only one bid was received, $300,000 by L. W. Jerome, but he was acting for the line.[14] Because they could not dispose of their chief assets, the owners decided to continue in 1867 and, once the decision was reached, planned aggressive action. To be successful, their sailings had to be as frequent as the French ones. The aid of the New York Mail Steamship Company was enlisted and a joint fortnightly service planned and widely advertised.[15] The latter company had two iron screw steamships available, the *Merrimack* and *Mississippi,* as well as the *Guiding Star* and the *Morning Star*. Further details of these vessels are given in chapters 16 and 19.

Although there was no January sailing in 1867, the *Arago* started the combined operation on February 16 and was followed by the *Mississippi* on March 2. Two weeks later came the *Fulton* and then the beam-engined paddle steamer *Guiding Star,* which replaced the *Merrimack* on a sailing scheduled for March 30. For the first time since the demise of the Collins Line, American steamships departed every two weeks for England and France and continued to do so through July of 1867. In June the *Morning Star,* another beam-engined steamship, was substituted on short notice for the *Mississippi.* All the New York Mail ships were splendid vessels that complemented the Havre ones.

The combined service met with some success, particularly during the summer months. The French Line, as a result, cut its fares to $160 first cabin and $100 second, as against the $120 and $70 charged by the American ships.[16] The *Fulton* left New York on May 11, 1867, with a full cargo of cotton and general merchandise, 125 passengers, and $341,700. Moreover, the *Guiding Star,* sailing from Havre on August 21 and Falmouth two days later, arrived with a passenger list of 472.

The New York Mail Steamship Company, however, was in a desperate financial state. After the *Guiding Star*'s arrival at New York on September 3, its sailings were canceled. The Havre Line held on for three more voyages, at four-week intervals, before closing down the service. The *Arago*'s last trip started from New York September 28, left Havre on October 30, and reached her home port on November 14. It was her fifty-fifth Atlantic voyage, of which thirteen were performed after the Civil War. The *Fulton,* from New York on October 26 and Havre on November 27, made the line's final trip, an unhappy one with her starboard wheelhouse stove in during a gale and her main rail damaged by seas coming aboard. She was further delayed by snowstorms outside New York and did not reach her destination until December 13. She had made forty-eight voyages in all, twelve of them in the postwar era. In addition to the line's twenty-five round trips since 1865, the *Mississippi, Guiding Star,* and *Morning Star* had added six more Havre voyages in 1867.

The *Arago* and *Fulton* were still in excellent condition and only eleven and twelve years old when they were, for the first time in their careers, laid up. In 1868 an intense rivalry on the California route developed between the long-established Pacific Mail Steamship Company and a more recent competitor, the North American Steamship Company, which had William H. Webb, the eminent shipbuilder, as its president.[17] In earlier years he had been a stockholder in, and had built many steamships for the Pacific Mail. The newer line was well-run and attracted large numbers of passengers. For a time it was sending vessels both to Aspinwall for the Panama crossing and to Greytown for the Nicaragua one. After March of 1868 only the Panama service was continued.

The *Arago* was chartered by Webb, fitted with hundreds of steerage berths, and left New York for Aspinwall on January 4, 1868. She put out from Aspinwall January 20 with 360 passengers but had difficulties with her forward cylinder when she was two days out, and carried on at reduced speed with the after one. Short of coal, she made Savannah on January 30.[18] Continuing her voyage two days later, she reached New York February 5.

The *Fulton* replaced her and left for Aspinwall February 15 under Captain Merry, who was transferred from the *Arago.* A second voyage was made in March and a third on May 5, with a return from Aspinwall on the 19th, arriving at New York after a rather long passage on the 28th.

For the second time the Havre Line offered its ships at auction on May 29, 1868, but received no satisfactory bids.[19] Again they lay idle, but in the fall the *Arago,* which always seems to have been the preferred steamer, was chartered by Ruger Brothers to make their last sailing that year to Bremen. She was sent out on September 5 and returned, via Southampton, with a load of immigrants. Departing from Bremen on October 6, it took eighteen days to reach New York; westerly gales slowed her the whole way.

A second Bremen voyage was undertaken, but who sponsored it is not clear. Godeffroy Branker and Company advertised a November 28 sailing with "room for cotton only."[20] The November 30 report of her clearance quoted the North American Steamship Company as her agent; that organization had ceased operating and it is surprising to see it listed for a steamship that it did not own and never had owned. She was next reported at Falmouth, England, having sailed from New York on December 2 with fifty passengers, a crew of ninety-eight, and a cargo of 2,700 bales of cotton.[21] Eleven days out she was running before a heavy gale under reefed foresail and fore topsail when a violent squall made her broach to and left her rolling uncontrollably. Boats were swept away, funnel guys torn loose (the Second Mate was carried unconscious from the deck when one struck him), skylights were broken in, dining tables torn loose, and water poured down into the hull. Winds of hurricane force heeled the ship on her beam ends, and the barometer dropped to a low of 28.49 inches! Her engines never stopped, and she was finally brought head to the sea until the wind abated and she could return to her course. Falmouth was made on the 19th to refill her empty coal bunkers. The account in the *New York Herald* was no exaggeration when it stated that "the opinion prevailed that the number of steamships was small indeed that would not have succumbed to the force of the elements had they been in a like position." George W. Browne, her master, proved himself a seaman in the very best tradition of the sea. Despite the mauling she received, she reached Bremen on December 24.

From there the *Arago* sailed on January 10, 1869, and was next heard from six days later at the English fishing port of Grimsby.[22] Because this North Sea port was a most unlikely one for any Atlantic steamship, one can only speculate that some new accident put her there. She must have wintered abroad, for she was next reported as sailing from Bremen on March 3, 1869, with a new master, A. G. Jones, and under a new charter to Ruger Brothers.[23] After a stop at Falmouth on the 11th and 12th, she sailed for Philadelphia, another unusual port. Leaking boilers forced her to shorten the trip; she changed course for New York and docked there on March 28, 1869.

Within a month she had been sold to undisclosed owners.[24] Later events showed her to be part of a comprehensive set of purchases by Peruvian authorities who were building up a fleet to use in a war with Spain.

Two American monitors left over from the Civil War, the *Catawba* and *Oneota*, had been built at Cincinnati for over $600,000 apiece.[25] Completed in the summer of 1865, they saw no service and had been rusting and rotting away near New Orleans until acquired by Peru. In order to tow them on the long voyage to the Pacific, two merchant steamships were acquired, the iron screw steamer *Havana*, formerly of the Brazil Line, and the *Arago*. Everything was renamed. The monitors became the *Atahualpa* and *Manco Capac*, the *Havana* the *Maranon*, and the *Arago* the *Pachitea*.

The last sailed from New York on May 1, 1869, still under her original name, to rendezvous with the others at St. Thomas in the Virgin Islands. By that time peace had been declared between the combatants, but a Spanish ironclad, the *Victoria*, was keeping watch when the four craft left the harbor on May 19, 1869, each under her own power. Once outside the harbor the monitors, because they could carry only a five-day coal supply, were taken in tow.[26]

By the time the fleet reached Rio de Janeiro it was in a sorry state. The wooden hulls of the ironclads flexed in a seaway and needed to be reinforced with extra beams and stanchions. A grounding damaged the *Manco Capac*. The *Maranon*, ex-*Havana*, suffered such severe boiler casualties that she could barely make four knots and was incapable of towing duty. The *Pachitea*, ex-*Arago*, therefore, had to tow one monitor at a time and then return for the second one. A three-month delay at Rio was needed to make the vessels seaworthy. Many of the American officers left in disgust as an excess number of inexperienced Peruvians arrived and joined the convoy.[27] The Strait of Magellan was reached in January of 1870, but not until May 11 did the squadron reach Callao.[28] The monitors saw actual service in 1880, when Peru and Chile were at war;[29] the *Maranon* became a school ship; but of the *Arago* nothing further is known.

As for the *Fulton*, which had been laid up since her return from Aspinwall in May of 1868, she had to endure a year's idleness before being taken up by Ruger Brothers. She left New York on May 29, 1869, under Captain A. G. Jones, for Copenhagen, via Cowes and Bremen. This was followed by a second voyage extending clear into the Baltic. From New York on August 21, 1869, the *Fulton* reached Swinemunde, the port for Stettin, on September 13. Five days later she left, stopping at both Copenhagen and Southampton, to deliver 650 passengers to New York on October 23. Stettin was substituted for the usual call at Bremen because she carried eastbound cargo for Russia.[30] It was transferred to a steamer named the *Hogbo* to take it on to Kronstadt.

The *Fulton* was laid up again, but by March of 1870 was at the Morgan Iron Works having her engines taken out while D. W. Edwards was simultaneously offering her hull for sale.[31] Thus, with the foreign purchase of the *Arago* in 1869 and the dismantling of the *Fulton* in 1870, both Havre steamships completed their careers at the age of fourteen. During their early years they were exceptional liners, the pride of their owners, of their masters, and of the entire American Merchant Marine. Their splendid workmanship and remarkable economy of operation were a tribute to all who assisted in their creation. The same features made them outstanding transports during the Civil War. But when they returned to the Atlantic at the end of 1865, European shipbuilding had forged ahead and left them behind as second-class steamships. Three years later they were reduced to the status of immigrant carriers but still found charters and managed to return profits to those who chose to use them under favorable circumstances. The *Arago* eventually logged fifty-seven voyages across the Atlantic and back while the *Fulton* did fifty. In addition, the *Arago* completed one round trip from New York to Aspinwall and made a delivery trip to a Peruvian port towing two monitors much of the way. The *Fulton* replaced the *Arago* on the Aspinwall route and completed three voyages thereon. Each made two trips in the Rugers' steerage service to the North Sea and the Baltic.

The Havre Line, if we include both its 1850-61 and 1865-67 operations, made 159 trips from New York to Havre, with stops at Cowes before the Civil War and at Falmouth after. But with monthly sailings instead of fortnightly ones, it was never able to equal the record of 185 voyages undertaken by the Collins Line. The Havre Line's stockholders were, nevertheless, consoled by excellent dividends, rather than gloriously unprofitable statistics. The Havre Line and its two final steamships deserve recognition as the first completely successful American steamship venture on the Atlantic. Moderate speeds and economical engines were the key to their success.

Notes

1. *New York Commercial Advertiser,* March 18, 1862.
2. House Executive Document 337.
3. Official Records, ser. 1, 7: 134-35, 169, 342-43.
4. *New York Herald,* November 9, 1863.
5. Official Records, ser. 1, 9: 264-68, 318-24.
6. *New York Herald,* November 24, 1863.
7. Official Records, ser. 1, 14: 399.
8. Ibid., 3: 392-93.
9. *New York Journal of Commerce,* December 17, 1864.
10. *New York Herald,* November 19, 1865.
11. *New York Journal of Commerce,* advertisement, June 4, 1866.
12. N.R.P Bosnor, *North Atlantic Seaway* (Prescott, Lancashire, 1955); pp. 207-9.
13. *New York Journal of Commerce,* advertisement, September 12, 1866.
14. Unidentified newspaper clipping, William H. Webb scrapbook.
15. *Boston Transcript,* advertisement, February 15, 1867.
16. *New York Herald,* advertisement, March 5, 1867.
17. John Haskell Kemble, *The Panama Route, 1848-1869* (Berkeley and Los Angeles, Calif., 1943), pp. 109-10.
18. *New York Journal of Commerce,* February 6, 1868.
19. Ibid., May 30, 1868.
20. Ibid., November 16, 1868.
21. *New York Herald,* January 7, 1869.
22. *Boston Shipping List,* January 1869.
23. *New York Journal of Commerce,* March 29, 1869.
24. Ibid., April 26, 1869.
25. Frank M. Bennett, *The Steam Navy of the United States* (Pittsburgh, Pa., 1896), pp. 348, 627.
26. *New York Herald,* June 5 and 6, 1869.
27. Ibid., September 23, 1869
28. John D. Alden, "Monitors 'Round Cape Horn," *U.S. Naval Institute Proceedings,* September 1974, pp. 80-82.
29. William Laird Clowes, *Four Modern Naval Campaigns* (London, 1902), pp. 77, 110, 113, 115, 122.
30. *New York Herald,* September 4, 1869.
31. *New York Maritime Register,* March 2, 1870.

19

The New York Mail Steamship Company, 1863-1867

ALTHOUGH the Civil War diverted many steamships from their normal routes, it did not entirely disrupt coastwise operations. The Army's endless demand for transports and the Navy's for merchant vessels that could be put to work as armed cruisers, gunboats, blockaders, and supply steamers encouraged owners to build new vessels to replace those already taken over. Moreover, once Admiral Farragut's fleet had fought its way past the Mississippi River forts below New Orleans and General Butler's troops had landed and taken the city, a thriving trade developed to that reopened port. Older firms, such as Spofford and Tileston, began to send steamships there and newer ones were formed to satisfy the growing demand for freight and passenger service.

One newcomer was the New York Mail Steamship Company, which had James A. Raynor as president. Raynor's intention was to own the largest and most luxurious passenger steamships on the East Coast. Two splendid side-wheel steamships were ordered from the Rosevelt and Joyce shipyard. The person directly engaged in their construction was Isaac L. Waterbury, who had been with Jacob Bell when the *Baltic*, *Pacific*, and *Pioneer* were built.[1] Since Vanderbilt's clear demonstration of the walking beam engine's economy on an ocean steamship, it had become the universal choice to power any sidewheeler. For Raynor's vessels, a reduction in cost and construction time was achieved by the use of secondhand engines taken from Greak Lakes steamers.

The pair were named *Morning Star* and *Evening Star* and, because their owner's corporate name was long, their sailings were usually advertised under the heading of "Star Line." The *Morning Star*'s official dimensions were 272'-0" by 39'-4" by 19'-8", with a tonnage of 2,022, while the *Evening Star* was rated a foot shorter and eight tons lower.

Waterbury gave the length on keel of the *Evening Star* as 275' and length overall as 288', with a light draft of 12'-6" forward and 13'-9" aft. She was strongly built with a keel 15" x 16" and a frame of white oak. Floors were 12" x 15", spaced 30", and the bottom was filled in solid for 200'. Yellow pine ceiling 7" thick was fitted inside the frames and oak planking 4 1/2" thick with 5" wales and 7" x 16" garboard strakes outside; the latter were bolted edgewise to the keel as well as vertically into the floors. Straps 4" x 5/8" were notched into the frames and their upper ends riveted to a 5" x 3/4" iron band extending around the ship just below the deck.

The *Morning Star*'s engine of 80" bore by 12'-0" stroke was built in 1853 by the Morgan Iron Works for the *Crescent City*. The *Evening Star*'s was identical in size, constructed by Cunningham and Belknap for a similar steamer, the *Queen of the West*. New return-tube boilers with forced draft blowers were installed. The paddle wheels were 33' in diameter and had 28 blades.[2]

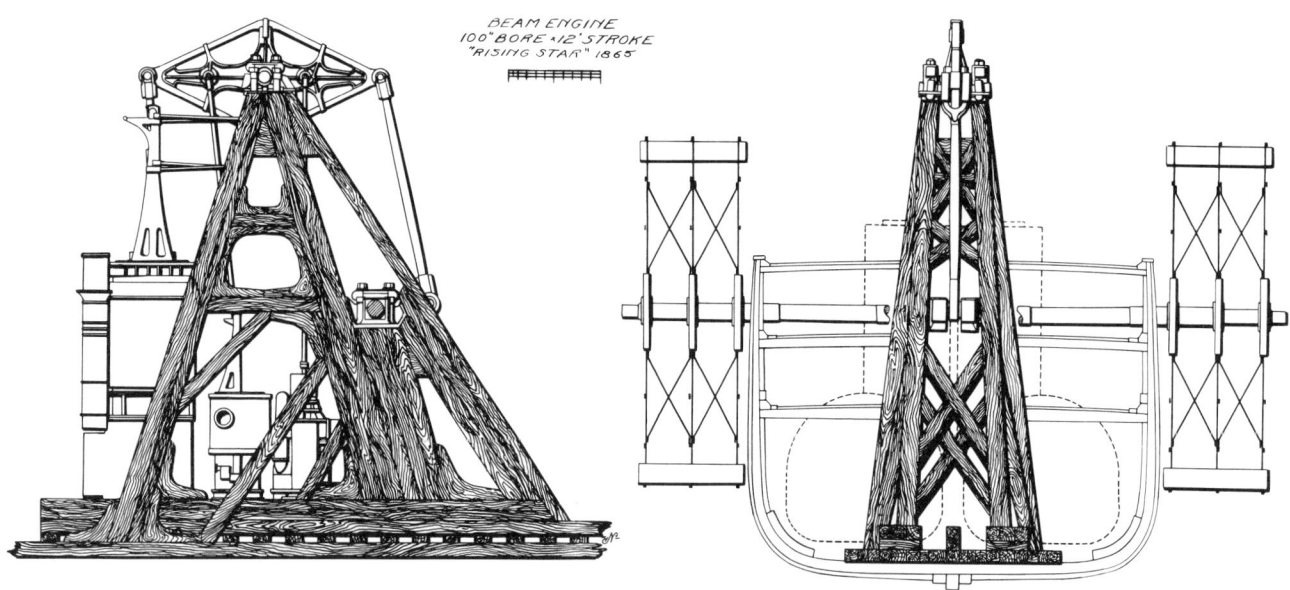

Walking beam engine 100″ bore by 12′-0″ stroke for the *Rising Star*.

The Star steamers came out as particularly handsome, single-stacked liners having black hulls to the rail line. A low forecastle was fitted forward. Above the hull the paddle boxes were white, to create an unbroken sheer line. A small two-masted rig had square courses and topsails on the foremast. Despite their large size, almost as long as the Havre Line's *Arago* and *Fulton*, they looked rather delicate. Cabins were installed for 223 first- and 24 second-cabin passengers; great comfort and elegance were stressed in their interior design. The few second-class berths were intended for servants traveling with their employers.

The *Morning Star* opened the line's operations to New Orleans by way of Havana on April 15, 1863, and made three sailings from New York before the *Evening Star* appeared. The latter left on a trial voyage to Havana on June 30, 1863, and, starting in July, the sisters sailed twice a month for New Orleans. At first Havana was a way stop, but in later years it was omitted.

A third and enlarged steamship, the *Guiding Star*, was ordered from the same builders, reorganized under the name of Rosevelt, Joyce, and Waterbury. Once more an existing beam engine, 81″ bore by 12′-0″ stroke, was obtained from another Great Lakes steamer, the *Mississippi*, which was being dismantled in Buffalo in 1863. It had come from Philadelphia's I. P. Morris works. The *Guiding Star* was 300′-6″ by 40′-6″ by 33′ depth of hold to the spar deck and had a tonnage of 2,385. Her first-class passenger capacity was increased to 250 and Raynor was to advertise her as "one of the very best as well as the most elegant steamships ever to be built in this country and for comfort or elegance, her saloons and staterooms are not excelled by those of any other steamer in the world."[3] For some reason she was given a larger sail plan than her predecessors, with square sails on both masts. The deck arrangement, with a half-height forecastle and a raised pilot house just aft of the foremast followed by a long deckhouse extending to the stern, was unchanged. The *Guiding Star* was launched August 13, 1864.[4] Only two months later she sailed for New Orleans.

As a wartime military base, that city prospered; the three Star liners were continuously filled with passengers and cargo. A fourth was laid down as the largest and grandest of the lot. She was to be 303′-6 1/2″ by 43′-8″ by 23″-0″ depth to the main deck with a draft of 17′-0″ and a tonnage (under the new system started in 1865) of 2,726. This made her considerably larger than the *Arago*, *Fulton*, or the last Vanderbilt liners, *Costa Rica* and *New York*; she was exceeded in size only by the surviving Collins liners, *Atlantic* and *Baltic*, and the *Vanderbilt* among the steamships on the Atlantic or Gulf Coast.

The beam engine was a new one designed by John Roach's Etna Iron Works. Of 100″ bore by 12′-0″ stroke, it had an unusual feature: a steam jacket surrounding the cylinder to prevent condensation inside when the exhaust valves opened to the cold condenser. A separate boiler of small size supplied steam to the jacket at 35 psi pressure;

from there it discharged into the main steam line at 25 psi pressure. There was, of course, a surface condenser. The paddle wheels were larger than the others, 36' in diameter.[5]

The *Herald* reported that "this vessel is the peer of anything before sailing from this port."[6] When completed, her cost ran to $800,000, $300,000 of it for the machinery. There were three decks, two masts with square sails, the usual single stack, and a large cargo capacity, the equivalent of 27,000 barrels. The dining saloon was in the after deckhouse and the best cabins directly below it. Although the Star quartet were built for coastal service, three of them would make Atlantic crossings; it is, therefore convenient to treat them as a group.

The *Morning Star* sailed from New York at four-week intervals, from April of 1863 through May of 1864, completing fifteen successive voyages. The only interruption was a brief one when, northbound, she ran aground on New Jersey's Deal Beach on July 31, 1863. Help was sent down from New York[7] and, with the aid of anchors planted off shore, she was kedged off and reached New York on August 2. The damage was slight; she cleared New York again on the 15th.

A year later she was off the run from June through September of 1864. It is likely that engine or boiler modifications were being made. Whatever the cause, there were alterations to the passenger quarters, for it was their incomplete state that delayed resumption of service until October 1.[8] Thereafter she ran regularly until the middle of June 1865. Her popularity is best described by another enthusiastic notice: "The unsurpassed accommodations of this magnificent vessel are so well known to travellers that it is rare that a vacant stateroom can be obtained on the day of sailing."[9]

The *Evening Star* became even more successful. Starting in June of 1863, she steamed along at four-week intervals for almost three years. But on her fortieth voyage she met a hurricane that overwhelmed and sank her on October 3, 1866, about 180 miles off the Georgia coast. A former Vanderbilt liner, the *Daniel Webster*, perished in the same storm. Of the 275 passengers and crew on the *Evening Star*, fully three-quarters were drowned.[10]

Despite the fact that her cylinder was only an inch larger in diameter than those of her two predecessors, the *Guiding Star* became the fastest of the three. Coming out in the fall of 1864, she left for New Orleans on October 22.

Her first trip was uneventful, but while returning from her second, her shaft fractured off the North Carolina coast. Although her engine was secondhand, the shafting, wheels, and boilers were all new. Many similar failures have been reported.

A small iron steamer, the *S. R. Spaulding*, towed the *Guiding Star* to Old Point Comfort. Her passengers landed on December 11 and were sent on by the Baltimore steamboat. A larger and more powerful transport, the *Arago*, took over and towed her into New York on December 16.[11] After six weeks at the Novelty Iron Works, she was ready to sail from New York on January 28, 1865.[12]

There were further engine problems in May when a crosshead shoe broke and her speed had to be kept below eight knots for most of the way to New Orleans. Once this was corrected, the *Guiding Star* began to perform with élan, each new passage time being shorter than the last. On her eighth voyage she sailed from New Orleans at 7:20 P.M., July 22, 1865, left the Mississippi by way of South West Pass at 6:30 the next morning, after a delay of four-and-one-half hours awaiting the tide. Sandy Hook was reached at 10:00 A.M. on the 28th and she docked at 11:30. The sea time was 5 days 1 hour 30 minutes and from wharf to wharf (after deducting the stops in the river) 5 days 10 hours.[13] The overall speed of 13.25 knots created a record that was to hold for ten years before it was surpassed by the screw steamers *Knickerbocker* and *Hudson*.[14]

The *Guiding Star* fared better than the *Evening Star*, even though her eleventh passage from New Orleans came close to disaster on two separate occasions. Leaving the Mississippi on October 14, 1865, under a river pilot, she ran hard aground just outside South West Pass. Five days elapsed before she could be got off, but the weather and sea remained calm and she was able to proceed on the 19th. Two days later a gale developed and turned into a hurricane. Her main spencer was set to try to head her into the wind and sea, but it was blown to ribbons. The *Guiding Star* began to roll so severely that her condition became perilous. Captain Berry had a heavy gun and its carriage made fast to a 10" hawser 80 fathoms long and dropped it over the bow to act as a drag. With its aid he was able to bring her head somewhat nearer to the wind and she managed to ride out the seas with safety.[15] The *Republic*, ex-*Tennessee*, was lost, and the *North Star* and *Northern Light*, both under charter to the Star Line, were badly damaged in nearby waters. The *Guiding Star* made for Port Royal to replace her depleted coal supplies, arriving there October 26. She took aboard eighteen passengers rescued from the *Republic*; five others elected to stay ashore rather than risk another sea voyage! New York was reached on November 2 and the *Guiding Star* had suffered so little from either her grounding or her laboring in tempestuous seas that she was able to head South four days later.

Her sixteenth and last New Orleans voyage began on March 17, 1866, and was completed at New York April 16.

Back to February of 1865. Once the *Guiding Star*'s

damaged shaft had been replaced, the Star Line had three magnificent steamers and excellent commercial prospects, but, although the *Rising Star* was on order, it still lacked enough vessels to undertake a weekly service. Raynor expanded its operations by chartering available craft while he looked around for suitable ones to purchase. In the former category were two wooden screw steamships of just over a thousand tons, the *Mariposa* and *Monterey*, which made weekly sailings possible after April of 1865. Fares of $60 were charged instead of the $80 ones for the more commodious paddlers.[16]

The end of the Civil War brought another venture to James Raynor's attention, that of expanding into ocean service. He went as far as to advertise that the *Guiding Star* would sail from New York for Southampton and Havre on June 15, 1865.[17] The passenger response must have been limited; no such sailing was made.

At the same time, when the *Golden Rule* was lost on Roncador Reef on May 30, 1865, the *Morning Star* was chartered to replace her for a single voyage to Greytown.[18] She left New York June 21, sailed from Nicaragua July 5, and docked at New York on the 13th.

As permanent additions to the fleet, the largest and most luxurious screw steamships available, the *Merrimack* and the *Mississippi*, were considered. Both had spent the war as transports. The *Mississippi* reached New York from New Orleans loaded with troops on May 20, 1865, only to steam back again, under charter to the Star Line, on June 27. After a second trip in July, the New York Mail was so favorably impressed with her that it bought both steamships in October of 1865.[19] The *Mississippi* left New York as a Star liner on October 19 and, along with many other craft already reported, ran into that month's devastating hurricane. Her decks were swept by seas that broke through the skylights and flooded the saloons below. The steering chains gave way, leaving her uncontrollable. Her purser's office was broken open and its cargo books, bills of lading, and mail bags swept overboard. But, like the *Guiding Star*, she survived.[20]

The *Merrimack*, after completing her last transport duties in May of 1865, was laid up until transferred to the New York Mail, whereupon she sailed for New Orleans on October 25. Her first commercial voyage was not without incident for, just after leaving the Mississippi on her way home, a severe leak suddenly developed near her stern tube. She turned back to New Orleans; several feet of water had to be pumped out of her on November 15. Extensive repairs took several weeks and she did not reach New York again until December 15.[21]

Simultaneously, the *Mississippi*, just after her hurricane damage, southbound from New York on her fourth voyage for the company, suffered a broken crankshaft on November 23 off the New Jersey Coast and had to be towed back to New York by the *Nevada*.[22] Because of these accidents, both were out of service for the winter of 1865-66.

The New York Mail Steamship Company now found itself in a difficult financial situation. Its success had led to the appearance of a number of competitors in the New Orleans trade and its construction and purchase of expensive steamships had put it badly in debt. Not only that, the wartime boom had ended. The first sign of overexpansion was the announcement in April 1866 of the sale of the *Rising Star*, which had been completed but not sent to sea, to the Pacific Mail Steamship Company.[23] The latter had recently taken over the New York to Aspinwall mail route and needed new ships to replace the inadequate Vanderbilt fleet it had inherited. The second sign of retrenchment was the continued use of its smaller, slower, more economical screw steamers, the *Mariposa* and *Monterey*. After repairs, the *Merrimack* made a single voyage to New Orleans at the end of March.

Instead of continuing, however, she and her sister were chartered to Ruger Brothers' North American Lloyd and sent to Bremen. The *Mississippi* left first, on May 10, 1866, under Captain Furber, stopping at Cowes on the way. Leaving Bremen on June 7 and Cowes on the 9th, she reached New York June 23 with a passenger list of 929. The *Merrimack*, with Captain Van Sice, sailed from New York on May 17 but did less well. She brought in 318 immigrants on July 2, fifteen days from Bremen, including a stop at Cowes on June 20.

After their European jaunts, both went back to the New Orleans route. But the Star Line, alas, had a new and very different management. Cornelius K. Garrison and Daniel B. Allen had acquired it. Garrison had replaced Raynor as the line's president and New York agent in July of 1866.[24] Under this far less imaginative regime, the *Evening Star* was the only one of the three side-wheelers immediately put to work. She made one round trip in September only to be lost on a second one in October. To replace her, the *Morning Star* was reactivated. The new management considered the side-wheelers too large and too costly; for them they substituted smaller screw steamers, the *Havana, Matanzas,* and *Missouri*, which had been under Garrison's control prior to the New York Mail acquisition. Because the *Havana* and *Matanzas* had iron hulls, the New York Mail Steamship Company now achieved the distinction of having the most modern fleet under the American flag; no one else could boast four first-rate iron screw steamships.

Another Garrison and Allen enterprise was the United States and Brazil Mail Steamship Company, which had been founded jointly with Thomas Asencio and Company. A subsidy of $150,000 per year had been offered by the United States and the Brazilian governments to bring about a monthly mail service from New

York to Rio de Janeiro. Large steamships of at least 2,000 tons were required.[25] The operation began in September of 1865, was completely taken over by Garrison and Allen in 1866, and eventually was operated by William R. Garrison, who succeeded his father, C. K. Garrison, as president of the corporation. Allen withdrew about 1868.

An immediate problem for the Brazil Line, as it was usually called, was that it owned only two steamships over the 2,000-ton requirement and could not, therefore, collect its 1865 fees on the September and December voyages made by the 1,801-ton *Havana*. To remedy the matter, it chartered the *Morning Star*. She departed from New York on March 29, 1866, stopping at St. Thomas, Para (now Belem), Pernambuco (Recife), and Bahia (Salvador) on the way. Returning, she sailed from Rio de Janeiro on May 4, reached St. Thomas, then a major center for Caribbean trade, on May 25, and arrived at New York May 31.

Later, when the Brazil and Star Lines came under the same management, the *Guiding Star* was transferred to the former line and made the Rio sailings from New York on September 22 and December 22, 1866. Because her running mates, the *North America*, ex-*Fort Jackson*, and the *South America*, ex-*Connecticut*, were both beam-engined steamships, she fitted nicely into the South American run. She was advertised to make the March 22, 1867, sailing as well, but was replaced by the *Merrimack* at the last moment; the latter left on the 23d, a day behind schedule.

Transatlantic Operations

The Garrison and Allen management not only transferred steamships between the New York Mail and the United States and Brazil Mail Line, but they also supplied ships to run with the Havre Line on fortnightly sailings to Falmouth and Havre. Their original plans contemplated using their two largest screw steamships to Havre in the summer of 1867.[26] The *Stars* were to be laid up or sent to Brazil and New Orleans was to have their smaller screw steamers. But nothing ever remained static with either Garrison or Allen; both had been steeped in the Vanderbilt tradition of day-to-day policy shifts.

True, the *Mississippi* sailed for Havre, as planned and as advertised, on March 2, 1867. Since she unloaded 674 passengers when she arrived home on April 20, it must have been a very satisfactory voyage, although a long one—fifteen days from Falmouth, slowed by bad weather. The *Merrimack* was to follow four weeks later, but, as we have just seen, she went to Brazil instead and the *Guiding Star* was shifted, at the last moment, to the Havre run. She left on March 30, made a better westward crossing than the *Mississippi*—thirteen days from Falmouth to New York—and arrived on May 15. The time was remarkably good considering that a gale slowed her and tore the woodwork off the paddle boxes on the 7th and 8th.[27] The *Guiding Star* continued on the Atlantic for two more trips from New York in May and July. Her size and speed made her especially suitable for the run. Her final crossing, from Havre August 21, arriving at New York on September 3 with 472 passengers, was the sixth and last transatlantic one by any New York Mail steamer.

After a second Havre trip by the *Mississippi* in April, she was replaced by the *Morning Star*, which began her career as an Atlantic liner on June 22, 1867. She was scheduled to return from Havre on July 24 via Falmouth on the 25th.

While the short-lived Atlantic venture was under way, the Star organization was both changing ownership and disintegrating. Its recent management had been far more interested in financial manipulations than in steamship operation. Garrison and Allen had discreetly withdrawn in April of 1867 and had taken with them the *Merrimack*, already on the way to Brazil. Furthermore, they retained title to the *Mississippi*, although she was not returned to them until September. Other vessels they had brought with them were taken back. The Star Line, with two small steamers, *Mariposa* and *Monterey*, and two large ones, *Guiding Star* and *Morning Star*, was left in the hands of George B. Hartson to see what he could salvage. Debts were also left; an action had been filed against the line in the Supreme Court of the State of New York for upwards of $350,000, including an unpaid coal bill of $108,139.34.[28]

A line that had at one time made two New Orleans sailings per week and had boasted five of the finest steamships afloat was about to founder. The *Mississippi* made one more New Orleans sailing, in August of 1867, for Hartson, but after that she went to the Brazil Line. The *Mariposa* and *Monterey* were all that Hartson had running. The former sailed from New York on November 30, 1867 and from New Orleans on December 15, and reached her home port two days before Christmas. That was the sad and sudden ending of what had been a splendid, well-run steamship line under James Raynor. In four and one-half years it had made 195 voyages from New York to New Orleans, one to Havana, and six to Havre. Moreover, it had supplied vessels to others for a voyage to Nicaragua, two to Bremen, and four to Rio de Janeiro.

Of its surviving *Stars*, the first, *Morning Star*, completed her active career in August of 1867 when she returned from Havre. She was only four years old and had traveled widely. In addition to thirty-four New Orleans

Rising Star docked at Erie Basin, Brooklyn. Her metal sheathing is being replaced and bottom planking caulked. *Courtesy of San Francisco Maritime Museum, Firemans Fund Collection.*

voyages she had made one to Nicaragua, one to Brazil, and one to France. She was laid up at Hoboken, Weehawken, and Red Hook until broken up at the Quintard Iron Works in the spring of 1872.[29]

The *Rising Star*

The *Rising Star* did much better as a Pacific Mail steamer than the other *Stars.* Although purchased in April of 1866, she was not put on their Aspinwall route until the late fall. Whether major modifications were needed is not known, but, obviously, several hundred steerage berths had to be installed. As a Star liner she was intended for only the highest class of passenger.

Her maiden voyage began at New York on November 21, 1866. At that time the Pacific Mail provided three sailings a month to California and had the *Ocean Queen*, from Vanderbilt; the *Rising Star*, from Raynor; and two steamships of their own design, the *Arizona* and *Henry Chauncey*, on the Atlantic. All were well-matched in size, speed, and capacity, and like all large, post-1860, coastwise vessels, they were reliably and, for the low steam pressures used, economically propelled by single walking beam engines. The *Rising Star* continued on the Aspinwall route through May 25, 1869, the completion date of her twenty-fourth voyage, when a new and even larger steamship, the *Alaska*, took her place.

Ruger Brothers chartered the *Rising Star* in 1870 for a European venture leaving New York March 24. She was at Havre from April 7 through the 20th, when she headed for Swinemunde, the port for Stettin. From Swinemunde on May 5 she sailed for Copenhagen, touching there on the 12th and Christiansand in Norway on the 15th, and, carrying a phenomenal passenger list of 1,267, arrived home on May 29.[30] A case of smallpox forced her to anchor off Quarantine until the passengers could be vaccinated and the ship fumigated.

After her release she made an Aspinwall voyage in July and was then extensively overhauled to replace the *Alaska*.[31] The latter left the East Coast to join the company's Pacific fleet. Returning to the Panama route, the *Rising Star* departed on February 4, 1871, and grounded off the Jersey Highlands but was got off, aided by two passing steamers, and proceeded on her way.[32] Her next trip was even more embarrassing to her master, A. G. Gray. Northbound from Aspinwall, she was coming into Kingston, Jamaica, for coal on March 27 and, while attempting to anchor, ran into and injured two vessels before she could be stopped. A third was driven ashore.[33] She continued with monthly sailings, together with the *Ocean Queen* and *Henry Chauncey*, for the rest of 1871 and most of 1872.

In October of the latter year she was sent to Brooklyn's Erie Basin for routine docking and overhaul. While the dry dock was being pumped out, the *Rising Star* slipped from the keel blocks and fell over on her side, collapsing the bilge blocks that should have held her upright. The hull was badly strained and leaking, and the whole engine was thrown out of alignment. She was refloated, towed to Hoboken, and put on a floating dock there.[34] This was not entirely a freak accident; the present-day naval architect is aware that a vessel with trim by the stern may lose stability during docking and that the worst condition occurs just before the keel comes to rest, throughout its length, on the blocks in the bottom of the dock. In the *Rising Star*'s day, no one had the necessary theoretical background to investigate the problem.

The hull was made tight, the cylinder, crosshead guides, and paddle shafts were realigned, and everything was put right in time for a November 21 sailing. By this time the days of the wooden side-wheeler were numbered; iron replacements were on order. The *Rising Star* ran regularly from November of 1872 through October of 1873. In 1874 she was relegated to freight service from New York to Aspinwall in January and February. Her last voyage, starting on February 7, 1874, ran into a winter gale so severe that her violent rolling shifted her boilers.[35] She put into Bermuda on February 15, leaking so badly that she was grounded to prevent her sinking. All cargo had to be discharged before she could go into the Admiralty dock there.[36] Although undocked on March 21, she could not continue to Aspinwall until April 5. There she was held for another month and did not reach New York until May 22. By that time there was no one who either wanted or needed a wooden beam-engined steamship over 300' long. The *City of Panama, Colon,* and *Acapulco,* three iron screw steamers, were running to Aspinwall. The *Ocean Queen* had been withdrawn in 1873; the *Henry Chauncey* was the only paddler still left

and she would be taken off the run in the spring of 1875.

The *Rising Star* was sold to John Roach for her scrap value, $40,000, in February of 1875.[37] She was to linger on for a while and was still afloat in January of 1876 at the Morgan Iron Works, on the East River at 9th Street, where her engine was being taken out.[38]

The *Rising Star* had a useful life of less than eight years, most of it from New York to Aspinwall, with calls for coal at Kingston on the return trip. She fitted in with the Pacific Mail Steamship Company's ideal of a large wooden paddler until that company belatedly realized that it could build and operate iron screw steamers at far less cost. As an Atlantic liner she saw a single voyage, for Ruger Brothers, to French, German, Danish, and Norwegian ports in 1870. For this, her large size was advantageous and she was able to bring back over twelve hundred sanguine immigrants.

Rather than continue with the final events in the *Guiding Star*'s career, the author will put them off until the chapter dealing with her last owner, the North American Steamship Company.

The Final Years of the *Mississippi* and the *Merrimack*

After their wartime transport service, the two largest iron steamers in the American merchant marine changed ownership rapidly. James Raynor acquired them for the New York Mail Steamship Company in October of 1865. Garrison and Allen took over the line about July of 1866 and then pulled out in April of 1867. At the time, the *Mississippi* was on the North Atlantic and the *Merrimack* on a Brazilian voyage. The latter was retained for South American service while the former was permitted to make a second trip to Havre, plus a final one to New Orleans, before being added to the Brazil Line fleet. During the latter part of 1867 both were out of service, being refitted to make them more suitable for the tropics. Their large passenger capacity was no longer needed, but more space for freight was. Passenger cabins and mess rooms were moved up to the spar deck houses, and the hurricane deck above them was extended to provide shelter.[39] Vacated tween deck spaces were used for cargo.

The *Mississippi* sailed for Rio de Janeiro late in December of 1867 and the *Merrimack* a month later. Together with the wood-hulled, beam-engined *South America*, a monthly service was maintained, each vessel making the 10,000-mile round trip every three months. The screw steamships were most satisfactory, but their partnership was soon severed. On her sixth voyage the *Mississippi* left New York March 23, 1869, and reached Rio without incident. She started North about April 25, stopping at Bahia, Pernambuco, and Para. Nearing the

307

Merrimack on the large balance dock at New York after her 1867 refit for Brazilian service. *Courtesy of San Francisco Maritime Museum, Firemans Fund Collection.*

French island of Martinique, on her way to St. Thomas, she struck on Grand Personnel Reef in the early morning of May 12. Part of her cargo of 14,000 bags of coffee was thrown overboard to lighten the ship, but leakage water caused the tightly packed beans in the hold to swell; within twenty four hours seams and decks were bursting open and she became a total loss.[40] The French steamer *Acheron* took the passengers and crew to Fort de France. As in the case of the *Evening Star*, Garrison and Allen had not insured the ship.[41]

In her short commercial life, after three busy years as a transport the *Mississippi* had completed twelve voyages to New Orleans for the Star Line and three transatlantic ones, one to Bremen for North American Lloyd in 1866 and two for the joint Havre Line-New York Mail operation of 1867. Her stranding prevented her from finishing six returns from Brazil.

The *Merrimack* carried on with four voyages a year for the duration of the Brazil Line's mail contract. New boilers were installed in 1870 within the normal one-month layover at New York.[42] Her splendid record for regularity was interrupted only once when, northbound, two days out of Para, on March 4, 1874, she suffered a broken propeller shaft.[43] Returning under sail, she reached Para on the 12th and resumed her voyage March 23. She must have carried a spare shaft on board. Although she arrived at New York over two weeks late, her next departure, April 23, 1874, was strictly on schedule. Later in the same year she suffered a broken cylinder about sixty miles south of St. Thomas.[44] Emergency repairs, made there on December 14, enabled her to continue at slow speed. She reached New York December 23, 1874, eight days from St. Thomas in lieu of her normal five or six. Once again there was time to put things right before her next scheduled sailing.

This, the first real test of an American service to South America, was unable to continue without a mail subsidy. The number of passengers carried was small; the largest published list for the *Merrimack* was seventy-six to New York in March of 1873; two years earlier, in the same month, it was a mere thirty-four. When the ten-year contract expired, the Garrisons (Allen having withdrawn) closed down the line. The *Merrimack* made her last sailing from New York on July 23, 1875, from Rio de Janeiro on August 25, and reached New York September 19. She was laid up for the next four years.

For a remarkably low price, $18,000, less than her scrap value, she was sold to Nehemiah Gibson of Boston on August 1, 1879.[45] After being towed to that port, she was rescued from destruction and so extensively rebuilt and reengined by her original builder, Harrison Loring, that she emerged in 1884 looking like a new vessel, with two masts instead of three; a tall, thin stack; and a long hurricane deck over an extensive deckhouse. Her appearance suffered. Only her handsome sheer and forward curving stem remained to show that Samuel Pook had designed her before the Civil War. A compound engine with a 38″ diameter high-pressure cylinder and a 70″ low-pressure one of 4′-0″ stroke and new boilers of higher pressure made her far more economical to run.[46]

Nine years after the original Brazil Line ended, Loring chartered his rejuvenated *Merrimack* to another Brazil line recently established by John Roach, the Chester, Pennsylvania, shipbuilder. She replaced the *Reliance*, which had been wrecked on the Brazilian coast.[47] After having her bottom cleaned and painted at Simpson's dry dock, the *Merrimack* left Boston on December 16, 1884, and departed from New York two days after Christmas for Newport News, Virginia; St. Thomas; Barbados; Para; Maranham; Pernambuco; Bahia; and Rio de Janeiro.[48] She was not quite so fast as the line's newer *Finance* and *Advance* and had trouble keeping up with their advertised schedules. Modifications were made to her engine and a different propeller installed to improve matters.[49]

In the course of a year she completed five trips to Rio, her last New York arrival being December 29, 1885. In January she made a single voyage to Havana and Vera Cruz, Mexico, for the Alexandre Line. After some modifications at New York she returned to Boston in June 1886. It has already been noted that F. W. Nickerson had been operating two services out of Boston, one North to Nova Scotia and Prince Edward Island and another South to Savannah. The *Merrimack* was used on both, and made five Charlottetown via Halifax sailings, along with the *Carroll* and the *Worcester*, and three to Savannah. From December of 1886 through March 1887 she was chartered for Weltch, Walter, and Company's Boston to Charleston line and then returned to the Savannah route for four more voyages. Finally, in July of 1887, she headed for Halifax and Charlottetown again. By this time Harrison Loring had taken over the Nickerson steamers.

Boston bound, the *Merrimack* left Halifax on July 9 with 150 passengers and crew. That night, shrouded in fog, she ran hard aground on Little Hope Island and came to rest surrounded by breakers. There was great difficulty in getting the passengers ashore; some passing fishermen sent their dories over to help.[50] The purser had to walk twenty-four miles to reach a telegraph office and summon aid; not until three days had passed did the passengers get back to Halifax. Meanwhile the wreck pounded on the rocks, broke in two, and disintegrated.

In a quarter of a century the *Merrimack* had started life as a Civil War transport, had made ten New Orleans voyages for the Star Line and thirty-two to Rio de Janeiro for the United States and Brazil Line, and had been rejuvenated for five more Brazil runs plus others to

Mexico, Nova Scotia, Prince Edward Island, Savannah, and Charleston. As an Atlantic liner she made one round trip to Bremen under charter to North American Lloyd in 1866.

The *Mississippi* and the *Merrimack* conclusively demonstrated that large iron screw steamships could be built in the United States and that they were stronger, faster, more economical, but just as comfortable as wooden side-wheelers. James Raynor, of the New York Mail Steamship Company, started by building magnificent paddlers but soon realized the merit of screw steamers and headed his company in the right direction when he acquired this splendid pair. Harrison Loring and his naval architect, Samuel Pook, were real innovators when they produced these remarkable steamships.

In a review of the Star Line's Atlantic sailings, it was directly responsible for a total of six voyages in 1867, all in conjunction with the Havre Line. The *Guiding Star* made three of these, the *Mississippi* two, and the *Morning Star* one. In addition, under charter to North American Lloyd, the *Merrimack* and *Mississippi* made one trip apiece to Bremen in 1866. The *Guiding Star* went to Copenhagen and Bremen in 1869, and the *Rising Star* sailed to Swinemunde in 1870, both for Ruger Brothers. Another Atlantic voyage by the *Guiding Star* was made in 1870 for William H. Webb. Thus the ships the Star Line had owned completed five Atlantic voyages under other house flags.

Notes

1. 39th Congress, 2d sess., House Ex. Doc. no. 51, pp. 2-5.
2. *Journal of the Franklin Institute* 45, 3d ser. (January-June, 1863): 379-80.
3. *New York Herald*, May 17, 1865.
4. Ibid., August 14, 1864.
5. T. Main, *The Progress of Marine Engineering* (New York, 1893), pp. 136-37.
6. *New York Herald*, January 26, 1866.
7. Ibid., August 1 and 3, 1863.
8. *New York Tribune*, advertisement, September 23, 1864.
9. *New York Herald*, January 20, 1865.
10. *New York Journal of Commerce*, October 9, 10, 15 and 17, 1866.
11. Ibid., December 14, 15 and 17, 1864.
12. *New York Herald*, January 23, 1865.
13. *New York Tribune*, July 29, 1865.
14. *New York Herald*, July 10 and August 15, 1875.
15. Ibid., November 3, 1865.
16. *New York Journal of Commerce*, February 3, 1866.
17. *New York Herald*, May 17, 1865.
18. John Haskell Kemble, *The Panama Route, 1848-1869* (Berkeley and Los Angeles, Calif., 1943), p. 228.
19. *New York Herald*, October 13, 1865.
20. Ibid., October 30, 1865.
21. *New York Journal of Commerce*, November 24 and December 16, 1865.
22. Ibid., November 23, 1865.
23. Ibid., April 30, 1866.
24. The *New York Journal of Commerce* lists Raynor as president for the last time in the Star Line advertisement for July 17, 1866.
25. Francis O. Braynard, "The First American Steam Passenger Line to South America," *The American Neptune* (1944), p. 139. Note that the title is incorrect. Nine years earlier the *Tennessee* ran regularly to Venezuela.
26. *Boston Transcript*, February 15, 1867.
27. *New York Journal of Commerce*, May 16, 1867.
28. Bill of Sale, GUIDING STAR, March 14, 1868. Webb Institute of Naval Architecture.
29. *New York Shipping and Commercial List*, January 1869-May 1872.
30. *New York Herald*, May 30 and 31, 1870.
31. Ibid., November 20, 1870.
32. Ibid., February 6, 1871.
33. Ibid., March 30, 1871.
34. Ibid., November 3 and 4, 1872.
35. Ibid., February 24, 28 and March 31, 1874.
36. *New York Maritime Register*, February 25 and March 4, 1874.
37. *New York Herald*, February 3, 1875.
38. *New York Maritime Register*, January 5, 1876.
39. *New York Herald*, March 27, 1874.
40. Ibid., June 7, 1869.
41. *New York Journal of Commerce*, May 25, 1869.
42. *New York Herald*, March 27, 1874.
43. *New York Maritime Register*, April 8, 1874.
44. *New York Journal of Commerce*, December 16 and 24, 1874.
45. *New York Maritime Register*, August 6, 1879.
46. American Shipmasters' Association, Record of American and Foreign Shipping, 1887.
47. *Nautical Gazette*, December 11 and 25, 1884.
48. *Boston Commercial and Shipping List*, December 3 and 20, 1884; *New York Maritime Register*, December 24 and 31, 1885.
49. *Nautical Gazette*, March 19 and November 19, 1885.
50. *Boston Post*, July 4, 12, 13 and 14, 1887.

20

Ruger Brothers, 1865-1870

ASIDE from their appearance as owners, charterers, or agents in the shipping columns of the daily newspapers, nothing is known about the individuals who operated under the name of Ruger Brothers. An inspection of New York City Directories shows that William Ruger first appeared with an office at 45 Beaver Street and a home on Staten Island in the 1860 issue.[1] By 1864 a second brother, Theodore, was added, even though he resided abroad. A third Ruger, Emil, appeared in 1868, and by 1874 all three had homes on Manhattan or Staten Island. They probably came from Germany, for their principal activity, as evidenced by 1863-65 advertisements, was as agents for sailing ships bound for Bremen and other North German ports. Vessels to European and, occasionally, South American destinations were also represented.

Ruger Brothers first managed an American steamship in August of 1865, when they acted for Arthur Leary during the *Circassian*'s first voyage to Bremen. Despite its abrupt termination, when she had to be run aground on Cape Breton Island, the expereince seems to have alerted the brothers to the potential sums that could be made by transporting steerage passengers from Bremen to the United States. Although that city's North German Lloyd operation was well established and the Hamburg American Line provided it with intense competition, the Rugers concluded that a third service could still profit from the rapidly expanding flow of westbound emigrants. They were right.

Steamships, once the Civil War ended, could be bought cheaply. The remaining Collins liners, *Atlantic* and *Baltic*, were purchased from the North Atlantic Steamship Company and a large beam-engined steamer, the *Western Metropolis*, from the firm of Benner and Brown. The last was built in 1863 and originally measured 2,269 tons. She had spent most of her life as an army transport.

In January of 1866 Ruger Brothers advertised that the *Western Metropolis* would leave New York for Bremen, via Cowes, on March 17 and would carry three classes of passengers at $105, $62.50, and $37.50 eastward.[2] Westward, however, was where the real passenger market lay, and fares were $112, $80, and $45. The name chosen for the service was North American Lloyd Steamship Company. The *Western Metropolis*, despite her recent completion, had a series of problems that would make her unavailable for any of the many sailing dates announced.

The *Atlantic*, after sixteen years of hard service, was still in good condition and made the line's first sailing on February 22, 1866. Her return justified the Ruger family's hopes; she had 900 passengers when she arrived at New York on April 9, eighteen days from Bremen and fourteen from Cowes. To replace the ailing *Western Metropolis*, the *Ericsson* was chartered and sent out in March. After her came the *Atlantic* and the *Baltic*, both in April.

Because the *Western Metropolis* was still not ready, the *Mississippi* and *Merrimack* were chartered from the New

York Mail Steamship Company for May departures. Their addition made weekly sailings possible, starting with the *Mississippi* on May 10. She was followed by the *Merrimack*, then the *Ericsson*, and, finally, the *Atlantic* on May 31. Subsequent fortnightly departures were scheduled but the *Western Metropolis* always refused to sail.

She was at long last deemed ready on June 28. During her extensive refit, unusual, patented paddle wheels had been installed. Once she was at sea their floats worked loose and fell off in anything but the smoothest weather.[3] Captain Sanders had to turn back and put into Boston on July 6. After temporary measures to make a short sea trip possible, she arrived back at New York on July 12. She was suposed to sail again on August 14, then August 30, but never did so.

After that fiasco only one more voyage was undertaken, by the *Atlantic* on July 19. Her entry into New York harbor on September 25, 1866, with 1,156 passengers, ended North American Lloyd's operation. Later sailings were advertised for August, September, and October, but nothing further transpired.

The ten completed voyages, four by the *Atlantic*, two by the *Baltic*, two by the *Ericsson*, and one each by the *Mississippi* and *Merrimack*, must have been financially successful. Of the six homeward trips for which statistics are available, the passenger load varied from 318 on the *Merrimack* to 1,158 on the *Atlantic*, with an average of 933. It may be that the continued work on the *Western Metropolis* was such a drain on the profits that the Rugers decided to abandon their service when they were fortunate enough to find a purchaser for their three steamships. Isaac Taylor and several associates bought them and planned to carry on to Bremen. They did so, in 1867, under the name of New York and Bremen Steamship Company.

Having disposed of their fleet, Ruger Brothers were almost inactive for a year with regard to steamships, although they continued to act as agents for sailing vessels in and out of New York. Their one exception was to charter the *Circassian*, which had been laid up since the September 1866 demise of Continental Mail Steamship Company, and send her off to Bremen about August 10, 1867. Continuing her usual record of misfortunes, she loaded 470 passengers but had to put into Falmouth October 9 with her engine damaged. She carried on to New York two days later; twenty-two days, with heavy gales all the way, passed before she arrived.[4]

The Rugers, undiscouraged by their second experience with this unfortunate steamer, set up another Bremen operation in 1868, after the New York and Bremen Steamship Company was dissolved. All their steamships from now on would be chartered, normally for a single voyage, and only summer sailings were made in order to take advantage of favorable weather and achieve low coal consumption. To further save fuel, all vessels were run at very moderate speeds.

The *Quaker City* opened the service with a May 29, 1868, sailing from New York, to be followed by the *Ariel* on June 13 and the *Circassian* on July 25. In August there were two sailings, by the *Ariel* and *Northern Light*, and on September 5 the *Arago*, the largest and finest steamer employed that year, set forth. Return sailings were scheduled four weeks after the New York departures, and Falmouth, rather than Cowes, was the English port of call. The passenger capacity of each steamer, aside from the *Arago*, was listed at 650[5] and on one occasion the smallest of the lot, the *Ariel*, did carry exactly that number when she arrived home September 23 after her second trip.

The six voyages in 1868 were so successful that a similar program was undertaken for 1869, but the route was extended to include the Danish capital of Copenhagen as well as Bremen. Up until this time there were no transatlantic steamships serving Denmark; it is not generally recognized that the Rugers were the first to provide anything beyond irregular departures of unscheduled sailing ships. The honor of opening the Scandinavian venture fell to the *Ariel*; she was making her 21st and final Atlantic voyage before being sent to the Pacific. She reached Copenhagen on April 5, 1869, after stopping at Cowes and Bremen, having left New York on March 13. After an eleven-day stay she came home, overcrowded with 750 passengers, but with the division between Danes and Germans unknown.

Following her came the *Northern Light*, *Ocean Queen*, *Guiding Star*, *Fulton*, and *Santiago de Cuba* at intervals of two to four weeks. The last listed was a beam-engined steamer built by Jeremiah Simonson in 1861 that had been purchased by the U.S. Navy and converted into an armed cruiser. Of 1,567 tons (1861) or 1,627 (1865), she had a reputation for speed and reliability. She will be discussed more fully in chapter 24. After the *Santiago de Cuba* departed on her maiden Atlantic crossing on June 16, it appeared that the Ruger service was ended for the year, but two months later the *Fulton* was sent out, not only to Copenhagen, but also to Swinemunde, the port for Stettin. Once again Ruger Brothers were breaking into new territory; part of her cargo was consigned to Russia and would be transshipped at Stettin. As a result, she sailed directly for the Baltic, arriving at Copenhagen on September 10 and Swinemunde on the 13th. Returning, she coaled at Southampton, left October 9 with 650 passengers, and arrived home on the 23d. This was her fiftieth and last Atlantic voyage.

Seven round trips were made by Ruger steamships in 1869, after six the year before. The vessels used, however, were generally superior. Certainly the *Ocean Queen*,

Guiding Star, and *Fulton* were the very best available. Once more results were successful enough to insure the service's continuing.

The 1870 operations began with an excellent pair of ships, both owned and admirably kept up by the Pacific Mail Steamship Company. Their itinerary was further expanded to include Dutch and Norwegian ports; the Rugers were showing far more enterprise than any other American operator. The *Ocean Queen* departed on March 3, 1870, and reached Havre on the 17th without an English stop. Five days later she left for Bremen, via Brouwershaven in the Netherlands, on the 26th. On April 1 she reached Swinemunde, her most distant port in the territory that was sometimes German and sometimes Polish. After a six-day stay she headed homeward via Christiansand, Norway, and, being a large steamer, was able to carry a large load, 1,137 passengers. Thus she justified the Rugers' selection of both ship and itinerary.

Following the *Ocean Queen* by three weeks, the *Rising Star* set out on roughly the same route but omitted Brouwershaven and added Copenhagen on the way back. Her passenger list, 1,267, was even better. Then, on May 18, the *Western Metropolis*, some four years late, finally got under way for the Rugers. She made a surprisingly fast crossing, thirteen days to Havre, and added one new port of call, Kiel, to the ones made by the *Rising Star*. With 954 passengers aboard, she was more crowded than her larger predecessors. Her return on July 7, 1870, marks the end of Ruger Brothers' operations to France, Germany, Norway, and Denmark. One week later, on July 15, France's Corps Legislatif declared war on Prussia.[6] Within three weeks, nine steamships and twenty-seven sailing vessels from North Germany[7] had taken refuge in the harbor of New York and all trade between the United States and Germany had halted. Faced with wartime uncertainties, the Rugers suspended their steamship operations, never to resume them.

In their earlier years they had managed steamships for others. Three cases have been mentioned for the *Circassian*: one was for Arthur Leary in 1866 and two for Ernest Fiedler in 1868. Another unusual operation was the *Arago*'s 1868-69 trip, which was advertised by Godeffroy and Brancker and was said to have the North American Steamship Company as agent, when she left New York on December 2, 1968, for Bremen. The steamship spent the winter abroad, probably undergoing repairs in England. But on her return, she was consigned to Ruger Brothers when she left Bremen, touched at Falmouth, and reached New York on March 28, 1869. If, as the author suspects, this was an actual charter for the westbound part of the trip, the *Arago*, rather than the *Ariel*, began the line's 1869 operations and the number of voyages for that year should be increased to six and one-half.

The firm of Ruger Brothers has been badly treated by Atlantic steamship historians, who always say that it was unsuccessful. Passenger lists prove the contrary. True, the Rugers did not operate on a year-round schedule, they held only minor mail contracts, and they used steamships belonging to others. There was no glamor to a steerage line, no matter how profitable. Yet theirs was the only American flag operation to stay on the Atlantic for five successive years after the Civil War. They were a group of very clever businessmen who chartered ships cheaply, operated them at moderate speeds and in good weather to keep their coal bills down, and regularly carried hordes of immigrants from Bremen and, later, from other ports that had never before had any steamships sailing to the United States. Counting in their one year of North American Lloyd activities, they were responsible for twenty-seven-and-a-half voyages that delivered to the New World over 20,000 souls seeking a new way of life. Sixteen different steamships were used from 1866 to 1870, the *Atlantic* being the only one to make four voyages. The *Ariel* completed three; the *Northern Light*, *Ocean Queen*, *Fulton*, *Circassian*, *Baltic*, and *Ericsson* two; and the *Arago* one round voyage and, perhaps, a westward crossing. All of this group had seen previous Atlantic service. In addition, the *Quaker City*, *Santiago de Cuba*, *Guiding Star*, *Rising Star*, *Western Metropolis*, *Mississippi*, and *Merrimack* made one trip each. The first was a prewar coaster, the next four were beam-engined steamships of Civil War vintage, and the last two were splendid iron screw steamers. The Ruger fleet was an interesting and varied one, practically a catalogue of ocean steamships from 1850 to 1865. Furthermore, the *Mississippi*, *Merrimack*, *Ocean Queen*, *Guiding Star*, and *Rising Star*, at their dates of charter, were in the best possible condition, well maintained by first-rate owners. Ruger Brothers regularly employed the finest steamships available. They were remarkably successful until the Franco-Prussian War cut off their North Sea and Baltic routes. But after 1870 the wooden side-wheeler could no longer continue as a profitable steamer. The Rugers, well aware of changing economic conditions, left the sea. Their profession, as listed in the New York Directories, changed from "merchants," in 1871, to "brokers" in succeeding years.

Notes

1. H. Nelson, compiler, *Trow's New York City Directory*, 1860-74.
2. *New York Herald*, January 2, 1866.
3. *New York Journal of Commerce*, July 7 & 8, 1866.
4. *New York Herald*, October 11 and November 3, 1867.
5. *New York Journal of Commerce*, July 14, 1868.
6. *New York Herald*, July 16, 1870.
7. Ibid., August 5, 1870.

21

The New York and Bremen Steamship Company, 1867

LATE in 1866, when Ruger Brothers chose to give up their North American Lloyd service from New York to Bremen, a new organization was formed, with Boston's Isaac Taylor as president, to purchase their three steamships and to continue them on the same route. Under the name of New York and Bremen Steamship Company, it provided a substantially improved service in 1867. The ships were kept in good condition, made regular trips on the dates announced, and, for most of the year, offered sailings every two weeks. In this respect the new line outdid the concurrent Baltimore and Liverpool Line and equaled the combined Havre-New York Mail service.

Because the *Atlantic*, *Baltic*, and *Western Metropolis*, the vessels acquired by the Bremen Line, were not enough to insure fortnightly sailings, the *Northern Light* was chartered from the Pacific Mail Steamship Company. First to set out was the *Atlantic*, under Captain Charles Hoyer, on February 7, 1867, for Bremen via Cowes. Eastbound fares started at $120, $70, and $37.50, all higher than those charged by North American Lloyd, but by April they had been reduced to $110, $65, and $35.[1] Coming back from Bremen with 500 passengers, the *Atlantic* ran into four days of winter gales that injured six crew members and damaged her forecastle, bulwarks, and companionways.[2] Waves washed her forward boats away and swept her deck. She arrived at New York on March 24, eighteen days from Bremen.

The *Baltic* followed the *Atlantic* two weeks later, and she, too, had a moderate passenger list, 408, when she arrived home on April 10. The *Western Metropolis*, so reluctant to sail in 1866, had been further altered over the winter and was sent off a fortnight after the *Baltic*. She became a success for her new proprietors when she reached New York April 24 loaded with 710 passengers, having taken twenty days from Bremen against head winds and gales. Eastbound she had done a little better, seventeen days to Cowes.

After two sailings each by the line's own steamships, the *Northern Light* was added to the fleet and departed from New York on May 18 and from Bremen June 9; she had no return cargo, but 944 passengers made up for its lack. Her presence gave the New York and Bremen Steamship Company a series of sailings every two weeks from April until the middle of September. But the *Western Metropolis* was again to prove ill-favored.

Eastbound with 249 Austrian troops, who were returning home from Mexico, she left New York August 24 and arrived at Southampton fifteen days later with a broken shaft.[3] It was so quickly replaced that she was able to make a return trip from Bremen on October 20 and reached New York in seventeen days, including the regular stop at Cowes.

The last sailing of the line, by the *Atlantic*, was almost a repetition of the first.[4] Winter gales carried away the forward boats, damaged the bulwarks, stove in the captain's quarters and houses on the guards, and broke nine deck beams. When she arrived on December 13, 1867, the last American line that could be relied upon for continuous, regularly scheduled, Atlantic sailings, came to an end.

Throughout the year the passenger loads had been most heartening. The *Atlantic* and *Baltic*, by May, were always carrying over 1,000 a trip, the former having reached a record number of 1,727. The minimum quoted was 410, in August, for the *Northern Light*'s last crossing. Based on incomplete figures, at least 12,000 westbound steerage passengers must have been transported in the ten-month period covered by the line's crossings. At times its passenger loads even exceeded those on the newer and faster North German Lloyd steamers.

Isaac Taylor, however, was well aware that his elderly wooden steamships would not continue indefinitely; twice the *Atlantic* had suffered damage from the sea and it is likely that both she and the *Baltic* needed new boilers. Hampered by low steam pressures and outmoded jet condensers, they consumed far more coal than the iron screw steamers sailing out of Bremen and Hamburg. Taylor, as a prudent businessman, chose to end the New York and Bremen Steamship Company while it was still profitable, rather than risk another year's operations.

The line's steamships were offered for sale, but as we have seen, the *Baltic*, whose hull was still in excellent condition, was eventually retained by Taylor and converted into a highly successful sailing ship. The *Atlantic* and *Western Metropolis* were auctioned off on June 30, 1868, for $41,000 and $57,000 respectively.[5] The low prices show how little demand there was for wooden side-wheelers three years after the Civil War ended.

The *Atlantic* made six voyages, the *Baltic* five, the *Western Metropolis* four, and the *Northern Light* two, for a total of seventeen for the New York and Bremen Steamship Company.

The year 1867 was the last for regular Atlantic service under the American flag. The Baltimore and Liverpool Steamship Company provided monthly sailings, the joint Havre-New York Mail operation left New York every two weeks for England and France, Taylor's line served Bremen at the same interval, and, late in the year, Boston would be linked to Liverpool by a splendid new steamship, the *Ontario*, especially built for the Atlantic. In all, forty-five round voyages were completed and many thousands of immigrants brought to the United States.[6] Every one of these lines ended in 1867. From 1868 through 1870, only sporadic crossings would take place. After 1870 there would be a few one-way crossings of steamers being delivered to the Orient.

Notes

1. *New York Journal of Commerce,* February 2, 1867.
2. Ibid., March 25, 1867.
3. Ibid., September 10, 1867.
4. Ibid., December 14, 1867.
5. Ibid., July 3, 1868.
6. Cedric Ridgely-Nevitt, "1867—A Forgotten Year in the History of the American Steamship," *Steamboat Bill* (1967), pp. 3-6.

22

The American Steamship Company, 1867-1871

LONG before the Civil War ended, a group of Boston merchants considered the establishment of a new transatlantic line from Boston to Liverpool. In July of 1864 they incorporated, under the name of American Steamship Company, with Edward S. Tobey as president, Joseph W. Balch as treasurer, and Hamilton Hill as secretary.[1] For technical advice they engaged Samuel H. Pook to design their steamships and Charles W. Copeland for corresponding work on the machinery. In October of 1865, the engineering work having been completed, a contract was made with George W. Jackman, Jr., former Mayor of Newburyport, Massachusetts, for the construction of two steamships of approximately 3,000 tons.[2] Although Jackman had little experience in the steamship field, he was well-known for the excellence of the many sailing ships he had built since 1849. The engines and boilers were to be supplied and installed by Harrison Loring's City Point Works in South Boston. Captain Eben Howes, of the Bureau Veritas, supervised the building of the hulls and Copeland the engines. These steamships, to be named *Ontario* and *Erie,* had wood hulls and screw propulsion, just the combination that had proved difficult to effect in vessels of large size and power. Nevertheless, the combination of Pook's and Jackman's theoretical and practical knowledge, together with Copeland's long years of engineering experience, actually did produce large steamships that were entirely successful.

Other contractors who contributed to this all-New England venture were Francis Low for the rigging, and Manson and Peterson for the cabin interiors.

The ships' sizes, as given on their official registers, differ considerably. The *Erie*'s official values are questionable and the 1868 edition of American Lloyd's Register shows dimensions almost identical to the *Ontario*'s. The depth, of course, was to the upper deck.

Steamer	Source	Length	Beam	Depth	Tonnage
Ontario	Custom House	323.5'	43.9'	20'	2,889
Erie	" "	319'	41'	21'	2,900
Erie	Am. Lloyd's	325'	44'	29'	2,900

The *Ontario* and *Erie* were a handsome pair, with sailing ship hulls. The bow was of the old-fashioned type, having a long overhang, beautiful curves, carved trail boards, and a feminine figurehead. In fact, if one looks at the launching photograph of the *Erie*, there is nothing to show that she is not a fine-lined clipper of extremely large size. There were high bulwarks and an unbroken sheer. Within the hull were three decks and four watertight bulkheads, the two adjacent to the machinery being made of iron.[3] The scantlings and details of her metal strapping and fastenings must have been planned with exceptional care in order to make the hull rigid enough for its long propeller shaft to operate smoothly.

On the spar deck was a continuous house for the full length of the ship, averaging twenty-two feet wide. Sixty feet of its length was taken up by a dining saloon seating all of the 125 first-class passengers. It was finished in satinwood and walnut and had a ceiling of light blue panels set in white moldings. Other public rooms consisted of a smoking room, a ladies' saloon, another for gentlemen, and a lounging room. The cabins were paneled in bird's-eye maple and black walnut.[4] In addition to these elegant quarters, there was space for 600 steerage passengers, 1,500 tons of freight, and 1,000 tons of coal.

An innovation was the use of wire rope for standing rigging. A spike bowsprit projected over the figurehead; three masts with considerable rake and two stacks forward of the mainmast produced an impression of power and speed. The fore and main had topgallant sails above double topsails and the mizzen was fore and aft rigged.

In their general concept, the *Ontario-Erie* pair were successors to the *Pioneer* of 1851; they had sailing ship hulls with steam machinery installed to drive them and a considerable sail plan as an emergency measure. The new steamers, however, corrected the mistakes made in the earlier craft. First, the steam plant was powerful enough to do its job. Second, the size and, particularly, the length were increased to carry sufficient fuel and at the same time keep the resistance down and the horsepower reasonable. Third, the bark rig was cut down to reduce its windage. Designers and builders had learned from past errors and had achieved a far better integration of a hull, engine, and rig to produce a superior total solution.

The engine output, about 2,500 indicated horsepower, was equal to the *Vanderbilt*'s, but well below the *Adriatic*'s 3,300. The hull, free of topside weights such as paddle wheels or walking beams, had a lower center of gravity. It could therefore be narrower and would make less waves. A second advantage was the lighter weight of a fast-turning screw engine, when compared with slow, cumbersome, paddle-wheel machinery. Finally, the combination of low beam, light machinery, and some improvement in fuel economy, all resulted in less coal and a marked reduction in displacement. Thus, for a longer, narrower, lighter vessel, the same horsepower could produce a very satisfactory speed. Furthermore, the American Steamship Company had neither the incentive nor the desire to make record passages; they would cost too much. The *Ontario* rarely, if ever, ran at her top speed.

The main engine had two cylinders 74" in diameter by 4'-0" stroke directly over the crankshaft. Steam was supplied at 30 psi by four boilers, two on each side of the ship with a centerline firing aisle.[5] For use in port a donkey boiler was available. The four-bladed propeller was very large, 18'-0" in diameter, having a pitch of 31'.

Such a big screw was excellent from an efficiency point of view, but undesirable in a seaway, for it tended to break water and race as the ship pitched. Vibration would develop as it speeded up. Because the designed draft was 21', the propeller was barely immersed, even in a fully loaded condition.

The launch of the *Ontario* on November 24, 1866, was a major event for Newburyport. The steamship was twice the size of anything ever built there and almost three times larger than any hull hitherto constructed by Jackman. Special trains brought officials, stockholders, guests, and the curious from Boston, Medford, Chelsea, and Portsmouth to help celebrate the occasion.

Her builder had a real problem to face: The Merrimack River was obstructed by Carr's Island, and he had to choose a slope that would be high enough for his launching ways to keep the steamer from sticking, yet not so great that she would cross the narrow channel and run ashore. As a precaution, both of the steamer's anchors were buried ashore with chains led to her bitts. Furthermore, a series of twelve-inch hawsers were run to large trees ashore. When the critical time came, she started nicely, pulled her anchors out of the ground, dragged them to the water's edge, uprooted several trees, and snapped three of the manila ropes before running hard aground on Carr's Island.[6] It was a spectacular launch.

At the next high tide, however, the *Ontario* was pulled off and faced her next ordeal, going through a narrow drawbridge barely wider than her beam. In order to achieve absolute control, blocks and tackles were rigged between the hull and the bridge, the hauling parts being led ashore, where they were pulled by yokes of oxen. After this critical maneuver was successfully carried out, she was towed to Boston by the steamers *American Eagle* and *Charles Pearson* for docking and engine installation. An examination on dock indicated no serious damage beyond a split rudder post.

Everything was completed by July of 1867. On the 22d, after two days of turning her screw at slow speed while she was tied up to a pier, she headed out for a twenty-six-hour sea trial.[7] She steamed past Cape Cod to the open sea but did not work up to full power. For twenty hours the shaft turned at 33 rpm instead of its intended 50. Coming home she was speeded up to make the run from Minots Ledge to Boston Light at eleven knots.[8]

Thereafter, a second, public trial was undertaken on July 29, complete with speeches, six-hundred guests, most of them stockholders in the company, and a catered luncheon. Like all such affairs, it was a convivial occasion and when the visitors were put ashore by tug in the late afternoon, no one bothered to report on technical matters. The ship had performed well and her officers were getting used to her. These included Frank Hallett as Captain, Richard Lavery as First Assistant Engineer, and

Charles Packard as Steward. Lavery had served in the engine room of the Loring-built *Mississippi* and Packard had been with both the Collins Line and the Pacific Mail. The American Steamship Company was making every effort to produce the best possible steamship and to man her with the best available men.

The *Ontario*, the first American steamship specifically designed and constructed as a transatlantic liner since the *Adriatic* some eleven years earlier, was advertised to sail from Boston on August 3, 1867. Fares of $125 in the first cabin and $80 second were listed; steerage was not mentioned. She departed two days late, on the 5th, and had only ten first-class passengers.[9] Her crossing took thirteen days eastward, but, coming home, she did much better, 10 days 6 hours at a speed of 11.8 knots. The weather was good, the wind light, and no sails were set. The passenger list recorded eighty-three in first and 228 steerage, rather low values when, in the same year, the *Atlantic* and *Baltic* averaged over 900 per westbound trip for the New York and Bremen Steamship Company.

On her second voyage, from Boston on October 5— again with a handful of passengers, thirteen first, and fifty-five steerage—the *Ontario* met with the accident to which screw steamers proved so prone.[10] She lost her propeller and made Liverpool under sail on the 16th. Despite the casualty, the 11-day crossing was an improvement over her first trip. A month passed before she could leave on November 13. In an attempt to obtain more passengers, she stopped at Queenstown the next day and then made the remarkably good time of 9 days 4 hours to Boston.[11] Her speed was at least 12.7 knots, an excellent figure against the prevailing wind and so late in the year. Only 130 steerage plus 31 cabin passengers continued to be disheartening numbers. A third and last eastbound departure had a full cargo of cotton and breadstuff, but only 17 cabin and 42 steerage passengers on December 7, 1867. She left Liverpool on the 29th, Queenstown on the 30th, and tied up at Boston January 11, 1868.

While building the *Erie*, George Jackman made a number of changes from the *Ontario* and employed another Newburyport master builder, W. C. Currier, to make the drawings needed. She was ready for launching on March 12, 1867, but because her engine was not yet ready for installation, that gala event was held up until April 13.[12] A lower inclination was chosen for the ways and two tugboats were waiting to slow her in the event that the anchors did not hold. This time all went smoothly and she stopped fifty feet from the opposite shore.[13] But her machinery and outfit were not completed in time for her to join the *Ontario*. It is a dismal fact that, by the end of 1867, the American Steamship Company had exhausted its financial resources. The cost of two steamships and the operation of the Ontario for three poorly patronized voyages, including a Liverpool propeller replacement at a high price, was said to be $1,500,000.[14] Cunard Line had for over a quarter of a century provided frequent sailings from Boston to Liverpool. The new service, starting too late in the year with a single ship, could do little to attract passengers or cargo from that formidable rival.

On June 3, 1868, an auction was held on board the *Erie*, which, with the *Ontario*, was moored at Grand Junction Wharf in East Boston. William H. Webb offered $720,000, J. J. Comstock, $730,000, and Nathaniel Winsor, of Boston, $750,000 for both. The last, naturally, was awarded the vessels. It appears that the full amount, to be paid in installments after an initial 10% deposit, was never paid. The Webb bid was the best actual offer; as president of the North American Steamship Company, Webb was actively engaged in the California trade while Winsor was planning to enter it. At a second auction, later in the year, a price of $512,000 for the pair was offered by unknown bidders.[15] This, too, must have fallen through, for the *Erie*'s first official register, dated January 15, 1870, still listed Edward Tobey, president of the American Steamship Company, as the nominal owner.

Both remained inactive at Grand Junction Wharf until the spring of 1870, when the *Ontario* was put on Simpson's dry dock to be inspected, recaulked, and, for the first time, sheathed with yellow metal.[16] The outbreak of the Franco-Prussian War sent her and the *Erie* abroad to carry arms and supplies to France. The *Ontario* left Boston October 23 for New York and sailed from there on November 7, 1870. She reached Cowes on the 22d and Havre on the 27th. In December she proceeded to Liverpool. The trip home was slowed by heavy weather that forced her to put into St. John's, Newfoundland, for coal on January 6, 1871. She reached Boston January 12 from Liverpool on December 23.

The *Erie*, making her debut as a working steamship, left Boston on November 10 and New York November 29, stopped at Cowes, and put into Havre on December 13. By that time Paris was besieged by Prussian troops and she was sent on to Brest, arriving there on the 16th. The *Erie* returned via London January 20, 1871, and Plymouth January 25, but at the latter port she entangled her propeller with a buoy's anchor chain. When the damaged propeller was lost eight days out, sail was hoisted and Captain Sears headed home under canvas. With light airs and fair winds, he was able to keep his topgallant sails up the whole way.[17] Despite her small bark rig, the *Erie* made 8 knots on the wind, 11 with it abeam, and passed all but one of the sailing ships sighted on the way. On February 25 she put into St. Thomas for badly needed provisions for both her passengers and a cargo of live animals consigned to P. T. Barnum, weighed anchor two days later, and disembarked thirty-

Ontario on the sectional dock at New York. Her bowsprit has been shortened to about half its original length. *Courtesy of San Francisco Maritime Museum, Firemans Fund Collection.*

three passengers at New York on March 13. A camel and a pony had been slaughtered to feed the carnivorous beasts.[18]

While these operations were under way, the United States and Brazil Mail Steamship Company had become interested in the ships. Its experience with the *Mississippi* and *Merrimack* on the very long run to Rio de Janeiro had been most satisfactory, but the loss of the former ship in May of 1869 had left it with two beam-engined paddlers, the *North America* and *South America*, as companions for their one remaining screw steamer.

As a result, the *Ontario* was sent from Boston to New York, arriving on January 29, 1871, consigned to Cornelius K. Garrison, president of the Brazil Line. Several months later she and her sister were bought, apparently from their original owners.[19] In commenting on the purchase, the *Herald* rated them as "probably the finest steamers owned in the United States at this time."[20] The sale price for both was $275,000, a fraction of their original cost. The New York firm of Lugar and Reid made alterations to fit them for South American operations. The first cabin spaces were increased to 150 passengers, the steerage was drastically reduced, and the cargo holds and tween decks were fitted out for 30,000 bags of coffee.[21] A new deck was added, presumably at the house top level. Near the end of the year, they were submitted to the United States authorities, inspected by a board of naval officers, and approved as meeting all the requirements of their owners' mail contract.

Under Captain George B. Slocum, the *Erie* was the first to steam from New York to Rio de Janeiro, via St. Thomas, Para, Pernambuco, and Bahia, her departure date being November 27, 1872. Homeward from Rio on December 26, she arrived at New York on January 20, after calling at the same ports in reverse order. The *South America* had left for Rio just four days ahead of the *Erie* but seems to have met unreported difficulties. The *Erie* took over the *South America*'s northbound sailing from Rio and the latter's December departure was put off for three weeks.

Neither of the new acquisitions was put into regular use until a year later; on November 23, 1872 the *Erie* replaced the *South America* and on December 23 the *Ontario* took the place of the *North America*. When the *Merrimack* headed South in January, an improved service entirely by screw steamer was finally achieved and the line's paddlers were laid up. Plans, however, do not always go as intended; the Brazil Line seems to have been particularly prone to misfortunes. The *Mississippi* had been wrecked in 1869. The *Erie*, under Captain E. L. Tinklepaugh, formerly with the Vanderbilt Line, made a successful trip to Rio de Janeiro and started North the day after Christmas. On January 1, 1873, loaded with twelve passengers and 24,676 bags of coffee, she left Pernambuco but, when about sixty miles out, smoke was discovered. At first it was thought the coal bunkers had caught fire, but it proved to be cargo in the tween deck below the galley. Hoses were put down ventilators, holes were cut in the deck to reach the fire, and steam was turned into the spaces to try to smother the flames, all to no avail. Eight boats were launched; in one the single woman passenger, her year-old child, and the stewardess were put, with the ship's doctor in charge. The eighty passengers and crew rowed ten miles to shore, leaving the ship ablaze from foremast to mizzen.[22] The official inquiry speculated that a newly installed galley stove may have caused the fire. It stated that the *Erie* was worth $400,000 and her cargo $750,000.

The Brazil service now had the *Merrimack*, survivor from the line's first pair of screw steamships, and the *Ontario*, from the second. Their better side-wheeler, the *South America*, was brought back from lay-up and the ill-assorted trio of one wood beam-engined steamer of 2,150 tons, one mostly iron propeller of 2,031 tons, and one unusually large wood screw steamer of 2,889 tons were used until the line ended its operations.

The *Ontario*, under Captain Slocum, began her South American career on December 23, 1872, and continued to depart from New York at three-month intervals. She was a model of regularity, sailing on the 23d, except when it fell on Sunday; in that event the departure was put off a day. A month later she left Rio, on either the 25th or 26th, and arrived at New York on the 20th of the following month. If conditions were good from St. Thomas up, she might arrive a day early; on two occasions she was a day late, but never more than one day off schedule. In all, she made twelve voyages for the Brazil Line as their most successful steamship. Nothing happened to her. There were no delays, no groundings, and but one reported case of damage when she shipped heavy seas, had some fore and aft sails carried away, and lost her fore spencer boom and gaff in November of 1873 in the general vicinity of stormy Cape Hatteras.[23]

In 1875, when the line's ten-year mail contract ended, the service could not survive without subsidy payments. The *Ontario* made the line's last voyage in September of that year and reached home November 20. The four steamships were laid up at various out-of-the-way points around New York harbor.

Two years later a pair of promoters from the unlikely city of Indianapolis, Indiana, issued a prospectus announcing a "scientific expedition" to sail around the world. James O. Woodruff was the director, and Daniel Macauley, former Mayor of that inland city, the secretary. The *Ontario* was to become a floating college and would carry a teaching staff to instruct those willing to take off for two years and pay $5,000 for the intellecually stimulating experience. For young men anxious to wear uniforms, drill as cadets, and travel in considerably less luxurious quarters, there was a special rate of $2,500.[24] The *Ontario*, however, spent the entire year of 1877 at Eagle Pier in Hoboken. She saw no preparations for an October 1 departure.[25]

The sailing was put off to the spring of 1878, with the *Ontario* superseded by other vessels; finally, in May of 1879, the whole thing was abandoned when Woodruff died.[26]

During 1878 and 1879, all the moldering Brazil Line steamships were disposed of. Only the *Merrimack* was to see further service. The others were sold for dismantling, the *Ontario* going to Nehemiah Gibson of Boston for $20,000.[27]

The *Ontario* and *Erie* were unique steamships, the only ones built in the United States specifically for transatlantic service after the Civil War. They demonstrated that wooden screw steamships above three hundred feet in length could be built successfully. For their original owners they were technical successes and commercial failures; they spent years laid up, both in Boston and New York, before they were finally put to work running to Rio de Janeiro rather than to Liverpool. The *Erie* made a single Atlantic voyage and was lost on her second one to South America. The *Ontario*, in a useful life of eight years, completed four Atlantic round trips and twelve to Rio de Janeiro. She was ill-conceived in being intended for a route well served by Cunard steamships. Yet, when the United States and Brazil Steamship Company purchased her for a fraction of her actual cost, she became their best and most reliable steamship.

Notes

1. John H. Morrison, *History of American Steam Navigation* (New York, 1854), pp. 403-4; *Boston Advertiser*, July 23, 1867.
2. *New York Herald*, October 21, 1865.
3. *Nautical Gazette*, July 15, 1871.
4. *Boston Advertiser*, July 24, 1867.
5. Robert K. Cheney, *Maritime History of the Merrimac* (Newburyport, Mass., 1864), pp. 79-80.
6. Ibid., pp. 81-83.
7. *New York Herald*, July 24, 1867.
8. *Boston Advertiser*, July 24 and 30, 1867.
9. *Boston Transcript*, August 3 and 6, 1867.
10. *New York Journal of Commerce*, October 22, 1867.
11. *Boston Transcript*, November 25, 1867.
12. Cheney, *Maritime History*, pp. 81-83, 333.
13. *Harper's Weekly*, April 20, 1867.
14. *New York Herald*, June 4, 1868.
15. *New York Maritime Register*, December 1, 1869.
16. *New York Journal of Commerce*, May 13, 1870.
17. Ibid., February 24 and March 14, 1871.
18. Cheney, *Maritime History*, pp. 83-84.
19. *New York Maritime Register*, July 6, 1871.
20. *New York Herald*, November 30, 1871.
21. *Nautical Gazette*, July 15 and November 26, 1871.
22. *New York Herald*, January 16, February 14, and March 17, 1873.
23. *New York Maritime Register*, November 26, 1873.
24. *New York Times*, July 30 and October 22, 1877.
25. *New York Shipping and Commercial List*, 1877.
26. *New York Times*, July 12, 1879.
27. *New York Journal of Commerce*, July 25 and August 1, 1879.

23

William H. Webb, 1868-1870

THE Nicaraguan route from New York to California was revived by Marshall O. Roberts in March of 1863 and continued under his direction until June of 1866. Advertisements had various headings; M. O. Roberts Line, Opposition Line, and Central American Transit Company. To either Vanderbilt or the Pacific Mail Steamship Company, Roberts was a minor competitor; he provided no more than monthly sailings and had just two steamships, one on the Atlantic and one on the Pacific. Late in 1866 there were signs of impending changes, the first being the adoption of an impressive title, North American Steamship Company. Additional steamships began to appear on the service, and advertisements listed William H. Webb as president, Charles Dana as vice-president, and D. N. Carrington as agent.[1] Roberts had sold his holdings to them. Webb, the New York shipbuilder who for many years had built the largest vessels for the Pacific Mail and had held a considerable block of its stock, was now dissatisfied with its management. Never a person to compromise, he cut his ties with the line and became the principal stockholder in, and leading spirit of, the North American Steamship Company.

Under his guidance its fleet was increased by both purchase and charter, fares were cut, and sailings as frequent as three times monthly were achieved. In addition to the Nicaragua service, alternate voyages were made to Aspinwall. Because the Panama route with its pleasant four-hour crossing by railroad proved more popular, Greytown, formerly San Juan del Norte, was eventually eliminated. Although the effort was intensive and the line well run, it was suddenly abandoned in October of 1868. Unlike Vanderbilt, Webb did not have unlimited funds to finance operations at a loss until he had to be bought out by a luckless competitor. As the chief creditor of the line, Webb was left as owner of most of the steamships it had purchased. On the Atlantic side were the *San Francisco*, ex-*Keystone State*, the *Dakota*, the *Santiago de Cuba*, and the *Guiding Star*.

Because he used the last two for transatlantic crossings, they are the ones of primary interest here. The *Santiago de Cuba* was a 1,627-ton beam-engined coaster, constructed in 1861, that spent the Civil War years as a unit of the U.S. Navy before being purchased by Roberts. When he sold out to the Webb group, the steamer passed into their hands but was mortgaged to Webb. The *Guiding Star*, 2,384 tons, had been purchased by Webb at a sheriff's sale in March of 1868 for the sum of $177,000.[2] She had been sold to settle some of the debts of the New York Mail Steamship Company. Her prior history appears in chapter 19.

Strangely enough, the North American Steamship Company's name first appears in connection with an Atlantic voyage when a former Havre liner, the *Arago*, departed on December 1, 1868, for Bremen carrying a cargo of cotton. After she had spent the winter abroad, Ruger Brothers took her over in March of 1869 for her return. Why Webb, or his company, should serve as agent for a steamship in which he no longer had any direct interest is inexplicable.

The *Santiago de Cuba*, after the abandonment of the California route, was chartered out for a few voyages to New Orleans in 1868 and 1869 and employed by Ruger brothers for a trip to Bremen and Copenhagen in the summer of the latter year. Then, in November of 1869, she left New York for New Orleans, loaded a cargo, and departed, under Webb's management, on November 29 for Havre. She called at Bermuda for coal on December 7 and reached Havre on the 29th after a protracted trip of nineteen days from Bermuda.[3] She arrived home at New York on February 2, 1870, with a damaged engine and a crushed stem resulting from a collision that sank the steamer *Brunette* the day before.[4]

After making good the bow damage and putting her engine in order, the *Santiago de Cuba* lay unused at New York until the fall, when Webb sent her out on a second Atlantic voyage to Havre, Cherbourg, Rotterdam, and Brouershaven, Holland. She departed from New York on September 3, took twenty-one days to Havre and even longer, twenty-four days, from Brouershaven home; she docked at New York on November 11, 1870.

The *Guiding Star*, immediately after her purchase for the North American Steamship Company, set out from New York on April 7, 1868, for Greytown. In May she was transferred to the Aspinwall route and made monthly sailings thereafter. She proved a most satisfactory addition to the line and, on each northbound trip, brought over 300 passengers to New York. Her record list occurred on September 12, 1868, when she unloaded 800 after a seven and one-half day run from Aspinwall.[5] Her next voyage, however, was her last for the line; its southbound sailing was September 19 and was followed by a return on October 11, 7 days 20 hours from the Isthmus, with 500 passengers.[6] The fact that the *Ocean Queen*, her Pacific Mail rival, had only 350 a few days earlier shows how successful the North American operation had become. It was a hollow victory, for the *Santiago de Cuba*'s arrival on October 28 marked the end of the line's sailing to and from the Isthmus of Panama.

Like the *Santiago de Cuba*, the *Guiding Star* was chartered to Ruger Brothers in 1869 for a single trip, from New York on May 1, to Bremen and Copenhagen and back to New York, with a stop at Southampton in both directions.

On August 25, 1870, Webb dispatched her to Havre. It is likely that she and the *Santiago de Cuba*, which followed a month later, carried military supplies to France for the Franco-Prussian War. The former reached Havre on September 9, spent six days there, then went on to Cherbourg. Thence she reached Cowes on September 26. From England she proceeded to the Belgian port of Antwerp on October 14. Leaving Antwerp November 3, she was forced to put into Southampton two days later, leaking badly. Her cargo had to be unloaded and her hull recaulked; the work held her up until November 20. From then on she met severe gales that slowed her so much that she changed course for Bermuda to replenish her coal. She arrived there December 12, twenty-two days from Southampton, sailed five days later, and reached New York on the 22d, bringing in 101 passengers, an infant born on the way, and the Bermuda mails.[7] It was a long, arduous trip and an unmemorable conclusion to the *Guiding Star*'s working career. She was a mere six years old and was still to endure a four-year lay-up during which Webb was unable to find anyone wanting to buy or charter what had been a large, fast, luxuriously fitted-out steamship. In October of 1874 she was broken up by John P. Dole on the beach at Cold Spring Harbor, Long Island.[8]

The *Guiding Star*, during her short life, had made five Atlantic voyages, three for New York Mail Steamship Company in 1867, one for Ruger Brothers in 1869, and one for Webb in 1870. The last is noteworthy as the final round trip in commercial service of any American-built side-wheel steamer. She had also made three voyages to South America for the Brazil Line, sixteen to New Orleans for the New York Mail Steamship Company, one to Greytown and Aspinwall, and five to Aspinwall for the North American Steamship Company. She was, in the author's opinion, the finest-looking ocean steamship built in the 1860 decade.

When she was withdrawn, her owner's activities on the North Atlantic ended. William H. Webb had never intended that his North American Steamship Company would be anything but a California operation; the two European trips by the *Santiago de Cuba* and the single one by the *Guiding Star* were attempts to achieve some return from steamships left over when their intended service ended. The eastbound trip of the *Arago* under the line's auspices remains unexplained.

Notes

1. *New York Journal of Commerce*, January 5, 1867.
2. *New York Herald*, March 15, 1868.
3. *New York Journal of Commerce*, November 8, 9, and December 4, 1869.
4. *New York Herald*, February 3 and 4, 1870.
5. *New York Journal of Commerce*, September 14, 1868.
6. *New York Herald*, October 12, 1868.
7. *New York Journal of Commerce*, November 28, December 21 and 23, 1870.
8. Ibid., October 19, 1874.

24
Occasional Atlantic Liners Part 2

DESPITE the somewhat chaotic condition of American steamship operations on the Atlantic from 1865, when the Civil War ended, up to 1870, when the wooden-hulled steamships made their last regular crossings, it has been possible to consider most of the voyages and describe a large majority of the steamships in the preceding chapters. These have been arranged chronologically, by steamship line, in the order of the first sailing from New York, Philadelphia, Baltimore, or Boston, to a European port. Only one minor line, the Continental Mail Steamship Company of 1866, has been omitted; this was done because its entire operation, a total of two voyages, had been covered by the individual histories of its chartered steamships, the *Ericsson* and *Circassian*.

There are still a number of unchronicled eastward crossings that transported passengers or freight, usually as part of a delivery voyage of a steamship bound for the Orient. A few of these will be considered here, in particular those made by steamships of a size or type representing important examples of the American shipbuilder's art. Two additional steamships, the *Santiago de Cuba* and the *Western Metropolis*, whose technical characteristics and extensive histories have not heretofore been considered, are covered in the following sections.

The *Santiago de Cuba*, 1861-1899

The name of Jeremiah Simonson has already appeared as the builder of Cornelius Vanderbilt's steamships. In between the *Vanderbilt* of 1856 and the *Costa Rica* of 1864, Simonson turned out a moderate-sized coastwise steamship for Valiente and Company's Cuban trade that was, appropriately, christened *Santiago de Cuba*. She was a shorter, wider, deeper version of the *Ariel*, with a somewhat higher tonnage. She was 227' long by 38' beam by 19' depth of hold and measured 1,567 tons, differing from the Vanderbilt series in that she had more freeboard forward, a bowsprit, and a curving stem, which made her more seaworthy. Overall she measured 240' on deck, had a beam over the guards of 52', and a depth of 27' to her spar deck.[1] She drew 14' and had a single stack and a half-brig rig. Although not absolutely necessary in a beam-engined ship, her floor timbers were filled in solid and, as usual with Simonson products, she was of superior construction. An engine of 66" bore by 11'-0" stroke was supplied by the Neptune Iron Works, together with two return-flue boilers 30' long. The paddle wheels were 29'-0" in diameter with twenty-two

Santiago de Cuba armed as a Civil War cruiser. *Courtesy of The Mariners Museum, Newport News, Va.*

widely spaced floats. After being launched at Greenpoint on April 2, 1861, the Cuba and New York Mail Steamship Company took delivery in July and sent her out on the 19th to Havana and Santiago. Returning from the latter port on August 10, she reached New York nine days later by way of Havana. The U.S. Navy bought her in September for $200,000. After being armed with two 20-pound rifled Parrott guns and eight 32-pound smooth bores, she became one of the most successful units of the fleet assembled to blockade Southern ports.[2] Captain O. S. Glisson, who commanded her in 1864, reported, "The *Santiago De Cuba* is the fastest on the blockade" after an overhaul.[3] She "worked well, running at a rate of 12 miles per hour and, at one time going 13 miles. As soon as our [untrained] firemen get a little more experienced, I am in hope we shall keep up this speed."

After commissioning at the New York Navy Yard, on November 5, 1861, she was ordered to Havana to cooperate with the local U.S. Consul, who provided information as to the presence and expected operations of suspicious vessels calling there. In April of 1862 she was cruising off Nassau, and, in a six-day period, intercepted and captured three prizes. The *Ella Warley*, formerly the *Isabel*, was chased for five hours on April 25, seized, and sent to Port Royal for coal and water. Her cargo included swords, rifles, and munitions. A day later the schooner *Mersey* was taken and towed to Port Royal. The *Santiago de Cuba* sailed in company with the *Warley* for New York on April 30 and, on the way, intercepted the schooner *Maria* bound for Charleston. The only dark spot on her record was that she had been unable to overtake a steamer thought to be the *Nashville*.[4] Commander Daniel B. Ridgely, U.S.N., her first naval commander, found himself extremely shorthanded after sending away fifty nine of his crew to man these and other captures.

After this brilliant performance, the *Santiago de Cuba* joined Commodore Charles Wilkes's "Flying Squadron" and was engaged in a fruitless search, under his misdirection, for the Confederate cruisers *Florida* and *Alabama*. In June of 1863 she was assigned to escort the Aspinwall mail steamers between Mayaguana Island in the Bahamas and Navassa in the Jamaica Channel. It was in these waters that the *Alabama* had captured the *Ariel* in December of 1862. But eighteen months of Naval service left the *Santiago de Cuba* in a rather bad state. She had been aground several times; her hull leaked badly—eight inches of water per hour when light, and double that when her coal bunkers were full. Her pumps were running day and night, her bottom was foul, and both the engine and the boilers needed attention. Worst of all, the galley was described as "worn out!" In December of 1863 she was sent to the Boston Navy Yard for an overhaul so extensive that it kept her out of service for six months.[5]

June of 1864 saw her steaming off Wilmington, North Carolina, and on that station she spent the remainder of her war service. During most of this period her third naval master, Captain Glisson, was the senior officer in command of a division of sixteen blockaders of the New Inlet Division. Another seventeen steamships were assigned to the Western Bar group. Even a fleet of this size never entirely closed off Wilmington, the last major port by which foreign supplies reached the Confederacy.[6]

When Fort Fisher, the guardian of Wilmington, was attacked in December of 1864, the *Santiago de Cuba* was part of the Admiral Porter's Third Division that fired on the fort for seven hours. Some of her seamen landed at Flag Pond Battery on Christmas Day and took sixty-five prisoners. These were delivered to Fortress Monroe.[7] She was back in time to lead the same Third Division into place during the second attack on Fisher and, after bombarding it, helped the Army to land its guns. After the surrender she gathered up wounded from the attacking force and took them to Norfolk.[8] Once Fort Fisher was eliminated, Wilmington surrendered and the end of the conflict became inevitable. The *Santiago de Cuba* remained off Wilmington but saw no further action.

She had served throughout the war and had been a most useful vessel. She was fast and reliable, and had acquired an enviable series of prizes. Her most valuable one was the *Victory*, taken on June 21, 1863, which brought in $330,000 prize money.[9] The latter had started life as the iron screw steamship *North Carolina*, built by the Novelty Works in 1861. Her numerous adventures and many names have been presented in chapter 16.

The Civil War began when South Carolina guns fired on Fort Sumter, in Charleston harbor, on April 12, 1861. When Major Anderson and its garrison surrendered, they were put aboard the Collins liner *Baltic*, which headed North flying the torn flag that had flown over the fort. Four years later, April 14, 1865, the *Santiago de Cuba* steamed up Charleston harbor to a battered pile of rubble that had been Fort Sumter. On board were Admiral J. A. Dahgren, whose ironclads had attacked it many times; Gustavus V. Fox, Assistant Secretary of the Navy; and, best of all, General Robert Anderson to raise the same tattered flag over what he had once commanded.[10]

The *Santiago de Cuba* was decommissioned at Philadelphia on June 17, 1865, then sold at auction on September 21 for $108,000. A number of bidders were eager to buy her and the successful one was Marshall O. Roberts, who needed a steamship for his Nicaraguan operations.[11] The *Santiago de Cuba* was brought to New York; her gunports were closed up, a new deckhouse was built, a light spar deck was added, and passenger quarters were installed. All this took only seven weeks.

November 11, 1865, saw her setting out from New York under the command of Jerry W. Smith, headed for Greytown. It was the first of a long series of monthly sailings. The *Santiago de Cuba* became the principal steamship on the Atlantic side of Roberts's California service. In 1865 he advertised under the name of Central American Transit Company, but starting in June of 1866 he substituted the name of North American Steamship Company. January of 1867 saw the entire operation sold to William H. Webb and others.

Until the fall of 1866 the *Santiago de Cuba* steamed placidly between New York and Nicaragua. On Saturday, September 29, 1866, loaded with five hundred passengers and a large cargo, she departed from New York; the ill-fated *Evening Star* left the same day. By Tuesday the former had run into a violent gale and was rolling so badly that her passengers were thrown out of their berths. Wednesday, the 3d, saw her in the midst of a hurricane that tore the cutwater off the bow, twisted the head off her rudder stock, and left her floundering uncontrollably. Her fore yard was broken, everything on deck ahead of the mast swept away, her forward deckhouse completely demolished, and her starboard paddle box stripped of its woodwork. The stem damage caused severe leaks and broken skylights rained water into the cabins. Two passengers and two crew members had been washed overboard. All passengers were ordered to a lower deck in case the remaining deckhouses went. Everyone bailed while the firemen staggered knee-deep in hot water swirling around the boilers. Then, to make matters worse, the hull flexed so badly that a crack appeared in the main steam line. The engine had to be worked by hand to keep it turning. Despite all these difficulties, a temporary tiller was rigged on Thursday and the ship brought under control. Two hours later it gave way. By good fortune, two of the passengers were shipwrights; they joined the ship's carpenter to cut away the rudder trunk and drill a hole through the rudder into which a temporary tiller could be inserted.[12]

In the meantime, the crew attacked the bow leaks by stuffing in pillows, blankets, and sheets. A sail was weighted along one side by spare grate bars for the boiler furnaces, put over the bow, and lashed in place. As the fury of the sea began to diminish, barrels of salt and pig iron were broken out from the forward hold and manhandled aft to raise the bow. Pumping out the forward fresh-water tanks helped lift it further.

The *Santiago de Cuba* headed for Norfolk but, on October 5, despite severe gales still present, Captain Smith decided that his ship could reach New York, where better repair facilities were available. She arrived on

October 7. Her survival with a damaged bow, a broken rudder, and a cracked steam line seems almost miraculous. Much of the credit must belong to Captain Jerry W. Smith, who never despaired, and to his unnamed Chief Engineer, who kept his terrified men at work while steam whistled overhead and hot water dashed over the floor plates on which they stood. The same hurricane sank the *Evening Star* and the *Daniel Webster*.

In the short period of three weeks the *Santiago de Cuba* was put back into order and resumed her interrupted voyage October 30. Her last trip for Roberts was from New York on November 27, 1866.

In 1867, under the North American Steamship Company's new management, she continued her monthly sailings until May 22. That day, on her way to New York, she ran hard aground in a dense fog just below Atlantic City.[13] While her 340 passengers were being rowed ashore, seven deaths resulted when a boat overturned in the breakers. It was ten days before the Coast Wrecking Company was able to pull her off and tow her to New York. The damage must have been severe; she did not return to service until September 20.

Starting in December of 1867, her southern terminus was shifted from Greytown to Aspinwall. The *Santiago de Cuba* continued routine calls there in 1868. It was she that made the North American Steamship Company's final trip from New York; when she passed Sandy Hook on October 28, she had completed eighteen voyages to Greytown and eight to Aspinwall. She was lucky enough to find immediate employment under charter to Livingston, Fox, and Company, who used her on monthly trips to New Orleans via Havana, sailing in November and December of 1868 and January of 1869. After finishing the last, on March 2, she was laid up for three months before Ruger Brothers took her over and sent her out, on her first Atlantic voyage, to Bremen and Copenhagen via Cowes. Departing on June 16, 1869, she made a slow eastward crossing of fourteen days to Cowes but a rather good return trip—sixteen days from Copenhagen on August 2 to New York. The Rugers were always more interested in fuel economy than speed.

Another three-month lay-up followed before she made a sailing to New Orleans, on November 6, as part of N. H. Brigham's Dispatch Line. There she loaded cotton and departed on the 29th, under North American Steamship management, for Havre. The eastbound crossing was a leisurely one from Bermuda, where she coaled on December 19, to Havre on the 29th. She left France January 16, 1870, and headed across the Atlantic by the southern route to avoid winter gales, but as she neared her destination she suffered a broken crank pin that slowed her down but did not immobilize her. Troubles multiplied on February 1, off Squan Beach, New Jersey, when she collided with the 274-ton iron steamer *Brunette*.[14] The latter sank, and two of her crew went down with her. The *Santiago de Cuba* picked up the rest and reached New York on the 2d. Because there was no immediate work in sight, repairs were delayed until August, when she was docked to replace her damaged stem and patch her torn sheathing.[15]

The Franco-Prussian War having created a sudden demand for supplies and munitions, she was scheduled to sail for Havre on September 3, 1870, with extremely low fares of $75, $60, and $30 quoted for those brave enough to enter a country at war.[16] From Havre and Cherbourg she sailed to the Netherlands ports of Rotterdam and Brouwershaven and arrived back at New York on November 11, 1870, after a long, slow crossing in heavy weather. Her three Atlantic voyages give no indication that the *Santiago de Cuba* was once famous for her speed.

For the next two years, William H. Webb, her owner, offered her for sale but found neither buyers nor operators until March of 1873, when O. I. Guilleaume and Company chartered her for two trips from New York to Matanzas and Havana.

Three idle years later she finally found a purchaser in the Philadelphia firm of William P. Clyde and Company, which was rapidly expanding its operations to East Coast and Gulf of Mexico ports. Shortly afterward, Clyde purchased a second paddler, the *Morro Castle*, which had been built in 1864 for the Havana trade. The two were completely gutted, their guards removed, their sterns rebuilt, new engines and boilers installed, and all quarters rearranged to become almost-new screw-propelled steamships. The whole operation was most unusual; they were the only American paddle steamers to undergo such a complete reconstruction.

The *Santiago de Cuba* was put on the sectional dock at New York in March of 1877, to have her underwater work done;[17] by September she was ready. The *Morro Castle* followed in 1878, and received a single-cylinder high-pressure engine of 50" diameter by 5'-0" diameter stroke, receiving steam at 60 psi from four cylindrical boilers 11'-0" diameter by 10'-0" long.[18] There was, of course, a surface condenser. Her new engine was built by Philadelphia's Neafie and Levy, a firm that had engined screw steamships for over a quarter of a century. It is probable that the *Santiago de Cuba*'s power plant was similar, if not identical.

Once the rejuvenated *Santiago de Cuba* was ready, she was chartered to the Pacific Mail Steamship Company; she left Philadelphia September 11, 1877, for Aspinwall.[19] Coming home she made an extremely slow fourteen-day passage, arriving at New York on October 24.[20] Another charter having been negotiated, she was turned over to James W. Quintard for Charleston service and left New York about November 1, on the first of a series of voyages that kept her employed for the next five

Deck view of the *Western Metropolis*. Photographed June 7, 1864, as a Civil War transport. *Photo by U.S. Army Signal Corps.*

months. Not until April 20, 1878, did she sail as part of Clyde Line's growing fleet; on that date she cleared New York for Havana. Thereafter, most of her sailings were to Cuba, although she was sent elsewhere from time to time when a profitable charter was offered. Quintard used her every so often for one or two Charleston voyages, and the Alexandre Line took her over, in February of 1881, for a far longer trip to the Mexican ports of Vera Cruz, Frontera, Campeachy, and Progreso.

Immediately after her Mexican venture she sailed for Charleston, under Quintard auspices, and had the misfortune to pound on the Charleston bar as she came out of the harbor on April 14. Both the sternpost and rudder were carried away. Her cargo was unloaded and put aboard Clyde's *Delaware*.[21]

Later in the year, under charter to Boston's F. Nickerson and Company, which owned the *Worcester, Carroll,* and *Somerset,* she was assigned to their Boston-Savannah branch in June of 1881. But when she reached Boston on December 14, 1881, after a dozen voyages, she was replaced by the *Worcester*. Because Clyde had no immediate use for her, she was laid up at Brooklyn's Erie Basin for the next five years.[22] A wood screw steamer powered by a one-cylinder single-expansion engine was, by 1882, obsolete; there were newer and better craft available.

In 1886 her engine was removed and she became a schooner-rigged coal barge under the name of *Marion*, enrolled at New London.[23] In this third transformation she continued for over a decade. She was still in good enough condition to warrant the installation of a new deck as late as 1897. Like the *New York, Costa Rica,* and *Vanderbilt,* she demonstrated that a Simonson steamer was so well built that it would long outlast its contemporaries. She was last listed by the American Bureau of Shipping in their 1899 Record.

In four decades of life the *Santiago de Cuba* had been a remarkably successful blockader during the Civil War, a Panama and Nicaragua steamer for twenty-six voyages; a coaster to New Orleans, Charleston, Boston, Savannah, and Vera Cruz; a coal barge; and, for three trips, an Atlantic liner. One of the last was for Ruger Brothers and two for William H. Webb. She had started life as a paddler; in middle age she became a screw steamer; and, as a dowager, although at the end of a tug's tow line, she could still set sails to help her on her way.

The *Western Metropolis*

The *Western Metropolis* was a large beam-engined coaster built in response to the Civil War's enormous demand for steamships. Her principal owners were George Griswold, A. Benner, and William Wall; others held smaller shares. Contemporary reports listed her as belonging to the firm of Benner and Brown. Her builder, Brooklyn's F. Z. Tucker, had produced steamboats but had no previous experience with oceangoing vessels.

She was large, 2,269 tons, when built in 1863, or 2,092, as remeasured in 1865; she had capacious holds, and made use of a fifteen-year-old engine of 75" bore and 12'-0" stroke that had already served two Great Lakes steamers, the *Empire State* in 1848 and the *Western Metropolis* in 1856. In had been built by Merrick and Towne in Philadelphia,[24] but the new boilers, wheels, and general reconditioning were by the Morgan Iron Works. The name of the second lake steamship carried over into the new hull; Benner and Brown did not bestir themselves to think up a new one.

She was as long as a Collins liner, 285'-4", and five feet narrower, 40'-8"; she had a depth of hold of 23'-0" and a draft of 16'-0". A single stack, a vertical stem, and a two-masted rig with a course and a topsail forward gave her an attractive appearance. A contemporary, Captain George H. Norton, described her as a "very slow, clumsy, unwieldy, hard steering steamer."[25]

The Quartermaster Corps chartered her, immediately upon completion, at $850 per day, and kept her in continuous use from December 1863 until late January of 1865. For her first voyage she steamed South to New Orleans and, on the way back, sighted a suspicious-looking steamer, the *Rosita*. Although his ship had no guns, Captain W. B. Hilton put after the *Rosita* in hot pursuit, overtook her, and sent aboard a boat's crew armed with pistols and cutlasses. They were assisted by two Navy and one Army officer, all of them passengers aboard the *Western Metropolis*.[26] The *Rosita* carried a cargo of munitions and liquor and had been sabotaged.

Her crew had removed the plug from a sea cock and waist-deep water was already rising in her hold. The safety valves had been tied shut and the boiler water-level blown down to induce an explosion. Prompt action prevented either of the imminent disasters and the *Western Metropolis* towed her prize into Key West on January 29, 1864. The event became the high point of her entire career; she never did anything else so well.

Most of 1864 was spent ferrying troops and supplies between New York and Hampton Roads in support of the Union Army's activities in Virginia. Northbound she carried hundreds of sick and wounded. In December she loaded troops for the attack on Fort Fisher.

February of 1865 saw her first commercial operation, under charter to M. O. Roberts, when she sailed from New York for Greytown on the 20th. After a single round trip, she was chartered to H. B. Cromwell and Company for a New Orleans voyage in April and then went back to the Quartermaster Corps in May. They also sent her to New Orleans, and, after completing governmental service in July, she made a third voyage, in August, to that port under a third sponsor, W. H. Robson and Company, bringing back a record cargo of 3,000 bales of cotton. Finally, on September 30, she steamed from New York to Apalachicola, Florida, on her only sailing for her original owners; after delivering cotton to New York on November 23, 1865, she was sold to Ruger Brothers.[27]

As reported in chapter 20, the *Western Metropolis* was a poor investment. It is not known just what was wrong with her. According to the Rugers' advertisements, she was "thoroughly refitted" and would sail for Bremen on March 17, 1866.[28] After that the sailing date was put off again; three months passed before she ever did get away, on June 28. She had been equipped with some new variety of paddle wheel that shed its floats in anything but a dead calm. The losses became so serious that the *Western Metropolis* had to turn around and head for Boston before all the paddles were lost.[29] She put in there on July 6, made temporary repairs, and headed back to New York on the 10th, unable to continue on to Bremen. It was a distressing situation; Ruger Brothers and their newest steamship must have been the laughing stock of the shipping industry. They did not try again, but sold her to the New York and Bremen Steamship Company.

The latter organization strengthened the wheels and sent her to sea with instructions not to come back until she had met heavy weather. After that trial she was in proved running condition and set out on March 7, 1867, under Captain William Weir for Cowes and Bremen. She took all of seventeen days to reach the former.[30] She continued making eastbound sailings at approximately eight-week intervals, but on her fourth voyage met with misfortune.[31] From New York on August 24, she arrived at Southampton September 8 with a broken shaft. She was repaired there, proceeded on to Bremen, and delivered 921 passengers to New York on November 6, 1867, having left Bremen October 20 and Cowes the 22d. Three deaths were recorded; an infant, a case of delirium tremens, and a case of apoplexy.[32]

The New York and Bremen Steamship Company, after a one-year trial, ended their operations and sold the *Western Metropolis* on June 30, 1868, for $57,000.[33] Her new owner advertised her for sale for the rest of 1868 and all of 1869 without success. In 1870, however, she was acquired by the Merchants' Steamship Company, Frederic Baker, Agent, which assembled a fleet of assorted steamships, some of them captured Civil War blockade runners, that required little cash outlay. Included were the *United States, Mississippi, Crescent City, General Meade, Sherman,* and *Emily B. Souder.* The familiar names *United States* and *Mississippi* do not refer to the steamships considered earlier; these were later and lesser vessels. Of this heterogeneous group of available steamships, the *Western Metropolis* stood out as the largest, and the only side-wheeler of the lot. From the spring of 1870 to that of 1875 the Merchants' Line provided regular sailings from New York to New Orleans.

The *Western Metropolis* made two trips, sailing on March 12 and April 9 of 1870, but was then chartered to Ruger Brothers. Now that someone else had straightened her out, they used her for a single voyage from New York to Havre, Bremen, Copenhagen, Swinemunde, Kiel, and Christiansand. She left New York May 18 under Captain H. S. Quick, and came home by the northern route. From the Shetlands to Newfoundland it was cold and foggy; a sailor died of pneumonia and three infants did not survive the rigors of the voyage. She delivered 954 immigrants of many nationalities on July 7.[34]

Two months later she was back with Baker's New Orleans steamers, sailing once a month. Once more bad luck descended. After she left New York on October 7, 1871, a crack appeared in her port shaft and began to lengthen; she was able to proceed at reduced speed and reached New Orleans October 16. When she left, without passengers or freight, she was partially towed by the *Sherman* and partially under her own power. Off Key West the crack increased so much that she could no longer use her engine.[35] The *Sherman,* unable to handle so large a craft, left her there. A new shaft was sent down from New York, together with skilled workmen and heavy hoisting gear. The replacement was completed and, going back to New Orleans for a cargo, she left February 8, 1872, taking seven days to New York.[36] The fact that the same vessel suffered two shaft breaks, one in 1867 and another in 1871, leads one to suspect that Civil War workmanship left much to be desired.

Aside from being laid up from February to August of 1873, she continued to New Orleans for the rest of her career. On her last voyage she left New York February 13, 1875, came out of South West Pass on March 2, and docked on the 9th. Within a month the Merchants' Steamship Company ceased operating. From then until 1878 she was laid up. During that time several changes in ownership were reported; the last sale was to Cornelius Delamater, who bought her for $15,000. In March of 1878 she was at the Delamater Iron Works on the North River side of Manhattan, where her engine was removed.[37]

During the useful portion of her life, from 1863 to 1875, the *Western Metropolis* seems to have been most successful as a large cotton carrier in the New Orleans trade. On the Atlantic she made four voyages for the New York and Bremen Steamship Company and one for Ruger Brothers. There her ample tween deck spaces provided berths for a thousand immigrants per trip. Nevertheless, two shafting casualties and an abortive European trip that had to be canceled when her paddle wheels spun off their buckets interrupted her planned services. She was a big, clumsy steamship, one that could be bought or chartered cheaply, but she could never endear herself to any owner.

Steamships for Chinese Waters: The *Suwonada* and the *Fung Shuey*

Within the two decades that saw the wood oceangoing steamship develop and decline, over fifty vessels were constructed in the United Sates for use in and around Chinese waters. Many others gravitated there and spent their later years in the Oriental trade.[38] Numbers of them were large, white, walking beam-engined craft of shallow draft, indistinguishable from their sisters on the Delaware, Chesapeake Bay, Long Island Sound, and other sheltered waters. Others were seagoing in type and had evolved from the coastwise steamship. Many made their way across the Atlantic by circuitous routes, in order to refuel on the way that would eventually take them around Africa's Cape of Good Hope. Even though deeply laden with coal, some took freight and paying passengers with them. It is not planned to cover here all the possible one-way crossings of the South Atlantic. Instead, two outstanding examples of these China coasters will be considered. Both were built during the Civil War; both served for a time in American waters; one was a very fast side-wheeler and the other a screw steamer of moderate size and speed.

The more glamorous of the two was a large beam-engined steamship with a rather low freeboard. Her two tall stacks, slightly curving stem, and a half-brig rig made her remarkably similar in appearance to the Collins liner

Suwonada at Hong Kong. *Courtesy of Peabody Museum of Salem.*

Adriatic, even though she had less than half the latter's tonnage.

The *Suwonada* or, as the name was often written, *Suwo Nada*, was ordered by Percival L. Everett of Boston for Augustine Heard and Company, one of the most important American trading houses in China. Her builders were John Englis and Son, whose yard was on Manhattan Island at 10th Street and the East River. John Englis, Sr., had been apprenticed to the Smith and Dimon shipyard; his own firm was to pass successively to his son, John, Jr., and his grandson, Charles M. Englis. All three generations were noted for their splendid river, bay and sound steamers. The *Suwonada*, an oceangoing vessel, was somewhat of an exception to the type usually turned out by the father-and-son partnership. Her dimensions were 258'-0" by 38'-2" by 21'-2", with a tonnage of 1,802 and a beam engine of 76" bore by 12'-0" stroke installed by the Neptune Iron Works.[39] There were four tubular boilers. It was a very large cylinder; when combined with a light hull it gave her a nice turn of speed. The long stroke resulted in large-diameter wheels whose paddle boxes towered high above the rather shallow hull. They were especially noticeable because her guards were located at the weather deck, far above the water.

Because of favorable Civil War charter rates, she

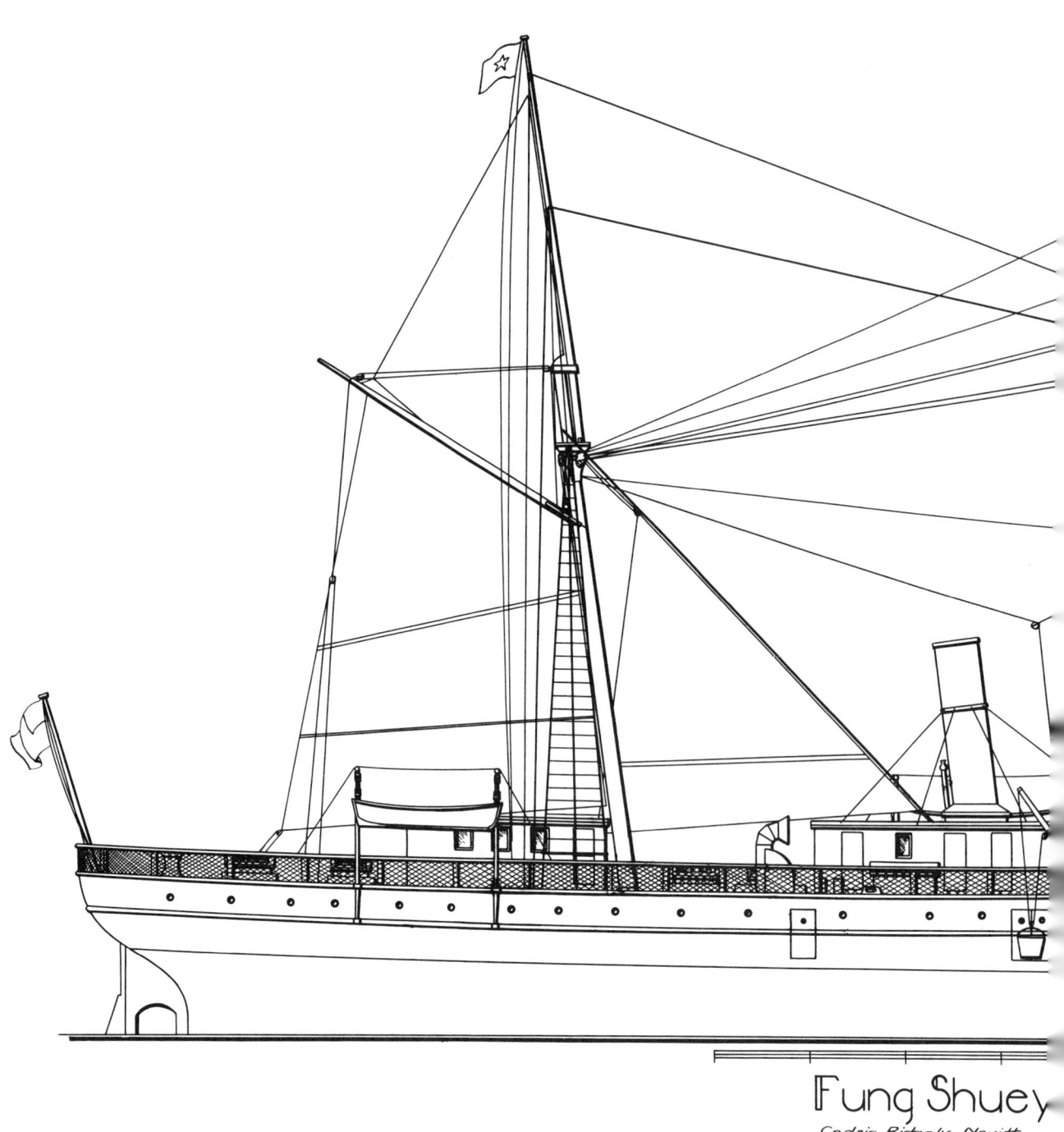

Fung Shuey, 1864, sail plan.

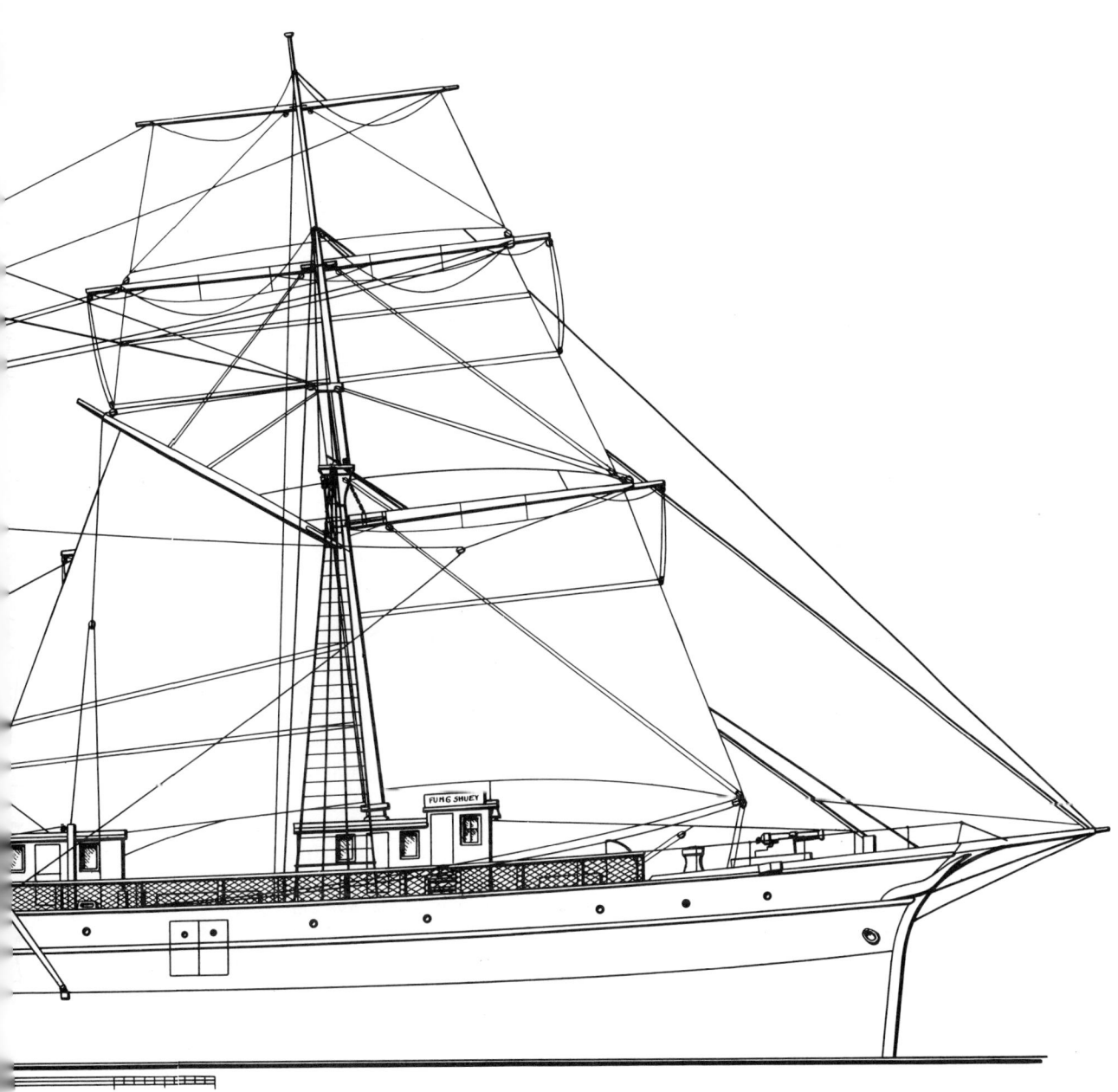

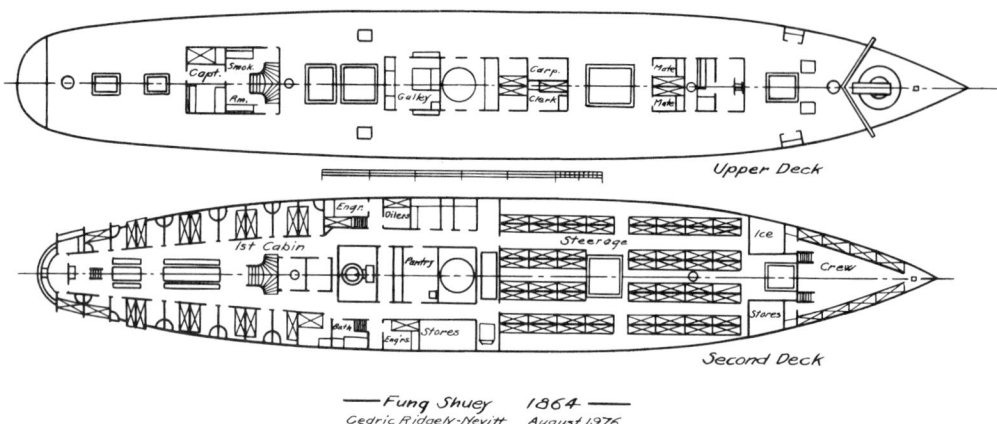

Deck plans of the *Fung Shuey*.

started work, in September of 1864, to New Orleans; her delivery to the Pacific was put off. Flying the New York Mail Steam Ship Company's house flag, she left New York on the 10th, New Orleans the 24th, and reached New York on the 30th with a fast northbound passage of less than six days.[40] Two further trips were made in October and November. Then, after the completion of the line's newest steamship, the *Guiding Star*, she was taken over by the U.S. Quartermaster Corps for Civil War transport work; this occupied her from January through April of 1865, when that long, tragic conflict came to an end.

At last she was made ready for the long voyage to Hong Kong. Clearing New York on May 23,[41] she went via Rio de Janeiro on June 23 and Mauritius, July 30, to Singapore, August 15, and Hong Kong, August 27. She steamed slowly to conserve her coal and took eighty-two or eighty-three days to Singapore.[42]

Despite the arduous trip, she arrived in excellent condition and was ready to leave Hong Kong for Shanghai on September 14, 1865. She continued on this route for the next four years. On June 25, 1869, under Captain Arthur H. Clark, one day out of Hong Kong, she struck an unchartered rock and began to leak badly. Captain Clark turned back and entered the empty Union Dock at Hong Kong on the 26th. The *Suwonada*, having won her race against time, sank, but no better location than a dry dock could have been chosen. Her master was praised by the underwriters. He is best known to modern readers as the author of that 1920 classic in marine history, *The Clipper Ship Era*. During his long career he commanded a variety of vessels that included the sailing ship *Verena* and the transatlantic liner *Indiana*.

In March and April of 1870, when the *Suwonada* was making a trip to Manila, she ran on a reef off Luzon Island but got off undamaged. Back in Hong Kong, Heard and Company chartered her to the China Sea, Saigon, and Straits Steamship Company, a firm whose name literally describes the route of its vessels. After a year under their auspices, she was overhauled at Hong Kong in April of 1871. Over the years she had proven to be the fastest and finest steamship on the China Coast, but her low-pressure engine of large horsepower had always led to large fuel bills. Despite this, she was put back on the Hong Kong-Shanghai route in May and continued there for the rest of her life.

Leaving Hong Kong on January 27, 1872, still commanded by Captain Clark, she was passing through Haitan Straits at her full speed of thirteen knots on the 29th, when she struck a submerged rock and was holed. Her master headed for a nearby sand bar, ran her aground, and kept his engine going ahead to keep all its attached pumps working. They could not cope with the

334

rising water and soon it drowned out her fires. Then, to make matters worse, local pirates attacked but were driven back by rifles and the ship's gun. Fortunately, the U.S.S. *Ashuelot* happened to pass the stranded steamer and frightened off the attackers. A rising tide caused the *Suwonada* to slide off her resting place and sink in deep water, but all hands were rescued.[43] The court of inquiry held at Hong Kong congratulated Captain Clark for his handling of the ship and saving all on board.

The *Suwonada* was a splendid example of a lightly built, yet structurally strong steamship of high power and high speed that was kept continuously busy, first to New Orleans, then as a troop transport, and, through much of her career, as a popular China coaster between Shanghai and Hong Kong. There were occasional trips as far away as Saigon or Manila. The seas in which she sailed were poorly charted, but her master was skilled in operating under adverse conditions and managed to bring her home or kedge her off on two occasions when she grounded. A third time, however, brought her to an end. Had it not done so, it is probable that she would have seen early replacement by a smaller, slower, far less handsome but more economical screw steamship with an iron hull.

The *Fung Shuey* was a wood propeller of 1,004 tons, built by Elisha P. Whitlock at Greenpoint in 1864.[44] She was a clean-looking craft 195'-6" long by 33' beam with a 24' depth of hold. The stem curved forward to support a short bowsprit, the sheer line was attractive to the eye, and the rake of the two masts and single stack gave her a yachtlike appearance. A one-cylinder engine of 44" bore by 3"-6" stroke was installed by the Delamater Iron Works, one of the first of the New York engine builders to make a specialty of engining screw steamers.

J. B. Hildreth seems to have been a man of many talents. He was not only the *Fung Shuey*'s master and part owner; he actually designed her. Two deck plans and a sail plan at the Peabody Museum of Salem show professional draftsmanship and carefully worked out details. The coal loading and ash handling, for example, indicate a twenty-three-foot boom mounted on the starboard side to hoist coal buckets from a barge to a side port. Inside was a truck on rails to move it inboard and dump it into the bunkers. The ashes were thrown overboard through a corresponding port on the port side. By sloping the boom forward instead of aft, it could serve to handle cargo through another side port.

The sail details are explicit. The two storm sails, the fore staysail and the mainsail, had removable bonnets such as Columbus and Magellan used, and the mainsail had reef points as well, as a means of setting the minimum possible amount of canvas in a typhoon. The former was of 19-ounce per yard weight, the heaviest possible grade. 13-ounce duck was used for the lightest sails, the topgallant and the gaff topsail. In 1867, just before she sailed for China, her three square yards were shortened four feet apiece and their sails cut down accordingly.

On the weather deck an essential feature was a pivot gun forward. We have seen that the *Suwonada* found hers necessary as a protection against pirates. The mates were housed in the forward deckhouse; the cooks, adjacent to the galley in the midship one; and the captain aft, next to the smoking room. Below decks there were sixteen crew berths at the bow. From there to the coal and ash gangway were two high bunks for one hundred and twenty steerage passengers. The first-class saloon was abaft the engine, with tables on the centerline and eighteen double staterooms along the sides. One more stateroom with an extrawide berth, conveniently located adjacent to the bath, was, obviously, the bridal suite.

July of 1864 is a difficult month to follow from the confusing newspaper reports that did not differentiate between two different steamships of almost the same name. One *Fung Shuey*—sometimes spelled *Foong Shuey*—sailed for Hong Kong on July 8 and returned on the 20th. She was a beam-engined steamship that had started under the name *Plymouth Rock* and, after arriving in China, would revert to that very American name. The screw steamer *Fung Shuey*, although intended for the Orient, sailed for New Orleans on July 2 under Captain Hildreth. After starting back on the 20th, the piston rod of her "pump" (probably the air pump) was broken and repaired twice at sea; long delays ensued before she limped into New York on July 29.[45]

Henry W. Hubbell, the latter *Fung Shuey*'s principal owner, had found wartime charter rates so high that his steamship could make more money at home than abroad. As a result she spent three years leased out, first to W. H. Robson, then to H. B. Cromwell, and, finally, to the Black Star Line; all kept her on the lucrative New York to New Orleans run.

By 1867, however, after thirty-one calls at New Orleans, the market for steamships had seriously declined; it was time to return to the original plans for the ship. She was advertised to leave New York, still under Captain Hildreth, on May 15, 1867,[46] via St. Thomas to Rio de Janeiro, where she would coal before heading across the South Atlantic, around the Cape of Good Hope to Mauritius off the east coast of Africa, and on to Singapore and Hong Kong. The *Suwonada* had followed the same route. Her actual departure was put off to May 18, J. S. Watson replaced Hildreth as master, and she reached Singapore on August 11, 1867. She was then put to work between Hong Kong and Shanghai. In September of 1868 she passed to Augustine Heard and Company, who used her between Hong Kong, Manila, and Amoy.[47] At the last-named port, on August 25,

1869, while being shifted from one berth to another, she broke free and drifted onto some rocks; there she remained hard and fast and, eventually, became a total loss.

The *Fung Shuey* was an admirable example of the wooden propeller of moderate size that had proved entirely successful wherever it was tried; there were neither structural nor vibration problems for a 1,000 tonner 200′ long. It seems unfair that a blunder in line handling ended her successful career so soon. Her name, which translates into "Good Luck," failed to protect her.

Pacific Mail Steamships

The master builders of the City of New York created the American oceangoing steamship between 1845 and 1850. They continued to perfect and enlarge their later vessels, eventually developing a limited number of splendid Atlantic liners in the 1850-60 decade. These culminated in two unique specimens, the *Vanderbilt* and *Adriatic*, both requiring extensions of the existing art of ship construction in order to achieve wooden hulls above 300′ long having the necessary strength and rigidity. From 1860 to 1869 it was still New York's master shipbuilders who launched a whole series of magnificent steamships, all beyond the 300′ mark, for the Pacific Mail Steamship Company. These were the largest and finest wood steamships ever built anywhere. After them would come a completely new type, from Delaware River shipyards, which would have iron hulls, compound engines, and screw propellers. The *George W. Clyde*, of 1872, would be the forerunner of a new era in American ships and shipbuilding.

Because three of the great Pacific Mail liners carried passengers across the Atlantic, and all of them represent the final development of the paddle-wheel steamship, a brief summary of the Pacific Mail's post-1860 fleet follows:

Date	Steamship	Length in Feet	1865 Tonnage	Builder
1861	Constitution	342	3,575	Webb
1863	Golden City	343	3,589	Webb
1864	Sacramento	304	2,682	Webb & Bell
1865	Henry Chauncey	319	2,656	Webb
1865	Colorado	314	3,728	Webb
1866	Arizona	324	2,793	Steers
1866	Montana	318	2,676	Webb & Bell
1867	Great Republic	360	3,881	Steers
1867	China	360	3,836	Webb
1868	Japan	362	4,351	Steers
1868	Alaska	346	4,011	Steers
1869	America	363	4,454	Steers

Three of the steamers, the *Henry Chauncey*, *Arizona*, and *Alaska*, started their careers on the Atlantic side of the Panama route. Four, the *Great Republic*, *China*, *Japan*, and *America*, were designed for transpacific operation from San Francisco to Yokohama and Hong Kong, a new service begun in 1867. The remaining five, *Constitution*, *Golden City*, *Sacramento*, *Colorado*, and *Montana* were intended for the 3,250 nautical mile passage from Panama to San Francisco. It was the *Colorado*, of this group, that, after alterations to her superstructure and additions to her masts and sails, made the first sailing from San Francisco to Japan and China on January 1, 1867.[48]

A common characteristic of all these steamships was an upper deck extending beyond the hull to provide extra space for the first-class passengers. Because much of the trip was in tropical waters, in an era before mechanical ventilation became practical, it was essential to have

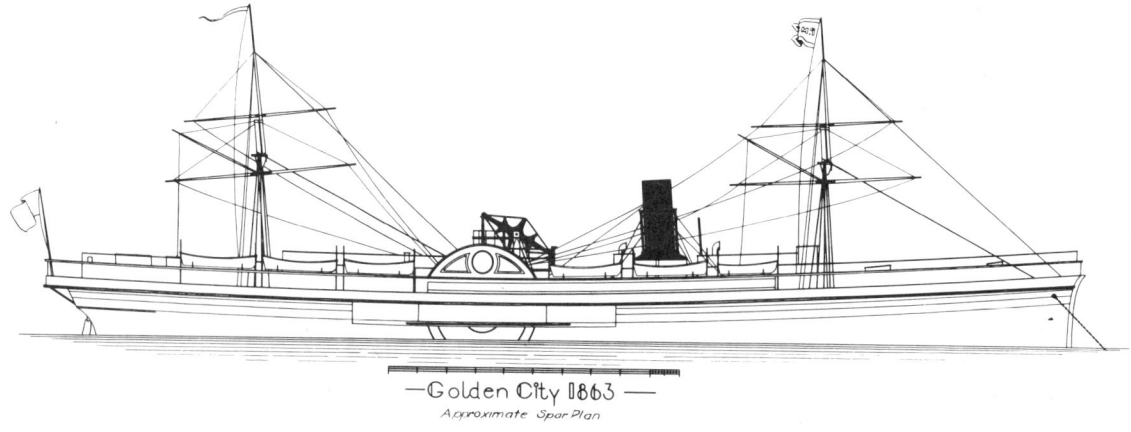

Golden City, approximate spar plan.

cabins shaded from the sun, and exposed to the wind but not to spray, if they were to be livable. Over everything was a lightly built hurricane deck providing a vast promenade. Above it, in good weather, an awning shaded everything. A second standard feature was a walking beam engine having an enormous single cylinder of 105" bore and 12'-0" stroke capable of producing around 1,800 IHP. In practice it would frequently be operated at a lower rating to conserve coal. The speed of the transpacific liners was restricted to ten knots or less, making their coal consumption forty-five tons per day, half that of a Collins liner. Fuel was a very costly commodity that had to be shipped vast distances by sailing ships. The Novelty Iron Works, which owned Pacific Mail stock, developed a standard engine that was automatically ordered for each successive ship.[49] The one exception was the *Sacramento*, which had a 100" cylinder instead of 105". All had 40' diameter wheels turning at a stately rate of nine or ten revolutions per minute.

These steamships had evolved from the *Golden Age*, completed for transpacific service by William H. Brown in 1853, and acquired by the Pacific Mail a year later. She contributed the rugged, long-lived, trouble-free walking beam engine. Both she and her contemporary, the *John L. Stephens*, had overhanging decks extending out to the guards, and had shown that these made possible increased numbers of passengers and increased comfort for them.

The *Constitution* incorporated both features and, in addition, adopted the *Stephens*'s split-boiler placement with separate fire rooms and stacks, one forward and one aft of the engine. As a faint relic from her sailing-ship ancestors, she had a tiny bowsprit supporting the topmast stay. Like the *California*, the first steamship for the line, she was designed and built by William H. Webb. He held considerable stock in the company and was on its board of directors.

In the following vessel, the *Golden City*, Webb did away with the bowsprit and moved all the boilers forward of the engine, where they could be served by one squat stack in place of two tall, thin ones. Boilers were reduced in size and their efficiency was increased by the use of return tubes rather than flues. All subsequent steamships retained the *Golden City*'s features. The coastwise ones had two masts with square sails; those for China and Japan had a three-masted bark rig with taller masts and more sails. In them, the after deck overhang was reduced; moreoever, it was completely eliminated forward of the foremast. The basic design, however, was established by Webb in his *Golden City*, and remained unchanged in the ten steamships following her.

Although it is possible to re-create the lines for all the Webb steamships, only two surviving drawings show any details. The Mariners Museum owns a cabin plan for the *Japan*.[50] The other is an outboard profile at Webb Institute of Naval Architecture, which shows the lines and the superstructure of the *Golden City*. The author has redrawn the latter, with two minor additions—the braces supporting the overhang and the walking beam.

The *Golden City*'s keel was laid on June 23, 1862, she was launched seven months later on January 24, 1863, and was completed August 7, 1863.[51] So short a building period was remarkable at the height of a war and for a steamship longer than the *Adriatic*. The Novelty Iron Works must have started work on the massive engine, its 22" diameter shafts, and its 40' wheels well before the keel was laid. Her 343' hull had a beam of 45'-1", a depth of hold of 30' to the spar deck, and a designed draft of 17'. She could carry almost two thousand passengers.[52]

On August 13, 1863, loaded down with 2,107 tons of coal, she cleared New York for San Francisco via the Strait of Magellan, and in November began the San Francisco-Panama sailings that kept her occupied until 1870. During that period she was considered the "most economical and best paying" steamship of the line's West Coast fleet.[53] Confirming the statement, an 1864 report shows that on a northbound trip she averaged 10.5 knots at 10.5 rpm with a steam pressure of 17.5 psi and a coal consumption of 39 tons per day. When slowed to 9 knots, she burned only a ton of coal an hour, a remarkable performance for so large a steamer.[54]

Departing from San Francisco on February 18, 1870, with $791,000 in gold and a $91,000 cargo, she ran aground in a dense fog on February 22 near Point Lazaro, Lower California. About four hundred passengers were landed and had to walk sixteen miles to Santa Maria Bay over barren wastes. Lacking food and water, nine strayed and were lost on the way. The survivors were picked up by the *Colorado*, which had passed the wreck on the 24th but was unable to land boats through the high surf. The *Golden City*, hammered by the sea, broke in two abaft the stack.

The *America*, 1869-1872

The first of these steamships to cross the Atlantic was the *America*, the last of the quartet built to cross the Pacific. Authorized by Congress in 1865, an agreement had been reached between the Post Office and the Pacific Mail to establish a monthly mail service from San Francisco to Yokohama and Hong Kong for a payment of $500,000 per year. A pair of steamships, the *Celestial Empire* and the *Great Republic*, were started at once, the former by William H. Webb, the latter by Henry Steers,

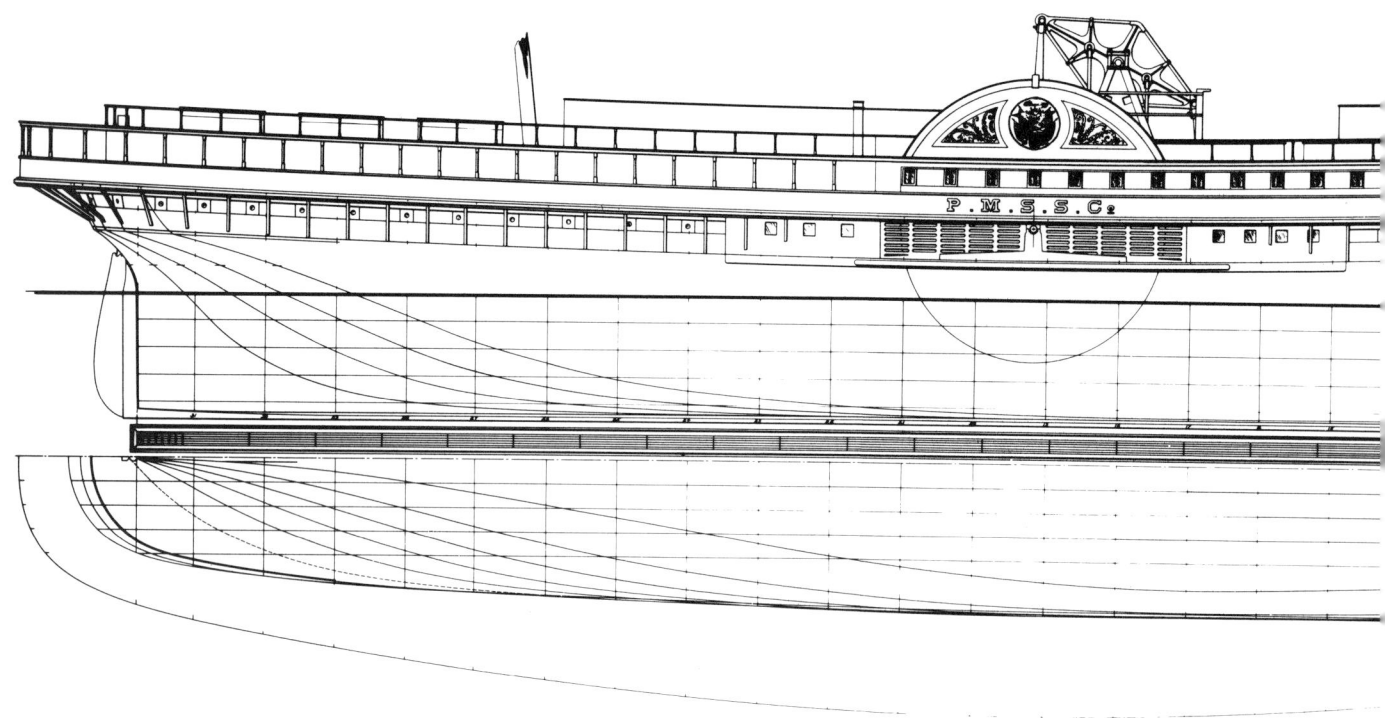

Golden City, 1863, lines.

son of James and nephew of George Steers. The younger Steers had been apprenticed to his uncle when the latter was foreman in William H. Brown's shipyard and had worked for his father while the *Adriatic* was being built. In 1859, after William H. Brown had retired, Steers rented a portion of his yard and began construction under his own name. He quickly acquired a reputation for the excellence of his sailing yachts, pilot boats, and fast steamers for the China trade. When William H. Webb fell out with the Pacific Mail Steamship Company and took over a competitor, the North American Steamship Company, Steers succeeded him as the regular builder of the Pacific Mail's steamships. As a result, the last two transpacific liners were Steers vessels. Named *Japan* and *America*, they were almost identical to the *China*—as the *Celestial Empire* was renamed after her trials—and the *Great Republic*, but with about three feet of length and twenty inches extra beam added. The hull construction and machinery details were very similar on all four. The *China* and *Japan* were extensively described in the New York newspapers; the information given is equally applicable to the *America*, last and largest— 4,454.69 tons—of any wooden paddler built in the United States for ocean service.[55] She was 363′ in length, 49.3′ in beam, with a depth to the upper deck of 31.3′.

The increase in length from 340′ for the *Golden City* to 360′ appears a small matter to modern readers. In a wood ship where 340′ represented an extreme development, the extra 20′ represented a major change. There would be greater stress and deflection problems, and as a result the hull construction had to be radically revised. In the late 1840s and early 1850s the use of diagonal iron strapping had made the *Adriatic* and, fifteen years later, the *Golden City* possible. For the four great Pacific steamships, it was decided that two sets of strapping would be necessary. The first was placed outside the frames in the usual manner. Over this yellow pine planking was laid, caulked, and payed with pitch to make it watertight. Then another set of straps was placed outside the planking and a second layer of 4 1/2″ oak installed below the waterline and a 5″ pine layer above. All this required exacting workmanship, for the bolts, spikes, and treenails had to be spaced to suit the frames, planks, and both sets of iron straps where they crossed. Obviously, the frames and planks had to be notched out around the strapping. All this was so expensive that each steamer cost just over $1,000,000, but the results justified the care taken. The hulls, once the fastenings tied together the two layers of planking, double sets of diagonal strapping, and the heavy ceiling inside the frames, were extremely strong. Not only was the shell of great strength, but the decks were especially tied into it; each orlop deck beam had two grown knees at each end

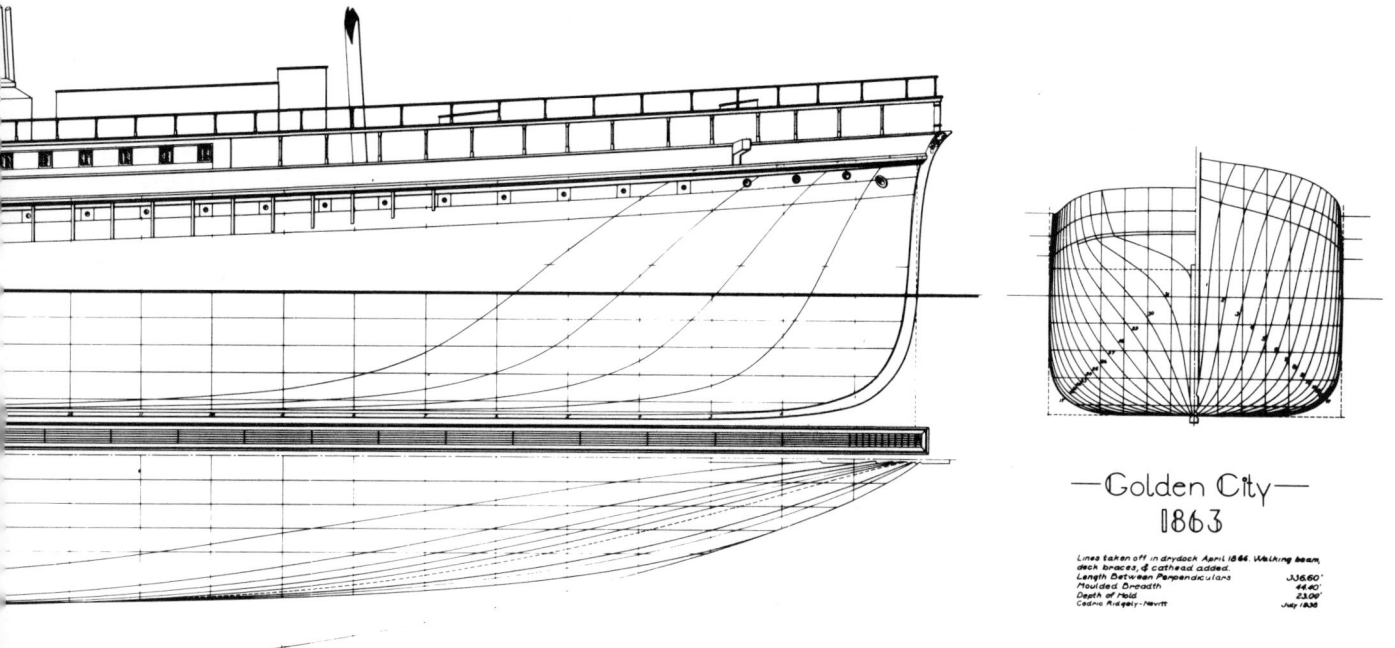

—Golden City—
1863

fastened with sixteen iron bolts. The berth and main deck beams had three knees per beam. Colossal amounts of timber were used and many thousands of holes had to be accurately bored by hand through wood of three to four feet in thickness.

The machinery, which took eighteen months to build, was a considerable advance over that built in the 1850s, which operated everything—the air pump, the feed pump, the bilge pump, and the circulating pump—directly from the main engine. In the Pacific liners, a separate two-cylinder engine drove both the air pump and a rotary pump circulating sea water through the surface condenser. An individual feed pump was fitted for each boiler; these, however, were driven off the main engine. Independently driven pumps could be used as bilge or fire pumps and could be fed steam by a donkey boiler when the main ones were cold. The main boilers, four in number, had six furnaces each, and supplied steam at 20 to 30 psi. On the *China*, and presumably on the *America*, they were 25' long, 13' wide, and 12'-6" high, with horizontal return tubes three inches in diameter.[56] Their outputs were led into a main steam pipe 28" in diameter and admitted to the cylinder by balanced poppet valves controlled by an Allen adjustable cut-off. The piston rod was 11", the connecting rod 13", the crank pin 14", and the early paddle shafts 22" in diameter. The *America*'s shafting was increased to 24" and weighed 67 tons. Each paddle wheel came to 74 tons and the walking beams weighed 25.[57] The sum of all the machinery weights was 880 tons.

The surface condenser had 4,464 brass tubes 3/4" diameter by 9'-4" long, and these fitted into joints with compressed-wood ferrules to make them water and vacuum tight. Its air pump was 60" bore by 6'-0" stroke.

The 40' paddle wheels had 34 oak buckets 3 1/2" thick by 24" deep with a 6" iron reinforcement top and bottom. Some of the paddles were weighted to balance the wheel against the effect of the piston and connecting rod; care was taken to leave some imbalance to prevent the wheels from stopping with the engine at dead center.

A photograph of the *China*'s wheel shows that the floats were installed at three different radial positions, some against the outer circular rim, some about two feet inside it, and the remainder two or three feet further in.

There was a broad expanse of spar deck having a pilot house and the master's cabin well forward, plus a small smoking room in the same deckhouse. Just abaft the wheels was a large social hall and an entry to the grand staircase leading down to the main deck. Aft on that deck was a dining saloon for one third of the ship's length, with fifteen large staterooms on each side. Outside was an open promenade. The forward third of the main deck consisted of steerage space with berths in three-high tiers. In the midship region, separating the classes, were the galleys, pantries, officers' mess, and service spaces. The port guard had a pen for livestock. Descending to the berth deck, a first-class ladies' saloon was aft with ten staterooms per side, the forecastle for the crew at the bow, and in the rest of the space were nine triple rooms for men, extensive steerage berthing, and the upper levels of the engine and boiler rooms. Below came a cargo deck and farther down, partial decks.

The *Japan*'s plans show berths for 122 cabin passengers and list 908 for steerage. The cargo capacity is quoted as 2,135 tons, obviously volume, not weight, and the coal bunkers hold 1,438 tons, weight, not volume. Fresh-water tanks contain 18,500 gallons. With so limited a capacity, additional amounts must have been distilled at sea.

The *San Francisco Alta*'s description of the *America*[58] is in complete agreement with the *Japan* drawings. Moreover, it adds more color than a modern reader would dare to imagine. The public rooms were painted "in the most exquisite style of fresco; peach blossom, lavender, purple and pea-green constitute the principal colors, with gold ornamental work." Not only that, the decks were uncarpeted so that the cabin passengers, when they lowered their gaze, could be dazzled by diagonal zebra stripes beneath their feet; alternate deck planks were of dark wood, black walnut, and light spruce.

If the safety of the passengers is considered, there were a dozen metallic lifeboats capable of carrying fifty persons apiece, and the captain's gig. Five rifled cannon, twenty-five rifles, twenty-five carbines, and twenty-five revolvers plus a dozen boarding pikes could be used to repel pirates or control a riot in the steerage spaces.

The *America*, when she was launched at Greenpoint, Long Island, on July 23, 1868, marked the end of New York's two decades of leadership in the construction of ocean steamships.[59] Simultaneously, the use of paddle wheels for propelling deep-water vessels came to a somewhat delayed conclusion.

After her launch, nine months were needed to install the *America*'s engine, finish her colorful cabins, and make her ready for sea. She was placed on the sectional dock at Hoboken to have her bottom coppered, was lowered into the water on April 17, 1869, and started running up and down the North River to try her engine.[60] On May 28 she went to sea with her cargo holds filled with coal, her paddle wheels deeply buried, and a passenger list of four. In command was Captain Seth Doane. The *America* was so big that she could carry enough fuel, over 2,000 tons, to take her all the way to the Orient, provided she steamed at a slow speed and set sail whenever the winds were fair. She arrived at St. Simons Bay, near the Cape of Good Hope, on July 5, sailed two days later, and anchored off Singapore on August 4, sixty-eight days out of New York.[61] The omission of coaling stops made her elapsed time much shorter than that of the *Fung Shuey* or the *Suwonada*.

Now that the holds were empty, the *America* loaded cargo and two hundred Chinese passengers at Singapore and delivered them to Hong Kong on August 21. Finally, on September 18, 1869, she began the transpacific service for which she was designed, sailing for Yokohama and San Francisco. On October 20, 1869, she entered the Golden Gate, having steamed seven-eighths of the way around the globe in less than five months. She brought in eight hundred steerage passengers and 7,000 packages of tea.[62]

For three years the *America* carried small numbers of first-class passengers, large numbers of Chinese, the United States mails, and many valuable cargoes across the Pacific. On her eleventh westbound crossing, she arrived at Yokohama just before noon on August 24, 1872, unloaded her passengers for Japan, and lay at anchor to take on coal. Her fires were drawn, there was no steam on her main boilers, and, with no demand for it, only 12 pounds of pressure in her donkey boiler. At 11:00 P.M. a stewardess smelled smoke, and a fire of incendiary origin was discovered on a lower deck aft that spread with great rapidity. There was no time to raise steam and, although the pumps were started, no water reached the hoses. Two first-cabin and 175 Chinese passengers were aboard. Many of the latter loaded themselves with the gold they were bringing home and drowned when they crowded the forward accommodation ladder and it collapsed under their weight. Others jumped overboard and sank. Around 60 were lost, including three of the crew and three Japanese workmen. Fruitless effort was made to extinguish the flames; a steam fire engine was loaded on a tug and brought out from shore but its hose streams were minute in comparison with the burning hull. An Italian corvette, the *Vettor Pisani*, sent a boat and a howitzer to sink the steamer; her thick sides were impervious to its shot. The blaze lasted a full day before a deluge of rain put it out.[63] The charred hulk was beached at Kanagawa on the 25th. About $400,000 in treasure was stowed in an iron-lined compartment. There were, however, wood stanchions passing through holes in its boundaries. When divers went down they found that the heat had melted the gold and part of it had run out through the holes; nevertheless, about $300,000 was recovered.[64] Thus perished the last great side-wheel liner built in the United States. It is difficult to say whether she or the *Adriatic* was the largest American example of this type ever to cross the Atlantic. Their tonnages were measured differently and cannot be compared. The *Adriatic* was shorter, wider, and deeper. If the length, beam, and depth to the upper deck are multiplied together and used as a measure of size, the *Adriatic* would become the largest. If, however, a figure representing the superstructure volume is added, the *America*'s total gives her first place. The *America*, because of her longer, shallower hull, was a more difficult structure to design and build, but she had the advantage of following a long series of steamships of ever-increasing size that climaxed the 1860-70 decade.

The *Arizona*, 1886-1882?

When the Pacific Mail Steamship Company finally eliminated Cornelius Vanderbilt from the California trade by buying both his steamships and his mail contract, they had already begun building replacements for the vessels, most of which were too small or too old for the most important passenger and mail service under the American flag. In 1864 orders for three large steamships, the *Henry Chauncey*, *Arizona*, and *Montana*, had been placed, two with William H. Webb and one with Henry Steers. Webb's Manhattan yard was so busy that he built only one, the *Henry Chauncey*, and subcontracted the *Montana* to his brother, Eckford Webb, a partner in the firm of Webb and Bell, whose yard was on the Long Island side of the East River. For Steers, the *Arizona* was his first really large steamship; she was twice the tonnage of any commercial vessel he had built previously; moreover, her success would lead to contracts for even larger ones.

These three were almost indistinguishable. The Webb brothers built theirs from the same set of lines; because the launching of the *Montana* was four months after the *Chauncey*'s, the molds from William's yard could be sent to Eckford's. Steers, however, added a few feet to length and increased the beam (if the official dimensions can be trusted) to make his hull 140 tons larger than the *Henry Chauncey*, the first to be completed. The one visible distinction was that the *Chauncey* had an open main deck forward of the foremast, while the *Montana*'s and *Arizona*'s were closed in.

The *Arizona* was 323.8' by 44.8' by 27', with a tonnage of 2,793, a slightly smaller version of the *Golden City* with less overhang to her decks. She was launched as early as January 19, 1865, but not completed until February 27, 1866.[65] On March 1 she joined the *Henry Chauncey* on the New York-Aspinwall route under the command of Jefferson Maury, former mate and later master of the *Adriatic*. This was the first of nine monthly voyages that continued until December of 1866. For the next five months she was tied up at the Novelty Iron Works, probably for engine or boiler work.[66]

From April of 1867 through November of 1870, the *Arizona* continued her Aspinwall sailings. The critical year of 1869 was marked by two epochal openings, of the first transcontinental railroad, in the United States, and of the Suez Canal, in Egypt. The first enabled those passengers whose primary interest was speed rather than comfort to cut their travel time from New York to San Francisco from three weeks to almost one. Slowly the new mode of transportation began to reduce the Pacific Mail's passenger lists. Instead of having too few steamships on the East Coast, as they did in 1865, they now had too many. It has already been reported that two somewhat smaller vessels, the *Costa Rica* and *New York*, were sent to the Orient in 1867. Now the *Arizona*, originally built as a replacement for them, was to be sent there, too.

For the first time the Pacific Mail took advantage of the Suez Canal's far shorter route to the East. The *Arizona* was advertised for a "Grand Oriental Trip" that would be "Interesting to tourists and others Contemplating an Escape from the Rigors of a Northern Winter."[67] Under Captain C. P. Seabury, she would leave on December 22, 1870, for Gibraltar, Malta, Port Said, the Suez Canal, Singapore, and Hong Kong. This was the first use of the canal by an American steamship.[68] Via Gibraltar on January 8, 1871, and Malta the 13th, she passed through that less-than-two-year-old waterway, stopped at Singapore, and finally anchored at Hong Kong on March 5.[69] In seventy-three days she completed a trip that had taken the *America* eighty-five.

At Yokohama the *Arizona* joined the *Ariel*, *Golden Age*, *New York*, and *Costa Rica* in the local service between Japanese ports and Shanghai. After three voyages in the spring and summer of 1871,[70] she was ordered across the Pacific to become a part of the San Francisco-Panama mail service; she was too big for the Shanghai trade. From Yokohama on September 13, she entered the Golden Gate October 6.[71]

Northbound from Panama, on October 27, 1872, she suffered a severe engine casualty. Her piston rod broke and the enormous 8'-9" diameter piston crashed through the lower cylinder head into the condenser.[72] Beam engines had always placed their condensers under the cylinder. The damage was extremely serious, for the Novelty Iron Works was no longer in business. It is not clear who produced the mammoth castings needed to rebuild the engine.

After drifting for five days, the helpless steamer was found by the *Salvador*, which took her to Acapulco. There she awaited the arrival of the *Constitution* to tow her North.[73]

Two years later the *Arizona* became the rescuer when she picked up the Pacific Mail's *Colima*, disabled by the loss of three of her four propeller blades, and towed her to San Francisco.[74] But the continued construction of iron-screw steamers like the *Colima*, smaller than the paddlers now that the railroads had made serious inroads into both the passenger and freight business, soon forced the side-wheelers to be withdrawn. The *Arizona* was laid up in 1876.[75] Some time later she was sold to Japanese

Alaska coaling at Malta. *Courtesy of Peabody Museum of Salem.*

owners and her later career is unknown, beyond a listing in Lloyd's Register up to 1882.

The *Alaska*, 1868-1885?

The *America* of 1869 represented the ultimate form of the transoceanic liner produced by American shipbuilders working in wood. The *Alaska*, completed one year earlier, was the ultimate paddle-wheel coastwise liner. Both came from the Henry Steers yard and were strikingly similar. The latter was 346' by 47.6' by 23.5', with a tonnage of 4,011. She was 17' shorter, 2'-0" narrower, and had the same depth as the *America*. Nothing approaching them would ever be built again. To go farther back to their common ancestor, the *Golden City*, the *Alaska* was almost the same length and depth but 2 1/2' beamier. Engines and paddle wheels were identical on all three. As a coastwise vessel the *Alaska* had a limited, two-masted rig, but followed the Atlantic and the China steamers in avoiding extensive overhanging decks. It is not clear why the Pacific Mail ordered so large a vessel so late; its management either ignored or underestimated the effect of the railroad competition that was about to begin.

The *Alaska* was launched on November 27, 1867,[76] had her machinery installed ready to make a trial trip on New York's Lower Bay on June 8, 1868, and was completed by July 10 at a cost of $964,138.[77] Under Captain A. G. Gray she set out for Aspinwall on August 8. It is worth noting that the passengers from the new *Alaska* would continue on to California aboard the venerable *Golden Age* of 1853. The bitter struggle between the North American Steamship Company and the Pacific Mail was then at its height and the latter had, to meet the competition, increased its sailings from three to four per month. The East Coast steamships were *Ocean Queen, Rising Star, Henry Chauncey*, and now, surpassing them all, the *Alaska*. The last made monthly sailings in 1868 and 1869. But once the North American Steamship Company had been eliminated, the frequency was reduced to three per month and, in October of 1869, to two. The Pacific Mail no longer needed all these vessels on the East Coast and, as passengers defected to the transcontinental railroad, could never hope to fill a mammoth steamship like the *Alaska*.

Her last Isthmus sailing took place on October 20, 1870. She left Aspinwall November 3 and reached New York on the 11th. She then joined the *Arizona* at the works of Cobanks and Weeks, where both were being overhauled for the Pacific.[78]

After the *Arizona* had departed, the *Alaska* was advertised to follow a month later. Once more it was a "Grand Oriental Trip" to Hong Kong via Gibraltar, Malta, Port Said, the Suez Canal, and Singapore.[79] Later Marseilles was added and the departure date put off until January 22, 1871. Under Captain Jefferson Maury, former master of the *Arizona*, the *Alaska* set out, the first American steamship offering to take passengers to Marseilles since the *William Penn* in 1854. She reached Gibraltar February 2 and Marseilles on the 16th. She was not only the second American vessel to transit the Suez Canal, but she was also the largest side-wheel steamship ever to do so. After reaching Hong Kong on April 14, she was added to the line's transpacific fleet and, starting in August, ran between San Francisco, Yokohama, and Hong Kong with the *Colorado*, *Great Republic*, *China*, *Japan*, and *America*. What a magnificent assemblage of steamships these were as they majestically covered thousands of miles at a stately speed of 9.5 knots! To save fuel, their masters were given twenty-two days to cross to Japan and were penalized if they arrived early.[80] As a result, there were no costly races between ships. To go all the way to Hong Kong, including the Yokohama stop, would take about thirty-three days. Best of all, their cargoes of tea, silk, treasure, and their many Chinese passengers in the steerage made them profitable steamships that were not totally dependent on mail contracts to keep them running.

The burning of the *America* in 1872 and the *Japan* two years later, along with an increased frequency of sailings, made the *Alaska* a welcome addition. But simultaneously, the Pacific Mail's rapidly changing management began to realize that it had an outmoded type of craft on its hands. It brought out two even larger iron steamships, the *City of Peking* and *City of Tokio* in 1874 and, within the next five years, replaced the sidewheelers.

The *Alaska*, when not needed on the Orient run, made trips to Panama. Upon occasion the China steamships, after arriving and coaling at San Francisco, headed South for Panama to avoid having to transship a particularly valuable cargo of tea and silk. The *Alaska* did so in June of 1873. She arrived on the 13th, thirty-two days from Hong Kong, and sailed six days later for Panama.[81] During her San Francisco stop she loaded further valuable cargo, 33,000 gallons of California wine.

Twice the *Alaska* suffered typhoon damage. On August 30, 1874, she arrived at Hong Kong with a broken shaft and was sent to Aberdeen, on Hong Kong Island, for repairs. During the night of September 22 a typhoon struck, tore loose her mooring lines, and sent her careening across the harbor, dragging her anchors after her.[82] She came to rest on a rock ledge, heeling heavily to starboard with her port wheel entirely out of water. She was eventually refloated, with the aid of pontoons, on December 22, after months of arduous effort.[83] Henry Steers, a master of ship construction, had produced an exeptional hull; despite its size and weight, it survived grinding on the rocks and was neither holed nor strained.

On her last Pacific crossing the *Alaska* sailed from San Francisco March 6, 1879, and met another typhoon, which forced her into Honolulu on April 1.[84] Seas had ripped up the port guards, stove in the bow, lifted a series of overhanging deck beams, and poured into the engine room. The engine stopped; steering was lost; she rolled wildly broadside to the waves. Eventually she was brought under control and the engine started again; then she headed toward the Hawaiian Isles where she spent ten days repairing before continuing on to Hong Kong.

The year 1879 was the last to see the giant paddlers crossing the Pacific. By that time only the *Alaska* and the *China* were left and they were rarely used. After the *Alaska*'s return, there was just one more trip; the *China* set out from San Francisco on June 2. From that date on, passengers had to make the crossing by a more up-to-date screw steamship. Unfortunately, iron hulls, heated by a tropical sun, take all night to cool off. One is led to wonder whether a first-class cabin at the stern, high in the air, well clear of the hot engine and its boilers, undisturbed by vibration from a thrashing propeller, and with far cooler wood boundaries, was not markedly more comfortable on one of the paddlers? The steerage spaces on them were clearly superior. But economics, alas, takes precedence over comfort.

The *Alaska* lay unused until 1882, when her engine was removed at San Francisco.[85] As a hulk, she was towed down to Acapulco in June of that year, and was still there as a stores ship in 1885.[86] The *China* was retained as a spare until 1883; she was used briefly as a floating hospital and then broken up three years later. The *Colorado* had been scrapped in 1881 and the *Great Republic*, under other owners, wrecked in 1879.

Although the Pacific Mail Steamship Company had been a part-owner of the North Atlantic Steamship Company, which provided a passenger service to Havre in 1860, their later Atlantic crossings, all in the eastbound direction, were incidental ones. It was sending its steamships from New York to the Orient, or to San Francisco, and chose the eastward route for some of them. The first two sailings, by the *Costa Rica* and the *New York* in 1867, have been earlier reported under the Vanderbilt Line heading. They went out via the Cape of Good Hope and were put on the Yokohama-Shanghai run. The much larger *America* followed in 1869 for the San Francisco-Hong Kong route. The other steamships built for Pacific service had gone around South America rather than Africa. In 1870 the *Arizona* headed East via Gibraltar and Suez. She was intended for the Shanghai route but, on second thought, was considered too large and continued on to San Francisco to be added to the company's Panama fleet. She was followed in 1871, again through the Suez Canal, by the *Alaska*, which made both Hong Kong and Panama voyages. In all, there were five eastward Atlantic crossings, three around the Cape of Good Hope and two through the Strait of Gibraltar. On at least three of these sailings, those of the *Costa Rica*, *Arizona*, and *Alaska*, advertisements appeared offering to take passengers. Four traveled in lonely splendor on the *America*'s South Atlantic crossing.

The *Kuroda* and the *Capron*

By 1870 the Japanese government had become interested in steamships for commercial and naval use and had begun to purchase some that were already operating in nearby waters, as well as ordering others from European and American sources. Two of the latter came from the Cornelius and Richard Poillon shipyard at the foot of Bridge Street in Brooklyn. The Poillon brothers were, like the Englis family, renowned for their river and sound steamers. They had completed the *Ocean Bird* after John Griffiths refused to make her into a practical steamship. Although classified as transports or training vessels, the Japanese craft were actually small passenger steamers.

The *Capron* was 136′-6″ in length over all, 125′-0″ on the waterline, with a 23′ beam, an 11′ depth of hold, and measured 374 tons. She was driven by a propeller, a single-cylinder engine of 30″ bore and 2′-6″ stroke furnished by the Delamater Iron Works, and had one return-tube boiler operating at 45 psi. Charles W. Copeland designed the engine and superintended the machinery installation. She was given a topsail schooner rig and had a large deckhouse containing eight staterooms. New developments, such as mechanical refrigeration that could make ice and an illuminating gas plant to light the cabin, were installed.[87] A journalist described her as suitable for a steam yacht.

The *Kuroda* was somewhat larger, 644 tons, with dimensions of 188′ by 28′ by 17′ depth of hold, and had two decks. Her load draft was 11′ forward and 13′ aft.[88] She was rigged as a half-brig and had wire rope shrouds and stays. Of particular interest was the innovation of a compound engine with a high-pressure cylinder of 24″ diameter and a low pressure one of 42″, both of 2″-10″ stroke.[89] It was designed by Copeland, built by Delamater, and, on trials, turned at 66 rpm when steam was supplied at 52 psi. It should be particularly emphasized that this, one of America's earliest marine compounds, was designed by an engineer whose career spanned the entire history of marine engineering for ocean steamships. Copeland had the enviable record of designing boilers for the U.S.S. *Fulton* in 1837; this country's first side-lever engines for the U.S.S. *Mississippi* in 1839; far larger engines of the same type for the Atlantic liners *Pacific* and *Baltic* in 1850; and the trunk engines for the *Pioneer* and *City of Pittsburg* in 1851.[90] After the Civil War came large direct-acting engines for the *Erie* and *Ontario*. He had been chief engineer of the West Point Foundry and the Allaire Iron Works, consultant to the U.S. Navy and to many steamship owners, Steamboat Inspector at New York, and Superintending Engineer for the U.S. Lighthouse Board.

The *Kuroda* made eleven knots on a trial conducted in New York harbor, on November 7, 1872, under the direct supervision of Messrs. R. Poillon, C. W. Copeland, and C. H. Delamater, three of America's outstanding members of the maritime profession.

The introduction of a compound engine would cut fuel consumption by about fifty percent and would, in turn, provide higher speeds, lower bunker capacities, and longer voyages for future steamships. A major technical problem with a single-expansion engine, the type used on every steamship considered to date, was that its cylinder was cooled once its exhaust valve opened to the condenser. Then, when a new supply of hot steam entered one-half revolutions later, part of it was wasted as it condensed on the cylinder walls.

In a compound, the steam was partially expanded in a small-diameter, high-pressure cylinder. It was then exhausted to a larger, low-pressure one and used again; finally, it escaped to the condenser. This permitted the use of higher steam pressure and at the same time kept the high-pressure cylinder at an elevated temperature, thereby avoiding the loss from premature condensation. The higher the steam pressure and the more the steam is expanded, the greater the efficiency of an engine and the lower its fuel consumption.

At increased pressures, however, any salt water in the boilers would produce severe scaling and damage. Not until a completely reliable condenser was available was it possible to go to high pressures and expand the steam successively in two or more cylinders. The compound engine was an old invention; so was the surface condenser. The latter had been used clear back to the *Massachusetts* in 1845. Leakage, however, was altogether too frequent until satisfactory metals for condenser tubes became available, together with a means of allowing them to expand and contract with changing temperatures. Real solutions were finally achieved, in the United States, around 1870. The general use of the compound engine and an increase in boiler pressure followed.

The *Kuroda*'s engine is, without question, the earliest of that type in any American steamship to cross the Atlantic. Westray and Gibbs, Shipping Agents, advertised that she would take passengers for Yokohama via the Suez Canal, leaving New York on December 21, 1872, under Captain Askins.[91] She reached Gibraltar January 6, Port Said the 27th, and Singapore on March 24, and arrived at Yokohama April 27, 1873.[92]

Similar advertisements appeared for the *Capron*, under Captain David Baker, in March of 1873.[93] This tiny steamer, little larger than the *Savannah* of 1819, must have sailed most of the way. She arrived at Bermuda on April 17, thirteen days out of New York, and did not reach Gibraltar until May 17. After a thirteen-day stay, she took another nineteen days to the entrance of the Canal. Hong Kong was made by August 9.[94]

It would be interesting to know whether any passengers were adventurous enough to sail on this 125' steamship. In neither sailing was a Japanese crew involved; a safe delivery was part of the contract.

Both steamers had long lives under the Japanese flag, mostly in commercial rather than governmental service. The *Kuroda* was renamed *Gembu Maru*, while the *Capron* became the *Karafuto Maru* and, later, the *Kyoryu Maru*.[95] In 1883 they were turned over by the Japanese Government to the Kiodo Unyu Kaisha, which three years later was merged with a Mitsubishi service to form Nippon Yusen Kaisha, the great Japanese steamship line that is still, in 1980, in operation.[96] Both were sold in 1897 but steamed on under other owners. The *Gembu Maru*, ex-*Kuroda*, was wrecked on February 17, 1907, near Higashi Tori. The *Capron*, again rechristened the *Kepuron Maru*, sank at Yesashi, Oshima, Hokkaido Island, on December 14, 1912.[97] These may have been very small vessels to take transatlantic passengers, but they were remarkable successful ones. The *Kuroda* is particularly interesting; her advanced power plant seems to have been completely overlooked by American historians. The *Pennsylvania*, built and engined by William H. Cramp and Sons in Philadelpha, has always been quoted as the first American steamship with a compound engine to cross the Atlantic. She, as an iron-screw steamship, inaugurated a new era when she left Philadelphia on May 22, 1873, bound for Liverpool. The *Kuroda*, however, anticipated her by six months. Her sixteen-day crossing to the Pillars of Hercules was extremely good for so small a vessel. The *Alaska*, many times her size, did no better, and the *Arizona* took a day longer.

Notes

1. *Journal of the Franklin Institute* 45, 3d ser., (January-June 1863), pp. 41-42.
2. Official Records, ser. 2, 1: 200.
3. Ibid., ser. 1, 10: 212.
4. Ibid., 17: 230-31.
5. Theodore P. Greene, letter book, June-December 1863. New York Historical Society.
6. Official Records, ser. 1, 10: 176, 243, 287, and 398.
7. Ibid., 11: 245-47, 258, and 260.
8. Ibid., 11: 426, 430, 570, and 723.
9. B. S. Osbon, *Handbook of the United States Navy* (New York, 1864), p. 218.
10. Official Records, ser. 1, 16: 372-73.
11. *New York Herald*, September 22, 1865.
12. Ibid., October 8, 1866.
13. Ibid., May 23 and 30, June 6 and 10, 1867.
14. Ibid., February 3 and 4, 1870.
15. Ibid., August 7, 1870.
16. Ibid., September 2, 1870.
17. *New York Journal of Commerce*, March 29, 1877.
18. Ibid., March 14 and July 6, 1878.
19. *New York Shipping and Commercial List*, September 15, 1877.
20. *New York Journal of Commerce*, October 25, 1877.
21. *New York Maritime Register*, April 20 and May 4, 1881.
22. *New York Shipping and Commercial List*, 1882-1886.
23. *New York Maritime Register*, June 23 and July 7, 1886; American Bureau of Shipping Record, 1899.
24. John H. Morrison, *History of American Steam Navigation* (New York, 1903), p. 373.
25. F. B. C. Bradlee notes. Peabody Museum.
26. *New York Tribune*, February 4, 1864.
27. *New York Herald*, January 5, 1866.
28. *New York Tribune*, January 4, 1866.
29. *New York Journal of Commerce*, July 7 & 8, 1866.
30. Ibid., February 14, March 7, and March 25, 1867.
31. Ibid., September 10, 1867.
32. *New York Herald*, November 7, 1867.
33. *New York Journal of Commerce*, July 3, 1868.
34. *New York Herald*, July 8, 1870.
35. *New York Journal of Commerce*, October 12, 17, and November 6, 1871.
36. *New York Herald*, January 27 and February 2, 1872.
37. *New York Maritime Register*, February 27, 1878.
38. Edward Kenneth Haviland, "American Steam Navigation in China," *The American Neptune*,(1958), pp. 68-73.
39. *New York Times*, June 6, 1865.
40. *New York Herald*, May 24, 1864.
41. *New York Tribune*, May 24, 1865.
42. Haviland, "American Steam Navigation," *The American Neptune*, (1957) pp. 49-51.
43. *New York Herald*, March 25 and April 3, 1872.
44. *New York Times*, June 20, 1865.
45. *New York Journal of Commerce*, July 30, 1864.
46. *New York Herald*, May 7 and 17, 1867.
47. Haviland, "American Steam Navigation," pp. 55-56, 135-36.
48. John Haskell Kemble, *The Panama Route, 1848-1869* (Berkeley and Los Angeles, Calif., 1943), p. 220.
49. John Haskell Kemble, "Side-Wheelers across the Pacific," *The American Neptune* (1942), p. 11.
50. Kemble, "Side-Wheelers," folding plate, pp. 242-43.
51. William H. Webb, Certificate Book, vol. 2. Webb Institute of Naval Architecture.
52. *Journal of the Franklin Institute* 50, 3d ser. (July-December, 1865): 53.
53. *New York Herald*, March 3, 1870.
54. *The Engineer*, July 15, 1864, p. 64.
55. *New York Herald*, December 6, 1866 and December 20, 1874.
56. F. B. C. Bradlee notes, Peabody Museum.
57. *San Francisco Alta*, October 21, 1869.
58. Ibid.
59. *New York Herald*, July 24, 1868.

60. Ibid., April 18 and May 29, 1869.
61. Kemble, "Side-Wheelers," p. 8.
62. *San Francisco Alta*, October 21, 1869.
63. *New York Herald*, September 5, 6, 14; October 3 and 11, 1872.
64. *New York Herald*, November 2, 1872.
65. Kemble, *Panama Route*, p. 216.
66. *New York Shipping and Commercial List*, November 1866 through April 1867.
67. *New York Journal of Commerce*, December 12, 1870.
68. *New York Maritime Register*, March 26, 1884.
69. *New York Maritime Reporter*, March 26, 1884.
70. Haviland, "American Steam Navigation," p. 311.
71. *New York Maritime Register*, January 24, 1872.
72. *New York Herald*, November 30, 1872.
73. *New York Maritime Register*, January 21 through May 3, 1871.
74. *New York Herald*, March 29 and 30, 1874.
75. Kemble, "Side-Wheelers," p. 20.
76. Kemble, *Panama Route*, p. 214.
77. Kemble, "Side-Wheelers," p. 14.
78. *New York Herald*, December 5, 1870.
79. Ibid., December 24, 1870.
80. August Mencken, *First-Class Passenger* (New York, 1938), pp. 59-60.
81. *San Francisco Alta*, June 14 and 20, 1873.
82. *New York Herald*, October 2, November 6, and December 2, 1874.
83. *New York Maritime Register*, October 7, November 1, December 2 and 30, 1874.
84. *New York Times*, April 23, 1879.
85. *New York Maritime Register*, February 1, April 26, and June 31, 1882.
86. Kemble, "Side-Wheelers," p. 38.
87. *New York Times*, March 19, 1873; *New York Herald*, November 21, 1872.
88. *New York Times*, November 8, 1872.
89. *Nautical Gazette*, March 9, April 13, and June 22, 1872.
90. *Seaboard Magazine*, February 14, 1895.
91. *New York Herald*, December 16, 1872.
92. *New York Maritime Register*, February 26 through April 30, 1873.
93. *New York Herald* advertisements, March 1, 10, and 15, 1873.
94. *New York Maritime Register*, April 30 through November 12, 1873.
95. *70th Year History of Nippon Yusen Kaisha* (in Japanese), (Tokyo, 1956).
96. *Golden Jubilee History of Nippon Yusen Kaisha* (Tokyo, 1935), pp. 8-11.
97. Havilland, "*American Steam Navigation*," p. 306.

25

The End of the Side-Wheel Era

THE histories of all the American-flag steamships that crossed the North or South Atlantic have been presented, with the exception of the shallow-draft vessels built for Chinese river and coastwise trades. These made one-way ocean passages en route to the Orient. A few early auxiliary steamships sent to China, two representative examples of oceangoing China coasters, and a pair of screw steamers built for Japanese waters have been included.

Starting with the *Savannah* in 1819, and ending with the *Capron* in 1873, sixty-six steamships have been described and their biographies narrated. These made a total of 683 eastward Atlantic sailings, mainly from New York. Baltimore saw thirty departures; Boston and Philadelphia accounted for ten. Almost half of the craft involved, thirty of them, made no more than one trip. The remainder, thirty-six in number, were the real Atlantic liners that accounted for 653 sailings between 1845 and 1870. This group comprised twenty-nine side-wheel steamships, all with wooden hulls, and seven screw steamships, two with iron hulls. Twenty-eight of the side-wheelers came from New York. Originating in Philadelphia, the *Quaker City* was the single outsider.

Three of the screw steamships, the *Massachusetts*, *Mississippi*, and *Ontario*, came from New England. The *Circassian* was built in Ireland and engined in Scotland. The three Baltimore and Ohio Railroad propellers were natives of New York. Thus, thirty-one of the thirty-six steamships that completed two or more voyages were products of New York shipyards and were engined by New York iron works at a time when the two combined to create the finest wooden steam vessels, of any size or type, ever seen on this side of the Atlantic.

If we count completed round trips, the *Atlantic*, with sixty-six, is the empress of them all. The *Baltic* and the *Arago* are tied for second place with fifty-seven. The *Washington* made fifty-four, the *Fulton* fifty, and the *Hermann* forty-seven. All were ordered by pioneering lines holding United States mail contracts that required year-round sailings. Moreover, with the exception of the *Washington*, all continued running after the Civil War. The *Hermann* was on the Pacific by that time, but the rest made further Atlantic voyages as immigrant carriers.

If we consider record speeds, rather than numbers, the *Baltic*'s name comes up again. She was the fastest Collins liner and, in 1851, with a Liverpool crossing of 9 days 18 hours 45 minutes, became the fastest steamship on the Atlantic. Three years later she reduced her time to 9 days 16 hours 53 minutes at an average speed of 13.23 knots.

Two magnificent steamships made maiden voyages in 1857, the *Vanderbilt* and the *Adriatic*. The former was a superbly built steamship with a pair of walking beam engines. The second was larger and far more expensive, had an elaborate interior decor, and was driven by two oscillating engines that had to be modified twice before they functioned properly.

The *Vanderbilt* made a faster crossing, 9 days 8 hours, Sandy Hook to the Needles, at 13.75 knots. She did not, however, hold her speed well against wind and waves. In 1860, when both were on the same route, the average eastward speed was 13.27 knots for the *Adriatic*, to the *Vanderbilt*'s 12.55. Both were splendid steamships but extremely expensive to run.

The shipyards, engine builders, and owners of all these vessels were located on lower Manhattan Island. When New York's East River waterfront became too crowded, some builders located their yards on the Long Island side of the channel. The owners had offices within easy walking distance of the piers, the shipyards, and the dry docks needed to repair their vessels. Much of America's maritime business was concentrated in this small area.

Two events made possible the rapid development of steamships and led to their construction in great numbers. First came Congressional action, in the late 1840s, authorizing mail contracts from New York to Charleston, Savannah, Havana, New Orleans, Aspinwall, Panama, and the then unsettled West Coast. Simultaneously, foreign routes to England, Germany, and France were created. Starting in 1847 and continuing to 1856, large, fast, seaworthy steamships were built to serve these routes.

Just as the first mail liners were being completed and setting out to sea, gold was discovered in California. The sudden demand for any means of reaching that distant shore led to an enormous expansion of every aspect of the maritime industry. New York, which had built all the Atlantic mail liners, rose to the occasion; its shipyards turned out hundreds of sailing ships and scores of steamships to take adventurers out and bring back the gold they found.

The seven years from 1850 to 1857, then, were years of excitement, expansion, progress, and profit to shipbuilders, ship owners, and financiers. Steamships were built to replace sailing packets in the coastwise trade. Bigger, faster, more palatial vessels were continuously demanded. The new Bremen liners doubled and trebled freight shipments to and from Germany. The Havre steamships brought luxury items from Paris to fill the best stores of New York. Various coastwise steamships took part of the imported silks, laces, and jewelry on to Charleston and New Orleans. The Collins liners made new speed records across the Atlantic and created a new standard of luxury to beguile the most demanding traveler.

All the large, fast steamships had wooden hulls and paddle wheels, and in most cases were propelled by enormous side lever engines using steam at low pressures and burning prodigious amounts of coal. The designs of the first coastwise steamships influenced the Atlantic liners, which in turn contributed to the development of later coasters. Neither was a unique, totally independent type: coasters were used on the ocean and Atlantic liners returned to take up coastwise routes.

The side-wheel steamship was the only realistic choice for a high-powered vessel, as long as its hull was wood. The screw propeller had been tried, had proved a highly successful means of driving a small steamer, but had too many disadvantages when applied to a large one. It needed a rigid iron hull before it could be successful in a major steamship. Philadelphia's shipyards and engine works concentrated on the latter means of propulsion but, wisely, kept the hull size down.

The Panic of 1857, later renamed a "depression," brought a halt to the rapid expansion of shipping that had been financed by California gold. Simultaneously, Congress reduced subsidy payments to the Collins Line and did not renew the Havre and Bremen Lines' mail contracts that expired that year. The next four years were therefore a period of retrenchment, with only the Havre Line providing year-round Atlantic sailings.

The year 1861 and the outbreak of the Civil War created a new demand for steamships. The Army's need for transports and the Navy's for gunboats, blockaders, and supply ships, not to mention warships, once again put enormous pressure upon every ship and engine builder in the United States. Quantity was the word, and simplicity in construction an asset. The walking beam engine (the simplest available) became the standard for any nonmilitary side-wheel steamer; a dozen engine builders in New York and lesser numbers in Philadelphia and Baltimore could supply them.

But when the war ended in 1865, there were all those transports, cruisers, and blockaders left over. There were besides additional captured blockade runners that were admitted to American registry. All could be bought cheaply. There were no new orders to employ the highly skilled shipwrights. And with no new steamships, the iron works had nothing to engine. Furthermore, completely new ways of designing and building ships had evolved in England and Scotland.

Faced by a shortage of timber, the British shipyards had developed the iron hull at the same time the American ones were extending and perfecting the art of building in wood. Iron was cheaper abroad; seamen's wages, uninflated by a war, were lower. A stronger, lighter ship, boasting many technical improvements, had been achieved. Screw propulsion had already arrived when the Inman Line, starting in 1850, sent a British-built iron steamship from Liverpool to Philadelphia. That successful organization never owned a side-wheeler!

Yet the New York shipbuilders still turned out enormous beam-engined steamships with wooden hulls for transpacific service in 1867, 1868, and 1869. These provided work for a few yards and one old-line engine works.

The 1865-1870 period, then, found the finest shipyards and iron works in New York with little to do beyond repair work. An extensive strike by the city's ship carpenters, caulkers, and joiners in 1866 was badly timed. They did not get the eight-hour day they had demanded.[1] Instead, repair work and sailing-ship construction continued to leave the city and go to other parts of the country where timber was readily available and wages lower. New England would continue to construct wooden sailing ships for decades.

The famous builders of steamships sold their shipyards and retired. William H. Webb turned his attention to steamship operation and, after severe financial losses, to nonmaritime investments. Jacob Westervelt became Superintendent of Docks for New York. Henry Steers left the city to assist John Roach in Chester, Pennsylvania. The Poillon brothers were able to get orders for small craft: yachts, pilot boats, ferries, and scows.

Webb and Bell continued on with one large contract, not for a ship, but for the enormous wood caissons that were needed to construct the foundations for the Brooklyn Bridge towers. The Englis yard turned out a number of moderate-sized coastwise propellers for the Alexandre Line in the early 1870s and river, bay, and sound steamers in later years, but finally became a specialist in building wood superstructures and fitting out the cabins of iron steamships built by others.[2]

Most of the engine works closed their gates. By October of 1869 the venerable Allaire establishment, after fifty-three years of engine manufacture at 466 Cherry Street, had become a horse stable.[3] The Neptune Iron Works was a saw mill and the Fulton Iron Works was selling off its tools. Although a number of small gunboats for Cuba were being outfitted at the Delamater Iron Works, not a single marine engine was being built in New York. Six months later, in April of 1870, the only hull on the stocks was a Brooklyn ferryboat at the Englis yard.[4]

Shipbuilding, however, was undergoing a drastic change. The wood era ended in New York and the iron one was about to succeed it. The shipyards needed had developed elsewhere, on the banks of the Delaware River. They had been nurtured, during the Civil War, by orders for warships with iron hulls and armor.

Harlan and Hollingsworth, a pioneer in iron starting as long ago as 1844, had the tools and the men to build and engine modern craft. Pusey and Jones, also in Wilmington, Delaware, while not so large, was a second iron specialist. Farther up the river, William Cramp, who had built a few iron hulls in Philadelphia, obtained an order for four Atlantic liners. The Pennsylvania Railroad was providing funds for their construction. Cramp and his sons acquired land and set up a new shipyard solely for iron construction. The *Pennsylvania*, a screw steamship with a compound engine, was launched there on August 16, 1872.[5]

John Roach, who had owned the Etna Works in New York, took over the larger Morgan Iron Works. In 1871 he purchased the bankrupt Reaney, Son, and Archbold shipyard in Chester, Pennsylvania. Four years later he had set up his own blast furnace and rolling mill, also at Chester, to supply iron plates and shapes to his shipyard.[6] The Morgan Works was used to build and install engines and boilers for his ships. A coastwise steamer, the *City of San Antonio*, was launched at the Roach yard, renamed the Delaware River Iron Shipbuilding and Engine Works, on April 10, 1872.[7]

The Delaware River was the ideal place for these enterprises. It was close to Pennsylvania's coal, iron, and rolling mills. The river and its connecting canal systems made delivery by water a simple, cheap, and rapid matter. Any other location in the United States would have had far higher transportation costs on the thousands of tons of material needed to build and engine a steamship.

To summarize, the wooden paddle steamship, its builders, and, with a few exceptions, its engine builders, vanished from New York in the years from 1865 to 1870. A new and different type of steamship was evolving and works were created to build it in the years from 1870 to 1875; new yards were established and old ones enlarged along the banks of the Delaware River. Engines became smaller, lighter, and more economical. Philadelphia, Camden, Chester, and Wilmington shipyards were now the builders of these new iron steamships.

The 1869 completion of that marvelous example of the wood shipbuilder's art, the side-wheeler *America*, ended the wood era. No oceangoing paddlers with wooden hulls ever followed her. New York, the center of shipbuilding in the United States, lost its eminence in that field. A widely different replacement, the iron screw steamship, was created on the lower Delaware River, a region that would dominate shipbuilding and engine building in the United States for the next seventy-five years. The past was dead and buried. Its handsome paddlers, rolling their thrashing wheels out of water in a storm, would be seen no more. It was time for them to give way to progress.

In 1870 the *Guiding Star* was the last wood side-wheeler to make an Atlantic round trip. In January of 1871 the *Alaska* was the last to set out on a one-way crossing.

Notes

1. John H. Morrison, *History of New York Ship Yards* (New York, 1909), p. 159.
2. *Master, Mate, and Pilot*, June 1911, pp. 56-61.
3. *New York Times*, October 23, 1869.
4. *New York Globe*, April 25, 1870.
5. David B. Tyler, *The American Clyde* (Newark, Del., 1958), p. 33.
6. Leonard Alexander Swann, Jr., *John Roach* (Annapolis, Md., 1965), pp. 53-59.
7. Souvenir of Roach's Shipyard, Philadelphia, 1895, pp. 7, 18.

Appendix A
Fulton-Livingston Steamboats

	LENGTH	BEAM	DEPTH HOLD	TONS	BUILT AT	BUILT BY	DATE
North River	142'	14'	4'	78	N.Y.C.	Charles Brownne	1807
North River	149'	17'-11"	7'	182	N.Y.C.		1808
Rariton	124'	21'	6'-8"	163	N.Y.C.	Charles Brownne	1809
Car of Neptune	169'	25'-2"	7'-3"	295	N.Y.C.	Charles Brownne	1809
Paragon	167'	26'-10"	7'-9"	331	N.Y.C.	Charles Brownne	1811
Fire-Fly	81'-6"	14'	4'-5"	47	N.Y.C.	Charles Brownne	1811
Fulton	134'	30'-9"	8'-9"	327	N.Y.C.	Adam & Noah Brown	1813
Richmond	154'-6"	28'-6"	9'	370	N.Y.C.	Charles Brownne	1814
Washington	130'	20'-6"	7'-4"	186	N.Y.C.	Charles Brownne	1815
Olive Branch	122'	30'	8'-1"	265	N.Y.C.	Noah Brown	1816
Connecticut	135'	31'	10'-7"	350	N.Y.C.	Adam & Noah Brown	1816
Chancellor Livingston	157'	33'-6"	10'-3"	495	N.Y.C.	Henry Eckford	1817
James Kent	135'	31'-6"	9'	346	N.Y.C.	Blossom, Smith & Dimon	1823

Appendix B
Fulton-Livingston Engines

	TYPE[1]	BORE	STROKE	DIA. WHEELS	ENGINE BUILDER	DATE
North River	Crosshead-Bell Crank	24"	4'-0"	15'-0"	Boulton & Watt & Fulton	1808
Car of Neptune	Crosshead-Bell Crank	33"	4'-4"	14'-0"	Fulton Iron Works	1809
Paragon	Crosshead	32"	4'-0"	16'-0"	Fulton Iron Works	1811
Fire-Fly	Crosshead	20"	3'-9"	12'-6"	Fulton Iron Works	1812
Fulton	Crosshead	36"	4'-0"	15'-6"	Fulton Iron Works	1813
Richmond	Crosshead	33"	4'-4"	15'-0"	Fulton Iron Works	1814
Washington	Crosshead	28"	4'-0"	14'-9"	Fulton Iron Works	1815
Olive Branch	Crosshead	36"	4'-0"	16'-6"	Fulton Iron Works	1816
Connecticut	Crosshead	40"	4'-6"	17'-0"	Fulton Iron Works	1816
Chancellor Livingston	Crosshead	40"	5'-0"	18'-0"	James P. Allaire	1817
Chancellor Livingston[2]	Crosshead	56"	6'-0"		James P. Allaire	1827
James Kent	Crosshead	48"			West Point Foundry	1823

1 All are single cylinder
2 Installed in *Portland* 1835

Appendix C
Early Steamboats

	LENGTH	BEAM	DEPTH HOLD	TONS	BUILT AT	BUILT BY	DATE
Phoenix	100'	16'	6'-4"		Hoboken, N.J.	Sayre, Morgan, Gailer, et al.	1808
Hope	149'	20'-8"	7'-7"	225	Albany, N.Y.	James Van Ingen	1811
Perseverance	147'	20'-6"	7'-6"	218	Albany, N.Y.	James Van Ingen	1811
Philadelphia	139'	21'	7'-8"	214	Kensington, Pa.	Nicholas Van Dusen	1812
Sea Horse	84'-6"	15'	5'-11"	71	Elizabeth, N.J.		1812
Stoudinger[1]	47'-7"	12'-5"	4'-10"	25	N.Y.C.	Tunis Bergh	1817
Patent	80'-2"	21'	6'-7"	98	Medford, Mass.	Thatcher Magoun	1821
Legislator	111'-8"	22'-10"	7'-3"	170	N.Y.C.	Noah Brown	1824
Washington	137'-6"	20'	9'	339	N.Y.C.	Brown & Bell	1825
MacDonough	129'-6"	26'	8'-9"	272	N.Y.C.	Brown & Bell	1826
New Philadelphia	133'	23'-6"	7'-6"	220	Kensington, Pa.	Sam'l. Grice	1826
Benjamin Franklin	142'-6"	31'-6"	6'	410	N.Y.C.	Brown & Bell	1828
President	158'	32'-4"	11'	518	N.Y.C.	Brown & Bell	1829
Portland	163'-6"	27'-2"	10'-7"	445	Portland, Me.	Nathan Dyer	1835

1 Renamed *Mouse of the Mountain*

Appendix D
Coastwise Steamboats

	LENGTH	BEAM	DEPTH HOLD	TONS	BUILT AT	BUILT BY	DATE
Enterprize[1]	80'-9"	21'	7'-6"	105	Hartford		1819
David Brown	130'-11"	18'-1"	8'-4"	190	N.Y.C.	Brown & Bell	1832
William Gibbons	140'	22'-6"	9'-10"	294	N.Y.C.	Brown & Bell	1834
Bangor[2]	156'	25'-10"	10'-1"	385	N.Y.C.	Brown & Bell	1834
Dolphin	115'-6"	16'	7'-6"	133	N.Y.C.	Bishop & Simonson	1835
South Carolina	172'	23'	12'-2"	466	Balt.		1835
Columbia	164'-6"	22'-6"	11'-10"	423	N.Y.C.	Brown & Bell	1835
Georgia	194'	24'	12'-2"	551	Balt.		1836
Charleston	201.4'	24.1'	12.0'	569	Phila.		1836
Home	210'-9"	22'-6"	11'-6"	537	N.Y.C.	Brown & Bell	1837
New York	160'-6"	22'-6"	10'-6"	365	N.Y.C.	Brown & Bell	1837
Pulaski	206'-1"	25'-2"	13'-7"	687	Balt.	Robb	1837
Neptune	215'	25'-4"	14'	745	N.Y.C.	Lawrence & Sneeden	1838
Natchez	206'-9"	18'-2"	14'-1"	792	Balt.	Rogers, Brown & Cully	1838

1 Renamed *Sea Gull* 1823
2 Renamed *Sudaver* 1842

Appendix E
Steamboat Engines

	TYPE	BORE	STROKE	APPROX. IHP	ENGINE BUILDER	DATE
Hope		30"	4'-0"	32	Robert McQueen	1811
Benjamin Franklin	2-Independent Walking Beam	41"	7'-0"		Allaire	1828
President	2-Independent Walking Beam	48"	7'-0"	320	Allaire	1829
Bangor	1-Crosshead	36"	9'-0"			1834
Columbia	1-Crosshead	56"	6'-0"		Allaire	1835
Home	1-Crosshead	56"	9'-0"		Allaire	1837
Pulaski	1-Crosshead	48"	12'-0"	200	Watchman & Bratt	1837
Neptune	1-Crosshead	50"	11'-6"	250	Allaire	1838
Natchez	1-Beam	56"	10'-0"		Chas. Reeder & Sons	1838

Appendix F
Auxiliary Steamships

	LENGTH	BEAM	DEPTH HOLD	TONS	BUILT AT	BUILT BY	DATE
Savannah[1]	98'-6"	25'-10"	14'-2"	319	N.Y.C.	Sam'l. Fickett & Wm. Crockett	1819
Fidelity[2]	88'	19'-4"	9'	139	N.Y.C.	Brown & Bell	1820
New York	105'	27'-6"	11'	281	Norfolk	Wm. Hunter	1822
Clarion[3,4]	91.9'	23.5'	11.75'	226	Phila.	Jos. P. Vogels	1838
Midas[4]	100'	21'-7"	9'-5"	186	Boston	Sam'l. Hall	1844
Edith	121'	26'-3"	14'	407	Boston	Sam'l. Hall	1844
Marmora[4]	143'-3"	24'-10 1/2"	11'-4"	380	Bath	Johnson Rideout	1844
Massachusetts	161'-3"	31'-8"	20'-0"	750	Boston	Sam'l. Hall	1845

1 Converted to sailing ship 1820
2 Converted to steam schooner 1821
3 Converted to steam bark 1841
4 Twin Screw

Appendix C
Early Steamboats

	LENGTH	BEAM	DEPTH HOLD	TONS	BUILT AT	BUILT BY	DATE
Phoenix	100'	16'	6'-4"		Hoboken, N.J.	Sayre, Morgan, Gailer, et al.	1808
Hope	149'	20'-8"	7'-7"	225	Albany, N.Y.	James Van Ingen	1811
Perserverance	147'	20'-6"	7'-6"	218	Albany, N.Y.	James Van Ingen	1811
Philadelphia	139'	21'	7'-8"	214	Kensington, Pa.	Nicholas Van Dusen	1812
Sea Horse	84'-6"	15'	5'-11"	71	Elizabeth, N.J.		1812
Stoudinger[1]	47'-7"	12'-5"	4'-10"	25	N.Y.C.	Tunis Bergh	1817
Patent	80'-2"	21'	6'-7"	98	Medford, Mass.	Thatcher Magoun	1821
Legislator	111'-8"	22'-10"	7'-3"	170	N.Y.C.	Noah Brown	1824
Washington	137'-6"	20'	9'	339	N.Y.C.	Brown & Bell	1825
MacDonough	129'-6"	26'	8'-9"	272	N.Y.C.	Brown & Bell	1826
New Philadelphia	133'	23'-6"	7'-6"	220	Kensington, Pa.	Sam'l. Grice	1826
Benjamin Franklin	142'-6"	31'-6"	6'	410	N.Y.C.	Brown & Bell	1828
President	158'	32'-4"	11'	518	N.Y.C.	Brown & Bell	1829
Portland	163'-6"	27'-2"	10'-7"	445	Portland, Me.	Nathan Dyer	1835

1 Renamed *Mouse of the Mountain*

Appendix D
Coastwise Steamboats

	LENGTH	BEAM	DEPTH HOLD	TONS	BUILT AT	BUILT BY	DATE
Enterprize[1]	80'-9"	21'	7'-6"	105	Hartford		1819
David Brown	130'-11"	18'-1"	8'-4"	190	N.Y.C.	Brown & Bell	1832
William Gibbons	140'	22'-6"	9'-10"	294	N.Y.C.	Brown & Bell	1834
Bangor[2]	156'	25'-10"	10'-1"	385	N.Y.C.	Brown & Bell	1834
Dolphin	115'-6"	16'	7'-6"	133	N.Y.C.	Bishop & Simonson	1835
South Carolina	172'	23'	12'-2"	466	Balt.		1835
Columbia	164'-6"	22'-6"	11'-10"	423	N.Y.C.	Brown & Bell	1835
Georgia	194'	24'	12'-2"	551	Balt.		1836
Charleston	201.4'	24.1'	12.0'	569	Phila.		1836
Home	210'-9"	22'-6"	11'-6"	537	N.Y.C.	Brown & Bell	1837
New York	160'-6"	22'-6"	10'-6"	365	N.Y.C.	Brown & Bell	1837
Pulaski	206'-1"	25'-2"	13'-7"	687	Balt.	Robb	1837
Neptune	215'	25'-4"	14'	745	N.Y.C.	Lawrence & Sneeden	1838
Natchez	206'-9"	18'-2"	14'-1"	792	Balt.	Rogers, Brown & Cully	1838

1 Renamed *Sea Gull* 1823
2 Renamed *Sudaver* 1842

Appendix E
Steamboat Engines

	TYPE	BORE	STROKE	APPROX. IHP	ENGINE BUILDER	DATE
Hope		30"	4'-0"	32	Robert McQueen	1811
Benjamin Franklin	2-Independent Walking Beam	41"	7'-0"		Allaire	1828
President	2-Independent Walking Beam	48"	7'-0"	320	Allaire	1829
Bangor	1-Crosshead	36"	9'-0"			1834
Columbia	1-Crosshead	56"	6'-0"		Allaire	1835
Home	1-Crosshead	56"	9'-0"		Allaire	1837
Pulaski	1-Crosshead	48"	12'-0"	200	Watchman & Bratt	1837
Neptune	1-Crosshead	50"	11'-6"	250	Allaire	1838
Natchez	1-Beam	56"	10'-0"		Chas. Reeder & Sons	1838

Appendix F
Auxiliary Steamships

	LENGTH	BEAM	DEPTH HOLD	TONS	BUILT AT	BUILT BY	DATE
Savannah[1]	98'-6"	25'-10"	14'-2"	319	N.Y.C.	Sam'l. Fickett & Wm. Crockett	1819
Fidelity[2]	88'	19'-4"	9'	139	N.Y.C.	Brown & Bell	1820
New York	105'	27'-6"	11'	281	Norfolk	Wm. Hunter	1822
Clarion[3],[4]	91.9'	23.5'	11.75'	226	Phila.	Jos. P. Vogels	1838
Midas[4]	100'	21'-7"	9'-5"	186	Boston	Sam'l. Hall	1844
Edith	121'	26'-3"	14'	407	Boston	Sam'l. Hall	1844
Marmora[4]	143'-3"	24'-10 1/2"	11'-4"	380	Bath	Johnson Rideout	1844
Massachusetts	161'-3"	31'-8"	20'-0"	750	Boston	Sam'l. Hall	1845

1 Converted to sailing ship 1820
2 Converted to steam schooner 1821
3 Converted to steam bark 1841
4 Twin Screw

Appendix G
Coastwise Steamships, 1820-1850

	LENGTH	BEAM	DEPTH HOLD	TONS	BUILT AT	BUILT BY	DATE
Robert Fulton	159'	33'-5"	17'-3"	702	N.Y.C.	Henry Eckford	1820
Southerner	191'-3"	30'-8"	14'-1"	785	N.Y.C.	Wm. H. Brown	1846
Northerner	203'-6"	32'-4"	21'-7 1/2"	1012	N.Y.C.	Wm. H. Brown	1847
California	199'-2"	33'-6"	20'	1057	N.Y.C.	Wm. H. Webb	1848
Panama	200'-4"	33'-10 1/2"	20'-2"	1087	N.Y.C.	Wm. H. Webb	1848
Oregon	202'-9"	33'-10"	20'	1099	N.Y.C.	Smith & Dimon	1848
Isabel	209'	33'-6"	16'-7"	1115	Balt.	S. H. Duncan	1848
Falcon	204'-2"	30'-2"	21'-5"	891	N.Y.C.	Wm. H. Brown	1848
Crescent City	233'-7"	33'-11"	22'-8"	1291	N.Y.C.	Wm. H. Brown	1848
Cherokee	210'-8"	35'-4"	22'[1]	1244	N.Y.C.	Wm. H. Webb	1848
Tennessee	211'-10"	35'-8"	22'	1275	N.Y.C.	Wm. H. Webb	1849
Empire City	238'-8"	39'-4"	24'-4"	1751	N.Y.C.	Wm. H. Brown	1849
Ohio	246'	46'	32'-9"	2432	N.Y.C.	Bishop & Simonson	1849
Georgia	248'	48'-8"[1]	33'-0"	2727	N.Y.C.	Smith & Dimon	1850

1 Corrected values differing from official documents

Appendix H
Coastwise Steamship Engines, 1820-1850

	TYPE	BORE	STROKE	STEAM PRESS psi	REV'S. PER MIN.	APPROX. IHP	WHEEL DIA.	FLOAT WIDTH	BOILERS	BUILT BY	DATE
Robert Fulton	1 crosshead	44"	5'-0"				18'-0"	6'-6"	1 ret. flue	Allaire	1820
Southerner	1 side lever	67"	8'-0"		15		31'-0"	7'-6"	2 ret. flue	Novelty	1846
Northerner	1 side lever	70"	8'-0"	18	14	430	31'-0"	7'-6"	2 ret. flue	Novelty	1847
California	1 side lever	70"	8'-0"				26'-0"	9'-0"	2 ret. flue	Novelty	1848
Panama	1 side lever	70"	8'-0"			550	26'-0"	8'-6"	2 ret. flue	Allaire	1848
Oregon	1 side lever	70"	8'-0"				26'-0"	9'-0"	2 ret. flue	Novelty	1848
Isabel	1 side lever	70"	8'-0"				30'-0"	7'-6"	2 ret. flue	Chas. Reeder	1848
Falcon	2 inclined[1]	60"	5'-0"	14	15	360	32'-0"	7'-9"	2 ret. flue	Hogg & Delamater	1848
Crescent City	1 side lever	80"	9'-0"			600	33'-6"	8'-0"	2 ret. flue	T. F. Secor & Co.	1848
Cherokee	1 side lever	75"	8'-0"	16	15	500	31'-4"	8'-0"	2 ret. flue	Novelty	1848
Tennessee	1 side lever	75"	8'-0"	16	15	500	31'-11"	8'-0"	2 ret. flue	Novelty	1949
Empire City	1 side lever	85"[2]	9'-0"						2 ret. flue	T. F. Secor & Co.	1849
Ohio	2 side lever	90"	8'-0"	15	12	1100	36'-0"	10'-6"	4 ret. flue	T. F. Secor & Co.	1849
Georgia	2 side lever	90"	8'-0"	15	12	1100	36'-0"	10'-6"	4 ret. flue	T. F. Secor & Co.	1850

1 From Hudson River Steamboat *Iron Witch*
2 Other sources quote a 75" bore

Appendix I
Coastwise Steamships, 1851-1860

	LENGTH	BEAM	DEPTH HOLD	TONS	BUILT AT	BUILT BY	DATE
Union	215'	34'	22'	1200	N.Y.C.	Wm. H. Webb	1851
Golden Gate	269'-6"	40'	22'	2067	N.Y.C.	Wm. H. Webb	1851
Illinois	266'-6"	40'-10"	22'	2123	N.Y.C.	Smith & Dimon	1851
James Adger	215'	33'-6"	21'-3"	1151	N.Y.C.	Wm. H. Webb	1852
John L. Stephens	274.3'	41'	17.3'	2182	N.Y.C.	Smith & Dimon	1852
Golden Age	272'-10"	41'-10"	25'-1"	2281	N.Y.C.	Wm. H. Brown	1853
George Law	278'-3"	40'	32'	2141	N.Y.C.	Wm. H. Webb	1853
San Francisco	281'-5"	41'	24'-10"	2272	N.Y.C.	Wm. H. Webb	1853
Nashville	215'-6"	34'-6"	21'-9"	1220	N.Y.C.	Wm. Collyer	1854
Tennessee	210'	33'-11"	16'-11 1/2"	1149	Balt.	John A. Robb	1854
Sonora	269'	35'-2"	24'	1619	N.Y.C.	Westervelt & Son	1854
St. Louis	270'-2"	35'-2"	25'-7"	1621	N.Y.C.	Westervelt & Son	1854
Quaker City	234.4'	35.7'	19.8'	1426	Phila.	Vaughan & Lynn	1854
Ocean Bird	229'-4"	36'-8"	25'	1467	Greenpoint	C. & R. Poillon[1]	1855
Columbia	234'	35'	23'	1347	Greenpoint	Wm. Collyer	1857

[1] Begun by John W. Griffiths. Renamed *Pajaro del Oceano* in 1856

Appendix J
Coastwise Steamship Engines, 1851-1860

	TYPE	BORE	STROKE	STEAM PRESS. psi	REV'S. PER MIN.	APPROX. IHP	WHEEL DIA.	FLOAT WIDTH	BOILERS	BUILT BY	DATE
Union	2 side lever	60″	7′-0″	17	16 1/2	830	30′	9′-0″	2	Allaire	1851
Golden Gate	2 oscillating	85″	9′-0″	12	14	1150	31′-0″	12′-0″	4 ret. tube	Novelty	1851
Illinois	2 oscillating	85″	9′-0″	18	11	1150	33′-6″	10′-6″	4 ret. rube	Allaire	1851
James Adger	1 side lever	75″	8′-0″							Allaire	1852
John L. Stephens	1 oscillating	85	9′-0″	20			32′-0″	10′-0″	2 ret. flue	Novelty	1852
Golden Age	1 beam	83″	12′-0″	27	16	1350	34′-0″	9′-6″	2 dbl. ended	Morgan	1853
George Law	2 inclined	65″	10′-0″							Morgan	1853
San Francisco	2 oscillating	65″	8′-0″	20			28′[1]	9′-0″	2 ret. flue	Morgan	1853
Nashville	1 side lever	86″	8′-0″	28			32′-0″	10′-0″	2 ret. flue	Novelty	1854
Tennessee	1 beam	72″	9′-0″	20		560	28′-0″	9′-0″	2 ret. flue	Chas. Reeder, Jr.	1854
Sonora	2 beam	50″	10′-0″	20	16	675	30′-0″	9′-6″	2 ret. flue	Morgan	1854
St. Louis	2 beam	50″	10′-0″	22	15	675	30′-0″	9′-0″	2 ret. flue	Morgan	1854
Quaker City	1 side lever	85″	8′-0″	21	14	750[2]	31′-0″	10′-0″	4 ret. flue	Merrick & Sons	1854
Ocean Bird	1 beam	65″	12′-0″	25	17	770	33′	7′-9″	4 ret. flue	Neptune	1855
Columbia	1 side lever	86″	9′-0″							Novelty	1857

1 14 feathering floats 9′-0″ x 4′-0″
2 Measured on trials, others computed by C.R.N.

Appendix K
Atlantic Liners, 1847-1857

	LENGTH	BEAM	DEPTH HOLD	TONS	BUILT AT	BUILT BY	DATE
Washington	230'-5"	38'-9 1/2"	31'[1]	1640	N.Y.C.	Westervelt & Mackay	1847
Hermann	234'-11"	39'-6"	31'-0"	1734	N.Y.C.	Westervelt & Mackay	1848
United States	244'-7"	40'1	30'-10"	1857	N.Y.C.	Wm. H. Webb	1848
Atlantic	284'	45'-11"	32'[1]	2845	N.Y.C.	Wm. H. Brown	1850
Pacific	281'	45'	32'-3"	2707	N.Y.C.	Jacob Bell	1850
Arctic	285'	45'-11"	32'[1]	2856	N.Y.C.	Wm. H. Brown	1850
Baltic	282'-6"	45'	32'[1]	2723	N.Y.C.	Jacob Bell	1850
Franklin	264'	41'-8"	25'-9"	2183	N.Y.C.	Westervelt & Mackay	1850
Humboldt	283'	40'	27'-2"	2181	N.Y.C.	Westervelt & Mackay	1851
Ericsson	253'-6"	39'-8"	26'-6"	1902	Greenpoint	Perrine, Patterson & Stack	1853
Arago	290'[1]	40'	31'-6"	2240	N.Y.C.	Westervelt & Sons	1855
Fulton	287'-6"	40'-10"	32'	2307	N.Y.C.	Smith & Dimon	1856
Adriatic	345'	50'	33'-2"	4145	N.Y.C.	James R. & George Steers	1856

1 Corrected values differing from official documents

Appendix L
Atlantic Liner Engines, 1847-1857

	TYPE	BORE	STROKE	STEAM PRESS. psi	REV'S. PER MIN.	APPROX. IHP	WHEEL DIA.	FLOAT WIDTH	BOILERS	BUILT BY	DATE
Washington	2 side lever	72″	10′-0″	12	11	700	39′-0″	7′-6″	2 ret. flue	Novelty	1847
Hermann	2 side lever	72″	10′-0″	12	11	821[2]	36′-0″	8′-0″	2 ret. flue	Novelty	1848
United States	2 side lever	80″	9′-0″	12	12	850	34′-8″	8′-4″	4 ret. flue	T. F. Secor & Co.	1848
Atlantic	2 side lever	95″	9′-0″	19	15	2100	35′-0″	12-4″	4 water tube	Novelty	1850
Pacific	2 side lever	95″	9′-0″	17	16	2122[2]	36′-0″	11′-8″	4 water tube	Allaire	1850
Arctic	2 side lever	95″	10′-0″	17	15.8	2298[2]	35′-6″[3]	12′-2″[3]	4 water tube	Novelty	1850
Baltic	2 side lever	95″	10′-0″	18	16	2300	36′-0″	12′-3″	4 water tube	Allaire	1850
Franklin	2 side lever	93″	8′-0″	16	14	1466[2]	32′-2″	11′-8″	4 ret. flue	Novelty	1850
Humboldt	2 side lever	95″	9′-0″	15	14	1600	34′-2″	12′-3″	4 ret. flue	Novelty	1851
Ericsson	4 hot air[1]	168″	6′-0″	8	10	275	32′-0″	10′-0″	—	Hogg & Delamater	1853
Arago	2 oscillating	65″	10′-0″	30	14	1450	33′-0″	10′-0″	2 ret. flue	Novelty	1855
Fulton	2 oscillating	65″	10′-0″	25	18	1600	31′-0″	9′0″	2 water tube	Morgan	1856
Adriatic	2 oscillating	100″	12′-0″	25	14	3300	40′-0″	12′-0″	8 water tube	Novelty	1856

1 Replaced by 2 inclined steam engines 62″ x 7′-8″.
2 Measured in service, others computed by C.R.N.
3 Wheels 34′-5″ x 11′-6″ when 2298 IHP measured.

Appendix M
Vanderbilt Steamships, 1850-1865

	LENGTH	BEAM	DEPTH HOLD	TONS	BUILT AT	BUILT BY	DATE
Prometheus	230'-6"	33'	20'-2"	1207	N.Y.C.	J. Simonson	1850
Daniel Webster	223'-4"	31'	18'-6"	1035	N.Y.C.	Wm. H. Brown	1851
Northern Light	253'-6"	38'-2"	28'[1]	1767	N.Y.C.	J. Simonson	1852
Star of the West	228'-4"	32'-8"	24'-6"	1172	N.Y.C.	J. Simonson	1852
North Star	262'-6"	38'-6"	28'	1867	N.Y.C.	J. Simonson	1853
Ariel	252'-6"	32'-6"	26'[1]	1295	N.Y.C.	J. Simonson	1855
Vanderbilt	331'[1]	47'-6"	31'-9"	3600[1]	Greenpoint	J. Simonson	1856
Ocean Queen	327'	42'	30'[1]	2801	N.Y.C.	S. G. Bogert	1859
Champion[2]	235'	35'	25'-4"[1]	1419	Wilmington	Harlan & Hollingsworth	1859
Costa Rica	269'	38'-10"	27'	1950	Greenpoint	J. Simonson	1864
New York	292.6'	41.7'	26.5'	2217[3]	Greenpoint	J. Simonson	1865

1 Corrected values differing from official documents
2 Iron Hull
3 1865 tonnage rules

Appendix N
Vanderbilt Steamship Engines, 1850-1865

	TYPE	BORE	STROKE	STEAM PRESS. psi	REV'S. PER MIN.	APPROX. IHP	WHEEL DIA.	FLOAT WIDTH	BOILERS	BUILT BY	DATE
Prometheus	2 beam	42"	10'-0"							T. F. Secor & Co.	1850
Daniel Webster	1 beam	56"	10'-0"						2 ret. flue	Allaire	1851
Northern Light	2 beam	60	10'-0"			900	33'-0"		4 ret. flue	Allaire	1852
Star of the West	2 beam	42"	10'-0"							Allaire	1852
North Star	2 beam	60"[1]	10'-0"	20	15	900	33'-0"[1]	8'-0"	4 ret. flue	Allaire	1853
Ariel	1 beam	75"	11'-0"			800	33'-0"	8'-0"	2 ret. flue	Allaire	1855
Vanderbilt	2 beam	90"	12'-0"	20	14	2500	41'-0"[2]	10'-0"	4 ret. tube	Allaire	1856
Ocean Queen	1 beam	90"	12'-0"	20	13	1160	38'-0"	10'-6"	3 ret. flue	Morgan	1859
Champion	2 beam	42"	10'-0"	30	16	620	30'-0"	6'-6"	2 ret. flue	Allaire	1859
Costa Rica	1 beam	81"	12'-0"			1330				Allaire	1864
New York	1 beam	90"	12'-0"			1680	35'-0"		2 tubular	Allaire	1865

1 Other sources quote 66" bore and 34'-0" wheel dia.
2 Other sources quote 42'-0" wheel dia.

Appendix O
Wood Screw Steamships

	LENGTH	BEAM	DEPTH HOLD	TONS	BUILT AT	BUILT BY	DATE
Lafayette	210'	32'-6"	19'	1059	Williamsburg	Perrine, Patterson & Stack	1851
S. S. Lewis	216'	32.6'	16.3'	1103	Phila.	T. Birely & Son	1851
William Penn	183'	26.4'	13.2'	613	Phila.	T. Birely & Son	1851
Pioneer	218'-4"	42'-6"	29'	1833	N.Y.C.	Jacob Bell	1851
City of Pittsburg	246'-8"	40'	31'-4"	1875	Williamsburg	Perrine, Patterson & Stack	1851
West Wind	144'	26.1'	13.5'	460	Wilmington	Thatcher	1851
South Carolina	200'-6"	37'-3"	30'	1301	Williamsburg	Jabez Williams	1851
Granite State	174.6'	26.4'	13.2'	582	Gloucester		1852
Star of the South	206.1'	31.2'	15.6'	960	Phila.	T. Birely & Son	1853
North Carolina	173.2'	31.1'	13.3'	672	Phila.	Vaughan & Fisher	1854
Worcester[1]	209'-6"	35'-6"	20'-8"[4]	1244	N.Y.C.	J. B. & J. D. Van Dusen	1863
Carroll[2]	209'	35'-6"	20'-8"	1244	N.Y.C.	J. B. & J. D. Van Dusen	1863
Somerset[3]	209'	34'-6"	20'-8"[4]	1244	N.Y.C.	J. B. & J. D. Van Dusen	1863
Fung Shuey	195-6"	33'	24'	1004	Brooklyn	Elisha S. Whitlock	1864
Ontario	323.5'	43.9'	20'	2889	Newburyport	G. W. Jackman, Jr.	1867
Erie	325'[4]	44'[4]	21'	2900	Newburyport	G. W. Jackman, Jr.	1867
Kuroda	188'	28'	17'	644	Brooklyn	C. & R. Poillon	1872
Capron	125'	23'	11'	374	Brooklyn	C. & R. Poillon	1873

1 Ex-*Glaucus*
2 Ex-*Proteus*
3 Ex-*Nereus*
4 Corrected values differing from official documents

Appendix P
Wood Screw Steamship Engines

	TYPE	BORE	STROKE	STEAM PRESS. psi	REV'S. PER MIN.	APPROX. IHP	DIA.	BOILERS	BUILT BY	DATE
Massachusetts	2 vee dir. acting	25"	3'-0"	37[1]	54[1]	241[1]	9'-6"	2 ret. tube	Hogg & Delamater	1845
Lafayette	2 vee dir. acting	50"	3'-8"				14'-0"	2 water tube	Hogg & Delamater	1851
S. S. Lewis	2 geared vee	60"	3'-4"	18	38	410	13'-0"	2	James T. Sutton	1851
William Penn	2 inv. dir. acting	34"	2'-10"				10'-0"	2 ret. flue	Reaney, Neafie & Co.	1851
Pioneer	2 vert. trunk	85 1/2" 39"	4'-3"	15	36	1100	16'-0"	2 ret. flue	West Point Foundry	1851
City of Pittsburg	2 vert. trunk	85 1/2" 39"	4'-3"	10[2]	32[2]	777[2]	16'-2"	3 ret. flue	West Point Foundry	1851
West Wind	2 inv. dir. acting	30"	2'-2"	15	48	135	8'-2"	1 ret. flue	Reaney, Neafie & Co.	1851
Star of the South	2 geared trunk	52"/-	3'-0"	36	90	850	10'-6"	2 vert. tube	James T. Sutton	1853
North Carolina	1 inv. geared	56"	4'-0"	23	69	290	9'-3"	2 ret. tube	Merrick & Son	1854
Worcester	2 inv. dir. acting	44"	3'-0"	15	35	290	12'-0"	2 tubular	Henry Esler & Co.	1863
Fung Shuey	1 inv. dir. acting	44"	3'-6"				11'-9"	1 tubular	Delamater Iron Works	1864
Ontario	2 inv. dir. acting	74"	4'-0"	30	50	2400	18'-0"	4	Harrison Loring	1867
Kuroda	2 cyl. compound D.A.	24" & 42"	2'-10"	52	66				Delamater Iron Works	1872
Capron	1 inv. dir. acting	30"	2'-6"	45	80	250	8'-4"	1 ret. tube	Delamater Iron Works	1873

1 Trial results with 10'-6" dia. propeller and return flue boilers
2 Average on Eastbound Atlantic crossing

Appendix Q
Iron Screw Steamships

	LENGTH	BEAM	DEPTH HOLD	TONS	BUILT AT	BUILT BY	DATE
Ashland[1]	97.5'	22.7'	9.1'	182	Wilmington	Betts, Harlan & Hollingsworth	1844
Ocean[1]	98.8'	23'	9.3'	191	Wilmington	Betts, Harlan & Hollingsworth	1844
Bangor[1]	118'-6"	24'-2"	8'-8 1/2"	231[4]	Wilmington	Betts, Harlan & Hollingsworth	1845
Circassian	242.1'	39'	22.4'	1457[5]	Belfast, Ireland	Robt. Hickson	1857
Matanzas	202'	29'-10"	20'-6"	862	N.Y.C.	Delamater Iron Works	1860
South Carolina	217'-10"	33'-5"	25'	1165	Boston	Harrison Loring	1860
Massachusetts	219'-8"	33'-2"	25'	1155	Boston	Harrison Loring	1860
North Carolina	196'-6"	29'-2"	13'-3"	618	N.Y.C.	Novelty Iron Works	1861
Oriental	218'	34'	17'	1202	Phila.	Neafie & Levy	1861
Mississippi	274.3'	39'	28'	2008	Boston	Harrison Loring	1862
Merrimack	272'	39'	28'	1991	Boston	Harrison Loring	1862
Havana	230'	35'		1347	Phila.	Neafie & Levy	1863
Crescent City[2]	265.4	34.0'	16.6'	2003		Lengthened in 1873	
Gulf Stream[3]	218.0'	30.0'	13.0'	998		Lengthened in 1874	

1 Twin-screw
2 Ex-*Massachusetts*
3 Ex-*North Carolina*
4 In 1846 after rebuilding at Bucksport, Maine
5 1865 tonnage

Appendix R
Iron Screw Steamship Engines

	TYPE	BORE	STROKE	STEAM PRESS. psi	PROP. REV'S. PER MIN.	APP- ROX. IHP	PROP. DIA.	BOILERS	BUILT BY	DATE
Ashland[1]	1 geared crosshead							1 drop flue	Betts, Harlan & Hollingsworth	1844
Bangor[1]	2 ind. dir. acting	22"	2'-0"	46	44	125	8'-6"	1 drop flue	Betts, Harlan & Hollingsworth	1845
Circassian	2 geared beam	60"	4'-0"	17				2 tubular	Randolph & Elder	1857
Matanzas	1 inv. dir. acting	56"	3'-9"	22			14'-0"	1 ret. tube	Delamater Iron Works	1860
South Carolina	1 inv. dir. acting	62"	3'-8"					2	Harrison Loring	1860
North Carolina	1 inv. dir. acting	42"	3'-6"	30	60	400	12'-0"	1 ret. tube	Novelty Iron Works	1861
Merrimack	2 inv. dir. acting	62"	4'-0"				15'-0"	4	Harrison Loring	1862
Havana	1 inv. dir. acting	60"	5'-0"				14'-5"		Neafie & Levy	1863
Merrimack	2 cyl. compound D.A.	38"&70"	4'-0"	Rebuilt with new engine & boilers					Harrison Loring	1884

1 Twin-screw

Appendix S
Coastwise Steamships, 1861-1865

	LENGTH	BEAM	DEPTH HOLD	TONS	BUILT AT	BUILT BY	DATE
Santiago de Cuba	227'	38'	19'	1567	Greenpoint	J. Simonson	1861
Morning Star	272'	39'-4"	19'-8"	2022	N.Y.C.	Rosevelt & Joyce	1863
Evening Star	271'	39'-4"	19'-8"	2014	N.Y.C.	Rosevelt & Joyce	1863
Western Metropolis	285'-4"	40'-8"	23'	2269	Brooklyn	F. Z. Tucker	1863
Suwonada	258'	38'-2"	21'-2"	1802	N.Y.C.	John Englis & Son	1864
Guiding Star	300'-6"	40'-6"	23'[1]	2384	N.Y.C.	Rosevelt, Joyce & Waterbury	1864
Morro Castle	262.2'	40'	23.1'	1680	N.Y.C.	J. A. & D. D. Westervelt	1864
Rising Star	303.54'	43.66'	23'	2726	N.Y.C.	Rosevelt, Joyce & Waterbury	1865

1 Corrected value differing from official document

Appendix T
Pacific Mail Steamships, 1861-1869

	LENGTH	BEAM	DEPTH HOLD	TONS	BUILT AT	BUILT BY	DATE
Constitution	342'-6"	44'-8"	30'[1]	3315	N.Y.C.	Wm. H. Webb	1861
Golden City	343'	45'-1"	30'[1]	3373	N.Y.C.	Wm. H. Webb	1863
Sacramento	304'	42'-6'	29'-2"	2647	Greenpoint	Webb & Bell	1864
Henry Chauncey	319'-5"	43'	28'[1]	2656	N.Y.C.	Wm. H. Webb	1865
Colorado	314'	45'	31'-9"	3357	N.Y.C.	Wm. H. Webb	1865
Arizona	323.8'	44.8'	27'[1]	2793	Greenpoint	Henry Steers	1866
Montana	318'	42.5'	27'[1]	2676	Greenpoint	Webb & Bell	1866
Great Republic	360'	47.4'	31.5'[1]	3881	Greenpoint	Henry Steers	1867
China	360'	47.4'	31.5'[1]	3836	N.Y.C.	Wm. H. Webb	1867
Japan	362'	49'	30.9'	4351	Greenpoint	Henry Steers	1868
Alaska	346'	47.6'	31.5'[1]	4011	Greenpoint	Henry Steers	1868
America	363'	49.3'	31.3'	4454	Greenpoint	Henry Steers	1869

1 Corrected values differing from official documents

Appendix U
Walking Beam Engines, 1861-1869

	BORE	STROKE	STEAM PRESS. psi	REV'S. PER MIN.	APPROX. IHP	WHEEL DIA. MIN.	FLOAT WIDTH	NO. FLOATS	BOILERS	BUILT BY REBUILT BY	DATE
1 *Santiago de Cuba*	66″	11′-0″				29′	9′-6″	22	2 ret. flue	Neptune Iron Works	1861
2 *Columbia*						32′-10″	11′-0″	28	2 ret. tube	Quintard Iron Works	1877
Constitution	105″	12′-0″				40′	12′-0″		4 ret. flue	Novelty Iron Works	1861
1 *Crescent City*						40′				Morgan Iron Works	1853
2 *Morning Star*	80″	12′-0″				33′		28	2 ret. tube	Morgan Iron Works	1863
1 *Queen of the West*										Cunningham & Belknap	1853
2 *Evening Star*	80″	12′-0″							2 ret. tube		1863
Golden City	105″	12′-0″	17.5	10.5	1100	40′	12′-0″		4 horiz. tube	Novelty Iron Works	1863
1 *Empire State*										Merrick & Towne	1848
2 *Western Metropolis*						38′				Buffalo Steam Engine Co.	1856
3 *Western Metropolis*	75″	12′-0″								Morgan Iron Works	1863
Suwonada	76″	12′-0″				31′			4 tubular	Neptune Iron Works	1864
1 *Mississippi*						38′				I. P. Morris	1853
2 *Guiding Star*	81″	12′-0″							2 ret. tube	Etna Iron Works	1864
1 *City of Buffalo*						38′				Morgan Iron Works	1857
2 *Morro Castle*	76″	12′-0″								Allaire Iron Works	1864
3 *Grand Republic*						36′	10′-6″	32	2 ret. tube	Quintard Iron Works	1878
Rising Star	100″	12′-0″	25			36′	11′-6″		2 flue & tube	Etna Iron Works	1865
China[1]	105″	12′-0″	30	10	1500	40′	12′-0″	34	4 ret. tube	Novelty Iron Works	1867

[1] *Great Republic, Japan, Alaska,* & *America,* 1867-1969, similar

Appendix V
Converted Steamships

	LENGTH	BEAM	DEPTH HOLD	TONS	ALTERED TO	ALTERED AT	DATE
Savannah	98'-6"	25'-10"	12-'11"	291	Sailing ship	N.Y.C.	1820
South Carolina	200'-6"	37'-3"	30'	1301	Sailing ship	N.Y.C.	1853
Pioneer	218'-4"	42'-6"	29'	1833	Bark	San Francisco	1854
Ericsson	248'	40.4'	27.6'	1645	Sailing ship	N.Y.C.	1868
Baltic	280.6'	45'	29.4[4]	2552	Sailing ship	N.Y.C.	1869
Oregon	200'	34'	19'	887	Bark	San Francisco	1865
Circassian	242.4'	39'	30.2[4]	1397	Sailing ship	N.Y.C.	1873
Three Brothers[1]	320'	48.8'	29.8'	2972	Sailing ship	San Francisco	1873
California	198.5'	33.8'	18'	794	Bark	San Francisco	1875
Santiago de Cuba	231'	38'	20'	1680	Screw Stmr.	N.Y.C.	1877
Morro Castle	255.2'	40'	23.1'	1714	Screw Stmr.	N.Y.C.	1878
Moro Castle[2]	258'	40'	16.2'	1006	3 Masted Schooner-Barge	N.Y.C.	1886
Marion[3]	231'	38'	20.2'	1008	Schooner-Barge	N.Y.C.	1886

1 Ex-*Vanderbilt*
2 Ex-*Morro Castle*
3 Ex-*Santiago de Cuba*
4 From unofficial sources

Appendix W
First Atlantic Sailings

NAME	SAILED FROM	DATE	DESTINATION	VIA	ARRIVED AT	DATE	TIME	EASTWARD CROSSINGS
Savannah	Savannah	5/22/19	St. Petersburg	Stockholm	Liverpool	6/20/19	29d11h	1
Bangor	Boston	8/16/42	Constantinople	Gibraltar	Malta	10/17/42	62d	1
Midas	New York	11/4/44	Hong Kong	Cape of Good Hope	Hong Kong	5/21/45	198d	1
Edith	New York	1/18/45	Hong Kong	Cape of Good Hope	Hong Kong	9/11/45	236d	1
Marmora	New York	9/2/45	Constantinople	Gibraltar	Liverpool	9/26/45	24d	1
Massachusetts	New York	9/15/45	Liverpool		Liverpool	10/3/45	17 1/2d	2
Washington	New York	6/1/47	Bremen	Cowes	Southampton	6/15/47	14d	54
Hermann	New York	3/21/48	Bremen	Halifax	Cowes	4/11/48	21d	47
United States	New York	4/8/48	Liverpool		Liverpool	4/22/48	13d20h	12[1]
Atlantic	New York	4/27/50	Liverpool		Liverpool	5/10/50	13d3h	66
Pacific	New York	5/25/50	Liverpool		Liverpool	6/7/50	12d13h	40
Franklin	New York	10/5/50	Havre	Cowes	Cowes	10/18/50	13d	22
Arctic	New York	10/26/50	Liverpool		Liverpool	11/7/50	11d7h	24
Baltic	New York	11/16/50	Liverpool		Liverpool	11/28/50	11d17h30m	57
Humboldt	New York	5/6/51	Havre	Cowes	Cowes	5/18/51	11d15h	16
Lafayette	Philadelphia	5/10/51	Liverpool	Cork	Liverpool	6/2/51	23d	1
S. S. Lewis	Boston	10/4/51	Liverpool		Liverpool	10/21/51	17 1/2d	1
Pioneer	New York	10/18/51	Liverpool		Liverpool	11/6/51	18d17h	1
City of Pittsburg	Philadelphia	10/26/51	Liverpool	Cork	Liverpool	11/16/51	19 1/2d	1
South Carolina	New York	8/11/52	Liverpool		Liverpool	8/29/52	18d	1
West Wind	New York	9/19/52	Sydney	Cape of Good Hope	Port Phillip	4/28/53	221d	1
North Star	New York	5/20/53	European Cruise		Southampton	6/1/53	12d	16
Golden Age	New York	9/30/53	Melbourne	Cape of Good Hope	Liverpool	10/12/53	11d18h	1
Nashville	New York	1/16/54	Havre	Cowes	Cowes	2/2/54	17d	4[2]
Union	New York	5/6/54	Havre	Cowes	Havre	5/20/54	14d	11
William Penn	New York	5/20/54	Marseilles	Gilbraltar	Gibraltar	6/7/54	18d	1

1 Includes 5 to Bremen & 2 to Galway under foreign flags
2 One as a Confederate Cruiser

NAME	SAILED FROM	DATE	DESTINATION	VIA	ARRIVED AT	DATE	TIME	EASTWARD CROSSINGS
St. Louis	New York	8/1/54	Havre	Cowes	Cowes	8/13/54	11d17h	6
North Carolina	Delaware Breakwater	3/23/55	Liverpool		Sunk in collision off Ireland	4/8/55		1
Ariel	New York	5/19/55	Havre		Havre	5/31/55	12d	21
Granite State	Philadelphia	5/29/55	Liverpool		Liverpool	6/16/55	18d	1
Arago	New York	6/2/55	Havre	Cowes	Cowes	6/15/55	13d5h	57
Ericsson	New York	6/16/55	Havre		Havre	6/30/55	14d4h	20
Tennessee	Baltimore	6/17/55	Havre		Havre			1
Star of the South	New York	6/20/55	Liverpool		Liverpool	7/7/55	17d	1
Fulton	New York	2/9/56	Havre	Cowes	Havre	2/25/56	16d	50
Quaker City	New York	2/16/56	Liverpool		Liverpool	3/2/56	14d12h	3
Pajaro del Oceano	New York	6/25/56	Cadiz	Vigo	Vigo	7/13/56	18d	2
Vanderbilt	New York	5/5/57	Havre	Cowes	Cowes	5/15/57	9d21h15m	23[3]
Columbia	New York	6/6/57	Liverpool		Liverpool	6/18/57	11d8h	2
Adriatic	New York	11/23/57	Liverpool		Liverpool	12/4/57	10d21h10m	13[4]
Northern Light	New York	6/12/58	Bremen	Cowes	Cowes	6/25/58	13d	6
Ocean Queen	New York	5/21/59	Havre	Cowes	Cowes	6/2/59	11d13h30m	7
Illinois	New York	4/7/60	Havre	Cowes	Cowes	4/23/60	15d16h	5
Annie Childs	Wilmington	2/5/62	Liverpool	Fayal & Madeira	Queenstown	3/8/62	31d	1[5]
Suwonada	New York	Abt. 5/24/65	Hong Kong	Cape of Good Hope	Singapore	8/15/65	82/83d	1
Circassian	New York[6]	8/19/65	Bremen	Cowes	Cowes	9/3/65	15d	17
Somerset	Baltimore	9/30/65	Liverpool		Liverpool	10/16/65	16d	10
Worcester	Baltimore	12/23/65	Liverpool		Liverpool	1/9/66	17d	12
Carroll	Baltimore	4/25/66	Liverpool		Liverpool	5/9/66	14d	7
Mississippi	New York	5/10/66	Bremen	Cowes	Cowes	5/25/66	15d	3
Merrimack	New York	5/17/66	Bremen	Cowes	Cowes	5/31/66	14d	1
Western Metropolis	New York	3/7/67	Bremen	Cowes	Cowes	3/24/67	17d	5
Guiding Star	New York	3/30/67	Havre	Falmouth	Brest	4/11/67	12d	5
Costa Rica	New York	4/1/67	Yokohama	Cape of Good Hope	Singapore	6/11/67	71d	1
Fung Shuey	New York	5/18/67	Hong Kong	Cape of Good Hope	Singapore	8/11/67	85d	1
Morning Star	New York	6/22/67	Havre	Falmouth	Havre			1
New York	New York	8/3/67	Yokohama	Cape of Good Hope	Capetown	9/9/67	37d	1
Ontario	Boston	8/5/67	Liverpool		Liverpool	8/18/67	13d	4
America	New York	5/28/69	Yokohama	Cape of Good Hope	Singapore	8/4/69	68d	1
Santiago de Cuba	New York	6/16/69	Copenhagen	Cowes	Cowes	6/30/69	14d	3
Rising Star	New York	3/24/70	Swinemunde	Havre	Havre	4/7/70	14d	1
Erie	New York	1/29/70	Brest	Cowes	Havre	12/13/70	14d	1
Arizona	New York	12/22/70	Hong Kong	Suez Canal	Gibraltar	1/8/71	17d	1
Alaska	New York	1/22/71	Hong Kong	Suez Canal	Gibraltar	2/7/71	16d	1
Kuroda	New York	12/21/72	Yokohama	Suez Canal	Gibraltar	1/6/73	16d	1
Capron	New York	4/4/73	Yokohama	Suez Canal	Gibraltar	5/17/73	43d	1

3 One as a Cruiser
4 Six as a British Steamship
5 Blockade Runner, Ex-*North Carolina*
6 First U.S. Flag Sailing. Had made 11 as a British Steamship.

Glossary of Marine Terms

ABAFT — aft of

AIR PUMP — a pump, normally driven by the main engine, that extracts air and condensed water from the condenser and discharges into the hot well

ANGLE — rolled iron shape in the form of an L. Used for deck beams and frames of iron hulls

BARK — three-masted vessel with square sails on the forward two masts

BARKENTINE — three-masted vessel with square sails on the foremast

BEAM — breadth of a hull; also, a deck beam

    EXTREME BEAM — breadth measured outside planking or wales

    MOLDED BEAM — breadth measured inside the planking

BEDPLATE — heavy iron casting forming the base of a marine engine

BELL CRANK — L-shaped lever turning about a pivot at the intersection of its two arms and serving a link between the piston rod, driving one arm, and the crank on the paddle shaft, turned by a connecting rod attached to the other arm

BILGE — the curved portion of the hull joining the bottom with the side. Descriptive of structure in the vicinity such as bilge ceiling, bilge keelson, bilge strake, or bilge planking. Also, drainage space where leakage, or bilge water, collects

BILGE PUMP — a pump used to discharge leakage water overboard

BLOWING DOWN — use of a boiler's steam pressure to force part of its salt-laden water overboard

BODY PLAN — the end view of a ship showing the shape of frames or transverse sections

BOILER — pressure vessel in which steam is generated. Early examples were made of copper, later ones of iron

    DONKEY BOILER — small auxiliary boiler for providing steam in port when the main boilers were cold

    DROP-FLUE BOILER — a boiler whose smoke and combustion products go down before returning through flues to the uptake and stack. This is always a return-flue boiler

    FIRE-TUBE BOILER — a boiler with tubes used to generate steam that have combustion gas inside and water outside

    RETURN-FLUE BOILER — a boiler whose smoke and combustion gases first flow away from the grates and then reverse direction and return through flues before escaping up the stack

    RETURN-TUBE BOILER — similar to a return-flue boiler but with many small tubes rather than a few large-diameter flues

    TUBULAR BOILER — indefinite description; a boiler may have vertical tubes, horizontal ones, fire tubes, or water tubes

    WATER-TUBE BOILER — unlike the fire-tube boiler, the tubes of this type of boiler used to generate steam have water inside and combustion gas outside

BOILER FLUES — large-diameter passages between a boiler furnace and the uptake

BOILER TUBES — small-diameter passages between a boiler furnace and the uptake

BOWSPRIT — spar extending forward of the stem

BREAK — a sudden change in deck height

BREAST HOOK — horizontal reinforcement, inside the

hull, tying the port and starboard sides together at the stem

BRIG—two-masted vessel with square sails on both

BRIGANTINE—two-masted vessel with square sails on the foremast but no sail on the fore yard. Despite modern pedants, the term was often used for hermaphrodite brig by the men who sailed both

BUCKET—paddle, or float, on a side wheel

BULWARK—planked extensions along the sides, above the weather deck, to keep water and spray from coming aboard

BUNKERS—spaces used for the stowage of coal

CABOOSE, CAMBOOSE—early term for galley, especially if located in a deckhouse

CAPSTAN—device with a vertical axis turned by crew members to haul in a line or lift an anchor chain. Portable bars were inserted into sockets so that many men could work at once

CATHEAD—heavy timber projecting from the bow to hoist the anchor aboard

CAULKING—the operation of driving oakum into seams between planks to make them watertight; also, the material used in the seams

CEILING—the interior planking from keelson to deck clamps

CHANNELS—horizontal platforms extending beyond the sides adjacent to the masts to increase the spread of the supporting rigging

CIRCULATING PUMP—a pump supplying sea water to a surface condenser for cooling purposes, and then forcing it overboard

CLAMPS—heavy lengthwise strakes inside the frames supporting the outer ends of the deck beams

CLIPPER BOW—an attractive sailing-ship feature comprising a curved cutwater forward of the stem and headrails and trailboards on each side, and featuring a carved figurehead or an ornamental billethead

COAMING—raised fore-and-aft boundary of a hatch

COMBUSTION CHAMBER—submerged chamber inside a boiler connecting one or more furnaces to a series of tubes or flues

COMPOUND ENGINE—one that uses the steam twice, once in a high-pressure cylinder and afterward in a low-pressure one

CONDENSER—enclosed chamber with a below-atmospheric pressure, which receives and cools exhaust steam from the engine and condenses it to water

JET CONDENSER—one that uses a jet of sea water to mix with and condense the steam

SURFACE CONDENSER—one in which the steam is condensed inside a series of tubes cooled on the outside by sea water. In later years the cooling water has been inside the tubes and steam outside.

CONDENSATE—water collecting in the bottom of a condenser

CONNECTING ROD—part of a reciprocating engine driving the crank from some other moving part such as a crosshead, crosstail, walking beam, or bell crank

COUNTER—that portion of the stern between the waterline and the lowest knuckle of the after overhang

COUNTER STERN—rounded stern, usually with one or more knuckles below the uppermost rail

COVERING BOARD—heavy lengthwise timber extending from the outside planking across the tops of the frames to the waterway of the deck. Sockets for the rail or bulwark stanchions were cut in it

CROSSHEAD—mechanical device attached to a piston rod sliding back and forth along one or more guides and transferring its force to some other linkage that will eventually turn linear into rotary motion at the crank shaft or paddle shaft

CROSSTAIL—part of a side-lever engine connecting the two side levers at their ends farthest from the cylinder. From its center a connecting rod turns the crank on the paddle shaft

CUT OFF—part of the valve gear that regulates the period during which steam is admitted to a cylinder

CUTWATER—extension of the stem forward and above the waterline

DAVITS—wood or iron cranes for hoisting boats

DEADWOOD—heavy timbers, where the vessel is very narrow, joining the keel to the stem or sternpost

DECK—horizontal platform

FORECASTLE DECK—raised deck forward

HURRICANE DECK—lightly built, exposed to the weather, often over a series of deckhouses

MAIN DECK—usually the heavily built deck just below the spar deck

ORLOP DECK—partial lower deck at the bow or stern

POOP DECK—full-height raised deck aft

QUARTER DECK—partially raised deck aft

SPAR DECK—uppermost deck to the hull extending from stem to stern

TWEEN DECK—space inside hull between two decks

UPPER DECK—same as spar deck

WEATHER DECK—one exposed to the wind and weather

DECK BEAM—transverse timber resting on the clamps at the sides and supporting the deck planking

DEPTH OF HOLD—depth, admidships, from the top of the ceiling to the upper side of the deck beam, sometimes measured to the uppermost deck, sometimes to the main deck below it. Official documents are inconsistent in their choice. Many published values, official or not, are incorrect. For tonnage purposes up to 1865, the depth was assumed to be one-half the beam. As a result, this value frequently appears in place of the actual depth

DISPLACEMENT—actual weight, in tons of 2,240 lbs., of a vessel plus its cargo, fuel, and stores

DRAFT—distance from waterline to bottom of ship

EXTREME DRAFT—maximum value to bottom of keel

MEAN DRAFT—average of bow and stern values

MOLDED DRAFT—a designer's term measured from waterline to the top of keel, amidships

DRAG—slope of keel with draft increasing from bow to stern

FAIRING—process of smoothing out irregularities in the lines, either on a drawing or on the mold loft floor

FALSE KEEL—addition below the main keel that can be crushed or torn off, without damage to the keel, should the vessel ground. Also called the shoe

FEATHERING WHEEL—paddle wheel with pivoted floats designed to enter and leave the water with the minimum of impact. It could be smaller in diameter and could rotate faster than a normal radial wheel with fixed floats

FLARE—outward slope of the sides of ship; the deck is wider than the waterline

FEED PUMP—a pump drawing condensed water from the hot well and forcing it into the boilers

FLOATS—paddles, usually wood, sometimes iron, mounted on the arms of a side wheel

FLOOR, FLOOR TIMBER—heaviest part of a ship's transverse frame that crosses the keel and supports the bottom planking and ceiling

FLUSH DECKED—description of a vessel whose upper deck has no discontinuities or breaks

FOREFOOT—region where the stem intersects the keel

FOULING—growth of sea grasses, weeds, and barnacles on a ship's bottom

FRAME—major transverse structural member; popularly, a rib. It is built up from a series of pieces, the floors, futtocks, and top timbers, and is two layers thick, so that every joint in one layer is backed up by a continuous piece in the other

FURNACE—the chamber inside a boiler in which grates are fitted and coal or wood burned

FUTTOCKS—portions of the transverse frames extending around the turn of the bilge from the floor to the deck and covering board. The uppermost piece is called the top timber

GARBOARD, GARBOARD STRAKE—one or more extra-heavy bottom planks on either side of the keel

GUARDS—extensions of the deck beyond the hull in a shallow-draft steamboat; also a separate pair of platforms with heavy boundaries extending out and around the side wheels on an oceangoing vessel. The latter not only protected the wheels from damage, but also supported the outboard paddle shaft bearing

HALF-BREADTH PLAN—bottom view of a hull showing waterlines, wales, knuckles, and rails. Deck outlines are shown for iron craft but not for wooden ones

HALF BRIG—two-masted vessel with square sails on the foremast, hermaphrodite brig

HALF MODEL—large-scale model, showing only one side of the hull, shaped by eye to establish the lines of a vessel, the method of design preferred by American shipbuilders working in wood

HAWSE HOLE, HAWSE PIPE—opening through the bow to lead an anchor chain aboard

HEEL—transverse inclination owing to wind or off-center weights

HERMAPHRODITE BRIG—two-masted vessel with square sails on the foremast, half brig

HOGGING—longitudinal bending in which the ends droop with respect to the midship region; the opposite of sagging. Many wood hulls without diagonal bracing hogged badly as they grew older

HORSEPOWER—unit, invented by James Watt, to measure the rate at which a steam engine does work

NOMINAL HORSE POWER (NHP)—fictitious rating used by the British Admiralty for comparing engines; it is well below the actual output and has no real meaning

INDICATED HORSE POWER (IHP)—actual output measured by a pressure-measuring device called an indicator. This is the only value ever quoted by the author

HOT WELL—tank to which the air pump discharges water from the condenser

INBOARD—toward the centerline

INVERTED ENGINE—engine with a cylinder over the crank shaft and its piston rod extending downward; used in screw steamships

JIB BOOM—spar extending forward beyond the bowsprit

KEEL—backbone of a ship, the first member laid down when a new ship is constructed; a major, centerline, lengthwise strength member projecting below the bottom planking

KEELSON—heavy longitudinal timber placed on top of the floors

CENTER, or MAIN, KEELSON—one or more timbers located directly over the keel

ENGINE KEELSON—keelson that acts as a foundation supporting the engine

SIDE KEELSON—off-center bottom reinforcement

KNEE—heavy reinforcement joining two portions of a ship's structure that intersect at an angle

HANGING KNEE—vertical knee below a deck beam connecting it to the side of the hull

LODGING KNEE—horizontal knee connecting a deck beam to the side

KNIGHTHEADS—timbers on each side of the stem that support a bowsprit, if one is fitted

KNOT—speed of one nautical mile (6,080 ft.) per hour. In the nineteenth century it was also used as a synonym for nautical mile

KNUCKLE—the intersection of two curved surfaces at an angle, usually around the stern

LAY DOWN—to draw the lines of a ship, full size, on the mold loft floor

LAY TO—to slow down, in a violent sea, and head in the direction that reduces the rolling and pitching of a vessel

LINES DRAWING, PLAN, or DRAUGHT—drawing showing the shape of a vessel, consisting of a profile, half-breadth, and body plan. If the designer worked from a half model, no such drawing had to be prepared. In a wooden vessel the shape delineated is inside the planking; in an iron one it is inside the plating

MIDSHIP SECTION—the largest transverse section of a ship, whether exactly at mid-length or not

MIDSHIPS—the center portion of a ship

MOLDED LINES or DIMENSIONS—drawings and half models were made to the outside of the framing (the inside of the planking). When the principal dimensions of a ship were measured from a drawing, or from the loft floor, rather than from the finished vessel, the values so determined were called molded

MOLDS—light wood patterns made in the mold loft and used in cutting, shaping, and beveling structural parts of the hull

MOLD LOFT—building, or portion of one, with a large, unobstructed, wood floor on which the lines of a ship were expanded to full size and faired. It did not have to be as long as the ship, for the bow half could be superposed on top of the stern

OAKUM—old rope picked apart and used for caulking seams

OUTBOARD—toward or beyond the side of the hull

PADDLE BEAMS—two extremely large transverse timbers, one forward and the other abaft the side wheels, which support the overhanging guards

PADDLE BOX—lightly constructed covering over the upper half of a side wheel, often with decorative ornaments; frequently called wheel house

PADDLE SHAFT—large-diameter axle with side wheels at its outboard ends

PARTNERS—extra-heavy deck beams and fore-and-aft timbers supporting a mast where it pierces a deck

PITCH—the distance a screw propeller would move forward in one revolution if it were working its way through a solid, rather than a liquid. Also, a black, tarry product used to fill seams and to coat the bottom of a wooden hull before its metal sheathing is nailed on

PITCHING—angular motion of a vessel in which the bow and stern alternately rise and fall

PLANKSHEER—uppermost strake of the side planking. Above it are the covering board and the bulwark

POPPET VALVE—the most common American means of controlling the entrance and exit of steam to a cylinder. It consists of a stem, and a circular top that is lifted and lowered onto a seat

PORT—to the left; modern replacement for larboard

PROFILE—side view of a hull that shows heights of rails, wales, planksheer, and buttock shapes

PROPELLER POST—in a screw steamer, the massive vertical timber on centerline just forward of the propeller

QUARTER GALLERIES—aft projections with windows on either side of a transom-sterned ship. Originally, in sailing warships, they contained the officers' water closets, but in smaller craft became nothing more than attractive ornaments

RAIL—fore-and-aft member forming the top of a bulwark. Also, the overall term for vertical stanchions, without planking, and the horizontal member they support

RAKE—inclination to the vertical. Masts, for example, raked aft

RUDDER—pivoted stern flap by which a vessel is steered

RUDDER STOCK—upward extension of the rudder through the counter into the hull

RUDDER POST—heavy vertical timber aft of the propeller on a screw steamer

SAGGING—longitudinal bending of a ship in which the midship region drops below the ends; the opposite of hogging

SCANTLINGS—sizes of the strength members of a hull

SCHOONER—a two- (or more) masted vessel with fore and aft sails; sometimes supplied with a square sail on the fore-topmast

SCUPPERS—holes cut through the sides and lined with lead to drain water overboard from a weather deck

SEAMS   longitudinal joints between planks

SHAPE—special rolled iron section with flanges for riveting to iron plates, usually L- or T-shaped

SHEATHING—thin sheets of copper or copper-bearing alloy (yellow metal) to protect a hull from wood borers and to prevent the growth of weeds and barnacles

SIDE LEVERS—two heavy levers low in the ship, one on each side of a cylinder, to transmit forces from the piston at one end to a connecting rod at the other. They are pivoted at their centers

SPENCER—a fore-and-aft sail set abaft the foremast or mainmast of a ship-rigged vessel. There is a gaff at the head but no boom at the foot of the sail

SQUARE STERN—same as transom stern

STABILITY—the ability of a floating body to right itself if pushed over to an angle. In a seaway, a vessel with high stability is extremely uncomfortable, and has a quick, jerky roll. Very low stability is dangerous, for then a gust of wind can overturn a vessel

STANCHION—vertical column or pillar supporting a deck or rail

STARBOARD—to the right

STEAM CHIMNEY—extension of a boiler's steam space up and around the base of the stack

STEM—heavy timber, vertical or raking, at the bow on centerline

STEP—the socket on top of the keelson, or on a deck, into which the lower end of a mast fits; also, a sudden change in deck level

STERNPORT—heavy, nearly vertical, centerline timber at the stern

STRAKE—one of a series of planks laid side by side

STUFFING BOX—devise with soft packing to prevent steam, air, or water leaking past a member that turns or slides

TEE—rolled-iron shape in the form of a T. Frequently used for deck beams of an iron vessel

TILLER—lever attached to, or inserted through, the rudder stock and used to turn it and the rudder

TONNAGE—legal measure of the interior volume, or size, of a ship. Prior to 1865 one ton equaled 95 cubic feet; after that date, 100 cubic feet. Before that date tonnage was computed by unrealistic formulae that made values for one-deck hulls out of line with two-deckers and rendered both results meaningless. Post-1865 American values were a better measure of a hull's actual interior volume plus that of any watertight structures extending above the upper deck. For freight, shippers charged by the cargo ton, of 40 cubic feet, or the weight ton, of 2,240 pounds, whichever produced the higher revenue

TRAIL BOARDS—curved supports connecting the cutwater to the stem and the bow planking. These were often carved, ornamented, and, in many cases, had the hawse holes for the anchor chains at their after ends

TRANSOM—wide, transverse, flat portion of the stern, usually pierced for windows to the cabin forward of it

TRANSOM STERN—stern with a transom, a holdover from the sailing ship that was replaced by rounded counter sterns; square stern

TRANSVERSE—side to side or athwartships

TREENAILS—large wood pins of oak or locust used to fasten planking and ceiling to the frames and to join the various futtocks of a frame together

TRUNNION—projection either side of an oscillating cylinder on which it rocks back and forth

TUMBLEHOME—inward slope of the sides of a ship; the opposite of flare

UPTAKE—the duct work above the boilers leading to the base of the stack to carry the smoke and other combustion products away

VALVE GEAR—series of levers, cams, eccentrics, and rods used to open and close the steam and exhaust valves of an engine. Inventors reveled in designing and patenting complex mechanisms for this purpose

WALKING BEAM—heavy, diamond-shaped, elevated lever pivoted at its center and transmitting the force from a vertical piston rod at one end to a connecting rod at the other. The name is a corruption of an earlier term, *working beam*

WALES—thick upper strakes projecting slightly beyond the outside planking and forming ornamental moldings along a ship's sides and around the stern. In later vessels, a series of thicker-than-normal strakes of side planking in the vicinity of the waterline

WATERWAYS—heavy timbers, port and starboard, forming deck boundaries at the intersection of deck beams and frames

WINDLASS—the means of hoisting an anchor, originally with a horizontal axis. In later ships it was replaced by a capstan

WORKING—continuous bending or deflecting of the hull in a seaway. Should this become excessive, the caulking would be forced out of the seams, with resultant leakage

Bibliography

Manuscripts

Cresson, J. C., and Alexander, J. H. *Report on Marine Condensers; and on Corrosion and Deposite in Steam Boilers*, Philadelphia, 1852 (Author's collection.)

John Ericsson Letters, 1831-1862. New York Historical Society.

Robert Fulton Papers. New York Historical Society.

Theodore P. Greene. Letter Book, June-December 1863. New-York Historical Society.

B. W. Ginsberg Records. A series of manuscript volumes in the collection of the late Frank C. Bowen.

Robert L. Livingston. Letter Book, 1817-1826. New-York Historical Society.

Richard W. Meade, II, U.S.N. Abstract Log of the *Massachusetts*. New-York Historical Society.

G. W. Murdock. Annotated picture albums. New York Historical Society.

Jeremiah Simonson. Notebook. Launches B. & C. S., Steamship Historical Society of America.

Stevens Family Papers. New Jersey Historical Society.

Herman Stryker. Diary. New York Historical Society.

U.S. National Archives. Naval Record Group 45, File AD, 1812-1859.

———. Record Group 41, Records of the Bureau of Marine Inspection and Navigation.

Virnelson, Marian. Summary of information about the *Ocean Bird*, ca. 1838.

Webb, William H. Certificate Books, vols. I & II.

———. Offset Book.

———. Ship Papers, Enrollments, Registers, Mortgages, and Bills of Sale. Webb Institute of Naval Architecture.

Newspapers—New York City

Commercial Advertiser

Commercial Pathfinder

Daily Advertiser

Daily Tribune

Evening Post

Gazette and General Advertiser

Globe

Herald

Journal of Commerce

Maritime Register

Mercantile Advertiser

Morning Courier and New York Enquirer

Shipping and Commercial List

Times

Other American Newspapers

Baltimore American

Baltimore American and Commercial Advertiser

Baltimore Sun

Boston Advertiser

Boston Courier

Boston Transcript

Boston Post

Boston Shipping List

Charleston Courier

Columbian Museum and Savannah Gazette

Frank Leslie's Illustrated Newspaper

Harper's Weekly

Lancaster (Pennsylvania) *Intelligencer*

New Orleans Picayune

New Orleans Price Current and General Intelligencer

Philadelphia Aurora and General Advertiser

Philadelphia Public Ledger

San Francisco Alta California

Foreign Newspapers

Belfast (Ireland) *Daily Mercury*
Illustrated London News
Willmer & Smith's European Times (Liverpool)
London Times
Mitchell's Maritime Register
Mitchell's Steam Shipping List

Periodicals

The American Neptune
Appleton's Mechanics' Magazine
The Artizan
California Historical Society Quarterly
Chambers Edinburgh Journal
The Engineer
Engineering
Gentleman's Magazine
Harper's New Monthly Magazine
The Historical Magazine
Journal of the Franklin Institute, 3d series
Master, Mate, and Pilot
Mechanics' Magazine
Merchant's Magazine and Commercial Review (Hunt's Merchants Magazine)
Monthly Nautical Magazine and Quarterly Review (U.S. Nautical Magazine), (U.S. Nautical Magazine and Naval Journal).
Nautical Gazette (Seaboard Magazine)
Pacific Marine Review
Scientific American
Scientific American Supplement
Steamboat Bill
United States National Museum Bulletin
U.S. Naval Institute Proceedings

Annual and Occasional Publications

American Bureau of Shipping Record
American Lloyds' Registry of American and Foreign Shipping
American Lloyd's Universal Register of Shipping
American Shipmasters' Association, Record of American and Foreign Shipping
House Executive Documents
Original American Lloyd's Register
Papers of the House of Commons
Transactions, Society of Naval Architects and Marine Engineers
Trow's New York City Directory

Books, Pamphlets, Articles

Alden, John D. *Monitors 'Round Cape Horn,* U.S. Naval Institute Proceedings, 1974.

Andriel, Pierre. *Coup d'oeil historique sur l'utilité des batimens-à-vapeur dans le royaume des Deux-Sicilies.* Naples, 1817.

Anon. Articles descriptive of the Caloric Ship *Ericsson* and of her Trial Excursion. . . . Washington, 1853.

———. [Carl C. Cutler]. *The Marine Collection at India House.* New York, 1935.

———. [Frederick Law Olmsted]. *Hospital Transports.* Boston, 1863.

———. *Golden Jubilee History of Nippon Yusen Kaisha.* Tokyo, 1935.

———. *70th Year History of Nippon Yusen Kaisha* (in Japanese). Tokyo, 1956.

———. *Schiffahrtsmuseum der Oldenburgischen Weserhafen.* Oldenberg, Germany, ca. 1968.

———. "The Old Shipbuilders of New York." *Harper's New Monthly Magazine*, 1882.

———. *Semi-Centennial Memoir of the Harlan & Hollingsworth Company.* Wilmington, Del., 1886.

———. *Souvenir of Roach's Shipyard.* Philadelphia, 1895.

Baker, William Avery. *A Maritime History of Bath, Maine.* Bath, 1973.

Bathe, Greville, and Bathe, Dorothy. *Oliver Evans.* Philadelphia, 1935; reprint ed. 1972.

Bartol, B. H. *A Treatise on the Marine Boilers of the United States.* Philadelphia, 1851.

Bauer, K. Jack. *Surfboats and Horse Marines.* Annapolis, Md., 1969.

Baughman, James P. *Charles Morgan.* Nashville, Tenn., 1968.

Beaufoy, Mark. *Nautical and Hydraulic Experiments.* London, 1834.

Bell, Malcolm, Jr. *Savannah Ahoy!* Savannah, Ga., 1959.

Bennett, Frank M. *The Steam Navy of the United States.* Pittsburgh, Pa., 1896.

Berthold, Victor M. *The Pioneer Steamer California, 1848-1849.* Boston and New York, 1932.

Bishop, J. Leander. *A History of American Manufacturers from 1608 to 1860.* 3d ed. Philadelphia, 1868.

Bonsor, N. R. P. *North Atlantic Seaway.* Prescott, Lancashire, 1955; 2d ed., vol. 1, Newton Abbott, London, Vancouver, 1975.

Bradlee, Francis B. C. *Blockade Running during the Civil War.* Salem, Mass., 1925.

———. *Piracy in the West Indies and Its Suppression.* Salem, Mass., 1923.

———. *Some Accounts of Steam Navigation in New England.* Salem, Mass., 1920.

Braynard, Francis O. "The First American Steam Passenger Line to South America." *The American Neptune,* 1944.

Braynard, Frank O. *S.S. Savannah.* Athens, Ga., 1963.

Brewington, M. V., and Brewington, Dorothy. *Marine Paintings and Drawings in the Peabody Museum.* Salem, Mass., 1968.

Brown, Alexander Crosby. "An Early American 'Neptune'." *The American Neptune,* 1952.

———. "James Rumsey—Steamboat Inventor." *Steamboat Bill,* 1969.

———. "The Sheet Iron Steamboat 'Codorus,' " *The American Neptune*, 1950.

———. "The Steamer *Vesta*," *The American Neptune, 1960*.

———. *Women and Children Last*. New York, 1961.

Bulloch, James D. *The Secret Service of the Confederate States in Europe*. New York, 1884.

Byrne, Alexander S. *Observations on the Best Means of Propelling Steamships*. 2d ed., New York, 1841.

Byrne, Oliver. *The American Engineer, Draftsman and Machinist's Assistant*. Philadelphia, 1853.

Chapelle, Howard I. "The Pioneer Steamship Savannah: A Study for a Scale Model." United States National Museum Bulletin 228, Washington, 1961.

Cheney, Robert K. *Maritime History of the* Merrimac. Newburyport, Mass., 1964.

Choules, John Overton. *The Cruise of the Steam Yacht* North Star. Boston and New York, 1854.

Church, William Conant. *The Life of John Ericsson*. New York, 1890.

Capt. Claxton, R.N. *Steamship* Great Britain. New York, 1845.

Clowes, William Laird. *Four Modern Naval Campaigns*. London, 1902.

Codman, John. *An American Transport in the Crimean War*. New York, 1896.

Colden, Cadwallader D. *The Life of Robert Fulton*. New York, 1817.

Cutler, Carl C. *Queens of the Western Ocean*. Annapolis, Md., 1961

Dayton, Fred Irving. *Steamboat Days*. New York, 1925.

Dickinson, H. W. *Robert Fulton, Engineer and Artist*. London, New York, and Toronto, 1913.

Emmerson, John C., Jr. *The Steamboat Comes to Norfolk Harbor*. Portsmouth, Va., 1949.

———. *Steam Navigation in Virginia and North Carolina Waters, 1826-1836*. Portsmouth, Va., 1949.

Ferguson, Eugene S. *John Ericsson and the Age of Caloric*. Contributions from the Museum of History and Technology, Paper 20. Washington, 1960.

Eugène Flachat. *Navigation à vapeur transocéanienne*. Paris, 1866.

Flexner, James T. *Steamboats Come True*. New York, 1944.

Forbes, Robert B. *Personal Reminiscences*. 2d ed. Boston, 1882.

Goodrich, Caspar F. *America's Part in Founding the German Navy*. U.S. Naval Institute Proceedings, 1924.

Griffiths, John W. *A Treatise on Marine and Naval Architecture*. new ed. London, 1856.

Hall, Henry. *Report on the Shipbuilding Industry of the United States*. Washington, D.C., 1882.

Haviland, Edward Kenneth. "American Steam Navigation in China." *The American Neptune*, 1956-1958.

Hodge, P. R. *The Steam Engine*. New York, 1840.

Holdcamper, Forrest R., ed.; William M. Lytle, comp. *Merchant Steam Vessels of the United States, 1807-1868*. The Lytle List. Mystic, Conn., 1952. Revised and edited by C. Bradford Mitchell, The Lytle Holdcamper List, Staten Island, N.Y., 1975.

Howe, Henry. *Memoirs of the Most Eminent American Mechanics*. New York, 1846.

Howland, S. A. *Steamboat Disasters and Railroad Accidents in the United States*. 2d ed. Worcester, Mass., 1840.

Hulls, Jonathan. *Description and Draft of a Newly-Invented Machine for Carrying Vessels or Ships out of and into any Harbor*. London, 1737.

Hungerford, Edward. *Story of the Baltimore and Ohio Railroad*. New York, 1928.

L'Illustration, Histoire de la Marine. Paris, 1934.

Isherwood, B. F. *Descriptions, Logs, Trial Reports of Steamships Contributed to the Journal of the Franklin Institute, 1852-1855*.

———. *Engineering Precedents for Steam Machinery*. New York, 1859.

Isherwood, J. H. "Collins Liner 'Adriatic,' " *Sea Breezes*, 1857.

Jacobus, Melancthon W. *The Connecticut River Story*. Hartford, 1956.

Johnson, Allen, ed. *Dictionary of American Biography*. New York, 1928.

Johnson, Robert Erwin. *Rear Admiral John Rogers, 1812-1882*. Annapolis, Md., 1967.

Kemble, John Haskell. "Genesis of the Pacific Mail Steamship Company." *California Historical Society Quarterly*, 1934.

———. *The Panama Route, 1848-1869*. Berkeley and Los Angeles, Calif., 1943.

———. *San Francisco Bay*. Cambridge, Md., 1957.

———. "Side-Wheelers across the Pacific." *The American Neptune*, 1942.

Klinkostrom, Axel. *Bref om de Forenta Staterna forfattade under en resa till Amerik*. Stockholm, 1824.

Lane, Wheaton J. *Commodore Vanderbilt*. New York, 1942.

Law, R. J. *The Steam Engine*. Science Museum, London, 1965.

Lawson, Will. *Pacific Steamers*. Glasgow, 1927.

Lee, Charles E. *The Blue Riband*. London, ca. 1930.

Lindsay, W. S. *History of Merchant Shipping*. London, 1876.

Lubbock, Basil. *The Down Easters*. Boston, 1929.

Macfarland, Robert. *History of Propellers and Steam Navigation*. New York, 1851.

McKay, Lauchlan. *The Practical Shipbuilder*. New York, 1839.

Maber, John M. *North Star to Southern Cross*. Prescot, Lancashire, 1967.

Main, T. *The Progress of Marine Engineering*. New York, 1893.

Marestier, Jean Baptiste. *Mémoire sur les bateaux à vapeur des Etats-Unis d'Amérique*. Paris, 1824. Partial English translation by Sidney Withington. Marine Historical Association, Mystic, Conn., 1957.

Mathews, F. C. "The Clipper Ship *Three Brothers*," *Pacific Marine Review*, 1921.

Mitchell, C. Bradfor, ed. *Merchant Steam Vessels of the United States, 1790–1868*. The Lytle-Holdcamper List. Staten Island, N.Y., 1975.

Mencken, August. *First-Class Passenger*. New York, 1938.

Mitman, Carl W. Catalogue of the Watercraft Collection in the United States National Museum, U.S. National Museum Bulletin 127, Washington, 1923.

Mone, Frederick. *Treatise on American Engineering*. New York, 1854.

Morrison, John H. *History of American Steam Navigation*. New York, 1903.

———. *History of New York Ship Yards*. New York, 1909.

———. "Iron and Steel Hull Steam Vessels of the United States, 1825-1905." *Scientific American Supplement*, 1905.

Murray, Andrew. *The Theory and Practice of Shipbuilding*. Edinburgh, 1861.

Murray, Marischal. *Ships and South Africa*. London, 1933.

Navy Department. Official Records of the Union and Confederate navies in the War of the Rebellion. Washington, 1894-1923.

———. Office of the Chief of Naval Operations, Naval History Division. *Civil War Chronology, 1861-1865*. Washington, 1962-1966.

———. *Dictionary of American Fighting Ships*. Washington, 1959-1976.

———. *History of Ships Named* Massachusetts, updated September 21, 1965.

———. "Summary of Dates from *Vanderbilt* Logs," n.d.

New York Board of Underwriters, Committee Report. "Causes of the Loss, USM Steamer *Central America*." 2d part. New York, 1858.

Olds, Edson B. Speech on the Appropriation for the Collins Steamers, Delivered in the House of Representatives, February 15, 1855.

Obson, B. S. *Handbook of the United States Navy*. New York, 1864.

Parker, H., and Bowen, Frank C. *Mail and Passenger Steamships of the Nineteenth Century*. Philadelphia, ca. 1930.

Partington, Charles Frederick. *An Historical and Descriptive Account of Steam Engines*. London, 1822.

Preble, Geo. Henry. *A Chronological History of the Origin and Development of Steam Navigation, 1543-1882*. Philadelphia, 1883.

Price, Marcus W. "Ships That Tested the Blockade of the Carolina Ports, 1861-1865." *The American Neptune*, 1948.

Pursell, Carroll W., Jr. *Early Stationary Engines in America*. Washington, 1969.

Rainey, Thomas. *Ocean Steam Navigation and the Ocean Post*. New York 1858.

Renwick, James. "Reminiscences of the First Introduction of Steam Navigation." *The Historical Magazine*, 1858.

Ridgely-Nevitt, Cedric. "The Baltimore and Liverpool Steamship Line." *Steamboat Bill*, 1965.

———. "The Steam Boat, 1807-1814." *The American Neptune*, 1967.

———. "1867—A Forgotten Year in the History of the American Steamship." *Steamboat Bill*, 1967.

Ringwald, Donald C. "First Steamboat to Albany." *The American Neptune*, 1964.

Robinson, William Morrison, Jr., *The Confederate Privateers*. New Haven, Conn., 1928.

Rumsey, James. *Explanation of a Steam Engine, and the Method of Applying It in a Boat*. Philadelphia, 1788.

———. *A Short Treatise on the Application of Steam*. Philadelphia, 1788.

Sanford, L. Cope. *A Century of Sea Trading, 1824-1924*. The General Steam Navigation Company Limited, London, 1924.

Scharf, J. Thomas. *History of the Confederate States Navy*. New York, 1887.

Spratt, H. Philip. *The Birth of the Steamboat*. London, 1958.

———. *Transatlantic Paddle Steamers*. 2d ed. Glasgow, 1967.

Stackpole, Edouard A. *The Wreck of the Steamer* San Francisco. Mystic., Conn., 1954.

Stanton, Samuel Ward. "A History of the First Century of Steam Navigation." *Nautical Gazette*, 1907-1910, and *Master, Mate and Pilot*, 1910-1912.

Stevens, John Austin, Jr. Memorial of the Chamber of Commerce of the State of New York to the Senate and House of Representatives. New York, 1864.

Stuart, Charles B. *Naval and Mail Steames of the United States*. 2d ed. New York, 1853.

Sutcliffe, Alice Crary. *Robert Fulton and the "Clermont."* New York, 1909.

Swann, Leonard Alexander, Jr. *John Roach*. Annapolis, Md., 1965.

Tredgold on the Steam Engine, Marine Engines and Boilers: Exemplified in Numerous Examples of British and American Steam Vessels. London, ca. 1851.

Turnbull, Archibald Douglas. *John Stevens*. New York and London, 1928.

Tyler, David B. *The American Clyde*, Newark, Del. 1958.

———. *Steam Conquers the Atlantic*. New York, 1939.

The Union Defense Committee of the State of New York, Minutes, Reports, and Correspondence, New York, 1885.

Ward, J. H. *Steam for the Millions*. New York, 1862.

Watkins, J. Elfreth. The Log of the Savannah. Report of the United States National Museum for 1890.

Webb, William H. *Plans of Wooden Vessels*. Privately printed and distributed by the author. New York, ca. 1895.

Westcott, Thompson. *Life of John Fitch*. Philadelphia, 1857.

Whittemore, Henry. *The Past and the Present of Steam Navigation on Long Island Sound*. Providence and Stonington Steamship Company, 1893.

Wiltsee, Ernest A. *Gold Rush Steamers*. San Francisco, Calif., 1938.

Wright, E. W. ed. *Lewis and Dryden's Marine History of the Pacific Northwest*. New York, 1895; reprint ed., 1961.

U.S. Congress. Mail Contracts by Steamships between New York and California, 32d Congress, 1st sess., House Executive Document 124.

———. Vessels Bought, Sold, and Chartered by the Quartermaster Corps since April 1861, 40th Congress, 2d sess., House Executive Document 337.

Index of Names of Vessels

A. Winants, 288
Acadia (Cunard), 99, 143
Acapulco, 241, 206
Adela, 277
Admiral (sail), 173
Adriatic (Collins), 161-70, 185, 230-32, 237, 250-53, 317, 336-41, 347-48, 357-58, 368
Adriatic (sail), 253
Advance, 309
Africa (Cunard), 156-57, 162, 176
Alabama (sail), 83
Alabama, U.S.S., 203
Alabama, C.S.S., 203, 238, 245, 247, 325
Alaska (ex-*Massachusetts*), 96, 364
Alaska (1868), 307, 336, 342-45, 349, 365, 368
Alexander Petion (ex-*Galatea*), 278-79, 292
Alhambra, 278
Allegany, 292
Alma (sail), 124
Alma (French), 184
Alps (Cunard), 184
America (Cunard), 154, 157-58
America (sailing yacht), 150, 270
America (towboat), 195
America (ex-*Coatzacoalcos*), 214
America (1869), 336-44, 349, 364-65, 368
American Eagle, 317
Andalusia, 278
Anna Maria (sail), 213
Annie Childs (ex-*North Carolina*), 285, 368
Arabia (Cunard), 160
Arago, 162, 167, 176-85, 213, 230, 244, 264, 296-303, 312-13, 322-23, 347, 357-58, 368
Arctic, 122, 150-56, 159-62, 168, 170, 213, 232-33, 357-58, 367
Ariel, 137, 213, 226-39, 241-42, 244, 250, 292, 312-13, 324-25, 341, 359, 368
Arizona, 241, 306, 336, 341-45, 364, 368
Ashland, 282, 362-63
Ashuelot, U.S.S., 335
Asia (Cunard), 157-59, 162, 173
Atahualpa, 299
Atalanta, 67
Atlantic, 122, 150-51, 154-61, 163, 169-70, 173, 213-14, 232, 244-46, 250-55, 261, 263-64, 269-70, 288, 302, 311-15, 318, 347, 357-58, 367
Augusta, 203, 277, 298
Australasian (Cunard), 252-53
Baltic, 150-63, 169-70, 192, 213-14, 232, 244, 250-56, 269, 302, 311-15, 318, 326, 344, 347, 357-58, 366-67
Baltimore (North German Lloyd), 293
Bangor (1835), 80-82, 96, 351-52, 367
Bangor (1845), 282, 285, 362-63
Bangor (1850), 155
Banshee, 297
Barcelone (French), 184
Bellona, 56

Benjamin Franklin (1826), 47-48, 99, 351-52
Benjamin Franklin (1851), 199-200
Big Brother (*Three Brothers*), 247
Blanco de Gray, 265
Borussia (Hamburg American), 137, 234
Bremen (North German Lloyd), 294
Britannia (Cunard), 63, 99, 131, 176
Brother Jonathan, 225
Brunette, 323, 327

C. Vanderbilt (*Vanderbilt*), 330
Cadmus (sail), 56
Caledonia (Cunard), 99
California, 102-10, 115-16, 167, 264, 337, 353-54, 366
Cambria (Cunard), 156, 173
Cambria (sail), 161
Canada (Cunard), 117, 158
Canada (National), 269
Capron, 344-47, 360-61, 368
Car of Neptune, 33-36, 43, 350
Carroll (ex-*Proteus*), 279, 292-95, 309, 328, 360, 368
Catawba, U.S.S., 299
Celestial Empire (China), 337
Central America (ex-*George Law*), 103, 124, 263, 275
Champion, 236, 238, 242, 282, 359
Chancellor Livingston, 39-49, 52, 56, 58, 62, 73, 80-81, 104, 111, 222, 350
Charles Pearson, 317
Charleston, 77, 351
Cherokee, 110-14, 120, 123, 125, 203, 353-54
Chicora, U.S.S., 277
China, 336, 339, 343, 364-65
Circassian, 214, 286-90, 292, 311-13, 324, 347, 362-63, 366, 368
City Ice Boat, 195
City of Buffalo, 365
City of Glasgow (Inman), 194-95
City of Manchester (Inman), 194-95
City of Panama, 307
City of Peking, 343
City of Philadelphia, 194
City of Pittsburg, 192, 194-95, 208, 344, 360-61, 367
City of San Antonio, 349
City of Tokio, 343
Clarion, 83-86, 96, 352
Clermont (*North River*), 23, 26
Codorus, 281
Colima, 341
Colon, 241, 307
Colorado, U.S.S., 246
Colorado (1865), 336-37, 343, 364
Columbia (1835), 75-76, 80, 351-52
Columbia (1857), 163, 170, 260, 263-64, 355-56, 368
Columbia (Cunard), 99
Columbia, C.S.S., 246
Columbia (Galway), 252
Columbia (ex-*Quaker City*), 278
Columbia (1877), 365
Commodore (Costa Rica), 243
Congress, U.S.S., 244
Connecticut (1816), 37-38, 43, 47-49, 56, 58, 71, 81, 350
Connecticut (1861), 278, 305
Constitution, 336-37, 341, 364-65
Corinthian (sail), 159
Costa Rica, 239-44, 248, 272, 302, 324, 328, 341, 343-44, 359, 368
Crescent City (1848), 108, 114-17, 120-24, 141, 146, 150, 230, 353-54
Crescent City (1853), 301, 365
Crescent City (ex-*Massachusetts*), 264, 283, 330, 362
Cruiser, 253
Cuba, 164
Cumberland, U.S.S., 244

378

Dakota, 322
Daniel Webster, 193, 223, 225, 303, 327, 359
David Brown, 74-75, 80, 351
Decatur, U.S.S., 95
Defiance (sail), 124
Delaware (transport), 297
Delaware (Clyde), 328
Demologos (steam battery), 66
Denmark (sail, ex-*Great Republic*), 256
Director, 223
Dolphin, 78-80, 351
Doras (sail), 293
Douro, 277-78
Dunbarton (sail), 277
Dunderberg (steam ram), 156

E. C. Knight, 286
Eagle (1813), 58, 71
Eagle (Spanish), 83, 150
Eagle (1860), 263
Edith, 88-89, 91, 94, 96, 352, 367
Ella Warley (ex-*Isabel*), 125, 239, 325
Ellen (sail), 124
Eliza Ann (sail), 102
Elizabeth (Russian), 65
Elizabeth, 292
Emily (sail), 274
Emily B. Souder, 330
Emma, 297
Emperor of Russia (*Connecticut*), 37
Empire City, 114-16, 120-25, 150, 213, 265, 268, 353-54
Empire State, 329, 365
Emu (Cunard), 184
England (National), 269
Enoch Train (sail), 220
Enterprize, 47, 54, 66, 351
Ericsson, 162-63, 170, 208-15, 244, 263, 288, 296, 312-13, 324, 357-58, 367-68
Erie, 316-21, 344, 360, 368
Erzherzog Johann (ex-*Acadia*), 146
Esmerelda (sail), 159
Etna (Cunard), 184
Europa (Cunard), 157
Europe (French), 297
Evening Star, 301-4, 307, 326-27, 363, 365
Experiment, 15

Falcon, 108, 116-17, 120-23, 146, 212, 269, 282, 353-54
Farralones (ex-*Massachusetts*), 96
Fidelity, 67, 352
Finance, 309
Fingal, 285
Fire-Fly, 35, 43-45, 350
Florida (1850), 203, 278-79
Florida, C.S.S., 325
Foong Shuey (ex-*Plymouth Rock*), 335
Fox (sail), 274
Francis Skiddy, 225
Franklin, 157, 172-76, 181, 184, 186, 201, 208, 261, 275, 357-58, 367
Friend (sail), 156
Friendship (sail), 274
Fulton, U.S.S. (*Demologos*), 66
Fulton, U.S.S. (1837), 344
Fulton (1813), 36-38, 43, 47-49, 58, 64, 71, 350
Fulton (1856), 137, 167, 182-85, 213, 232, 264, 296-99, 302, 312-13, 347, 357-58
Fung Shuey (ex-*Plymouth Rock*), 335
Fung Shuey (1864), 330, 332-36, 340, 360-61, 368

Galatea, U.S.S., 278, 291-92
Gazelle (sail), 194
Gembu Maru (ex-*Kuroda*), 345

General Admiral (Russian), 156
General Meade, 330
Genil, 141
Genkai Maru (ex-*Costa Rica*), 244
George Canning (sail), 65
George Law, 103, 120, 124, 275, 335-56
George W. Clyde, 336
Georgia (1836), 79, 351
Georgia (1850), 115-26, 152, 175, 206-7, 222, 265, 269, 271, 353-54
Georgia, C.S.S., 245, 247
Germania (sail), 110
Germania (ex-*Acadia*), 146
Gertrude, 245
Glaucus, 291, 360
Glory of the Seas (sail), 215
Golden Age, 244, 269-72, 337, 341-42, 355-56, 367
Golden City, 336-39, 341-42, 364-65
Golden Eagle (sail), 254
Golden Gate, 167, 178-80, 272, 355-56
Golden Rule, 214, 304
Goliah, 191
Gorgona, 189
Governor Dudley, 123
Granada, 268
Grand Gulf, U.S.S., 297
Grand Republic, 365
Granite State, 202, 360, 368
Grapeshot (sail), 164
Great Britain, 116, 154, 281
Great Eastern, 156
Great Republic (sail), 164, 256
Great Republic, 244, 336, 343, 364-65
Great Western, 99
Guiding Star, 297-98, 302-7, 312-13, 322-23, 334, 349, 363, 365, 368
Gulf Stream (ex-*North Carolina*), 285-86, 362

Habana, 276
Hammonia (Hamburg American), 137, 234
Hansa (ex-*United States*), 141-43, 146-47
Harriet Lane, 254
Hartford, U.S.S., 274
Harvey Birch (sail), 261
Havana, 282, 299, 304-5, 362-63
Havre (*Humboldt*), 172
Heath Park (sail), 289
Henry Chauncey, 241, 254, 306-7, 336, 341-42, 364
Hercules, 192
Hermann, 132-41, 143, 147, 151, 157, 159, 162, 172-75, 189, 213, 232-33, 242, 261, 347, 357-58, 367
Hibernia (Cunard), 129, 141
Hibernia (Galway), 253
Hiroshima Maru (ex-*Golden Age*), 244, 272
Home, 76-78, 80, 98, 351-52
Hope, 35, 46, 35-52
Hoqua (Sail), 150
Hornet, U.S.S., 55
Housatonic, U.S.S., 277
Howquah, U.S.S., 296
Hudson, 303
Humboldt, 172-75, 178-79, 184, 186, 208, 260-61, 357-58, 367
Huron (sail), 161

Illinois, 120, 167, 178-79, 236-37, 244, 251, 264-69, 276-77, 296, 355-56, 368
Independence, 191
Indian Empire (ex-*United States*), 147-48
Indiana (British), 175-76
Indiana (1873), 334
Iron Witch, 86, 123, 212, 282, 354
Isaac Mead (sail), 102
Isabel, 112, 125, 157, 239, 254, 260, 272, 325, 353-54
Isthmus, 108

James Adger, 103, 260, 263, 273, 355-56
James Kent, 46, 49, 104, 184, 350
Jane (sail), 189
Jane, 274
Japan, 336-37, 340, 343, 364-65
Johanna Elizabeth (sail), 245
John L. Stephens, 181, 206, 272, 337, 355-56
John M. Clayton, 223
John Marshall, 130
Josephine Thompson, 286
Julie Usher (ex-*North Carolina*), 285
Juniata (ex-*South Carolina*), 283
Jupiter, 292
Jura (Cunard), 184

Kamschatka (Russian), 98
Karafuto Maru (ex-*Capron*), 345
Kennebec, 195
Kepuron Maru (ex-*Capron*), 345
Keystone State, 206, 277, 296
Khersonese, 286-87
Kite (sail), 65
Knickerbocker, 303
Kuroda, 344-46, 360-61, 368
Kyoryu Maru (ex-*Capron*), 345

Lady Chapman (sail), 121
Lafayette (*Franklin*, 1848), 131, 172
Lafayette (1851), 188-89, 208, 360-61, 367
Lafayette (French), 297
Lebanon (sail), 161
Lebanon (Cunard), 184
Legislator, 48, 351
Leviathan, 175
Lion, 83, 131, 150
Little Juliana, 27, 206
Lively (sail), 36
Louis (sail), 285
Louisa (sail), 274
Lyonnais (French), 184

MacDonough (*Macdonough*), 49, 351
Madawaska, U.S.S., 212
Magdalena, 272
Manco Capac (ex-*Oneota*), 299
Maranon (ex-*Havana*), 299
Maratanza, U.S.S., 278
Marcia (sail), 123
Margaret and Jessie, 296
Marine (sail), 124
Marion (1851), 103, 174, 260, 263
Marion (ex-*Santiago de Cuba*), 328-29, 366
Mariposa, 304-5
Marmora, 86-87, 96, 352, 367
Massachusetts (1845), 89-96, 117, 134, 141, 180, 193, 212, 289, 344, 347, 352, 361, 367
Massachusetts (1860), 283, 362
Matanzas, 282, 304, 362-63
Mercedita, U.S.S., 277, 285
Merrimack, U.S.S., 244
Merrimac (*Merrimack*, 1862), 284
Merrimack (1862), 283-84, 297-98, 304-13, 320-21, 362-63, 368
Mersey (sail), 325
Mexican, 293-94
Michigan, U.S.S., 281-82
Midas, 86-87, 96, 352, 367
Minna, 287
Minnesota, 246
Mississippi, U.S.S., 100, 152, 344
Mississippi (1853), 302, 365
Mississippi (1862), 283-84, 297-98, 304-13, 320, 347, 362, 368
Mississippi (ex-U.S.S. *Memphis*), 330
Missouri, U.S.S., 100

Missouri (1862), 304
Mobile (ex-*Tennessee*, 1854), 274
Mohawk, U.S.S., 124
Monadnock, U.S.S., 246
Montana, 336, 341, 364
Montauk, U.S.S., 263
Monterey, 304-5
Montgomery, 274
Morning Star, 298, 301-5, 363, 365, 368
Moro Castle (ex-*Morro Castle*), 366
Morro Castle, 263-64, 327, 363, 365-66
Moses Taylor, 214, 236-37
Mount Vernon (1820), 71
Mount Vernon, U.S.S., 284
Mouse of the Mountain, (ex-*Stoudinger*), 222, 351

Nagoya Maru (ex-*Oregonian*), 244
Nansemond, U.S.S., 278, 296
Napoleon III (French), 297
Nashville, 103, 161, 170, 174, 186, 260-64, 272-73, 275-76, 325, 355-56, 367
Natchez, 80, 351-52
Nautilus, 56
Nemeha, 292
Neptune (1838), 77-78, 80, 351-52
Neptune (1863), 291-92
Neptune (tug), 247
Nereus, U.S.S., 291-92, 360
New Jersey, 58
New Orleans, 123
New Philadelphia, 46, 351
New World, 155
New York (1822), 48, 67-74, 83, 86, 352
New York (1837), 76, 80, 351
New York (1865), 241-44, 248, 272, 302, 328, 341, 343, 359, 368
Niagara (Cunard), 157
Niagara (1865), 220
North America (1827), 74
North America (ex-*Fort Jackson*), 305, 322
North Carolina (sail), 277
North Carolina (1854), 204-7, 360-61, 368
North Carolina (1861), 282-86, 362-63, 368
North River (Steamboat), 23-27, 31-36, 38, 40-42, 49, 58-59, 100, 156, 350
North Star, 137, 213, 223-34, 237-41, 250, 254, 264, 271-72, 274, 278, 359, 367
Northern Light, 191, 223-25, 234-42, 250, 254, 256, 264, 274, 278, 312-15, 359, 368
Northerner, 101-3, 121-24, 257, 260, 353-54, 368
Nuevo Pajaro del Oceano (ex-*Niagara*), 220

Ocean, 282, 362
Ocean Bird, 215-21, 344, 355-56
Ocean Queen, 213, 236-37, 241-42, 306-7, 312-13, 342, 359, 368
Ohio, U.S.S., 109
Ohio (1848), 115-17, 120-23, 126, 152, 175, 222-23, 265, 353-54
Olive Branch, 38-39, 46, 350
Oliver Ellsworth, 56
Oneota, U.S.S., 299
Onondaga, U.S.S., 246
Ontario, 315-21, 344, 347, 360-61, 368
Oregon, 102-5, 108-10, 115, 122, 125, 167, 264, 353-54, 366
Oriental, 282, 362
Oregonian, 244, 272
Orizaba, 235
Orus, 108

Pachitea (ex-*Arago*), 299
Pacific (Collins), 150-59, 161-63, 170, 179, 213, 225, 232, 261, 276, 344, 357-58, 367
Pacific (Lowry and Jarvis), 223, 225
Pajaro del Oceano (ex-*Ocean Bird*), 219-20, 355, 368
Palm (Sail), 156

380

Palmetto State, U.S.S., 277
Panama, 102-3, 105, 108-10, 115, 167, 264, 353-54
Paragon, 34-36, 43, 350
Patent, 48, 67, 70, 75, 351
Pawnee, U.S.S., 254
Peerless, 203
Pennsylvania, 345, 349
Periere (French), 297
Perseverance (Fitch), 15
Perseverance (1811), 35, 46, 351
Persia (Cunard), 162-64, 232-33, 251-52
Peterhoff, 245
Petersburg, 71
Philadelphia (1812), 38
Philadelphia (1849), 120, 123
Phoenix (1808), 26-28, 32-33, 38, 58, 351
Phoenix (1815), 35
Pioneer, 177, 191-97, 223, 317, 344, 360-61, 366-67
Portland, H.M.S., 195
Portland (1835), 49, 350-51
Potomac, 71
Powhatan (1817), 67
Powhatan, U.S.S., 246
President, 48, 351-52
Prince Albert, 198
Princeton, U.S.S., 189
Prometheus, 223-25, 359
Proteus, U.S.S., 291-92, 360
Pulaski, 79-80, 351-52

Quaker City, 162, 170, 275-80, 288, 292, 312-313, 347, 355-56, 368
Queen, U.S.S. (ex-*North Carolina*), 285
Queen of the Pacific (*Ocean Queen*), 335
Queen of the West, 301, 365
Queen of the South, 137

Rainbow (sail), 117
Rapidan, 293
Rariton, 28-35, 38, 46, 75, 350
Rattlesnake (ex-*Nashville*), 263
Red Jacket (sail), 283
Reliance, 309
Relief, 288
Republic (1849), 167
Republic (ex-*Tennessee*), 239, 264, 274, 278, 303
Rescue, 247
Richmond (1814), 35-37, 41, 43, 350
Richmond (1818), 71
Rising Star, 302-7, 313, 342, 363, 365, 368
Robert (sail), 207
Robert Fulton, 52-55, 64, 68, 71, 73-74, 76, 78, 80, 98-99, 103, 111, 353-54
Robert Gilfillan, 261
Roscius (sail), 150, 192, 225
Roscoe (sail), 123
Rosita, 329
Royal William, 99
Ruby, 292

S. R. Spaulding, 303
S. S. Lewis, 189-92, 195, 199, 202, 224-25, 360-61, 367
Sacramento, 336-37, 364
St. Lawrence, 218
St. Louis (sail), 156
St. Louis (1854), 164, 176-80, 184, 186, 274-75, 355-56, 368
Salnave (ex-*Maratanza*), 278-79
Salvador, 341
Samson, 192
San Francisco (1853), 182-83, 260, 355-56
San Francisco (ex-*Keystone State*), 322
San Jacinto, U.S.S., 253
Santiago de Cuba, 214, 285, 312-13, 322-29, 363, 365-66, 368

Sarah A. Smith (sail), 123
Sarah Ann (sail), 159
Sarah Sands, 101, 103
Saranac, U.S.S., 43, 235
Savannah, 57-67, 71, 86, 88-89, 289, 345, 347, 352, 366-67
Saxon (sail), 244
Scourge (ex-*Bangor*), 282
Screw Steamer (ca. 1855), 196-98
Sea Gull (ex-*Enterprize*), 54, 66, 351
Sea Horse, 351
Sea Witch (sail), 109, 117
Shenandoah (sail), 155
Sherman, 330
Sierra Nevada, 235
Sir Henry Bulwer, 223
Sirius, 99
Somerset, U.S.S., 287
Somerset (1863), 279, 292-95, 328, 360, 368
Sonora, 138, 176, 178, 272, 274, 355-56
South America (ex-*Connecticut*), 305, 307, 320
South Boston (sail), 76
South Carolina (1835), 79, 351
South Carolina (1851), 195-96, 360, 366-67
South Carolina (1860), 283, 285, 362-63
Southerner, 93-103, 121, 124, 129, 150, 163, 257, 260, 353-54
Star of the South, 202-3, 206, 360-61, 368
Star of the West, 224-25, 228, 250, 268, 359
Steam Boat (*Steamboat, North River*), 24-27
Stoudinger, 222, 351
Suchil, 276
Sudaver (ex-*Bangor*), 82, 351
Superb, H.M.S., 36
Suwo Nada (*Suwonada*), 331
Suwonada, 330-35, 340, 363, 365, 368

T. D. Wagner (ex-*North Carolina*), 285
Tell-Tale (sail), 68
Tennessee (1848), 110-14, 117, 125, 179, 192, 203, 265, 273, 353-54
Tennessee (1854), 239, 264, 272-78, 285, 355-56, 368
Tennessee, C.S.S., 274
Thomas L. Wragg (ex-*Nashville*), 262
Three Brothers (ex-*Vanderbilt*), 246-47, 255-56, 366
Tokio Maru (ex-*New York*), 244
Trent, 245, 253
Trumpeter, 292
Tuscarora, U.S.S., 246, 271

Uncle Sam, 227
Union, 157, 175, 177, 184, 186, 257-61, 264, 275, 355-56, 367
United States (1821), 47
United States (1848), 115, 125, 140-48, 173, 357-58, 367
United States (1864), 330

Vanderbilt, 169, 185, 213, 228-41, 244-47, 251-55, 268-69, 278, 296, 302, 317, 324, 328, 336, 347-48, 359, 366, 368
Varuna, U.S.S., 274
Verena (sail), 334
Vermont, U.S.S., 203
Vesta, 160-61
Vettor Pisani, 340
Vicksburg, 285
Victoria (sail), 255
Victory (ex-*North Carolina*), 285, 326
Vigo (French), 184
Ville de Paris (French), 297
Virginia (1817), 71
Virginia (ca. 1820), 71
Virginia, C.S.S., 244, 296
Virginia (National), 269
Vixen, U.S.S., 234

Wabash, U.S.S., 246
Wasdale (sail), 215

Washington (1815), 37, 350
Washington (1825), 47-48, 71, 99, 351
Washington (sail), 234
Washington (1847), 101, 128-43, 147, 150-51, 155, 159, 162-63, 172, 174-75, 184, 213, 233, 294, 347, 357-58, 367
Washington (French), 297
Waterloo, 65
West Wind (1851), 198, 360-61, 367
Western Metropolis (1856), 329, 365
Western Metropolis (1863), 214, 254-55, 311-14, 329-30, 363, 365, 368
William Gibbons, 74-76, 80, 98, 351
William H. Webb, 169, 278
William Norris (*Ocean Bird*), 217
William Penn, 199-202, 206, 343, 360-61, 367
Winfield Scott, 260
Wissahickon, U.S.S., 263
Winchester (sail), 136
Worcester (ex-*Glaucus*), 279, 292-95, 309, 328, 360-61, 368

Yankee Blade, 227
Yorkshire (sail), 94
Young America (sail), 215, 247

Index of Owners, Operators, Shipbuilders, and Engine Builders

Abrahams and Sons, John J., 292
Accessory Transit Company, 224, 227
African Steamship Company, 253
Alexander, James, 196
Alexandre Line, 309, 328
Allaire, James P., 33, 39, 59, 63, 66, 74, 76, 78, 232
Allaire Iron Works, 104, 126-27, 150, 152, 154, 223, 225, 229, 232, 236, 260, 264, 344, 349
Allen, Daniel B., 237, 304-5, 309
Allen, Horatio, 128, 153, 168
Allmand, John, 68
American Atlantic and Ship Canal Company, 138
American Steamship Company (1854), 275
American Steamship Company (1864), 316-21
Anchor Line, 247
Asencio and Company, Thomas, 304
Aspinwall, George A., 198
Aspinwall, William H., 104, 176, 275
Atlantic and Pacific Steamship Company, 235, 237
Atlantic Mail Steamship Company, 238, 263
Atlantic Royal Mail Steam Navigation Company, 252
Atlantic Steam Navigation Company, 147

Badell, J. and S., 203
Baird, James, 284
Baker, Frederic, 330
Baker, M. V., 275

Balch, Joseph W., 316
Baltimore and Liverpool Steamship Company, 291-95, 314-15
Baltimore and Ohio Railroad, 292, 294, 347
Baltimore and Southern Packet Company, 272
Bangor Steam Navigation Company, 282
Barnes, Joseph, 17
Bates and Company, 253
Bell, Isaac, 297
Bell, Jacob, 150-51, 156, 191, 301
Bender, Wright, and Company, 198
Benner, A., 329
Benner and Brown, 311, 329
Bennet and Henderson, 99
Bethlehem Steel Company, 282
Betts, Harlan, and Hollingsworth, 282
Birely, Theodore, 189, 199, 202, 207
Birely and Son, Theodore, 189, 199
Bishop, Joseph, 115, 125
Bishop and Simonson, 115, 222-23
Black Ball Line, 140
Black Star Line, 335
Blossom, Smith, and Dimon, 46
Boardman, 218
Bogert, Stephen G., 235
Boston and Bangor Steamship Company, 80
Boulton, Mathew, 13
Boulton and Watt, 20, 23, 33, 35, 42
Brazil Line, 305, 307, 309, 320, 323
Bremen Line, 128-39, 147, 159, 174, 178, 185, 187, 233, 348
Brigham, N. H., 288, 327
Brooks, Davis, 260
Brown, Adam, 150
Brown, Adam and Noah, 36-37, 150
Brown, David, 150
Brown, George, 80
Brown, James, 147, 163, 170
Brown, Noah, 150
Brown, Stewart H., 147, 170
Brown, William H., 98-103, 114, 117, 121-22, 124-25, 150-51, 155, 223, 244, 272, 337
Brown and Bell, 67, 74-76, 80, 150
Brown Brothers, 147, 154, 170
Brownell, Thomas, 129
Brownne, Charles, 20-21, 33
Bunker, Elihu S., 35-37, 47, 53, 58

Caldwell, Robert, 195
California, New York, and European Steamship Company, 138
Calla, Etienne, 19
Carrington, D. N., 322
Ceballos, J. M., 219
Central American Transit Company, 214, 322, 326
Chauncey, Henry, 104, 140
China Sea, Saigon, and Straits Steamship Company, 334
Churchward, Richard, 68
City Point Works, 316
Clark, James, 292
Clyde, William P., 286
Clyde and Company, William P., 264, 286, 327-28
Coast Wrecking Company, 288-89
Cobanks and Weeks, 343
Colden, Cadwallader D., 36-37, 53
Collins, Edward K., 128, 140, 147, 149-50, 154, 156, 158, 160-61, 163, 168, 170, 232, 269
Collins Line, 137, 149-73, 178, 180-81, 187, 210, 213-14, 224, 230, 232, 246, 251, 253-55, 260-61, 263, 265, 269-70, 276, 279, 298-99, 311, 329-30, 337, 347-48
Collyer, William, 261, 263
Compagnie Franco Américaine, 184
Compagnie Générale Transatlantique, 297
Confederated German States, 146
Continental Mail Steamship Company, 214, 288-89, 312, 324
Coombs and Taylor, 247

Copeland, Charles W., 100, 152, 192, 316, 344
Coryell, Miers, 182, 251
Cramp and Sons, William H., 345, 349
Crockett, Samuel, 59
Cromwell and Company, H. B., 274, 282, 285, 329, 335
Croskey and Company, 175
Cross, Amos H., 48
Cunard Line, 129, 131, 135-37, 142, 154, 161-62, 184, 192, 197, 212, 233, 251, 318, 321
Cunningham and Belknap, 301
Currier, W. C., 318

Dale, Mary A., 188
Dana, Charles, 322
Davis, Gartner, and Webb, 282
Day and Ballock, 174
De Agreda, Jove, and Company, S., 273
De Wolf, James R., 289
De Wolf, John S., 289
Dejois, Admiral, 279
Delamater, Cornelius H., 188, 282, 330, 344
Delamater Iron Works, 282, 330, 344, 349
Delaware River Iron Shipbuilding and Engine Works, 349
Dickerson, E. N., 168
Dimon, John, 46, 104
Dispatch Line, 327
Dod, Daniel, 59, 67, 70
Dole, John P., 323
Dramatic Line, 140, 150
Duncan, S. H., 125
Dunham, David, 52, 55-57
Dunham and Company, 56-57, 213-14

Eckford, Henry, 39, 46, 52, 104, 111
Edwards, D. W., 299
Empire City Line, 116, 123, 195, 270
Englis, Charles M., 331
Englis, John, Jr., 331
Englis, John, Sr., 331
Englis and Son, John, 331, 349
Ericsson, John, 83, 86-89, 123, 134, 162, 189, 197, 202, 208-12, 232
Etna Iron Works, 302, 349
Evans, Oliver, 13, 43
Everett, Percival L., 331

Faron, John, Jr., 152
Fickett, William, 59
Fiedler, Ernest, 288-89, 313
Fitch, John, 13, 17-18, 20, 58, 67
Forbes, J. M., 86, 89
Forbes, Robert B., 86-89, 128
Foster and Company, J. H., 284
Fox and Livingston, 172
Fox, 176, 275
Franklin Iron Works, 189, 202
Fraser, Trenholm, and Company, 262
French Line, 297-98
Fritze, W. A., 146
Fritze, Lemkuhl, and Company, 147
Fuller, Dudley, 170
Fulton, Robert, 13, 18-31, 33-36, 39, 46, 59, 63, 100, 212
Fulton Iron Works, 219, 349

Galway Line, 147-48, 252-53, 287, 289
Gardner, James S., 89
Garrison, Cornelius K., 227, 240, 304-5, 309, 320
Garrison, William R., 305, 309
Garrison and Allen, 263, 282, 304-5, 307
Gibbons, Thomas, 46
Gibson, Nehemiah, 309
Gillilan, Edward, 192
Glover, Russell, 83

Glover, Stephen, 83
Godeffroy, Branker, and Company, 298, 313
Goodall, Nelson, and Perkins, 110
Graham, John, 217-18
Griffiths, John W., 21, 117, 121-22, 150, 179, 206-7, 215-17, 228, 230, 344
Griswald, George, 329
Guilleaume and Company, O. I., 327
Guion, Archer, 123
Guion and Boardman, 218

Hackstaff, William G., 140, 142
Haggerty, John, 75-76
Hall, Samuel, 86-89, 283
Hallowell, Rufus, 87
Hamburg American Line, 137, 233, 311
Hammill, Henry F., 240
Hargous and Company, 276
Hargous Brothers, 276
Harlan and Hollingsworth, 223, 236, 282, 349
Harland and Wolff, 286
Harris, Arnold, 104
Hart, Samuel, 150
Hartson, George B., 263, 305
Harnden Express Company, 189
Haswell, Charles H., 132, 212
Havre Line, 128, 159, 167, 172-87, 229-30, 232, 260-61, 264, 273, 275-76, 286, 296-300, 302, 305, 314-15, 348
Heard and Company, Augustine, 331, 334
Henderson, John, 292
Heron, Alexander, Jr., 206
Heron and Martin, 276
Hickson and Company, Robert, 286
Hildreth, J. B., 335
Hill, Hamilton, 316
Hogg, Peter, 188
Hogg and Delamater, 86, 188, 210-11, 282
Holdridge, Nathan A., 66-67
Holladay and Flint, 110
Hooper, James, 272-73, 292
Howard, John, 122
Howard, Joseph, 122
Howard and Son, 123
Howes, Eben, 316
Howes, George, 246
Howland, Gardiner G., 104
Howland and Aspinwall, 104
Hubbell, Henry W., 335
Hulls, Jonathan, 13
Hunter, W. C., 86
Hunter, William F., 68

Inman, William, 194
Inman Line, 194, 348
Iselin, William, 177

Jackman, George W., 316-18
Jouffroy, Marquis de, 13

Kenyon, Moses, 35
King, Edward, 89
Kitching, John B., 162, 208-14

Lasala and Company, J. B., 80
Latrobe, Benjamin, 18
Law, George, 114, 124
Law Line, 114, 116, 121
Lawrence and Sneeden, 78
Leary, Arthur, 278, 288-99, 311, 313
Leary Brothers, 278
Leese, R. H., 198
Lemkuhl, Carl, 146
Lenthall, John, 150

Lever, John O., 147
Lewis, G. S., 175
Lincoln, E., 189, 200-201
Lincoln, J. B., 200-201
Livingston, Edward P., 39
Livingston, John R., 28, 46
Livingston, Mortimer, 128, 172, 177, 185, 273, 276
Livingston, Robert J., 28
Livingston, Robert L., 39
Livingston, Robert R., 18-19, 23, 26-29, 35-36, 39, 46
Livingston, Fox, and Company, 203, 240, 278, 327
Loper, Richard F., 189, 192, 198-99, 202, 207, 282
Loring, Harrison, 283, 294-95, 309-10, 316, 318
Louisiana and Tehuantepec Company, 276
Lowber, Alfred B., 211-12, 214
Luling, Charles, 255
Lynch, Dominick, Jr., 39

Macauley, Daniel, 320
McCready and Company, N. L., 201
McCready, Mott and Company, 201
McIlvaine, Bowes R., 114
McKinne, Joseph P., 59
McQueen, Robert, 29, 33, 35
Mallory and Company, C. H., 287
Marshall, Charles H., 140
Merchant's Steamship Company, 330
Merrick and Sons, 206, 276
Merrick and Towne, 189, 329
Mills, Edward, 128, 227
Mitchill, S. L., 103, 111, 203
Mitsubishi Mail Steamship Company, 272
Mordecai and Company, 257, 291, 293
Morey, Samuel, 13
Morgan, Charles, 75-76, 122-23, 140, 227, 235, 240, 263, 273
Morgan, Joseph, 27
Morgan Iron Works, 176, 182, 235, 270, 275, 297, 299, 301, 307, 329, 349
Morris Works, I. P., 199, 302
Mott and Ayres, 134
Murray and Hazelhurst, 167

National Line, 269
Neafie and Levy, 282, 327
Neptune Iron Works, 218, 324, 331, 349
New England Ocean Steamship Co., 189
New York and Australia Steam Navigation Company, 270
New York and Bremen Steamship Company, 240-42, 312, 314-15, 329-30
New York and Charleston Steam Packet Company, 74
New York and Havre Steam Navigation Company, 172
New York and Havre Steamship Company, 297
New York and Liverpool United States Mail Steamship Company, 149-71
New York and Matanzas Steamship Company, 282
New York and San Francisco Steamship Line, 260
New York and Venezuela Line, 285
New York Mail Steamship Company, 238-40, 288, 297-311, 314-15, 322-23, 334
New York, Nuevitas, and Cuba Steamship Company, 282
Newton, Isaac, 122
Nicholas and Whittle, 196
Nickerson and Company, F. W., 295, 309, 328
Norris, William, 215
North American Lloyd Steamship Company, 214, 254-55, 304, 309-10, 314-14
North American Steamship Company, 214, 298, 307, 313, 318, 322-23, 326-27, 342
North Atlantic Steam Navigation Company, 287, 289
North Atlantic Steamship Company, 250-53, 255, 343
North German Lloyd Line, 236, 294, 311, 315
North River Steamboat Company, 38, 40
Novelty Iron Works, 101, 103-4, 111, 121, 126, 129, 132-33, 150, 152, 162, 168, 173, 180, 239, 251, 261, 263, 282, 285, 326, 337, 341

Ocean Steam Navigation Company, 130-39, 172, 233
Oelrichs, Herman, 128
Ogden, Aaron, 46
Ogden, John, 196
Opposition Line, 322
Oregon and Mexico Steamship Company, 110
Ormsbee, Elijah, 13

Pacific Mail Steamship Company, 104, 110, 112, 114, 116, 120, 123, 138-40, 149, 167, 176-78, 181-82, 222, 235, 237-42, 247, 250-52, 254, 260, 264, 272, 274-76, 298, 304, 306, 314, 322-23, 327, 336-44
Page, William R., 87-88
Panama Railroad, 250, 254
Papin, Denis, 13
Patterson, Ariel, 188
Pearson, Z. C., 248, 287
Penn and Sons, John, 192
Penn Works, 198, 202
Pennoyer, James, 74-75, 78-80
Perrine, William, 188
Perrine, Patterson, and Stack, 188, 194, 208, 212, 215
Phoenix Foundry, 87
Plummer, George, 215
Poillon, Cornelius, 218, 344, 349
Poillon, Richard, 218, 344, 349
Pook, Samuel H., 283, 309, 316
Porter, Samuel, 48
Porter, Seward, 48
Porter, William, 49
Potter, Charles H., 170
Pusey and Jones, 349

Quartermaster Corps, 124, 213-14, 238, 241, 253, 269, 284, 296, 329, 334
Quintard, George W., 176, 270, 275
Quintard, James W., 286, 327-28
Quintard and Merritt, 270
Quintard Iron Works, 305

Randolf, Elder, and Company, 287
Raynor, James A., 301, 304-7, 310
Reaney, Neafie, and Company, 198-99, 282
Reaney, Son, and Archbold, 349
Reeder, Charles, 126
Reeder, Charles, Jr., 273
Renwick, H. B., 251
Renwick, James, 22-23
Revere, Paul, 29
Reynolds, Samuel, 189, 200
Richard, Nicholas, 110
Richardson, Thomas, 195
Richardson and Company, Thomas, 206
Richardson Brothers, 194
Roach, John, 241, 302, 307, 309, 349
Robb, John A., 79, 272
Roberts, Marshall O., 114, 124, 139, 214, 236-37, 269, 322, 326, 329
Roberts Line, M. O., 214, 322
Robson, William H., 274, 329, 335
Roche Brothers and Coffee, 203
Rogers, Moses, 28, 35, 38, 58-59, 63-64, 66-67
Roosevelt, Nicholas, 13, 18, 33
Roosevelt and Joyce, 301
Roosevelt, Joyce, and Waterbury, 302
Ross, James G., 289
Rowland, George, 68
Royal Mail Steam Packet Company, 270
Ruger, Emil, 311
Ruger, Theodore, 311

Ruger, William, 311
Ruger Brothers, 238, 240-41, 254, 278, 288-89, 298-99, 304, 306, 310-14, 322, 327, 329-30
Rumsey, James, 13, 16-18, 20, 58, 67

Sand, Christian H., 128, 213
Savannah Steamship Company, 59, 63, 66-67
Sayre, Nathaniel, 27
Scarborough, William, 59
Secor, Theodosius F., 140, 223
Secor and Braisted, 232
Secor and Company, T. F., 115-16, 122-23, 126, 140-41, 176, 223
Sewell, William, 152, 168
Sickels, F. E., 168
Simonson, Jeremiah, 115, 125, 222-23, 230, 236, 269, 312, 324
Skiddy, William, 115, 172, 176, 178, 274-75
Sloo, Albert G., 114
Smallman, James, Jr., 29
Smith, Erastus W., 133-34, 178-79
Smith, Stephen, 46, 104
Smith, Thomas, 198
Smith and Ashby, 143
Smith and Dimon, 104, 115, 120, 125, 178, 181, 264, 269, 331
Smith and Dunning, 285
Sound Steamboat Line, 38
Southern Steamship Company, 283
Speedwell Iron Works, 59
Spofford, Paul, 98
Spofford, Tileston, and Company, 98, 103, 163, 174, 193, 201, 257, 261, 263-64, 301
Stack, Thomas, 188
Stanton, Miss F. S., 196
Stanton, J. W., 202
Stanton, T. P., 202
Star Line, 301-6, 309-10
Steers, George, 150, 164, 166, 270, 338
Steers, Henry, 337, 341, 349
Steers, James R., 164, 166, 338
Stevens, Francis Bowes, 154
Stevens, John, 13, 26-27, 38
Stevens, Robert L., 27-28, 38, 99, 154
Stillman and Allen, 101, 155
Stoudinger, Charles, 33, 39
Sturgis, Russell, 274
Suckley, George, 67
Sutton, George D., 255
Sutton and Company, J. T., 189, 202
Symington, William, 13

Taylor, Isaac, 255-56, 312, 314-15
Taylor, Richard, 287
Tennent, W. M., 198
Tileston, Thomas, 98, 101
Thatcher, William A., 198, 282
Thornton, William, 15

Trenholm, George A., 196
Tobey, Edward S., 316, 318
Torrance, Daniel, 229-30
Tucker, F. Z., 329

Union Defense Committee of New York, 277
Union Steamship Company, 283
United States and Brazil Mail Steamship Company, 204, 321
United States Mail Steamship Company, 114, 116, 120-23, 149, 167, 178, 222-23, 227-28, 234, 236, 240, 250, 264, 268, 275

Vail, Stephen, 59
Valiente and Company, 324
Van Dusen, J. B. and J. D., 291, 295
Vanderbilt, Cornelius, 46, 48, 123, 137, 159, 167, 191, 193-94, 222-29, 234-37, 244, 247-51, 263, 268-69, 273, 301, 324
Vanderbilt European Line, 222-49, 269
Vanderbilt Line, 185, 213, 222-49, 251, 268-69, 282, 304, 320, 322, 341-43
Vaughan and Fisher, 206
Vaughan and Lynn, 276
Vogels, Joseph, 83
Voight, Henry, 14, 17, 28

Wall, William, 329
War Department, 269
Watchman and Bratt, 79
Waterbury, Isaac L., 301
Waterman, Robert H., 194
Watt, James, 13, 89, 168, 212
Webb, Eckford, 341
Webb, Isaac, 111
Webb, William H., 104, 111, 114, 117, 125, 140, 146, 156, 163, 164, 178, 260, 269, 273, 298, 310, 318, 322-23, 326-27, 329, 337, 341, 349
Webb and Bell, 341, 349
Weltch, Walter, and Company, 309
West Point Foundry, 46, 192, 344
Westervelt, Jacob A., 176-77, 235, 349
Westervelt and Company, Jacob A., 176, 178, 274
Westervelt and Mackay, 128, 131, 135, 172, 176
Westervelt and Son, Jacob, 178
Westray and Gibbs, 345
Wheeler, S. G., 263
Whitlock, Elisha P., 335
Williams, E. F., 194
Williams, Jabez, 195
Williams, John, 247
Williams, William, 77
Williams, William P., 291
Winsor, Nathaniel, 318
Wright, George, 138
Woodruff, James O., 320-21

Young, John H., 83